Fundamentals of Building Construction

FUNDAMENTALS OF BUILDING CONSTRUCTION

MATERIALS AND METHODS

THIRD EDITION

Edward Allen

Drawings *by*

Joseph Iano

JOHN WILEY & SONS. INC.

New York / Chichester / Weinheim / Brisbane / Singapore / Toronto

Photo on page ii courtesy of Kawneer Company, Inc.

This publication is designed to provide accurate and authoritative
information in regard to the subject matter covered. It is sold with the
understanding that the publisher is not engaged in rendering professional
services. If professional advice or other expert assistance is required,
the services of a competent professional person should be sought.

Library of Congress Cataloging-in-Publication Data:
Allen, Edward, 1938-
    Fundamentals of building construction : materials and methods /
    Edward Allen : drawings by Joseph Iano. – 3rd ed.
        p.   cm.
    Includes bibliographical references and index.
    ISBN 0-471-18349-0 (alk. paper)
    1. Building.   2. Building materials.   I. Title.
    TH145.A417 1998
    690–dc21   98-8019

Printed in the United States of America.
10 9 8 7 6 5 4 3 2

CONTENTS

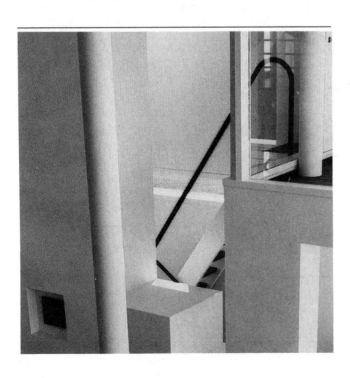

Fireproofing of Steel Framing 389

Longer Spans in Steel 394

Composite Columns 406

Industrialized Systems in Steel 406

Steel and the Building Codes 407

The Uniqueness of Steel 407

12 Light Gauge Steel Frame Construction 417

The Concept of Light Gauge Steel Frame Construction 418

Framing Procedures 419

Other Common Uses of Light Gauge Steel Framing 426

Advantages and Disadvantages of Light Gauge Steel Framing 429

Light Gauge Steel Framing and the Building Codes 430

Finishes for Light Gauge Steel Framing 431

13 Concrete Construction 437

History 438

Cement and Concrete 439

Making and Placing Concrete 443

Formwork 447

Reinforcing 448

Prestressing 461

ACI 301 467

14 Sitecast Concrete Framing Systems 471

Casting a Concrete Slab on Grade 473

Casting a Concrete Wall 476

Casting a Concrete Column 481

One-Way Floor and Roof Framing Systems 483

Two-Way Floor and Roof Framing Systems 493

Concrete Stairs 499

Sitecast Posttensioned Framing Systems 499

Selecting a Sitecast Concrete Framing System 501

Innovations in Sitecast Concrete Construction 504

Architectural Concrete 507

Longer Spans in Sitecast Concrete 516

Designing Economical Sitecast Concrete Buildings 519

Sitecast Concrete and the Building Codes 519

The Uniqueness of Sitecast Concrete 521

15 Precast Concrete Framing Systems 527

Precast, Prestressed Concrete Structural Elements 530

Assembly Concepts for Precast Concrete Buildings 532

The Manufacture of Precast Concrete Structural Elements 535

Joining Precast Concrete Elements 538

The Construction Process 552

Precast Concrete and the Building Codes 552

The Uniqueness of Precast Concrete 557

16 Roofing 565

Low-Slope Roofs 567

Steep Roofs 592

Roofing and the Building Codes 604

17 Glass and Glazing 609

History 610

The Material Glass 612

Glazing 623

PREFACE

There are two motivations for a new edition of this work. One is a need to keep up with the continuing evolution of the construction industry, which has brought with it changes in both construction materials and construction practices. The other is a desire to make the book clearer and more readable, following advice that was offered by the many teachers who responded to a request for suggestions concerning the third edition.

In the years that have elapsed since publication of the second edition, many new building products have been introduced, many older products have assumed new and more important roles in construction, and the social and political climate in which we build has changed in significant ways. New wood products have captured a substantial portion of the market for construction lumber. Insulating concrete forms and light-gauge steel framing have found increasing use in residences and small commercial buildings. A new grade of steel, produced from scrap in mini-mills, has begun to predominate the structural market, and the last of the big steel mills has closed. The newest building to hold the title of world's tallest has been constructed on columns of high-strength concrete rather than structural steel. Roofing and glass technologies have advanced at an astonishing pace. Construction workplace safety has become more and more important, and environmental regulations have altered the composition of many building materials. Energy conservation is increasingly reflected through regulations that affect the building process. A single, unified model building code is near adoption in the United States. All of these developments are reflected in this new edition.

In reorganizing the book for clarity and readability, care has been taken to preserve its fundamental organization. Materials and construction systems continue to be considered simultaneously. The basic order of the book remains unchanged, but two large chapters from the second edition have been broken down into smaller ones. Masonry, which formerly ran nearly 100 pages, is now reorganized into three shorter, more easily digested chapters. Cladding has also been divided into three chapters, the first of them covering basic theory of cladding, the second treating masonry and concrete cladding systems, and the third devoted exclusively to aluminum and glass systems. Light-gauge steel framing has been given expanded coverage in a chapter of its own, reflecting its greater role in small-scale construction. Windows and doors are discussed in more detail in a new chapter that includes comprehensive information on energy efficiency testing and rating programs. A new sidebar on fabric structures has been added to reflect the growing importance of these miraculous buildings. Hundreds of new illustrations, both photographs and drawings, have been added. Some of these replace earlier illustrations that have become obsolete or needed clarification, and some represent new information.

The author wishes to thank the many photographers and representatives of construction organizations who furnished information and illustrations for this edition; their contributions were indispensable. He thanks especially his colleague and good friend Joseph Iano, whose skillful and revealing drawings, the newest of them done by computer, continue to grace these pages.

The preparation of this new edition was initiated and directed by Senior Editor Amanda Miller at John Wiley & Sons, Inc. Her wisdom, professionalism, and patience are greatly appreciated. Karin Kincheloe applied her genius to the design for this third edition, as she did for the first two, and followed through with her characteristic attention to every detail. Donna Conte deftly managed the intricate, seemingly impossible process of production with efficiency and good humor.

Developmental Editor Matthew Van Hattem, Editorial Assistant Katherine Gayle, and Publisher Robert Garber helped in innumerable ways to keep the publication process moving and on track. The author offers his profuse thanks to all of these professionals, without whom the book would not have found its way into print.

This book is dedicated with admiration to Ruth Dietz and to the memory of Professor Albert G. H. Dietz.

E.A.

South Natick, Massachusetts
August 1998

DISCLAIMER

The drawings, tables, descriptions, and photographs in this book have been obtained from many sources, including trade associations, suppliers of building materials, governmental organizations, and architectural firms. They are presented in good faith, but the author, illustrator, and publisher, while they have made every reasonable effort to make this book accurate and authoritative, do not warrant, and assume no liability for, its accuracy or completeness or its fitness for any particular purpose. It is the responsibility of users to apply their professional knowledge in the use of information contained in this book, to consult the original sources for additional information when appropriate, and to seek expert advice when appropriate.

FUNDAMENTALS OF BUILDING CONSTRUCTION

MAKING BUILDINGS

An ironworker connects a steel wide-flange beam to a column.

(*Courtesy of Bethlehem Steel Company*)

We build because little that we do can take place outdoors. We need shelter from sun, wind, rain, and snow. We need dry, level platforms for our activities. Often we need to stack these platforms to multiply available groundspace. On these platforms, and within our shelter, we need air that is warmer or cooler, more or less humid, than outdoors. We need less light by day, and more by night, than is offered by the natural world. We need services that provide energy, communications, and water, and dispose of wastes. So we gather materials and assemble them into the constructions we call buildings in an attempt to satisfy these needs.

DESIGNING BUILDINGS

A building begins as an idea in someone's mind, a desire for new and ample accommodations for a family, many families, an organization, or an enterprise. For any but the smallest of buildings, the next step for the owner of the prospective building is to engage, either directly or through a hired construction manager, the services of building design professionals. An architect helps to consolidate the owner's ideas about the new building, develops the form of the building, and assembles a group of engineering specialists to help work out concepts and details of foundations, structural support, and mechanical, electrical, and communications services. This team of designers, working with the owner, then develops the scheme for the building in progressively finer degrees of detail. Drawings and written specifications are produced by the architect–engineer design team to document how the building is to be made and of what. A general contractor is selected, either by negotiation or by competitive bidding. The general contractor hires subcontractors to carry out many specialized portions of the work. The drawings and specifications are submitted to the municipal inspector of buildings, who checks them for conformance with zoning ordinances and building codes before issuing a permit to build. Construction may then begin, with the building inspector, the architect, and the engineering consultants inspecting the work at frequent intervals to be sure that it is carried out according to plan.

CHOOSING BUILDING SYSTEMS: CONSTRAINTS

Although a building begins as an abstraction, it is built in a world of material realities. The designers of a building—the architects and engineers—work constantly from a knowledge of what is possible and what is not. They are able, on the one hand, to employ any of a limitless palette of building materials and any of a number of structural systems to produce a building of almost any desired form and texture. On the other hand, they are inescapably bound by certain physical limitations: how much land there is with which to work; how heavy a building the soil can support; how long a structural span is feasible; what sorts of materials will perform well in the given environment. They are also constrained by a construction budget and by a complex web of legal restrictions. Those who work in the building professions need a broad understanding of many things, including people, climate, the physical principles by which buildings work, the technologies available for utilization in buildings, the legal restrictions on buildings, and the contractual arrangements under which buildings are built. This book is concerned primarily with the technologies of construction materials—what the materials are, how they are produced, what their properties are, and how they are crafted into buildings. But these must be studied with reference to many other factors that bear on the design of buildings, some of which require explanation here.

Zoning Ordinances

The legal restrictions on buildings begin with local *zoning ordinances,* which govern what types of activities may take place on a given piece of land, how much of the land may be covered by the building or buildings, how far buildings must be set back from each of the property lines, how many parking spaces must be provided, how large a total floor area may be constructed, and how tall the building may be. In many cities, the zoning ordinances establish special center-city fire zones in which buildings must be built of noncombustible materials. Copies of the zoning ordinance for a municipality are available for purchase or reference at the office of the building inspector or the planning department, or they may be consulted at public libraries.

Building Codes

In addition to its zoning ordinances, each local government also regulates building activity by means of a *building code.* The intent of a building code is to protect public health and safety by setting a minimum standard of construction quality. Most building codes in the United States are based on a *model building code,* a standardized code prepared by a national organization of local building code officials. Beginning in the year 2000, more and more local codes throughout the United States will conform to

Table 503
HEIGHT AND AREA LIMITATIONS OF BUILDINGS
Height limitations of buildings (shown in upper figure as stories and feet above grade plane)[m], and area limitations of one- or two-story buildings facing on one street or public space not less than 30 feet wide (shown in lower figure as area in square feet per floor[m]). See Note a.

Use Group			Type 1 Protected Note b 1A	Type 1 Protected Note b 1B	Type 2 Protected 2A	Type 2 Protected 2B	Type 2 Unprotected 2C	Type 3 Protected 3A	Type 3 Unprotected 3B	Type 4 Heavy timber 4	Type 5 Protected 5A	Type 5 Unprotected 5B
A-1	Assembly, theaters		Not limited	Not limited	5 St. 65' 19,950	3 St. 40' 13,125	2 St. 30' 8,400	3 St. 40' 11,550	2 St. 30' 8,400	3 St. 40' 12,600	1 St. 20' 8,925	1 St. 20' 4,200
A-2	Assembly, nightclubs and similar uses		Not limited	Not limited 7,200	3 St. 40' 5,700	2 St. 30' 3,750	1 St. 20' 2,400	2 St. 30' 3,300	1 St. 20' 2,400	2 St. 30' 3,600	1 St. 20' 2,550	1 St. 20' 1,200
A-3	Assembly	Lecture halls, recreation centers, terminals, restaurants other than nightclubs	Not limited	Not limited	5 St. 65' 19,950	3 St. 40' 13,125	2 St. 30' 8,400	3 St. 40' 11,550	2 St. 30' 8,400	3 St. 40' 12,600	1 St. 20' 8,925	1 St. 20' 4,200
A-4	Assembly, churches	Note c	Not limited	Not limited	5 St. 65' 34,200	3 St. 40' 22,500	2 St. 30' 14,400	3 St. 40' 19,800	2 St. 30' 14,400	3 St. 40' 21,600	1 St. 20' 15,300	1 St. 20' 7,200
B	Business		Not limited	Not limited	7 St. 85' 34,200	5 St. 65' 22,500	3 St. 40' 14,400	4 St. 50' 19,800	3 St. 40' 14,400	5 St. 65' 21,600	3 St. 40' 15,300	2 St. 30' 7,200
E	Educational	Note c	Not limited	Not limited	5 St. 65' 34,200	3 St. 40' 22,500	2 St. 30' 14,400	3 St. 40' 19,800	2 St. 30' 14,400	3 St. 40' 21,600	1 St. 20' 15,300 Note d	1 St. 20' 7,200 Note d
F-1	Factory and industrial, moderate		Not limited	Not limited	6 St. 75' 22,800	4 St. 50' 15,000	2 St. 30' 9,600	3 St. 40' 13,200	2 St. 30' 9,600	4 St. 50' 14,400	2 St. 30' 10,200	1 St. 20' 4,800
F-2	Factory and industrial, low	Note h	Not limited	Not limited	7 St. 85' 34,200	5 St. 65' 22,500	3 St. 40' 14,400	4 St. 50' 19,800	3 St. 40' 14,400	5 St. 65' 21,600	3 St. 40' 15,300	2 St. 30' 7,200
H-1	High hazard, detonation hazards	Notes e, i, k, l	1 St. 20' 16,800	1 St. 20' 14,400	1 St. 20' 11,400	1 St. 20' 7,500	1 St. 20' 4,800	1 St. 20' 6,600	1 St. 20' 4,800	1 St. 20' 7,200	1 St. 20' 5,100	Not permitted
H-2	High hazard, deflagration hazards	Notes e, i, j, l	5 St. 65' 16,800	3 St. 40' 14,400	3 St. 40' 11,400	2 St. 30' 7,500	1 St. 20' 4,800	2 St. 30' 6,600	1 St. 20' 4,800	2 St. 30' 7,200	1 St. 20' 5,100	Not permitted
H-3	High hazard, physical hazards	Notes e, l	7 St. 85' 33,600	7 St. 85' 28,800	6 St. 75' 22,800	4 St. 50' 15,000	2 St. 30' 9,600	3 St. 40' 13,200	2 St. 30' 9,600	4 St. 50' 14,400	2 St. 30' 10,200	1 St 20' 4,800
H-4	High hazard, health hazards	Notes e, l	7 St. 85' Not limited	7 St. 85' Not limited	7 St. 85' 34,200	5 St. 65' 22,500	3 St. 40' 14,400	4 St. 50' 19,800	3 St. 40' 14,400	5 St. 65' 21,600	3 St. 40' 15,300	2 St. 30' 7,200
I-1	Institutional, residential care		Not limited	Not limited	9 St. 100' 19,950	4 St. 50' 13,125	3 St. 40' 8,400	4 St. 50' 11,550	3 St. 40' 8,400	4 St. 50' 12,600	3 St. 40' 8,925	2 St. 35' 4,200
I-2	Institutional, incapacitated		Not limited	Not limited	4 St. 50' 17,100	2 St. 30' 11,250	1 St. 20' 7,200	1 St. 20' 9,900	Not permitted	1 St. 20' 10,800	1 St. 20' 7,650	Not permitted
I-3	Institutional, restrained		Not limited	Not limited	4 St. 50' 14,250	2 St. 30' 9,375	1 St. 20' 6,000	2 St. 30' 8,250	1 St. 20' 6,000	2 St. 30' 9,000	1 St. 20' 6,375	Not permitted
M	Mercantile		Not limited	Not limited	6 St. 75' 22,800	4 St. 50' 15,000	2 St. 30' 9,600	3 St. 40' 13,200	2 St. 30' 9,600	4 St. 50' 14,400	2 St. 30' 10,200	1 St. 20' 4,800
R-1	Residential, hotels		Not limited	Not limited	9 St. 100' 22,800	4 St. 50' 15,000	3 St. 40' 9,600	4 St. 50' 13,200	3 St. 40' 9,600	4 St. 50' 14,400	3 St. 40' 10,200	2 St. 35' 4,800
R-2	Residential, multiple-family		Not limited	Not limited	9 St. 100' 22,800	4 St. 50' 15,000 Note f	3 St. 40' 9,600	4 St. 50' 13,200 Note f	3 St. 40' 9,600	4 St. 50' 14,400	3 St. 40' 10,200	2 St. 35' 4,800
R-3	Residential, one- and two-family and multiple single-family		Not limited	Not limited	4 St. 50' 22,800	4 St. 50' 15,000	3 St. 40' 9,600	4 St. 50' 13,200	3 St. 40' 9,600	4 St. 50' 14,400	3 St. 40' 10,200	2 St. 35' 4,800
S-1	Storage, moderate		Not limited	Not limited	5 St. 65' 19,950	4 St. 50' 13,125	2 St. 30' 8,400	3 St. 40' 11,550	2 St. 30' 8,400	4 St. 50' 12,600	2 St. 30' 8,925	1 St. 20' 4,200
S-2	Storage, low	Note g	Not limited	Not limited	7 St. 85' 34,200	5 St. 65' 22,500	3 St. 40' 14,400	4 St. 50' 19,800	3 St. 40' 14,400	5 St. 65' 21,600	3 St. 40' 15,300	2 St. 30' 7,200
U	Utility, miscellaneous		Not limited	Not limited	5 St. 65' 19,950	4 St. 50' 13,125	2 St. 30' 8,400	3 St. 40' 11,550	2 St. 30' 8,400	4 St. 50' 12,600	2 St. 30' 8,925	1 St. 20' 4,200

Note a. See the following sections for general exceptions to Table 503:
Section 504.2 Allowable height increase due to automatic sprinkler system installation.
Section 506.2 Allowable area increase due to street frontage.
Section 506.3 Allowable area increase due to automatic sprinkler system installation.
Section 506.4 Allowable area reduction for multistory buildings.
Section 507.0 Unlimited area one-story buildings.
Note b. Buildings of Type 1 construction permitted to be of unlimited tabular heights and areas are not subject to special requirements that allow increased heights and areas for other types of construction (see Section 503.1.4).
Note c. For height exceptions for auditoriums in occupancies in Use Groups A-4 and E, see Section 504.3.
Note d. For height exceptions for day care centers in buildings of Type 5 construction, see Section 504.4.
Note e. For exceptions to height and area limitations for buildings with occupancies in Use Group H, see Chapter 4 governing the specific use groups.
Note f. For exceptions to height of buildings with occupancies in Use Group R-2 of Types 2B and 3A construction, see Sections 504.6 and 504.7.
Note g. For height and area exceptions for open parking structures, see Section 406.0.
Note h. For exceptions to height and area limitations for special industrial occupancies, see Section 503.1.1.
Note i. Occupancies in Use Groups H-1 and H-2 shall not be permitted below grade.
Note j. Rooms and areas of Use Group H-2 containing pyrophoric materials shall not be permitted in buildings of Type 3, 4 or 5 construction.
Note k. Occupancies in Use Group H-1 are required to be detached one-story buildings (see Section 707.1.1).
Note l. For exceptions to height for buildings with occupancies in Use Group H, see Section 504.5.
Note m. 1 foot = 304.8 mm; 1 square foot = 0.093 m^2.

FIGURE 1.1

Height and area limitations of buildings of various types of construction, as defined in the BOCA National Building Code/1996.

Table 506.4
REDUCTION OF AREA LIMITATIONS

Number of stories	Type of construction		
	1A & 1B	2A	2B, 2C, 3A, 3B, 4, 5A, 5B
1	None	None	None
2	None	None	None
3	None	5%	20%
4	None	10%	20%
5	None	15%	30%
6	None	20%	40%
7	None	25%	50%
8	None	30%	60%
9	None	35%	70%
10	None	40%	80%

FIGURE 1.2

For a building more than two stories in height, the area permitted by the BOCA Code for each story of the building must be reduced in accordance with this table.

the new *International Building Code (IBC)*, the first unified model code in U.S. history. It will take many years for this new code to become fully adopted. In the meantime, in the western United States and parts of the Midwest, most codes are modeled after the *Uniform Building Code (UBC)*. In the East and other areas of the Midwest, the *BOCA National Building Code (BOCA)* is the model in most states. The *Standard Building Code (SBC)* has been adopted by many southern and southeastern states. Canada publishes its own model code, the *National Building Code of Canada*. While these model codes differ in detail, they are similar in their approach. Each begins by defining *use groups* for buildings: buildings of assembly, business, industry, high-hazard industry (such as plants working with highly flammable or explosive substances), institutional buildings, mercantile buildings, residential buildings, and simple buildings used only for storage or agricultural purposes.

These definitions are followed by a set of definitions of *construction types*. At the head of this list are highly fire-resistive types of construction such as masonry, reinforced concrete, and fire-protected steel. At the foot of it are types of construction that are relatively combustible

because they are framed with small wood members. In between are a range of construction type with varying levels of resistance to fire.

With use groups and construction types carefully defined, the code proceeds to match the two, setting forth in a table which use groups may be housed in which types of construction, and under what limitations. Figure 1.1 is reproduced from a model code prepared by the Building Officials and Code Administrators International, Incorporated (BOCA). This table concentrates a great deal of useful information into a very small space. A designer may enter it with a particular use group in mind—an electronics plant, for example—and find out very quickly what types of construction will be permitted and what shape the plant may take. Under the BOCA classification scheme, an electronics plant belongs to Use Group F-1, Factory and Industrial, moderate. Reading across the chart from left to right, one finds immediately that this factory may be built to any desired size, without limit, using Type 1 construction. (For buildings that are taller than two stories, the area limitations given in Figure 1.1 must be reduced in accordance with the table in Figure 1.2).

Type 1 construction is defined in a nearby table in the BOCA Code,

reproduced here as Figure 1.3. Looking down this table under Type 1 construction, one finds a rather detailed listing of the required *fire resistance ratings*, measured in hours, of the various parts of either a Type 1A or a Type 1B building. In a Type 1A building, for example, one finds on line 10 that floor beams must be rated at 3 hours, and on line 9 that lower-story columns, which support more than one floor, must be rated at 4 hours. For the required fire resistances of exterior walls (line 1), one is referred to another section of the BOCA Code, not reproduced in this book, that specifies fire resistance in relation to the distance to surrounding buildings. Line 2 refers to a Table 707.1 for fire walls and party walls, which is reproduced as Figure 22.6 of this book.

Fire resistance ratings of actual construction components are not found in the BOCA Code (although the Standard Building Code and Uniform Building Code do contain partial listings). Instead, they are tabulated in a variety of catalogs and handbooks issued by building material manufacturers, construction trade associations, and organizations concerned with fire protection of buildings. In each case, the ratings are derived from full-scale laboratory fire tests of building components carried out in accordance with Standard E119 of the American Society for Testing and Materials, to assure uniformity of results. (This fire test is described more fully in Chapter 22 of this book.) Figures 1.4 through 1.6 reproduce small sections of tables from catalogs and handbooks to illustrate how this type of information is presented.

It is not possible in this volume to reproduce a comprehensive listing of fire resistance ratings for every type of building component, but what can be said in a very general way (and with many exceptions) is that the higher the degree of fire resistance, the higher the cost. In general, therefore, buildings are built with the least level of fire resistance that is permitted by the applicable building code. The hypothetical electronics plant could be built using Type 1 construction, but does it really need to be constructed to this high standard?

Let us suppose that the owners want the electronics plant to be a two-story building with 10,000 square feet on each floor. The table in Figure 1.1 makes it clear that it cannot be built of unprotected metal (Type 2C), unprotected joisted construction (Type 3B), or unprotected wood frame (Type 5B), because none of these types will allow construction of floors as large as 10,000 square feet. But it can be built of steel with a relatively small amount of applied fire protective material (Type 2B), of protected wood joists with exterior walls either of masonry or of wood (Type 3A), of Heavy Timber construction (Type 4), or of Protected Combustible construction (Type 5A). (The names associated with the various construction types are defined in later chapters of this book.)

Often the situation is more complicated than this. Figure 1.1 applies only to one- and two-story buildings; for taller buildings, the BOCA Code applies the reduction factors shown in Figure 1.2 (each of the model building codes does this slightly differently). The presence or absence of

an automatic sprinkler system for suppression of fires also affects the choice of a construction system under all the building codes. The BOCA Code is typical in this regard: A fully sprinklered building belonging to a nonhazardous use group is allowed to have floor areas three times those shown in Figure 1.1 if it is one or two stories tall, and twice the tabulated areas if it is more than two stories tall. In addition, BOCA allows the building to be a story taller than the height limitation indicated in Figure 1.1 if it is fully sprinklered. Further increases are granted by all the model codes for buildings that front on streets or open spaces on more than one side. Additionally, if a building is divided by fire walls having the fire resistance ratings specified in Table 707.1 (Figure 22.6 of this book), each portion of the building that is separated from the remainder of the building by fire walls may be considered as a separate building for purposes of computing its allowable area, which effectively permits the architect to create a building many times larger than Figure 1.1 would indicate.

A building code goes far beyond what is illustrated here. A typical code also establishes standards for natural light, ventilation, means of emergency egress, structural design, floor, wall, ceiling, and roof construction, chimney construction, fire protection systems, accessibility of the building to disabled persons, and energy efficiency.

The building code is not the only code with which a new building must comply. There are also health codes,

---

. . . the architect should have construction at least as much at his fingers' ends as a thinker his grammar.

Le Corbusier, *Towards a New Architecture,* **1927**

---

fire codes, plumbing codes, and electrical codes in force in most communities. Some of these are locally written, but most are based on model national codes.

Other Legal Constraints

Other types of legal restrictions must also be observed in the design and construction of buildings. *Access standards* regulate the design of entrances, stairs, doorways, elevators, and toilet facilities to assure that they are accessible by physically handicapped members of the population. The *Americans with Disabilities Act (ADA)* makes accessibility to buildings a civil right of all Americans. The U.S. *Occupational Safety and Health Act (OSHA)* controls the design of workplaces to minimize hazards to the health and safety of workers. OSHA sets safety standards under which a building must be constructed and also has an important effect on the design of industrial and commercial buildings. Many states have established comprehensive standards of *energy efficiency* for buildings; these affect primarily a designer's choices of thermal insulation, windows, and heating and cooling systems. An increasing number of states have limitations on the amount of volatile organic compounds such as solvents for paints and construction adhesives that building products can release into the atmosphere. States and localities have conservation laws that protect wetlands and other environmentally sensitive areas from encroachment by buildings. Fire insurance companies exert a major influence on construction standards through their testing and certification organizations (Underwriters Laboratories and Factory Mutual, for example) and through their rate structures for building insurance coverage, which offer strong financial incentives for more fire-resistant construction. Building contractors and construction labor unions also have standards, both formal and informal,

Table 602
FIRERESISTANCE RATINGS OF STRUCTURE ELEMENTS[k]

Structure element Note a		Type of construction Section 602.0									
		Noncombustible					Noncombustible/Combustible		Combustible		
		Type 1 Section 603.0		Type 2 Section 603.0			Type 3 Section 604.0		Type 4 Section 605.0	Type 5 Section 606.0	
		Protected		Protected		Unprotected	Protected	Unprotected	Heavy timber Note c	Protected	Unprotected
		1A	1B	2A	2B	2C	3A	3B	4	5A	5B
1 Exterior walls	Loadbearing	4	3	2	1	0	2	2	2	1	0
		Not less than the fireresistance rating based on fire separation distance (see Section 705.2)									
	Nonloadbearing	Not less than the fireresistance rating based on fire separation distance (see Section 705.2)									
2 Fire walls and party walls (Section 707.0)		4	3	2	2	2	2	2	2	2	2
		Not less than the fireresistance rating required by Table 707.1									
3 Fire separation assemblies (Section 709.0)	Fire enclosure of exits (Sections 1014.11, 709.0 and Note b)	2	2	2	2	2	2	2	2	2	2
	Shafts (other than exits) and elevator hoistways (Sections 709.0, 710.0 and Note b)	2	2	2	2	2	2	2	2	1	1
	Mixed use and fire area separations (Section 313.0)	Not less than the fireresistance rating required by Table 313.1.2									
	Other separation assemblies (Note i)	1	1	1	1	1	1	1	1	1	1
		Note d									
4 Fire partitions (Section 711.0)	Exit access corridors (Note g)	Not less than the fireresistance rating required by Section 1011.4									
		Note d									
	Tenant spaces separations (Note f)	1	1	1	1	0	1	0	1	1	0
		Note d									
5 Dwelling unit and guestroom separations (Sections 711.0, 713.0 and Notes f and j)		1	1	1	1	1	1	1	1	1	1
		Note d									
6 Smoke barriers (Section 712.0 and Note g)		1	1	1	1	1	1	1	1	1	1
7 Other nonloadbearing partitions		0	0	0	0	0	0	0	0	0	0
		Note d									
8 Interior loadbearing walls, loadbearing partitions, columns, girders, trusses (other than roof trusses) and framing (Section 716.0)	Supporting more than one floor	4	3	2	1	0	1	0	see Sec. 605.0	1	0
	Supporting one floor only or a roof only	3	2	1½	1	0	1	0	see Sec. 605.0	1	0
9 Structural members supporting wall (Section 716.0 and Note g)		3	2	1½	1	0	1	0	1	1	0
		Not less than fireresistance rating of wall supported									
10 Floor construction including beams (Section 713.0 and Note h)		3	2	1½ Note l	1	0	1	0	see Sec. 605.0 Note c	1	0
11 Roof construction, including beams, trusses and framing, arches and roof deck (Section 715.0 and Notes e, m)	15' or less in height to lowest member	2	1½	1	1	0	1	0	see Sec. 605.0 Note c	1	0
				Note d							
	More than 15' but less than 20' in height to lowest member	1	1	1	0	0	0	0	see Sec. 605.0	1	0
				Note d							
	20' or more in height to lowest member	0	0	0	0	0	0	0	see Sec. 605.0	0	0
				Note d							

Note a. For fireresistance rating requirements for structural members and assemblies which support other fireresistance rated members or assemblies, see Section 716.1.

Note b. For reductions in the required fireresistance rating of exit and shaft enclosures, see Sections 1014.11 and 710.3.

Note c. For substitution of other structural materials for timber in Type 4 construction, see Section 2304.2.

Note d. For fireretardant-treated wood permitted in roof construction and nonloadbearing walls where the required fireresistance rating is 1 hour or less, see Sections 603.2 and 2310.0.

Note e. For permitted uses of heavy timber in roof construction in buildings of Types 1 and 2 construction, see Section 715.4.

Note f. For reductions in required fireresistance ratings of tenant separations and dwelling unit separations, see Sections 1011.4 and 1011.4.1.

Note g. For exceptions to the fireresistance rating of construction supporting exit access corridor walls, tenant separation walls in covered mall buildings, and smoke barriers, see Sections 711.4 and 712.2.

Note h. For buildings having habitable or occupiable stories or basements below grade, see Section 1006.3.1.

Note i. Not less than the rating required by this code.

Note j. For Use Group R-3, see Section 310.5.

Note k. Fireresistance ratings are expressed in hours.

Note l. In buildings which are required to comply with the provisions of Section 403.3, the required fireresistance rating for floor construction, including beams, shall be 2 hours, see Section 403.3.3.1.

Note m. 1 foot = 304.8 mm.

FIGURE 1.3
Fire resistance of structure elements, as required by the BOCA National Building Code. (Figures 1.1–1.3 are from the BOCA National Building Code/1996, © 1996, Building Officials and Code Administrators International, Inc. Published by arrangements with author. All rights reserved. No parts of the BOCA Code may be reproduced or transmitted in any form or by any means, electronic or mechanical, including photocopying, recording or by an information storage and retrieval system without advance permission in writing from Building Officials and Code Administrators International, Inc. For information, address: BOCA, Inc., 4051 West Flossmoor Road, Country Club Hills, IL 60477)

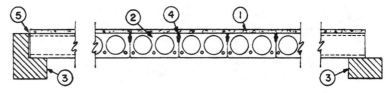

Design No. J921
Restrained Assembly Ratings—2 and 3 Hr. (See Item 1)
Unrestrained Assembly Rating—2 Hr.

Restrained
End Detail

Unrestrained
End Detail

1. **Concrete Topping**—3000 psi compressive strength. Normal weight 150 ± 3 pcf unit weight or lightweight, 112 ± 3 pcf unit weight.

Rating	Thickness-In.
2	0
3	1¼

2. **Precast Concrete Units***—Light-weight aggregate. Cross-section similar to the above illustration. Units 8, 10 or 12 in. thick, 16, 20 or 24 in. wide with 2 or 3 core holes.

3. **End Details**—Restrained and unrestrained. Min bearing 3 in.
4. **Joint**—Clearance between slabs at bottom, full length, min ¹⁄₁₆ in., max ⁵⁄₁₆ in. grouted full length with sand-cement grout, 3500 psi min.
   Note—A ³⁄₄-in. lateral expansion joint to be provided the full length and depth of the slabs every 14 ft. Expansion should be obtained with noncombustible, compressible material, for example; 12 sheets of 1/16 in. thick asbestos paper.
5. **End Clearance**—Clearance for expansion at each end of slabs shall be equal to L/17 (¼ ± 1/16) in., where "L" equal to length of span in feet.

*Bearing the UL Classification Marking

FIGURE 1.4
Examples of fire resistance ratings for concrete and masonry structure elements. The upper detail, taken from the Underwriters Laboratory Fire Resistance Directory, is for a precast concrete hollow-core plank floor with poured concrete topping. "Restrained" and "unrestrained" refer to whether or not the floor is prevented from expanding longitudinally when exposed to the heat of a fire. The lower detail is from literature published by the Brick Institute of America. (*Reprinted by permission of Underwriters Laboratory, Inc., and Brick Institute of America, respectively*)

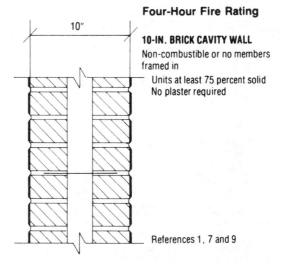

Four-Hour Fire Rating

10"

10-IN. BRICK CAVITY WALL
Non-combustible or no members framed in

Units at least 75 percent solid
No plaster required

References 1, 7 and 9

Design No. A814
Restrained Assembly Rating—3 Hr.
Unrestrained Assembly Rating—3 Hr.
Unrestrained Beam Rating—3 Hr.

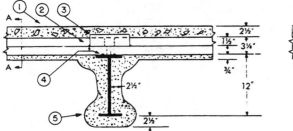

Section A-A

Beam—W12×27, min size.
1. **Sand-Gravel Concrete**—150 pcf unit weight 4000 pcf compressive strength.
2. **Steel Floor and Form Units***—Non-composite 3 in. deep galv units. All 24 in. wide, 18/18 MSG min cellular units. Welded to supports 12 in. O.C. Adjacent units button-punched or welded 36 in. O.C. at joints.

3. **Cover Plate**—No. 16 MSG galv steel.
4. **Welds**—12 in. O.C.
5. **Fiber Sprayed***—Applied to wetted steel surfaces which are free of dirt, oil or loose scale by spraying with water to the final thickness shown above. The use of adhesive and sealer and the tamping of fiber are optional. The min ind density of the finished fiber should be 11 pcf and the specified fiber thicknesses require a min fiber density of 11 pcf. For areas where the fiber density is between 8 and 11 pcf, the fiber thickness shall be increased in accordance with the following formula:

$$\text{Thickness, in.} = \frac{(11)\ (\text{Design Thickness, in.})}{\text{Actual Fiber Density, pcf.}}$$

Fiber density shall not be less than 8 pcf. For method of density determination refer to General Information Section.

*Bearing the UL Classification Marking.

Design No. X511
Rating—3 Hr.

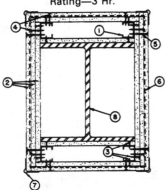

1. **Steel Studs**—1⅝ in. wide with leg dimensions of 1-5/16 and 1-7/16 in. with a ¼-in. folded flange in legs, fabricated from 25 MSG galv steel, ¾- by 1¾-in. rectangular cutouts punched 8 and 16 in. from the ends. Steel stud cut ½ in. less in length than assembly height.
2. **Wallboard, Gypsum***—½ in. thick, three layers.

3. **Screws**—1 in. long, self-drilling, self-tapping steel screws, spaced vertically 24 in. O.C.
4. **Screws**—1⅝ in. long, self-drilling, self-tapping steel screws, spaced vertically 24 in. O.C., except on the outer layer of wallboard on the flange, which are spaced 12 in. O.C.
5. **Screws**—2¼ in. long, self-drilling, self-tapping steel screws, spaced vertically 12 in. O.C.
6. **Tie Wire**—One strand of 18 SWG soft steel wire placed at the upper one-third point, used to secure the second layers of wallboard only.
7. **Corner Beads**—No. 28 MSG galv steel, 1¼ in. legs or 27 MSG uncoated steel, 1¾ in. legs,

FIGURE 1.5
Fire resistance ratings for a steel floor structure and column, respectively, taken from the Underwriters Laboratory Fire Resistance Directory. Figure 11.73 is a photograph of the type of column fireproofing shown here. (*Reprinted by permission of Underwriters Laboratory, Inc.*)

Fire Rating	Sound Rating STC	GA File No.	DETAILED DESCRIPTION	SKETCH AND DESIGN DATA	
				Fire	Sound
1 HR	30 to 34	WP 3620	**Construction Type: Gypsum-Veneer Base, Veneer Plaster, Wood Studs** One layer ½″ type X gypsum veneer base applied at right angles to each side of 2 x 4 wood studs 16″ o.c. with 5d etched nails, 1¾″ long, 0.099″ shank, ¼″ heads, 8″ o.c. Minimum 1/16″ gypsum-veneer plaster over each face. Stagger vertical joints 16″ and horizontal joints each side 12″. Sound tested without veneer plaster. (LB)	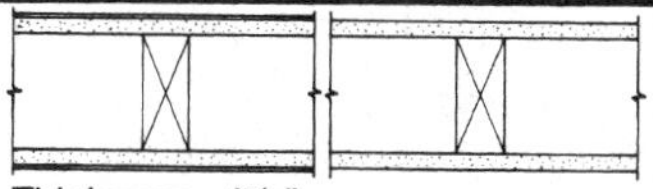 Thickness: 4⅞″ Approx. Weight: 7 psf Fire Test: UC, 1-12-66 Sound Test: G & H IBI-35FT. 5-26-64	

FIGURE 1.6

A sample of fire resistance ratings published by the Gypsum Association, in this case for an interior partition. (*Courtesy of Gypsum Association*)

that affect the ways in which buildings are built. Unions have work rules and safety rules that must be observed; contractors have particular types of equipment, certain kinds of skills, and customary ways of going about things. All of these vary significantly from one place to another.

CHOOSING BUILDING SYSTEMS: INFORMATION RESOURCES

The tasks of the architect and the engineer would be impossible without the support of dozens of organizations that produce and disseminate information on materials and methods of construction. Several organizations and types of organizations whose work has the most direct impact on day-to-day building design and construction operations are discussed in the sections that follow.

ASTM, CSA, and ANSI

The *American Society for Testing and Materials (ASTM)* establishes standard specifications for commonly used materials of construction. These specifications are accepted throughout the United States, and are generally referred to by number (ASTM C150, for example, is a specification for

portland cement). These numbers are frequently used in construction specifications for specific buildings as precise shorthand designations for the quality of material that is required. Throughout this book, ASTM specification numbers will be provided for the major building materials. Should you wish to examine the contents of any ASTM specification, they can be found in the ASTM publication listed at the end of this chapter. In Canada, corresponding standards are set by the *Canadian Standards Association (CSA)*.

The *American National Standards Institute (ANSI)* is an organization that develops standards for many industrial products, such as aluminum windows and many of the mechanical components of buildings. Some government agencies, most notably the U.S. Department of Commerce, also establish standards for certain building products. For the most part, the standards set by the various organizations are complementary and do not overlap or conflict.

Construction Trade and Professional Associations

Building professionals, building materials manufacturers, and building contractors have formed a large number of organizations that work toward the development of technical

standards and the dissemination of information with relation to their respective fields of interest. The *Construction Specifications Institute (CSI)*, whose *Masterformat* standard is described in the following section, is one example. It is composed both of independent building professionals, such as architects and engineers, and of members from industry. The *Western Wood Products Association*, to choose an example from among a hundred or more *trade associations*, is made up of producers of lumber and wood products. It carries out programs of research on wood products, establishes uniform standards of product quality, certifies mills and products that conform to its standards, and publishes authoritative technical literature concerning the use of lumber and related products. Associations with a similar range of activities exist for virtually every material and product used in building. All of them publish technical data relating to their fields of interest, and many of these publications are indispensable references for the architect or engineer. A considerable number are incorporated by reference into various building codes and standards. Selected publications from professional and trade associations are identified in the reference lists at the ends of the chapters in this book. The reader is encouraged to obtain these publications and to request complete publication lists from the various organizations.

Masterformat

The Construction Specifications Institute of the United States and *Construction Specifications Canada (CSC)* have evolved over a period of many years a standard outline called *Masterformat* for organizing information about construction materials and components. Masterformat is used as the outline for construction specifications for nearly all large construction projects in the two countries. It also forms the basis on which trade associations' and manufacturers' technical literature is cataloged and filed. Its 16 primary divisions are as follows:

Division 1	General Requirements
Division 2	Site Work
Division 3	Concrete
Division 4	Masonry
Division 5	Metals
Division 6	Wood and Plastics
Division 7	Thermal and Moisture Protection
Division 8	Doors and Windows
Division 9	Finishes
Division 10	Specialties
Division 11	Equipment
Division 12	Furnishings
Division 13	Special Construction
Division 14	Conveying Systems
Division 15	Mechanical
Division 16	Electrical

Within these broad divisions, several levels of subdivision are established to allow the user to reach any desired degree of detail. A five-digit code, in which the first two digits correspond to the division numbers above, gives the exact reference to any category of information. Within Division 5, Metals, for example, some standard reference codes are

05120	Structural Steel
05210	Steel Joists
05310	Steel Deck
05400	Cold-Formed Metal Framing
05725	Ornamental Metal Castings

Each chapter of this book gives the major Masterformat designations for the information presented, to help the reader know where to look in construction specifications and in the technical literature for further information. The full Masterformat system is contained in the volume referenced at the end of this chapter.

CHOOSING BUILDING SYSTEMS: THE WORK OF THE DESIGN PROFESSIONAL

The designers of a building must make many choices before its design is complete and ready for construction. In making these choices, they confront four basic questions:

1. What will give the required functional performance?

2. What will give the desired aesthetic result?

3. What is possible legally?

4. What is most economical?

Increasingly, engineers and architects are adding a fifth question to this list: What is best for the environment? In other words, what building materials are obtained with least damage to the global environment? What building practices are least disruptive of the natural world? How may a building be designed so that it uses as little fossil fuel and nuclear energy as possible?

This book examines primarily the first of these questions, the func-

tional performance of the materials and methods of building construction most commonly employed in North America. But reference is made constantly to the other four questions, and the reader should always have them uppermost in mind.

The chapters that follow describe many alternative ways of doing things: different structural systems, different systems of enclosure, different systems of interior finish. Each system has characteristics that distinguish it from the alternatives. Sometimes a system is distinguished chiefly by its visual qualities, as one might acknowledge in choosing one type of granite over another, one color of paint over another, or one type of tile pattern over another. However, visual distinctions can extend beyond surface qualities; a designer may prefer the massive appearance of a masonry bearing wall building to the slender look of an exposed steel frame on one project, yet would choose the steel for another building whose situation is different. Again, one may choose for purely functional reasons, as in selecting terrazzo flooring instead of carpet or wood in a restaurant kitchen. One could choose on purely technical grounds, as, for example, in electing to posttension a long concrete beam rather than merely to reinforce it. A designer is often forced into a particular choice by some of the legal constraints identified earlier in this chapter. And frequently the selection is made on purely economic grounds. The economic criterion can mean any of several things: Sometimes one system is chosen over another because its first cost is less; sometimes the entire *life-cycle costs* of competing systems are compared by means of formulas that include first cost, maintenance cost, energy consumption cost (if any), the useful lifetime and replacement cost of the system, and interest rates on invested money; and, finally, a system may be chosen because there is keen competition among local suppliers and/or installers that keeps the price at the lowest possible level. This is often a reason to specify a very standard type of roofing material, for example, that can be furnished and installed by any of a number of companies, instead of a new system that is theoretically better from a functional standpoint, but can only be furnished by a single company that has the new equipment and skills required to install it.

One cannot gain all the knowledge needed to make such decisions from a textbook. It is incumbent upon the reader to go far beyond what can be presented here—to other books, to catalogs, to trade publications, to professional periodicals, and especially to the design office, the workshop, and the building site. There is no other way to gain much of the required information than to get involved in the art and business of building. One must learn how materials feel in the hand; how they look in a building; how they are manufactured, worked, and put in place; how they perform in service; how they deteriorate with time. One must become familiar with the people and organizations that produce buildings—the architects, engineers, materials suppliers, contractors, subcontractors, workers, inspectors, managers, and building owners—and learn to understand their respective methods, problems, and points of view. In the meantime, this long and

> **Go into the field where you can see the machines and methods at work that make the modern buildings, or stay in construction direct and simple until you can work naturally into building-design from the nature of construction.**
>
> **Frank Lloyd Wright, *To the Young Man in Architecture*, 1931**

hopefully enjoyable process of education in the materials and methods of building construction can begin with the information presented in the chapters that follow.

Recurring Concerns

Certain themes are woven throughout this book and surface again and again in different, often widely varying contexts. These represent a set of concerns that fall into two broad categories: one that has to do with building *performance,* and one that has to do with building *construction.*

The performance concerns relate to the inescapable problems that must be confronted in every building: fire; building movement of every kind, including foundation set-

tlement, structural deflections, and expansion and contraction due to changes in temperature and humidity; heat flow through building assemblies; water vapor migration and condensation; water leakage; acoustical privacy; deterioration, cleanliness, and building maintenance.

The construction concerns are associated with the everyday problems of getting a building built safely, on time, within budget, and to the required standard of quality: division of work between the shop and the field; optimum use of the various building trades; sequencing of construction operations for maximum productivity; convenient worker access to construction operations; dealing with inclement weather; making building components fit together; and quality assurance in construction materials and components, through grading, testing, and inspection.

To the novice, these matters may seem of minor consequence when compared to the larger and often more interesting themes of building form and function. To the experienced building professional, who has seen buildings fail both aesthetically and physically for want of attention to one of more of these concerns, they are issues that must each be resolved as a matter of course before the work of a building project can be allowed to proceed.

SELECTED REFERENCES

1. Allen, Edward. *How Buildings Work* (2nd ed.). New York, Oxford University Press, 1995.

What do buildings do, and how do they do it? This book sets forth in easily understandable terms the physical principles by which buildings stand up, enclose a piece of the world, and modify it for human use.

2. American Society for Testing and Materials. *ASTM Standards in Building Codes* (31st ed.). Philadelphia, 1994.

In this volume are found the ASTM standards for all the materials used in the construction of buildings. (Address for ordering: 1916 Race Street, Philadelphia, PA 19103.)

3. The Construction Specifications Institute and Construction Specifications Canada. *Masterformat—Master List of Section Titles and Numbers.* Alexandria, VA, and Toronto, 1995.

This book gives the full set of numbers and titles under which construction infor-

mation is filed and utilized. (Address for ordering: 601 Madison Street, Alexandria, VA 22314, or One St. Clair Avenue West, Suite 1206, Toronto, Ontario M4V 1K6.)

4. Building Officials and Code Administrators International, Inc. *The BOCA National Building Code.* (Address for ordering: 4051 West Flossmoor Road, Country Club Hills, IL 60477.)

5. Associate Committee on the National Building Code, National Research

Council of Canada. *National Building Code of Canada.* (Address for ordering: NRCC, Ottawa, Ontario K1A 0R6.)

6. Southern Building Code Congress International, Inc. *Standard Building Code.* (Address for ordering: 900 Montclair Road, Birmingham, AL 35213.)

7. International Conference of Building Officials. *Uniform Building Code.* (Address for ordering: 5360 South Workman Mill Road, Whittier, CA 90601.)

8. Allen, Edward, and Joseph Iano. *The Architect's Studio Companion* (2nd ed.). New York, John Wiley & Sons, Inc., 1995.

This reference book contains precalculated tables that allow an instant determination of the allowable heights and areas for any building under any of the model building codes. It also explains clearly what each construction type means, relating it to actual construction materials and structural systems, and it gives extensive rules of thumb for structural systems, mechanical systems, and egress planning.

9. Guise, David. *Design and Technology in Architecture.* New York, John Wiley & Sons, Inc., 1985.

Sixteen illustrated case studies of actual buildings are the core around which this book is built. It deals in a very practical way with the influence of codes, zoning, materials, structure, and mechanical systems on building design.

KEY TERMS AND CONCEPTS

zoning ordinance	use group	CSA
building code	construction type	ANSI
model building code	fire resistance rating	CSI
IBC	access standards	Masterformat
UBC	ADA	CSC
BOCA	OSHA	trade association
SBC	energy efficiency	life-cycle costs
National Building Code of Canada	ASTM	

REVIEW QUESTIONS

1. Who are the members of the typical team that designs a major building? What are their respective roles?

2. What are the major constraints under which the designers of a building must work?

3. What types of subjects are covered by zoning ordinances? By building codes?

4. In what units is fire resistance measured? How is the fire resistance of a building assembly determined?

5. If you are designing a five-story office building (Use Group B) with 11,300 square feet per floor, what types of construction will you be permitted to use under the BOCA Code if you do not install sprinklers? (Be sure to correct the area limitations in Figure 1.1 for the height of the building, using the reduction factors in Figure 1.2.) How does the situation change if you install sprinklers?

EXERCISES

1. Have each class member write to two or three trade associations at the beginning of the term to request their lists of publications, and then have each send for some of the publications. Display and discuss the publications.

2. Repeat the above exercise for manufacturers' catalogs of building materials and components.

3. Obtain copies of your local zoning ordinance and building code (they may be in your library). Look up the applicable provisions of these documents for a specific site and building. What setbacks are required? How large a building is permitted? What construction types may be employed? What types of roofing materials are permitted? What are the restrictions on interior finish materials? Outline in complete detail the requirements for emergency egress from the building.

FOUNDATIONS

2

- **Foundation Loads**
- **Foundation Settlement**
- **Soils**
  Types of Soils
  Subsurface Exploration and Soil Testing
- **Excavation**
  Slope Support
  Sheeting
  Soil Mixing
  Bracing
  Dewatering
- **Foundations**
  Shallow Foundations
  Deep Foundations

- **Seismic Base Isolation**
- **Underpinning**
- **Retaining Walls**
- **Waterproofing and Drainage**
- **Basement Insulation**
- **Shallow Frost-Protected Foundations**
- **Backfilling**
- **Up–Down Construction**
- **Designing Foundations**
- **Foundation Design and the Building Codes**

Foundation work in progress for a midrise hotel and apartment building in Boston. The earth surrounding the excavation is retained with steel sheet piling supported by steel walers and tiebacks. Equipment enters and leaves the site via the earth ramp at the bottom of the picture. Although a large backhoe at the right continues to dig around old piles from a previous building on the site, the installation of pressure-injected concrete pile footings is already well underway, with two piledrivers at work in the near and far corners, and clusters of completed piles visible in the center of the picture. Concrete pile caps and column reinforcing are under construction in the center of the excavation. (*Courtesy of Franki Foundation Company*)

The function of a *foundation* is to transfer the structural loads from a building safely into the ground. Every building needs a foundation of some kind: A backyard toolshed will not be damaged by slight movements of its foundations and may need only wooden skids to spread its load across an area of the surface of the ground sufficient to support its weight. A house needs greater stability than a toolshed, so its foundation reaches through the unstable surface to underlying soil that is free of organic matter and unreachable by winter's frost. A larger building of masonry, steel, or concrete weighs many times more than a house, and its foundations probe into the earth until they reach soil or rock that is competent to carry its massive loads; on some sites, this means going a hundred feet or more below the surface. Because of the variety of soil, rock, and water conditions that are encountered below the surface of the ground and the unique demands that many buildings make upon their foundations, foundation design is a highly specialized field of geotechnical engineering, a branch of civil engineering, that can be sketched here only in its broad outlines.

1. The foundation, including the underlying soil and rock, must be safe against a structural failure that could result in collapse.

2. During the life of the building, the foundation must not settle in such a way as to damage the structure or impair its function.

3. The foundation must be feasible both technically and economically, and practical to build without adverse effects to surrounding property.

FOUNDATION SETTLEMENT

All foundations settle to some extent as the soil around and beneath them adjusts itself to these loads. Foundations on bedrock settle a negligible amount. Foundations on certain types of clay, such as that found in Mexico City, may settle to an alarming degree, allowing buildings to subside by amounts that are measured in feet or meters. *Foundation settlement* in most buildings is measured in millimeters or fractions of inches. If settlement occurs at roughly the same rate from one side of the building to the other (*uniform settlement*), no harm is likely to be done to the building, but if large amounts of *differential settlement* occur, in which the various columns and loadbearing walls of the building settle by substantially different amounts, the frame of the building may become distorted, floors may slope, walls and glass may crack, and doors and windows may refuse to work properly (Figure 2.1).

FOUNDATION LOADS

A foundation supports a number of different kinds of loads:

• The *dead load* of the building, which is the sum of the weights of the frame; the floors, roofs, and walls; the electrical and mechanical equipment; and the foundation itself

• The *live load*, which is the sum of the weights of the people in the building; the furnishings and equipment they use; and snow, ice, and water on the roof

• *Wind loads*, which can apply lateral, downward, and uplift loads to a foundation

• Horizontal pressures of earth and water against basement walls

• In some buildings, horizontal or inclined *thrusts* from arches, rigid frames, domes, vaults, or tensile structures

• On some building sites, buoyant uplift forces from underground water, identical to the forces that cause a boat to float

• During earthquakes, horizontal and vertical forces caused by the motion of the ground relative to the building

A satisfactory foundation for a building must meet three general requirements:

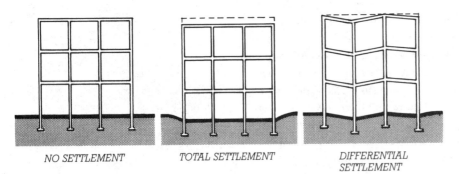

NO SETTLEMENT   TOTAL SETTLEMENT   DIFFERENTIAL SETTLEMENT

FIGURE 2.1
Uniform settlement is usually of little consequence in a building, but differential settlement can cause severe structural damage.

Accordingly, a primary objective in foundation design is to minimize differential settlement by loading the soil in such a way that equal settlement occurs under the various parts of the building. This is not difficult when all parts of the building rest on the same kind of soil, but can become a problem when a building occupies a piece of ground that is underlain by two or more areas of different types of soil with very different loadbearing capacities. Most foundation failures are attributable to excessive differential settlement; gross failure of a foundation, in which the soil fails completely to support the building, is extremely rare.

SOILS

Types of Soils

The following distinctions may be helpful in acquiring an initial understanding of how soils are classified for engineering purposes:

• *Rock* is a continuous mass of solid mineral material, such as granite or limestone, that can only be removed by drilling and blasting. Rock is not completely monolithic, but is crossed by a system of joints that divide it into irregular blocks. Despite these joints, rock is generally the strongest and most stable material on which a building can be founded.

• *Soil* is a general term referring to earth material that is particulate.

• If an individual particle of soil is too large to lift by hand, or requires two hands to lift, it is known as a *boulder.*

• If it takes the whole hand to lift a particle, it is a *cobble.*

• If a particle can be lifted easily with thumb and forefinger, the soil is *gravel.* In the Unified Soil Classification System (Figure 2.2), gravels are classified visually as having more than half their particles larger than 0.25 inch (6.5 mm) in diameter.

			Group Symbols	Typical Names
Coarse-grained Soils	Gravels	Clean Gravels	GW	Well-graded gravels, gravel-sand mixtures, little or no fines
			GP	Poorly graded gravels, gravel-sand mixtures, little or no fines
		Gravels with Fines	GM	Silty gravels, poorly graded gravel-sand-silt mixtures
			GC	Clayey gravels, poorly graded gravel-sand-clay mixtures
	Sands	Clean Sands	SW	Well-graded sands, gravelly sands, little or no fines
			SP	Poorly graded sands, gravelly sands, little or no fines
		Sands with Fines	SM	Silty sands, poorly graded sand-silt mixtures
			SC	Clayey sands, poorly graded sand-clay mixtures
Fine-grained Soils	Silts and Clays	(Liquid limit greater than 50)	ML	Inorganic silts and very fine sands, rock flour, silty or clayey fine sands with plasticity
			CL	Inorganic clays of low to medium plasticity, gravelly clays, sandy clays, silty clays, lean clays
			OL	Organic silts and organic silt-clays of low plasticity
		(Liquid limit less than 50)	MH	Inorganic silts, micaceous or diatomaceous fine sandy or silty soils, elastic silts
			CH	Inorganic clays of high plasticity, fat clays
			OH	Organic clays of medium to high plasticity
Highly Organic Soils			Pt	Peat and other highly organic soils

FIGURE 2.2

A soil classification chart based on the Unified Soil Classification System. The group symbols are a universal set of abbreviations for soil types, as used in Figure 2.7.

• If the individual particles can be seen but are too small to be picked up individually, the soil is *sand*. Sand particles range in size from about 0.25 to 0.002 inch (6.5 to 0.06 mm). Sand and gravel are considered to be *coarse-grained soils*.

• *Silt* particles are approximately equidimensional, and range in size from 0.002 to 0.00008 inch (0.06 to 0.002 mm). Because of their low surface-area-to-volume ratio, which approximates that of sand and gravel, the behavior of silt is controlled by the same mass forces that control the behavior of the coarse-grained soils.

• *Clay* particles are plate shaped rather than equidimensional (Figure 2.3) and smaller than silt particles, less than 0.00008 inch (0.002 mm). Clay particles, because of their

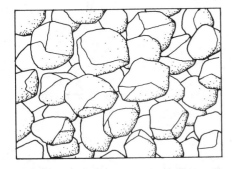

FIGURE 2.3
Silt particles (top) are approximately equidimensional granules, while clay particles (bottom) are platelike. Notice that the clay particles are much smaller than the silt particles; a circular area of clay particles has been magnified to make their structure easier to see.

smaller size and flatter shape, have a surface-area-to-volume ratio hundreds or thousands of times greater than that of silt. As particle size decreases, the size of the pores between the particles also decreases, and soil behavior increasingly depends on surface forces. As particle shape becomes more platelike, the *fabric* (arrangement of particles and pores) becomes more important: The volume of the pores and the amount of water in the pores greatly influences the properties of a clay soil.

• Peat, topsoil, and other organic soils are not suitable for the support of building foundations. Because of their high content of organic matter, they are spongy and compress easily, and their properties can change over time due to changing water content or biological activity in the soil.

Clay soils are generally referred to as being *cohesive*, which means that they retain a measurable shear resistance in the absence of confining forces. To put this more directly, a clay soil sticks together. In contrast, in a *frictional* or *cohesionless* soil such as sand or silt, the shear resistance is directly proportional to the confining force pushing the particles together. In the absence of a confining force, the shear resistance disappears and the soil cannot stand with a vertical face (Figure 2.4). Thus, sand confined by surrounding soil within the earth can support a heavy building, whereas a conical pile of sand on the surface of the earth can support nothing, because there is little or no shear resistance between the particles.

A building site is usually underlain by a number of superimposed layers (*strata*) of different soils. These strata were deposited one after another in very ancient times by volcanic action and by the action of water, wind, and ice. Most of the soils in these strata are mixtures of several different sizes and types of particles

and bear such names as silty gravels, gravelly sands, clayey sands, silty clays, and so on (Figure 2.2). The distribution of particle sizes and types in a soil is important to know when designing a foundation because it is helpful in predicting the loadbearing capacity of the soil, its stability, and its drainage characteristics.

Figure 2.5 gives some typical ranges of loadbearing capacities for various types of soil. These values give only a general idea of the relative strengths of different soils; the strength of an actual soil is also dependent on factors such as the presence or absence of water, the depth at which the soil lies beneath the surface, and the size of the foundation that applies the load to the soil. The designer may also arbitrarily choose to reduce the pressure of the

EXCAVATION IN FRICTIONAL SOIL

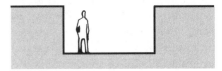

EXCAVATION IN HIGHLY COHESIVE SOIL

FIGURE 2.4
Excavations in highly cohesive and frictional soils.

foundations on the soil to well below these values in order to reduce the anticipated amount of settlement of the building. Rock is generally the best material on which to found a building; the proliferation of tall office buildings in Manhattan is attributable in part to the bedrock that lies a short distance beneath the surface. When rock is too deep to be reached by foundations, the designer must choose from the strata of different soils that lie closer to the surface, and design a foundation to perform satisfactorily in the selected soil.

The *stability* of a soil is its ability to retain its engineering properties under the varying conditions that may occur during the lifetime of the building. Rock, gravels, sands, and many silts tend to be stable soils. Many clays are dimensionally unstable under changing subsurface moisture conditions. Clay may swell considerably as it absorbs water and shrink as it dries. When wet clay is put under pressure, water can be squeezed out of it, with a corresponding reduction in volume. Taken together, these properties make clays that are subject to changes in water content the least stable and least predictable soils for supporting buildings.

The drainage characteristics of a soil are important in predicting how water will flow on and under building sites and around building substructures. Water passes readily through clean gravels and sands, slowly through very fine silts and sands, and almost not at all through many clays. A building site with clayey or silty soils near the surface drains poorly and is likely to be muddy and covered with puddles during rainy periods, whereas a gravelly site is likely to remain virtually dry. Underground, water passes quickly through strata of gravel and sand, but accumulates above strata of clay and fine silt. An excellent way to keep a basement dry is to surround it with a thick layer of clean gravel or crushed stone. Water passing through the soil toward the building cannot reach the basement without first falling to the bottom of the gravel layer, from where it can be drawn off in perforated pipes before it accumulates (Figures 2.57, 2.58). It does little good to place perforated drainage pipes directly in clay or silt because water cannot flow easily toward the pipes.

Subsurface Exploration and Soil Testing

Prior to designing a foundation for any building larger than a single-family house, it is necessary to determine the soil and water conditions beneath the site. This can be done by digging test pits or by making test borings. Test pits are useful when the foundation is not expected to extend deeper than about 8 feet (2.5 m), which is the maximum practical reach of small excavating machines. The strata of the soil can be observed and measured in the pit, and soil samples can be taken for laboratory testing. The level of the *water table* (the elevation at which the pressure of the ground water is atmospheric) will be readily apparent in coarse-grained soils if it falls within the depth of the pit because water will seep through the walls of the pit up to the level of the water table. On the other hand, if a test pit is excavated below the water table in clay, free water will not seep into the pit because the clay is relatively impermeable. In this case, the level of the water table must be determined by means of an observation well or special electronic devices that are installed to measure water pressure. If desired, a load test can be performed on the soil in the bottom of a test pit to determine the stress the soil can safely carry and the amount of settlement that should be anticipated under load.

If a pit is not dug, borings with standard penetration tests can give an indication of the bearing capacity of the soil by the number of blows of a standard driving hammer required to advance a sampling tube into the soil by a fixed amount. Test boring

Table 1804.3
PRESUMPTIVE LOADBEARING VALUES OF FOUNDATION MATERIALS

Class of material	Loadbearing pressure (pounds per square foot)[a]
1. Crystalline bedrock	12,000
2. Sedimentary rock	6,000
3. Sandy Gravel	5,000
4. Sand, silty sand, clayey sand, silty gravel and clayey gravel	3,000
5. Clay, sandy clay, silty clay & clayey silt	2,000

Note a. 1 psf = 47.9 Pa.

FIGURE 2.5
Presumptive surface bearing values of various soils, as given in the BOCA National Building Code/1996. (*Reproduced by permission*)

(a)   *(b)*   *(c)*

FIGURE 2.6
Examples of soil test boring equipment. (*a*) A portable cathead drilling rig. A small gasoline engine on the leg of the tripod next to the operator spins a metal drum called a "cathead." By alternately tightening and loosening the loops of rope around the cathead, the operator lifts and drops the cylindrical weight to drive a casing or a sampling tube into the earth. (*b*) A trailer-mounted hydraulic feed core drill. (*c*) A truck-mounted hydraulic drill rig with core augers. (*Courtesy of Acker Drill Company, Inc.*)

(Figure 2.6) extends the possible range of exploration much deeper into the earth than test pits and returns information on the thickness and locations of the soil strata and the depth of the water table. Laboratory-quality soil samples can also be recovered for testing. Usually a number of holes are drilled across the site; the information from the holes is coordinated and interpolated in the preparation of drawings that document the subsurface conditions for the use of the engineer who will design the foundation (Figure 2.7).

Laboratory testing of soil samples is an important preliminary to foundation design. By passing a dried sample of coarse-grained soil through a set of sieves with graduated mesh sizes, the particle size distribution in the soil can be determined. Further tests on fine-grained soils assist in their identification and provide information on their engineering properties. Important among these are tests

SOIL DESCRIPTION	DEPTH
Topsoil	0.5'
Loose silt, some fine sand and clay (ML)	6.5'
Loose to medium dense fine to coarse sand, some silt, trace of fine gravel (SM)	20.5'
Medium dense fine to coarse sand, some silt (SM)	30.5' 34.5'
Medium dense silt, some fine sand (ML)	40' 43'
Medium dense fine to coarse sand, some silt, trace fine gravel (SP-ML)	52.5'
Medium dense silt, some fine sand (ML)	
Firm to stiff clay, some silt (CL)	
Very dense fine to coarse sand, some silt, trace fine to coarse gravel (SP-SM)	84'

FIGURE 2.7
A typical log from a soil test boring, indicating the type of soil in each stratum and the depth in feet at which it was found. The abbreviations in parentheses refer to the Unified Soil Classification System, and are explained in Figure 2.2.

that establish the *liquid limit,* the water content at which the soil passes from a plastic state to a liquid state, and the *plastic limit,* the water content

at which the soil loses its plasticity and begins to behave as a solid. Additional tests can determine the water content of the soil, its perme-

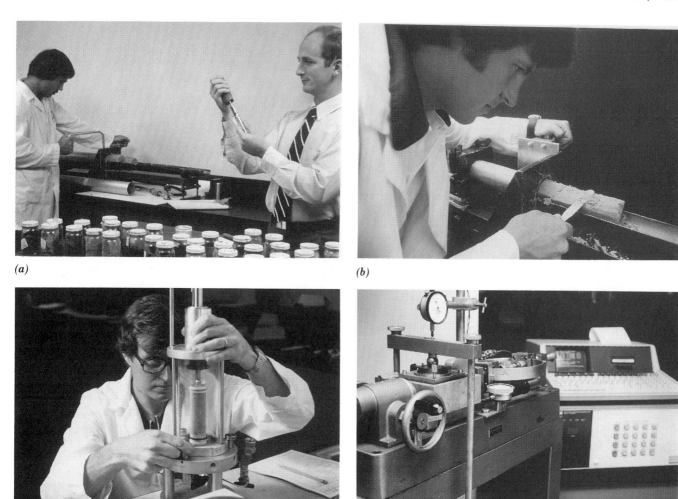

(a)

(b)

(c)

(d)

FIGURE 2.8

Some laboratory soil testing procedures. (*a*) To the right, the hardness of split spoon samples is checked with a penetrometer to be sure that they come from the same stratum of soil. To the left, a soil sample from a Shelby sampling tube is cut into sections for testing. (*b*) A section of undisturbed soil from a Shelby tube is trimmed to examine the stratification of silt and clay layers. (*c*) A cylindrical sample of soil is set up for a triaxial load test, the principal method for determining the shear strength of soil. The sample will be loaded axially by the piston in the top of the apparatus, and also circumferentially by water pressure in the transparent cylinder. (*d*) A direct shear test, used to measure the shear strength of cohesionless soils. A rectangular prism of soil is placed in a split box and sheared by applying pressure in opposite directions to the two halves of the box.

(continued)

ability to water, its shrinkage when dried, its shear and compressive strengths, the amount by which the soil can be expected to consolidate under load, and the rate at which the consolidation will take place (Figure 2.8). These latter two qualities are helpful in predicting the rate and magnitude of foundation settlement in a building.

The information gained through subsurface exploration and soil test-

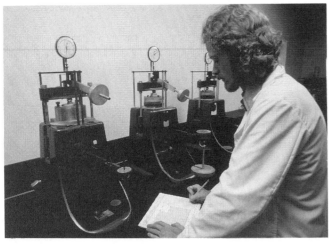

(e)

(f)

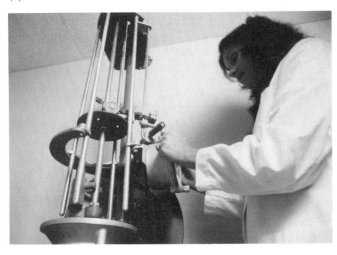

(g)

FIGURE 2.8 *(continued)*

(e) **One-dimensional consolidation tests in progress on fine-grained soils, to determine their compressibility and expected rate of settlement. Each sample is compressed over an extended period of time to allow water to flow out of the sample. (*f*) A panel for running 30 simultaneous constant head permeability tests, to determine the rate at which a fluid, usually water, moves through a soil. (*g*) A Proctor compaction test, in which successive layers of soil are compacted with a specified tamping force. The test is repeated for the same soil with varying moisture content, and a curve is plotted of density achieved versus moisture content of the soil, to identify an optimum moisture content for compacting the soil in the field. Not shown here are some common testing procedures for grain size analysis, liquid limit, plastic limit, specific gravity, and unconfined compression.** (*Courtesy of Ardaman and Associates, Inc., Orlando, Florida*)

ing is summarized in a written soils report. This report includes the results of both the field tests and the laboratory tests, recommended types of foundations for the site, recommended depths and bearing stresses for the foundations, and an estimate of the expected rate of foundation settlement. This information can be used directly by foundation and structural engineers in the design of the excavations, dewatering, and slope support systems, foundations, and substructure.

EXCAVATION

At least some *excavation* is required for every building. Foundations should not be placed on organic soils, which are subject to decomposition and to shrinking and swelling with changing moisture content. Organic topsoil, which is excellent for growing lawns and landscape plants, is scraped away from the building area and stockpiled for redistribution over the site after construction of the building is complete. When the topsoil has been removed, further digging is necessary to place the footings out of reach of water and wind erosion and, in colder climates, to place them below the level to which the ground freezes in winter. Soil expands slightly as it freezes, with a force that can lift and break buildings. Under certain soil and temperature conditions, upward migration of water vapor from the pores in the soil can result in the formation of *ice lenses*, thick layers of frozen water

crystals than can lift buildings by larger amounts, unless the foundations are placed below the frost line or are insulated in such a way that the soil beneath them cannot freeze.

Excavation is required on many sites to place the footings at a depth where soil of the appropriate bearing capacity is available. Excavation is frequently undertaken so one or more levels of basement space can be added to a building, whether for additional habitable rooms, for parking, or for mechanical equipment and storage. Where footings must be placed deep to get below the frost line or reach competent soil, a basement is often bargain-rate space, adding little to the overall cost of the building.

In particulate soils, a variety of machines can be used to loosen and lift the soil from the ground: bulldozers, shovel dozers, backhoes, bucket loaders, scrapers, trenching machines, and power shovels of every type. If the soil must be moved more than a short distance, dumptrucks come into use.

In rock, excavation is slower and many times more costly. Weak or decomposing rock can sometimes be fractured and loosened with power shovels, tractor-mounted rippers, pneumatic hammers, or drop balls such as those used in building demolition. Blasting, in which explosives are placed and detonated in lines of closely spaced holes drilled deep into the rock, is often necessary. On tight urban sites where blasting is impractical, rock can be fractured with hydraulic splitters.

If . . . solid ground cannot be found, but the place proves to be nothing but a heap of loose earth to the very bottom, or a marsh, then it must be dug up and cleared out and set with piles made of charred alder or olive wood or oak, and these must be driven down by machinery, very closely together . . .

Marcus Vitruvius Pollio,
first century B.C.

Slope Support

If the site is sufficiently larger than the area to be covered by the building, the edges of the excavation can be sloped back or *benched* at an angle such that the soil will not slide back into the hole. This angle can be steep for cohesive soils such as the stiffer clays, but must be shallow for frictional soils such as sand and gravel. On constricted sites, the soil around an excavation must be held back by temporary walls called *sheeting*. The sheeting, in turn, must be supported against the pressure of the earth and water it retains (Figure 2.9). If the sheeting is less than one story tall, this can often be done by driving it deeply enough into the soil that it can act as a vertical cantilever beam. For taller walls, and in less stable soils, some form of *bracing* is required for support.

f.02_09.TIF

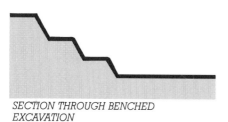

SECTION THROUGH BENCHED EXCAVATION

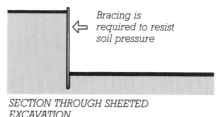

Bracing is required to resist soil pressure

SECTION THROUGH SHEETED EXCAVATION

FIGURE 2.9
On a spacious site, an excavation can be benched. On an urban site, lack of space often requires the use of sheeting to retain the soil around an excavation.

Sheeting

Sheeting can take any of several forms, depending on the qualities of the soil and circumstances such as the equipment and preferences of the contractor, the proximity of surrounding buildings, the level of the water table, or a desire to retain the sheeting as a permanent part of the substructure of the building. Among the most common forms of sheeting are soldier beams and lagging, sheet piling, and slurry walls. *Soldier beams* are steel wide-flange sections driven vertically into the earth at close intervals around an excavation site before digging begins. As earth is removed, *lagging,* usually consisting of heavy wood planks, is placed against the flanges of the soldier beams to retain the soil outside the excavation (Figures 2.10, 2.11). The soldier beams and lagging may be removed after construction of the substructure has been completed.

Sheet piling consists of vertical planks of wood, steel, or precast concrete that are placed tightly against one another and driven into the earth to form a solid wall before excavation begins (Figures 2.12, 2.13). Steel and precast concrete sheet piling are often left in place as part of the substructure of the building, or, like wood sheet piling, they may be pulled from the soil when their work is finished.

A *slurry wall* is a more complicated and expensive type of sheeting that is usually economical only if it becomes the permanent foundation wall of the building. The first steps in creating a slurry wall are to lay out the wall location on the surface of the ground with surveying instruments and to define the location and thickness of the wall with shallow poured concrete *guide walls* (Figures 2.14, 2.15). When the formwork has been removed from the guide walls, a special narrow *clamshell bucket,* mounted on a crane, is used to excavate the soil from between the guide walls. As the narrow trench deepens, the tendency of its earth walls to collapse is counteracted by filling the trench with a slurry of bentonite clay and water, which exerts a pressure back against the earth. The clamshell bucket is lowered and raised through the slurry to continue excavating the soil from the bottom of the trench until the desired depth has been reached,

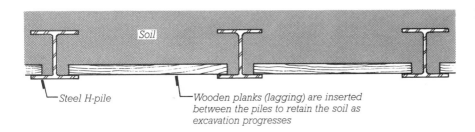

FIGURE 2.10
Soldier beams and lagging, seen in horizontal section.

Soil

Steel H-pile

Wooden planks (lagging) are inserted between the piles to retain the soil as excavation progresses

FIGURE 2.11
Soldier beams and lagging. Lagging planks are added at the bottom as excavation proceeds. The drill rig is boring a hole for a tieback to support the soldier beams. (*Courtesy of Franki Foundation Company*)

often a number of stories below the ground.

Meanwhile, workers have welded together cages of steel bars designed to reinforce the concrete wall that will replace the slurry in the trench. Steel tubes whose diameter corresponds to the width of the trench are driven vertically into the trench at predetermined intervals to divide it into sections of a size that can be rein-

forced and concreted conveniently. The concreting of each section begins with the lowering of a cage of reinforcing bars into the slurry. Then concrete is poured into the trench to fill it from the bottom up, using a funnel-and-tube arrangement called a *tremie*. As the concrete rises in the trench, it displaces the slurry, which is pumped out into holding tanks, where it is stored for reuse. After the

concrete hardens sufficiently, the vertical pipes on either side are withdrawn from the trench. This process is repeated for each section of the wall. When the concrete in the trenches has cured to its intended strength, earth removal begins inside the wall, which serves as sheeting for the excavation.

In addition to the *sitecast* slurry wall described in the preceding para-

TIMBER SHEET PILING

STEEL SHEET PILING

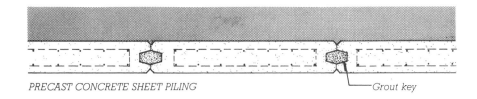
PRECAST CONCRETE SHEET PILING ——Grout key

FIGURE 2.12
Horizontal sections through three types of sheet piling. The shading represents the retained earth.

FIGURE 2.13
Drilling tieback holes for a wall of steel sheet piling. Notice the completed tieback connections to the waler in the foreground and background. The hole in the top of each piece of sheet piling allows it to be pulled from the ground by a crane. (*Courtesy of Franki Foundation Company*)

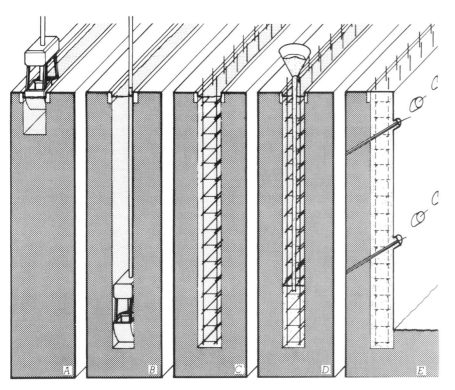

FIGURE 2.14
Steps in constructing a slurry wall. (*a*) The concrete guide walls have been installed, and the clamshell bucket begins excavating the trench through a bentonite clay slurry. (*b*) The trench is dug to the desired depth, with the slurry serving to prevent collapse of the walls of the trench. (*c*) A welded cage of steel reinforcing bars is lowered into the slurry. (*d*) The trench is concreted from the bottom up with the aid of a tremie. The displaced slurry is pumped from the trench, filtered, and stored for reuse. (*e*) The reinforced concrete wall is tied back as excavation progresses.

(a)

(b)

(c)

(d)

(e)

FIGURE 2.15
Constructing a slurry wall. (*a*) The guide
walls are formed and poured in a shallow
trench. (*b*) The narrow clamshell bucket
discharges a load of soil into a waiting
dump truck. Most of the trench is covered with wood pallets for safety. (*c*) A
detail of the narrow clamshell bucket
used for slurry wall excavation. (*d*)
Hoisting a reinforcing cage from the area
where it was assembled, getting ready to
lower it into the trench. The depth of the
trench and height of the slurry wall are
equal to the height of the cage, which is
about four stories for this project. (*e*)
Concreting the slurry wall with a tremie.
The pump just to the left of the tremie
removes slurry from the trench as concrete is added. (*Photos* b, c, *and* d *courtesy
of Franki Foundation Company. Photos* a
and e *courtesy of Soletanche*)

graphs, *precast* concrete slurry walls
are also built. The slurry for precast
walls is a mixture of water, bentonite
clay, and portland cement. The wall is
cast in sections and pretensioned
(see Chapter 15) in a precasting
plant, then trucked to the construction site. Before each section is lowered by a crane into the slurry, its face
is coated with a compound that prevents the clay–cement slurry from
adhering to it (Figure 2.16). The sections are installed side by side in the
trench, joined by tongue-and-groove
edges or synthetic rubber gaskets.
After the cement has caused the
slurry to harden to a soil-like consistency, excavation can begin, with the
hardened slurry on the inside face of
the wall dropping away from the
coated surface as soil is removed. The
primary advantages of a precast slurry
wall over a sitecast one are better surface quality, more accurate wall alignment, a thinner wall (due to the
structural efficiency of prestressing),
and improved watertightness of the

FIGURE 2.16
**Workers apply a nonstick coating to a
section of precast concrete slurry wall as
it is lowered into the trench.** (*Courtesy of
Soletanche*)

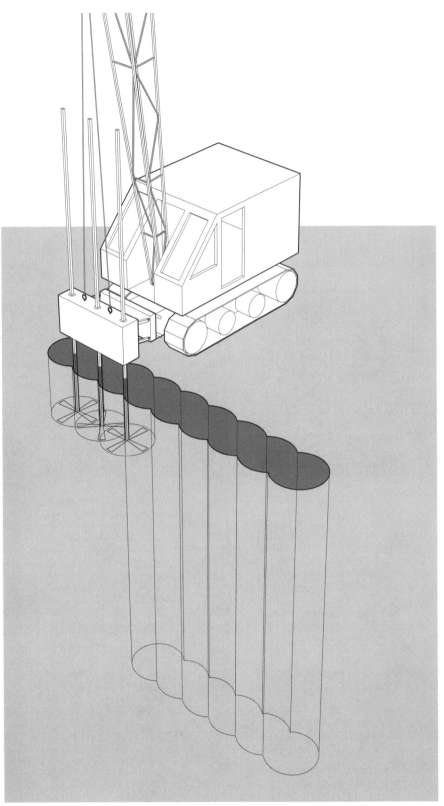

FIGURE 2.17
Soil mixing.

wall because of the continuous layer of solidified clay outside.

Soil Mixing

Soil mixing is a technique of adding a modifying substance to soil and blending it in place by means of paddles rotating on the end of a shaft (Figure 2.17). This technique has several different applications, one of which is to remediate soil contaminated with a chemical or biological substance by blending it with a chemical that renders it harmless. Another is to mix portland cement and water with a soil to create a cylinder of low-strength concrete in the ground. A series of these cylinders in a line becomes a concrete wall that can serve as a cutoff wall against water penetration and, in some cases, as sheeting for an excavation.

Bracing

Soldier beams, sheet piling, and slurry walls all need to be braced against soil and water pressures as the excavation deepens (Figure 2.18). *Crosslot bracing* utilizes temporary steel wide-flange columns that are driven into the earth at points where braces will cross. As the earth is excavated down around the sheeting and the columns, tiers of horizontal bracing struts, usually of steel, are added to support *walers,* which are beams that span across the face of the sheeting. Where the excavation is too wide for crosslot bracing, sloping *rakers* are used instead, bearing against *heel blocks* or other temporary footings.

Both rakers and crosslot bracing, especially the latter, are a hindrance to the excavation process. A clamshell bucket on a crane must be used to remove the earth between the braces, which is much less efficient and much more costly than removing soil with a shovel dozer or backhoe in an open excavation.

Where subsoil conditions permit, *tiebacks* can be used instead of braces to support the sheeting while maintaining an open excavation. At each level of walers, holes are drilled at intervals through the sheeting and the surrounding soil into rock or a stratum of stable soil. Steel cables or tendons are then inserted into the holes, grouted to anchor them to the rock or soil, and stretched tight with hydraulic jacks (*posttensioned*) before they are fastened to the walers (Figures 2.19, 2.20).

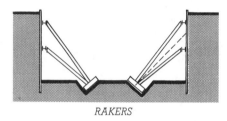

CROSSLOT BRACING

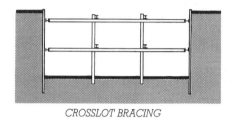

RAKERS

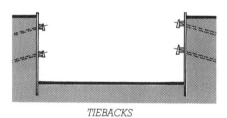

TIEBACKS

FIGURE 2.18
Three methods of bracing sheeting, drawn in cross section. The connection between the brace, raker, or tieback and the waler needs careful structural design. The broken line between rakers indicates the mode of excavation: The center of the hole is excavated first with sloping sides as indicated by the broken line. The heel blocks and uppermost tier of rakers are installed. As the sloping sides are excavated deeper, more tiers of rakers are installed. Notice how the tiebacks leave the excavation totally free of obstructions.

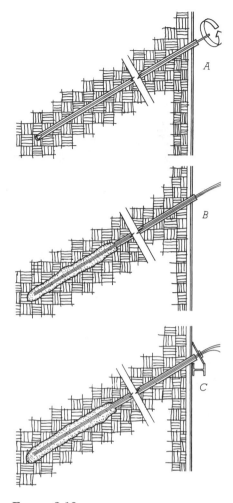

FIGURE 2.19
Three steps in the installation of a tieback to a soil anchor. (*a*) A rotary drill bores a hole through the sheeting and into stable soil or rock. A steel pipe casing keeps the hole from caving in where it passes through noncohesive soils. (*b*) Steel prestressing tendons are inserted into the hole and grouted under pressure to anchor them to the soil. (*c*) After the grout has hardened, the tendons are tensioned with a hydraulic jack and anchored to a waler.

(a)

FIGURE 2.20
Installing tiebacks. (a) Drilling through a slurry wall for a tieback. The ends of hundreds of completed tiebacks protrude from the wall. (b) Inserting prestressing tendons into the steel casing for a tieback. The apparatus in the center of the picture is for pressure-injecting grout around the tendons. (c) After the tendons have been tensioned, they are anchored to a steel chuck that holds them under stress, and the cylindrical hydraulic jack is moved to the next tieback. (d) Slurry walls and tiebacks used to support historic buildings around deep excavations for a station of the Paris Metro.
(*Photos* a, b, *and* c *courtesy of Franki Foundation Company. Photo* d *courtesy of Soletanche*)

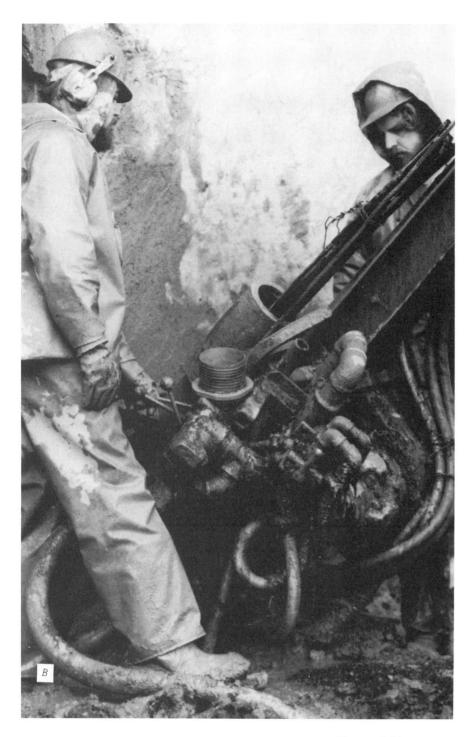

FIGURE 2.20 *(continued)*

FIGURE 2.20 (continued)

C

D

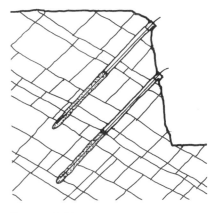

FIGURE 2.21
Rock anchors are similar to tiebacks, but are used to hold jointed rock formations in place around an excavation.

Excavations in fractured rock can often avoid sheeting altogether, either by injecting grout into the joints of the rock to stabilize it or by drilling into the rock and inserting *rock anchors* that fasten the blocks together (Figure 2.21).

Bracing and tiebacks in excavations are usually temporary. Their function is taken over permanently by the floor structure of the basement levels of the building, which is designed specifically to resist lateral loads from the surrounding earth as well as ordinary floor loads.

Dewatering

Specifications, codes, and common sense require that foundation construction be performed "in the dry," meaning an excavation that is free of water. Therefore, when excavation work is carried out below the water table in the surrounding soil, the site must be *dewatered*. The most common method of dewatering is to pump the water from *sumps*, which are pits in the bottom of the excavation. In certain types of soils, particularly sands and silt, the seepage into the sumps may soften the undisturbed soil on which the foundations bear. Therefore, it may be necessary to keep ground water from entering the

FIGURE 2.22
Two methods of keeping an excavation dry, viewed in cross section. The water sucked from well points draws the water table in the immediate vicinity below the level of the excavation. Watertight barrier walls work only if their bottom edges are inserted into an impermeable stratum that prevents water from working its way under the walls.

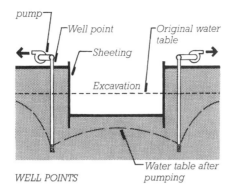

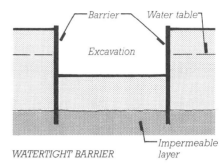

excavation, either by pumping water from the surrounding soil to depress the water table below the level of the bottom of the excavation or by erecting a watertight barrier around the excavation (Figure 2.22).

Well points are commonly used to depress the water table. These are vertical pieces of pipe with screened openings at the foot that keep out soil particles while allowing water to enter. Closely spaced well points are driven into the soil around the entire circumference of the excavation, and are connected to horizontal header pipes leading to pumps that continu-

ally suck water from the system and discharge it away from the building site. Once pumping has drawn down the water table in the area of the excavation, work can continue "in the dry" (Figure 2.23). For excavations deeper than the 20 feet (6 m) or so that can be drained by a suction

FIGURE 2.23
The pump in the foreground, drawing water through a header that connects to well points spaced every meter or so around the perimeter of the excavation, allows work to continue in the dry despite a high water table maintained by the lake just visible at the upper right. (*Courtesy of Griffin Dewatering Corporation*)

pump stationed at ground level, two rings of well points may be required, the inner ring being driven to a deeper level than the outer ring. Else a single ring of deep wells with submersible force pumps may have to be installed.

In some cases, the lowering of the water table by well points or deep wells can have serious adverse effects on neighboring buildings by causing consolidation and settling of soil under their foundations or by exposing untreated wood foundation piles, previously protected by total immersion in water, to decay. In these cases, a *watertight barrier wall* is used as an alternative to well points. A slurry wall makes an excellent watertight barrier. Sheet piling can also work, but tends to leak. A watertight barrier must resist the hydrostatic pressure of the surrounding water, which grows greater as the foundation grows deeper, so a strong system of bracing or tiebacks is required. A watertight barrier only works if it reaches into an impermeable stratum that lies beneath the water table; otherwise, water will flow beneath the barrier and up into the excavation.

FOUNDATIONS

It is convenient to think of a building as consisting of three major parts: the *superstructure,* which is the above-ground portion of the building; the *substructure,* which is the habitable below-ground portion; and the *foundations,* which are the components of the building that transfer its loads into the soil (Figure 2.24).

There are two basic types of foundations: shallow and deep. *Shallow foundations* are those that transfer the load to the earth at the base of the column or wall of the substructure. *Deep foundations* transfer the load at a point some distance below the substructure. Shallow foundations are generally less expensive than deep ones and can be used where suitable soil is found at the level at the bottom

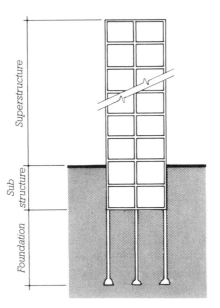

FIGURE 2.24
Superstructure, substructure, and foundation. The substructure in this example contains two levels of basements, and the foundation consists of bell caissons.

of the substructure, whether this be a few feet or several stories below the surface. Deep foundations, either piles or caissons, penetrate through upper layers of incompetent soil in order to reach competent bearing soil or rock deeper within the earth.

The primary factors that affect the choice of a foundation type for a building are:

• Subsurface soil and ground water conditions

• Structural requirements, including foundation loads, configurations, and depth

Secondary factors that may be important include:

• Construction methods, including access and working room

• Environmental factors, including noise, traffic, and disposal of earth and water

• Codes and regulations

• Impact on adjacent property

• Time available for construction

• Construction risks

The foundation engineer is responsible for assessing these factors and, working together with other members of the design and construction team, selecting the most suitable foundation system.

Shallow Foundations

Most shallow foundations are simple concrete *footings.* A *column footing* is a square block of concrete, with or without steel reinforcing, that accepts the concentrated load placed on it from above by a building column and spreads this load across an area of soil large enough that the allowable bearing stress of the soil is not exceeded. A *wall footing* or *strip footing* is a continuous strip of concrete that serves the same function for a loadbearing wall (Figures 2.25, 2.26).

To minimize settlement, footings are placed on undisturbed soil. (The only exception to this rule is that some buildings are built on *engineered fill,* which is earth that has been deposited in thin layers at a controlled moisture content and compacted in accordance with detailed procedures that assure a known degree of long-term stability.) The last few inches of soil in the excavation for a footing are removed with hand tools to avoid the loosening of the bearing layer of soil that would be caused by digging machinery.

If the horizontal projection of the footing beyond the face of the wall or column is less than half the vertical dimension of the footing block, steel reinforcing is not required. If the footing needs to project farther than this to achieve the necessary bearing area, reinforcing bars are placed near the bottom of the footing to accept the tensile stresses that occur in this area, as illustrated in Chapter 14.

Footings appear in many forms in different foundation systems. In climates with little or no ground frost, a concrete *slab on grade* with thickened edges is the least expensive foundation and floor system one can use and is applicable to one- and two-story buildings of any type of construction (see Chapter 14 for further information on slabs on grade). For floors raised above the ground, either over a *crawlspace* or a *basement,* support is provided by concrete or masonry foundation walls supported on concrete strip footings (Figure 2.27). When building on slopes, it is often necessary to step the footings to maintain the required depth of foot-

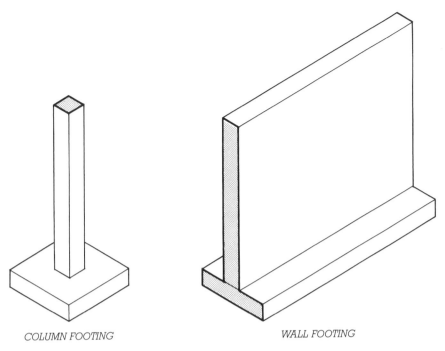

COLUMN FOOTING   WALL FOOTING

FIGURE 2.25
A column footing and a wall footing of concrete. The steel reinforcing bars have been omitted for clarity.

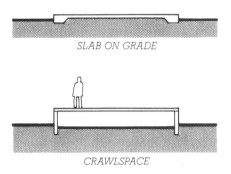

SLAB ON GRADE

CRAWLSPACE

BASEMENT

FIGURE 2.27
Three types of substructures using simple wall footings. The slab on grade is most economical under many circumstances, especially where the water table lies near the surface of the ground. A crawlspace is often used under a floor structure of wood or steel, and gives much better access to underfloor piping and wiring than a slab on grade.

FIGURE 2.26
These concrete foundation walls for an apartment building, with their steel formwork not yet stripped, rest on wall footings. For more extensive illustrations of wall and column footings, see Figures 14.5, 14.11, and 14.13. (*Courtesy of Portland Cement Association, Skokie, Illinois*)

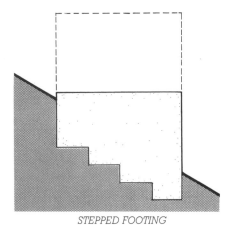

STEPPED FOOTING

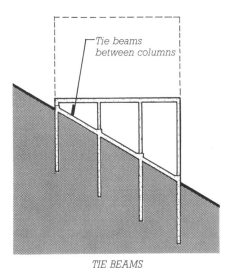

TIE BEAMS

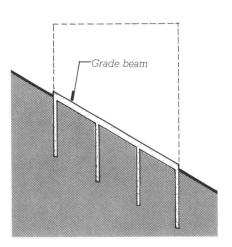

ing at all points around the building (Figure 2.28). If soil conditions or earthquake precautions require it, column footings on steep slopes are linked together with reinforced concrete *tie beams* to avoid possible differential slippage between footings.

Footings cannot legally extend beyond a property line, even for a building built tightly against it. If the outer toe of the footing were simply cut off at the property line, the footing would be eccentrically loaded by the column or wall and would tend to rotate and fail. *Combined footings* and *cantilever footings* solve this problem by tying the footings for the outside row of columns to those of the next row in such a way that any rotational tendency is neutralized (Figure 2.29).

In situations where the allowable bearing capacity of the soil is low in relation to the weight of the building,

column footings may become large enough that it is more economical to merge them into a single *mat* or *raft foundation* that supports the entire building. Mats for very tall buildings are often 6 feet (1800 mm) or more thick and are heavily reinforced (Figure 2.30).

Where the bearing capacity of the soil is low and settlement must be carefully controlled, a *floating foundation* is sometimes used. A floating foundation is essentially the same as a mat, but is placed beneath a building at a depth such that the weight of the soil removed from the excavation is equal to the weight of the building. Roughly speaking, one story of excavated soil weighs about the same as five to eight stories of superstructure, depending on the density of the soil and the construction of the building (Figure 2.31).

FIGURE 2.28
Foundations on sloping sites, viewed in a cross section through the building. The broken line indicates the outline of the superstructure. Wall footings are stepped to maintain the necessary distance between the bottom of the footing and the surface of the ground. Separate column foundations, whether caissons, as shown here, or column footings, are often connected with reinforced concrete tie beams to reduce differential movement between the columns. A grade beam differs from a tie beam by being reinforced to distribute the continuous load from a loadbearing wall to separate foundations.

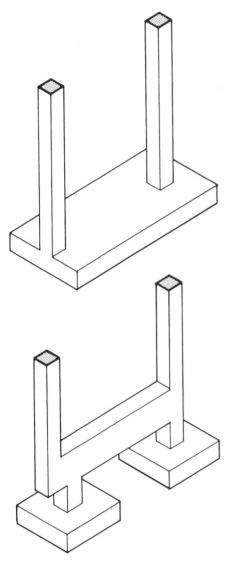

FIGURE 2.30
Pouring a large foundation mat. Six truck-mounted pumps receive concrete from a continuous procession of transit-mix concrete trucks and deliver this concrete to the heavily reinforced mat. Concrete placement continues nonstop around the clock until the mat is finished to avoid "cold joints" between hardened concrete and fresh concrete. The soil around this excavation is supported with a sitecast concrete slurry wall. Most of the slurry wall is tied back, but a set of rakers is visible at the lower right. (*Courtesy of Schwing America, Inc.*)

FIGURE 2.29
Either a combined footing (top) or a cantilevered footing (bottom) is used when columns must abut a property line. By combining the foundation for the column against the property line, at the left, with the foundation for the next interior column to the right, in a single structural unit, a balanced footing design can be achieved. The concrete reinforcing steel has been omitted from these drawings for the sake of clarity.

FIGURE 2.31
A cross section through a building with a floating foundation. The building weighs approximately the same as the soil excavated for the substructure, so the stress in the soil beneath the building has not changed.

Deep Foundations

Caissons

A *caisson* (Figure 2.32) is similar to a column footing in that it spreads the load from a column over a large enough area of soil that the allowable stress in the soil is not exceeded. It differs from a column footing in that it reaches through strata of unsatisfactory soil beneath the substructure of a building until it reaches a satisfactory bearing stratum, such as rock, dense sands and gravels, or firm clay. A caisson is constructed by drilling or hand-digging a hole, belling (flaring) the hole out at the bottom as necessary to achieve the required bearing area, and filling the hole with concrete. Large auger drills (Figures 2.33, 2.34) are used for drilling caissons; hand excavation is used only if the soil is too bouldery for the drill. A temporary cylindrical steel casing is usually lowered around the drill as it progresses, to support the soil around the hole. When firm bearing is reached, the bell, if required, is created at the bottom of the shaft either by hand excavation or by a special

FIGURE 2.34
For cutting through hard material, the caisson drill is equipped with a carbide-toothed coring bucket. (*Courtesy of Calweld, Inc.*)

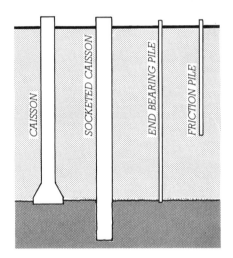

FIGURE 2.32
Deep foundations. The caissons are concrete cylinders poured into drilled holes. They reach through weaker soil (light shading) to bear on competent soil beneath. The end bearing caisson at the left is belled as shown when additional bearing capacity is required. The socketed caisson is drilled into a hard stratum and transfers its load primarily by friction between the soil or rock and the sides of the caisson. Piles are driven into the earth. End bearing piles act in the same way as caissons. The friction pile derives its load-carrying capacity from friction between the soil and the sides of the pile.

FIGURE 2.33
A 6-foot (1828-mm) diameter auger on a telescoping 70-foot (21-m) bar brings up a load of soil from a caisson hole. The auger will be rotated rapidly to spin off the soil before being reinserted in the hole. (*Courtesy of Calweld, Inc.*)

belling bucket on the drill (Figure 2.35). The bearing surface of the soil at the bottom of the hole is then inspected to be sure it is of the anticipated quality, and the hole is filled with concrete, withdrawing the casing as the concrete rises. Reinforcing is seldom used in the concrete except near the top of the caisson, where it joins the columns of the superstructure.

Caissons are large, heavy-duty units. Their shaft diameters range up to 8 feet (2400 mm) and more, from a minimum of 18 inches (460 mm) for a mechanical drill with a belling bucket, or 32 inches (800 mm) for hand excavation. *Belled caissons* are practical only where the bell can be excavated in a cohesive soil (such as clay) that can retain its shape until concrete is poured, and where the bearing stratum is impervious to the passage of water, to prevent flooding of the hole.

A *socketed caisson* (Figure 2.32) is drilled into rock at the bottom, rather than belled. Its bearing capacity comes not only from its end bearing, but from the frictional forces between the sides of the caisson and

> **We must never trust too hastily to any ground. . . . I have seen a tower at Mestre, a place belonging to the Venetians, which, in a few years after it was built, made its way through the ground it stood upon . . . and buried itself in earth up to the very battlements.**
>
> **Leon Battista Alberti, 1404–1472**

FIGURE 2.35
The bell is formed at the bottom of the caisson shaft by a belling bucket with retractable cutters. The example shown here is for an 8-foot (2.44-m) shaft, and makes a bell 21 feet (6.40 m) in diameter. (*Courtesy of Calweld, Inc.*)

the rock as well. Figure 2.36 shows the installation of a *rock caisson* or *drilled-in caisson,* a special type of socketed caisson with a steel H-section core.

The term "caisson" originally referred to a watertight chamber within which foundation work could be carried out underwater or below the water table. In common usage, it has also come to mean the type of deep foundation unit described in the preceding paragraphs. To avoid confusion, the term *drilled pier* is sometimes used in the literature to describe the foundation unit.

Piles

A *pile* is distinguished from a caisson by being driven into the earth rather than drilled and poured. The simplest kind of pile is a *timber pile,* a tree trunk with its branches and bark removed; it is supported small end down in a *piledriver* and beaten into the earth with repeated blows of a very heavy mechanical hammer (called a *pile hammer*).

If a pile is driven until its tip encounters firm resistance from rock

(a)

(b)

FIGURE 2.36
Installing a rock caisson foundation. (*a*) The shaft of the caisson has been drilled through softer soil to the rock beneath, and cased with a steel pipe. A churn drill is being lowered into the casing to begin advancing the hole into the rock. (*b*) When the hole has penetrated the required distance into the rock stratum, a heavy steel H-section is lowered into the hole and suspended on steel channels across the mouth of the casing. The space between the casing and the H-section is then filled with concrete, producing a caisson with a very high load-carrying capacity because of the composite structural action of the steel and the concrete. (*Courtesy of Franki Foundation Company*)

or dense sands and gravels, it is an *end bearing pile.* If it is driven only into softer material, without encountering a firm bearing layer, it will still develop a considerable load-carrying capacity through the frictional resistance between the sides of the pile and the soil through which it is driven; in this case it is known as a *friction pile* (Figure 2.32).

Piles are generally driven closely together in clusters that contain from two to twenty-five piles. The piles in each cluster are later joined at the top by a reinforced concrete *pile cap,* which distributes the load of the column or wall equally among the piles (Figures 2.37, 2.38).

End bearing piles are used on sites where a firm bearing stratum exists at a depth that can be reached by piles. End bearing piles work exactly the same as caissons, but can be used when noncohesive soils or subsurface water conditions preclude the installation of caissons. Each pile is driven "to refusal," the point at which little additional penetration is made with continuing blows of the hammer, indicating that the pile has

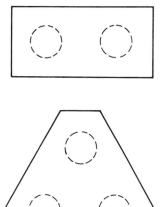

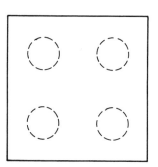

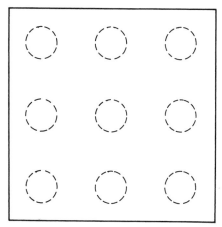

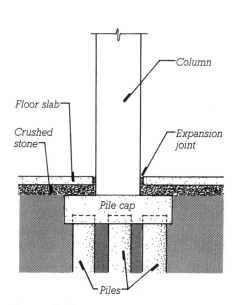

FIGURE 2.38
An elevation view of a pile cap, column, and floor slab.

FIGURE 2.37
Clusters of two, three, four, and nine piles with their concrete caps, viewed from above. The caps are reinforced to transmit column loads equally into all the piles in the cluster, but the reinforcing steel has been omitted here for the sake of clarity.

penetrated the bearing layer and is firmly embedded in it.

Friction piles are used when no firm bearing stratum exists at a reasonable depth, or when a high water table would make footings or caissons prohibitively expensive. Friction piles work best in silty, clayey, and sandy soils, and are driven to a predetermined depth or resistance rather than to refusal. Clusters of friction piles have the effect of distributing a column load to a large volume of soil around and below the cluster, at stresses that lie safely within the capability of the soil (Figure 2.39).

The loadbearing capacities of piles are calculated in advance based on soil test results and the properties of the piles and piledriver. To verify the correctness of the calculation, test piles are often driven and loaded on the building site before foundation work begins.

Where piles are used to support loadbearing walls, reinforced concrete *grade beams* are constructed between the pile caps to transmit the wall loads to the piles (Figure 2.40). Grade beams are also used with caisson foundations for the same purpose.

Pile Driving

Pile hammers are massive weights lifted by the energy of steam, compressed air, compressed hydraulic fluid, or a diesel explosion, then dropped against a block that is in firm contact with the top of the pile. Single-acting hammers fall by gravity alone, while double-acting hammers are forced downward by reverse application of the energy source that lifts the hammer. The hammer travels on tall vertical rails called *leads* (pronounced "leeds") at the front of a piledriver (Figure 2.41). It is first

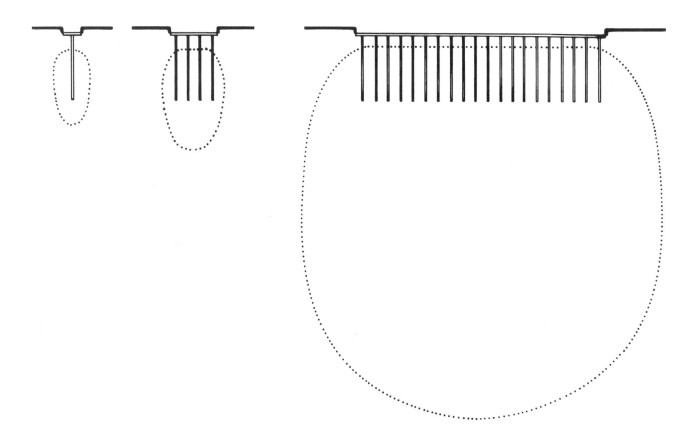

FIGURE 2.39
A single friction pile (left) transmits its load into the earth as an equal shear pressure along the bulb profile indicated by the dotted line. As the size of the pile cluster increases, the piles act together to create a single, larger bulb of higher pressure that reaches deeper into the ground. A building with many closely spaced clusters of piles (right) creates a very large, deep bulb. Care must be taken to ensure that large pressure bulbs do not overstress the soil or cause excessive settlement of the foundation. The settlement of a large group of friction piles in clay, for example, will be considerably greater than that for a single pile.

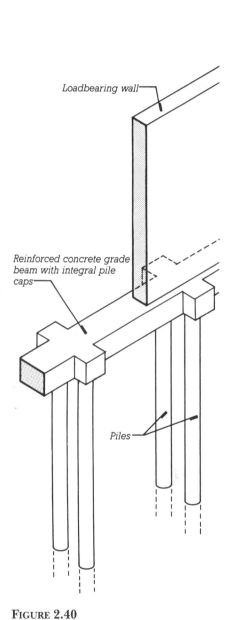

FIGURE 2.40
If they must support a loadbearing wall, pile caps are joined by a grade beam. The reinforcing in the grade beam is similar to that in any ordinary continuous concrete beam, and has been omitted for clarity. A concrete loadbearing wall can generally be reinforced to act as its own grade beam.

FIGURE 2.41
A steam piledriver hammers a precast concrete pile into the ground. The pile is supported by the vertical structure (leads) of the piledriver, and driven by a heavy piston mechanism that follows it down the leads as it penetrates deeper into the soil.
(*Courtesy of Lone Star/San-Vel Concrete*)

hoisted up the leads to the top of each pile as driving commences, then follows the pile down as it penetrates the earth. The piledriver mechanism includes lifting machinery to raise each pile into position before driving.

In certain types of soils, piles can be driven more efficiently by vibration than by a heavy hammer, using a special vibratory hammer mechanism. When piles must be driven beneath an existing building to increase the capacity of its foundations, as is often necessary when increasing the height of a building, they are forced into the soil by hydraulic jacks acting against the weight of the building.

Pile Materials

Piles may be made of timber, steel, sitecast or precast concrete, and various combinations of these materials (Figure 2.42). Timber piles have been used since Roman times, at which time they were driven by large mechanical hammers hoisted by muscle power. Their main advantage is that they are economical for lightly loaded foundations. On the minus side, they cannot be spliced during driving, and are therefore limited in length to the length of available tree trunks, approximately 60 feet (18 m) maximum. Unless pressure treated with a wood preservative or completely submerged below the water

table, they will decay. Relatively small hammers must be used in driving timber piles to avoid splitting them. Capacities of timber piles lie in the range of 10 to 35 tons each (9000 to 32,000 kg).

Two forms of steel piles are used, H-piles and pipe piles. *H-piles* are special hot-rolled, wide-flange sections, 8 to 14 inches deep (200 to 360 mm), that are approximately square in cross section. They are used mostly in end bearing applications. H-piles displace relatively little soil during driving. This minimizes upward heaving of adjacent structures, which is sometimes a problem on urban sites with large numbers of piles. They can be

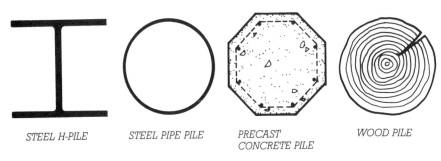

STEEL H-PILE   STEEL PIPE PILE   PRECAST CONCRETE PILE   WOOD PILE

FIGURE 2.42
Cross sections of common types of piles. Precast concrete piles may be square or round instead of the octagonal cross section shown, and may be hollow in the larger sizes.

brought to the site in any convenient lengths, welded together as driving progresses to form any necessary length of pile, and cut off with an oxy-acetylene torch when the required depth is reached. Cut-off ends can then be welded onto other piles to avoid waste. Corrosion can be a problem in some soils, however, and, unlike closed pipe piles and hollow precast concrete piles, H-piles cannot be inspected after driving to be sure they are straight and undamaged. Allowable loads on H-piles run from 30 to 120 tons (27,000 to 110,000 kg).

Steel *pipe piles,* with diameters of 8 to 16 inches (200 to 400 mm), may be driven with the lower end either open or closed with a heavy steel plate. An open pile is easier to drive than a closed one, but its interior must be cleaned of soil and inspected before being filled with concrete, whereas a closed pile can be inspected and concreted immediately after driving. Pipe piles are stiff and can carry heavy loads (50 to 150 tons, or 45,000 to 125,000 kg). They displace relatively large amounts of soil during driving, which can lead to upward heaving of nearby soil and buildings. The larger sizes of pipe piles require a very heavy hammer for driving.

Precast concrete piles are square, octagonal, or round in section, and in the larger sizes often have open cores to allow inspection (Figures 2.42–2.44). Most are prestressed, but some for smaller buildings are merely reinforced. Sizes range typically from 10 to 16 inches (250 to 400 mm), and bearing capacities from 60 to 120 tons (55,000 to 110,000 kg). Advantages of precast piles include high load capacities, an absence of corrosion or decay problems, and, in most situations, a relative economy of cost. Precast piles must be handled carefully to avoid bending and cracking before installation. Splices between lengths of precast piling can be made effectively with mechanical fastening devices that are cast into the ends of the sections.

FIGURE 2.43
Precast, prestressed concrete piles. Lifting loops are cast into the sides of the piles as crane attachments for hoisting them into a vertical position. (*Courtesy of Lone Star/San-Vel Concrete*)

FIGURE 2.44
A driven cluster of six precast concrete piles, ready for cutting off and capping. (*Photo by Alvin Ericson*)

A *sitecast concrete pile* is made by driving a hollow steel shell into the ground and filling it with concrete. The shell is sometimes corrugated to increase its stiffness; if the corrugations are circumferential, a heavy steel *mandrel* (a tight-fitting liner) is inserted in the shell during driving to protect the shell from collapse, then withdrawn before concreting. Some shells with longitudinal corrugations are stiff enough that they do not require mandrels. Some types of mandrel-driven piles are limited in length, and the larger diameters of sitecast piles (up to 16 inches, or 400 mm) can cause ground heaving. Load capacities range from 50 to 120 tons (45,000 to 110,000 kg). The primary reason to use sitecast concrete piles is their economy.

There are many proprietary systems of sitecast concrete piles, each with various advantages and disadvantages (Figures 2.45, 2.46). *Pressure-injected footings* (Figure 2.46) are interesting because they share characteristics of piles, piers, and footings and because they are highly resistant to uplift forces.

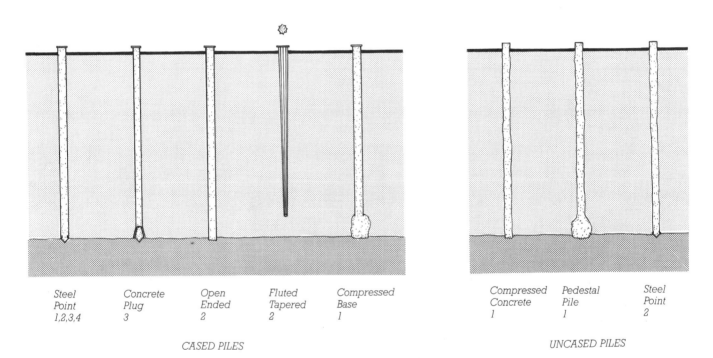

| Steel Point 1,2,3,4 | Concrete Plug 3 | Open Ended 2 | Fluted Tapered 2 | Compressed Base 1 |

CASED PILES

| Compressed Concrete 1 | Pedestal Pile 1 | Steel Point 2 |

UNCASED PILES

FIGURE 2.45

Some proprietary types of sitecast concrete piles. All are cast into steel casings that have been driven into the ground; the uncased piles are made by withdrawing the casing as the concrete is poured, and saving it for subsequent reuse. The numbers refer to the methods of driving that may be used with each: 1. Mandrel driven. 2. Driven from the top of the tube. 3. Driven from the bottom of the tube to avoid buckling it. 4. Jetted. Jetting is accomplished by advancing a high-pressure water nozzle ahead of the pile to wash the soil back alongside the pile to the surface. Jetting has a tendency to disrupt the soil around the pile, so it is not a favored method of driving under most circumstances.

(a)   (b)   (c)

(d)   (e)   (f)

FIGURE 2.46
Steps in the construction of a proprietary pressure-injected, bottom-driven concrete pile footing. (*a*) A charge of a concrete mix that is very low in moisture is inserted into the bottom of the steel drive tube at the surface of the ground and compacted into a sealing plug with repeated blows of a drop hammer. (*b*) As the drop hammer drives the sealing plug into the ground, the drive tube is pulled along by the friction between the plug and the tube. (*c*) When the desired depth is reached, the tube is held and a bulb of concrete is formed by adding small charges of concrete and driving the concrete out into the soil with the drop hammer. The bulb provides an increased bearing area for the pile and strengthens the bearing stratum by compaction. (*d, e*) The shaft is formed of additional compacted concrete as the tube is withdrawn. (*f*) Charges of concrete are dropped into the tube from a special bucket supported on the leads of the driving equipment. (*Courtesy of Franki Foundation Company*)

SEISMIC BASE ISOLATION

In areas where strong earthquakes are common, buildings are sometimes founded on *base isolators* that allow the ground to move laterally back and forth beneath the building while the substructure and superstructure remain more or less at rest and free from damage. A frequently used type of base isolator is a multilayered sandwich of rubber and steel plates (Figure 2.47). The rubber layers deform in shear to allow the rectangular isolator to become a parallelogram in cross section in response to relative motion between the ground and the building. A lead core deforms enough to allow this motion to occur, provides damping action, and keeps the layers of the sandwich aligned.

UNDERPINNING

Underpinning is the process of strengthening and stabilizing the foundations of an existing building. It may be required for any of a number of reasons: The existing foundations may never have been adequate to carry their loads, leading to excessive settlement of the building. A change in building use or additions to the building may overload the existing foundations. New construction near a building may disturb the soil around its foundations or require that its foundations be carried deeper. Whatever the cause, underpinning is a slow, expensive, highly specialized task that is seldom the same for any two buildings. Three different alternatives are available when foundation capacity needs to be increased: The foundations may be enlarged; new, deep foundations can be inserted under shallow ones to carry the load to a deeper, stronger stratum of soil; or the soil itself can be strengthened by grouting or by chemical treatment. Figures 2.48 and 2.49 illustrate in diagrammatic form some concepts of underpinning.

RETAINING WALLS

A *retaining wall* holds soil back to create an abrupt change in the elevation of the ground. A retaining wall must resist the pressure of the earth that bears against it on the uphill side. Retaining walls may be made of masonry, preservative-treated wood, coated or galvanized steel, precast concrete, or, most commonly, sitecast concrete.

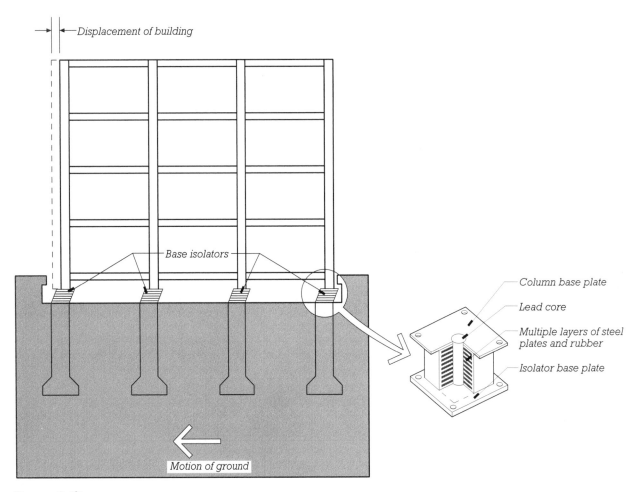

FIGURE 2.47
Base isolation.

The structural design of retaining walls is complicated by such factors as the height of the wall, the character of the soil behind the wall, the presence or absence of ground water behind the wall, any structures whose foundations apply pressure to the soil behind the wall, and the character of the soil beneath the base of the wall, which must support the footing that keeps the wall in place. The rate of structural failure in retaining walls is high relative to the rate of failure in other types of structures, and may occur through fracture of the wall, overturning of the wall due to soil failure, lateral sliding of the wall, or undermining of the wall by flowing

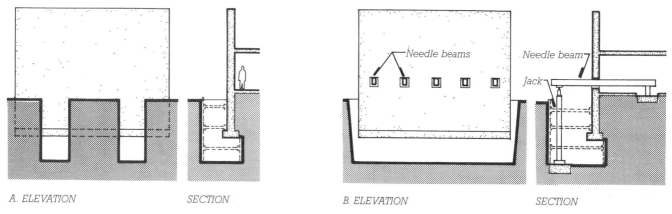

A. ELEVATION   SECTION   B. ELEVATION   SECTION

FIGURE 2.48
Two methods of supporting a building while carrying out underpinning work beneath its foundation, each shown in both elevation and section. (*a*) Trenches are dug beneath the existing foundation at intervals, leaving the majority of the foundation supported by the soil. When new footings have been completed in the trenches, another set of trenches may be dug to add new footings to more of the foundation, and so on. (*b*) The foundations of an entire wall can be exposed at once by needling, in which the wall is supported temporarily on *needle beams* threaded through holes cut in the wall. After underpinning has been accomplished, the jacks and needle beams are removed and the trench is backfilled.

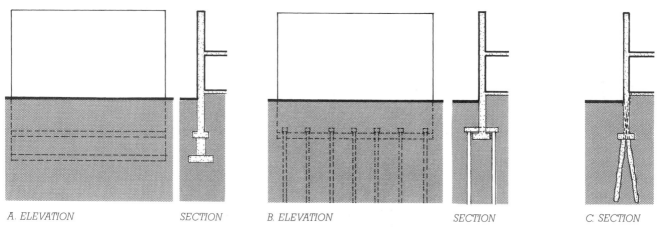

A. ELEVATION   SECTION   B. ELEVATION   SECTION   C. SECTION

FIGURE 2.49
Three types of underpinning. (*a*) A new foundation wall and footing are constructed beneath the existing foundation. (*b*) New piles or caissons are constructed on either side of the existing foundation. (*c*) Concrete mini-piles are cast into holes drilled diagonally through the existing foundation. Mini-piles do not generally require excavation or temporary support of the building.

ground water (Figure 2.50). Careful engineering design and site supervision are crucial to the success of a retaining wall.

There are many ways of building retaining walls. For walls less than 3 feet (900 mm) in height, simple, unreinforced walls of various types are often appropriate (Figure 2.51). For taller walls, and ones subjected to unusual loadings or ground water, the type most frequently employed today is the *cantilevered concrete retaining wall,* two examples of which are shown in Figure 2.52. The shape and reinforcing of a cantilevered wall can be custom designed to suit almost any situation.

Earth reinforcing (Figure 2.53) is an economical alternative to conventional retaining walls in many situations. Soil is compacted in thin layers, alternating with strips or meshes of galvanized steel, polymer fibers, or glass fibers, which stabilize the soil in much the same manner as the roots of plants.

WATERPROOFING AND DRAINAGE

The substructure of a building is subject to penetration of ground water, especially if it lies below the water table. Even if the concrete walls of a basement are perfectly constructed, water can work its way to the inside through the microscopic pores in the concrete, and basement walls are usually far from perfect, being riddled with shrinkage cracks, form tie holes, and joints between pours of concrete.

There are two fundamental approaches to waterproofing: waterproof membranes and drainage. In theory, either will work independently of the other, but, in practice, they are usually combined. *Waterproof membranes* may be made of plastic, asphaltic, or synthetic rubber sheets; of asphaltic mastics or coatings; of special cementitious plasters; or of bentonite clay. Cementitious plasters

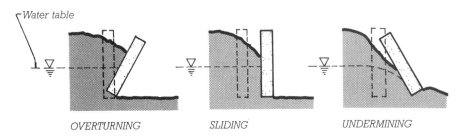

OVERTURNING   SLIDING   UNDERMINING

FIGURE 2.50
Three failure mechanisms in retaining walls.

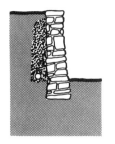
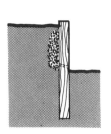
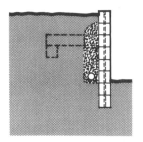

STONE GRAVITY WALL   VERTICAL TIMBER CANTILEVERED WALL   HORIZONTAL TIMBER WALL WITH DEADMEN

FIGURE 2.51
Three types of simple retaining walls, usually used for heights not exceeding 3 feet (900 mm). The deadmen in the horizontal timber wall are timbers embedded in the soil behind the wall and connected to it with timbers inserted into the wall at right angles. The timbers, which should be pressure treated with a wood preservative, are held together with very large spikes or with steel reinforcing bars driven into drilled holes. The crushed stone drainage trench behind each wall is important as a means of relieving water pressure against the wall to prevent wall failure. With proper engineering design, any of these types of construction can also be used for taller retaining walls.

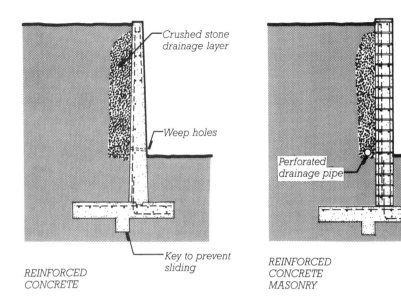

REINFORCED
CONCRETE

REINFORCED
CONCRETE
MASONRY

FIGURE 2.52
Cantilevered retaining walls of concrete and concrete masonry. The footing is shaped to resist sliding and overturning, and drainage behind the wall reduces the likelihood of undermining. The pattern of steel reinforcing (broken lines) is designed to resist the tensile forces in the wall.

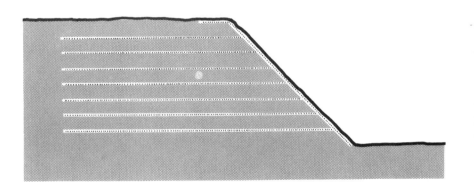

FIGURE 2.53
Earth reinforcing. The embankment in the top section is placed in layers of earth alternated with layers of synthetic mesh fabric. The retaining wall in the lower section is made of precast concrete panels fastened to long galvanized steel straps that run back into the soil.

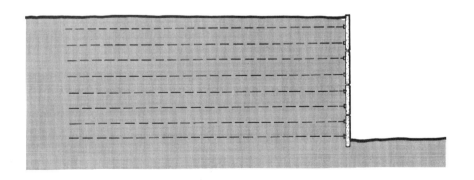

can generally be applied either inside or outside the foundation, with some preference for the inside because in this location the membrane may be inspected and repaired as needed. The major problem with plasters is that they are brittle and will crack and leak if the foundation cracks. Sheets, mastics, coatings, and clay membranes must generally be applied to the outside of the wall so that they can be supported against water pressure by the wall (Figure 2.54). Membranes outside the wall cannot be reached for repair, so the installation must be inspected with extreme care before the excavation around the foundation wall is backfilled with soil. Once successfully installed and backfilled, such membranes are shielded from weather and will last for a very long time, provided they are not subsequently torn by foundation cracking. The safest of all membranes is perhaps *bentonite clay,* which, when wetted, swells to several times its dry volume and becomes practically impervious to the passage of water. Moist bentonite clay is sufficiently flexible to adjust even to fairly large movements in the wall it protects. The clay is either sprayed onto the wall as a slurry, or applied in the form of corrugated cardboard sheets whose cells are filled with dry clay. When the panels are wetted by ground water, the clay expands to form a continuous membrane, and the cardboard disintegrates.

At movement joints and joints between pours of concrete in basement walls, structural movement is likely to cause cracking and leakage in most types of waterproof membranes. Preformed synthetic rubber *waterstops* are cast into the mating concrete edges at these joints to maintain a seal despite subsequent movement (Figures 2.55, 2.56).

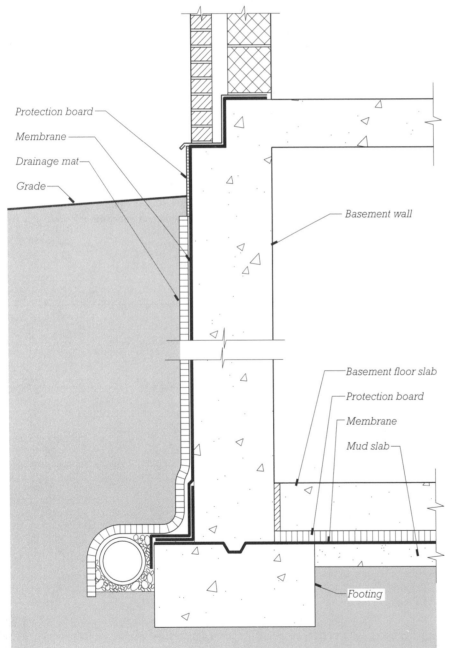

Protection board
Membrane
Drainage mat
Grade
Basement wall
Basement floor slab
Protection board
Membrane
Mud slab
Footing

FIGURE 2.54
A diagrammatic representation of the placement of sheet membrane waterproofing around a basement.

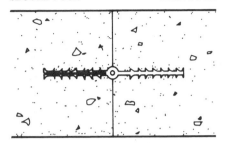

SECOND POUR   FIRST POUR

FIGURE 2.55
A synthetic rubber waterstop is used to seal against water penetration at movement joints and at joints between pours of concrete in a foundation. The type shown here is split on one side so its halves can be placed flat against the end of the formwork. After the formwork has been removed from the first pour, the split halves are reunited before the next pour.

FIGURE 2.56
Installing waterstops in a building foundation. (*a*) Workers unroll the waterstop around a concrete form, where it will be inserted in the top of the footing. (*b*) The top half of the waterstop is left projecting above the concrete of the footing and will seal the joint between the footing and the concrete wall that will be poured atop the footing. (*c*) A waterstop ready for the next pour of a concrete wall, as diagrammed in Figure 2.55.
(*Courtesy of Vulcan Metal Products, Inc., Birmingham, Alabama*)

(*c*)

GEOTEXTILES

Flexible fabrics made of chemically inert plastics are highly resistant to deterioration in the soil and are increasingly used for a variety of purposes relating to foundations of buildings. One use of these *geotextiles* is soil reinforcement. The accompanying text describes soil reinforcement with a plastic mesh fabric or grid being used to create a retaining wall. The same mesh may be utilized in layers to stabilize engineered fill beneath a shallow footing (Figure A). In a similar way, mesh geotextiles are employed to stabilize marginal soils under driveways, roads, and airport runways, acting very much as the roots of plants do in preventing the movement of soil particles. Another geotextile that was introduced earlier in this chapter is drainage matting, an open matrix of plastic filaments with a feltlike filter fabric on one side to keep soil particles from entering the matrix. In addition to providing free drainage around foundation walls, drainage matting is often used beneath the soil in the bottoms of planter boxes and under heavy paving tiles on rooftop terraces, where it maintains a free passage for the drainage of water above a waterproof membrane. Synthetic filter fabrics are used alone to wrap over and around subterranean crushed stone drainage layers such as the one frequently used around a foundation drain. In this position, they keep the stone and pipe from becoming clogged with soil particles. Similar fabrics are supported vertically above ground on stakes as temporary barriers to filter soil out of water that runs off a construction site, thus preventing contamination of lakes and streams. Special geotextiles are manufactured to stake down on freshly cut slopes to prevent soil erosion and encourage revegetation; some of these are designed to decay and disappear into the soil as plants take over the function of slope stabilization. Another type of geotextile is used for weed control in landscaped beds, where it allows rainwater to penetrate the soil but blocks sunlight, preventing weeds from sprouting.

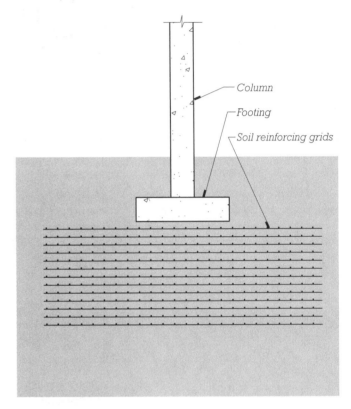

Column
Footing
Soil reinforcing grids

FIGURE A

Drainage is more secure than membranes in keeping a basement dry and has the advantage of preventing the buildup of potentially destructive water pressure against basement walls and slabs. A slab, in particular, is many times less expensive if it does not have to be reinforced against hydrostatic pressure. Several schemes for basement drainage are shown in Figures 2.57 and 2.58. The perforated pipes are at least 4 inches (100 mm) in diameter, and serve to maintain an open channel in the crushed stone bed through which water can flow by gravity either "to daylight" below the building on a sloping site or to a sump pit that is periodically pumped dry. With a properly installed drainage system, the only function of membrane waterproofing is to keep soil humidity from permeating the basement walls.

Small residential basements in dry locations are often coated on the outside with an inexpensive asphalt *dampproofing* that is not a reliable barrier to the passage of water, but will prevent most moisture from coming through the wall. Every residential basement should have an external drain of the type shown in Figure 2.57, with the perforated pipe running to daylight or to a sump and pump.

BASEMENT INSULATION

Comfort requirements, heating fuel efficiency, and building codes often

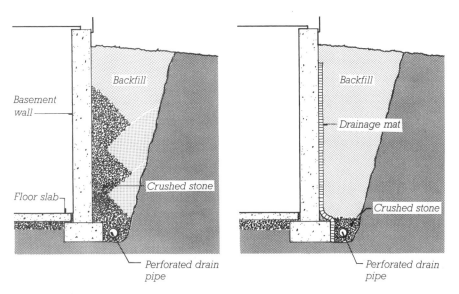

FIGURE 2.57
Two methods of relieving water pressure around a building substructure. The gravel drain (left) is the standard method, but it is hard to do well because of the difficulty of depositing the crushed stone and backfill soil in neatly separated, alternating layers. The drainage mat or drainage panel (right) is much easier and often more economical to install. It is a manufactured component that may be made of a loose mat of inert fibers, a plastic egg-crate structure, or some other very open, porous material. It is faced on the outside with a filter fabric that prevents fine soil particles from entering and clogging the drainage passages in the mat or panel. Water approaching the wall falls through the porous material to the drain pipe at the footing.

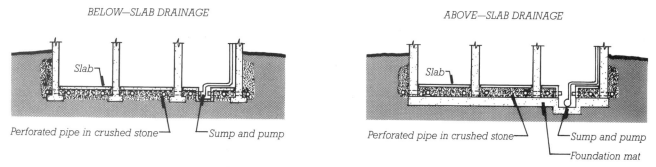

FIGURE 2.58
For a high degree of security against substructure flooding, drainage both around and under the basement is required. Above-slab drainage is used in buildings with mat foundations.

require that basement walls be thermally insulated to retard the loss of heat from basement rooms to the soil outside. Thermal insulation may be applied either inside or outside the basement wall. Inside the wall, mineral batt or plastic foam insulation may be installed between furring strips, as shown in Figure 23.5. Alternatively, polystyrene foam or glass fiber insulation boards, typically 2 to 4 inches thick (50 to 100 mm), may be placed on the outside of the wall, held there either by adhesive or the pressure of the soil. Proprietary products are available that combine insulation board and drainage mat in a single assembly.

SHALLOW FROST-PROTECTED FOUNDATIONS

Because polystyrene foam insulation boards do not deteriorate in the soil, they can be used in cold climates to construct footings that lie above the normal frost line in the soil, resulting in lower excavation costs. Continuous layers of insulation board are placed around the perimeter of the building in such a way that heat flowing into the soil in winter from the interior of the building maintains the soil beneath the footings at a temperature above freezing (Figure 2.59). Even beneath unheated buildings, properly installed thermal insulation can trap enough geothermal heat around shallow foundations to prevent freezing.

BACKFILLING

After the basement walls have been waterproofed or dampproofed, insulating boards have been applied, drainage features installed, and internal supporting constructions such as walls and floors have been completed, the area around a substructure in an open excavation is *backfilled* to restore the level of the ground. (A substructure built in a sheeted excavation needs little or no backfilling.) The backfilling operation involves placing soil back against the outside of the basement walls and compacting it there in layers, taking care not to damage drainage or waterproofing components or to place excessive pressure against the walls. An open, fast-draining soil such as sand is preferred for backfilling

because it allows the perimeter drainage system around the basement to do its work. Compaction must be sufficient to minimize subsequent settling of the backfilled area.

In some situations, especially in backfilling utility trenches under either roadways or floor slabs, settling can virtually be eliminated by backfilling with *controlled low-strength material (CLSM),* which is made from portland cement and/or fly ash (a byproduct of coal-burning power plants), sand, and water. CLSM, sometimes called "flowable fill," is brought in concrete mixer trucks and poured into the excavation, where it compacts and levels itself, then hardens into soil-like material. The strength of CLSM is matched to the situation: For a utility trench, for example, CLSM is proportioned so that it is weak enough to be excavated easily by ordinary digging equipment when the pipe needs servicing, yet as strong as good-quality compacted backfill. CLSM has many other uses in and around foundations. It is often used to pour *mud slabs,* which are weak concrete slabs used to create a level, dry base in an irregular, wet excavation. The mud slab serves as a working surface for the reinforcing and pouring of a foundation mat or basement floor slab. CLSM is also used to replace pockets of unstable soil that may be encountered beneath a substructure, and it may be used to create a stable volume of backfill around a basement wall.

UP–DOWN CONSTRUCTION

Sometimes speed of construction is crucial, especially in expensive building projects, where the interest costs on the money invested in the incomplete building are high; early occupancy is desirable so that the building will begin generating income as soon as possible. Normally, the substruc-

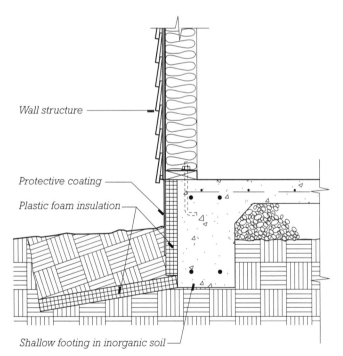

Wall structure

Protective coating

Plastic foam insulation

Shallow footing in inorganic soil

FIGURE 2.59
A typical detail for a shallow frost-protected footing.

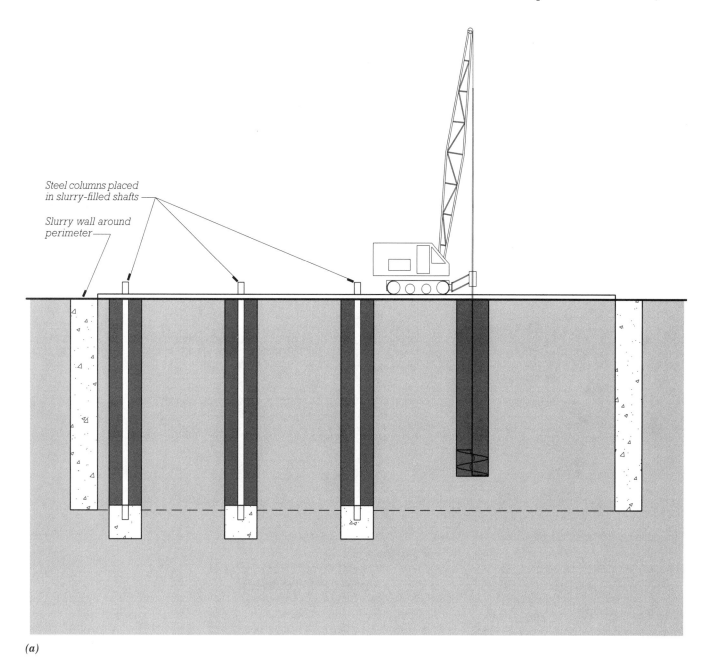

Steel columns placed in slurry-filled shafts

Slurry wall around perimeter

(a)

FIGURE 2.60
Up–down construction.

(continued)

ture of a building is completed before work begins on its superstructure. If the building has several levels of basements, however, substructure work can take many months. In such a case, *up–down construction* is sometimes an economical option, even though it costs somewhat more than the normal procedure, because it can save considerable construction time. As diagrammed in Figure 2.60, up–down construction begins with installation of a perimeter slurry wall. Internal steel columns for the substructure are lowered into drilled, slurry-filled holes, and concrete footings are tremied beneath them. Once the ground-floor slab is in place and connected to the substructure columns, erection of the superstructure may begin. Construction continues simultaneously on the substructure, largely through the use of mining machinery: A story of soil is

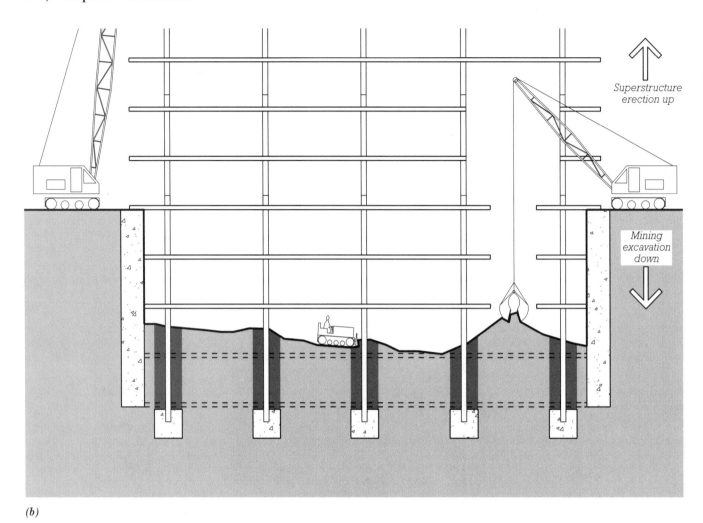

(b)

Figure 2.60 *(CONTINUED)*

excavated from beneath the ground-floor slab and a level mud slab of CLSM is poured. Working on the mud slab, workers reinforce and pour a concrete structural slab for the floor of the topmost basement level, again connecting it to the columns. When the slab is sufficiently strong, another story of soil is removed from beneath it, along with the mud slab, and the process is repeated until the substructure is complete.

DESIGNING FOUNDATIONS

It is a good idea to begin the design of the foundations of a building at the same time as architectural design work commences. Subsurface condi-tions beneath a site can strongly influ-ence several fundamental decisions about a building—its location, its size and shape, its weight, and the required degree of flexibility of its construction. On a large building project, at least three designers are involved in these decisions: the archi-tect, who has primary responsibility for the location and form of the building; the structural engineer, who has primary responsibility for its physical integrity; and the foundation engineer, who must decide, on the basis of site exploration and labora-tory reports, how best to support it in the earth. More often than not, it is possible for the foundation engineer to design foundations for a building design that was dictated entirely by the architect, but, in some cases, the cost of the foundations may consume a much larger share of the construc-tion budget than the architect has anticipated, unless certain compro-mises can be reached on the form and location of the building. It is safer and more productive for the architect to work with the foundation engineer from the outset, seeking alternative site locations and building configurations that will result in the fewest foundation problems and the lowest foundation cost.

In designing a foundation, a number of different design thresh-olds need to be kept in mind. If the designer crosses any of these thresh-olds, foundation costs take a sudden jump. Some of these thresholds are:

• *Building below the water table.* If the substructure and foundations of a building are above the water table, little or no effort will be required to keep the excavation dry during construction. If the water table is penetrated, whether by an inch or a hundred feet, expensive steps will have to be taken to dewater the site, strengthen the slope support system, waterproof the foundation, and either strengthen the basement floor slab against hydrostatic uplift pressure or provide for adequate drainage to relieve this pressure. For an extra inch or foot of depth, the expense would probably not be justified; for another story or two of useful building space, it might be.

• *Building close to an existing structure.* If the excavation can be kept well away from adjacent structures, the foundations of these structures can remain undisturbed and no effort and expense are required to protect them. When digging close to an existing structure, and especially when digging deeper than its foundations, the structure will have to be braced temporarily against structural collapse, and may require permanent underpinning with new foundations. Furthermore, an excavation at a distance from an existing structure may not require sheeting, while one immediately adjacent almost certainly will.

• *Increasing the column or wall load from a building beyond what can be supported by a shallow foundation.* Shallow foundations are far less expensive than piles or caissons under most conditions. If the building grows too tall, however, a shallow foundation may no longer be able to carry the load, and a threshold must be crossed into the realm of deep foundations. If this has happened for the sake of an extra story or two of height, the designer should consider reducing the height by broadening the building. If individual column loadings are too high for shallow foundations, perhaps they can be reduced by increasing the number of columns in the building and decreasing their spacing.

For very small buildings, at the scale of one- and two-family dwellings, foundation design is usually much simpler than for large buildings because foundation loadings are low. The uncertainties in foundation design can be reduced with reasonable economy by adopting a large factor of safety in calculating the bearing capacity of the soil. Unless the designer has reason to suspect poor soil conditions, the footings are usually designed using rule-of-thumb allowable soil stresses and standardized footing proportions. The designer then examines the actual soil when the excavations have been made. If it is not of the quality that was expected, the footings can be hastily redesigned using a revised estimate of soil-bearing capacity before construction continues. If unexpected ground water is encountered, better drainage provisions may have to be provided around the foundation, or the depth of the basement decreased.

FOUNDATION DESIGN AND THE BUILDING CODES

Because of the public safety considerations that are involved, building codes contain numerous provisions relating to the design and construction of excavations and foundations. The BOCA National Building Code, as one example, defines which soil types are considered satisfactory for bearing the weight of buildings and establishes a set of requirements for subsurface exploration, soil testing, and submission of soil reports to the local building inspector. It goes on to

C.S.I./C.S.C. Masterformat Section Numbers for Foundations	
02010	**SUBSURFACE INVESTIGATION**
02050	**DEMOLITION**
02100	**SITE PREPARATION**
02110	Site Clearing
02120	Structure Moving
02140	**DEWATERING**
02150	**SHORING AND UNDERPINNING**
02160	**EXCAVATION SUPPORT SYSTEMS**
02162	Cribbing and Walers
02164	Soil and Rock Anchors
02168	Slurry Wall Construction
02200	**EARTHWORK**
02210	Grading
02220	Excavation, Backfilling, and Compacting
02350	**PILES AND CAISSONS**
02355	Pile Driving
02360	Driven Piles
02380	Caissons
02500	**PAVING AND SURFACING**

specify the methods of engineering design that may be used for the foundations. It sets forth maximum load-bearing values for soils that may be assumed in the absence of detailed test procedures (Figure 2.5). It establishes minimum dimensions for footings, caissons, piles, and foundation walls, and contains lengthy discussions relating to the installation of piles and the drainage and water-proofing of substructures. The BOCA Code also requires engineering design of retaining walls. Through these or similar provisions, each building code attempts to ensure that every building will rest upon secure foundations and a dry substructure.

SELECTED REFERENCES

1. Ambrose, James E. *Simplified Design of Building Foundations* (2nd ed.). New York, John Wiley & Sons, Inc., 1988.

After an initial summary of soil properties, this small book covers simplified foundation computation procedures for both shallow and deep foundations.

2. Liu, Chen, and Jack B. Evett. *Soils and Foundations.* Englewood Cliffs, New Jersey, Prentice–Hall, Inc., 1981.

This is a fairly detailed discussion of the engineering properties of soils, subsurface exploration techniques, soil mechanics, and shallow and deep foundations, but is well suited to the beginner.

3. Schroeder, W. L. *Soils in Construction.* New York, John Wiley & Sons, Inc., 1984.

A well-illustrated, clearly written, moderately detailed survey of soils, soil testing, subsurface construction, and foundations.

KEY TERMS AND CONCEPTS

foundation	ice lenses	deep foundations	friction pile
dead load	sheeting	footing	pile cap
live load	bracing	column footing	grade beam
wind load	soldier beams	wall or strip footing	leads
thrust	lagging	engineered fill	H-pile
foundation settlement	sheet piling	slab on grade	pipe pile
uniform settlement	slurry wall	crawlspace	precast concrete pile
differential settlement	guide wall	basement	sitecast concrete pile
rock	clamshell bucket	tie beam	mandrel
soil	tremie	combined footing	pressure-injected footing
boulder	sitecast slurry wall	cantilever footing	base isolator
cobble	precast slurry wall	mat or raft	underpinning
gravel	soil mixing	foundation	needle beam
sand	crosslot bracing	floating foundation	retaining wall
coarse-grained soil	waler	caisson	cantilevered concrete retaining wall
silt	rakers	belling bucket	earth reinforcing
clay	heel block	belled caisson	geotextiles
fabric	tiebacks	socketed caisson	waterproof membrane
cohesive	rock anchors	rock or drilled-in	bentonite clay
frictional or cohesionless	dewatering	caisson	waterstop
strata	sump	drilled pier	drainage
stability	well points	pile	dampproofing
water table	watertight barrier wall	timber pile	backfill
liquid limit	superstructure	piledriver	controlled low-strength material (CLSM)
plastic limit	substructure	pile hammer	mud slab
excavation	shallow foundations	end bearing pile	up–down construction

REVIEW QUESTIONS

1. What is the nature of the most common type of foundation failure?

2. Explain in detail the differences among fine sand, silt, and clay, especially as they relate to foundations of buildings.

3. List three different ways of sheeting an excavation. Under what circumstances would sheeting not be required?

4. Under what conditions would you use a watertight barrier instead of well points when digging below the water table?

5. If shallow foundations are substantially less costly than deep foundations, why do we use deep foundations?

6. What soil conditions favor the use of belled caissons?

7. Which types of friction piles can carry the heaviest load per pile?

8. List and explain some cost thresholds frequently encountered in foundation design.

EXERCISES

1. Obtain the foundation drawings for a nearby building. Look first at the log of test borings. What sorts of soils are found beneath the site? How deep is the water table? What type of foundation do you think should be used in this situation (keeping in mind the relative weight of the building)? Now look at the foundation drawings. What type is actually used? Why?

2. What type of foundation and substructure is normally used for houses in your area? Why?

3. Look at several excavations for major buildings under construction. Note carefully the arrangements made for slope support and dewatering. How is the soil being loosened and carried away? What is being done with the excavated soil? What type of foundation is being installed? What provisions are being made for keeping the substructure permanently dry?

3

WOOD

(*Weyerhaeuser Company Photo*)

Wood is probably the best loved of all the materials that we use for building. It delights the eye with its endlessly varied colors and grain patterns. It invites the hand to feel its subtle warmth and varied textures. When it is fresh from the saw, its fragrance enchants. We treasure its natural, organic qualities and take pleasure in its genuineness. Even as it ages, bleached by the sun, eroded by rain, worn by the passage of feet and the rubbing of hands, we find beauty in its transformations of color and texture.

Wood earns our respect as well as our love. It is strong and stiff, yet by far the least dense of the materials used for the beams and columns of buildings. It is worked and fastened easily with small, simple, relatively inexpensive tools. It is readily recycled from demolished buildings for use in new ones, and, when finally discarded, it biodegrades rapidly to become natural soil. It is our only renewable building material, one that will be available to us for as long as we manage our forests with an eye to the perpetual production of wood.

But wood, like a valued friend, has its idiosyncrasies. A piece of lumber is never perfectly straight or true, and its size and shape can change significantly with changes in the weather. Wood is peppered with defects that are relics of its growth and processing. Wood can split; wood can warp; wood can give splinters. If ignited, wood burns. If left in a damp location, it decays and harbors destructive insects. But the skillful designer and the seasoned carpenter know all these things and understand how to build with wood to bring out its best qualities, while neutralizing or minimizing its problems.

TREES

Because wood comes from trees, an understanding of tree physiology is essential to knowing how to build with wood.

Tree Growth

The trunk of a tree is covered with a protective layer of dead *bark* (Figure 3.1). Inside the dead bark is a layer of living bark composed of hollow longitudinal cells that conduct nutrients down the trunk from the leaves to the roots. Inside this layer of living bark lies a very thin layer, the *cambium,* which creates new bark cells toward the outside of the trunk and new wood cells toward the inside. The thick layer of living wood cells inside the cambium is the *sapwood.* In this zone of the tree, nutrients are stored and sap is pumped upward from the roots to the leaves and distributed laterally in the trunk. At the inner edge of this zone, sapwood dies progressively and becomes *heartwood.* In many species of trees, heartwood is easily distinguished from sapwood by its darker color. Heartwood no longer participates in the life processes of the tree but continues to contribute to its structural strength. At the very center of the trunk, surrounded by heartwood, is the *pith* of the tree, a small zone of weak wood cells that were the first year's growth.

An examination of a small section of wood under a low-powered microscope shows that it consists primarily of tubular cells whose long axes are parallel to the long axis of the trunk. The cells are structured of tough *cellulose* and are bound together by a softer cementing substance called *lignin.* The direction of the long axes of the cells is referred to as the *direction of grain* of the wood. Grain direction is important to the designer of wooden buildings because the properties of wood parallel to grain and perpendicular to grain are very different.

In temperate climates, the cambium begins to manufacture new sapwood cells in the spring, when the air is cool and ground water is plentiful, conditions that favor rapid growth. Growth is slower during the heat of the summer when water is scarce. *Springwood* or *earlywood* cells are therefore larger and less dense in substance than the *summerwood* or *latewood* cells. Concentric bands of springwood and summerwood make up the annual growth rings in a trunk that can be counted to determine the age of a tree. The relative proportions of springwood and summerwood also have a direct bearing on the structural properties of the wood a given tree will yield because summerwood is stronger and stiffer than springwood. A tree grown under continuously moist, cool conditions grows faster than another tree of the same species grown under warmer, drier conditions, but its wood is not as dense or as strong.

Softwoods and Hardwoods

Softwoods come from coniferous trees and *hardwoods* from broad-leafed trees. The names can be deceptive because many coniferous trees produce harder woods than many broad-leafed trees, but the distinction is nevertheless a useful one. Softwood trees have a relatively simple microstructure, consisting mainly of large longitudinal cells (*tracheids*) together with a small percentage of radial cells (*rays*), whose function is the storage and radial transfer of nutrients (Figure 3.2). Hardwood trees are more complex in structure, with a

FIGURE 3.1
Summerwood rings are prominent and a few rays are faintly visible in this cross section of an evergreen tree, but the cambium, which lies just beneath the thick layer of bark, is too thin to be seen, and heartwood cannot be distinguished visually from sapwood in this species.
(*Courtesy of Forest Products Laboratory, Forest Service, USDA*)

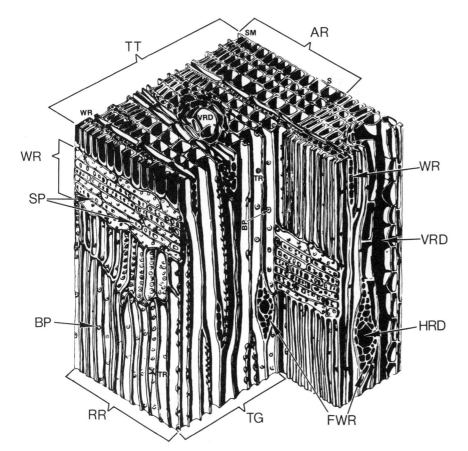

Cell structure of a softwood

FIGURE 3.2
Vertical cells (tracheids, labeled TR) dominate the structure of a softwood, seen here greatly enlarged, but rays (WR), which are cells that run radially from the center of the tree to the outside, are clearly in evidence. An annual ring (labeled AR) consists of a layer of smaller summerwood cells (SM) and layer of larger springwood cells (S). Simple pits (SP) allow sap to pass from ray cells to longitudinal cells and vice versa. Resin is stored in vertical and horizontal resin ducts (VRD and HRD), with the horizontal ducts centered in fusiform wood rays (FWR). Border pits (BP) allow the transfer of sap between longitudinal cells. The face of the sample labeled RR represents a radial cut through the tree, and TG, a tangential cut. (*Courtesy of Forest Products Laboratory, Forest Service, USDA*)

much larger percentage of rays and two different types of longitudinal cells: small-diameter *fibers* and large-diameter *vessels* or *pores*, which transport the sap of the tree (Figure 3.3).

When cut into lumber, softwoods generally have a coarse and relatively uninteresting grain structure, while many hardwoods show beautiful patterns of rays and vessels (Figure 3.4). Most of the lumber used today for frames of building comes from softwoods, which are comparatively plentiful and inexpensive. For fine furniture and interior finish details, hardwoods are often chosen. A few softwood and hardwood species widely used in North America are listed in Figure 3.5, along with the principal uses of each; however, it should be borne in mind that literally thousands of species of wood are used in construction around the world and that the available species

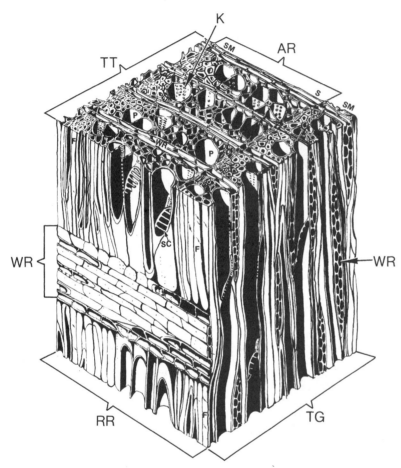

Cell structure of a hardwood

FIGURE 3.3

Rays (WR) constitute a large percentage of the mass of a hardwood, as seen in this sample, and are largely responsible for the beautiful grain figures associated with many species. The vertical cell structure is more complex than that of a softwood, with large pores (P) to transport the sap, and smaller wood fibers (F) to give the tree structural strength. Pore cells in some hardwood species end with crossbars (SC), while those of other species are entirely open. Pits (K) pass sap from one cavity to another.

(*Courtesy of Forest Products Laboratory, Forest Service, USDA*)

(a)

(b)

FIGURE 3.4

The grain figures of two softwood species (left) and two hardwoods (right) demonstrate the difference in cellular structure between the classes of woods. From left to right: (a) The cells in Sugar pine are so uniform that the grain structure is almost invisible except for scat-

vary considerably with geographic location. The major lumber-producing forests in North America are in the western and eastern mountains of both the United States and Canada, and the southeastern United States, but other regions also produce significant quantities.

(c)

(d)

tered resin ducts. (*b*) Vertical-grain Douglas fir shows very pronounced dark bands of summerwood. (*c*) Red oak exhibits large open pores amid its fibers. (*d*) This quarter-sliced Mahogany veneer has a pronounced "ribbon" figure caused by varying light reflections off its fibers. (*Photos by the author*)

SOFTWOODS	HARDWOODS
Used for Framing, Sheathing, Paneling	*Used for Moldings, Paneling, Furniture*
Alpine fir	Ash
Balsam fir	Beech
Douglas fir	Birch
Eastern hemlock	Black walnut
Eastern spruce	Butternut
Eastern white pine	Cherry
Englemann spruce	Lauan
Idaho white pine	Mahogany
Larch	Pecan
Loblolly pine	Red oak
Lodgepole pine	Rosewood
Longleaf pine	Teak
Mountain hemlock	Tupelo gum
Ponderosa pine	White oak
Red spruce	Yellow poplar
Shortleaf pine	*Used for Finish Flooring*
Sitka spruce	Pecan
Southern yellow pine	Red oak
Western hemlock	Sugar maple
White spruce	Walnut
Used for Moldings, Window and Door Frames	White oak
Ponderosa pine	
Sugar pine	
White pine	
Used for Finish Flooring	
Douglas fir	
Longleaf pine	
Decay-Resistant Woods, Used for Shingles, Siding, Outdoor Structures	
California redwood	
Southern cypress	
Western red cedar	
White cedar	

FIGURE 3.5
Some species of wood commonly used in construction in North America, listed alphabetically in groups according to end use. All are domestic except Lauan (Asia), Mahogany (Central America), Rosewood (South America and Africa), and Teak (Asia).

LUMBER

Sawing

The production of *lumber*, lengths of squared wood for use in construction, begins with the felling of trees and the transportation of the logs to a sawmill (Figure 3.6). Sawmills range in size from tiny family operations to giant, semiautomated factories, but the process of lumber production is much the same regardless of scale. Each log is first passed repeatedly through a large *headsaw*, which may be either a circular saw or a bandsaw, to reduce it to untrimmed slabs of lumber (Figure 3.7). The *sawyer* (with the aid of a computer in the larger mills) judges how to obtain the maximum marketable wood from each log and uses hydraulic machinery to rotate and advance the log in order to achieve the required succession of cuts. As the slabs fall from the log at each pass, a conveyor belt carries them away to smaller saws that reduce

them to square-edged pieces of the desired widths (Figure 3.8). The sawn pieces at this stage of production have rough-textured surfaces and may very slightly in dimension from one end to the other.

Most lumber intended for use in the framing of buildings is *plainsawed,* a method of dividing the log that produces the maximum yield and therefore the greatest economy (Figure 3.9). In plainsawed lumber, some pieces have the annual rings running virtually perpendicular to the faces of the piece, some have the rings on various diagonals, and some have the rings running almost parallel to the faces. These varying grain orientations cause the pieces to distort differently during seasoning, to have very different surface appearances from one another, and to erode at different rates when used in applications such as flooring and exterior siding. For uses where any of these variations will cause a problem, especially for finish flooring, interior trim,

and furniture, hardwoods are often *quartersawn* to produce pieces of lumber that have the annual rings running more nearly perpendicular to the face of the piece. Boards with this *vertical grain* orientation tend to remain flat despite changes in moisture content. The visible grain on the surface of a quartersawn piece makes a tighter and more pleasing *figure*. The wearing qualities of the piece are improved because there are no broad areas of soft springwood exposed in the face, as there are when the annual rings run more nearly parallel to the face.

Seasoning

Growing wood contains a quantity of water that can vary from about 30 percent to as much as 300 percent of the oven-dry weight of the wood. After a tree is cut, this water slowly starts to evaporate. First to leave the wood is the *free water* that is held in the cavities of the cells. When the free water is

FIGURE 3.6
Loading Southern pine logs onto a truck for their trip to the sawmill. (*Courtesy of Southern Forest Products Association*)

FIGURE 3.7
In a large, mechanized mill, the operator controls the high-speed bandsaw from an overhead booth. (*Courtesy of Western Wood Products Association*)

FIGURE 3.8
Sawn lumber is sorted into stacks according to its cross-sectional dimensions and length. (*Courtesy of Western Wood Products Association*)

PLAINSAWING

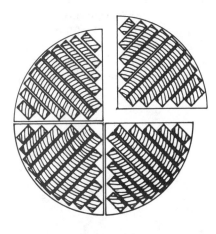

QUARTERSAWING

TYPICAL SAWING OF A LARGE LOG

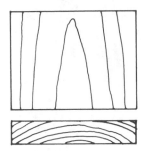

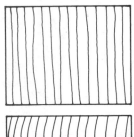

FIGURE 3.9
Plainsawing produces boards with a broad grain figure, as seen in end and top views below the plainsawed log. Quartersawing produces a vertical grain structure, which is seen on the face of the board as tightly spaced parallel summerwood lines. A large log of softwood is typically sawn to produce some large timbers, some plainsawed dimension lumber, and, in the horizontal row of small pieces seen just below the heavy timbers, some pieces of vertical-grain decking.

gone, the wood still contains about 26 to 32 percent moisture, and the *bound water* held within the cellulose of the cell walls begins to evaporate. As the first of the bound water disappears, the wood starts to shrink, and the strength and stiffness of the wood begin to increase. The shrinkage, stiffness, and strength increase steadily as the moisture content decreases. Wood can be dried to any desired moisture content, but framing lumber is considered to be *seasoned* when its moisture content is 19 percent or less. For framing applications that require closer control of wood shrinkage, lumber seasoned to a 15 percent moisture content, labeled "MC 15," is available. It is of little use to season ordinary framing lumber to a moisture content below about 13 percent because wood is hygroscopic and will take on or give off moisture, swelling or shrinking as it does so, in order to stay in equilibrium with the moisture in the air.

Most lumber is seasoned at the sawmill, either by *air drying* in loose stacks for a period of months or,

> **Nor do I ever come to a lumber yard with its citylike, graduated masses of fresh shingles, boards and timbers, without taking a deep breath of its fragrance, seeing the forest laid low in it by processes that cut and shaped it to the architect's scale of feet and inches . . .**
>
> **Frank Lloyd Wright**

more commonly, by *kiln drying* under closely controlled conditions of temperature and humidity for a period of days (Figures 3.10, 3.11). Seasoned lumber is stronger and stiffer than unseasoned (*green*) lumber, more dimensionally stable, and lighter and more economical to ship. Kiln drying is generally preferred to air drying because it is much faster and produces better quality lumber.

Wood does not shrink and swell uniformly with changes in moisture content. Moisture shrinkage along the length of the log (*longitudinal shrinkage*) is negligible for practical purposes. Shrinkage in the radial direction (*radial shrinkage*) is very large by comparison, and shrinkage around the circumference of the log (*tangential shrinkage*) is about half again greater than radial shrinkage (Figure 3.12). If an entire log is seasoned before sawing, it will shrink very little along its length, but it will grow noticeably smaller in diameter, and the difference between the tangential and radial shrinkage will cause it to *check*, that is, to split open all along its length (Figure 3.13).

These differences in shrinkage rates are so large that they cannot be ignored in building design. In constructing building frames of plain-sawed lumber, a simple distinction is made between *parallel-to-grain shrinkage*, which is negligible, and *perpendicular-to-grain shrinkage*, which is considerable. The difference between radial and tangential shrink-

FIGURE 3.10
For proper air drying, lumber is supported well off the ground. The stickers, which keep the boards separated for ventilation, are carefully placed above one another to avoid bending the lumber, and a watertight roof protects each stack from rain and snow. (*Courtesy of Forest Products Laboratory, Forest Service, USDA*)

FIGURE 3.11
Measuring moisture content in boards in a drying kiln. (*Courtesy of Western Wood Products Association*)

age is not considered because the orientation of the annual rings in plainsawed lumber is random and unpredictable. As we will see in Chapter 5, wood building frames are carefully designed to equalize the amount of wood loaded perpendicular to grain from one side of the structure to the other in order to avoid the noticeable tilting of floors and tearing of wall finish materials that would otherwise occur.

The position in a log from which a piece of lumber is sawn determines in large part how it will distort as it dries. Figure 3.14 shows how the differences between tangential and radial shrinkage cause this to happen. These effects are pronounced and are readily predicted and observed in everyday practice.

FIGURE 3.12
Shrinkage of a typical softwood with decreasing moisture content. Longitudinal shrinkage, not shown on this graph, is so small by comparison to tangential and radial shrinkage that it is of no consequence in wood buildings.
(*Courtesy of Forest Products Laboratory, Forest Service, USDA*)

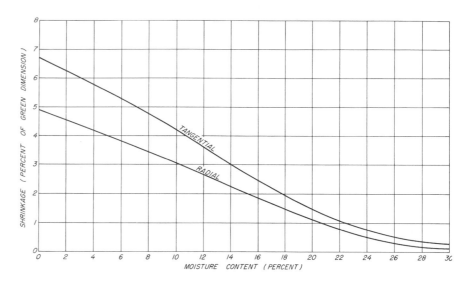

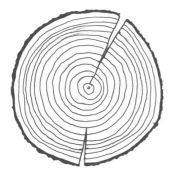

FIGURE 3.13
Because tangential shrinkage is so much greater than radial shrinkage, high internal stresses are created in a log as it dries, resulting inevitably in the formation of radial cracks called checks.

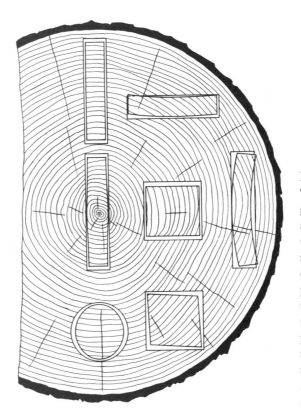

FIGURE 3.14
The difference between tangential and radial shrinkage also produces seasoning distortions in lumber. The nature of the distortion depends on the position the piece of lumber occupied in the tree. The distortions are the most pronounced in plainsawed lumber (upper right, extreme right, lower right). (*Courtesy of Forest Products Laboratory, Forest Service, USDA*)

Surfacing

Lumber is *surfaced* to make it smooth and more dimensionally precise. Rough (unsurfaced) lumber is often available commercially and is used for many purposes, but surfaced lumber is easier to work with because it is more square and uniform in dimension and less damaging to the hands of the carpenter. Surfacing is done by high-speed automatic machines whose rotating blades plane the surfaces of the piece and round the edges slightly. Most lumber is surfaced on all four sides (*S4S*), but hardwoods are often surfaced on only two sides (*S2S*), leaving the two edges to be finished by the craftsman.

Lumber is usually seasoned before it is surfaced, which allows the planing process to remove some of the distortions that occur during seasoning, but for some framing lumber this order of operations is reversed. The designation *S-DRY* in a lumber gradestamp indicates that the piece was surfaced (planed) when in a seasoned (dry) condition, and *S-GRN* indicates that it was planed when green.

Lumber Defects

Almost every piece of lumber contains one or more discontinuities in its structure caused by *growth characteristics* of the tree from which it came, or *manufacturing characteristics* that were created at the mill (Figures 3.15, 3.16). Among the most common

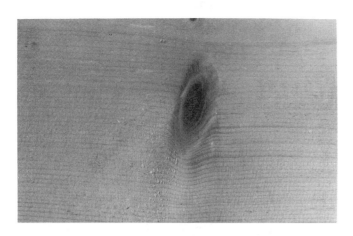

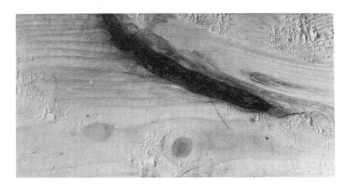
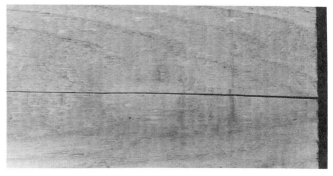

growth characteristics are *knots,* which are places where branches joined the trunk of the tree; *knotholes,* which are holes left by loose knots dropping out of the wood; *decay;* and *insect damage.* Knots and knotholes reduce the structural strength of a piece of lumber, make it more difficult to cut and shape, and are often considered to be detrimental to its appearance. Decay and insect damage that occurred during the life of the tree may or may not affect the useful properties of the piece of lumber, depending on whether the organisms are still alive in the wood and the extent of the damage they have done.

Manufacturing characteristics arise largely from changes that take place during the seasoning process because of the differences in rates of shrinkage with varying orientations to the grain. *Splits* and checks are usually caused by shrinkage stresses. *Crooking, bowing, twisting,* and *cupping* all occur because of nonuniform shrinkage. *Wane* is an irregular rounding of edges or faces that is caused by sawing pieces too close to the perimeter of the log. Experienced carpenters judge the extent of these defects and distortions in each piece of lumber and decide accordingly where and how to use the piece in the building. Checks

and shakes are of little consequence in framing lumber, but a joist or rafter with a crook in it is usually placed with the convex edge (the "crown") facing up, to allow the floor or roof loads to straighten the piece. Carpenters straighten badly bowed joists or rafters by sawing or planing away the crown before they apply subflooring or sheathing over them. Badly twisted pieces are put aside to be cut up for blocking. The effects of cupping in flooring and interior baseboards and trim are usually minimized by using quartersawn stock and by shaping the pieces so as to reduce the likelihood of distortion (Figure 3.17).

FIGURE 3.15
Surface features often observed in lumber include, in the left-hand column from top to bottom, a knot cut crosswise, a knot cut longitudinally, and a bark pocket. To the right are a grade-stamp, wane on two edges of the same piece, and a small check. The gradestamp indicates that the piece was graded according to the rules of the American Forest Products Association, that it is #2 grade Spruce–Pine–Fir, and that it was surfaced after drying. The 27 is a code number for the mill that produced the lumber. (*Photos by the author*)

FIGURE 3.17
The effects of seasoning distortions can often be minimized through knowledgeable detailing practices. As an example, this wood baseboard, seen in cross section, has been formed with a relieved back, which is a broad, shallow groove that allows the piece to lie flat against the wall even if it cups (broken lines). The sloping bottom on the baseboard assures that it can be installed tightly against the floor despite the cup. The grain orientation in this piece is the worst possible with respect to cupping. If quartersawn lumber were not available, the next best choice would have been to mill the baseboard with the center of the tree toward the room rather than toward the wall.

FIGURE 3.16
Four types of seasoning distortions in dimension lumber.

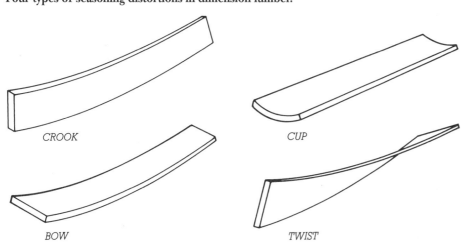

CROOK

BOW

CUP

TWIST

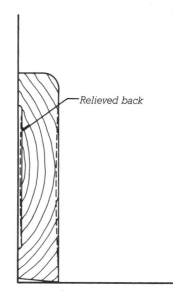

Relieved back

Lumber Grading

Each piece of lumber is graded either for appearance or for structural strength and stiffness, depending on its intended use, before it leaves the mill. Lumber is sold by species and grade; the higher the grade, the higher the price. Grading offers the architect and the engineer the opportunity to build as economically as possible by using only as high a grade as is required for a particular use. In a specific building, the main beams or columns may require a very high structural grade of lumber, while the remainder of the framing members will perform adequately in an intermediate, less expensive grade. For blocking, the lowest grade is perfectly adequate. For finish trim that will be coated with a clear finish, a high appearance grade is desirable; for painted trim, a lower grade will suffice.

Structural grading of lumber may be done either visually or by machine. In *visual grading,* trained inspectors examine each piece for ring density and for growth and manufacturing characteristics, then judge it and stamp it with a grade in accordance with industry-wide grading rules (Figure 3.18). In *machine grading,* an automatic device assesses the structural properties of the wood and stamps a grade automatically on the piece. This assessment is made either by flexing each piece between rollers and measuring its resistance to bending, or by scanning the wood electronically to determine its density. Appearance grading, naturally enough, is done visually. Figure 3.19 outlines a typical grading scheme for framing lumber, and Figure 3.20 outlines the appearance grades for nonstructural lumber. Light framing lumber for houses and other small buildings is usually ordered as "#2 and better" (a mixture of #1 and #2 grades) for floor joists and roof rafters, and as "Stud" grade for wall framing.

The Structural Strength of Wood

The strength of a piece of wood depends chiefly on its species, its grade, and the direction in which the load acts with respect to the direction of grain of the piece. Wood is several times stronger parallel to grain than perpendicular to grain. With its usual assortment of defects, it is stronger in compression than in tension. *Allowable strengths* (structural stresses that include factors of safety) vary tremendously with species and grade. Allowable compressive strength parallel to grain, for example, varies from 325 to 1700 pounds per square inch (2.24 to 11.71 MPa) for commercially available grades and species of framing lumber, a difference of more than five times. Figure 3.21 compares the structural properties of an "average" framing lumber to those of the other common structural materials—brick masonry, steel, and concrete. Of the four materials, only wood and steel have useful tensile strength. Defect-free wood is comparable to steel on a strength-per-unit-weight basis, but with the ordinary run of defects, an average piece of lumber is somewhat inferior to steel by this yardstick.

When designing a wooden structure, the architect or engineer determines the maximum stresses that are likely to occur in each of the structural members and selects an appropriate species and grade of lumber for each. In a given locale, a limited number of species and grades are usually available in retail lumberyards, and it is from these that the selection is made. It is common practice to use a stronger but more expensive species (Douglas fir or Southern pine, for example) for highly stressed major members, and to use a weaker, less expensive species (such as Eastern hemlock) or species group (Hem–Fir, Spruce–Pine–Fir) for the remainder of the structure. Within each species, the designer selects grades based on published tables of allowable stresses. The higher the grade, the higher the allowable stress, but the lower the grade, the less costly the lumber.

There are many factors other than species and grade that influence the useful strength of wood: These include the length of time the wood will be subjected to its maximum load, the temperature and moisture conditions under which it serves, and the size and shape of the piece. Certain fire-retardant treatments also reduce the strength of wood slightly. All these factors are taken into account when engineering a building structure of wood.

Lumber Dimensions

Lumber sizes in the United States are given as *nominal dimensions* in inches. A piece nominally 1 by 2 inches in cross section is a 1 × 2 ("one by two"), a piece 2 by 10 inches is a 2 × 10, and so on. At the time a piece of lumber is sawn, it may approach these dimensions. Subsequent to sawing, however, seasoning and surfacing diminish its size substantially. By the time a kiln-dried 2 × 10 reaches the lumberyard,

FIGURE 3.18
A grader marks the grade on a piece with a lumber crayon, preparatory to applying a gradestamp. (*Courtesy of Western Wood Products Association*)

DIMENSION LUMBER GRADES

Table **2.1**

Product	Grades	WWPA Western Lumber Grading Rules Section Reference	Uses
Structural Light Framing (SLF) 2″ to 4″ thick 2″ to 4″ wide	SELECT STRUCTURAL NO.1 NO.2 NO.3	(42.10) (42.11) (42.12) (42.13)	Structural applications where highest design values are needed in light framing sizes.
Light Framing (LF) 2″ to 4″ thick 2″ to 4″ wide	CONSTRUCTION STANDARD UTILITY	(40.11) (40.12) (40.13)	Where high-strength values are not required, such as wall framing, plates, sills, cripples, blocking, etc.
Stud 2″ to 4″ thick 2″ and wider	STUD	(41.13)	An optional all-purpose grade designed primarily for stud uses, including bearing walls.
Structural Joists and Planks (SJ&P)	SELECT STRUCTURAL NO.1 NO.2 NO.3	(62.10) (62.11) (62.12) (62.13)	Intended to fit engineering applications for lumber 5″ and wider, such as joists, rafters, headers, beams, trusses, and general framing.

STRUCTURAL DECKING GRADES

Table **2.2**

Product	Grades	WWPA Western Lumber Grading Rules Section Reference	Uses
Structural Decking 2″ to 4″ thick 4″ to 12″ wide	SELECTED DECKING	(55.11)	Used where the appearance of the best face is of primary importance.
	COMMERCIAL DECKING	(55.12)	Customarily used when appearance is not of primary importance.

TIMBER GRADES

Table **2.3**

Product	Grades	WWPA Western Lumber Grading Rules Section Reference	End Uses
Beams and Stringers 5″ and thicker, width more than 2″ greater than thickness	DENSE SELECT STRUCTURAL* DENSE NO. 1* DENSE NO. 2* SELECT STRUCTURAL NO.1 NO.2	(53.00 & 170.00) (53.00 & 170.00) (53.00 & 170.00) (70.10) (70.11) (70.12)	Grades are designed for beam and stringer type uses when sizes larger than 4″ nominal thickness are required.
Post and Timbers 5″ × 5″ and larger, width not more than 2″ greater than thickness	DENSE SELECT STRUCTURAL* DENSE NO. 1* DENSE NO. 2* SELECT STRUCTURAL NO.1 NO.2	(53.00 & 170.00) (53.00 & 170.00) (53.00 & 170.00) (80.10) (80.11) (80.12)	Grades are designed for vertically loaded applications where sizes larger than 4″ nominal thickness are required.

*Douglas Fir or Douglas Fir-Larch only.

FIGURE 3.19

Standard structural grades for western softwood lumber. For each species of wood, the allowable structural stresses for each of these grades are tabulated in the structural engineering literature. (*Courtesy of Western Wood Products Association*)

APPEARANCE LUMBER GRADES Table **2.5**

	Product	Grades[1]	Equivalent Grades in Idaho White Pine	WWPA Grading Rules Section Number
Highest Quality Appearance Grades	Selects *(all species)*	B & BTR SELECT C SELECT D SELECT	SUPREME CHOICE QUALITY	10.11 10.12 10.13
	Finish *(usually available only in Doug Fir and Hem-Fir)*	SUPERIOR PRIME E		10.51 10.52 10.53
	Special Western Red Cedar Pattern Grades	CLEAR HEART A GRADE B GRADE		20.11 20.12 20.13
General Purpose Grades	Common Boards (WWPA Rules) *(primarily in pines, spruces, and cedars)*	1 COMMON 2 COMMON 3 COMMON 4 COMMON 5 COMMON	COLONIAL STERLING STANDARD UTILITY INDUSTRIAL	30.11 30.12 30.13 30.14 30.15
	Alternate Boards (WCLIB Rules) *(primarily in Doug Fir and Hem-Fir)*	SELECT MERCHANTABLE CONSTRUCTION STANDARD UTILITY ECONOMY		**WCLIB[3]** 118-a 118-b 118-c 118-d 118-e
	Special Western Red Cedar Pattern[2] Grades	SELECT KNOTTY QUALITY KNOTTY		**WCLIB[3]** 111-e 111-f

[1]Refer to WWPA's *Vol. 2, Western Wood Species* book for full-color photography and to WWPA's *Natural Wood Siding* for complete information on siding grades, specification, and installation.

FIGURE 3.20
Nonstructural boards are graded according to appearance. (*Courtesy of Western Wood Products Association*)

Material	Allowable Tensile Strength	Allowable Compressive Strength	Density	Modulus of Elasticity
Wood (average)	700 psi (4.83 MPa)	1,100 psi (7.58 MPa)	30 pcf (480 kg/m^3)	1,200,000 psi (8,275 MPa)
Brick masonry (average)	0	250 psi (1.72 MPa)	120 pcf (1,920 kg/m^3)	1,200,000 psi (8,275 MPa)
Steel (ASTM A36)	22,000 psi (151.69 MPa)	22,000 psi (151.69 MPa)	490 pcf (7,850 kg/m^3)	29,000,000 psi (200,000 MPa)
Concrete (average)	0	1,350 psi (9.31 MPa)	145 pcf (2,320 kg/m^3)	3,150,000 psi (21,720 MPa)

FIGURE 3.21
Comparative physical properties of the four common structural materials: wood, masonry, steel, concrete.

its actual dimensions are $1\frac{1}{2}$ by $9\frac{1}{4}$ inches (38×235 mm). The relationship between nominal lumber dimensions (which are always written without inch marks) and *actual dimensions* (which are written *with* inch marks) is given in simplified form in Figure 3.22 and in complete form in Figure 3.23. Anyone who designs or constructs wooden buildings soon commits the simpler of these relationships to memory. Because of changing moisture content and manufacturing tolerances, however, it is never wise to assume that a piece of lumber will conform precisely to its intended dimension. The experi-

FIGURE 3.22

The relationship between nominal and actual dimensions for the most common sizes of kiln-dried lumber is given in this simplified chart, which is extracted from the complete chart in Figure 3.23.

Nominal Dimension	Actual Dimension
1″	$\frac{3}{4}$″ (19 mm)
2″	$1\frac{1}{2}$″ (38 mm)
3″	$2\frac{1}{2}$″ (64 mm)
4″	$3\frac{1}{2}$″ (89 mm)
5″	$4\frac{1}{2}$″ (114 mm)
6″	$5\frac{1}{2}$″ (140 mm)
8″	$7\frac{1}{4}$″ (184 mm)
10″	$9\frac{1}{4}$″ (235 mm)
12″	$11\frac{1}{4}$″ (286 mm)
over 12″	$\frac{3}{4}$″ less (19 mm less)

STANDARD SIZES—FRAMING LUMBER   Table **2.4**
Nominal & Dressed (Based on *Western Lumber Grading Rules*)

Product	Description	Nominal Size — Thickness (inches)	Nominal Size — Width (inches)	Dressed Dimensions — Thicknesses & Widths (inches) — Surfaced Dry	Dressed Dimensions — Thicknesses & Widths (inches) — Surfaced Unseasoned	Length (feet)
DIMENSION	S4S	2 3 4	2 3 4 5 6 8 10 12 over 12	$1\frac{1}{2}$ $2\frac{1}{2}$ $3\frac{1}{2}$ $4\frac{1}{2}$ $5\frac{1}{2}$ $7\frac{1}{4}$ $9\frac{1}{4}$ $11\frac{1}{4}$ off $\frac{3}{4}$	$1\frac{9}{16}$ $2\frac{9}{16}$ $3\frac{9}{16}$ $4\frac{5}{8}$ $5\frac{5}{8}$ $7\frac{1}{2}$ $9\frac{1}{2}$ $11\frac{1}{2}$ off $\frac{1}{2}$	6′ and longer, generally shipped in multiples of 2′
TIMBERS	Rough or S4S (shipped unseasoned)	5 and larger		**Thickness (unseasoned)** — 1/2 off nominal (S4S). See 3.20 of WWPA Grading Rules for Rough.	**Width (unseasoned)**	6′ and longer, generally shipped in multiples of 2′
DECKING	2″ (Single T&G)	**Thickness** 2	**Width** 5 6 8 10 12	**Thickness (dry)** $1\frac{1}{2}$	**Width (dry)** 4 5 $6\frac{3}{4}$ $8\frac{3}{4}$ $10\frac{3}{4}$	6′ and longer, generally shipped in multiples of 2′
	3″ and 4″ (Double T&G)	3 4	6	$2\frac{1}{2}$ $3\frac{1}{2}$	$5\frac{1}{4}$	

Abbreviations: FOHC—Free of Heart Center T&G—Tongued and grooved Rough Full Sawn—Unsurfaced lumber cut to full specified size
S4S—Surfaced four sides

FIGURE 3.23

A complete chart of nominal and actual dimensions for both framing lumber and finish lumber. (*Courtesy of Western Wood Products Association*)

A NATURALLY GROWN BUILDING MATERIAL

by

JOSEPH IANO

Wood is the only major building material that is organic in origin. This accounts for much of its uniqueness as a structural material. Because trees grow naturally, wood is a renewable resource, and most of the work of "manufacturing" wood is done for us by the processes of life and growth within the tree. The consequences of this are both economic and practical. Wood is cheap to produce. Most lumber need only be harvested, cut to size, and dried before it is ready for use. However, because wood is produced by trees, we have little control over the product. Unlike other structural materials such as steel or concrete, we can do little to adjust or refine wood's properties to suit our needs. Rather, we must accept its natural strengths and limitations.

Though we use wood much as we find it in the living tree, it is remarkably well suited to our building needs. This is because a tree is itself a structure made from wood. From a mechanical point of view, a tree is a tower erected for the purpose of displaying leaves to the sun.* Thus a tree is subjected to many of the same forces as the buildings we erect: It supports itself against the pull of gravity. It withstands the forces of wind, and the accumulation of snow and ice on its limbs. And it resists the natural stresses of our environment, including temperature and moisture extremes, attack by other organisms, and physical abuse. Since wood is the material that the tree uses to resist these forces, it is not surprising that we find wood a suitable material for our structures as well.

One of wood's natural strengths is its high resistance to loads or forces that act only for short periods of time.

---

* Brayton F. Wilson and Robert R. Archer, "Tree Design: Some Biological Solutions to Mechanical Problems," *BioScience*, Vol. 29, No. 5, p. 293.

The greatest forces that trees experience in nature are from the wind, particularly in combination with ice or heavy snow. Because of this, wood has evolved to be more capable of withstanding forces such as the wind that are applied over shorter periods of time than forces applied over much longer periods.

When we build with wood, we can take advantage of this unusual short-term strength. In the engineering design of any wooden structure, the maximum forces that the structural members may carry are increased as the length of time these forces are expected to act decreases. This adjustment is called the *duration of load factor*. Increases from 15 to 100 percent in the allowable stresses are applied to wooden structures when considering the effects of wind, snow, and earthquake—the types of forces also resisted best by the tree.

The structural form of the tree is also well suited to its environment. Tree branches are supported at one end only—an arrangement called a *cantilever*. The joint supporting a cantilevered branch is a strong, stiff, efficient connection that utilizes the tree's material resources to the utmost. Despite the stiffness of this joint, however, the branch it supports can still deflect relatively great amounts since it is fastened at one end only. This high deflection can be advantageous for the tree. A branch that droops under a heavy load of snow can drop that snow and relieve its load. Similarly, a tree that sways and bends in the wind can shed much of the force acting upon it. This capacity to deflect helps a tree to survive forces that might otherwise break it.

When we build with wood, we cannot directly exploit the efficient structural form of the tree. The high deflections characteristic of cantilevered beams are unsuitable for our buildings because such large movements would cause discomfort to the occupants and undue distress to

---

enced detailer of wooden buildings knows not to treat wood as if it had the dimensional accuracy of steel. The designer working with an existing wooden building will find a great deal of variation in lumber sizes. Wood members in hot, dry locations such as attics often will have shrunk to dimensions substantially below their original measurements.

Members in older buildings may have been manufactured to full nominal dimensions or to earlier standards of actual dimensions such as $1\frac{5}{8}$ inch or $1\frac{3}{4}$ inch (41 or 44 mm) for a nominal 2-inch member.

Pieces of lumber less than 2 inches in nominal thickness are called *boards*. Pieces ranging from 2 to 4 inches in nominal thickness are

referred to collectively as *dimension lumber*. Pieces nominally 5 inches and more in thickness are termed *timbers*.

Dimension lumber is usually supplied in 2-foot (610-mm) increments of length. The most commonly used lengths are 8, 10, 12, 14, and 16 feet (2.44, 3.05, 3.66, 4.27, and 4.88 m), but most retailers stock rafter material in lengths to 24 feet (7.32 m).

other building components. Nor can we take advantage of the naturally strong joints that support these cantilevers, though they could offer benefits in their stiffness and economy of materials. In preparing wood for our uses, we saw the tree into pieces of convenient size and shape. This destroys the structural continuity between the tree's parts. When we reassemble these pieces, we must devise new ways to join the wood. Despite the many methods developed for fabricating wood connections, joints with strength and stiffness comparable to the tree's are impossible to make on the building site. This is why most of our wood buildings rely on the much simpler type of connections that suffice when a beam is supported at both ends simultaneously.

Wood in a tree performs many functions that wood in a building does not, because wood is fundamentally involved with all aspects of the tree's life processes and growth. In order to meet numerous and specialized tasks, wood has evolved a complex internal structure that is highly directional. The fact that wood is not a uniform and "isotropic" material (the same in all directions) places basic limitations on how it can be used. Virtually all of wood's physical properties vary greatly, depending on whether they are measured parallel to the direction of the grain, or across it. This directional quality of wood is often taken for granted, even though no other major building material is like this. In fact, the direction of the grain in a piece of wood affects every aspect of how we can use it, including its production, shaping, and fastening, its structural capacity, its durability, and its beauty.

An example that illustrates this is a comparison of two early American wood building types: the log cabin and the heavy timber frame. The log cabin was a building system appropriate to the most primitive technology. The logs were found on or near the site and prepared for use with a minimum of labor and tools. In their final form they served many functions, including structural support, interior finish, exterior cladding, and insulation. But because the logs were stacked with their grain running horizontally, the wood was being used inefficiently from a struc-

tural point of view. Not only did this arrangement of the timbers produce a relatively weak wall, it also maximized the effects of vertical shrinkage and expansion in the structure, resulting in a high degree of dimensional instability. Where more sophisticated building techniques and a wider range of building materials were available, the log cabin was never used. Rather, buildings were framed with heavy timbers, utilizing a structural system that more closely resembled the structure of the tree. This more natural configuration of timbers in which a wall was framed with vertical posts and spanned with horizontal beams could be stronger, required less wood, was more dimensionally stable, and permitted the incorporation of other materials better suited to the wall's nonstructural requirements. A simple difference in orientation of the timbers produced a radically different building system that was more efficient, durable, and comfortable.

Many of the modern developments in the wood industry reflect a desire to overcome the natural limitations of wood. Plywood sheets are larger than could normally be obtained from trees, are dimensionally more stable than natural wood, and minimize the adverse effects of grain and natural defects in the wood. Newer panel products such as particleboard, waferboard, and hardboard use scraps and waste that would otherwise be discarded. Glue-laminated and veneer-based beams are larger in size and of higher quality than can be obtained from nature, and successfully utilize techniques of joinery that are as strong as those of the original tree. Chemical treatments can increase wood's resistance to fire and decay. Thus the trend is toward using wood less in its natural state, and more as a raw material in sophisticated manufacturing processes. These new products share more of the qualities of being highly refined and carefully controlled that are characteristic of other man-made building materials. However, despite such technological changes the tree remains not only a valuable resource but also an inspiration in its grace and strength, and in the lessons it offers in understanding our own building materials and methods.

Actual lengths are usually a fraction of an inch longer than nominal lengths.

Lumber in the United States is priced by the *board foot*. Board foot measurement is based on nominal dimensions, not actual dimensions. A board foot of lumber is defined as a solid volume 12 square inches in nominal cross-sectional area and 1 foot long. A 1 × 12 or 2 × 6 10 feet long contains 10 board feet. A 2 × 4 10 feet long contains $[(2 \times 4)/12] \times 10 = 6.67$ board feet, and so on. Prices of dimension lumber and timbers in the United States are usually quoted on the basis of dollars per thousand board feet. In other parts of the world, lumber is sold by the cubic meter.

The architect and engineer specify lumber for a particular construction use by designating its species, grade, seasoning, surfacing, nominal size, and chemical treatment, if any. When ordering lumber, the contractor must additionally give the required lengths of pieces and the required number of pieces of each length.

WOOD PRODUCTS

Much of the wood used in construction has been processed into laminated wood, structural composite wood, or wood panel products in order to overcome various of the shortcomings of solid wood structural members.

Laminated Wood

Large structural members are often produced by joining many small strips of wood together with glue to form *laminated wood* (called *glulam* for short). There are three major reasons to laminate: size, shape, and quality. Any desired size of structural member can be laminated, up to the capacities of the hoisting and transportation machinery needed to deliver and erect it, without having to search for a tree of sufficient girth and height. Wood can be laminated into shapes that cannot be obtained in nature: curves, angles, varying cross sections (Figure 3.24). Quality can be specified and closely controlled in laminated members because defects can be cut out of the wood before laminating. Seasoning is carried out before the wood is laminated (largely eliminating the checks and distortions that characterize solid timbers), and the strongest, highest-quality wood can be placed in the parts of the member that will be subjected to the highest structural stresses. The fabrication of laminated members obviously adds to the cost per board foot, but this is often overcome by the smaller size of the laminated member that can replace a solid timber of equal load-carrying capacity. In many cases, solid timbers are simply not available at any price in the required size, shape, or quality.

Individual laminations are usually 1½ inches (38 mm) thick except in curved members with small bending radii, where ³/₄-inch (19-mm) stock is used. End joints between individual pieces are either *finger jointed*

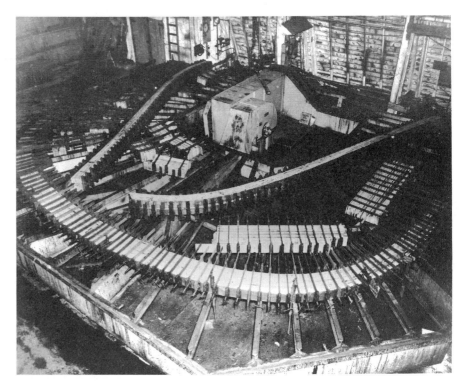

FIGURE 3.24
Glue laminating a U-shaped timber for a ship. Several smaller members have also been glued and clamped and are drying alongside the larger timber. (*Courtesy of Forest Products Laboratory, Forest Service, USDA*)

or *scarf jointed*. These types of joints allow the glue to transmit tensile and compressive forces longitudinally from one piece to the next within a lamination (Figure 3.25). Adhesives are chosen according to the moisture conditions under which the member will serve. Any size member can be laminated, but standard depths range from 3 to 75 inches (76 to 1905 mm). Standard widths range from 2⅛ to 14¼ inches (54 to 362 mm).

Structural Composite Lumber

A number of wood producers have developed *structural composite lumber* products that are made up of ordinary plywood veneers. Unlike plywood, however, in these products the grains of all the veneers are oriented in the longitudinal direction of the

piece of lumber to achieve maximum bending strength. One type of product, *laminated veneer lumber (LVL)*, uses the veneers in sheets and looks like a thick plywood with no crossbands (Figure 3.46). In another type, *parallel strand lumber (PSL)*, the veneers are sliced into narrow strands that are coated with adhesive, oriented longitudinally, pressed into a rectangular cross section, and cured under heat and pressure (Figure 3.26). LVL is generally produced in smaller sizes that approximate those of dimension lumber, whereas PSL is manufactured in a wide range of sizes including large dimensions like those of glue-laminated timbers. Structural composite lumber products offer much the same benefits as laminated wood products, but they can be produced more economically through a higher degree of mechanization.

JOINTS IN LAMINATIONS

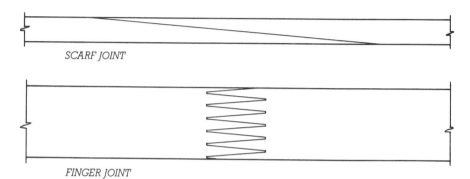

SCARF JOINT

FINGER JOINT

Figure 3.25
Joints within a lamination of a glue-laminated beam, seen in the upper drawing in a small-scale elevation view, must be scarf jointed or finger jointed to transmit the tensile and compressive forces from one piece of wood to the next. The individual pieces of wood are prepared for jointing by high-speed machines that mill the scarf or fingers with rotating cutters of the appropriate shape.

"Your chessboard, sire, is inlaid with two woods: ebony and maple. The square on which your enlightened gaze is fixed was cut from the ring of a trunk that grew in a year of drought: you see how its fibers are arranged? Here a barely hinted knot can be made out: a bud tried to burgeon on a premature spring day, but the night's frost forced it to desist. . . . Here is a thicker pore: perhaps it was a larvum's nest. . . ." The quantity of things that could be read in a little piece of smooth and empty wood overwhelmed Kublai; Polo was already talking about ebony forests, about rafts laden with logs that come down the rivers, of docks, of women at the windows . . .

Italo Calvino, in *Invisible Cities,* New York, Harcourt Brace Jovanovich, 1978

Figure 3.26
This parallel strand lumber, made up of strands of wood veneer, is largely free of defects, making it far stronger and stiffer than ordinary dimension lumber. (*Courtesy of Trus Joist MacMillan*)

FIGURE 3.27
Plywood is made of veneers selected to give the optimum combination of economy and performance for each application. This sheet of roof sheathing plywood is faced with a D veneer on the underside and a C veneer on the top side. These veneers, though unattractive, perform well structurally and are much less costly than the higher grades incorporated into plywoods made for uses where appearance is important. (*Courtesy of APA–The Engineered Wood Association*)

Wood Panel Products

Wood in panel form is advantageous for many building applications (Figure 3.27). The panel dimensions are usually 4 by 8 feet (1.22 by 2.44 m). Panels require less labor for installation than boards because fewer pieces must be handled, and wood panel products are fabricated in such a way that they minimize many of the problems of boards and dimension lumber: Panels are more nearly equal in strength in their two principal directions than solid wood; shrinking, swelling, checking, and splitting are greatly reduced. Additionally, panel products make more efficient use of forest resources than solid wood products through

FIGURE 3.28
Five different wood panel products, from top to bottom: plywood, composite panel, waferboard, oriented strand board, and particleboard. (*Courtesy of APA–The Engineered Wood Association*)

less wasteful ways of reducing logs to building products and through utilization in some types of panels of material that would otherwise be thrown away—branches, undersized trees, and mill wastes.

Structural wood panel products fall into three general categories (Figure 3.28):

1. *Plywood panels,* which are made up of thin wood *veneers* glued together. The grain on the front and back face veneers runs in the same direction, while the grain in one or more interior *crossbands* runs in the perpendicular direction. There is always an odd number of layers in plywood, which equalizes the effects of moisture movement, but an interior layer may be made up of a single veneer or of two veneers with their grains running in the same direction.

2. *Composite panels,* which have two parallel face veneers bonded to a core of reconstituted wood fibers.

3. *Nonveneered panels,* which are of three different types:

  a. *Oriented strand board (OSB),* which is made of long, strandlike wood particles compressed and glued into three to five layers. The strands are oriented in the same manner in each layer as the grains of the veneer layers in plywood. Because of the length and controlled orientation of the strands, oriented strand board is generally stronger and stiffer than the other two types of nonveneered panels.

  b. *Waferboard,* which is composed of large, waferlike flakes of wood compressed and bonded into panels.

  c. *Particleboard,* which is manufactured in several different classes, all of which are made up of smaller wood particles than oriented strand board or waferboard, compressed and bonded into panels.

APA–The Engineered Wood Association has established performance standards that allow considerable interchangeability among these

various types of panels for many construction uses. The standards are based on structural adequacy for a specified use, dimensional stability under varying moisture conditions, and the durability of the adhesive bond that holds the panel together. Because it can be produced from small trees and even branches, oriented strand board is generally more economical than plywood and is gradually replacing it in many construction applications.

Plywood Production

Veneers for structural panels are *rotary sliced:* Logs are soaked in hot water to soften the wood, then each is rotated in a large lathe against a stationary knife that peels off a continuous strip of veneer, much as paper is unwound from a roll (Figures 3.29, 3.30). The strip of veneer is clipped into sheets that pass through a drying kiln where in a few minutes their moisture content is reduced to roughly 5 percent. The sheets are then assembled into larger sheets, repaired as necessary with patches glued into the sheet to fill open defects, and graded and sorted according to quality (Figure 3.31). A machine spreads glue onto the veneers as they are laid atop one another in the required sequence and grain orientations. The glued veneers are transformed into plywood in presses that apply elevated temperatures and pressures to create dense, flat panels. The panels are trimmed to size, sanded as required, graded and gradestamped before shipping. Grade B veneers and higher are always sanded smooth, but panels intended for sheathing are always left unsanded because sanding slightly reduces the thickness, which diminishes the structural strength of the panel. Panels intended for subfloors and floor underlayment are lightly *touch sanded* to produce a more flat and uniform surface without seriously affecting their structural performance.

(a)

(b)

(c)

(d)

FIGURE 3.29

Plywood manufacture. (*a, b*) A 250-horsepower lathe spins a softwood log as a knife peels of a continuous sheet of veneer for plywood manufacture. (*c*) An automatic clipper removes unusable areas of veneer and trims the rest into sheets of the proper size for plywood panels. (*d*) The clipped sheets are fed into a continuous forced-air dryer, along whose 150-foot (45-m) path they will lose about half their weight in moisture. (*e*) Leaving the dryer, the sheets have a moisture content of about 5 percent. They are graded and sorted at this point in the process. (*f*) The higher grades of veneer are patched on this machine that punches out defects and replaces them with tightly fitted wood plugs. (*g*) In the layup line, automatic machinery applies glue to one side of each sheet of veneer, and alternates the grain direction of the sheets to produce loose plywood panels. (*h*) After layup, the loose panels are prepressed with a force of 300 tons per panel to consolidate them for easier handling. (*i*) Following prepressing, panels are squeezed individually between platens heated to 300 degrees

Fahrenheit (150°C) to cure the glue. (*j*) After trimming, sanding, or grooving as specified for each batch, the finished plywood panels are sorted into bins by grade, ready for shipment. (*Photos* b *and* i *courtesy of Georgia-Pacific; others courtesy of APA–The Engineered Wood Association*)

(e)

(f)

(g)

(h)

(i)

(j)

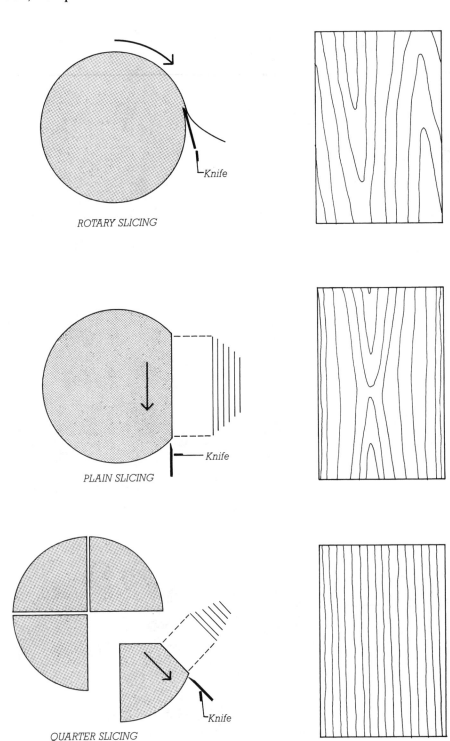

ROTARY SLICING

PLAIN SLICING

QUARTER SLICING

FIGURE 3.30
Veneers for structural plywood are rotary sliced, which is the most economical method. For better control of grain figure in face veneers of hardwood plywood, flitches are plain sliced or quarter sliced. The grain figure produced by rotary slicing, as seen in the detail to the right, is extremely broad and uneven. The finest figures are produced by quarter slicing, which results in a very close grain pattern with prominent rays.

Veneers for hardwood plywoods intended for interior paneling and cabinetwork are usually sliced from square blocks of wood called *flitches* in a machine that moves the flitch vertically against a stationary knife (Figure 3.30). Flitch-sliced veneers are analogous to quartersawn lumber: They exhibit a much tighter and more interesting grain figure than rotary-sliced veneers. They can also be arranged on the plywood face in such a way as to produce symmetrical grain patterns.

Standard plywood panels are 4 by 8 feet (1220 × 2440 mm) in surface area and range in thickness from $\frac{1}{4}$ to 1 inch (6.4 to 25.4 mm). Longer panels are manufactured for siding and industrial use. Actual surface dimensions of the structural grades of plywood are slightly less than nominal. This permits the panels to be installed with small spaces between them to allow for moisture expansion. Composite panels and nonveneered panels are manufactured by analogous processes to the same set of standard sizes as plywood panels and to some larger sizes as well.

Specifying Structural Wood Panels

For structural uses such as subflooring and sheathing, wood panels may be specified either by thickness or by *span rating*. The span rating is determined by laboratory load testing and is given on the gradestamp on the back of the panel, as shown in Figures 3.32 and 3.33. The purpose of the span rating system is to permit the use of many different species of woods and types of panels to achieve the same structural objectives. A panel with a span rating of $\frac{32}{16}$ may be used as roof sheathing over rafters spaced 32 inches (813 mm) apart, or as subflooring over joists spaced 16 inches (406 mm) apart. The long dimension of the sheet must be placed perpendicular to the length of the supporting members. A $\frac{32}{16}$ panel may be plywood, composite, or OSB, may be composed of any accepted wood species, and may be any of several thicknesses, so long as it passes the structural tests for a $^{32}/_{16}$ rating.

The designer must also select from three *exposure durability classifications* for structural wood panels: *Exterior, Exposure 1,* and *Exposure 2.* Panels marked "Exterior" are suitable for use as siding or in other permanently exposed applications.

"Exposure 1" panels have fully waterproof glue but do not have veneers of as high a quality as those of "Exterior" panels; they are suitable for structural sheathing and subflooring, which must often endure repeated wetting during construction. "Exposure 2" is suitable for panels that will be fully protected from weather and will be subjected to a minimum of wetting during construction. About 95 percent of structural panel products are classified "Exposure 1."

For panels intended as finish surfaces, the quality of the face veneers is of obvious concern and should be specified by the designer. For some types of work, fine flitch-sliced hardwood face veneers may be selected rather than rotary-sliced softwood veneers, and the matching pattern of the veneers specified.

FIGURE 3.31
Veneer grades for softwood plywood.
(*Courtesy of APA–The Engineered Wood Association*)

TABLE 1

VENEER GRADES

A Smooth, paintable. Not more than 18 neatly made repairs, boat, sled, or router type, and parallel to grain, permitted. Wood or synthetic repairs permitted. May be used for natural finish in less demanding applications.

B Solid surface. Shims, sled or router repairs, and tight knots to 1 inch across grain permitted. Wood or synthetic repairs permitted. Some minor splits permitted.

C Plugged. Improved C veneer with splits limited to 1/8-inch width and knotholes or other open defects limited to 1/4 x 1/2 inch. Wood or synthetic repairs permitted. Admits some broken grain.

C Tight knots to 1-1/2 inch. Knotholes to 1 inch across grain and some to 1-1/2 inch if total width of knots and knotholes is within specified limits. Synthetic or wood repairs. Discoloration and sanding defects that do not impair strength permitted. Limited splits allowed. Stitching permitted.

D Knots and knotholes to 2-1/2-inch width across grain and 1/2 inch larger within specified limits. Limited splits are permitted. Stitching permitted. Limited to Exposure 1 or Interior panels.

FIGURE 3.32
Typical gradestamps for structural wood panels. Gradestamps are found on the back of each panel. (*Courtesy of APA–The Engineered Wood Association*)

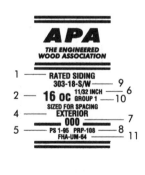

1 Panel grade
2 Span Rating
3 Tongue-and-groove
4 Exposure durability classification
5 Product Standard
6 Thickness
7 Mill number
8 APA's Performance Rated Panel Standard

9 Siding face grade
10 Species group number
11 HUD/FHA recognition
12 Panel grade, Canadian standard
13 Panel mark – Rating and end-use designation, Canadian standard.
14 Canadian performance-rated panel standard
15 Panel face orientation indicator

TABLE 2

GUIDE TO APA PERFORMANCE RATED PANELS[a][b]
FOR APPLICATION RECOMMENDATIONS, SEE FOLLOWING PAGES.

APA RATED SHEATHING Typical Trademark	APA THE ENGINEERED WOOD ASSOCIATION RATED SHEATHING 40/20 19/32 INCH SIZED FOR SPACING EXPOSURE 1 000 PS 2-92 SHEATHING PRP-108 HUD-UM-40C	APA THE ENGINEERED WOOD ASSOCIATION RATED SHEATHING 24/16 7/16 INCH SIZED FOR SPACING EXPOSURE 1 000 PRP-108 HUD-UM-40C	Specially designed for subflooring and wall and roof sheathing. Also good for a broad range of other construction and industrial applications. Can be manufactured as plywood, as a composite, or as OSB. EXPOSURE DURABILITY CLASSIFICATIONS: Exterior, Exposure 1, Exposure 2. COMMON THICKNESSES: 5/16, 3/8, 7/16, 15/32, 1/2, 19/32, 5/8, 23/32, 3/4.
APA STRUCTURAL I RATED SHEATHING[c] Typical Trademark	APA THE ENGINEERED WOOD ASSOCIATION RATED SHEATHING STRUCTURAL I 32/16 15/32 INCH SIZED FOR SPACING EXPOSURE 1 000 PS 1-95 C-D PRP-108	APA THE ENGINEERED WOOD ASSOCIATION RATED SHEATHING 32/16 15/32 INCH SIZED FOR SPACING EXPOSURE 1 000 STRUCTURAL I RATED DIAPHRAGMS-SHEAR WALLS PANELIZED ROOFS PRP-108 HUD-UM-40C	Unsanded grade for use where shear and cross-panel strength properties are of maximum importance, such as panelized roofs and diaphragms. Can be manufactured as plywood, as a composite, or as OSB. EXPOSURE DURABILITY CLASSIFICATIONS: Exterior, Exposure 1. COMMON THICKNESSES: 5/16, 3/8, 7/16, 15/32, 1/2, 19/32, 5/8, 23/32, 3/4.
APA RATED STURD-I-FLOOR Typical Trademark	APA THE ENGINEERED WOOD ASSOCIATION RATED STURD-I-FLOOR 24 oc 23/32 INCH SIZED FOR SPACING T&G NET WIDTH 47-1/2 EXPOSURE 1 000 PS 2-92 SINGLE FLOOR PRP-108 HUD-UM-40C	APA THE ENGINEERED WOOD ASSOCIATION RATED STURD-I-FLOOR 20 oc 19/32 INCH SIZED FOR SPACING T&G NET WIDTH 47-1/2 EXPOSURE 1 000 PRP-108 HUD-UM-40C	Specially designed as combination subfloor-underlayment. Provides smooth surface for application of carpet and pad and possesses high concentrated and impact load resistance. Can be manufactured as plywood, as a composite, or as OSB. Available square edge or tongue-and-groove. EXPOSURE DURABILITY CLASSIFICATIONS: Exterior, Exposure 1, Exposure 2. COMMON THICKNESSES: 19/32, 5/8, 23/32, 3/4, 1, 1-1/8.
APA RATED SIDING Typical Trademark	APA THE ENGINEERED WOOD ASSOCIATION RATED SIDING 24 oc 19/32 INCH SIZED FOR SPACING EXPOSURE 1 000 PRP-108 HUD-UM-40C	APA THE ENGINEERED WOOD ASSOCIATION RATED SIDING 303-18-S/W 16 oc 11/32 INCH GROUP 1 SIZED FOR SPACING EXTERIOR 000 PS 1-95 PRP-108 FHA-UM-64	For exterior siding, fencing, etc. Can be manufactured as plywood, as a composite or as an overlaid OSB. Both panel and lap siding available. Special surface treatment such as V-groove, channel groove, deep groove (such as APA Texture 1-11), brushed, rough sawn and overlaid (MDO) with smooth- or texture-embossed face. Span Rating (stud spacing for siding qualified for APA Sturd-I-Wall applications) and face grade classification (for veneer-faced siding) indicated in trademark. EXPOSURE DURABILITY CLASSIFICATION: Exterior. COMMON THICKNESSES: 11/32, 3/8, 7/16, 15/32, 1/2, 19/32, 5/8.

(a) Specific grades, thicknesses and exposure durability classifications may be in limited supply in some areas. Check with your supplier before specifying.

(b) Specify Performance Rated Panels by thickness and Span Rating. Span Ratings are based on panel strength and stiffness. Since these properties are a function of panel composition and configuration as well as thickness, the same Span Rating may appear on panels of different thickness. Conversely, panels of the same thickness may be marked with different Span Ratings.

(c) All plies in Structural I plywood panels are special improved grades and panels marked PS 1 are limited to Group 1 species. Other panels marked Structural I Rated qualify through special performance testing. Structural II plywood panels are also provided for, but rarely manufactured. Application recommendations for Structural II plywood are identical to those for APA RATED SHEATHING plywood.

FIGURE 3.33
A guide to specifying structural panels. Plywood panels for use in paneling, furniture, and other applications where appearance is important are graded by the quality of their face veneers. (*Courtesy of APA–The Engineered Wood Association*)

Other Wood Panel Products

Several types of nonstructural or semistructural panels of wood fiber are often used in construction. *Hardboard* is a thin, dense panel made of highly compressed wood fibers. It is available in several thicknesses and surface finishes, and in some formulations it is durable against weather exposure. Hardboard is produced in configurations for residential siding and roofing as well as in general-purpose panels of standard dimension. *Cane fiber board* is a thick, low-density panel with some thermal insulating value; it is used in wood construction chiefly as a nonstructural or semistructural wall sheathing. Panels made of recycled paper are low in cost and are useful for wall sheathing, carpet underlayment, and incorporation in certain proprietary types of roof decking and insulating assemblies.

CHEMICAL TREATMENT

Various chemical treatments have been developed to counteract two major weaknesses of wood: its combustibility and its susceptibility to attack by decay and insects. Fire-retardant treatment is accomplished by placing lumber in a pressure vessel and impregnating it with certain chemical salts that greatly reduce its combustibility. The cost of fire-retardant-treated wood is such that it is little used in single-family residential construction. Its major uses are roof sheathing in attached houses, and nonstructural partitions and other interior components in buildings of fire-resistant construction.

Decay and insect resistance is very important in wood that is used in or very near to the ground, and in wood used for exposed outdoor structures such as marine docks, fences, decks, and porches. Decay-resistant treatment is accomplished by *pressure impregnation* with any of several types of preservatives. *Creosote* is an oily derivative of coal that is widely used

> I shall always remember how as a child I played on the wooden floor. The wide boards were warm and friendly, and in their texture I discovered a rich and enchanting world of veins and knots. I also remember the comfort and security experienced when falling asleep next to the round logs of an old timber wall; a wall which was not just a plain surface but had a plastic presence like everything alive. Thus sight, touch, and even smell were satisfied, which is as it should be when a child meets the world.
>
> **Christian Norberg-Schulz**

in engineering structures, but because of its odor, toxicity, and unpaintability, it is unsuitable for most purposes in building construction. *Pentachlorophenol* is also impregnated as an oil solution, and, as with other oily preservatives, wood treated with it cannot be painted. The most widely used preservatives in building construction are *waterborne salts.* The most common of these is chromated copper arsenate (CCA), which imparts a greenish color to the wood but permits subsequent painting or staining. A newer formulation is based on copper and quaternary ammonia and causes wood to weather to a warm brown color. Preservatives based on boron compounds are also available. Preservatives of any of these types can be brushed or sprayed onto wood, but long-lasting protection (30 years and more) can only be accomplished by pressure impregnation. Wood treated with waterborne salts is often sold without drying, which is appropriate for use in the ground, but for interior use, it must be kiln dried after treatment.

Because of the poisonous nature of most types of wood preservatives, their use is often controlled by environmental regulations.

The heartwood of some species of wood is naturally resistant to decay and insects and can be used instead of preservative-treated wood. The most commonly used decay-resistant species are Redwood, Bald cypress, and Red and White cedars. The sapwood of these species is no more resistant to attack than that of any other tree, so "All-Heart" grade should be specified.

Wood–polymer composite planks intended for use in outdoor construction are a recent addition to the lumber market. A typical product of this type is made from wood fiber and recycled polyethylene. Its advantages are decay resistance and easy workability. Most such products are not as stiff as sawn lumber and require special attention to structural considerations in design.

Most wood-attacking organisms need both air and moisture to live. Most can therefore be kept out of wood by constructing and maintaining a building so that its wood components are kept dry at all times. This includes keeping all wood well clear of the soil, ventilating attics and crawlspaces to remove moisture, using good construction detailing to keep wood dry, and fixing roof and plumbing leaks as soon as they occur.

WOOD FASTENERS

Fasteners have always been the weak link in wood construction. The interlocking timber connections of the past, laboriously mortised and pegged, were weak because much of the wood in a joint had to be removed to make the connection. In today's wood connections, it is usually impossible to insert enough nails, screws, or bolts in a connection to develop the full strength of the members being joined. Adhesives and toothed plates are often capable of achieving this strength, but are

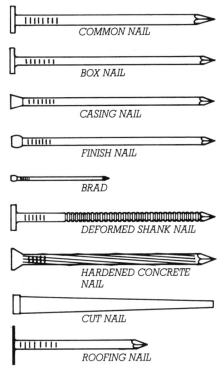

FIGURE 3.34
All nailed framing connections are made with common nails or their machine-driven equivalent. Box nails, which are made of lighter-gauge wire, do not have as much holding power as common nails; they are used in construction for attaching wood shingles. Casing nails, finish nails, and brads are used for attaching finish components of a building. Their heads are set below the surface of the wood with a steel punch, and the holes are filled before painting. Deformed shank nails, which are very resistant to withdrawal from the wood, are used for such applications as attaching gypsum wallboard and floor underlayment, materials that cannot be allowed to work loose in service. Concrete nails can be driven short distances into masonry or concrete for attaching furring strips and sleepers. Cut nails, once widely used for framing connections, are now used mostly for attaching finish flooring because their square tips punch through the wood rather than wedge through, minimizing splitting of brittle woods. The large head on roofing nails is needed to apply sufficient holding power to the soft material of which asphalt shingles are made.

largely limited to factory installation. Fortunately, most connections in wood structures depend primarily on the direct bearing of one member on another for their strength, and a variety of simple fasteners are adequate for the majority of purposes.

Nails

Nails are sharpened metal pins that are driven into wood with a hammer or a mechanical nail gun. Common nails and finish nails are the two types most frequently used. *Common nails* have flat heads and are used for most structural connections in light frame construction. *Finish nails* are virtually headless and are used to fasten finish woodwork, where they are less obtrusive than common nails (Figure 3.34).

In the United States, the size of a nail is measured in *pennies*. This strange unit probably originated long ago as the price of a hundred nails of a given size and persists in use despite the effects of inflation on nail prices. Figure 3.35 shows the dimensions of the various sizes of common nails. Finish nails are the same length as common nails of the corresponding penny designation.

Nails are ordinarily furnished *bright*, meaning that they are made of plain, uncoated steel. Nails that will be exposed to the weather should be of a corrosion-resistant type, either *hot-dip galvanized*, aluminum, or stainless steel. (The zinc coating on *electrogalvanized* nails is very thin, and is often damaged during driving.) Corrosion resistance is particularly important in nails for exterior siding, trim, and decks, which would be stained by rust leaching from bright nails.

The three methods of fastening with nails are shown in Figure 3.36. Each of these methods has its uses in building construction, as illustrated in Chapters 5 and 6. Nails are the favored means of fastening wood because they require no predrilling of holes under most conditions and they are extremely fast to install.

FIGURE 3.35
Standard sizes of common nails, reproduced full size. The abbreviation "d" stands for "penny." The length of each nail is given below its size designation. The three sizes of nail most used in light frame construction, 16d, 10d, and 8d, are shaded.

FIGURE 3.36
Face nailing is the strongest of the three methods of nailing. End nailing is relatively weak and is useful primarily for holding framing members in alignment until gravity forces and applied sheathing make a stronger connection. Toe nailing is used in situations where access for end nailing is not available. Toe nails are surprisingly strong—load tests show them to carry about five-sixths as much load as face nails of the same size.

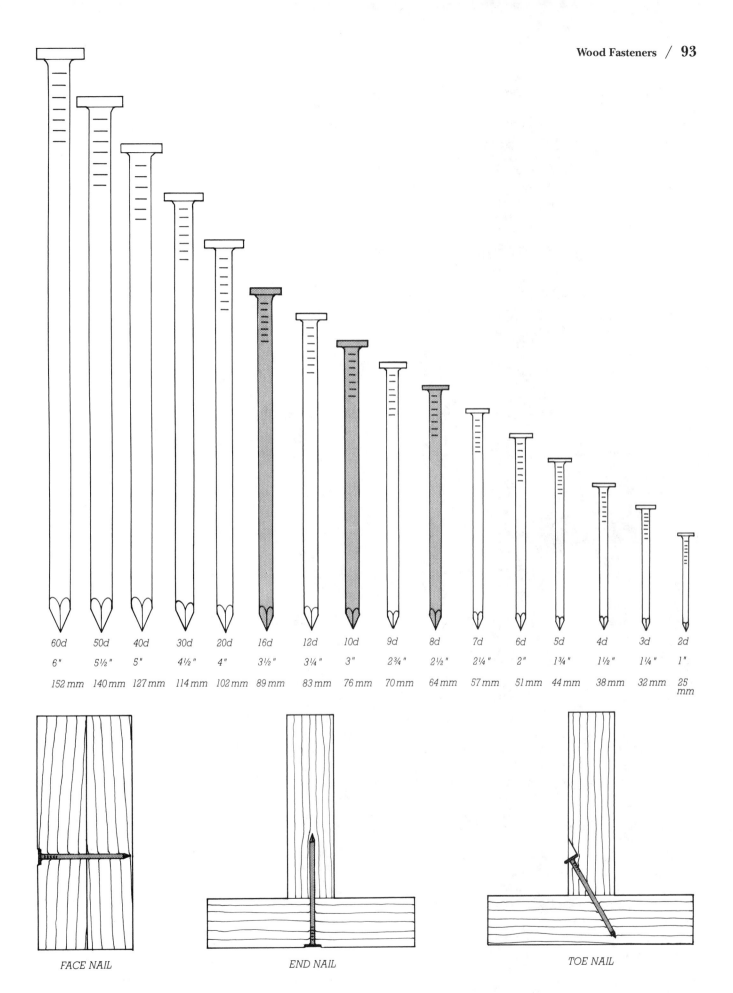

60d	50d	40d	30d	20d	16d	12d	10d	9d	8d	7d	6d	5d	4d	3d	2d
6"	5½"	5"	4½"	4"	3½"	3¼"	3"	2¾"	2½"	2¼"	2"	1¾"	1½"	1¼"	1"
152 mm	140 mm	127 mm	114 mm	102 mm	89 mm	83 mm	76 mm	70 mm	64 mm	57 mm	51 mm	44 mm	38 mm	32 mm	25 mm

FACE NAIL

END NAIL

TOE NAIL

Wood Screws and Lag Screws

Screws are inserted into drilled holes and turned into place with a screwdriver or wrench (Figure 3.37). They are little used in ordinary wood light framing because they cost more and take much longer to install than nails, but they are often used in cabinetwork, in furniture, and for mounting hardware such as hinges. Screws form tighter, stronger connections than nails and can be backed out and reinserted if a component needs to be adjusted or remounted. Very large screws for heavy structural connections are called *lag screws.* They have square or hexagonal heads and are driven with a wrench rather than a screwdriver. Small *drywall screws,* which can be driven by a power screwdriver, without first drilling holes in most cases, are made for use in attaching gypsum board to wood

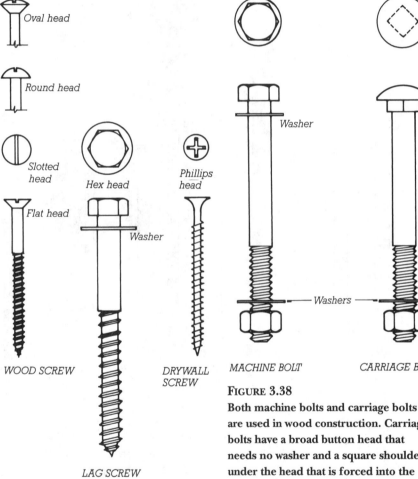

Oval head

Round head

Slotted head

Hex head

Phillips head

Flat head

Washer

WOOD SCREW

DRYWALL SCREW

LAG SCREW

FIGURE 3.37
Flat-head screws are used without washers and are driven flush with the surface of the wood. Round-head screws are used with flat washers, and oval-head with countersunk washers. The drywall screw does not use a washer, and is the only screw shown here that does not require a predrilled hole. Slotted, hex, and Phillips heads are all common.

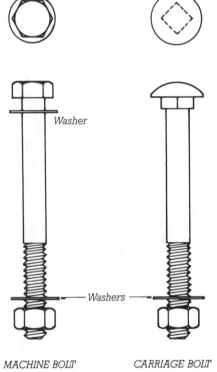

Washer

Washers

MACHINE BOLT

CARRIAGE BOLT

FIGURE 3.38
Both machine bolts and carriage bolts are used in wood construction. Carriage bolts have a broad button head that needs no washer and a square shoulder under the head that is forced into the drilled hole in the wood to prevent the bolt from turning as the nut is tightened.

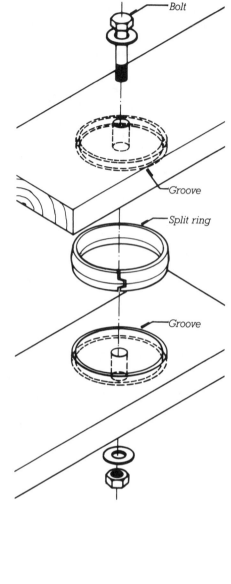

Bolt

Groove

Split ring

Groove

FIGURE 3.39
Split rings are high-capacity connectors used in heavily loaded joints of timber frames and trusses. After the center hole has been drilled through the two pieces, they are separated and the matching grooves are cut with a special rotary cutter driven by a power drill. The joint is then reassembled with the ring in place.

framing members and can be at least as fast to install as nails. They are weak and extremely brittle, however, so they cannot be used for structural connections. Machine-driven screws are also finding increasing use for attaching subflooring to joists in wood light frame construction; they result in a tighter floor construction that is less likely to squeak.

Bolts

Bolts are used for major structural connections in heavy timber framing. Commonly used bolts range in diameter from ³⁄₈ to 1 inch (9.53 to 25.4 mm), in any desired length. Flat steel disks called *washers* are inserted under heads and nuts of bolts to distribute the compressive force from the bolt across a greater area of wood (Figure 3.38).

Timber Connectors

Various types of timber connectors have been developed to increase the load-carrying capacity of bolts. The most widely used of these is the *split ring connector* (Figure 3.39). The split ring is used in conjunction with a bolt and is inserted in matching circular grooves in the mating pieces of wood. Its function is to spread the load across a much greater area of wood than can be done with a bolt alone. The split permits the ring to adjust to wood shrinkage. Split rings are useful primarily in heavy timber construction.

Toothed Plates

Toothed plates (Figure 3.40) are used in factory-produced lightweight roof and floor trusses. They are inserted into the wood with hydraulic presses, pneumatic presses, or mechanical rollers and act as metal splice plates, each with a very large number of built-in nails (Figures 3.41–3.43). They are extremely effective connectors because their multiple, closely spaced points interlock tightly with the fibers of the wood.

Sheet Metal and Metal Plate Framing Devices

Dozens of ingenious sheet metal and metal plate devices are manufactured for strengthening common connections in wood framing. The most frequently used is the *joist hanger,* but all of the devices shown in Figures 3.44 and 3.45 find extensive use. There

are two parallel series of this type of device, one made of sheet metal for use in light framing and one of thicker metal plate for heavy timber and laminated wood framing. The devices for light framing are attached with nails, and the heavier devices with bolts or lag screws.

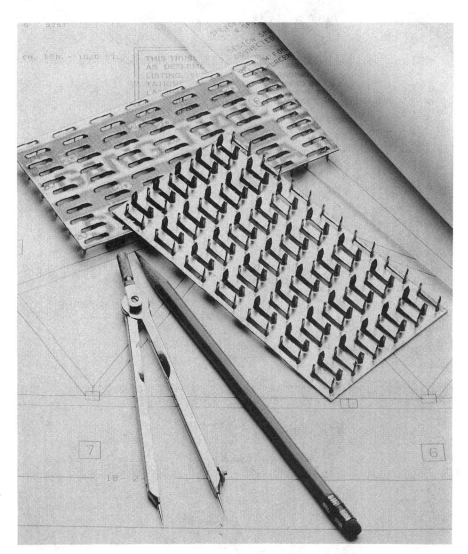

FIGURE 3.40
Manufacturers of toothed plate connectors also manufacture the machinery to install them and provide computer programs to aid local truss fabricators in designing and detailing trusses for specific buildings. The truss drawing in this photograph was generated by a plotter driven by such a program. The small rectangles on the drawing indicate the positions of toothed plate connectors at the joints of the truss. (*Courtesy of Gang-Nail System, Inc.*)

FIGURE 3.41
Factory workers align the wood members of a roof truss and position toothed plate connectors over the joints, tapping them with a hammer to embed them lightly and keep them in place. The roller marked "Gantry" then passes rapidly over the assembly table and presses the plates firmly into the wood. (*Courtesy of Wood Truss Council of America*)

FIGURE 3.42
The trusses are transported to the construction site on a special trailer. (*Courtesy of Wood Truss Council of America*)

FIGURE 3.43
Trusses can be designed and produced in almost any configuration. (*Courtesy of Wood Truss Council of America*)

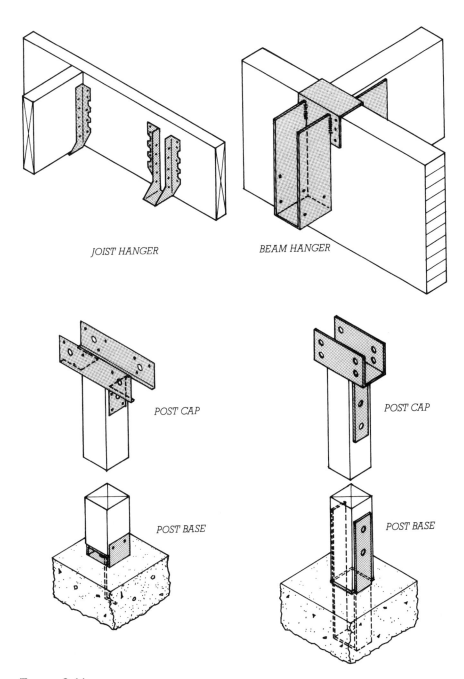

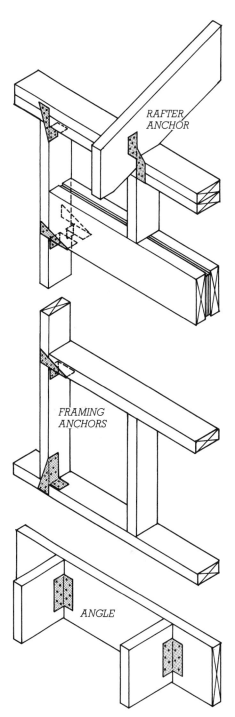

JOIST HANGER

BEAM HANGER

POST CAP

POST CAP

POST BASE

POST BASE

RAFTER ANCHOR

FRAMING ANCHORS

ANGLE

FIGURE 3.44
Joist hangers are used to make strong connections in floor framing wherever wood joists bear on one another at right angles. The heavier steel beam hangers are used primarily in heavy timber construction. Post bases serve the twofold function of preventing water from entering the end of the post and anchoring the post to the foundation. The bolts and lag screws used to connect the wood members to the heavier connectors are omitted from this drawing.

FIGURE 3.45
The sheet metal connectors shown in this diagram are less commonly used than those in Figure 3.44, but are invaluable in solving special framing problems and in reinforcing frames against wind uplift and earthquake forces.

Machine-Driven Staples and Nails

The speed of wood frame construction can be increased significantly through the use of pneumatically operated *nail guns* and *staple guns*. The nails or staples are of special design and are prepackaged by the manufacturer in self-feeding strips, each containing a large number of fasteners. These are loaded into guns that are powered by air fed through a hose from a small air compressor. They can be driven one by one as fast as the operator can pull the trigger. Several uses of machine-driven nails and staples are illustrated in Chapters 5 and 6.

Adhesives

Adhesives are widely employed in the factory production of plywood and panel products, laminated wood, wood structural components, and cabinetwork, but have relatively few uses on the construction job site. This disparity is explained by the need to clamp and hold adhesive joints under controlled environmental conditions until the adhesives have cured, which is easy to do in the shop but much more cumbersome and unreliable in the field. The area in which job site adhesives are most common is in securing subflooring panels to wood framing. Adhesives for these purposes are applied from a sealant gun in mastic form, and are clamped by simply nailing the panels to the framing. The nails usually serve as the primary structural connection, with the adhesive acting to eliminate squeaks in the floor.

WOOD MANUFACTURED BUILDING COMPONENTS

Dimension lumber, structural panel products, mechanical fasteners, and adhesives are used in combination to manufacture a number of highly efficient structural components that offer certain advantages to the designer of wooden buildings.

Trusses

Trusses for both roof and floor construction are manufactured in small, highly efficient plants in every part of North America. Most are based on 2 × 4s and 2 × 6s joined with toothed plate connectors. The designer or builder needs to specify only the span, the roof pitch, and the desired overhang detail. The truss manufacturer either uses a preengineered design for the specified truss or employs a sophisticated computer program to engineer the truss and develop the necessary cutting patterns for its constituent parts. The manufacture of trusses is shown in Figures 3.41 through 3.43, and several uses of trusses are depicted in Chapter 5.

Roof trusses use less wood than a comparable frame of conventional rafters and ceiling joists. Like floor trusses, they span the entire width of the building in most applications, allowing the designer complete freedom to locate interior partitions anywhere they are needed. The chief disadvantages of roof trusses as they are most commonly used are that they make the attic space unusable and that they generally restrict the designer to the spatial monotony of a flat ceiling throughout the building. Truss shapes can be designed and manufactured to overcome both these limitations, however (Figure 3.43).

Wood I-Joists

Manufactured wood I-shaped members (*I-joists*) are useful for framing of both roofs and floors (Figure 3.46). The flanges of the members may be made from solid lumber or laminated veneer lumber. The webs consist of plywood or OSB. Like trusses, these

FIGURE 3.46
The I-joist in the front of this photograph consists of two flanges of laminated veneer lumber (LVL) glued to a web of oriented strand board (OSB). The third member from the front is a beam made of LVL. The remainder of the beams and posts in this photograph are made of parallel strand lumber (PSL). (*Courtesy of Trus Joist MacMillan*)

components use wood more efficiently than conventional rafters and joists, and they can span farther between supports than dimension lumber. Further advantages include lighter weight than corresponding members made of dimension lumber, freedom from crooks and bows, and availability in lengths to 40 ft (12.2 m). Chapter 5 shows the use of I-joists in wood framing.

Panel Components

Dimension lumber and wood structural panels lend themselves readily to the prefabrication of many kinds of floor, wall, and roof panels. At their simplest, such panels consist of a section of framing, usually 4 feet (1219 mm) wide, sheathed with a sheet of plywood, OSB, or waferboard. In light frame construction, these are trucked to the construction site and rapidly nailed together into a complete frame, sheathed and ready for finishing. Panels can also be larger, encompassing whole walls or floors in a single component. They can become more sophisticated, including insulation, wiring, windows, doors, and exterior and interior finishes, if the thorny problems of joinery and code acceptance can be resolved.

For greatest structural efficiency and economy, wall, floor, and roof panels can utilize top and bottom sheets of plywood or OSB as the primary load-carrying members of the panel, if they are joined firmly together by a stiff plastic foam core to make a *sandwich panel* or by dimension lumber spacers to make a *stressed-skin panel* (Figure 3.47). In common usage, the name "stressed skin" is applied to both these panel types.

Dimension lumber and structural wood panels are the favored materials of the *manufactured housing* industry, which factory-builds entire houses as finished boxes, often complete with furnishings, and trucks them to prepared foundations where they are set in place and made ready for occupancy in a matter of hours. If the house is 14 feet (4.27 m) or less in width, it is constructed on a rubber-tired frame, is completely finished in the factory, and is known as a *mobile*

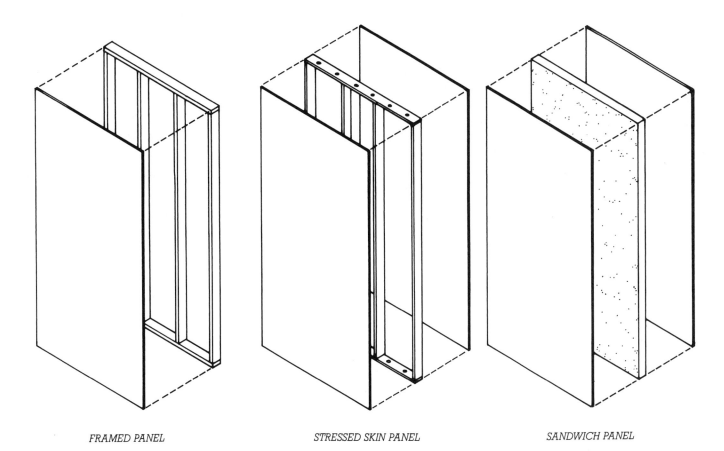

FRAMED PANEL   *STRESSED SKIN PANEL*   *SANDWICH PANEL*

FIGURE 3.47
Three types of prefabricated wood panels. The framed panel is identical to a segment of a conventionally framed wall, floor, or roof. The facings on the stressed-skin panel are adhesive bonded to thin wood spacers to form a structural unit in which the facings carry the major stresses. A sandwich panel functions structurally in the same way as a stressed-skin panel, but its facings are bonded to a core of insulating foam instead of wood spacers.

home. If wider than this, or more than one story high, it is built in two or more completed sections that are joined at the site, and is known as a *sectional home* or *modular home*. Mobile homes are sold at a fraction of the price of conventionally constructed houses. This is due in part to the economies of factory production and mass marketing, and in part to the use of components that are lighter and less costly, and therefore of substantially shorter life expectancy. But at prices that more closely approach the cost of conventional on-site construction, many companies manufacture modular housing to the same standards as conventional construction.

C.S.I./C.S.C. Masterformat Section Numbers for Wood	
06050	**FASTENERS AND ADHESIVES**
06100	**ROUGH CARPENTRY**
06110	Wood Framing
06115	Sheathing
06120	Structural Panels
06130	**HEAVY TIMBER CONSTRUCTION**
06170	**PREFABRICATED STRUCTURAL WOOD**
06180	Glued-Laminated Construction
06190	Wood Trusses
06195	Prefabricated Wood Beams and Joists
06200	**FINISH CARPENTRY**
06300	**WOOD TREATMENT**
06310	Preservative Treatment
06320	Fire-Retardant Treatment
06330	Insect Treatment
06400	**ARCHITECTURAL WOODWORK**

TYPES OF WOOD CONSTRUCTION

Wood construction has evolved into two major systems of on-site construction: The hand-hewn frames of centuries past have become the heavy timber frames of today, used primarily for structures larger than single-family houses. And from the heavy timber frame has sprung wood light frame construction, which is the dominant system for houses, small apartment buildings, and small commercial structures. These two systems of framing are detailed in Chapters 4 and 5.

SELECTED REFERENCES

1. Hoadley, R. Bruce. *Understanding Wood.* Newtown, Connecticut, The Taunton Press, 1980.

Beautifully illustrated and produced, this volume vividly explains and demonstrates the properties of wood as a material of construction.

2. Western Wood Products Association. *Western Woods Use Book.* (Address for ordering: 522 SW Fifth Avenue, Suite 400, Portland, OR 97204-2122.)

This looseleaf binder houses a complete reference library on the most common species of dimension lumber, timbers, and finish lumber. It includes the *National Design Specification for Wood Construction,* the standard for engineering design of wood structures. Updated frequently.

3. APA–The Engineered Wood Association. *Products and Applications.* (Address for ordering: P.O. Box 11700, Tacoma, WA 98411. Ask also for the catalog of APA publications.)

Updated frequently, this is a complete looseleaf guide to manufactured wood products including plywood, oriented strand board, waferboard, structural insulated panels, and glue-laminated wood.

4. Canadian Wood Council. *Wood Reference Handbook.* Ottawa, 1995.

A superbly illustrated encyclopedic reference on wood materials. (Telephone number for ordering: 800-463-5091.)

5. Callahan, Edward E. *Metal Plate Connected Wood Truss Handbook* (2nd ed.). Madison, Wisconsin, Wood Truss Council of America, 1997.

This is the standard reference on every phase of design, manufacturing, and installation of wood light trusses. (Address for ordering: Wood Truss Council of America, 5937 Meadowood Drive, Suite 14, Madison, WI 53711-4125.)

KEY TERMS AND CONCEPTS

bark
cambium
sapwood
heartwood
pith
cellulose
lignin
direction of grain
springwood
(earlywood)
summerwood
(latewood)
softwood
hardwood
tracheids
rays
fibers
vessels or pores
lumber
headsaw
sawyer
plainsawed
quartersawn
vertical grain
figure
free water
bound water
seasoning

air drying
kiln drying
green lumber
longitudinal, radial, tangential shrinkage
check
parallel-to-grain shrinkage
perpendicular-to-grain shrinkage
surfacing
S4S, S2S
S-DRY
S-GRN
growth characteristics
manufacturing characteristics
knot, knothole
decay
insect damage
split
crooking
bowing
twisting
cupping
wane
visual grading
machine grading
allowable strengths
nominal dimension
actual dimension
boards

dimension lumber
timbers
board foot
laminated wood (glulam)
finger joint, scarf joint
structural composite lumber
laminated veneer lumber (LVL)
parallel strand lumber (PSL)
structural wood panel
plywood panel
veneer
crossband
composite panel
nonveneered panel
oriented strand board (OSB)
waferboard
particleboard
rotary sliced
touch sanded
flitch
span rating
exposure durability classification
hardboard
cane fiber board
pressure impregnation
creosote
pentachlorophenol
waterborne salts

wood–polymer composite
nail
common nail
finish nail
penny
bright nail
hot-dip galvanized nail
electrogalvanized nail
screw
lag screw
drywall screw
bolt
washer
split ring connector
toothed plate
joist hanger
nail gun
staple gun
adhesive
truss
I-joist
sandwich panel
stressed-skin panel
manufactured housing
mobile home
sectional or modular home

REVIEW QUESTIONS

1. Discuss the changes in moisture content of the wood and their effects on a piece of dimension lumber, from the time the tree is cut, through its processing, until it has been in service in a building for an entire year.

2. Give the actual cross-sectional dimensions of the following pieces of kiln-dried lumber: 1×4, 2×4, 2×6, 2×8, 4×4, 4×12.

3. Why is wood laminated?

4. What is meant by a span rating of $^{32}/_{16}$? What types of wood products are rated in this way?

5. For what reasons might you specify preservative-treated wood?

6. Which common species of wood have decay- and insect-resistant heartwood?

7. Why are nails the fasteners of choice in wood construction?

EXERCISES

1. Visit a nearby lumberyard. Examine and list the species, grades, and sizes of lumber carried in stock. For what uses are each of these intended? While at the yard, look also at the available range of fasteners.

2. Pick up a number of scraps of dimension lumber from a shop or construction site. Examine each to see where it was located in the log before sawing. Note any drying distortions in each piece: How well do these correspond to the distortions you would have predicted? Measure accurately the width and thickness of each scrap and compare your measurements to the specified actual dimensions for each.

3. Assemble samples of as many different species of wood as you can find. Learn how to tell the different species apart, by color, odor, grain figure, ray structure, relative hardness, and so on. What are the most common uses for each species?

4. Visit a construction site and list the various types of lumber and wood products being used. Look for a gradestamp on each, and determine why the given grade is being used for each use. If possible, look at the architect's written specifications for the project and see how the lumber and wood products were specified.

HEAVY TIMBER FRAME CONSTRUCTION

Architects Moore, Lyndon, Turnbull, and Whitaker and structural engineer Patrick Morreau designed a rough-sawn heavy timber frame both as a means of supporting this seaside condominium dwelling and as a major feature of its interior design. For further photographs of this project, see Figures 4.31 and 4.32. (*Photo by Morley Baer*)

Wood beams have been used to span roofs and floors of buildings since the beginning of civilization. The first timber-framed buildings were crude pit houses, lean-tos, teepees, and basketlike assemblages of bent saplings. In earliest historic times, roof timbers were combined with masonry load-bearing walls to build houses and public buildings. In the Middle Ages, braced wall frames of timber were built for the first time (Figures 4.1, 4.2). The British carpenters who emigrated to North America in the 17th and 18th centuries brought with them a fully developed knowledge of how to build efficiently braced frames, and for two centuries North Americans lived and worked almost exclusively in buildings framed with hand-hewn wooden timbers joined by interlocking wood-to-wood connections (Figures 4.3, 4.4). Nails were rare and expensive, so they were used only in door and window construction and, sometimes, for fastening siding boards to the frame.

Until two centuries ago, logs could be converted to boards and timbers only by human muscle power. To make timbers, axemen skillfully scored and hewed logs to reduce them to a rectangular profile. Boards were produced slowly and laboriously with a long, two-man pit saw, one man standing in a pit beneath the log pulling the saw down and the other standing above, pulling it back up. But at the beginning of the 19th century, water-powered sawmills began to take over the work of transforming tree trunks into lumber, squaring timbers and slicing boards in a fraction of the time that it took to do the same work by hand.

Most of the great industrial mills of the 19th century, which manufactured textiles, shoes, machinery, and all the goods of civilization, consisted of heavy, sawn timber floors and roofs supported by masonry exterior walls (Figures 4.5, 4.6). The house builders and barn raisers of the early 19th century switched from hand-hewn to sawn timbers as soon as they became available. Many of these mills, barns, and houses still survive, and with them survives a rich tradition of heavy timber building that continues to the present day.

FIGURE 4.1
Braced wall framing was not developed until the late Middle Ages and early Renaissance, when it was often exposed on the face of the building in the style of construction known as *halftimbering*. The space between the timbers was filled with brickwork, or with wattle and daub, a crude plaster of sticks and mud, as seen here in Wythenshawe Hall, a 16th-century house near Manchester, England. (*Photo by James Austin, Cambridge, England*)

FIGURE 4.2
European timber house forms generally followed a progression of development from crude pit dwellings, made of earth and tree trunks, to cruck frames, to braced frames. The *crucks* (curved timbers), hewn by hand from appropriately shaped trees, were precursors of the laminated wood arches and rigid frames that are widely used today.

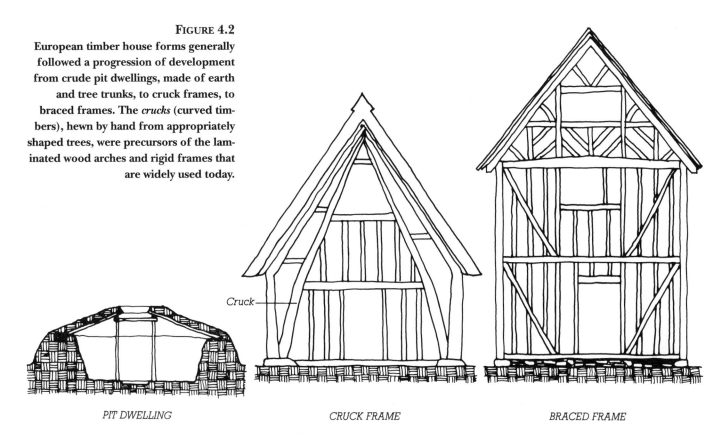

PIT DWELLING   CRUCK FRAME   BRACED FRAME

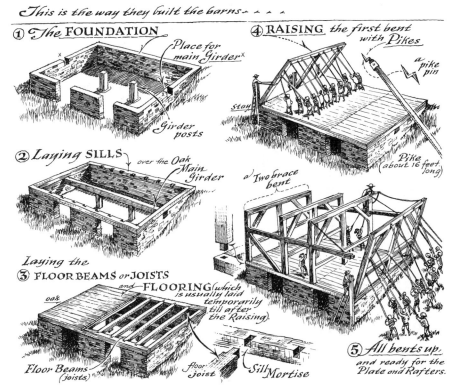

FIGURE 4.3
The European tradition of Heavy Timber framing was brought to North America by the earliest settlers and was used for houses and barns until well into the 19th century. (*Drawing by Eric Sloane, courtesy of the artist*)

(a)

(b)

(c)

FIGURE 4.4
Traditional timber framing has been revived in recent years by a number of builders who have learned the old methods of joinery and updated them with the use of modern power tools and equipment. (*a*) Assembling a bent (a plane of columns, beams, rafters, and braces). (*b*) The completed bents are laid out on the floor, ready for raising. (*c*) Raising the bents, using a truck-mounted crane, and installing floor framing and roof purlins. (*d*) The completed frame. (*e*) Enclosing the timber-framed house with sandwich panels consisting of waferboard faces bonded to an insulating foam core. (*Courtesy of Benson Woodworking, Alstead, New Hampshire*)

(d)

(e)

The old country builder, when he has to get out a cambered beam or a curved brace, goes round his yard and looks out the log that grew in the actual shape, and taking off two outer slabs by handwork in the sawpit, chops it roughly to shape with his side-axe and works it to the finished face with the adze, so that the completed work shall for ever bear the evidence of his skill . . .

Gertrude Jekyll, English garden
designer, writing in 1900

FIGURE 4.5
Most 19th-century industrial buildings in America were constructed of heavy timber roofs and floors supported at the perimeter on masonry loadbearing walls. This impressive group of textile mills stretches for a distance of 2 miles (3 km) along the Merrimac River in Manchester, New Hampshire. (*Photo by Randolph Langenbach*)

FIGURE 4.6
The generous windows in the mills provided plenty of daylight to work by. Columns were of wood or cast iron. Most New England mills, like this one, were framed very simply, with decking carried by beams running at right angles to the exterior walls, supported at the interior on two lines of columns. Notice that the finish flooring runs at right angles to the structural decking. Overhead sprinklers add a considerable measure of fire safety. (*Photo by Randolph Langenbach*)

HEAVY TIMBER (MILL) CONSTRUCTION

The Fire Resistance of Heavy Timbers

Large timbers, because of their greater capacity to absorb heat, are much slower to catch fire and burn than smaller lumber. When exposed to fire, a heavy timber beam, though deeply charred by gradual burning, will continue to support its load long after an unprotected steel beam would have collapsed. If the fire is not prolonged, a heavy timber beam or column can often be sandblasted afterward to remove the surface char and can continue in service. For these reasons, building codes recognize heavy timber construction that meets certain specific code requirements as having fire-resistive properties.

	Supporting Floor Loads	Supporting Roof and Ceiling Loads Only
Columns	8 × 8 (184 × 184 mm)	6 × 8 (140 × 184 mm)
Beams and Girders	6 × 10 (140 × 235 mm)	4 × 6 (89 × 140 mm)
Trusses	8 × 8 (184 × 184 mm)	4 × 6 (89 × 140 mm)
Decking	3″ + 1″ finish (64 mm + 19 mm finish)	2″, or 1⅛″ plywood (38 mm, or 29 mm plywood)

FIGURE 4.7
Minimum member sizes for Heavy Timber construction, as typically specified in a building code.

Heavy Timber construction with exterior walls of masonry or concrete is listed in the building code table in Figure 1.1 as Type 4 construction, Heavy Timber, and is often referred to as *Mill construction* or *Slow-burning construction*. Minimum sizes for the timbers and decking of Heavy Timber construction are specified in another section of the code; these requirements are summarized in Figure 4.7. Either sawn or laminated timbers are permitted. Traditionally, the long edges of the timbers are *chamfered* (beveled at 45°) to eliminate the thin corners of wood that catch fire most easily, but many codes no longer require this.

Wood Shrinkage in Heavy Timber Construction

The perimeter of the floors and roof of a Heavy Timber building must be supported on concrete or masonry [unless the building is at least 30 feet (9 m) from any surrounding buildings], whereas the interior may be supported on heavy wood columns. Wood, unlike masonry, is subject to large amounts of expansion and contraction caused by seasonal changes in moisture content, particularly in the direction perpendicular to its grain. A Heavy Timber building is detailed to minimize the effects of this differential shrinkage by eliminating cross-grain wood from the interior lines of support. In the tradi-

tional Mill construction shown in Figures 4.8 through 4.12, iron *pintles* or steel caps carry the column loads past the cross-grain of the beams at each floor, so that the beams and girders can shrink without causing the floors and roof to sag. In current practice, a laminated column may be fabricated as a single piece running the entire height of the building, with the beams connected to the column by wood bearing blocks or welded metal connectors; or columns may be butted directly to one another at each floor with the aid of metal connectors (Figures 4.11, 4.13, 4.14, 4.15).

Anchorage of Timber Beams and Masonry Walls

Where heavy timber beams join masonry or concrete walls, three problems must be solved: First, the beam must be protected from decay that might otherwise be caused by moisture seeping through the wall. This is done by leaving a ventilating airspace of at least ½ inch (13 mm) between the masonry and all sides of the beam except the bottom, unless the beam is chemically treated to resist decay. The second and third problems have to be solved together: The beam must be securely anchored to the wall so it cannot pull away under normal service, yet must be able to rotate freely so it does not pry the wall apart if it burns through during a severe fire (Figure 4.10). Two

methods of accomplishing these dual needs are shown in Figure 4.9, and another in Figure 4.13.

Floor and Roof Decks for Heavy Timber Buildings

Building codes require that Heavy Timber buildings have floors and roofs of solid wood construction, without internal cavities. Figure 4.16 shows several different types of *decking* used for these purposes. Minimum permissible thicknesses of decking are given in Figure 4.7. Floor decking must be covered with a finish floor consisting of nominal 1-inch (25-mm) *tongue-and-groove* boards laid at right angles or diagonally to the structural decking, or with ½-inch (13-mm) plywood or particleboard.

Lateral Bracing of Heavy Timber Buildings

A Heavy Timber building with an exterior masonry or concrete bearing wall is normally braced against wind and seismic forces by the shear resistance of its walls, working together with the diaphragm action of its roof and floor decks. In areas of high seismic risk, the walls must be reinforced both vertically and horizontally, and the decks may have to be specially nailed, or overlaid with plywood, to increase their shear resistance. In buildings with framed exterior walls, diagonal bracing or shear panels must be provided.

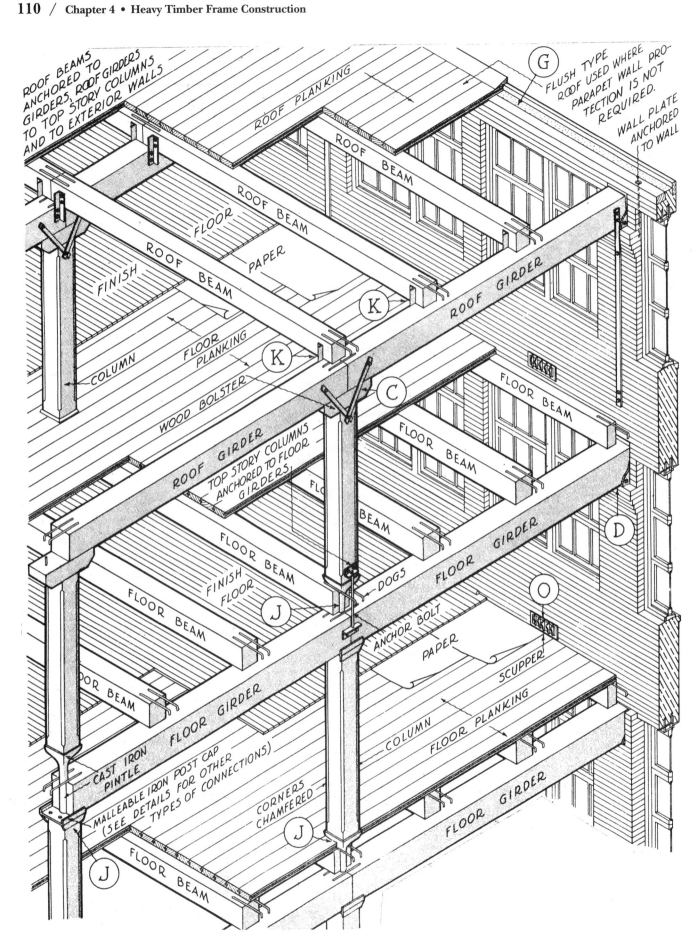

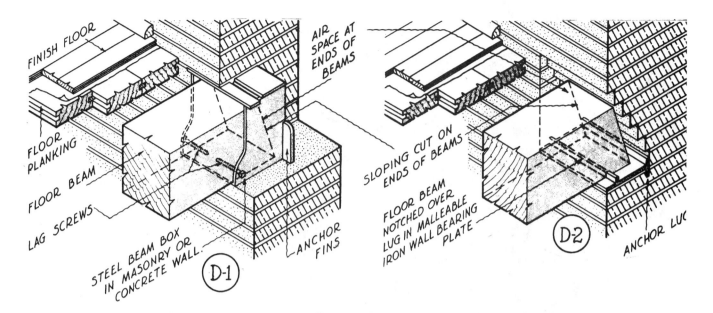

FINISH FLOOR

FLOOR PLANKING

FLOOR BEAM

LAG SCREWS

AIR SPACE AT ENDS OF BEAMS

STEEL BEAM BOX IN MASONRY OR CONCRETE WALL

ANCHOR FINS

D-1

SLOPING CUT ON ENDS OF BEAMS

FLOOR BEAM NOTCHED OVER LUG IN MALLEABLE IRON WALL BEARING PLATE

ANCHOR LUG

D-2

FIGURE 4.9
Two alternative details for the bearing of a beam on masonry in traditional Mill construction. In each case, the beam end is firecut to allow it to rotate out of the wall if it burns through (Figure 4.10), but anchored against pulling away from the wall, either with lag screws or a lug on the iron bearing plate. (*Courtesy of National Forest Products Association*)

FIGURE 4.8
Traditional Mill construction bypasses problems of wood shrinkage at the interior lines of support by using cast iron pintles to transmit the column loads through the beams and girders. Iron dogs tie the beams together over the girders. A long steel strap anchors the roof girder to a point sufficiently low in the outside wall that the weight of the masonry above the anchorage point is sufficient to resist wind uplift on the roof. (*Courtesy of National Forest Products Association*)

FIGURE 4.10
A timber beam that burns through in a prolonged fire is likely to topple its supporting masonry wall (left) unless its end is *firecut* (right). The beam is anchored to the wall in each case by a steel strap anchor.

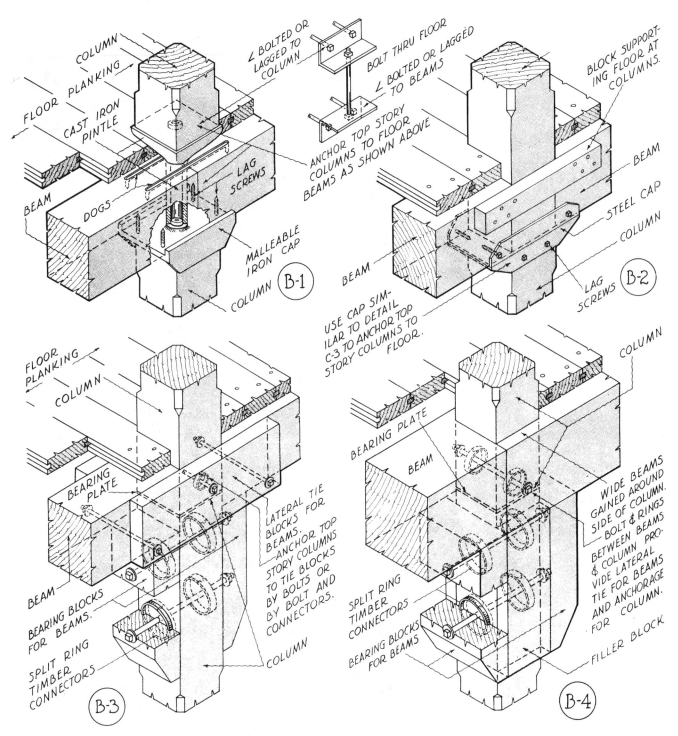

FIGURE 4.11

Four alternative details for interior girder–column intersections. Detail B-1 avoids
wood shrinkage problems with an iron pintle, while the other three details bring the
columns through the beams with only a steel bearing plate between the ends of the column sections. Split ring connectors are used in the lower two details to form a strong
enough connection between the bearing blocks and the columns to support the loads
from the beams; it would take a much larger number of bolts to do the same job.

(*Courtesy of National Forest Products Association*)

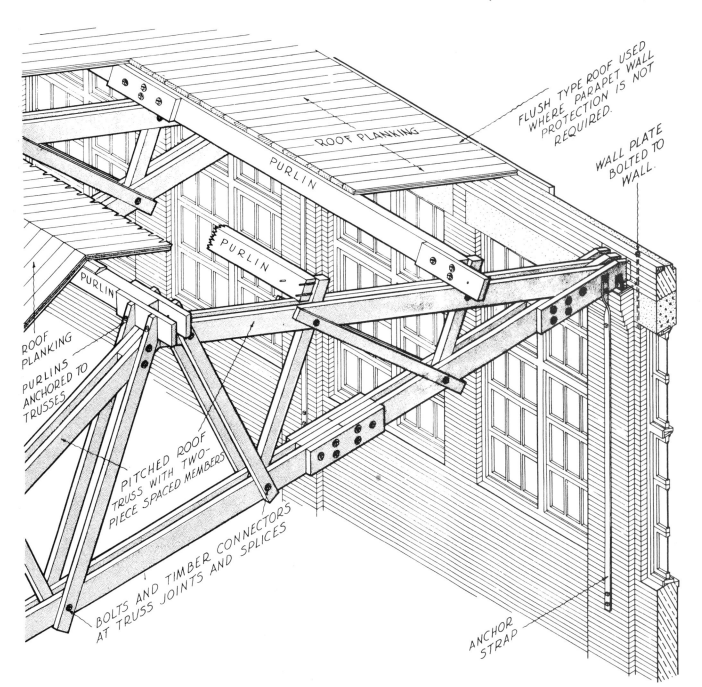

FIGURE 4.12
Heavy timber roof trusses for Mill construction. Split ring connectors are used to transmit the heavy forces between the overlapping members of the truss. A long anchor strap is again used at the outside wall, as explained in the legend to Figure 4.8.

(*Courtesy of National Forest Products Association*)

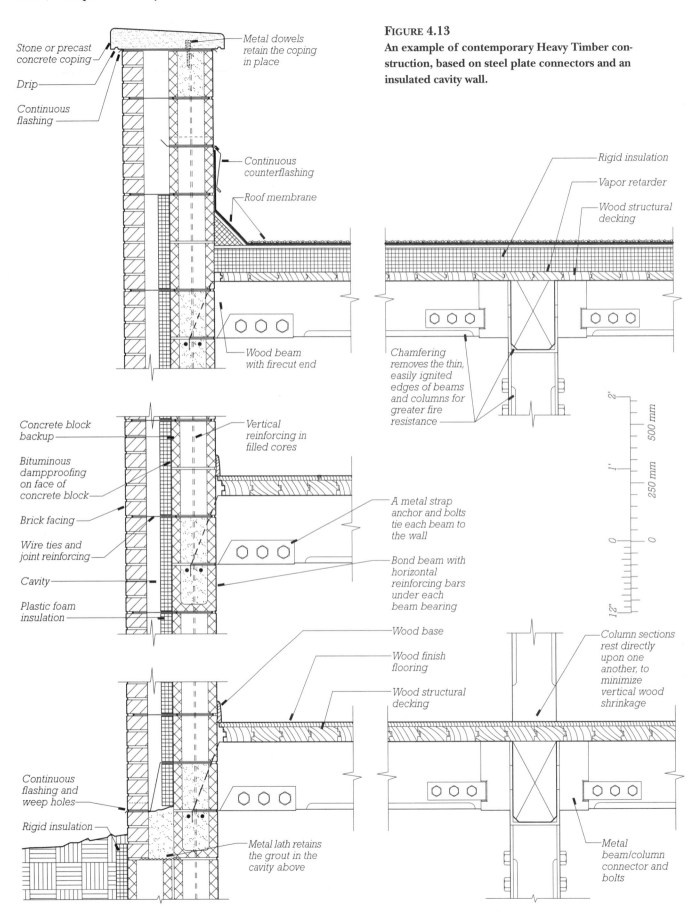

Stone or precast concrete coping

Drip

Continuous flashing

Metal dowels retain the coping in place

Continuous counterflashing

Roof membrane

Wood beam with firecut end

Concrete block backup

Bituminous dampproofing on face of concrete block

Brick facing

Wire ties and joint reinforcing

Cavity

Plastic foam insulation

Vertical reinforcing in filled cores

A metal strap anchor and bolts tie each beam to the wall

Bond beam with horizontal reinforcing bars under each beam bearing

Continuous flashing and weep holes

Rigid insulation

Metal lath retains the grout in the cavity above

Wood base

Wood finish flooring

Wood structural decking

Rigid insulation

Vapor retarder

Wood structural decking

Chamfering removes the thin, easily ignited edges of beams and columns for greater fire resistance

Column sections rest directly upon one another, to minimize vertical wood shrinkage

Metal beam/column connector and bolts

2'

500 mm

1'

250 mm

0   0

12"

FIGURE 4.13
An example of contemporary Heavy Timber construction, based on steel plate connectors and an insulated cavity wall.

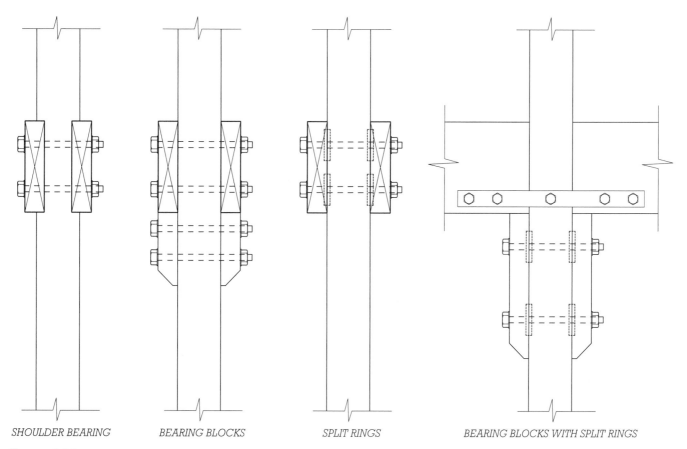

SHOULDER BEARING   BEARING BLOCKS   SPLIT RINGS   BEARING BLOCKS WITH SPLIT RINGS

FIGURE 4.14
Some typical beam–column connections for Heavy Timber construction. The first
three examples are for doubled beams sandwiched on either side of the column, and
the fourth shows single beams in the same plane as the column. In the shoulder bearing connection, the beams are recessed into the column by an amount that allows the
load to be transferred safely by wood-to-wood bearing; the bolts serve only to keep the
beams in the recesses. *Bearing blocks* allow more bolts to be inserted in a connection
than can fit through the beams, and each bolt in the bearing blocks can hold several
times as much load as one through the beam because it acts parallel to the grain of the
wood rather than perpendicular. It is generally impossible to place enough bolts in a
beam–column joint to transfer the load successfully without bearing blocks unless split
rings are used, as shown in the third example. The steel straps and bolts in the fourth
example hold the beams on the bearing blocks.

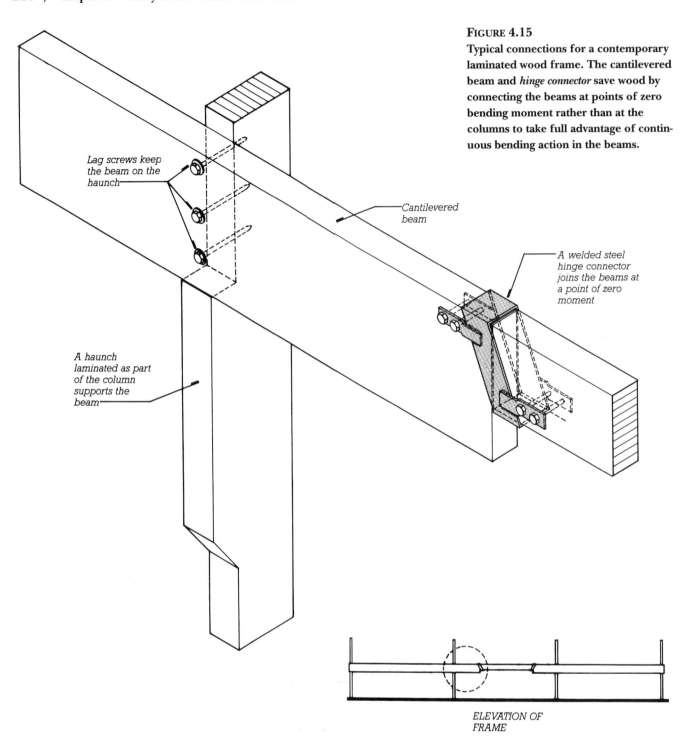

Lag screws keep
the beam on the
haunch

A haunch
laminated as part
of the column
supports the
beam

Cantilevered
beam

A welded steel
hinge connector
joins the beams at
a point of zero
moment

ELEVATION OF
FRAME

FIGURE 4.15
Typical connections for a contemporary
laminated wood frame. The cantilevered
beam and *hinge connector* save wood by
connecting the beams at points of zero
bending moment rather than at the
columns to take full advantage of continuous bending action in the beams.

FIGURE 4.16

Large-scale cross sections of four types of heavy timber decking. Tongue-and-groove is the most common, but the other three types are slightly more economical of lumber because wood is not wasted in the milling of the tongues. Laminated deck is a traditional type for longer spans and heavier loads; it consists of ordinary dimension lumber nailed together. Glue-laminated decking is a modern type. In the example shown here, five separate boards are glued together to make each piece of decking. Decking of any type is usually furnished and installed in random lengths, with end joints staggered in the roof to avoid zones of structural weakness. The splines, tongues, or nails allow the narrow strips of decking to share concentrated structural loads as if they constituted a continuous sheet of solid wood.

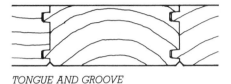

TONGUE AND GROOVE

LAMINATED DECK

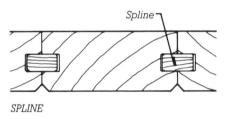

Spline

SPLINE

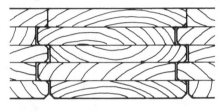

GLUE LAMINATED DECKING

cannot be simply inserted between ceiling joists but must be placed on top of the roof deck. If the roof is nearly flat, it can be insulated and roofed in the manner shown for low-slope roofs in Chapter 16, but if it is steeply pitched, a nailing surface for the shingles must be installed on the outside of the insulation. Electrical wiring for lighting fixtures on the ceiling will have to be run either through exposed metal conduits, which may be visually unsatisfactory, or through the insulation above the deck. If the walls and partitions are made of masonry or stressed-skin panels instead of light framing, special arrangements are necessary for installing wiring, plumbing, and heating devices in the walls as well.

LONGER SPANS IN HEAVY TIMBER

For buildings that require spans longer than the 20 feet (6 m) or so that is the maximum usually associated with framing of sawn timbers, the designer may select from among several different types of timber structural devices.

Large Beams

With large, old-growth trees no longer readily available, very large timber beams are usually glue laminated rather than sawn. Laminated beams are stronger and more dimensionally stable than sawn wood beams and can be made in the exact size and shape desired (Figures 4.17–4.19).

Glue-laminated wood beams that incorporate layers of fiber-reinforced plastic near their top and bottom faces are now becoming available. The layers of stiff, strong fibers (either glass, aramid, or steel embedded in a plastic binder) impart additional strength and rigidity to the beam, often reducing its required depth.

COMBUSTIBLE BUILDINGS FRAMED WITH HEAVY TIMBER

Heavy timbers are often used in combination with smaller wood framing members to construct buildings that do not meet all the fire-resistive requirements for Heavy Timber construction. These buildings, classified under the BOCA Code as Type 5 (Wood Light Frame) construction, bring the appearance and structural performance of beam-and-decking framing to small freestanding residential, commercial, religious, and institutional buildings. In these applications, there are no restrictions on minimum size of timber, minimum thickness of decking, or exterior wall material, and light framing of nominal 2-inch (51-mm) lumber can be incorporated as desired, so long as the code requirements for Type 5 construction are met.

The use of exposed beams and decking in buildings that are normally made of light framing poses some new problems for the designer because a heavy timber structure does not have the concealed cavities that are normally present in the light frame structure. Thermal insulation

FIGURE 4.17
Installing tongue-and-groove roof decking over laminated beams and girders. Note the hinge connectors in the beams at the extreme lower left corner of the photograph. (*Courtesy of American Institute of Timber Construction*)

FIGURE 4.18
This roof uses glue-laminated beams to support proprietary long-span trusses made of wood with steel tube diagonals. Notice that the beams are hinged in the manner shown in Figure 4.16. The trusses, roof joists, and plywood roof deck are prefabricated into panels to reduce installation time. (*Courtesy of Trus Joist Corporation*)

FIGURE 4.19
Installing a roof deck of prefabricated panels, each consisting of a beam, joists, and plywood, with hangers attached to the beam. Plywood sheets are installed in the gaps over the glue-laminated girders after the panels are in place to create a fully continuous plywood diaphragm to stiffen the building against wind and seismic forces. Note the hinge connectors in the girders. (*Courtesy of American Institute of Timber Construction*)

Rigid Frames

The cruck (Figure 4.2), cut from a bent tree, was a form of *rigid frame* or *portal frame*. Today's rigid frames are glue laminated to shape, and find wide use in longer-span buildings. Standard configurations are readily available (Figures 4.20–4.22), or the designer may order a custom shape. Rigid frames exert a horizontal thrust, so they must be tied together at the base with steel tension rods. In laminated wood construction, rigid frames are often called *arches*, acknowledging that the two structural forms act in very nearly the same manner.

Trusses

The majority of wood *trusses* built each year are light roof trusses of nominal 2-inch (51-mm) lumber joined by toothed plates (see Figures 3.40–3.43, 5.54, and 5.55), but heavy timber trusses are often used in larger buildings. Their joints are made with steel bolts and welded steel plate connectors, or with split ring connectors. Both sawn and laminated timbers are used, sometimes in combination with steel rod tension members. Many shapes of heavy timber truss are possible, and spans of over 100 feet (30 m) are common (Figures 4.12, 4.23–4.26).

Arches and Domes

Long curved timbers for making *vaults* and *domes* are easily fabricated in laminated wood and are widely used in athletic arenas, auditoriums, suburban retail stores, warehouses, and factories (Figures 4.27, 4.28, 4.29). Arched structures, like rigid frames, exert lateral thrusts that must be countered by tie rods or suitably designed foundations.

HEAVY TIMBER AND THE BUILDING CODES

The table in Figure 1.1 shows the range of building types that may be built of Heavy Timber construction, and of light frame construction utilizing heavy timbers. Notice that the allowable heights and areas for Heavy Timber construction are greater than those for unprotected steel (Type 2C construction under the BOCA National Building Code), and are comparable to those for lightly protected noncombustible construction (Type 2B). The references in the Heavy Timber column of Figure 1.3 to "Section 605.0" are to several paragraphs of the BOCA Code that are

FIGURE 4.20
Three-hinged arches of glue-laminated wood carry laminated wood roof *purlins*. The short crosspieces of wood between the purlins are temporary ladders for workers. (*Courtesy of American Institute of Timber Construction*)

FIGURE 4.21
Typical details for three-hinged arches of laminated wood. The *tie rod* is later covered by the floor slab.

FOR PRELIMINARY DESIGN OF A HEAVY TIMBER STRUCTURE

• Estimate the depth of **wood roof decking** at $\frac{1}{40}$ of its span. Estimate the depth of **wood floor decking** at $\frac{1}{30}$ of its span. Standard nominal depths of wood decking are 2, 3, 4, 6, and 8 inches.

• Estimate the depth of **solid wood beams** at $\frac{1}{15}$ of their span, and the depth of **glue-laminated beams** at $\frac{1}{20}$ of their span. Add a nominal 6 inches to these depths for girders. The width of a solid wood beam or girder is usually $\frac{1}{3}$ to $\frac{1}{2}$ of its depth. The width of a glue-laminated beam typically ranges from $\frac{1}{3}$ to $\frac{1}{4}$ of its depth.

• Estimate the depth of **timber triangular roof trusses** at $\frac{1}{2}$ to $\frac{1}{5}$ of their span.

• To estimate the size of a **wood column,** add up the total roof and floor area supported by the column. A nominal 6-inch column can support up to about 600 square feet (56 m²) of area, and an 8-inch column 1000 square feet (93 m²), a 10-inch column 2000 square feet (186 m²), a 12-inch column 3000 square feet (280 m²), and a 14-inch column 4000 square feet (372 m²). Wood columns are usually square or nearly square in proportion.

For actual sizes of solid timbers in both conventional and metric units, see Figures 3.22 and 3.23. Standard sizes of glue-laminated timbers are given in Chapter 3. For a building that must qualify as Heavy Timber or Mill construction under the building code, minimum timber sizes are given in Figure 4.7.

These approximations are valid only for purposes of preliminary building layout, and must not be used to select final member sizes. They apply to the normal range of building occupancies such as residential, office, commercial, and institutional buildings. For manufacturing and storage buildings, use somewhat larger members.

For more comprehensive information on preliminary selection and layout of a structural system and sizing of structural members, see Allen, Edward, and Joseph Iano, *The Architect's Studio Companion* (2nd ed.), New York, John Wiley & Sons, Inc., 1995.

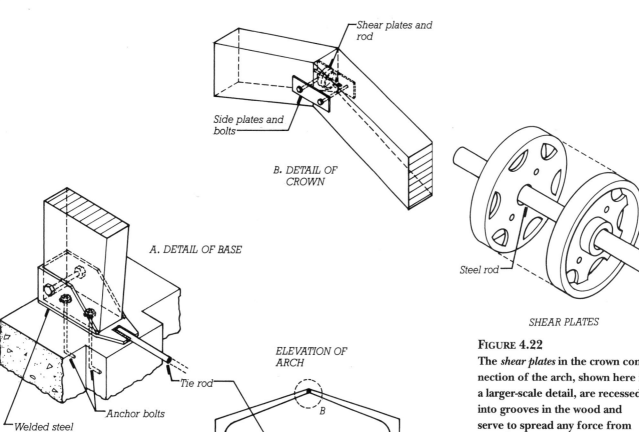

B. DETAIL OF CROWN

A. DETAIL OF BASE

Shear plates and rod

Side plates and bolts

Steel rod

SHEAR PLATES

Welded steel shoe

Anchor bolts

Tie rod

ELEVATION OF ARCH

FIGURE 4.22
The *shear plates* in the crown connection of the arch, shown here in a larger-scale detail, are recessed into grooves in the wood and serve to spread any force from the steel rod across a much wider surface area of wood to avoid crushing and splitting.

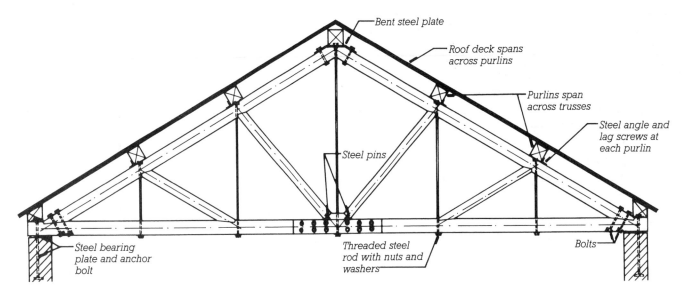

Bent steel plate

Roof deck spans across purlins

Purlins span across trusses

Steel angle and lag screws at each purlin

Steel pins

Steel bearing plate and anchor bolt

Threaded steel rod with nuts and washers

Bolts

FIGURE 4.23
A heavy timber roof truss with steel rod tension members. This type of truss is easy to construct but cannot be used if predicted wind uplift is sufficiently strong to cause the forces in the tension members to reverse. The center splice in the lower chord of the truss is required only if it is impossible to obtain lumber long enough to reach in one piece from one end of the truss to the other. Compare this mode of truss construction with that shown in Figure 4.12; still another common mode is to form the truss of a single layer of heavy members connected by steel side plates and bolts, as shown in Figure 4.26.

FIGURE 4.24
Laminated wood roof trusses with steel connector plates. The steel rods are for lateral stability, to keep the bottom chords of the trusses from moving sideways. (*Architects: Woo and Williams. Photo by Richard Bonarrigo. Courtesy of the architects*)

FIGURE 4.25
Roof trusses for the Montville, New Jersey, Public Library consist of a single plane of glue-laminated members joined by steel side plates and bolts. An extensive scheme of diagonal bracing resists wind and seismic forces in both axes of the building.

FIGURE 4.26
The Montville Public Library timbers are joined with custom-designed connectors that are welded together from steel plates. (*Design and photographs by Eliot Goldstein, The Goldstein Partnership, Architects*)

FIGURE 4.27
Semicircular laminated wood arches support the timber roof of the new Back Bay Station in Boston. Notice the use of horizontal steel tie rods to resist the thrust of the arches at the base. (*Architects: Kallmann, McKinnell & Wood. Photo © Steve Rosenthal*)

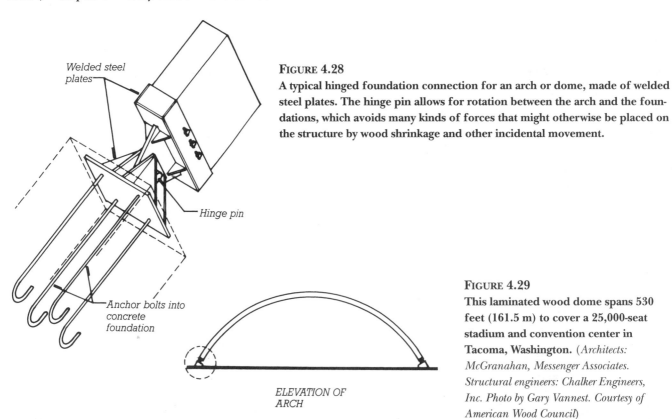

Welded steel plates

Hinge pin

Anchor bolts into concrete foundation

FIGURE 4.28
A typical hinged foundation connection for an arch or dome, made of welded steel plates. The hinge pin allows for rotation between the arch and the foundations, which avoids many kinds of forces that might otherwise be placed on the structure by wood shrinkage and other incidental movement.

ELEVATION OF ARCH

FIGURE 4.29
This laminated wood dome spans 530 feet (161.5 m) to cover a 25,000-seat stadium and convention center in Tacoma, Washington. (*Architects: McGranahan, Messenger Associates. Structural engineers: Chalker Engineers, Inc. Photo by Gary Vannest. Courtesy of American Wood Council*)

summarized here in Figure 4.7. The allowable floor areas can be increased by installing an automatic fire suppression sprinkler system in the building: Sprinklers triple the allowable area under this code for a one- or two-story building and double it for taller buildings.

THE UNIQUENESS OF HEAVY TIMBER FRAMING

Heavy timber cannot span as far or with such delicacy as steel, and it cannot mimic the structural continuity or smooth shell forms of concrete, yet many people respond more positively to the idea of a timber building than they do to one of steel or concrete. To some degree, this response may stem from the color, grain figure, and warmer feel of wood. In larger part, it probably comes from the pleasant associations that people have with the

FIGURE 4.30
Architects Greene and Greene of Pasadena, California, were known for their carefully wrought timber frame houses such as this one, built for David B. Gamble in 1909. (*Photo by Wayne Andrews*)

FIGURE 4.31
Each of the attached dwellings in Sea Ranch Condominium #1 in northern California, built in 1965, is framed with a simple cage of unplaned timbers sawn from trees taken from another portion of the site. The diagonal members are wind braces. (*Architects: Moore, Lyndon, Turnbull, and Whitaker. Photo by the author*)

(a)

(b)

FIGURE 4.32
(*a*) The exterior of Sea Ranch Condominium #1, an interior view of which is shown in the opening photograph of this chapter, is sheathed with vertical 2-inch (51-mm) unplaned tongue-and-groove decking, and clad with ¾-inch (19-mm) tongue-and-groove Redwood siding. (*b*) Another view of the exposed timbers and connectors inside a dwelling in this building. The flooring is vertical grain Douglas fir. (*Photos by Morley Baer*)

sturdy, satisfying houses our ancestors erected from hand-hewn timbers only a few generations ago. Most people today live in dwellings where none of the framing is exposed. Exposed ceiling beams in one's house or apartment have become a much-desired amenity, and shopping or restaurant dining in a converted mill building of heavy timber and masonry construction is generally thought to be a pleasant experience. There is something in all of us that derives satisfaction from seeing wood beams at work.

In economic reality, heavy timber must compete successfully on the basis of price with other materials of construction, and, as our forests have diminished in quality and shipping costs have risen, timber is no longer an automatic choice for building a mill or any other type of structure. But for many buildings, heavy timber is an economic alternative to steel and concrete, particularly in situations where the appearance and feel of a heavy timber structure will be highly valued by those who use it, in regions close to commercial forests, or where code provisions or fire insurance premiums create a financial incentive.

SELECTED REFERENCES

1. American Institute of Timber Construction. *Timber Construction Manual* (4th ed.). New York, John Wiley & Sons, Inc., 1994.

This is a comprehensive design handbook for timber structures, including detailed engineering procedures as well as general information on wood and its fasteners.

2. American Institute of Timber Construction. *Typical Construction Details.* AITC 104-84, 1984.

The 32 pages of this pamphlet contain dozens of examples of how to detail connections in Heavy Timber buildings. Especially instructive are the bad examples that are presented as lessons in what to avoid. (Address for ordering: AITC, 7102 South Revere Parkway, Suite 140, Englewood, CO 80112.)

KEY TERMS AND CONCEPTS

halftimbering	chamfer	bearing block	rigid or portal frame
cruck	pintle	hinge connector	arch
Heavy Timber construction	firecut	purlin	truss
Mill construction	decking	tie rod	vault
Slow-burning construction	tongue-and-groove	shear plate	dome

REVIEW QUESTIONS

1. Why does Heavy Timber construction receive relatively favorable treatment from building codes and insurance companies?

2. What are the important factors in detailing the junction of a wood beam with a masonry loadbearing wall? Draw several ways of making this joint.

3. Draw from memory one or two typical details for the intersection of a wood column with a floor of a building of Heavy Timber construction.

EXERCISES

1. Determine from the code table in Figure 1.1 whether a building you are currently designing could be built of Heavy Timber construction, and what modifications you might make in your design so that it will conform to the requirements of Heavy Timber construction.

2. Find a barn or mill that was constructed in the 18th or 19th century and sketch some typical connection details. How is the structure stabilized against horizontal wind forces?

3. Obtain a book on traditional Japanese construction from the library and compare Japanese timber joint details with 18th- or 19th-century American practice.

5

WOOD LIGHT FRAME CONSTRUCTION

A school of arts and crafts is housed in a cluster of small buildings of Wood Light Frame construction that cling to a dramatic mountainside site overlooking the ocean in New England. (*Architect: Edward Larrabee Barnes. Photograph by Joseph W. Molitor*)

Wood Light Frame construction is the most flexible of all building systems. There is scarcely a shape it cannot be used to construct, from a plain rectilinear box to cylindrical towers or to complex foldings of sloping roofs with dormers of every description. During the century and a half since it first came into use, wood light framing has served to construct buildings in styles ranging from reinterpretations of nearly all the historical fashions to uncompromising expressions of every 20th-century architectural philosophy. It has assimilated without difficulty during this same period a bewildering and unforeseen succession of technical improvements in building: central heating, air conditioning, gas lighting, electricity, thermal insulation, indoor plumbing, prefabricated components, and electronic communications.

Light frame buildings are easily and swiftly constructed with a minimal investment in tools. Many observers of the building industry have criticized the supposed inefficiency of light frame construction, which is carried out largely by hand methods on the building site, yet it has successfully fought off competition from industrialized building systems of every sort, partly by incorporating their best features, to remain the least expensive form of durable construction. It is the common currency of small residential and commercial buildings in North America today.

Wood Light Frame construction has its deficiencies: If ignited, it burns rapidly; if exposed to dampness, it decays. It expands and contracts by significant amounts in response to changes in humidity, sometimes causing chronic difficulties with cracking plaster, sticking doors, and buckling floors. The framing itself is so unattractive to the eye that it is seldom left exposed in a building. But these problems can be controlled by clever design and careful workmanship, and there is no arguing with success: Frames made by the monotonous repetition of wooden joists, studs, and rafters are likely to remain the number one system of building in North America for a long time to come.

FIGURE 5.1
Carpenters apply plywood roof sheathing to a platform-framed apartment building. The ground floor is a concrete slab on grade. The edge of the wooden platform of the upper floor is clearly visible between the stud walls of the ground and upper floors. Most of the diagonal bracing is temporary, but permanent let-in diagonal braces occur between the two openings at the lower left, and immediately above in the rear building. The openings have been framed incorrectly, without supporting studs for the headers. (*Courtesy of Southern Forest Products Association*)

HISTORY

Wood Light Frame construction was the first uniquely American building system. It was developed in the first half of the 19th century when builders recognized that the closely spaced vertical members used to infill the walls of a heavy timber building frame were themselves sufficiently strong that the heavy posts of the frame could be eliminated. Its development was accelerated by two technological breakthroughs of the period: Boards and small framing members of wood had recently become inexpensive for the first time in history because of the advent of the water-powered sawmill; additionally, machine-made nails had also become remarkably cheap compared to the hand-forged nails that preceded them.

The earliest version of wood light framing, the *balloon frame,* was a building system framed solely with slender, closely spaced wooden members: *joists* for the floors, *studs* for the walls, *rafters* for the sloping roofs. Heavy posts and beams were completely eliminated, and, with them, the difficult, expensive, mortise-and-tenon joinery they required. There was no structural member in a frame that could not be handled easily by a single carpenter, and each of the hundreds of joints was made with lightning rapidity using two or three nails. The impact of this new building system was revolutionary: In 1865, G. E. Woodward could write that "A man and a boy can now attain the same results, with ease, that twenty men could on an old-fashioned frame . . . the Balloon Frame can be put up for forty percent less money than the mortise and tenon frame."

The balloon frame (Figure 5.2) used full-length studs that ran continuously for two stories from foundation to roof. In time, these were recognized as being too long to erect efficiently, and the tall hollow spaces between studs acted as multiple chim-

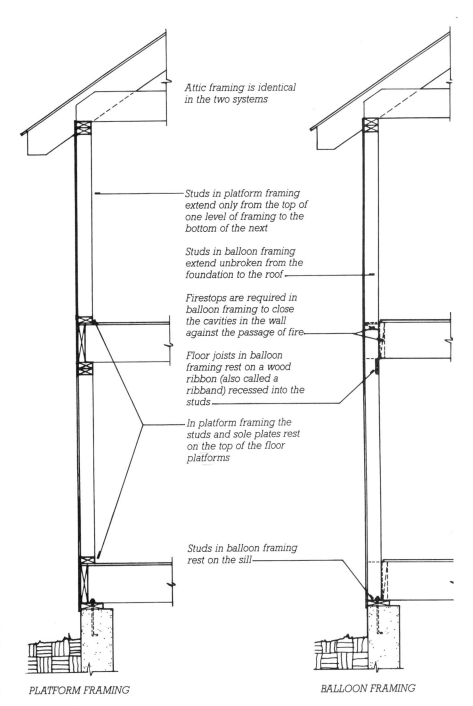

Attic framing is identical in the two systems

Studs in platform framing extend only from the top of one level of framing to the bottom of the next

Studs in balloon framing extend unbroken from the foundation to the roof

Firestops are required in balloon framing to close the cavities in the wall against the passage of fire

Floor joists in balloon framing rest on a wood ribbon (also called a ribband) recessed into the studs

In platform framing the studs and sole plates rest on the top of the floor platforms

Studs in balloon framing rest on the sill

PLATFORM FRAMING

BALLOON FRAMING

FIGURE 5.2
Comparative framing details for platform framing (left) and balloon framing (right). Platform framing is easier to erect but settles considerably as the wood dries and shrinks. If nominal 12-inch (300-mm) joists are used to frame the floors in these examples, the total amount of loadbearing cross-grain wood between the foundation and the attic joists is 33 inches (838 mm) for the platform frame, and only 4½ inches (114 mm) for the balloon frame.

neys in a fire, spreading the blaze rapidly to the upper floors, unless they were closed off with wood or brick *firestops* at each floor line. Several modified versions of the balloon frame were subsequently developed in an attempt to overcome these difficulties, and the most recent of these, the *platform frame,* is now the universal standard.

THE PLATFORM FRAME

While complex in its details, the platform frame is very simple in concept. A floor platform is built. Loadbearing walls are erected upon it. A second floor platform is built upon these walls, and a second set of walls upon this platform. The attic and roof are then built upon the second set of walls. There are, of course, many variations: A concrete slab that lies directly on the ground is sometimes substituted for the ground-floor platform; a building may be one or three stories tall instead of two; and several types of roofs are frequently built that do not incorporate attics. But the essentials remain: A floor platform is completed at each level, and the walls bear upon the platform rather than directly upon the walls of the story below.

The advantages of the platform frame over the balloon frame are that is uses short, easily handled lengths of lumber for the wall framing; its vertical hollow spaces are automatically firestopped at each floor; and its platforms are convenient working surfaces for the carpenters who build the frame. The major disadvantage of the platform frame is that each platform constitutes a thick layer of wood whose grain runs horizontally. This leads inevitably to a relatively large amount of vertical shrinkage in the frame as the excess moisture dries from the wood, which can lead to distress in the exterior and interior finish surfaces.

A conventional platform frame is made entirely of nominal 2-inch members, which are actually $1\frac{1}{2}$ inches (38 mm) in thickness. These are ordered and delivered cut to the nearest 2-foot (600-mm) length. They are measured and sawn to exact length on the building site. All connections are made with nails, using either face nailing, end nailing, or toe nailing (Figures 3.36, 5.18) as required by the characteristics of each joint. Nails are driven either by hammer or by hand-held pneumatic nailing machines (nail guns). In either case, the connection is quickly made because the nails are installed without drilling holes or otherwise preparing the joint.

Each plane of structure in a platform frame is made by aligning a number of pieces of framing lumber parallel to one another at specified intervals, nailing these to crosspieces that head off the parallel framing pieces at either end to maintain their spacing and flatness, then covering the plane of framing with *sheathing,* a facing layer of boards or panels that join and stabilize the pieces into a single structural unit, ready for the application of finish materials inside and out. In a floor structure, the parallel pieces are the floor joists, and the crosspieces at the ends of the joists are called *headers, rim joists,* or *band joists.* The sheathing on a floor is known as the *subfloor.* In a wall structure, the parallel pieces are the studs, the crosspiece at the bottom of the wall is the *sole plate,* and the crosspiece at the top (which is doubled for strength if the wall bears a load from above) is called the *top plate.* In a sloping roof, the rafters are headed off by the top plates at the lower edge of the roof and by the *ridge board* at the peak.

Openings are required in all these planes of structure: for windows and doors in the walls, for stairs and chimneys in the floors, and for chimneys, skylights, and dormers in the roofs. In each case, these are made by

heading off the opening: Openings in floors are framed with *headers* and *trimmers* (Figure 5.14). Headers and trimmers must be doubled to support the higher loads placed on them by the presence of the opening. In walls, *sills* head off the bottoms of openings, with strong window headers and door headers across the tops (Figure 5.29).

Sheathing is a key component of platform framing. The end nails that connect the plates to the studs have little holding power against uplift of the roof by wind, but the sheathing connects the frame into a single, strong unit from foundation to roof. The rectilinear geometry of the parallel framing members has no useful resistance to *wracking* by lateral forces such as wind, but rigid sheathing panels or diagonal sheathing boards brace the building effectively against these forces. Sheathing also furnishes a surface to which shingles, boards, and flooring are nailed for finish surfaces. In buildings without sheathing, or with sheathing materials that are too weak to tie and brace the frame, such as insulating plastic foam, diagonal bracing must be applied to the walls to impart lateral stability.

FOUNDATIONS FOR LIGHT FRAME STRUCTURES

Foundations for light framing, originally made of stone or brick, are now made in most cases of sitecast concrete, concrete block masonry, or preservative-treated wood (Figures 5.3–5.13). Concrete and masonry foundations are highly conductive of heat and usually must be insulated to meet code requirements concerning energy conservation (Figures 5.6, 5.7, 5.12, 5.13). A wood foundation is easily insulated in the same manner as the frame of the house it supports. Further advantages claimed for wood foundations are that they can be constructed in any weather by the same crew of carpenters that will frame the building, and that they allow for easy installation of electrical wiring, plumbing, and interior finish materials in the basement. A basement of any material needs to be carefully dampproofed and drained to avoid flooding with ground water and to prevent the buildup of water pressure in the surrounding soil that could cave in the walls (Figure 5.4).

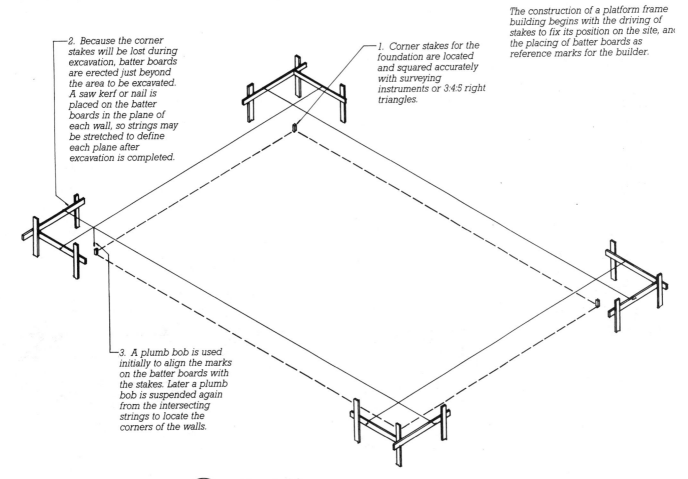

The construction of a platform frame building begins with the driving of stakes to fix its position on the site, and the placing of batter boards as reference marks for the builder.

2. Because the corner stakes will be lost during excavation, batter boards are erected just beyond the area to be excavated. A saw kerf or nail is placed on the batter boards in the plane of each wall, so strings may be stretched to define each plane after excavation is completed.

1. Corner stakes for the foundation are located and squared accurately with surveying instruments or 3:4:5 right triangles.

3. A plumb bob is used initially to align the marks on the batter boards with the stakes. Later a plumb bob is suspended again from the intersecting strings to locate the corners of the walls.

FIGURE 5.3
Step One in the construction of a simple platform frame building: establishing the position, shape, and size of the building on the site. This drawing begins a series of isometric drawings that will follow the erection of a building step by step throughout the course of this chapter.

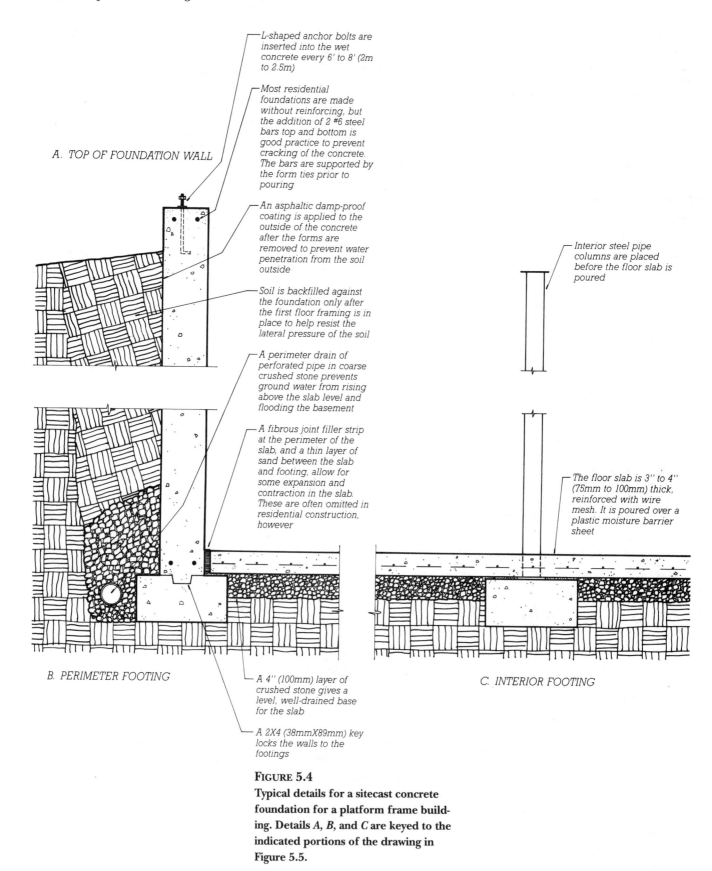

A. TOP OF FOUNDATION WALL

L-shaped anchor bolts are inserted into the wet concrete every 6' to 8' (2m to 2.5m)

Most residential foundations are made without reinforcing, but the addition of 2 #6 steel bars top and bottom is good practice to prevent cracking of the concrete. The bars are supported by the form ties prior to pouring

An asphaltic damp-proof coating is applied to the outside of the concrete after the forms are removed to prevent water penetration from the soil outside

Soil is backfilled against the foundation only after the first floor framing is in place to help resist the lateral pressure of the soil

A perimeter drain of perforated pipe in coarse crushed stone prevents ground water from rising above the slab level and flooding the basement

A fibrous joint filler strip at the perimeter of the slab, and a thin layer of sand between the slab and footing, allow for some expansion and contraction in the slab. These are often omitted in residential construction, however

B. PERIMETER FOOTING

A 4" (100mm) layer of crushed stone gives a level, well-drained base for the slab

A 2X4 (38mmX89mm) key locks the walls to the footings

Interior steel pipe columns are placed before the floor slab is poured

The floor slab is 3" to 4" (75mm to 100mm) thick, reinforced with wire mesh. It is poured over a plastic moisture barrier sheet

C. INTERIOR FOOTING

FIGURE 5.4
Typical details for a sitecast concrete foundation for a platform frame building. Details *A*, *B*, and *C* are keyed to the indicated portions of the drawing in Figure 5.5.

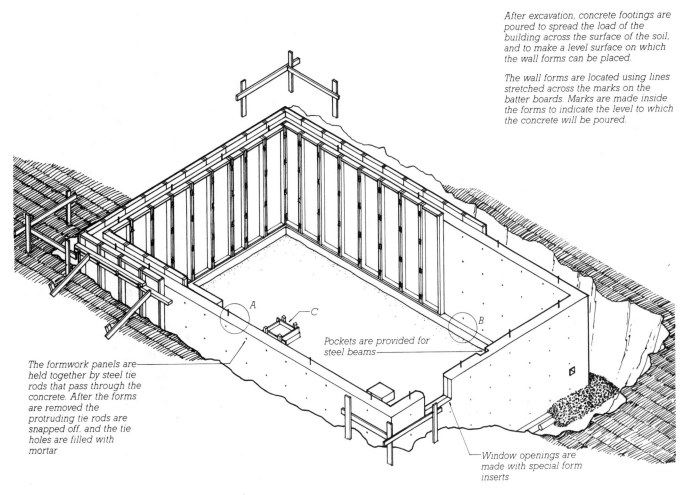

After excavation, concrete footings are poured to spread the load of the building across the surface of the soil, and to make a level surface on which the wall forms can be placed.

The wall forms are located using lines stretched across the marks on the batter boards. Marks are made inside the forms to indicate the level to which the concrete will be poured.

Pockets are provided for steel beams

The formwork panels are held together by steel tie rods that pass through the concrete. After the forms are removed the protruding tie rods are snapped off, and the tie holes are filled with mortar

Window openings are made with special form inserts

FIGURE 5.5
Step Two in the construction of a typical platform frame building: excavation and foundations. The letters *A*, *B*, and *C* indicate portions of the foundation that are detailed in Figure 5.4.

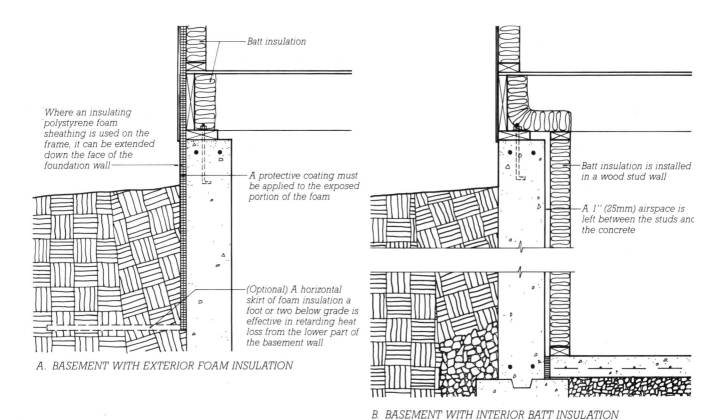

Batt insulation

Where an insulating polystyrene foam sheathing is used on the frame, it can be extended down the face of the foundation wall

A protective coating must be applied to the exposed portion of the foam

(Optional) A horizontal skirt of foam insulation a foot or two below grade is effective in retarding heat loss from the lower part of the basement wall

A. BASEMENT WITH EXTERIOR FOAM INSULATION

Batt insulation is installed in a wood stud wall

A 1'' (25mm) airspace is left between the studs and the concrete

B. BASEMENT WITH INTERIOR BATT INSULATION

Cantilevered joists allow use of foam on the basement portion of the house only

C. BASEMENT WITH EXTERIOR FOAM INSULATION

Batt insulation is stapled to the header joist and extended down the concrete wall and 2' (600mm) onto the floor of the crawlspace

A plastic moisture barrier sheet keeps the crawlspace and insulation dry

D. CRAWLSPACE WITH INTERIOR BATT INSULATION

FIGURE 5.6
Most building codes require thermal insulation of the foundation. Here are three different ways of adding insulation to a poured concrete or concrete masonry basement, and one way of insulating a crawlspace foundation. The crawlspace may alternatively be insulated on the outside with panels of plastic foam. The interior batt insulation shown in *B* is commonly used but raises unanswered questions about how to avoid possible problems arising from moisture accumulating between the insulation and the wall.

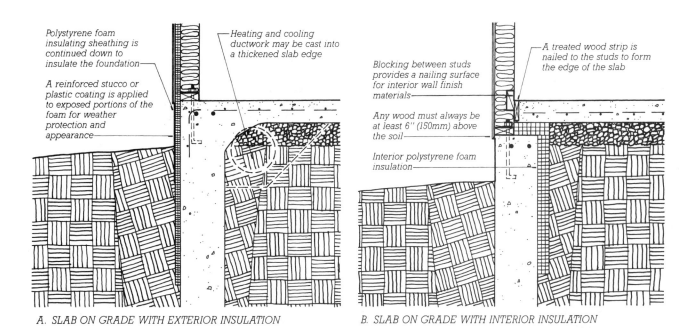

Polystyrene foam insulating sheathing is continued down to insulate the foundation—

A reinforced stucco or plastic coating is applied to exposed portions of the foam for weather protection and appearance—

—Heating and cooling ductwork may be cast into a thickened slab edge

A. SLAB ON GRADE WITH EXTERIOR INSULATION

Blocking between studs provides a nailing surface for interior wall finish materials—

Any wood must always be at least 6'' (150mm) above the soil—

Interior polystyrene foam insulation—

—A treated wood strip is nailed to the studs to form the edge of the slab

B. SLAB ON GRADE WITH INTERIOR INSULATION

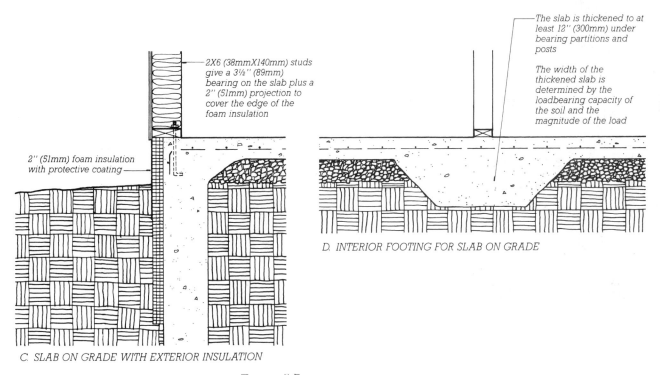

—2X6 (38mmX140mm) studs give a 3½'' (89mm) bearing on the slab plus a 2'' (51mm) projection to cover the edge of the foam insulation

2'' (51mm) foam insulation with protective coating—

C. SLAB ON GRADE WITH EXTERIOR INSULATION

—The slab is thickened to at least 12'' (300mm) under bearing partitions and posts

The width of the thickened slab is determined by the loadbearing capacity of the soil and the magnitude of the load

D. INTERIOR FOOTING FOR SLAB ON GRADE

FIGURE 5.7
Some typical concrete slab on grade details with thermal insulation.

FIGURE 5.8
Erecting formwork for a sitecast concrete foundation wall. The footing has already been cast and its formwork removed. It is visible in front of the worker at the left. (*Photo by Joseph Iano*)

FIGURE 5.9
Masons construct a foundation of concrete masonry. The first coat of parging, portland cement plaster used to help dampproof the foundation, has already been applied on the outside of the wall, and the drainage layer of crushed stone has been put in place. The projecting *pilaster* in the center of the wall will support a beam under the center of the main floor. After a second coat of parging, the outside of the foundation will be coated with an asphaltic dampproofing compound. (*Courtesy of Portland Cement Association, Skokie, Illinois*)

FIGURE 5.10
Erecting a preservative-treated wood foundation. One worker applies a bead of sealant to the edge of a panel of preservative-treated wood components, as another prepares to push the next panel into position against the sealant. The panels rest on a horizontal preservative-treated plank, which, in turn, rests on a drainage layer of crushed stone. A major advantage of a wood foundation is that it can be insulated in the same way as the superstructure of the building. (*Courtesy of APA–The Engineered Wood Association*)

FIGURE 5.11
The difference in color makes it clear where the preservative-treated wood foundation leaves off and the untreated superstructure of the building begins. (*Courtesy of APA–The Engineered Wood Association*)

FIGURE 5.12
External polystyrene foam insulation applied to the outside of a concrete masonry foundation. (*Courtesy of Dow Chemical Company*)

FIGURE 5.13
Following completion of the exterior siding, a worker staples a glass fiber reinforcing mesh to the foam insulation on the exposed portions of the basement wall. Next he will trowel onto the mesh two thin coats of a cementitious, stucco-like material that will form a durable, attractive finish coating over the foam. (*Courtesy of Dow Chemical Company*)

BUILDING THE FRAME

Planning the Frame

While it is true that an experienced carpenter can frame a simple building from the most minimal of drawings, a platform frame of wood for a custom-designed building should be planned as carefully as a frame of steel or concrete for a larger building. The architect or engineer should determine an efficient layout and the appropriate sizes for joists and rafters, and communicate this information to the carpenters by means of *framing plans* (Figures 5.14, 5.44, 5.49). For most purposes, member sizes can be determined using the standardized structural tables in Reference 4 at the end of this chapter. Detailed section drawings, similar to those seen throughout this chapter, are also prepared for the major connections in the building. The *architectural floor plans* serve to indicate the locations and dimensions of all the walls, partitions, and openings, and the *exterior elevations* are drawings that show the outside faces of the building, with vertical framing dimensions indicated as required. For most buildings, *sections* are also drawn that cut completely through the building, showing the dimensional relationships of the various floor levels and roof planes, and the slopes of the roof surfaces. *Interior elevation* views are often prepared for kitchens, bathrooms, and other rooms with elaborate interior features.

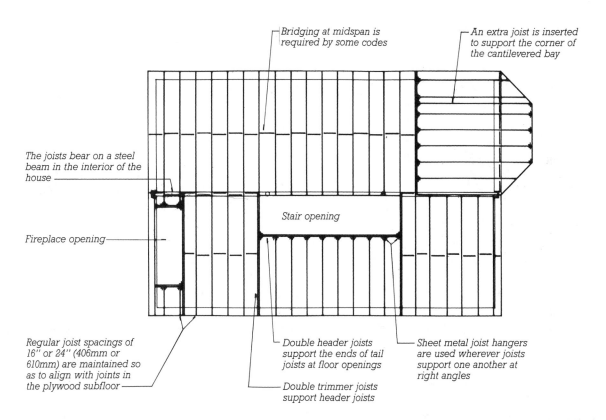

Bridging at midspan is required by some codes

An extra joist is inserted to support the corner of the cantilevered bay

The joists bear on a steel beam in the interior of the house

Stair opening

Fireplace opening

Regular joist spacings of 16" or 24" (406mm or 610mm) are maintained so as to align with joints in the plywood subfloor

Double header joists support the ends of tail joists at floor openings

Double trimmer joists support header joists

Sheet metal joist hangers are used wherever joists support one another at right angles

FLOOR FRAMING PLAN

FIGURE 5.14

A framing plan for the ground-floor platform of the building shown in Figure 5.16. Where joists are cantilevered to create an overhanging bay, the cantilever distance should not exceed one-quarter of the total length of the joist.

Erecting the Frame

The erection of a typical platform frame (referred to, imprecisely, as *rough carpentry* in architects' specifications) can best be understood by following the sequential isometric diagrams that begin with Figure 5.16. Notice the basic simplicity of the building process: A platform is built, walls are assembled horizontally on the platform and tilted up into place, and another platform or a roof is built on top of the walls. Most of the work is accomplished without the use of ladders or scaffolding, and temporary bracing is needed only to support the walls until the next level of framing is installed and sheathed.

The details of a platform frame are not left to chance. While there are countless local and personal vari- ations in framing details and tech- niques, the sizes, spacings, and con- nections of the members in a platform frame are closely regulated by building codes, down to the sizes and spacings of the studs, the size and number of nails for each connection (Figure 5.18), and the thicknesses and nailing of the sheathing panels.

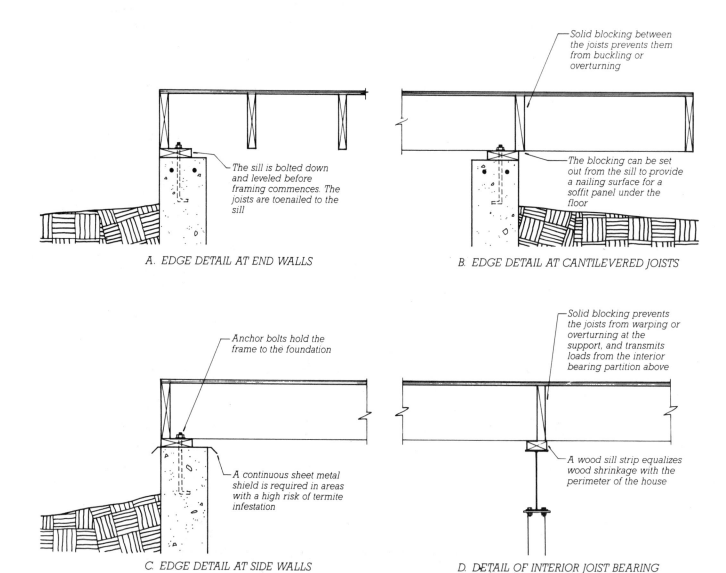

A. EDGE DETAIL AT END WALLS

The sill is bolted down and leveled before framing commences. The joists are toenailed to the sill

B. EDGE DETAIL AT CANTILEVERED JOISTS

Solid blocking between the joists prevents them from buckling or overturning

The blocking can be set out from the sill to provide a nailing surface for a soffit panel under the floor

C. EDGE DETAIL AT SIDE WALLS

Anchor bolts hold the frame to the foundation

A continuous sheet metal shield is required in areas with a high risk of termite infestation

D. DETAIL OF INTERIOR JOIST BEARING

Solid blocking prevents the joists from warping or overturning at the support, and transmits loads from the interior bearing partition above

A wood sill strip equalizes wood shrinkage with the perimeter of the house

FIGURE 5.15
Ground-floor framing details, keyed to the lettered circles in Figure 5.16. A fibrous sill sealer material should be installed between the sill and the top of the foundation to reduce air leakage, but the sealer material is not shown on these diagrams. Using accepted architectural drafting conventions, continuous pieces of lumber are drawn with an X inside, and intermittent blocking with a single slash.

Attaching the Frame to the Foundation

The *sill,* preferably preservative-treated or naturally decay-resistant wood, is bolted to the foundation as a base for the wood framing. A single sill, as shown in the details here, is all that is required by most codes, but in better-quality work, the sill is often doubled for greater stiffness. The top of the foundation is usually somewhat uneven, so at low spots the sill must be shimmed with wood shingle wedges to be able to transfer loads from the frame to the foundation. A compressible, fibrous *sill sealer* should be inserted between the sill and the foundation to reduce air infiltration through the gap (Figure 5.19). The normal foundation bolts are sufficient to hold most buildings on their foundations, but tall frames in areas subject to high winds or earthquakes may require more elaborate attachments (Figure 5.42).

When the foundation is complete, basement beams are placed, sills are bolted to the foundation, and the first floor joists and subfloor are installed.

Plywood sheets are considerably stiffer along their length than across their width, so they must be laid with their long dimension perpendicular to the joists. The end joints are staggered to avoid lines of weakness

Joist bridging is required by some building codes

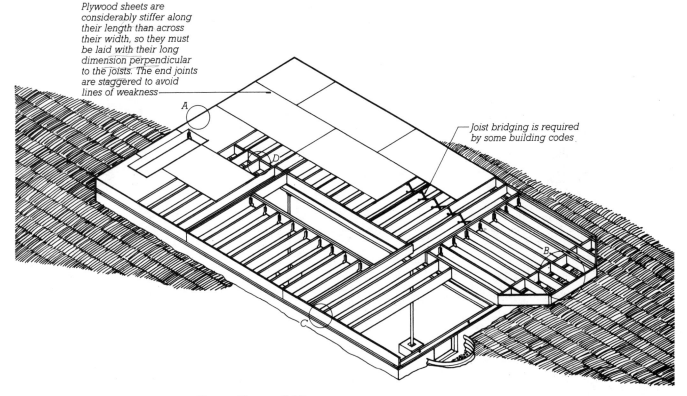

FIGURE 5.16

Step Three in erecting a typical platform frame building: the ground-floor platform. Compare this drawing with the framing plan in Figure 5.14. Notice that the direction of the joists must be changed to construct the cantilevered bay on the end of the building. A cantilevered bay on a long side of the building could be framed by merely extending the floor joists over the foundation.

FIGURE 5.17
Alternative ways of constructing an interior line of support for ground-floor joists. Space is left over the tops of the beams in details *B* and *D* to allow for drying shrinkage in the joists.

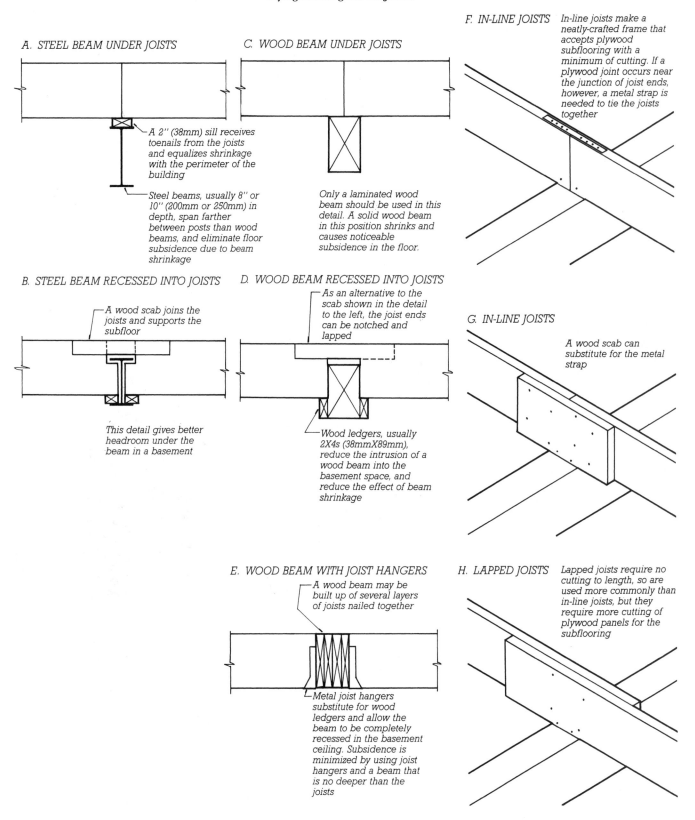

A. STEEL BEAM UNDER JOISTS

A 2" (38mm) sill receives toenails from the joists and equalizes shrinkage with the perimeter of the building

Steel beams, usually 8" or 10" (200mm or 250mm) in depth, span farther between posts than wood beams, and eliminate floor subsidence due to beam shrinkage

C. WOOD BEAM UNDER JOISTS

Only a laminated wood beam should be used in this detail. A solid wood beam in this position shrinks and causes noticeable subsidence in the floor.

F. IN-LINE JOISTS   In-line joists make a neatly-crafted frame that accepts plywood subflooring with a minimum of cutting. If a plywood joint occurs near the junction of joist ends, however, a metal strap is needed to tie the joists together

B. STEEL BEAM RECESSED INTO JOISTS

A wood scab joins the joists and supports the subfloor

This detail gives better headroom under the beam in a basement

D. WOOD BEAM RECESSED INTO JOISTS

As an alternative to the scab shown in the detail to the left, the joist ends can be notched and lapped

Wood ledgers, usually 2X4s (38mmX89mm), reduce the intrusion of a wood beam into the basement space, and reduce the effect of beam shrinkage

G. IN-LINE JOISTS

A wood scab can substitute for the metal strap

E. WOOD BEAM WITH JOIST HANGERS

A wood beam may be built up of several layers of joists nailed together

Metal joist hangers substitute for wood ledgers and allow the beam to be completely recessed in the basement ceiling. Subsidence is minimized by using joist hangers and a beam that is no deeper than the joists

H. LAPPED JOISTS   Lapped joists require no cutting to length, so are used more commonly than in-line joists, but they require more cutting of plywood panels for the subflooring

Connection	Common Nail Size
Stud to sole plate	4-8d toe nails, or
	2-16d end nails
Stud to top plate	2-16d end nails or toe nails
Double studs	10d face nails 12″ apart
Corner studs	16d face nails 24″ apart
Sole plate to joist or blocking	16d face nails 16″ apart
Double top plate	10d face nails 16″ apart
Lap joints in top plate	2-10d face nails
Rafter to top plate	3-8d toe nails
Rafter to ridge board	2-16d end nails or toe nails
Jack rafter to hip rafter	3-10d toe nails, or
	2-16d end nails
Floor joists to sill or beam	3-8d toe nails
Ceiling joists to top plate	3-16d toe nails
Ceiling joist lap joint	3-10d face nails
Ceiling joist to rafter	3-10d face nails
Collar tie to rafter	3-10d face nails
Bridging to joists	2-8d face nails each end
Let-in diagonal bracing	2-8d face nails each stud or plate
Tail joists to headers	Use joist hangers
Headers to trimmers	Use joist hangers
Ledger to header	3-16d face nails per tail joist

FIGURE 5.18
Platform framing members are fastened according to this nailing schedule, which framing carpenters know by memory, and which is incorporated into most building codes.

FIGURE 5.19
Carpenters apply a preservative-treated wood sill to a sitecast concrete foundation. Fluffy glass fiber sill sealer has been placed on the top of the concrete wall, and the sill has been drilled to fit over the projecting *anchor bolts*. Before each section of sill is bolted tightly, it is leveled as necessary with wood shingle shims between the concrete and the wood. As the anchor bolts are tightened, the sill sealer squeezes down to a negligible thickness. A length of completed sill is visible at the upper right. The basement windows were clamped into reusable steel form inserts and placed in the formwork before the concrete was cast, which causes the window to become integral with the basement wall. After the concrete was cast and the formwork was stripped, the steel inserts were removed, leaving a neatly formed concrete frame around each window. A pocket in the top of the wall for a steel beam can be seen at the upper left. (*Photo by the author*)

Floor Framing and Bridging

Floor framing (Figure 5.16) is usually laid out in such a way that the ends of uncut subflooring panels will fall directly over joists; otherwise, many panels will have to be cut, wasting both materials and time. The standard joist spacings are 16 or 24 inches o.c. (406 or 610 mm o.c.; "o.c." stands for "on center," meaning that the spacing is measured from center to center of the joists). Occasionally, a joist spacing of 19.2 inches (486 mm) is used. Any of these spacings automatically provides a joist at every panel end.

Subflooring should be glued to the joists to prevent squeaking and increase floor stiffness (Figure 5.24).

Plywood and OSB panels must be laid with the grain of their face layers perpendicular to the direction of the joists because these panels are considerably stiffer in this orientation. Sheathing and subflooring panels are normally manufactured ⅛ inch (3 mm) short in each face dimension so that they may be spaced slightly apart at all their edges to prevent floor buck-

FIGURE 5.20
Installing floor joists. *Blocking* will be inserted between the joists over the two interior beams to prevent overturning of the joists. (*Photo by the author*)

FIGURE 5.21
Various types of manufactured joists and floor trusses are often used instead of dimension lumber. This I-joist has laminated veneer lumber (LVL) flanges and a plywood web. I-joists are manufactured in very long pieces, and tend to be straighter, stronger, stiffer, and lighter in weight than sawn joists. See also Figure 3.46. (*Courtesy of Trus Joist MacMillan*)

ling during construction from the expansion of storm-wetted panels.

Bridging, which is crossbracing or solid blocking between joists at midspan, is a traditional feature of floor framing (Figure 5.25). Its function is to hold the joists straight and to help them share concentrated loads. Although many building codes no longer require it, bridging should be used on better-quality buildings because it makes a noticeable difference in the rigidity of a floor.

Where ends of joists butt into supporting headers, as around stair openings and at changes of joist direction for projecting bays, end nails and toe nails cannot carry the full weight of the joists, and sheet metal *joist hangers* must be used. Each provides a secure pocket for the end of the joist and punched holes into which a number of special short nails are driven to make a safe connection.

Manufactured I-joists and floor trusses are increasingly used in place of sawn joists because they can span farther between supports and they tend to be straighter (Figures 5.21, 5.22).

FIGURE 5.22

These floor trusses (shown here being set up for a demonstration house in a parking lot) are made of sawn lumber members joined by toothed plate connectors. The oriented strand board (OSB) web at each end of the truss allows workers to shorten the truss with a saw if necessary. Trusses are deeper than sawn joists or I-joists, but can span farther between supports and offer large passages for ductwork and pipes. (*Courtesy of Wood Truss Council of America*)

FIGURE 5.23
Applying OSB subflooring. (*Courtesy of APA–The Engineered Wood Association*)

FIGURE 5.24
For stiffness and squeak resistance, sub-flooring should be glued to the joists. The adhesive is a thick mastic that is squeezed from a sealant gun. (*Courtesy of APA–The Engineered Wood Association*)

FIGURE 5.25
Bridging between joists may be solid blocks of joist lumber, diagonal wood crossbridging, or, as seen here, diagonal steel crossbridging. During manufacture, the thick steel strip was folded across its width into a V-shape section to make it stiff. Steel crossbridging requires only one nail per piece and no cutting, so it is the fastest to install. (*Courtesy of American Plywood Association*)

Wall Framing, Sheathing, and Bracing

Wall framing, like floor framing, is laid out in such a way that a framing member, in this case a stud, occurs under each vertical joint between sheathing panels. The lead carpenter initiates wall framing by marking the stud locations on the top plate and sole plate of each wall (Figure 5.30). Other carpenters follow behind to cut the studs and headers and assemble the walls in a horizontal position on the subfloor. As each wall frame is completed, the carpenters tilt it up and nail it into position, bracing it temporarily as needed (Figures 5.31–5.34).

Sheathing made of rigid panels or diagonal boards acts as permanent bracing for the walls and is applied as soon as possible after the wall is framed. Many builders apply sheathing to each wall while it still lies flat on the floor platform. Some types of sheathing panels are intended only as insulation and have no structural value. Where these are used, *let-in diagonal bracing* is recessed into the outer face of the studs of each wall before it is erected (Figure 5.35). The let-in brace is customarily a wood 1 × 4, but steel braces are sometimes substituted.

Corners and partition intersections must furnish nailing surfaces for the edge of each plane of exterior and interior finish materials. This requires a minimum of three studs at each intersection, unless special metal clips are used to reduce the number to two (Figure 5.29).

Long studs for tall walls often must be larger than a 2 × 4 in order to resist wind forces. Because long pieces of sawn dimension lumber tend not to be perfectly straight, studs manufactured of laminated veneer lumber, parallel strand lumber, or finger-jointed lumber are sometimes used instead. Some building codes require blocking to be installed at the midheight of the studs in wall frames that are taller than 8 feet (2.4 m). The purpose of this blocking is to stop off the cavities between studs to restrict the spread of fire.

Headers over window and door openings must be sized in accordance with building code criteria. Typically, a header consists of two nominal 2-inch members standing on edge, separated by a plywood spacer that serves to make the header as thick as the depth of the wall studs (Figure 5.29). Headers for long spans and/or heavy loads are often made of laminated veneer lumber or parallel strand lumber, either of which is stronger and stiffer than dimension lumber because of an absence of knots.

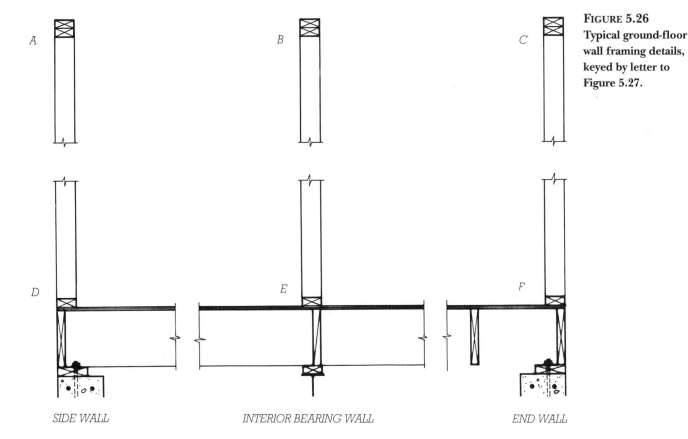

FIGURE 5.26
Typical ground-floor wall framing details, keyed by letter to Figure 5.27.

SIDE WALL

INTERIOR BEARING WALL

END WALL

The upper top plate
overlaps the lower top
plate at corners to join the
walls

The subfloor makes a convenient
platform on which to assemble the first
floor wall frames. The assembled
frames are tilted up into place, nailed
to the floor and to one another, and
supported by temporary braces.

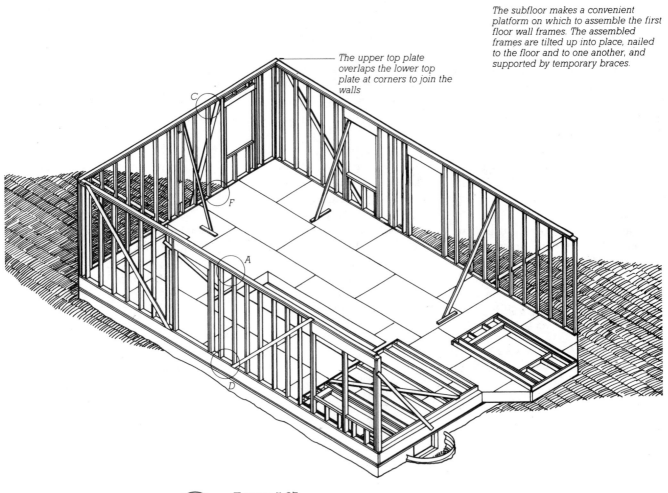

4

FIGURE 5.27
Step Four in erecting a platform frame
building: The ground-floor walls are
framed.

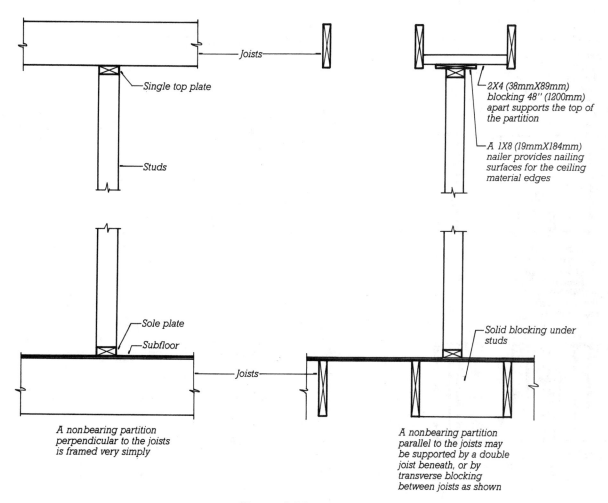

FIGURE 5.28
Framing details for nonloadbearing interior partitions.

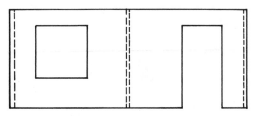

1. This is the layout of a typical exterior wall. It meets two other exterior walls at the corners, and a partition in the middle. It has two rough openings, one for a window and one for a door.

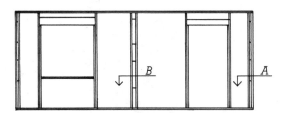

2. The framer begins by marking all the stud and opening locations on the sole plate and top plate. The "special" studs are cut and assembled first: two corner posts, a partition intersection, and full-length studs and supporting studs for the headers over the openings.

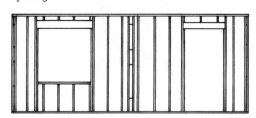

3. The wall is next filled with studs on a regular 16" (400mm) or 24" (600mm) spacing, to provide support for edges of sheathing panels.

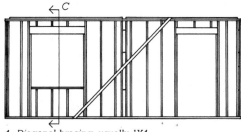

4. Diagonal bracing, usually 1X4 (19mmX89mm), is let into the face of the frame if the building will not have rigid sheathing. The second top plate may be added before the wall is tilted up, or after.

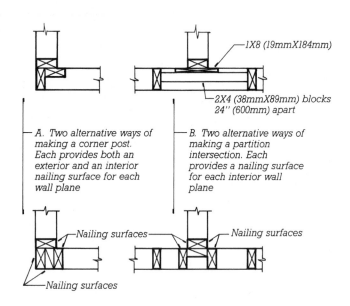

A. Two alternative ways of making a corner post. Each provides both an exterior and an interior nailing surface for each wall plane

1X8 (19mmX184mm)

B. Two alternative ways of making a partition intersection. Each provides a nailing surface for each interior wall plane

2X4 (38mmX89mm) blocks 24" (600mm) apart

Nailing surfaces

Nailing surfaces

Nailing surfaces

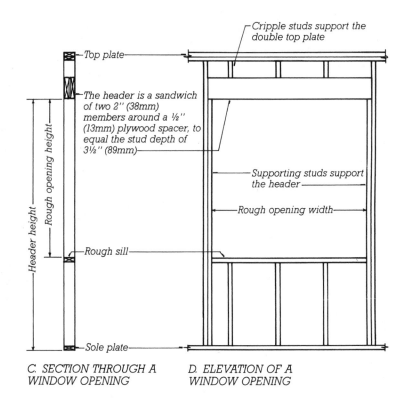

Top plate

The header is a sandwich of two 2" (38mm) members around a ½" (13mm) plywood spacer, to equal the stud depth of 3½" (89mm)

Rough sill

Sole plate

Header height

Rough opening height

C. SECTION THROUGH A WINDOW OPENING

Cripple studs support the double top plate

Supporting studs support the header

Rough opening width

D. ELEVATION OF A WINDOW OPENING

FIGURE 5.29
Procedure and details for wall framing.

FIGURE 5.30
The lead carpenter aligns the top plate and sole plate side by side on the floor platform and marks each stud location on both of them simultaneously. This is the first step in constructing a wall frame. (*Photo by the author*)

FIGURE 5.31
Assembling studs to a plate, using a pneumatic nail gun. The triple studs are for a partition intersection. (*Courtesy of Senco Products, Inc.*)

FIGURE 5.32
Tilting an interior partition into position. The gap in the upper top plate will receive the projecting end of the upper top plate from another partition that intersects at this point. (*Courtesy of APA–The Engineered Wood Association*)

FIGURE 5.33
Fastening a wall to the floor platform. The horizontal blocks between the studs will receive horizontal lines of nails used to attach vertical wood siding. (*Courtesy of Senco Products, Inc.*)

FIGURE 5.34
Ground-floor wall framing is supported by temporary diagonal bracing. After the upper-floor framing is in place and wall bracing or sheathing is complete, the frame becomes completely self-bracing and the temporary bracing is removed. The outer walls of this building are framed with 2 × 6s (38 × 140 mm) to allow for a greater thickness of thermal insulation, whereas the interior partitions are made of 2 × 4s (38 × 89 mm).
(*Photo by Joseph Iano*)

FIGURE 5.35
Applying a panel of insulating foam sheathing. Because this type of sheathing is too weak to brace the frame, diagonal bracing is inserted into the outside faces of the studs at the corners of the building. Steel bracing, nailed at each stud, is used in this frame and is visible just to the right of the carpenter's leg. (*Courtesy of The Celotex Corporation*)

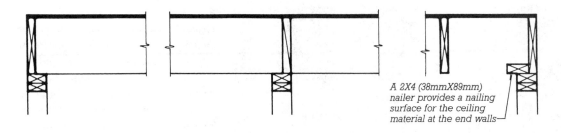

*A 2X4 (38mmX89mm)
nailer provides a nailing
surface for the ceiling
material at the end walls*

A. *SIDE WALL*   B. *INTERIOR BEARING WALL*   C. *END WALL*

FIGURE 5.36
Details of the second-floor platform, keyed to the letters on Figure 5.37. The extra
piece of lumber on top of the top plate in detail *C*, End Wall, is continuous blocking
whose function is to provide a nailing surface for the edge of the finish ceiling material, which is usually either gypsum board or veneer plaster base.

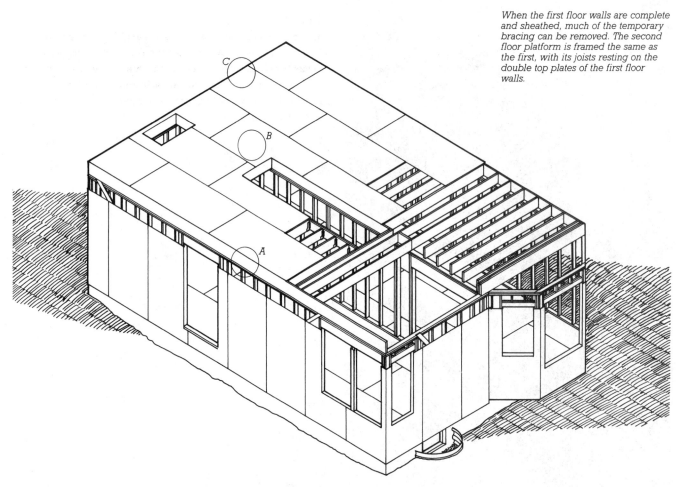

*When the first floor walls are complete
and sheathed, much of the temporary
bracing can be removed. The second
floor platform is framed the same as
the first, with its joists resting on the
double top plates of the first floor
walls.*

(5)

FIGURE 5.37
Step Five in erecting a two-story platform
frame building: Building the upper-floor
platform.

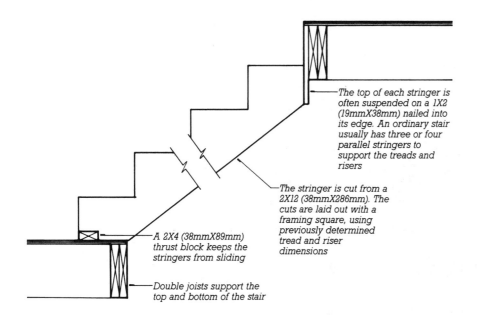

The top of each stringer is often suspended on a 1X2 (19mmX38mm) nailed into its edge. An ordinary stair usually has three or four parallel stringers to support the treads and risers

The stringer is cut from a 2X12 (38mmX286mm). The cuts are laid out with a framing square, using previously determined tread and riser dimensions

A 2X4 (38mmX89mm) thrust block keeps the stringers from sliding

Double joists support the top and bottom of the stair

FIGURE 5.38
Interior stairways are usually framed as soon as the upper-floor platform is completed. This gives the carpenters easy up-and-down access during the remainder of the work. Temporary treads of joist scrap or plywood are nailed to the *stringers*. These will be replaced by finish treads after the wear and tear of construction are finished.

Wall framing procedures for the second floor are identical to those for the first floor.

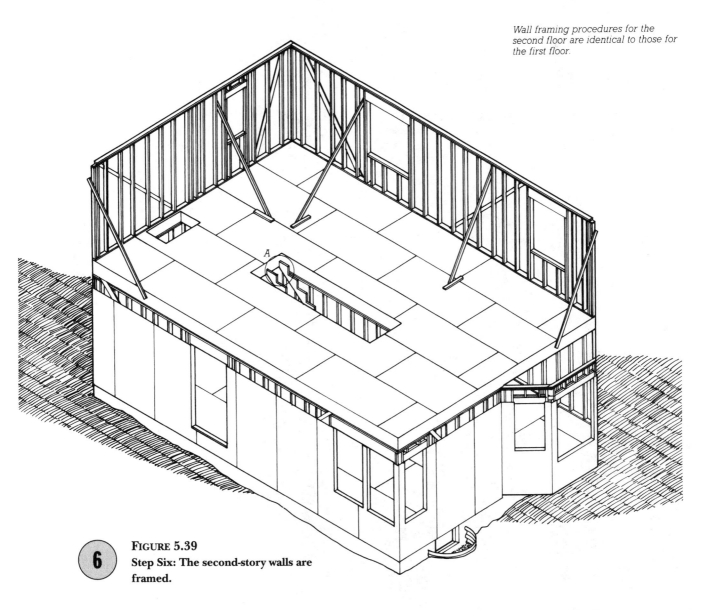

6 **FIGURE 5.39**
Step Six: The second-story walls are framed.

FIGURE 5.40
Nailing upper-floor joists to the top plates. (*Photo by the author*)

FIGURE 5.41
Installing upper-floor subflooring. The grade of the plywood panels used in this building is C-C Plugged, in which all surface voids are filled and the panel is lightly sanded to allow carpeting to be installed directly over the subfloor without additional underlayment. The long edges of the panels have interlocking tongue-and-groove joints to prevent excessive deflection of the edge of a panel under a heavy concentrated load such as a standing person or the leg of a piano. (*Courtesy of APA–The Engineered Wood Association*)

Framing for Increased Thermal Insulation

The 2 × 4 (38 × 89 mm) has been the standard wall stud since light framing was invented. In recent years, however, pressures for heating fuel conservation have led to building codes that often require more thermal insulation than can be inserted in the cavities of a wall framed with 2 × 4s. Some designers and builders have simply adopted the 2 × 6 (38 × 140 mm) as the standard stud, usually at a spacing of 24 inches (610 mm). Others have stayed with the 2 × 4 stud but cover the wall either inside or out with insulating plastic foam sheathing, thus reaching an insulation value about the same as that of a 2 × 6 wall. Many use both 2 × 6 studs and insulating sheathing to achieve even better thermal performance. Some designers in very cold climates provide still more space for thermal insulation with exterior walls that are framed in two separate layers or with vertical truss studs made up of two ordinary studs joined at intervals by plywood plates. Some of these constructions are illustrated in Figures 7.11 and 7.14.

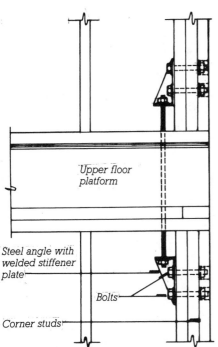

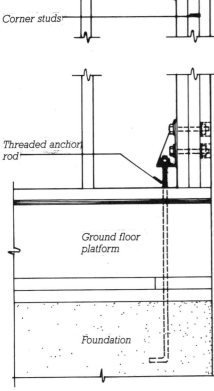

FIGURE 5.42
Tall, narrow platform frame buildings in areas subject to high winds or earthquakes must sometimes be reinforced against uplift at the corners. Hold-downs of the type shown here tie the entire height of the building securely to the foundation. To compensate for wood shrinkage, the nuts should be retightened after the first heating season, which can mean that access holes must be provided through the interior wall surfaces.

Roof Framing

The generic roof shapes for Wood Light Frame buildings are shown in Figure 5.43. These are often com-bined to make roofs that are suited to the covering of more complex plan shapes and building volumes.

For structural stability, gable and hip roofs must be securely tied by

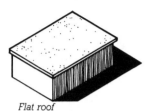

Flat roof

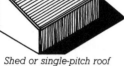

Shed or single-pitch roof

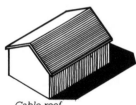
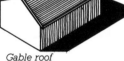

Gable roof

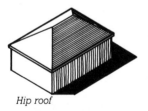

Hip roof

Gambrel roof

Mansard roof

Flat and shed roofs exert no lateral thrust

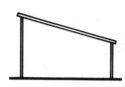

Ceiling joist

Ridge beam

Gable and hip rafters must be either tied with ceiling joists, or supported by a structural ridge beam

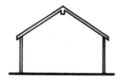

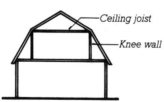

Ceiling joist

Knee wall

Gambrel and mansard roofs require both knee walls and ceiling joists for structural stability

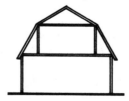

FIGURE 5.43
Basic roof shapes for Wood Light Frame buildings.

well-nailed ceiling joists to make what is, in effect, a series of triangular trusses. If the designer wishes to eliminate the ceiling joists to expose the sloping underside of the roof as the finished ceiling surface, a beam or bearing wall must be inserted at the ridge unless a system of exposed ties is designed to replace the ceiling joists.

While a college-graduate architect or engineer would find it difficult to lay out the rafters for a sloping roof using trigonometry, a carpenter, without resorting to mathematics, has little problem making the layout if the *pitch* (slope) is specified as a ratio of *rise* to *run*. Rise is the vertical dimension and run is the horizontal. In the United States, pitch is usually given on the architect's drawings as inches of rise per foot (12 inches) of run. An old-time carpenter uses these two figures on the two edges of a framing square to lay out the *rafter* as shown in Figures 5.45 and 5.51. The actual

> **The balloon frame is closely connected with the level of industrialization which had been reached in America [in the early 19th century]. Its invention practically converted building in wood from a complicated craft, practiced by skilled labor, into an industry. . . . This simple and efficient construction is thoroughly adapted to the requirements of contemporary architects . . . elegance and lightness [are] innate qualities of the balloon-frame skeleton.**
>
> **Sigfried Giedion, historian**

length of the rafter is never figured, nor does it need to be, because the square is used to make all the measurements as horizontal and vertical distances. Today, many carpenters prefer to do rafter layout with the aid of tables that give actual rafter lengths for various pitches and horizontal distances; these tables are stamped on the framing square itself or printed in pocket-size booklets.

Hips and *valleys* introduce another level of trigonometric complexity in rafter layout, but the experienced carpenter has little difficulty even here: Again, he or she can use published tables for hip and valley rafters, or do the layout the traditional way, as illustrated in Figure 5.49. The head carpenter lays out only one rafter of each type by these procedures. This then becomes the *pattern rafter* from which other carpenters trace and cut the remainder of the rafters (Figure 5.52).

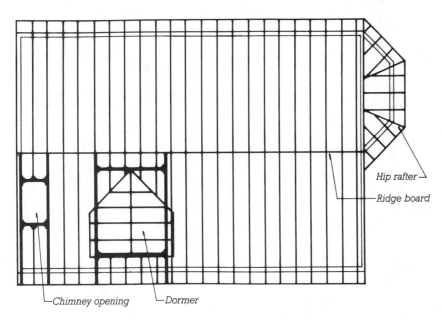

ROOF FRAMING PLAN

FIGURE 5.44
A roof framing plan for the building illustrated in Figure 5.46. The dormer and chimney openings are framed with doubled header and trimmer rafters. The dormer is then built as a separate structure that is nailed to the slope of the main roof.

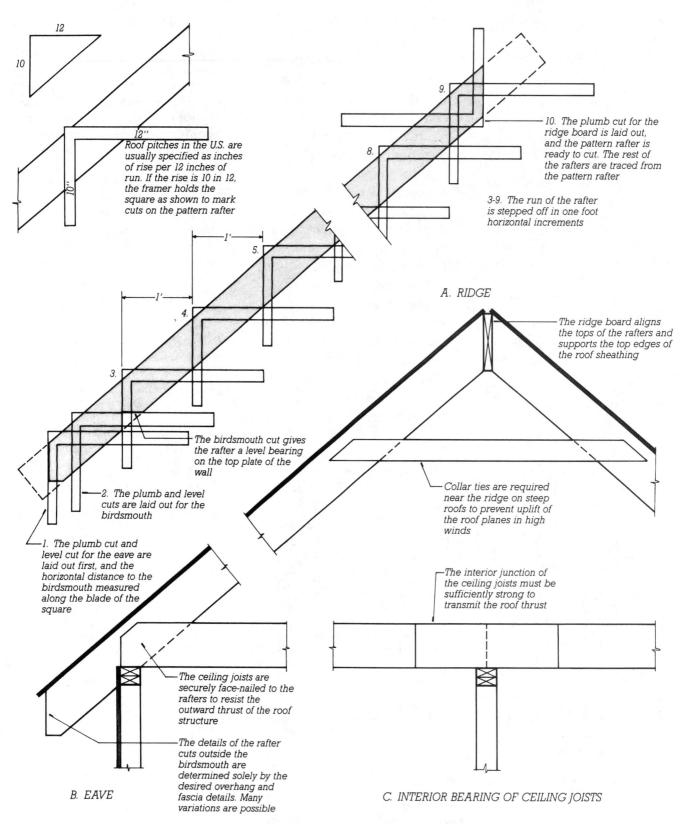

Roof pitches in the U.S. are usually specified as inches of rise per 12 inches of run. If the rise is 10 in 12, the framer holds the square as shown to mark cuts on the pattern rafter

10. The plumb cut for the ridge board is laid out, and the pattern rafter is ready to cut. The rest of the rafters are traced from the pattern rafter

3-9. The run of the rafter is stepped off in one foot horizontal increments

5.

4.

3.

A. RIDGE

The birdsmouth cut gives the rafter a level bearing on the top plate of the wall

2. The plumb and level cuts are laid out for the birdsmouth

1. The plumb cut and level cut for the eave are laid out first, and the horizontal distance to the birdsmouth measured along the blade of the square

The ridge board aligns the tops of the rafters and supports the top edges of the roof sheathing

Collar ties are required near the ridge on steep roofs to prevent uplift of the roof planes in high winds

The interior junction of the ceiling joists must be sufficiently strong to transmit the roof thrust

The ceiling joists are securely face-nailed to the rafters to resist the outward thrust of the roof structure

The details of the rafter cuts outside the birdsmouth are determined solely by the desired overhang and fascia details. Many variations are possible

B. EAVE

C. INTERIOR BEARING OF CEILING JOISTS

FIGURE 5.45
Roof framing: The lettered details are keyed to Figure 5.46. The remainder of the page shows how a framing square is used to lay out a pattern rafter, reading from the first step at the lower end of the rafter to the last step at the top.

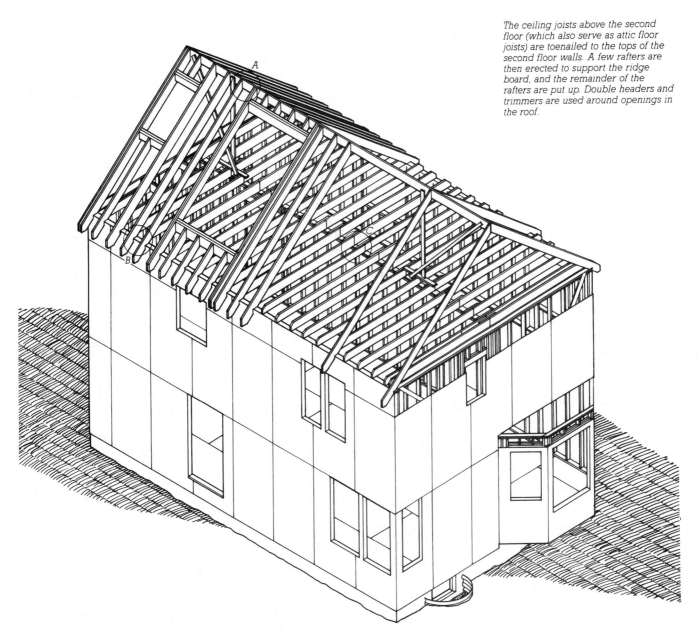

The ceiling joists above the second floor (which also serve as attic floor joists) are toenailed to the tops of the second floor walls. A few rafters are then erected to support the ridge board, and the remainder of the rafters are put up. Double headers and trimmers are used around openings in the roof.

7

FIGURE 5.46
Step Seven: Framing the attic floor and roof.

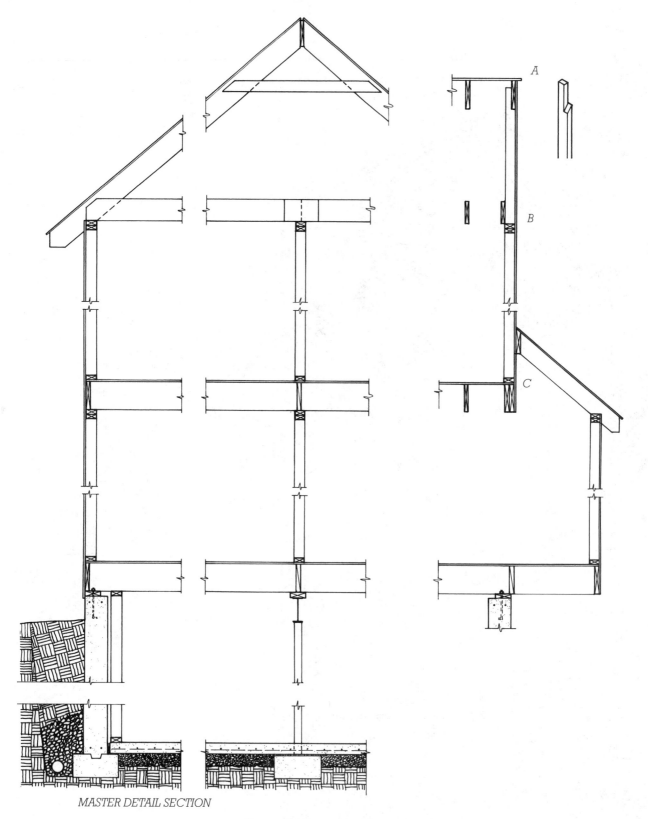

MASTER DETAIL SECTION

FIGURE 5.47

A summary of the major details for the structure shown in Figure 5.48, aligned in relationship to one another. The lettered details are keyed to Figure 5.48. The gable end studs are cut as shown in detail A and face nailed to the end rafter.

The framing of the building is completed with installation of the roof sheathing, the gable end walls, and the dormer.

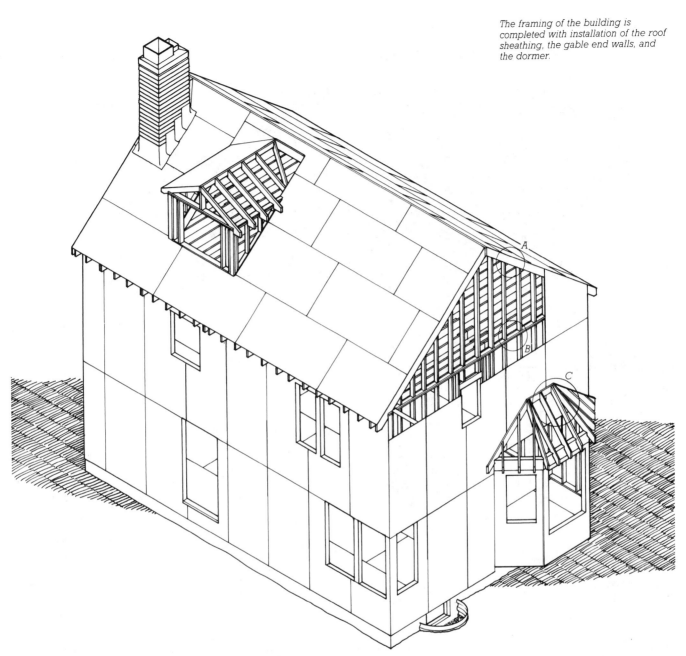

FIGURE 5.48
Step Eight: The frame is completed.

8

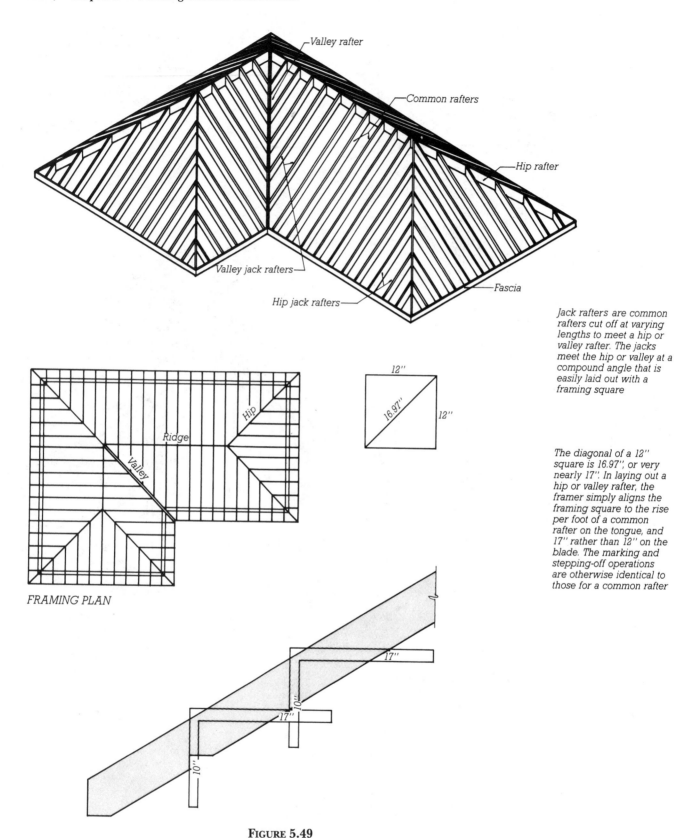

Jack rafters are common rafters cut off at varying lengths to meet a hip or valley rafter. The jacks meet the hip or valley at a compound angle that is easily laid out with a framing square

The diagonal of a 12" square is 16.97", or very nearly 17". In laying out a hip or valley rafter, the framer simply aligns the framing square to the rise per foot of a common rafter on the tongue, and 17" rather than 12" on the blade. The marking and stepping-off operations are otherwise identical to those for a common rafter

FRAMING PLAN

FIGURE 5.49

Framing for a hip roof. The difficult geometric problem of laying out the diagonal hip rafter is solved easily by using the framing square in the manner shown.

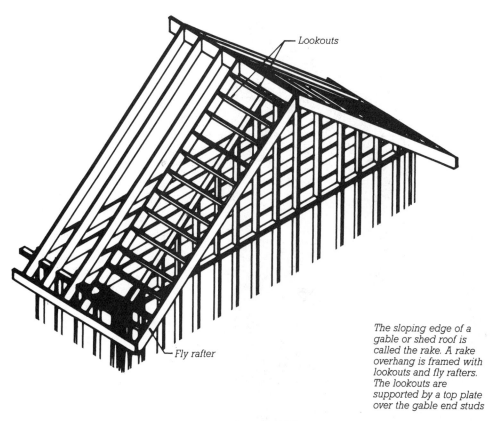

Lookouts

Fly rafter

The sloping edge of a gable or shed roof is called the rake. A rake overhang is framed with lookouts and fly rafters. The lookouts are supported by a top plate over the gable end studs

FIGURE 5.50
Framing for an overhanging rake.

FIGURE 5.51
A framing square being used to mark rafter cuts. The run of the roof, 12 inches, is aligned with the edge of the rafter on the blade of a square, and the rise, 7 inches in this case, is aligned on the tongue of the square. A pencil line along the tongue will be perfectly vertical when the rafter is installed in the roof, and one along the blade will be horizontal. True horizontal and vertical distances can be measured on the blade and tongue, respectively. (*Photo by the author*)

FIGURE 5.52
Tracing a pattern rafter to mark cuts for the rest of the rafters. The corner of the building behind the carpenters has let-in corner braces on both floors, and most of the rafters are already installed. (*Courtesy of Southern Forest Products Association*)

FIGURE 5.53
I-joists may be used as rafter material instead of solid lumber. (*Courtesy of Trus Joist MacMillan*)

Prefabricated Framing Assemblies

Roof trusses (and, to a lesser extent, *floor trusses*) find widespread use in platform frame buildings because of their speed of erection, economy of material usage, and long spans. Most are light enough to be lifted and installed by two carpenters (Figures 5.54, 5.55).

Manufactured wall panels have been adopted more slowly than roof and floor trusses, except by large builders who mass-market hundreds or thousands of houses per year. For the smaller builder, wall framing can be done on site with the same amount of material as with panels and with little or no additional overall expenditure of labor, especially when the building requires walls of varying heights and shapes.

FIGURE 5.54
Roof framing with prefabricated trusses. Sheet metal clips and nails anchor the trusses to the top plate. (*Courtesy of Gang-Nail Systems, Inc.*)

FIGURE 5.55
Prefabricated trusses are delivered to the building site in bundles. (*Courtesy of Gang-Nail Systems, Inc.*)

FIGURE 5.56
Applying plywood roof sheathing to a
half-hipped roof. Blocking between
rafters at the wall line has been drilled
with large holes for attic ventilation. The
line of horizontal blocking between studs
is to support the edges of plywood siding
panels applied in a horizontal orientation. (*Courtesy of APA–The Engineered
Wood Association*)

FIGURE 5.57
Fastening roof sheathing with a
pneumatic nail gun. (*Courtesy of
Senco Products, Inc.*)

FIGURE 5.58
A house frame sheathed with OSB
panels. (*Courtesy of APA–The
Engineered Wood Association*)

FOR PRELIMINARY DESIGN OF A WOOD LIGHT FRAME STRUCTURE

- Estimate the depth of **rafters** on the basis of the *horizontal* (not slope) distance from the outside wall of the building to the ridge board in a gable or hip roof, and the horizontal distance between supports in a shed roof. A 2 × 4 rafter spans approximately 7 feet (2.1 m), a 2 × 6 10 feet (3.0 m), a 2 × 8 14 feet (4.3 m), and a 2 × 10 17 feet (5.2 m).

- The depth of **wood light roof trusses** is usually based on the desired roof pitch. A typical depth is one-quarter of the width of the building, which corresponds to a 6/12 pitch in a gable truss. Trusses are generally spaced 24 inches (600 mm) o.c.

- Estimate the depth of **wood floor joists** as follows: 2 × 6 joists span up to 8 feet (2.4 m), 2 × 8 joists 11 feet (3.4 m), 2 × 10 joists 14 feet (4.3 m), and 2 × 12 joists 17 feet (5.2 m).

- Estimate the depth of manufactured wood I-joists as follows: 9.5-inch (240-mm) joists span 16 feet (4.9 m), 11⅞-inch (300-mm) joists span 19 feet (5.8 m), 14-inch (360-mm) joists span 22 feet (6.7 m), and 16-inch (400-mm) joists span 25 feet (7.6 m).

- Estimate the depth of **wood floor trusses** as 1/18 of their span. Typical depths of floor trusses range from 12 to 28 inches (305 to 710 mm) in 2-inch (51-mm) increments.

- 2 × 4 studs 16 inches o.c. can support one floor plus attic and roof. 2 × 6 studs 16 inches o.c. can support two floors plus attic and roof.

Framing members in light frame buildings are usually spaced either 16 or 24 inches (400 or 600 mm) o.c.

For actual sizes of dimension lumber in both conventional and metric units, see Figure 3.22.

These approximations are valid only for purposes of preliminary building layout, and must not be used to select final member sizes. They apply to the normal range of building occupancies such as residential, office, commercial, and institutional buildings. For manufacturing and storage buildings, use somewhat larger members.

For more comprehensive information on preliminary selection and layout of a structural system and sizing of structural members, see Allen, Edward, and Joseph Iano. *The Architect's Studio Companion* (2nd ed.), New York, John Wiley & Sons, Inc., 1995.

WOOD LIGHT FRAME CONSTRUCTION AND THE BUILDING CODES

It can be seen in the table in Figure 1.1 that the BOCA National Building Code, like all the model codes, allows buildings of every use group to be constructed with wood platform framing, which is classified as Type 5 construction, but with relatively severe restrictions on height and floor area. For some use groups, higher fire resistance ratings are required for some components of the building. (These are usually achieved by using gypsum board or plaster finishes to add fire resistance to the components.) Additionally, zoning ordinances in most cities identify certain densely built areas in which

wood platform frame construction is simply not permitted. And in some buildings, higher fire insurance premiums may discourage the owner from considering platform framing even if it is legally permitted.

Several examples serve to illustrate the limits within which wood platform framing can be used for different sorts of buildings. A nursing home would be classified under the BOCA Code as Use Group I-2, Institutional, incapacitated; Type 5B construction would not be permitted, and Type 5A could be used for a single-story building up to 7650 square feet (710 m²) in floor area. A meeting hall for a summer camp, Use Group A-3, can be only one story or 20 feet (6.1 m) tall, and 8925 square feet (830 m²) in area if protected (Type 5A) or 4200 square feet (390 m²) if

unprotected (Type 5B). Retail stores (Use Group M) can be one or two stories tall with as much as 10,200 square feet (950 m²) of area per floor. Hotels, apartments, and single-family dwellings may be as tall as three stories, the maximum permitted for wood platform frame construction in any use group.

Buildings larger than these limits are permitted under two additional provisions of a typical building code that relate to automatic sprinklers and fire walls. The BOCA Code allows floor areas in a fully sprinklered building to be increased beyond the limits set forth in Figure 1.1 by 200 percent for one- and two-story buildings and by 100 percent for taller buildings. With certain exceptions for hazardous and institutional buildings, the allowable height may also be

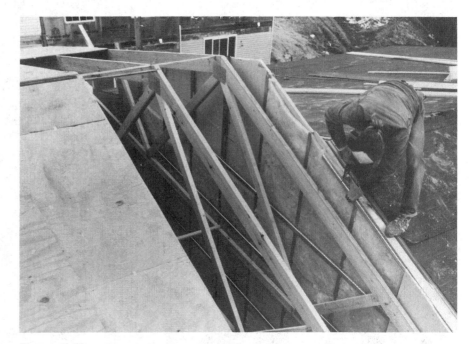

FIGURE 5.59
This proprietary fire wall consists of light-gauge metal framing, noncombustible insulation, and gypsum board. The metal framing is attached to the wood structure on either side with special clips (not shown here) that break off if the wood structure burns through and collapses, leaving the wall supported by the undamaged clips on the adjacent building. (*Courtesy of Gold Bond Building Products, Charlotte, North Carolina*)

increased by one story, thus allowing a sprinklered platform frame building of some use groups to rise as high as four stories.

The area of a building may also be increased by subdividing the building with *fire walls*. Under the BOCA Code, each portion of a building that is separated from the remainder of the building by fire walls may have a floor area as large as that permitted by Figure 1.1. Every building code has a similar provision. The required fire resistance of the fire wall is given in a separate table in the code; the BOCA table is reproduced in this book as Figure 22.6. A fire wall must extend from the foundation through the roof and must be constructed so that it remains standing even if the wooden construction on one side burns completely away. The

traditional fire wall is made of brick or concrete masonry, but proprietary systems using metal framing and gypsum board are also accepted in most areas (Figures 5.59, 5.60). This latter type of fire wall is especially useful in constructing attached dwellings. Building codes allow access through fire walls by means of self-closing fire doors; Figure 22.7 gives the BOCA requirements for the fire resistance of such doors. (It should be noted that the permitted increases in height and area relating to automatic sprinkler systems and fire walls apply to all construction systems, not just to wood platform framing.)

These limits are broad enough that a surprisingly high percentage of all building projects can be built using wood platform frame construction. And its economies are such that

most building owners will choose it over more fire-resistant types of construction if given the opportunity.

It would be a mistake, however, to assume from these facts that building codes are permissive when dealing with platform frame construction. On the contrary, wood platform framing is probably regulated more closely by codes than any other construction type. Building codes include detailed specifications for framing members and fasteners, and detailed drawings that show exactly how certain parts of the structure must be assembled. Emergency exit requirements for residential structures generally include minimum size and maximum sill height dimensions for bedroom windows to allow occupants to escape and firefighters to enter. An automatic fire alarm system

FIGURE 5.60
This townhouse has collapsed completely as the result of an intense fire, but the houses on either side, protected by fire walls of the type illustrated in Figure 5.59, are essentially undamaged. (*Courtesy of Gold Bond Building Products, Charlotte, North Carolina*)

including smoke detectors is almost universally required in residential buildings to awaken the occupants and get them moving toward the exits before the building becomes fully involved in a fire. It seems likely, too, that automatic sprinkler systems may soon become a requirement in many types of wood frame buildings, especially as simpler, cheaper sprinkler systems are developed.

THE UNIQUENESS OF WOOD LIGHT FRAME CONSTRUCTION

Wood light framing is popular because it is an extremely flexible and economical way of constructing small buildings. Its flexibility stems from the ease with which carpenters with ordinary tools can create build-ings of astonishing complexity in a variety of geometries. Its economy can be attributed in part to the relatively unprocessed nature of the materials from which it is made, and in part to mass-market competition among suppliers of components and materials and local competition among small builders.

Platform framing is the one truly complete and open system of construction that we have. It incorporates structure, enclosure, thermal insulation, mechanical installations, and finishes into a single constructional concept. Thousands of products are made to fit it: dozens of competing brands of windows and doors; a multitude of interior and exterior finish materials; scores of electrical, plumbing, and heating products. For better or worse, it can be dressed up to look like a building of wood or of masonry in any architectural style from any era of history.

Architects have failed to exhaust its formal possibilities and engineers have failed to invent a new environmental control system that it cannot assimilate.

At the other end of the scale, platform frame construction has led through sinister logic to the creation of countless horrors around the perimeters of our cities. To create the minimum-cost detached single-family house, one can begin by reducing the number of exterior wall corners to four because extra corners cost extra dollars for framing and finishing. Then the roof slope can be made as shallow as asphalt shingles will permit, to allow roofers to work without scaffolding, and roof overhangs can be eliminated to save material and labor. Windows can be made as small as building codes will allow, because a square foot of window costs more than a square foot of wall. The thinnest, cheapest finish materials

FIGURE 5.61
The W. G. Low house, built in Bristol, Rhode Island, in 1887 to the design of architects McKim, Mead, and White, illustrates both the essential simplicity of wood light framing and the complexity of which it is capable. (*Photo by Wayne Andrews*)

can be used inside and out, even if they will begin to look shabby after a short period of time, and all ceilings can be made flat and as low as is legally permitted, to save materials and labor. The result is an uninteresting, squat, dark residence that weathers poorly, has little meaningful relation to its site, and contributes to a lifeless, uninspiring neighborhood. The same principles, with roughly the same result, can be applied to the construction of apartments and commercial buildings. In most cases, things only get worse if one tries to dress up this minimal box building with inexpensive shutters, gingerbread, window boxes, and cheap imitations of fancy materials. The single most devastating fault of Wood Light Frame construction is not its com-

FIGURE 5.62
In the late 19th century, the stick-like qualities of wood light framing often found expression in the exterior ornamentation of houses.
(*Photo by the author*)

FIGURE 5.63
The Village Corner shopping center in San Diego, California (SPGA Planning & Architecture, Architects), makes skillful use of the easy informality of wood framing. (*Photo by Wes Thompson. Courtesy of American Wood Council*)

FIGURE 5.64
In this contemporary New England cottage, designer Dennis Wedlick has exploited the sculptural possibilities of platform framing. (*Photo by Michael Moran*)

FIGURE 5.65
The Thorncrown Chapel in Eureka Springs, Arkansas, designed by Fay Jones and Associates, Architects, combines large areas of glass with special framing details to create a richly inspiring space. To avoid damage to the sylvan site, all materials were carried in by hand rather than on trucks. Thus, all framing was done with nominal 2-inch (38-mm) lumber rather than heavy timbers. (*Photo by Christopher Lark. Courtesy of American Wood Council*)

bustibility or its tendency to decay, but its ability to be reduced to the irreducibly minimal—scarcely human—building system. Yet one can look to the best examples of the Carpenter Gothic, Queen Anne, Shingle Style, and Craftsman Style buildings of the 19th century, or the Bay Region and Modern styles of our own time, to realize that wood light framing gives to the designer the freedom to make a finely crafted building that nurtures life and elevates the spirit.

C.S.I./C.S.C. **Masterformat Section Numbers for Wood Light Framing**		
06100	**ROUGH CARPENTRY**	
06105	**Treated Wood Foundations**	
06110	**Wood Framing**	
	Preassembled components	
06115	**Sheathing**	
06120	**Structural Panels**	
06125	**Wood Decking**	
06170	**PREFABRICATED STRUCTURAL WOOD**	
06190	**Wood Trusses**	
06195	**Prefabricated Wood Beams and Joists**	

SELECTED REFERENCES

1. Thallon, Rob. *Graphic Guide to Frame Construction*. Newtown, Connecticut, The Taunton Press, 1991.

Unsurpassed for clarity and usefulness, this is an encyclopedic collection of details for wood platform frame construction.

2. Council of American Building Officials. *CABO One and Two Family Dwelling Code*. Falls Church, Virginia, updated at frequent intervals.

All the U.S. model building codes refer to this document as the definitive legal guide for platform frame residential construction.

3. Dietz, Albert G. H. *Dwelling House Construction* (5th ed.). Cambridge, Massachusetts, M.I.T. Press, 1991.

In his classic text, Dietz presents over 400 pages of lucid text and illustrations concerning light frame construction.

4. American Forest and Paper Association. *Span Tables for Joists and Rafters*. Washington, D.C., updated frequently.

This is the standard reference for the structural design of platform frame apartments and single-family residences. For use groups with heavier floor loadings, joist tables are available in *Wood Structural Design Data*, published by the same organization. (Address for ordering: American Wood Council, Publication Order Department, P.O. Box 5364, Madison, WI 53705-5364.)

KEY TERMS AND CONCEPTS

Wood Light Frame construction	trimmer	let-in diagonal bracing	mansard
balloon frame	sill	stringer	knee wall
joist	pilaster	supporting stud	hip rafter
stud	wracking	cripple stud	eave
rafter	framing plan	pitch	plumb cut
firestop	architectural floor plan	rise	level cut
platform frame	exterior elevations	run	collar tie
sheathing	sections	rafter	dormer
header	interior elevation	hip	fascia
rim joist	rough carpentry	valley	framing square
band joist	batter board	pattern rafter	hip rafter
subfloor	sill sealer	roof truss	lookout
sole plate	anchor bolt	floor truss	fly rafter
top plate	blocking	shed	jack rafter
ridge board	bridging	gable	birdsmouth
header	joist hanger	gambrel	fire wall

REVIEW QUESTIONS

1. Draw a series of very simple section drawings to illustrate the procedure for erecting a platform frame building, starting with the foundation and continuing with the ground floor, the ground-floor walls, the second floor, the second-floor walls, and the roof. Do not show details of connections but simply represent each plane of framing as a heavy line in your section drawing.

2. Draw from memory the standard detail sections for a two-story platform frame dwelling. *Hint:* The easiest way to draw a detail section is to draw the pieces in the order in which they are put in place during construction. If your simple drawings from Question 1 are correct, and if you follow this procedure, you will not find this question so difficult.

3. What are the differences between balloon framing and platform framing? What are the advantages and disadvantages of each? Why has platform framing become the method of choice?

4. Why is firestopping not usually required in platform framing?

5. Why is a steel beam or glue-laminated wood beam preferred to a solid wood beam at the foundation level?

6. How is a platform frame building braced against wind and earthquake forces?

7. Light framing of wood is highly combustible. In what different ways does a typical building code take this fact into account?

EXERCISES

1. Visit a building site where a wood platform frame is being constructed. Compare the details that you see on the site with the ones shown in this chapter. Ask the carpenters why they do things the way they do. Where their details differ from the ones illustrated, make up your own mind about which is better, and why.

2. Develop floor framing and roof framing plans for a building you are designing. Estimate the approximate sizes of the joists and rafters using the rules of thumb on page 171.

3. Make thumbnail sketches of 20 or more different ways of covering an L-shaped building with combinations of sloping roofs. Start with the simple ones (a single shed, two intersecting sheds, two intersecting gables) and work into the more elaborate ones. Note how the varying roof heights of some schemes could provide room for a partial second-story loft, or for high spaces with clerestory windows. How many ways do you think there are of covering an L-shaped building with sloping roofs? Look around you as you travel through areas with wood-frame buildings, especially older areas, and see how many ways designers and framers have roofed simple buildings in the past. Build up a collection of sketches of ingenious combinations of sloping roof forms.

4. Build a scale model of a platform frame from basswood or pine, reproducing accurately all its details, as a means of becoming thoroughly familiar with them. Better yet, build a small frame building for someone at full scale, perhaps a toolshed, playhouse, or garage.

## 6

EXTERIOR FINISHES FOR WOOD LIGHT FRAME CONSTRUCTION

Architects Hartman–Cox clad this small church with an economical but attractive siding of plywood and vertical wood battens, all painted white. The roofs are of asphalt shingles and the windows of wood. The rectangular windows are double hung. (*Photo by Robert Lautman. Courtesy of American Wood Council*)

As the rough carpentry of a platform frame building nears completion, a logical sequence of exterior finishing operations begins. First, the eaves and rakes of the roof are finished, which permits the roof to be shingled so as to offer as much weather protection as possible to subsequent operations. When the roof has been completed, the windows and doors are installed. Then the siding is applied. The building is now "tight to the weather," allowing interior finishing work to take place completely sheltered from sun, rain, snow, and wind. The outside of the building is ready for exterior painting or staining. Finish grading, landscaping, and paving work may commence as electricians, plumbers, sheet metal workers, plasterers, finish carpenters, and flooring installers swarm about the interior of the building.

siding, which is not installed until after the eaves and rakes, should be easy to join to them. The edges of the roof shingles should be positioned and supported in such a way that water flowing over them will drip free of the trim and siding below. The eaves must be ventilated to allow free circulation of air beneath the roof sheathing. And provision must be made to drain rainwater and snow melt from the roof without damaging the structure below.

Roof Drainage

Gutters and *downspouts* (downspouts are also called *leaders*) are often installed along the eaves of a sloping roof to remove rainwater and snow melt without wetting the walls or causing splashing or erosion on the ground below. Gutters that are

ROOFING

Finishing the Eaves and Rakes

Before a roof can be shingled, the *eaves* (horizontal roof edges) and

rakes (sloping roof edges) must be completed. Several typical ways of doing this are shown in Figures 6.1 through 6.3. When designing eave and rake details, the designer should keep several objectives in mind: The

FIGURE 6.1
Two typical details for the rakes (sloping edges) of sloping roofs. One detail has no rake overhang, and the other has an overhang supported on lookouts.

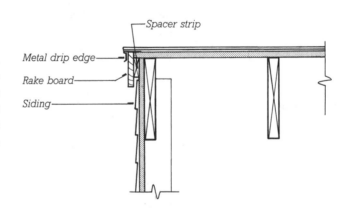

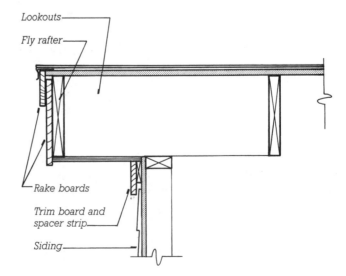

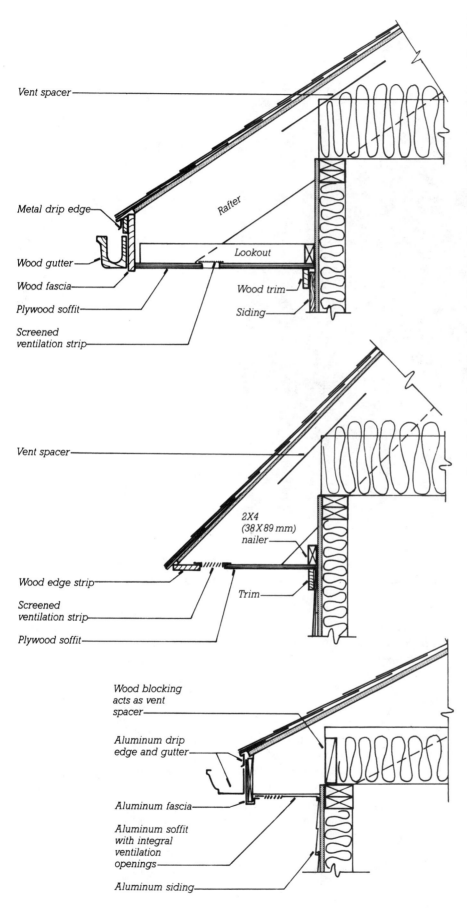

Vent spacer

Metal drip edge

Rafter

Wood gutter

Lookout

Wood fascia

Plywood soffit

Wood trim

Screened
ventilation strip

Siding

Vent spacer

2X4
(38 X 89 mm)
nailer

Wood edge strip

Trim

Screened
ventilation strip

Plywood soffit

Wood blocking
acts as vent
spacer

Aluminum drip
edge and gutter

Aluminum fascia

Aluminum soffit
with integral
ventilation
openings

Aluminum siding

FIGURE 6.2
Three ways, from among many, of finishing the eaves of a Light Frame Wood building. The top detail has a wood fascia and a wood gutter. The gutter is spaced away from the fascia on wood blocks to help prevent decay of the gutter and fascia. The width of the overhang may be varied by the designer, and a metal or plastic gutter may be substituted for the wood one. The sloping line at the edge of the ceiling insulation indicates a vent spacer as shown in Figure 6.5. The middle detail has no fascia or gutter; it works best for a steep roof with a sufficient overhang to drain water well away from the walls below. The bottom detail is finished entirely in aluminum. It shows wood blocking as an alternative to vent spacers for maintaining free ventilation through the attic. Sometimes designers entirely eliminate eave overhangs and gutters from their buildings for the sake of a "clean" appearance, but this is ill advised because the water from the roof washes over the walls, which leads to staining, leaking, decay, and premature deterioration of the windows, doors, and siding.

FIGURE 6.3
A simple wood eave detail with an aluminum gutter and downspout. The wood bevel siding, corner boards, and eave are all painted white. The carpentry at the intersection of the eave and the rake shows some poorly fitted joints and badly finished ends, and the soffit is not ventilated at all, which may lead to the formulation of ice dams on the roof (see **Figure 6.4**). (*Photo by the author*)

recessed into the surface of the roof, once popular with architects, are difficult to construct sufficiently well to prevent water from entering the building when the gutters become clogged with debris. They have been superseded almost completely by external gutters of wood, aluminum, or plastic (Figures 6.2, 6.3). These are sloped toward the points at which vertical downspouts drain away the collected water. Spacings and flow capacities of downspouts are determined by formulas found in the building codes. At the bottom of each downspout, a means must be provided for getting the water away from the building in order to prevent soil erosion and basement flooding. A simple precast concrete *splash block* can spread the water and direct it

away from the foundation, or a system of underground piping can collect the water from all the downspouts and conduct it to a storm sewer, a *dry well* (an underground seepage pit filled with broken rock), or a drainage ditch.

Some codes allow rainwater gutters to be omitted if the overhang of the roof is sufficient, thus avoiding the problems of clogging and ice buildup commonly associated with gutters. Typical minimum overhangs for gutterless buildings are 1 foot (300 mm) for single-story buildings, and twice this for two-story buildings. To prevent soil erosion and mud spatter from dripping water, the drip line at ground level below the gutterless eave must be protected with a bed of crushed stone.

Ice Dams and Roof Ventilation

Snow on a sloping roof tends to be melted by heat escaping through the roof from the heated space below. At the eave, however, the shingles and gutter, which are not directly above heated space, can be very cold, and the snow melt can freeze and begin to build up layers of ice. Soon an *ice dam* forms, and melt water accumulates above it, seeping between the shingles, through the roof sheathing, and into the building, causing damage to walls and ceilings (Figure 6.4). Ice dams can be prevented by keeping the entire roof cold enough so that snow will not melt. This is done by ventilating the roof continuously with cold outside air through vents at the eave and ridge. In buildings with an

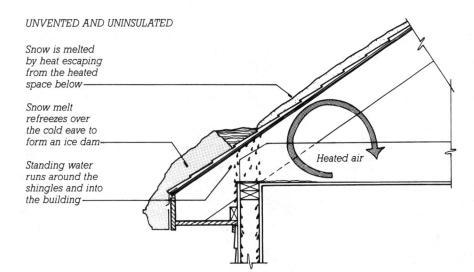

UNVENTED AND UNINSULATED

Snow is melted by heat escaping from the heated space below

Snow melt refreezes over the cold eave to form an ice dam

Standing water runs around the shingles and into the building

Heated air

FIGURE 6.4

Ice dams form because of inadequate insulation combined with a lack of ventilation, as shown in the top diagram. The lower two diagrams show how insulation, attic ventilation, and vent spacers are used to prevent snow from melting on the roof. Many building codes also require the installation of a strip of water-resistant sheet material under the shingles in the area of the roof where ice dams are likely to form—see Chapter 16 for further information.

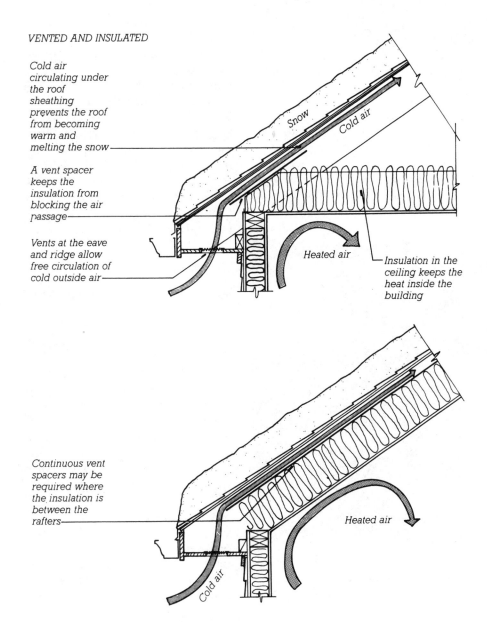

VENTED AND INSULATED

Cold air circulating under the roof sheathing prevents the roof from becoming warm and melting the snow

A vent spacer keeps the insulation from blocking the air passage

Vents at the eave and ridge allow free circulation of cold outside air

Snow

Cold air

Heated air

Insulation in the ceiling keeps the heat inside the building

Continuous vent spacers may be required where the insulation is between the rafters

Heated air

Cold air

FIGURE 6.5
To maintain clear ventilation passages where thermal insulation materials come between the rafters, wood blocking may be inserted, or *vent spacers* of plastic foam or wood fiber may be installed as shown here. The positioning of the blocking or vent spacers is shown in Figures 6.2 and 6.4. Vent spacers are generally more economical than blocking and can be installed along the full slope of a roof to maintain free ventilation between the sheathing and the insulation where the interior ceiling finish material is applied to the bottom of the rafters. (*Courtesy of Poly-Foam, Inc.*)

FIGURE 6.6
A continuous soffit vent made of perforated sheet aluminum permits generous airflow to all the rafter spaces but keeps out insects. Both roof and walls of this building are covered with cedar shingles. (*Photo by the author*)

attic, the attic itself is ventilated and kept as cold as possible. Insulation is installed in the ceilings below to retain the heat in the building. Where there is no attic and the insulation is installed between the rafters, the spaces between the rafters should be ventilated by means of air passages between the insulation and the roof sheathing (Figure 6.5).

Soffit vents create the required ventilation openings at the eave; these usually take the form of a continuous slot covered either with insect screening or with a perforated aluminum strip made especially for the purpose (Figure 6.6). Ventilation openings at the ridge can be either *gable vents* just below the roof peak in each of the end walls of the attic, or a *continuous ridge vent*, a screened cap that covers the ridge of the roof and draws air through gaps in the roof sheathing on either side of the ridge board (Figure 6.7). Building codes establish minimum area requirements for roof ventilation, based on the floor area of the building. Roof ventilation also keeps a structure cooler in summer by dissipating solar heat that has come through the shingles and roof sheathing.

Roof Shingling

Wood Light Frame buildings can be roofed with any of the systems described in Chapter 16. Because asphalt *shingles* are much less expensive than any other type of roofing, and because they are highly resistant to fire, they are used in the majority of cases. They are applied either by the carpenters who have built the frame or by a roofing subcontractor. Figures 6.8 and 16.41 through 16.44 show how an asphalt shingle roof is applied. Wood shingle and shake roofs (Figures 16.38–16.40), clay and concrete tile roofs (Figure 16.47), and architectural sheet metal roofs (Figures 16.49–16.53, 16.55) are also fairly common. Low slope roofing, either a built-up or a single-ply membrane, may also be applied to a light frame building (Figures 16.4–16.34).

FIGURE 6.7
This building has both a louvered gable vent and a continuous ridge vent to discharge air that enters through the soffit vents. The gable vent is of wood and has an insect screen on the inside. The aluminum ridge vent is designed to prevent snow or rain from entering even if blown by the wind. Vertical square-edged shiplap siding is used on this building. The roofing is a two-layered asphalt shingle designed to mimic the rough texture of a wood shake roof. (*Photo by the author*)

FIGURE 6.8
Applying asphalt shingles to a roof. Shingles and other roof coverings are covered in detail in Chapter 16.
(*Photo by the author*)

WINDOWS AND DOORS

Windows and doors are covered in detail in Chapter 18. Figure 6.9 demonstrates how a conventional wood window unit is installed in a Wood Light Frame building. Plastic-clad wood windows, aluminum windows, and plastic windows are even simpler to mount, because they include fins around the perimeter through which nails are driven into the sheathing and studs to fasten the unit securely in place.

Figures 6.10 and 6.11 relate to residential exterior wood doors. The installation procedure for a door is essentially identical to that for a window.

(a)

(b)

(c)

(d)

(e)

(f)

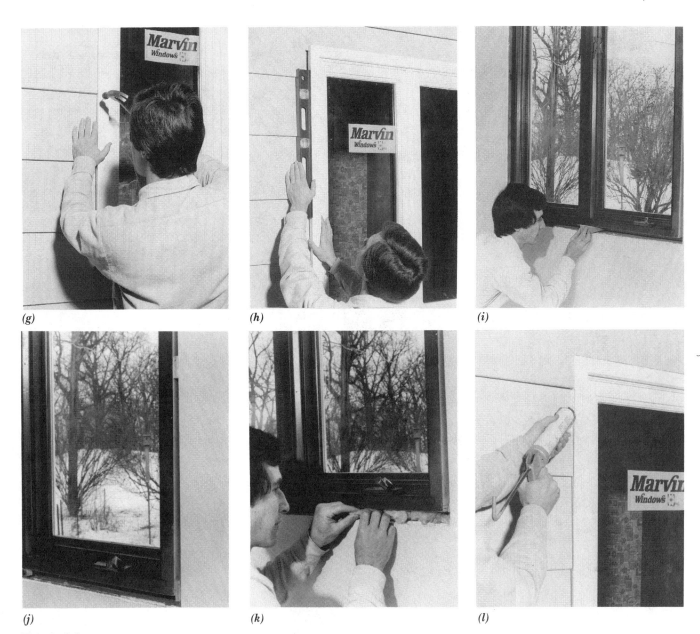

(g) (h) (i)

(j) (k) (l)

FIGURE 6.9

Installing a wood double-casement window unit in an existing building. (Installation in new construction is identical, except that the siding is installed after the window.) (*a*) The carpenter checks the dimensions and squareness of the rough opening to be sure the window will fit. Small inaccuracies in the rough opening are usually of no consequence because of the large dimensional tolerance between the window unit and the rough opening. (*b*) The window unit, with its factory-applied exterior wood casing flange, is placed in the opening and centered. (*c*) A single finish nail is driven through a lower corner of the casing and into the wall framing. (*d*) The sill is checked with a spirit level to see if it is level. If it is not, it will be shimmed temporarily to make it level. (*e*) When the sill is level, a nail is driven in the second corner of the sill to maintain the window in a level position. (*f*) The level is used again to be sure that the jamb of the window unit is plumb. This is important because the unit may have been distorted during shipping or installation and will not operate properly unless it is brought back into square. (*g*) A nail halfway up the jamb holds the unit square in the opening. (*h*) Before inserting nails around the remainder of the unit, the jamb is checked again for verticality near the top. (*i*) Inside the building, wedges are used to support the sill snugly at several points. Most carpenters prefer to use pairs of wedges (usually pieces of wood shingle), laid over one another in opposing directions, to create a flat support at each point. (*j*) The jambs on both sides are shimmed at top, center, and bottom with wood shingle wedge pairs to keep the unit square. (*k*) Thermal insulation is stuffed into the gaps between the window unit and the framing to reduce air leakage and heat loss. Plastic foam is often injected from a pressure can to fill these gaps, instead of fibrous insulation. (*l*) Sealant is applied between the siding and the exterior casing of the unit to reduce water leakage. The installation is now complete except for the installation of interior casings, which will be shown in the next chapter. Installation procedures for other types of wood windows are similar.

(*Courtesy of Marvin Windows, Warroad, Minnesota*)

FIGURE 6.10
A six-panel wood entrance door with flanking sidelights and a fan-light above. A number of elaborate traditional entrance designs such as this are available from stock for use in light frame buildings. (*Courtesy of Morgan Products, Ltd., Oshkosh, Wisconsin*)

Everybody loves window seats, bay windows, and big windows with low sills and comfortable chairs drawn up to them. . . . A room where you feel truly comfortable will always contain some kind of window place.

Christopher Alexander et al., in *A Pattern Language*, New York, Oxford University Press, 1977

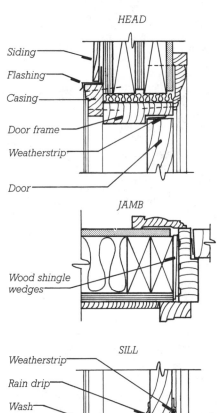

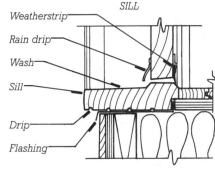

FIGURE 6.11
Details of an exterior wood door installation. The door opens toward the inside of the building in this detail.

PAINTS AND COATINGS

Paints and other architectural coatings (stains, varnishes, lacquers, sealers) protect and beautify the surfaces of buildings. A good coating job begins with thorough surface preparation to make the surface (called the *substrate*) ready to receive the coating. The coating materials must be carefully chosen and skillfully applied using the proper tools and techniques. And, to finish the job, environmental conditions must be right for the drying or curing of the coating.

Coating Materials

Most architectural coating materials are formulated to include four basic types of ingredients: vehicles, solvents, pigments, and additives. The *vehicle* provides adhesion to the substrate and forms a film over it; it is often referred to as a *film-former*. *Solvents* are volatile liquids used to improve the working properties of the paint or coating. The most common solvents used in coating materials are water and hydrocarbons, but turpentine, alcohols, ketones, esters, and ethers are also used. *Pigments* are finely divided solids that add color, opacity, and gloss control to the coating material. They also impart hardness, abrasion resistance, and weatherability to the coating. *Additives* modify various properties of the coating material. *Driers,* for example, are additives that hasten the curing of the coating. Other additives may relate to easy of application, resistance to fading, and other functions.

Coating materials fall into two major groups, solvent based and water based. *Solvent-based coatings* use solvents other than water. Their vehicles are most commonly alkyd resins, or sometimes natural oils or polyurethane resins. These coatings cure by such mechanisms as solvent evaporation, oxidation of the vehicle, or moisture curing from reaction of the vehicle with the humidity in the air. Cleanup after painting is done with mineral spirits or turpentine for most solvent-based formulations. *Water-based coatings* use water as a solvent. Most vehicles in water-based coatings are vinyl or acrylic resins. Cleanup is done with soap and water. Solvent-based and water-based coatings each have many uses for which they are preferred, and some uses for which they are more or less interchangeable. In everyday use, the common solvent-based paints are often referred to as "oil paints" or "alkyd paints," and the water-based as "latex paints."

As a component of their air pollution control programs, many states now place stringent controls on the amount of *volatile organic compounds* (*VOCs*) that coatings may release into the atmosphere. Consequently, coating manufacturers have undertaken the development and promotion of new types of water-based coatings, including water-based clear coatings, and the production of solvent-based coatings has declined significantly.

Types of Coating Materials

The various types of architectural coatings can be defined by the relative proportions of vehicle, solvent, pigment, and additives in each.

Paints contain relatively high amounts of pigments. The highest pigment content is in *flat* paints, those that dry to a completely matte surface texture. Flat paints contain a relatively low proportion of film-forming vehicle.

Paints that produce more glossy surfaces are referred to as *enamels*. A high-gloss enamel contains a very high proportion of vehicle and a relatively low proportion of pigment. The vehicle cures to form a hard, shiny film in which the pigment is fully submerged. A semigloss enamel has a somewhat lower proportion of vehicle, though still more than a flat paint.

Stains range from transparent stains to semitransparent and solid stains. Transparent stains are intended only to change the color of the substrate, usually wood and sometimes concrete. They contain little or no vehicle or pigment, a very high proportion of solvent, and a dye additive. Excess stain is wiped off with a rag a few minutes after application, leaving only the stain that has penetrated the substrate. Usually a surface stained with a transparent stain is subsequently coated with a clear finish such as varnish to bring out the color and figure of the wood and produce a durable, easily cleaned surface.

Semitransparent stains have more pigment and vehicle than transparent ones. They are not wiped after application. They are intended for exterior application in two coats and do not require a clear top coat. Also self-sufficient are the solid stains, which are usually water based. These contain much more pigment and vehicle than the other two types of stains, and resemble a dilute paint more than they do a true stain. They are intended for exterior use.

The *clear coatings* are high in vehicle and solvent content, and contain little or no pigment. Their purpose is to protect the substrate, make it easier to keep clean, and bring out its inherent beauty, whether it be wood, metal, stone, or brick. *Lacquers* are clear coatings that dry extremely rapidly by solvent evaporation. They are based on nitrocellulose or acrylics, and are employed chiefly in factories and shops for rapid finishing of cabinets and millwork. A slower-drying clear coating is known as a *varnish*. Most varnishes harden either by oxidation of an oil vehicle or by moisture curing. Varnishes are useful for on-site finishing. Varnishes and lacquers are available in gloss, semigloss, and flat formulations.

Shellac is a clear coating for interior use that is made from secretions of a particular Asian insect that have been dissolved in alcohol. Shellac dries rapidly and gives a very fine finish, but it is highly susceptible to damage by water or alcohol.

There are a number of materials that are designed specifically to prepare a surface to receive paints or clear coatings. Paste *fillers* are used to fill the small pores in open-grained woods such as oak, walnut, and mahogany prior to finishing. Various patching and caulking compounds serve to fill larger holes in the substrate. A *primer* is a pigmented coating especially formulated to make a surface more paintable. A wood primer, for example, improves the adhesion of paint to wood. It also hardens the surface fibers of the wood so it can be sanded smooth after priming. Other primers are designed as first coats for various metals or masonry materials. A *sealer* is a thin, unpigmented liquid that can be thought of as a primer for a clear coating. It seals the pores in the substrate so that the clear coating will not be absorbed.

There are many finishes, used primarily for furniture and indoor woodwork, that are based on simple formulations of natural oils and waxes. A mixture of boiled linseed oil and turpentine, rubbed into wood in many successive coats, gives a soft, water-resistant finish that is attractive to sight, smell, and touch. Waxes such as beeswax and carnauba can be rubbed over sealed (and sometimes unsealed) surfaces of wood and masonry to give a pleasingly lustrous finish. Finishes based on waxes and oils usually require reapplication at frequent intervals to maintain the character of the surface.

There are countless specialized coatings formulated for particular purposes, such as the intumescent coatings that can add fire resistance to steel (see Chapter 11), and various chemical-resistant and heat-resistant coatings for industrial use. There are also specialized impermeable coatings based on various polymer vehicles, and on materials such as asphalt (for painting roofs) and portland cement (for painting masonry and concrete).

Many of the newest and most durable architectural coatings are designed to be applied in the factory, where controlled environmental conditions and customized machines permit the use of many types of materials and techniques that would be difficult or impossible to use in the field. These include powder coatings that are sprayed on dry and fused into continuous films by the application of heat, two-component coating formulations such as

epoxies, fluoropolymer finishes (see Chapter 21), and many others.

Field Application of Architectural Coatings

There is no aspect of painting and finishing work that is more important than surface preparation. Unless the substrate is clean, dry, smooth, and sound, no paint or clear coating will perform satisfactorily. Normal preparation of wood surfaces involves scraping and sanding of any previous coatings, patching and filling holes and cracks, and sanding smooth. Preparation of metals may involve solvent cleaning to remove oil and grease, scraping and wire brushing to remove corrosion and mill scale, sandblasting if the corrosion and scale are tenacious, and, on some metals, chemical etching of the bare metal to improve the paint bond. New masonry, concrete, and plaster surfaces generally require some aging before coating to assure that the chemical curing reactions within the materials themselves are complete.

Most finish coatings adhere better if the substrate is first primed. Specialized primers are produced for both interior and exterior surfaces of wood, various metals, plaster, gypsum board, and a range of masonry and concrete surfaces. Wood trim and casings in high-quality work are *back primed* by applying primer to their back surfaces before they are installed; this helps equalize the rate of moisture change on both sides of the wood during periods of changing humidity, which reduces cupping and other distortions.

Coatings should be applied only to dry surfaces, or they may not adhere. To prevent premature drying, freshly painted surfaces should not be exposed to direct sunlight, and air temperatures during application should be between 50 and 90 degrees Fahrenheit (10 to 32°C). Wind speeds should not exceed 15 miles per hour (7 m/s). The paint materials themselves should be at normal room temperature.

Paint and other coatings may be applied by brush, roller, pad, or spray. Brushing is the slowest and most expensive, but best for detailed work and for applying many types of stains and varnishes. Spraying is the fastest and least expensive, but also the most difficult to control. Roller application is economical and effective for large expanses of flat surface. Many painters prefer to apply transparent stains to smooth surfaces by rubbing with a rag that has been saturated with stain.

A single coat of paint or varnish is usually insufficient to cover the substrate and build the required thickness of film over it. A typical requirement for a satisfactory paint coating is one coat of primer plus two coats of finish material. Two coats of varnish are generally required over raw wood, sanding the surface lightly to produce a smooth surface after the first coat has dried.

Different coatings need curing periods of lengths that range from minutes for lacquers to days for most paints. During this time, environmental conditions similar to those required for the application of the coating need to be maintained to prevent premature deterioration of the finish.

Deterioration of Paints and Finishes

Coatings are the parts of a building that are exposed to the most wear and weathering, and they deteriorate with time, requiring recoating. The ultraviolet component of sunlight is particularly damaging, causing fading of paint colors and chemical decomposition of the paint film. Clear coatings are especially susceptible to ultraviolet (UV) damage, often lasting no more than a year before discoloring and peeling, which is why they are generally avoided in exterior locations. Some clear coatings are manufactured with special UV-blocking ingredients so that they may be used on exterior surfaces. The other major force of destruction for paints and other coatings is water. Most peeling of paint is caused by water getting behind the paint film and lifting it off. The most common sources of this water in wood sidings are lumber that is damp at the time it is coated, rainwater leakage at joints, and water vapor migrating from damp interior spaces to the outdoors during the winter. Good construction practices and airtight vapor retarders can eliminate these problems.

Other major forces that cause deterioration of architectural coatings are oxygen, air pollutants, fungi, dirt, degradation of the substrate through rust or decay, and mechani-cal wear. Most exterior paints are designed to "chalk" slowly in response to these forces, allowing the rain to wash the surface clean at frequent intervals.

Typical Architectural Coating Systems

The accompanying table summarizes some typical specifications of coating systems for new surfaces of buildings.

Selected Reference

The Sherwin–Williams Company. "Painting and Coating Systems for Specifiers and Applicators." Cleveland, Ohio, published annually.

Some Typical Coating Specifications for New Surfaces

Exterior

Substrate	Surface Preparation	Primer and Topcoats
Wood siding, trim, window, doors	Sand smooth, spot-prime knots and pitch streaks. Fill and sand surface blemishes after prime coat.	Exterior alkyd primer and 2 coats latex paint, *or* 2 coats semitransparent or solid stain
Concrete masonry	Must be clean and dry, at least 30 days old	Block filler primer, 2 coats latex paint
Concrete walls	Must be clean and dry, at least 30 days old	2 coats latex paint
Stucco	Must be clean and at least 7 days old	2 coats latex paint
Iron and steel	Remove rust, mill scale, oil, and grease	Metal primer, 2 coats alkyd or latex paint
Aluminum	Clean with a solvent to remove oil, grease, and oxide	Zinc chromate primer, 2 coats enamel

Interior

Substrate	Surface Preparation	Primer and Topcoats
Plaster	Conventional plaster must be 30 days old (veneer plaster may be coated with latex paint immediately after hardening)	Interior primer, 2 coats alkyd or latex paint
Gypsum board	Must be clean, dry, and free from dust	Latex primer, 1 or 2 coats alkyd or latex paint
Wood doors, windows, and trim	Sand smooth. Fill and sand small surface blemishes after prime coat. Sand lightly between coats. Remove sanding dust before coating.	Alkyd enamel primer, 1 or 2 coats alkyd or latex enamel
Concrete masonry	Must be clean and dry, at least 30 days old	Latex primer, 2 coats alkyd or latex paint
Hardwood floors	Fill surface blemishes and sand before coating. Sand lightly between coats.	Apply oil stain if color change is desired. 2 coats oil or polyurethane varnish.

SIDING

The exterior cladding material applied to the walls of a Wood Light Frame building is referred to as *siding*. Many different types of materials are used as siding: wood boards with various profiles, applied either horizontally or vertically; plywood; wood shingles; metal or plastic sidings, brick or stone; and stucco (Figures 6.13–6.32).

Just before siding is applied, the wall sheathing is covered with a layer of paper or felt, which acts as an air barrier and backup waterproofing layer. This layer should allow water vapor to pass freely so that it does not accumulate in the wall. The traditional material for this layer is asphalt-saturated felt (commonly known as "tarpaper"). A universal concern with heating fuel efficiency has led to the development of airtight, vapor-permeable papers made of synthetic fibers as higher-performance substitutes for felt. These are stapled to the sheathing in as large sheets as possible, in order to minimize seams. For best performance in reducing air leakage, the seams in the paper should be cemented together or sealed with tape, and the edges around windows and doors should be carefully sealed.

An alternative approach to achieving airtightness is represented by Figure 6.12. Given that most sheathing panel products are impervious to the passage of air, air leakage can occur only through the seams between panels and around doors and windows. By conscientiously tapping all these gaps, an airtight envelope can be achieved without having to take any special care in the application of paper or felt.

Board Sidings

Horizontally applied board sidings, made either of solid wood or of wood composition board, are usually nailed in such a way that the nails pass through the sheathing and into the studs, giving a very secure attachment. This procedure also allows horizontal sidings to be applied directly over insulating sheathing materials without requiring a *nail-base sheathing*, a material such as plywood or waferboard that is dense enough to hold nails. *Siding nails*, whose heads are intermediate in size between those of common nails and those of

FIGURE 6.12
Using a tool supplied by the manufacturer, a carpenter applies a proprietary air barrier tape to all the seams in the sheathing of a house. This will eliminate most sources of air infiltration through the walls, saving heating and cooling costs and increasing comfort. (*Gap Wrap is a registered trademark of Benjamin Obdyke, Inc., Warminster, Pennsylvania*)

finish nails, are used to give the best compromise between holding power and appearance when attaching horizontal siding. Siding nails should be hot-dip galvanized, or made of aluminum or stainless steel, to prevent corrosion staining. Nailing is done in such a way that the individual pieces of siding may expand and contract freely without damage (Figure 6.13). The ends of horizontal siding boards are butted tightly to corner boards and window and door trim, usually with a sealant material applied to the joint during assembly.

More economical horizontal siding boards made of various types of wood composition materials are readily available. Caution should be exercised in selecting these materials, as some have experienced problems in service with excessive absorption of water, decomposition, or fungus growth. Composition sidings made from wood fiber and portland cement appear to be the most reliable.

In North America, horizontal siding boards are customarily nailed tightly against the sheathing. In Europe, and increasingly on this side of the Atlantic, siding is nailed to vertical wood spacers that are aligned over the studs (Figure 6.14). This "rainscreen" detail creates a pressure equalization chamber behind the sid-

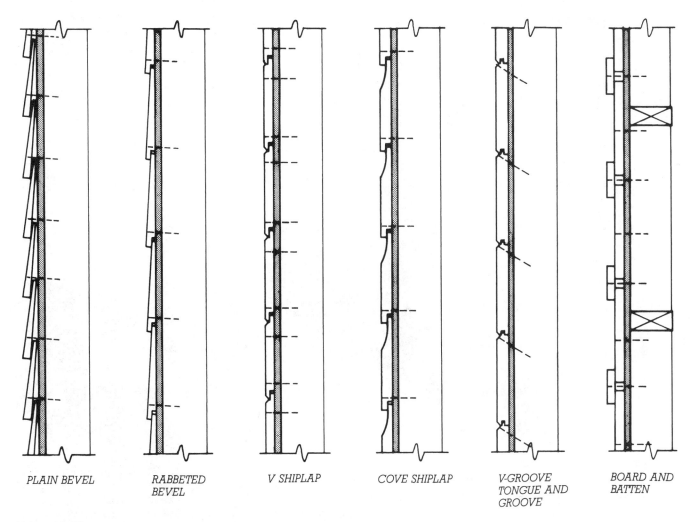

PLAIN BEVEL   RABBETED BEVEL   V SHIPLAP   COVE SHIPLAP   V-GROOVE TONGUE AND GROOVE   BOARD AND BATTEN

FIGURE 6.13

Six types of wood siding, from among many. The four *bevel* and *shiplap sidings* are designed to be applied in a horizontal orientation. *Tongue-and-groove siding* may be used either vertically or horizontally. *Board-and-batten siding* may only be applied vertically. The nailing pattern (shown with broken lines) for each type of siding is designed to allow for expansion and contraction of the boards. Nail penetration into the sheathing and framing should be a minimum of $1\frac{1}{2}$ inches (38 mm) for a satisfactory attachment.

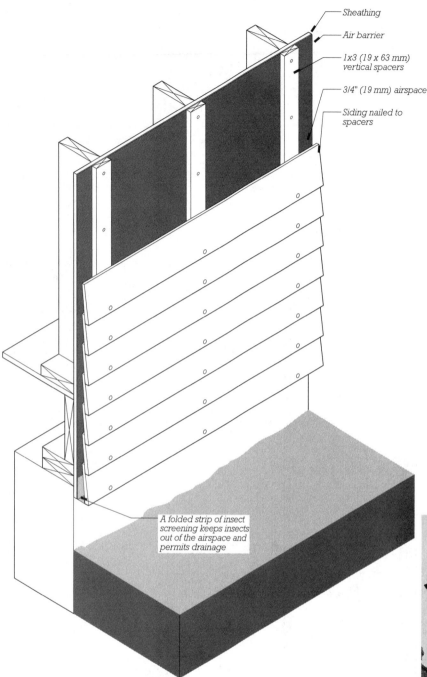

Sheathing

Air barrier

1x3 (19 x 63 mm) vertical spacers

3/4" (19 mm) airspace

Siding nailed to spacers

A folded strip of insect screening keeps insects out of the airspace and permits drainage

FIGURE 6.14
Rainscreen siding application. Any water that penetrates through joints or holes in the siding drains away before it reaches the sheathing.

ing that acts to prevent water penetration in a wind-driven rain, as explained in Chapter 19. It also provides a free drainage path for any water that leaks through the siding, and it permits rapid drying of the siding if it should become soaked with water, advantages that usually justify the additional expense of this procedure. Special attention must be given to corner boards and window and door casings to take account of the additional thickness of the wall.

Vertically applied sidings (Figures 6.16, 6.17) are nailed at the top and bottom plates of the frame, and at one or more intermediate horizontal lines of wood blocking

FIGURE 6.15
Bevel siding on this late-18th-century house is combined with wood quoins at the corners of the house and elaborately molded casings and brackets around the windows **and bay.** (*Photo by the author*)

installed between the studs. Rainscreen detailing of vertical sidings is inherently difficult because the horizontal spacers would not allow free drainage of water from the cavity.

Heartwood redwood, cypress, and cedar sidings, which are decay-resistant, may be left unfinished if desired, to weather to various shades of gray. The bare wood will erode gradually over a period of decades and will eventually have to be replaced. Other woods must be either stained or painted to prevent weathering and decay. If these coatings are renewed faithfully at frequent intervals, the siding beneath will last indefinitely.

FIGURE 6.16
A carpenter applies V-groove tongue-and-groove redwood siding to an eave soffit, using a pneumatic nail gun. (*Courtesy of Senco Products, Inc.*)

FIGURE 6.17
A completed installation of tongue-and-groove siding, vertically applied in the foreground of the picture and diagonally in the area seen behind the chairs. The lighter-colored streaks are sapwood in this mixed heartwood/sapwood grade of redwood. The windows are framed in dark-colored aluminum. (*Architect: Zinkhan/Tobey. Photo by Barbeau Engh. Courtesy of California Redwood Association*)

Plywood Sidings

Plywood siding materials (Figures 6.18, 6.19) have become popular for their economy. The cost of the material per unit area of wall is usually somewhat less than for other siding materials, and labor costs tend to be relatively low because the large sheets of plywood are more quickly installed than equivalent areas of boards. In many cases, the sheathing can be eliminated from the building if plywood is used for siding, leading to further cost savings. All plywood sidings must be painted or stained, even those made of decay-resistant heartwoods, because their veneers are too thin to withstand weather erosion for more than a few years. The most popular plywood sidings are those that are grooved to imitate board sidings. The grooves also serve to conceal the vertical joints between sheets.

The largest problem in using plywood sidings for multistory buildings is how to detail the horizontal end joints between sheets. A Z-flashing of aluminum (Figure 6.19) is the usual solution, and is clearly visible. The designer should include in the con-

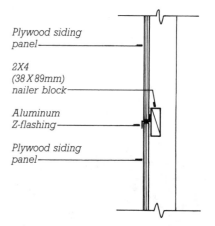

Plywood siding panel

2X4 (38 X 89mm) nailer block

Aluminum Z-flashing

Plywood siding panel

FIGURE 6.19
A detail of a simple Z-flashing, the device most commonly used to prevent water penetration at horizontal joints in plywood siding.

FIGURE 6.18
Grooved plywood siding is used vertically on this commercial building by Roger Scott Group, Architects. The horizontal metal flashings between sheets of plywood are purposely emphasized here with a special flashing detail that casts a dark shadow line.
(*Courtesy of American Plywood Association*)

struction drawings a sheet layout for the plywood siding that organizes the horizontal joints in an acceptable manner. This will help avoid a random, unattractive pattern of joints on the face of the building.

Shingle Sidings

Wood shingles and shakes (Figures 6.20–6.26) require a nail-base sheathing material such as OSB, plywood, or waferboard. Either corrosion-resistant box nails or air-driven staples may be used for attachment. Most shingles are of cedar or redwood heartwood and need not be coated with paint or stain unless such a coating is desired for cosmetic reasons. Figures 6.22 and 6.23 show two different ways of turning corners with wood shingles.

The application of wood shingle siding is labor intensive, especially around corners and openings, where many shingles must be cut and fitted. Several manufacturers produce shingle siding in panel form by stapling and/or gluing shingles to wood backing panels. A typical panel size is approximately 2 feet high and 8 feet long (600 by 2450 mm). Several different shingle application patterns are available in this form, along with prefabricated woven corner panels. The application of panelized shingles is much more rapid than that of individual shingles, which results in sharply lower labor costs and in most cases a lower overall cost.

FIGURE 6.20
Applying wood shingle siding over asphalt-saturated felt building paper. The corners are woven as illustrated in Figure 6.22. (*Courtesy of Red Cedar Shingle and Handsplit Shake Bureau*)

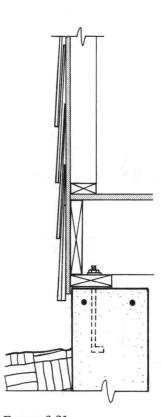

FIGURE 6.21
A detail of wood shingle siding at the sill of a wood platform frame building. The first course of shingles projects below the sheathing to form a drip, and is doubled so that all the open vertical joints between the shingles of the outside layer are backed up by the undercourse of shingles. Succeeding courses are single, but are laid so that each course covers the open joints in the course below.

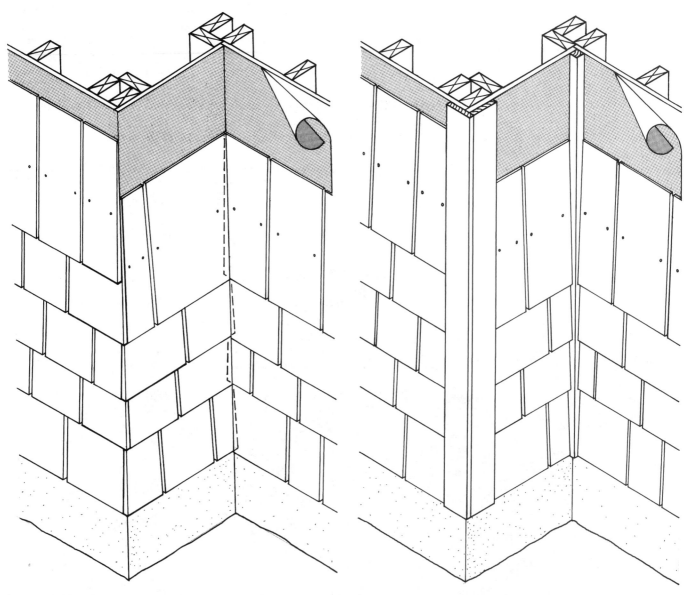

FIGURE 6.22
Wood shingles can be woven at the corners to avoid corner boards. Each corner shingle must be carefully trimmed to the proper line with a block plane, which is time consuming and relatively expensive, but the result is a more continuous, sculptural quality in the siding.

FIGURE 6.23
Corner boards save time when shingling walls, and become a strong visual feature of the building. Notice in this and the preceding diagram how the joints and nailheads in each course of shingles are covered by the course above.

FIGURE 6.24
Fancy cut wood shingles were often a featured aspect of shingle siding in the late 19th century. Notice the fish-scale shingles in the gable end, the serrated shingles at the lower edges of walls, and the sloping double shingle course along the rakes. Corners are woven. (*Photo by the author*)

FIGURE 6.25
Fancy cut wood shingles stained in contrasting colors are used here on a contemporary restaurant. (*Courtesy of Shakertown Corporation*)

FIGURE 6.26
Both roof and walls of this New England house by architect James Volney Righter are covered with wood shingles. Wall corners are woven. (*Photo © Nick Wheeler/Wheeler Photographics*)

Metal and Plastic Sidings

Painted wood sidings become unattractive and are apt to decay unless they are carefully scraped and repainted every 3 to 6 years. Sidings formed of prefinished sheet aluminum or molded of vinyl plastic, usually designed to imitate wood sidings, will not decay and are generally guaranteed against needing repainting for long periods, typically 20 years (Figures 6.27, 6.28). Such sidings do have their own problems, however, including the poor resistance of aluminum sidings to denting and the tendency of plastic sidings to shatter on impact, especially in cold weather. Although plastic and aluminum sidings bear a superficial similarity to the wood sidings that they mimic, their details around openings and corners in the wall are sufficiently different that they are usually inap-

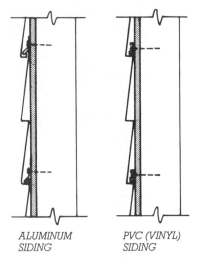

ALUMINUM
SIDING

PVC (VINYL)
SIDING

FIGURE 6.27
Aluminum and vinyl sidings are both intended to imitate horizontal lap siding in wood. Their chief advantage in either case is low maintenance. Nails are completely concealed in both systems.

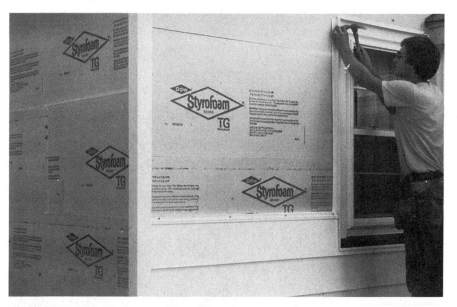

FIGURE 6.28
Installing aluminum siding over insulating foam sheathing on an existing residence. Special aluminum pieces are provided for corner boards and window casings; each has a shallow edge channel to accept the cut ends of the siding. (*Courtesy of Dow Chemical Company*)

FIGURE 6.29
Applying exterior stucco over wire netting. The workers to the right hold a hose that sprays the stucco mixture onto the wall, while the man at the left levels the surface of the stucco with a straightedge. The small rectangular opening at the base of the wall is a crawlspace vent. (*Courtesy of Keystone Steel and Wire Company*)

propriate for use in historic restoration projects.

Stucco

Stucco, a portland cement plaster, is a strong, durable, economical, fire-resistant material for siding light frame buildings. Figure 6.29 shows the application of exterior stucco with a spray apparatus, and Figure 6.30 displays an unusual use of the material. Stucco is discussed in greater detail in Chapter 23.

FIGURE 6.30
This small office building in Los Angeles demonstrates the plasticity of form that is possible with stucco siding. Vertical and horizontal joints in the stucco serve to minimize its tendency to crack as it dries out after curing. Although it is supported below on concrete columns and steel girders, this is actually a Wood Light Frame building. (*Eric Owen Moss Architects. Photo by Tom Bonner*)

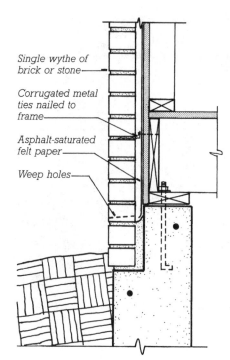

Single wythe of brick or stone

Corrugated metal ties nailed to frame

Asphalt-saturated felt paper

Weep holes

FIGURE 6.31
A detail of masonry veneer facing for a platform frame building. The weep holes drain any moisture that might collect in the cavity between the masonry and the sheathing. The cavity should be at least 1 inch (25 mm) wide. A 2-inch (50-mm) cavity is better because it is easier to keep free of mortar droppings that can clog the weep holes.

Masonry Veneer

Light frame buildings can be faced with a single wythe of brick or stone in the manner shown in Figure 6.31. The corrugated metal ties support the masonry against falling away from the building, but allow for differential movement between the masonry and the frame. Masonry materials and detailing are covered in Chapters 8 and 9.

FIGURE 6.32
Careful detailing is evident in every aspect of the exterior finishes of this commercial building. Notice especially the cleanly detailed window casings, the purposeful use of both vertical boards and wood shingles for siding, and the neat junction between the sidewall shingles and the rake boards. (*Architects: Woo and Williams. Photo by Richard Bonarrigo. Courtesy of the architects*)

Exterior Construction

Wood is often used outdoors for porches, decks, stairs, stoops, and retaining walls (Figure 6.33). Decay-resistant heartwoods, wood that is pressure-treated with preservatives, and certain wood/plastic composite planks are suitable for these exposed uses. If nondurable woods are used, they will soon decay at the joints, where water is trapped and held by capillary action. Wood decking that is exposed to the weather should always have open, spaced joints to allow for drainage of water through the deck and for expansion and contraction of the decking.

FIGURE 6.33
An exterior deck of redwood. The decking boards have spaces between to allow water to drain through. (*Designer: John Matthias. Photo by Ernest Braun. Courtesy of California Redwood Association*)

EXTERIOR PAINTING, FINISH GRADING, AND LANDSCAPING

The final steps in finishing the exterior of a light frame building are painting or staining of exposed wood surfaces; finish grading of the ground around the building; installation of paving for drives, walkways, and terraces; and seeding and planting of landscape materials. By the time these operations take place, interior finishing operations are well underway, having begun as soon as the roofing, sheathing, and windows and doors were in place.

C.S.I./C.S.C. Masterformat Section Numbers for Exterior Finishes for Light Frame Construction in Wood	
07300	SHINGLES AND ROOFING TILES
07400	MANUFACTURED ROOFING AND SIDING
07460	Siding
08200	WOOD AND PLASTIC DOORS
08500	METAL WINDOWS
08510	Steel Windows
08520	Aluminum Windows
08600	WOOD AND PLASTIC WINDOWS
08610	Wood Windows
08630	Plastic Windows
09900	PAINTING
09910	Exterior Painting
09920	Interior Painting
09930	Transparent Finishes

SELECTED REFERENCES

In addition to consulting the reference works previously listed at the end of Chapter 5, the reader should acquire current catalogs from a number of manufacturers of residential windows. Most lumber retailers also distribute unified catalogs of windows, doors, and millwork that can be an invaluable part of the designer's reference shelf.

KEY TERMS AND CONCEPTS

eave	shingles	paint	back primed
rake	substrate	enamel	siding
gutter	vehicle	stain	nail-base sheathing
downspout or leader	film-former	clear coating	siding nail
splash block	solvent	lacquer	bevel siding
dry well	pigment	varnish	shiplap siding
ice dam	additive	shellac	tongue-and-groove siding
soffit vent	drier	filler	board-and-batten siding
gable vent	solvent-based coating	primer	stucco
continuous ridge vent	water-based coating	sealer	masonry veneer
vent spacer	volatile organic compound (VOC)	primed	

REVIEW QUESTIONS

1. In what order are exterior finishing operations carried out on a platform frame building, and why?

2. At what point in exterior finishing operations can interior finishing operations begin?

3. Which types of siding require a nail-base sheathing? What are some nail-base

sheathing materials? Name some sheathing materials that cannot function as nail bases.

4. What are the reasons for the relative economy of plywood sidings? What special precautions are advisable when designing a building with plywood siding?

5. How are corners made when siding a building with wood shingles?

6. Specify two alternative exterior coating systems for a building clad in wood bevel siding.

7. What are the usual reasons for premature paint failure on a wood-sided house?

EXERCISES

1. For a completed wood frame building, make a complete list of the materials used for exterior finishes and sketch a set of details of the eaves, rakes, corners, and windows. Are there ways in which each could be improved?

2. Visit a lumberyard and look at all the alternative choices of sidings, windows, doors, trim lumber, and roofing. Study one or more systems of gutters and downspouts. Look at eave vents, gable vents, and ridge vents.

3. For a Wood Light Frame building of your design, list precisely and completely

the materials you would like to use for the exterior finishes. Sketch a set of typical details to show how these finishes should be applied to achieve the appearance you desire, with special attention to the roof edge details.

## 7

INTERIOR FINISHES FOR WOOD LIGHT FRAME CONSTRUCTION

- **Thermal Insulation and Vapor Retarder**

  Thermal Insulating Materials

  Increasing Levels of Thermal Insulation

  Radiant Barriers

  Vapor Retarders

  Air Infiltration and Ventilation

  Airtight Drywall Approach

- **Wall and Ceiling Finish**

- **Millwork and Finish Carpentry**

  Interior Doors

  Window Casings and Baseboards

  Cabinets

  Finish Stairs

  Miscellaneous Finish Carpentry

- **Flooring and Ceramic Tile Work**

- **Finishing Touches**

In this Wood Light Frame dwelling, architect Richard Meier has produced uncommon results from the most common, least expensive interior wall finish material: gypsum board. (*Photo by Ezra Stoller, © 1971 ESTO*)

As the exterior roofing and siding of a platform frame building approach completion, the framing carpenters and roofers are joined by workers from a number of other building trades. Masons commence work on fireplaces and chimneys (Figures 7.1, 7.2). Plumbers begin *roughing in* their piping ("roughing in" refers to the process of installing the components of a system that will not be visible in the finished building). First to be installed are the large *DWV* (*drain–waste–vent*) *pipes,* which drain by gravity and must therefore have right of way, then the small *supply pipes,* which bring hot and cold water to the fixtures, and the *gas pipes* (Figures 7.3, 7.4). If the building is to have a *hydronic* (forced hot water) *heating system,* the plumbers put in the boiler and rough in the heating pipes and convectors at this time. If the building will have central warm-air heating and/or air conditioning, sheet metal workers install the furnace and ductwork (Figures 7.5, 7.6). The electricians are usually the last of the mechanical and electrical trades to complete their roughing in because their wires are flexible and can generally be routed around the pipes and ducts without difficulty (Figure 7.7). When the plumbers, sheet metal workers, and electricians have completed their rough work, which consists of everything except the plumbing fixtures, electrical outlets, and air registers and grills, inspectors from the municipal building department check each of the systems for compliance with the plumbing, electrical, and sheet metal codes.

Once these inspections have been passed, connections are made to external sources of water, gas, electricity, and communications services, and to a means of sewage disposal, either a sewer main or a septic tank and leaching field. Thermal insulation and a vapor retarder are added to the exterior ceilings and walls. At this point, the general building inspector of the municipality is called in to examine the framing, exterior finish, and insulation of the building. Any deficiencies must be made good before the project is allowed to proceed further.

Now a new phase of construction begins, the interior finishing operations, during which the inside of the building undergoes a succession of radical transformations. The elaborate tangle of fram-

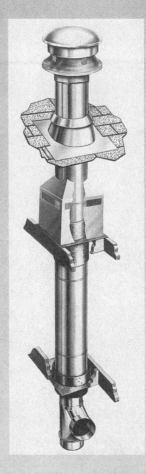

ing members, ducts, pipes, wires, and insulation rapidly disappears behind the finish wall and ceiling materials. The interior millwork—doors, finish stairs, railings, cabinets, shelves, closet interiors, and door and window casings—is installed by finish carpenters who specialize in this exacting work. The finish flooring materials are installed as late in the process as possible to save them from damage by the passing armies of workers, and the finish carpenters follow behind the flooring installers to add the baseboards that cover the last of the rough edges in the construction.

Finally, the building hosts the painters who prime, paint, stain, varnish, and paper its interior surfaces. The plumbers, electricians, and sheet metal workers make brief return appearances on the heels of the painters to install the plumbing fixtures, the electrical receptacles, switches, and lighting fixtures, and the air grills and registers. At last, following a final round of inspections and a last-minute round of repairs and corrections to remedy lingering defects, the building is ready for occupancy.

FIGURE 7.1
Insulated metal flue systems are often more economical than masonry chimneys for furnaces, boilers, water heaters, package fireplaces, and solid fuel stoves, and can be installed by carpenters.
(*Courtesy of Selkirk Metalbestos*)

FIGURE 7.2
A mason adds a section of clay flue liner to a chimney. The large flue is for a fireplace, and the three smaller flues are for a furnace and two woodburning stoves.
(*Photo by the author*)

FIGURE 7.3
At the basement ceiling, the plumber installs the plastic waste pipes first to be sure that they are properly sloped to drain. The copper supply pipes for hot and cold water are installed next.
(*Photo by the author*)

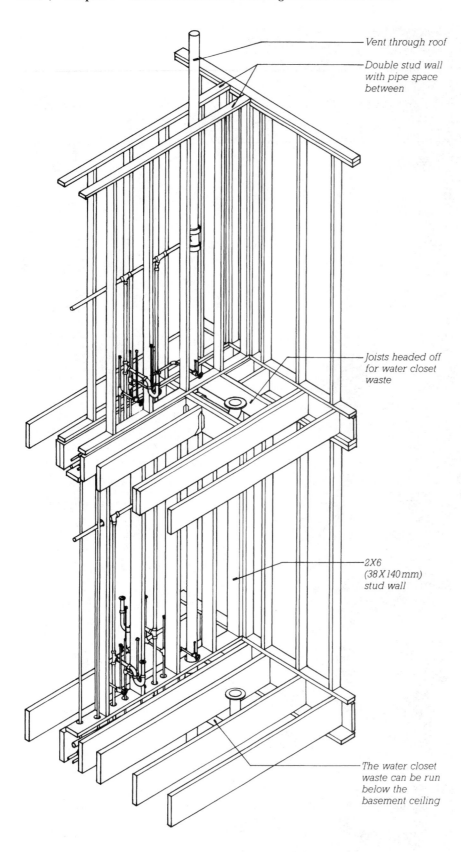

Vent through roof

Double stud wall with pipe space between

Joists headed off for water closet waste

2X6 (38 X 140 mm) stud wall

The water closet waste can be run below the basement ceiling

FIGURE 7.4
The plumber's work is easier and less expensive if the building is designed to accommodate the piping. The "stacked" arrangement shown here, in which a second-floor bathroom and a back-to-back kitchen and bath on the first floor share the same vertical runs of pipe, is economical and easy to rough in, as compared with plumbing that does not align vertically from one floor to the next. The double wall framing on the second floor allows plenty of space for the waste, vent, and supply pipes. The second-floor joists are located to provide a slot through which the pipes can pass at the base of the double wall, and the joists beneath the water closet (toilet) are headed off to house its waste pipe. The first floor shows an alternative type of wall framing using a single layer of deeper studs, which must be drilled to permit horizontal runs of pipe to pass through.

FIGURE 7.5
The installation of this warm air furnace and air conditioning unit is almost complete, needing only electrical connections. The metal pipe running diagonally to the left carries the exhaust gases from the oil burner to the masonry chimney. The ductwork is insulated to prevent moisture from condensing on it during the cooling season and to prevent excessive losses of energy from the ducts.
(*Photo by the author*)

FIGURE 7.6
Ductwork and electrical wiring are installed conveniently through the openings in these floor trusses, making it easy to apply a finish ceiling if desired. The 2 × 6 that runs through the trusses in the center of the photograph is a bridging member that restrains the trusses from buckling. (*Courtesy of Trus Joint Corporation*)

FIGURE 7.7
The electrician begins work by nailing plastic fixture boxes to the framing of the building in the locations shown on the electrical plan. Then holes are drilled through the framing, and the plastic-sheathed cable, which houses two insulated copper conductors and a bare ground wire, is pulled through the holes and into the boxes, where it is held by insulated staples driven into the wood. After the interior wall materials are in place, the electrician returns to connect outlets, lighting fixtures, and switches to the wires and the boxes and to affix cover plates to finish the installation. (*Photo by the author*)

THERMAL INSULATION AND VAPOR RETARDER

Thermal insulation helps keep a building cooler in summer and warmer in winter by retarding the passage of heat through the exterior surfaces of the building. It helps keep the occupants of the building more comfortable by moderating the temperatures of the interior surfaces of the building and reducing convective drafts, and it reduces the energy consumption of the building for heating and cooling to a fraction of what it would be without insulation.

Thermal Insulating Materials

Figure 7.8 lists the most important types of thermal insulating materials for Wood Light Frame buildings, and gives some of their characteristics. Glass fiber batts are the most popular type of insulation for wall cavities in new construction. They are also widely used as attic and roof insulation. Some details of their installation are shown in Figures 7.9 and 7.10.

Increasing Levels of Thermal Insulation

Until a couple of decades ago, the maximum resistance to heat flow (*R-value*) that could be achieved in a platform frame building was limited by the thickness of the studs and ceiling joists. With the rapid rise in energy costs in the mid-1970s, however, economic analyses indicated the strong desirability of insulating buildings to R-values well beyond those that could be reached by merely filling the available voids in the framing. At about the same time, in an effort to reduce consumption of nonrenewable fossil fuel resources, building codes began to specify higher minimum insulating values for walls, ceilings, windows, and basement walls. These developments led to experimentation with thicker walls, insulating sheathing materials, and various

schemes for insulating the concrete and masonry walls of basements. Figures 7.11 through 7.14 show some of the results of this experimentation. Insulating sheathing materials and 2 × 6 (38 × 140 mm) exterior wall studs have become widely accepted in the building industry in North America, as have exterior and interior insulation for basements. Treated wood foundations (Figures 5.10, 5.11), which are easily insulated to high R-values, are gaining increasing acceptance. The other schemes illustrated here for adding extra insulation to platform frame buildings are not yet standard practice but are representative of current experimentation by designers and building contractors in the colder regions of the United States and Canada.

Manufacturers of glass fiber insulating batts have responded to the call for better-insulated buildings by producing high-density batts whose R-value per inch of thickness is substantially higher than that of conventional batts (Figure 7.8). A 3.5-inch (89-mm) batt of this material has a total R-value of 15 (104). When this resistance is added to that of the interior wall finish, sheathing, and siding, it is often possible to meet a legal requirement for an R19 wall while using ordinary 2 × 4 studs.

Ceiling insulation batts usually must be compressed somewhat to fit them into the diminishing space under the roof sheathing, over the top of the exterior wall. This necessity, which can be observed in Figures 7.11 and 7.14, reduces the thermal resistance of the insulating layer of the building at this point. A *raised-heel roof truss* (Figure 7.15) overcomes this problem.

Radiant Barriers

In warmer regions, *radiant barriers* are increasingly used in roofs and walls to reduce the flow of solar heat into the building. These are thin sheets or panels faced with a bright metal foil that reflects infrared radiation. Most types of radiant barriers are made to be installed over the rafters or studs and beneath the sheathing, so they must be put in place during the framing of the building. They are effective only if the bright surface of the barrier faces a ventilated airspace; this allows the reflective surface to work properly, and it provides for convective removal of solar heat that has passed through the outer skin of the building. Radiant barrier panels are configured with folds that provide this airspace automatically. Sheet materials are draped loosely over rafters so that their sag creates the airspaces. Radiant barriers are used in combination with conventional insulating materials to achieve the desired overall thermal performance.

Vapor Retarders

Vapor retarders (often called, less accurately, *vapor barriers*) have also received increased attention during this period of renewed interest in fuel economy. A vapor retarder is a membrane of metal foil, plastic, or treated paper placed on the warm side of thermal insulation to prevent water vapor from entering insulation and condensing into liquid. The function of a vapor retarder is explained in detail on pages 572–574. Its role increases in importance as thermal insulation levels increase.

Many batt insulation materials are furnished with a vapor retarder

Type	Material(s)	Method of Installation	R-value[a]	Combustibility	Advantages and Disadvantages
Batt or blanket	Glass wool; rock wool	The batt or blanket is installed between framing members and is held in place either by friction or by a facing stapled to the framing	3.2–3.7 *22–26*	The glass wool or rock wool is incombustible, but paper facings are combustible	Low in cost, fairly high R-value, easy to install
High-density batt	Glass wool	Same as above	4.3 *30*	Same as above	Same as above
Loose fill	Glass wool; rock wool	The fill is blown onto attic floors, and into wall cavities through holes drilled in the siding	2.5–3.5 *17–24*	Noncombustible	Good for retrofit insulation in older buildings. May settle somewhat in walls
Loose fibers with binder	Treated cellulose; glass wool	As the fill material is blown from a nozzle, a light spray of water activates a binder that adheres the insulation in place and prevents settlement	3.1–4.0 *22–27*	Noncombustible	Low in cost, fairly high R-value
Foamed in place	Polyurethane	The foam is mixed from two components and sprayed or injected into place, where it adheres to the surrounding surfaces	5–7 *35–49*	Combustible, gives off toxic gases when burned	High R-value, high cost, good for structures that are hard to insulate by conventional means
Foamed in place	Polyicynene	Two components and sprayed or injected into place, where they react chemically and adhere to the surrounding surfaces	3.6–4.0 *25–27*	Resistant to ignition, combustible, self-extinguishing	Fairly high R-value. Seals against air leakage.
Rigid board	Polystyrene foam	The boards are applied over the wall framing members, either as sheathing on the exterior, or as a layer beneath the interior finish material	4–5 *27–35*	Combustible but self-extinguishing in most formulations	High R-value, can be used in contact with earth, moderate cost
Rigid board	Polyurethane foam, Polyisocyanurate foam	same	5.6 *39*	Combustible, gives off toxic gases when burned	High R-value, high cost
Rigid board	Glass fiber	same	3.5 *24*	Noncombustible	Moderate cost, vapor permeable
Rigid board	Cane fiber	same	2.5 *17*	Combustible	Low cost

[a] R-values are expressed first in units of hr-ft²-°F/Btu-in., then, in italics, in m-°K/W.

(a)

(b)

(c)

(d)

(e)

FIGURE 7.9

(*a*) Installing a polyethylene vapor retarder over glass fiber batt insulation using a staple hammer, which drives a staple each time it strikes a solid surface. The batts are unfaced and stay in place between the studs by friction. (*b*) Stapling faced batts between roof trusses. The R-value of the insulation is printed on the facing. (*c*) Placing unfaced batts between ceiling joists in an existing attic. Vent spacers should be used at the eaves (see Figure 6.5). (*d*) Insulating a floor over a crawlspace. Batts in this type of installation are usually retained in place by pieces of stiff wire cut slightly longer than the distance between joists and sprung into place at frequent intervals below the insulation. (*e*) Insulating crawlspace walls with batts of insulation suspended from the sill. The header space between the joist ends has already been insulated. (See also Figure 5.6.) (*Photos courtesy of Owens–Corning Fiberglass Corporation*)

(a)

(b)

FIGURE 7.10

(*a*) **Spraying polyicynene foam insulation between studs of a Wood Light Frame building. At the time of spraying, the components are dense liquids, but they react immediately with one another to produce a low-density foam, as has already occurred at the bottom of the cavity. The foam in the cavity to the right has already been trimmed off flush with the studs.** (*b*) **Trimming off excess foam with a special power saw.** (*Courtesy of Icynene, Incorporated, Toronto, Canada*)

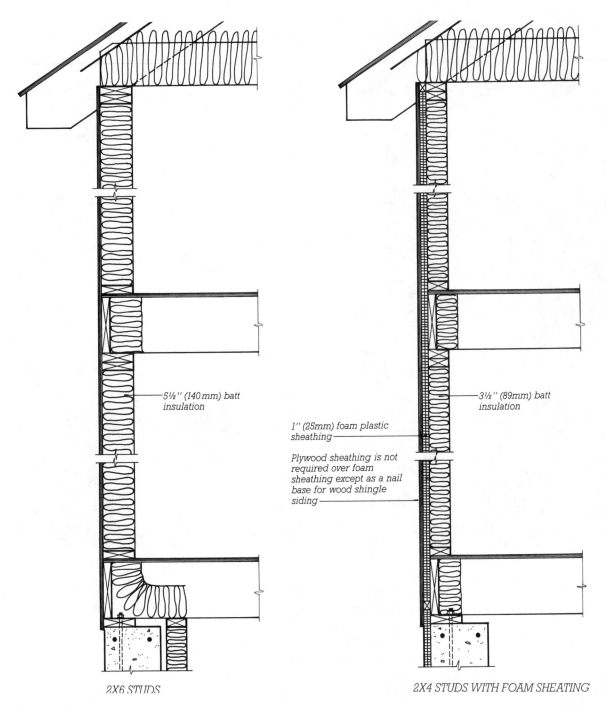

5½" (140 mm) batt insulation

1" (25mm) foam plastic sheathing

Plywood sheathing is not required over foam sheathing except as a nail base for wood shingle siding

3½" (89mm) batt insulation

2X6 STUDS

2X4 STUDS WITH FOAM SHEATING

FIGURE 7.11
Insulation levels in walls of light frame buildings can be increased from the R12 (83 in SI units) of a normal stud wall to R19 (132) by using either 2 × 6 framing and thicker batt insulation (left) or 2 × 4 framing with plastic foam sheathing in combination with batt insulation (right). The foam sheathing insulates the wood framing members as well as the cavities between, but can complicate the process of installing some types of siding.

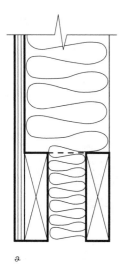

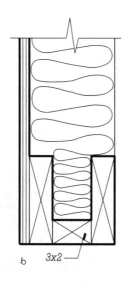

a.   b.   3x2

FIGURE 7.12
Window and door headers in 2 × 6 framing require special detailing.
Two alternative header details are shown here in section view: (*a*)
The header members are installed flush with the interior and exterior surfaces of the studs, with an insulated space between. This
detail is thermally efficient, but may not provide sufficient nailing for
interior finish materials around the window. (*b*) A 3 × 2 spacer provides full nailing around the opening.

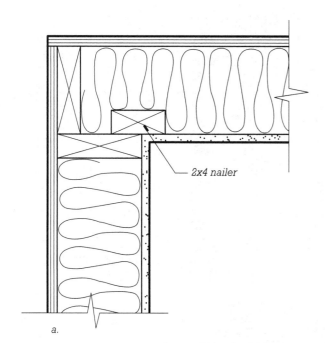

2x4 nailer

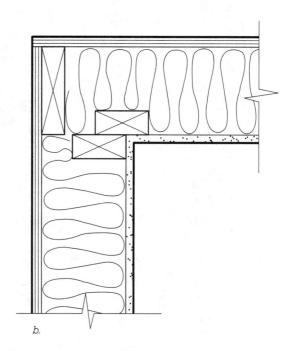

a.   b.

FIGURE 7.13
Two alternative corner post details for 2 × 6 stud walls, shown in plan view: (*a*) Each
wall frame ends with a full 2 × 6 stud, and a 2 × 4 nailer is added to accept fasteners
from the interior wall finish. (*b*) For maximum thermal efficiency, one wall frame
ends with a 2 × 4 stud flush with the interior surface, which eliminates any thermal
bridging through studs.

layer of treated paper or aluminum foil already attached. But most designers and builders in cold climates prefer to use unfaced batts and to apply a separate vapor retarder of polyethylene sheet, because a vapor retarder attached to batts has a seam at each stud that can leak significant quantities of air and vapor, while the separately applied sheet has few seams.

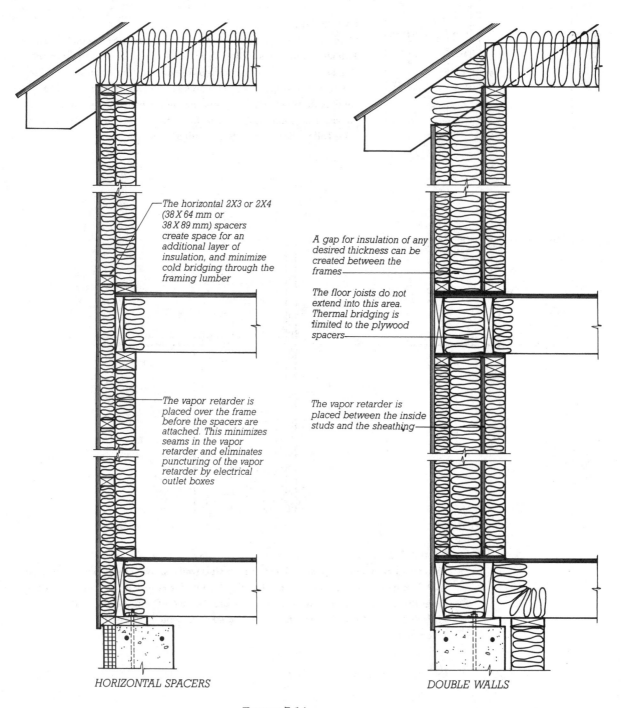

The horizontal 2X3 or 2X4 (38 X 64 mm or 38 X 89 mm) spacers create space for an additional layer of insulation, and minimize cold bridging through the framing lumber

The vapor retarder is placed over the frame before the spacers are attached. This minimizes seams in the vapor retarder and eliminates puncturing of the vapor retarder by electrical outlet boxes

A gap for insulation of any desired thickness can be created between the frames

The floor joists do not extend into this area. Thermal bridging is limited to the plywood spacers

The vapor retarder is placed between the inside studs and the sheathing

HORIZONTAL SPACERS

DOUBLE WALLS

FIGURE 7.14
With the two framing methods shown here, walls can be insulated to any desired level of thermal resistance.

Air Infiltration and Ventilation

Much attention is given to reducing air infiltration between the indoors and outdoors in a residential building because this leakage often accounts for the major portion of the fuel burned to heat the building in winter. This has led to higher standards of window and door construction, increased use of vapor-permeable air barrier papers under the exterior siding, and greater care in sealing around electrical outlets and other penetrations of the exterior wall. This trend, coupled with the use of continuous vapor retarders inside the walls and ceilings, has sometimes resulted in houses and apartments that exchange so little air with the outdoors that indoor moisture, odors, and chemical pollutants build up to intolerable and often unhealthful levels. Opening a window to ventilate the dwelling in cold weather, of course, wastes heating fuel, so many designers and builders install forced ventilation systems that employ *air-to-air heat exchangers,* devices that recover most of the heat from the air exhausted from the building and add it to the outside air that is drawn in. Nonresidential buildings of Wood Light Frame construction generally require higher rates of ventilation than residential structures, and here, too, low-infiltration construction and air-to-air heat exchangers are of value in reducing heating and cooling costs.

Airtight Drywall Approach

As an alternative to conventional vapor retarders, an increasing number of designers and builders are using the *airtight drywall approach* (*ADA*). ADA is based on the theory that most condensation problems are caused by moisture that is carried into the cavities of walls and ceilings by air leaking from the interior of the building toward the outdoors. Accordingly, meticulous attention is given to sealing all the joints in the gypsum board (drywall) panels that are used to finish the interior walls and ceilings. Compressible foam gaskets or sealants are used to eliminate air leakage around the edges of the floor platforms (Figure 7.16). Gypsum board is applied to all the interior surfaces of the outside walls before the interior partitions are framed, thus eliminating potential air leaks where partitions join the outside walls. Gaskets, sealants, or special airtight boxes are used to seal air leaks around electrical fixtures. In addition to providing good control of water vapor, ADA also reduces heat losses that occur because of air infiltration, which can result in substantial fuel savings.

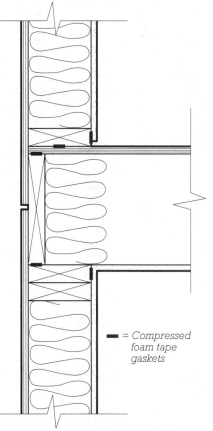

= Compressed foam tape gaskets

FIGURE 7.16
In the airtight drywall approach, foam tape gaskets or beads of sealant are applied to joints before assembly, greatly reducing air leakage through the exterior walls.

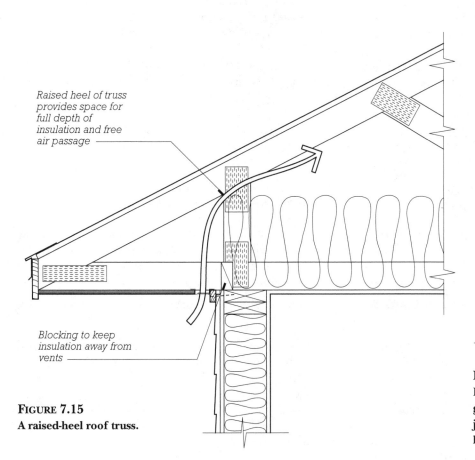

Raised heel of truss provides space for full depth of insulation and free air passage

Blocking to keep insulation away from vents

FIGURE 7.15
A raised-heel roof truss.

PROPORTIONING FIREPLACES

Ever since fireplaces were first developed in the Middle Ages, people have sought formulas for their construction that would ensure that the smoke from the fire would go up the chimney and the heat into the room, rather than the other way around, which is all too often the case. To this day, there is little scientific information on how fireplaces work and how best to design them, except for some measurements taken from fireplaces that seem to work reasonably well. These measurements have been arranged into tables of dimensions that enable designers to reproduce the critical features of these fireplaces as closely as possible.

Several general principles are clear: The chimney should be as tall as possible. The cross-sectional area of the *flue* should be about one-tenth the area of the front opening of the fireplace. A *damper* should be installed to close off the chimney when there is no fire burning and to regulate the passage of air through the firebox when the fire *is* burning (Figure A). A *smoke shelf* above the damper reduces fireplace malfunctions caused by cold downdrafts in the chimney. Sloping sides and a sloping back in the firebox reduce smoking and throw more heat into the room.

There are also a number of restrictions on fireplace and chimney construction in building codes. Typically these call for a 2-inch (51-mm) clearance between wood framing and the masonry of a chimney or fireplace, and clearances to combustible materials around the opening of the fireplace as shown on the accompanying diagram. Building codes specify also the minimum thicknesses of masonry around the firebox and flue, the minimum size of flue, and the minimum extension of the chimney above the roof. For an extended and knowledgeable discussion of fireplace design, see Albert G. H. Dietz, *Dwelling House Construction,* Cambridge, Massachusetts, M.I.T. Press, 1991, pages 167–175. For ready reference in proportioning a fireplace, use the values in the accompanying Figures B and C. In most cases, the designer need not detail the internal construction of the fireplace beyond the information given in these dimensions, because masons are well versed in the intricacies of assembling a fireplace.

There are several alternatives to the conventional masonry fireplace. One is the steel or ceramic fireplace liner that replaces the firebrick lining, damper, and smoke chamber, and which has internal passages to warm air drawn from the room and return it to the room. The liner is set in place on the underhearth, and built into a masonry facing and chimney by the mason. Another alternative is the "package" fireplace, a self-contained, fully insulated unit that needs no masonry whatsoever. It is set directly on the subfloor and fitted with a chimney of prefabricated, insulated sections of metal pipe. It can be faced with any desired ceramic or masonry materials. A third alternative is a metal stove that burns wood, coal, or both. Stoves are available in hundreds of styles and sizes. Their principal functional advantage is that they provide more heat to the interior of the building per unit of fuel burned than a fireplace. A stove requires a hearth and a fire-protected wall that are rather large in extent—the designer should consult the local building code at an early stage of design to be sure the room is large enough to hold a stove of the desired dimensions.

FIGURE A
A cutaway view of a conventional masonry fireplace. Concrete block masonry is used wherever it will not show, to reduce labor and material costs. The damper and ash doors are prefabricated units of cast iron. The flue liners are made of fired clay and are highly resistant to heat. The flue from the furnace or boiler in the basement slopes as it passes the firebox until it adjoins the fireplace flue, to keep the chimney as small as possible.

WALL AND CEILING FINISH

Plaster-type finishes have always been the most popular for walls and ceilings in wood frame buildings. Their advantages include substantially lower installed costs than any other types of finishes, adaptability to either painting or wallpapering, and, most importantly, a degree of fire resistance that offers considerable protection to a combustible frame. Three-coat *plaster* applied to wood strip *lath* was the prevalent wall and ceiling finish system until the Second World War, when *gypsum board* came into increasing use because of its lower cost, more rapid installation, and utilization of less skilled labor. More recently, *veneer plaster* systems

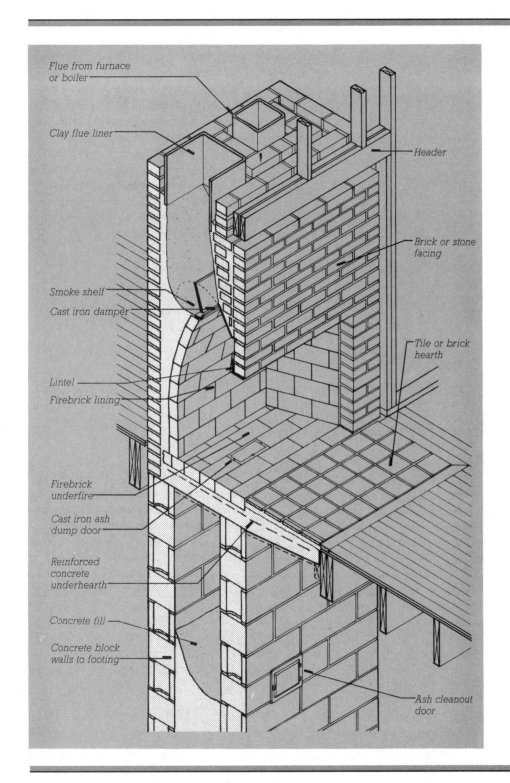

Flue from furnace or boiler

Clay flue liner

Header

Brick or stone facing

Smoke shelf

Cast iron damper

Tile or brick hearth

Lintel

Firebrick lining

Firebrick underfire

Cast iron ash dump door

Reinforced concrete underhearth

Concrete fill

Concrete block walls to footing

Ash cleanout door

have been developed that offer surfaces whose quality and durability are superior to those of gypsum board at comparable prices.

Plaster, veneer plaster, and gypsum board finishes are presented in detail in Chapter 23. Gypsum board remains the favored material for small builders who do all the interior finish work in a building themselves because the skills and tools it requires are largely those of carpenters rather than plasterers. In geographic areas where there are plenty of skilled plasterers, veneer plaster captures a substantial share of the market. Almost everywhere in North America there are subcontractors who specialize in gypsum board installation and finishing and who are able to finish the interior surfaces of larger projects,

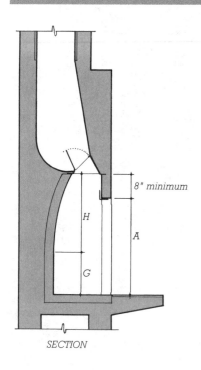

SECTION

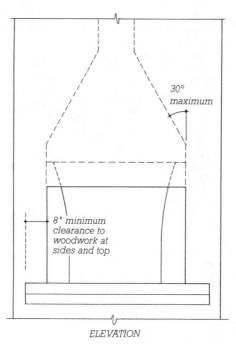

ELEVATION

8" minimum

30° maximum

8" minimum clearance to woodwork at sides and top

8" minimum

FIGURE B

The critical dimensions of a conventional masonry fireplace, keyed to the table in Figure C. Dimension *D*, the depth of the hearth, is commonly required to be 16 inches (405 mm) for fireplaces with openings up to 6 square feet (0.56 m²) and 20 inches (510 mm) if the opening is larger. The side extension of the hearth, *E*, is usually fixed at 8 inches (203 mm) for smaller fireplace openings, and 12 inches (305 mm) for larger ones.

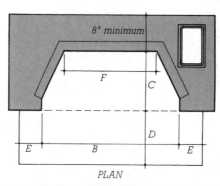

PLAN

FIGURE C

Recommended proportions for conventional masonry fireplaces, based largely on figures given in Ramsey/Sleeper, *Architectural Graphic Standards* (9th ed.), John Ray Hoke, Jr., F.A.I.A., Editor, New York, John Wiley & Sons, Inc., 1994.

Fireplace Opening				Vertical Backwall Height (G)	Inclined Backwall Height (H)	Flue Lining	
Height (A)	Width (B)	Depth (C)	Minimum Backwall Width (F)			Rectangular (outside dimensions)	Round (inside diameter)
24″ (610 mm)	28″ (710 mm)	16 to 18″ (405 to 455 mm)	14″ (355 mm)	14″ (355 mm)	16″ (405 mm)	8½ × 13″ (216 × 330 mm)	10″ (254 mm)
28 to 30″ (710 to 760 mm)	30″ (760 mm)	16 to 18″ (405 to 455 mm)	16″ (405 mm)	14″ (355 mm)	18″ (455 mm)	8½ × 13″ (216 × 330 mm)	10″ (254 mm)
28 to 30″ (710 to 760 mm)	36″ (915 mm)	16 to 18″ (405 to 455 mm)	22″ (560 mm)	14″ (355 mm)	18″ (455 mm)	8½ × 13″ (216 × 330 mm)	12″ (305 mm)
28 to 30″ (710 to 760 mm)	42″ (1065 mm)	16 to 18″ (405 to 455 mm)	28″ (710 mm)	14″ (355 mm)	18″ (455 mm)	13 × 13″ (330 × 330 mm)	12″ (305 mm)
32″ (815 mm)	48″ (1220 mm)	18 to 20″ (455 to 510 mm)	32″ (815 mm)	14″ (355 mm)	24″ (610 mm)	13 × 13″ (330 × 330 mm)	15″ (381 mm)

such as apartments, retail stores, and rental office buildings, as well as individual houses, at highly competitive prices.

In most small buildings, all wall and ceiling surfaces are covered with plaster or gypsum board. Even wood paneling should be applied over a gypsum board backup layer for increased fire resistance. In buildings that require fire walls between dwelling units, or fire separation walls between different uses, a gypsum board wall of the required degree of fire resistance can be installed, eliminating the need to employ masons to put up a wall of brick or concrete masonry (Figures 5.59, 5.60).

MILLWORK AND FINISH CARPENTRY

Millwork (so named because it is manufactured in a planing and molding mill) includes all the wood interior finish components of a building. Millwork is generally produced from much higher quality wood than that used for framing: The softwoods used are those with fine, uniform grain structure and few defects, such as Sugar pine and Ponderosa pine. Flooring, stair treads, and millwork intended for transparent finish coatings such as varnish or shellac are customarily made of hardwoods and hardwood plywoods: Red and White oak, Cherry, Mahogany, Walnut.

Moldings are also produced from high-density plastic foams, parallel strand lumber, and particleboard as a way of reducing costs and minimizing moisture expansion and contraction. All these materials must be painted.

Millwork is manufactured and delivered to the building site at a very low moisture content, typically in the range of 10 percent, so it is important to protect it from moisture and high humidity before and during installation to avoid swelling and distortions. The humidity within the building is high at the conclusion of plastering or gypsum board work. The framing lumber, concrete work, masonry mortar, plaster, drywall joint compound, and paint are still diffusing large amounts of excess moisture into the interior air. As much of this moisture as possible should be ventilated to the outdoors before *finish carpentry* (the installation of millwork) commences. Windows should be left open for a few days, and in cool or damp weather the building's heating system should be turned on to raise the interior air temperature and help drive off excess water. In hot, humid weather, the air conditioning system should be activated to dry the air.

Interior Doors

Figure 7.17 illustrates five doors that fall into three general categories: Z-brace, panel, and flush. *Z-brace doors,* built on site, are used infrequently because they are subject to distortions and large amounts of moisture expansion and contraction in the broad surface of boards whose grain runs perpendicular to the width of the door. *Panel doors* were developed centuries ago to minimize dimensional changes and distortions caused by the seasonal changes in the moisture content of the wood. They are widely available in ready-made form from millwork dealers. *Flush doors* may be either *solid core* or *hollow core.* Solid core doors consist of two

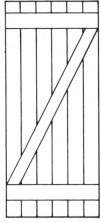

Z-BRACE

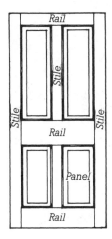

FOUR PANEL

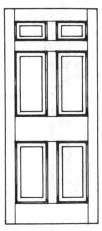

SIX PANEL

FLUSH SOLID CORE

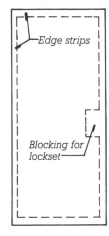

FLUSH HOLLOW CORE

FIGURE 7.17
Types of wood doors.

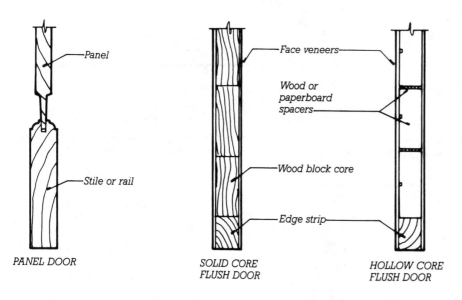

PANEL DOOR

Panel

Stile or rail

SOLID CORE
FLUSH DOOR

Face veneers

Wood or
paperboard
spacers

Wood block core

Edge strip

HOLLOW CORE
FLUSH DOOR

FIGURE 7.18
Edge details of three types of wood
doors. The panel is loosely fitted to the
stiles and rails in a panel door to allow
for moisture expansion of the wood. The
spacers and edge strips in hollow core
doors have ventilation holes to equalize
air pressures inside and outside the door.

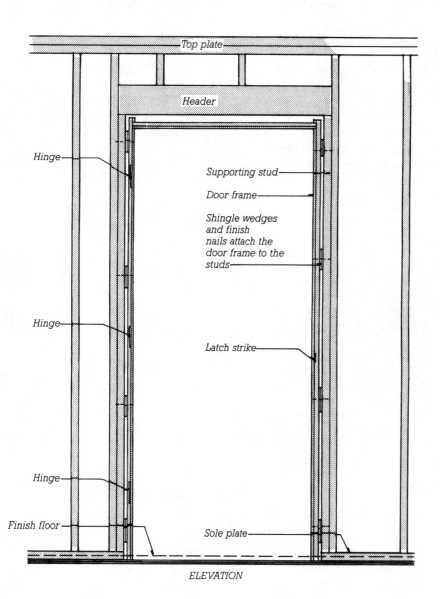

Top plate

Header

Hinge

Hinge

Hinge

Finish floor

Supporting stud

Door frame

Shingle wedges
and finish
nails attach the
door frame to the
studs

Latch strike

Sole plate

ELEVATION

FIGURE 7.19
Installing a door frame in a rough opening. The wood shingle wedges at each
nailing point are paired in opposing
directions to create a flat, precisely
adjustable shim to support the frame.

veneered faces glued to a core of wood blocks or bonded wood chips (Figure 7.18). They are much heavier, stronger, and more resistant to the passage of sound than hollow core doors, and are also more expensive. In residential buildings, their use is usually confined to entrance doors, but they are frequently installed throughout commercial and institutional buildings, where doors are subject to greater abuse. Hollow core doors have an interior grid of wood or paperboard spacers to which the veneers are bonded. Flush doors

of either type are available in a variety of veneer species, the least expensive of which are intended to be painted.

For speed and economy of installation, most interior doors are furnished *prehung,* meaning that they have been hinged and fitted to their frames at the mill. The carpenter on the site merely tilts the prehung door and frame unit up into the rough opening, plumbs it carefully with a spirit level, shims it with pairs of wood shingle wedges between the finish and rough jambs, and nails it to the studs with finish nails through the

jambs (Figure 7.19). *Casings* are then nailed around the frame on both sides of the partition to close the ragged gap between the door frame and the wall finish (Figure 7.20). To save the labor of applying casings, door units can also be purchased with *split jambs* that enable the door to be cased at the mill. At the time of installation, each door unit is separated into halves, and the halves are installed from opposite sides of the partition to telescope snugly together before being nailed in place (Figure 7.21).

(a)

(b)

(c)   *(d)*

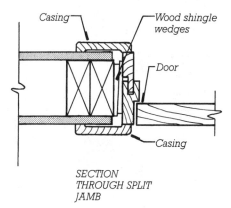

SECTION
THROUGH SPLIT
JAMB

FIGURE 7.21
A split-jamb interior door arrives on the construction site prehung and precased. The halves of the frame are separated and installed from opposite sides of the partition.

FIGURE 7.20
Casing a door frame. (*A*) The finish nails in the frame are set with a steel nail set. (*B*) The top piece of casing, mitered to join the vertical casings, is ready to install, and glue is spread on the edge of the frame. (*C*) The top casing is nailed into place. (*D*) The nails are set below the surface of the wood, ready for filling. (*Photos by Joseph Iano*)

(a)   (b)   (c)

(d)   (e)   (f)

(g)   (h)   (g)

FIGURE 7.22
Casing a window. (*A*) Marking the length of a casing. (*B*) Cutting the casing to length
with a power miter saw. (*C*) Nailing the casing. (*D*) Coping the end of a molded edge of
an apron with a coping saw, so that the molding profile will terminate neatly at the end
of the apron. (*E*) Planing the edge of the apron, which has been ripped (sawn lengthwise, parallel to the grain of the wood) from wider casing stock. (*F*) Applying glue to the
apron. (*G*) Nailing the apron, which has been wedged temporarily in position with a
stick. (*H*) The coped end of the apron, in place. (*I*) The cased window, ready for filling,
sanding, and painting. (*Photos by Joseph Iano*)

Window Casings and Baseboards

Windows are cased in much the same manner as doors (Figures 7.22, 7.23). After the finish flooring is in place, *baseboards* are installed at the junction of the floor and the wall to close the ragged gap between the flooring and the wall finish, and to protect the wall finish against damage by feet, furniture legs, and cleaning equipment (Figure 24.31).

The carpenter recesses the heads of finish nails below the surface of the wood, using a hammer and a *nail set,* a hardened steel punch. Later, the painters will fill these nail holes with a paste filler and sand the surface smooth after the filler has dried so that the holes will be invisible in the painted woodwork. For transparent wood finishes, either the filler must be selected carefully to match the color of the wood, or the nail holes may be left unfilled.

The nails in casings and baseboards must reach through the plaster or gypsum board to penetrate the framing members beneath in order to make a secure attachment. Eight- or ten-penny finish or casing nails are customarily used.

FIGURE 7.23
Simple but carefully detailed and skillfully crafted window casings in a restaurant. (*Architects: Woo and Williams. Photo by Richard Bonarrigo. Courtesy of the architects*)

Cabinets

Cabinets for kitchens, bathrooms, and workrooms are nearly always fabricated and coated in specialty cabinet shops and are brought to the construction site fully finished. They are installed by shimming against wall and floor surfaces as necessary to make them level, and screwing through the backs of the cabinet units into the wall studs (Figures 7.24, 7.25). The tops are then attached with screws driven up from the cabinets beneath. Kitchen and bath countertops are cut out for built-in sinks and lavatories, which are subsequently installed by the plumber.

Finish Stairs

Finish stairs are either constructed in place (Figures 7.26–7.28) or shop built (Figure 7.29). The shop-built stairs tend to be more tightly constructed and to squeak less in use, but site-built stairs can be fitted more closely to the walls and are more

FIGURE 7.24
Prepainted wood kitchen cabinets installed, but lacking shelves, drawers, doors, and countertops. (*Photo by the author*)

FIGURE 7.25
Custom-designed cabinets brighten a remodeled kitchen in an older house. (*Architects: Dirigo Design of Belmont, Massachusetts. Photo © 1989 by Lucy Chen*)

STAIR PLANS

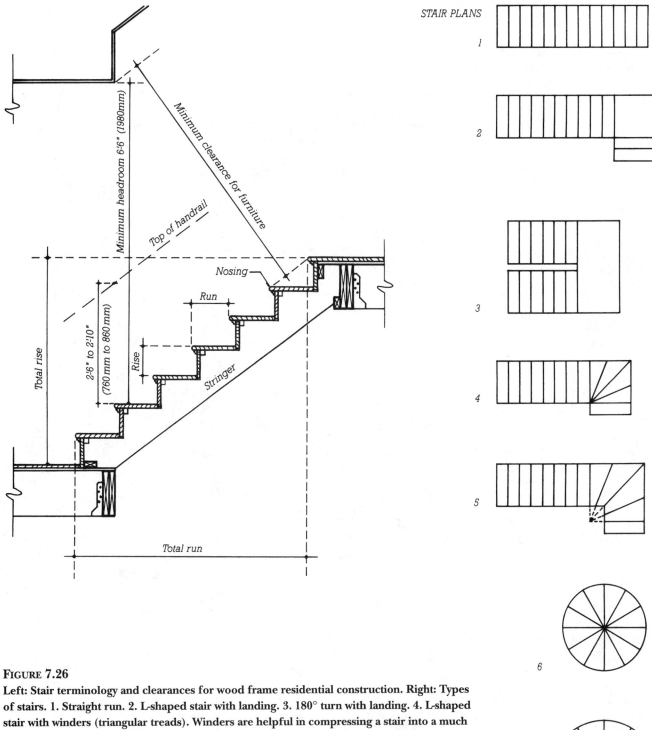

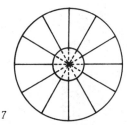

FIGURE 7.26

Left: Stair terminology and clearances for wood frame residential construction. Right: Types of stairs. 1. Straight run. 2. L-shaped stair with landing. 3. 180° turn with landing. 4. L-shaped stair with winders (triangular treads). Winders are helpful in compressing a stair into a much smaller space but are perilously steep where they converge, and their treads become much too shallow for comfort and safety. Many codes do not permit winders. 5. L-shaped stair whose winders have an offset center. The offset center can increase the minimum tread dimension to within legal limits. 6. A spiral stair (in reality a helix, not a spiral) consists entirely of winders and is generally illegal for any use but a secondary stair in a single-family residence. 7. A spiral stair with an open center of sufficient diameter can have its treads dimensioned to legal standards. (*Adapted by permission from Albert G. H. Dietz,* Dwelling House Construction *(4th ed.), Cambridge, Massachusetts, M.I.T. Press, © 1974 Massachusetts Institute of Technology*)

adaptable to special situations and framing irregularities. Stair treads are usually made of wear-resistant hardwoods such as Oak and Maple. Risers and stringers may be made of any reasonably hard wood, such as Oak, Maple, or Douglas fir.

FIGURE 7.27
Constructing a finished stair in place. The joint between the riser and the open stringer is a miter. The balusters, posts, and handrail are purchased ready made from millwork suppliers and cut to fit. (*Adapted by permission from Albert G. H. Dietz,* Dwelling House Construction *(4th ed.), Cambridge, Massachusetts, M.I.T. Press, © 1974 Massachusetts Institute of Technology*)

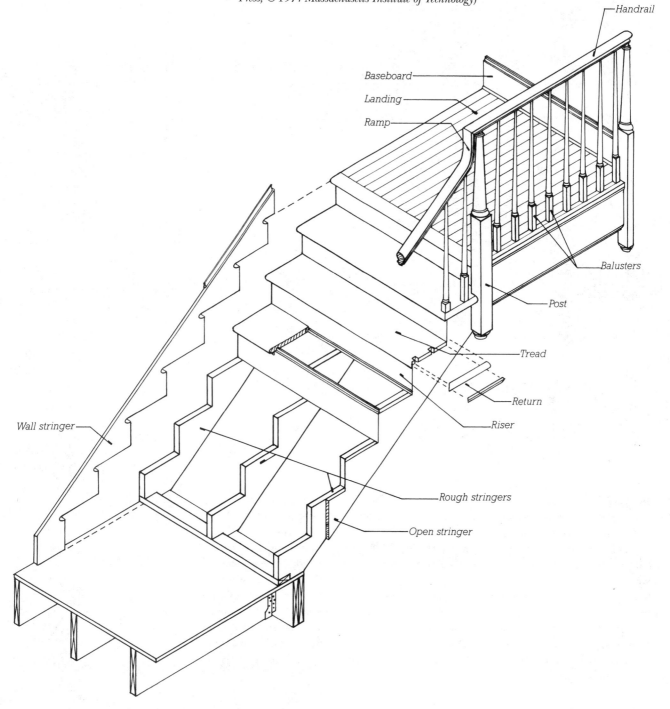

FIGURE 7.28
Millwork suppliers offer several different stair "packages" of coordinated parts in traditional designs. (*Courtesy of Morgan Products, Ltd.*)

FIGURE 7.29
A shop-built stair. All the components are glued firmly together in the shop and the stair is installed as a single piece.

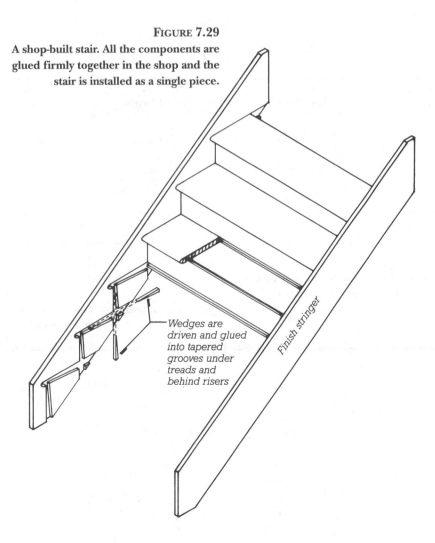

Wedges are driven and glued into tapered grooves under treads and behind risers

Finish stringer

Three openings are required in stair-cases; the first is the door thro' which one goes up to the stair-case, which the less it is hid to them that enter into the house, so much the more it is to be commended. And it would please me much, if it was in a place, where before that one comes to it, the most beautiful part of the house was seen; because it makes the house (even tho' small) seem very large; but however, let it be manifest, and easily found. The second opening is the windows that are necessary to give light to the steps; they ought to be in the middle, and high, that the light may be spread equally every where alike. The third is the opening thro' which one enters into the floor above; this ought to lead us into ample, beautiful, and adorned places.

Andrea Palladio, 1508–1580

PROPORTIONING STAIRS

Building codes encourage the design of safe, comfortable stairs through a number of detailed requirements. The overall dimensional limitations for stairs as given in the BOCA National Building Code are summarized in Figure A. Beyond these restrictions, some codes further specify that the risers and treads shall be proportioned so that twice the riser dimension added to the tread dimension shall equal 24 to 25 inches (610 to 635 mm). This formula was derived in France two centuries ago from measurements of actual dimensions of comfortable stairs. Figure B gives an example of how it is used in designing a new stair, in this case for a single-family dwelling. Because the code does not allow variations greater than $\frac{3}{16}$ inch (5 mm) between successive treads or risers, the floor-to-floor dimension should be divided equally into risers to an accuracy of 0.01 inch (0.5 mm) to avoid cumulative errors. The framing square used by carpenters in the United States to lay out stair stringers has a scale of hundredths of an inch, and riser dimensions should be given in these units rather than fractions to achieve the necessary accuracy.

Most building codes do not allow a rise of more than 12 feet (3660 mm) between landings in a stair. Landings contribute to the safety of a stair by providing a moment's rest to the legs between flights of steps. (Architects also generally avoid designing flights of less than three risers because short flights, especially in public buildings, sometimes go unnoticed, leading to dangerous falls.) The width and depth of the landing must each be equal to the width of the stair. The width of a required exit stairway is calculated according to the number of occupants served by the stair, in accordance with formulas given in building codes.

Monumental outdoor stairs, such as those that lead to entrances of public buildings, are designed with lower risers and deeper treads than indoor stairs. Many designers ease the $2R + T$ formula a bit for outdoor stairs, raising the sum to 26 or 27 inches (660 or 685 mm), but it is always best to make a full-scale mockup of a section of such a stair to be sure that it is comfortable underfoot.

	Minimum Width	Maximum Riser Height	Minimum Tread Depth	Minimum Headroom
Stair within a residence	36″ (915 mm)	7³⁄₄″ (197 mm)	10″ (254 mm)	6′8″ (2032 mm)
Nonresidential stair, occupancy of 50 persons or fewer	36″ (915 mm)	7″ (178 mm)	11″ (279 mm)	6′8″ (2032 mm)
Nonresidential stair, occupancy load greater than 50 persons	44″ (1120 mm)	7″ (178 mm)	11″ (279 mm)	6′8″ (2032 mm)

FIGURE A
Dimensional limitations for stairs as established by the BOCA National Building Code.

FIGURE B
A sample calculation for proportioning a residential stair.

Procedure	English	Metric
1. Determine the height (H) from finish floor to finish floor.	$H = 9'4\frac{3}{8}''$, or 112.375″	$H = 2854$ mm
2. Divide H by the approximate riser height desired, and round off to obtain a trial number of risers for the stair.	$\frac{H}{7''} = \frac{112.375''}{7''} = 16.05$ Try 16 risers	$\frac{H}{180 \text{ mm}} = \frac{2854 \text{ mm}}{180 \text{ mm}} = 15.86$ Try 16 risers
3. Divide H by the trial number of risers to obtain an exact trial riser height. Work to the nearest hundredth of an inch, or the nearest millimeter, to avoid any cumulative error that would result in one riser being substantially lower or higher than the rest. Check to make sure this trial riser height falls within the limits set by the building code.	$\frac{H}{16} = \frac{112.375''}{16} = 7.02''$ $R = 7.02''$ $7.02'' < 7.75''$ *OK*	$\frac{H}{16} = \frac{2854 \text{ mm}}{16} = 178$ mm $R = 178$ mm $178 \text{ mm} < 197 \text{ mm}$ *OK*
4. Substitute this trial riser height into the given formula and solve for the tread depth. The depth can be rounded down somewhat if desired, as long as $2R + T \geq 24$. Check the tread depth against the code minimum.	$2R + T = 25''$ $2(7.02) + T = 25''$ $T = 25'' - 14.04''$ $T = 10.96''$, say 10.9″ $10.9'' > 10''$ *OK*	$2R + T = 635$ mm $2(178 \text{ mm}) + T = 635$ mm $T = 635 \text{ mm} - 356 \text{ mm}$ $T = 279$ mm, say 275 mm $275 \text{ mm} > 254 \text{ mm}$ *OK*
5. Summarize the results of these calculations. There is always one fewer tread than risers in a flight of stairs.	16 risers @ 7.02″ 15 treads @ 10.9″ Total run = (15) (10.9″) = 163.5″ = 13′7½″	16 risers @ 178 mm 15 treads @ 275 mm Total run = (15) (275 mm) = 4125 mm
6. If desired, a steeper or shallower stair can be tried as an alternative by subtracting or adding one riser and tread and recalculating riser and tread dimensions. Subtracting a riser and tread can reduce the overall run of the stair dramatically, which is helpful when designing a stair for a limited amount of space.	Try 15 risers: $R = \frac{112.375''}{15} = 7.49''$ $7.49'' < 7.75''$ *OK* $2(7.49) + T = 25''$ $T = 10.02''$, say 10″ $10'' = 10''$ *OK* 15 risers @ 7.49″ 14 treads @ 10″ Total run = (14) (10″) = 140″ = 11′8″ Saves almost 2′ of run	Try 15 risers: $R = \frac{2854 \text{ mm}}{15} = 190$ mm $190 \text{ mm} < 210 \text{ mm}$ *OK* $2(190) + T = 635$ mm $T = 255$ mm $255 \text{ mm} > 254 \text{ mm}$ *OK* 15 risers @ 190 mm 14 treads @ 255 mm Total run = (14) (255 mm) = 3570 mm Saves 555 mm of run

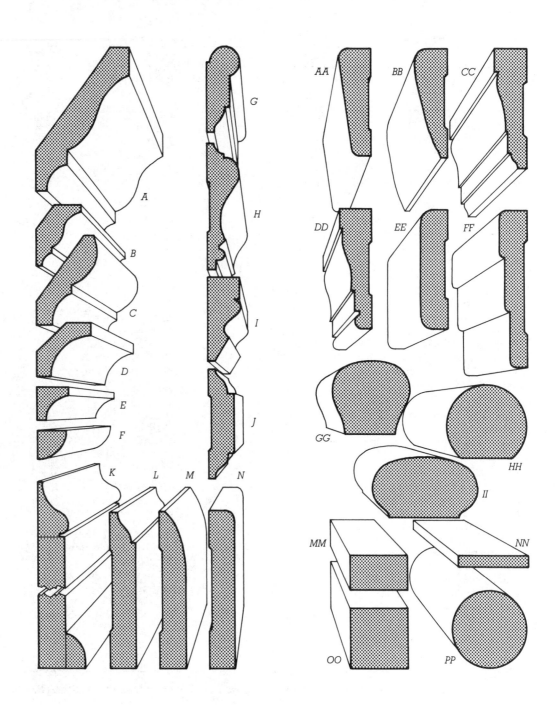

FIGURE 7.30

Some common molding patterns for wood interior trim. *A* and *B* are *crowns, C* is a *bed, D* and *E* are *coves.* All are used to trim the junction of a ceiling and a wall. *F* is a *quarter-round,* for general-purpose trimming of inside corners. Moldings *G* through *J* are used on walls—*G* is a *picture molding,* applied near the top of a wall so that framed pictures can be hung from it at any point on special metal hooks that fit over the rounded portion of the molding. *H* is a *chair rail,* installed around dining rooms to protect the plaster from damage by the backs of chairs. *I* is a *panel molding* and *J* a *batten,* both used in traditional paneled wainscoting. Baseboards (*K* through *N*) include three single-piece designs, and one traditional design (*K*) using a separate *cap molding* and *shoe* in addition to a piece of S4S stock that is the baseboard itself (see also Figure 24.31). Notice the shallow groove, also called a *relieved back,* in the single-piece baseboards and many other flat moldings on this page—this serves to reduce cupping forces on the piece, and makes it easier to install even if it is slightly cupped.

Designs *AA* through *FF* are standard *casings* for doors and windows. *GG, HH,* and *II* are handrail stock. *MM* is representative of a number of sizes of S4S material available to the finish carpenter for miscellaneous uses. *NN* is *lattice* stock, also used occasionally for flat trim. *OO* is square stock, used primarily for balusters. *PP* represents several available sizes of round stock for balusters, handrails, and closet poles. Wood moldings are furnished in either of two grades: N Grade, for transparent finishes, must be of a single piece. P Grade, for painting, may be finger jointed or edge glued from smaller pieces of wood. P Grade is less expensive because it can be made up of short sections of lower-grade lumber with the defects cut out. Once painted, it is indistinguishable from N Grade. The shapes shown here represent a fraction of the moldings that are generally available from stock. Custom molding patterns can be easily produced because the molding cutters used to produce them can be ground quickly to the desired profiles, working from the architect's drawings.

Miscellaneous Finish Carpentry

Finish carpenters install dozens of miscellaneous items in the average building—closet shelves and poles, pantry shelving, bookshelves, wood paneling, chair rails, picture rails, ceiling moldings, mantelpieces, laundry chutes, door hardware, weather stripping, doorstops, and bath accessories (towel bars, paper holders, and so on). Many of these items are available ready made from millwork and hardware suppliers (Figures 7.30, 7.31), but others have to be crafted by the carpenter (Figures 7.32, 7.33).

FIGURE 7.31
Fireplace mantels are available from millwork suppliers in a number of designs. Each is furnished largely assembled but is detailed in such a way that it can easily be adjusted to fit any fireplace within a wide range of sizes. (*Courtesy of Morgan Products, Ltd.*)

FIGURE 7.32
An existing room of a Victorian house before interior refinishing.

FIGURE 7.33
The same room finished in redwood paneling and casework. (*Architect: Marshall Roath.*
Photos by Karl Riek. Courtesy of California Redwood Association)

FLOORING AND CERAMIC TILE WORK

Before finish flooring can be installed, the subfloor is scraped free of plaster droppings and swept thoroughly. *Underlayment* panels of C-C Plugged plywood or particleboard (in areas destined for resilient flooring materials and carpeting) are glued and nailed over the subfloor, their joints offset from those in the subfloor to eliminate weak spots. The thicknesses of the underlayment panels often are chosen to make the finished floor surfaces as nearly equal in level as possible at junctions between different flooring materials.

In multistory Wood Light Frame commercial and apartment buildings, a floor underlayment specially formulated of poured gypsum or lightweight concrete is often poured over the subfloor. This has a three-fold function: It provides a smooth, level surface for finish floor materials; it furnishes additional fire resistance to the floor construction; and it reduces the transmission of sound through the floor to the apartment or office below. The gypsum or concrete is formulated with superplasticizer admixtures that make it virtually self-leveling as it is applied (Figure 7.34). A minimum thickness is ¾ inch (19 mm). Poured underlayments are also used to level floors in older buildings, to add fire and sound resistance to precast concrete floors, and to embed plastic tubing or electric resistance wires for in-floor radiant heat.

Floor finishing operations require cleanliness and freedom from traffic, so other trades are banished from the area as the flooring materials are applied. Hardwood flooring is sanded level and smooth after installation, then vacuumed to remove the sanding dust. The finish coatings are applied in as dust-free an atmosphere as possible to avoid embedded specks. Resilient-flooring installers vacuum the underlayment meticulously so that particles of dirt will not become trapped beneath the thin flooring and cause bumps in the surface. The finished floors are often covered with sheets of heavy paper or plastic to protect them during the final few days of construction activity. Carpet installation is less sensitive to dust, and installed carpets are less prone to damage than hardwood and resilient floorings, but temporary coverings are applied as necessary to protect the carpet from paint spills and water stains.

The application of ceramic tile to a shower stall is illustrated in Figure 7.35. The installation of ceramic tile and finish flooring materials is covered in more detail in Chapters 23 and 24.

FIGURE 7.34
Workers apply a gypsum underlayment to an office floor. The gypsum is pumped through a hose and distributed with a straightedge tool. Because the gypsum seals against the bottom of the interior partitions, it can also help reduce sound transmission from one room to another. (© *Gyp-Crete Corporation, Hamel, Minnesota*)

FIGURE 7.35
Installing sheets of ceramic tile in a shower stall. The base coat of portland cement plaster over metal lath has already been installed. Now the tilesetter applies a thin coat of tile adhesive with a trowel and presses a sheet of tiles into it, taking care to align the tiles individually around the edges. A day or two later, after the adhesive has hardened sufficiently, the joints will be grouted to complete the installation. (*Photos by Joseph Iano*)

FINISHING TOUCHES

When flooring and painting are finished, the plumbers install and activate the lavatories, water closets, tubs, sinks, and shower fixtures. Gas lines are connected to appliances and the main gas valve is opened. The electricians connect the wiring for the water heater and heating and air conditioning equipment; mount the receptacles, switches, and lighting fixtures; and put metal or plastic cover plates on the switches and receptacles. The electrical circuits are energized and checked to be sure they work. The smoke alarms and heat alarms, required by most codes in residential structures, are also connected and tested by the electricians, along with any communications, entertainment, and security system wiring. The heating and air conditioning system is completed with the installation of air grills and registers, or with the mounting of metal convector covers, then turned on and tested. Last-minute problems are identified and corrected through cooperative effort by the contractors, the owner of the building, and the architect. The building inspector is called in for a final inspection and issuance of an occupancy permit. After a thorough cleaning, the building is ready for use.

FIGURE 7.36
Ceramic tile is used for the floor, countertops, and backsplash in this kitchen. The border was made by selectively substituting tiles of four different colors for the white tiles used for the field of the floor. (*Designer: Kevin Cordes. Courtesy of American Olean Tile*)

FIGURE 7.37
Varnished oak flooring, millwork, and casework.
(*Architects: Woo and Williams. Photo by Richard Bonarrigo. Courtesy of the architects*)

C.S.I./C.S.C.
Masterformat Section Numbers for Interior Finishes for Light Frame Wood Construction

06200	FINISH CARPENTRY
06220	Millwork
	Cabinets
	Closet and Storage Shelving
	Molding and Trim
06240	Laminates
06260	Board Paneling
06400	ARCHITECTURAL WOODWORK
06410	Custom Casework
06420	Paneling
06430	Stairwork and Handrails
06440	Wood Ornaments
06450	Standing and Running Trim
07190	VAPOR RETARDERS
07200	INSULATION
07210	Building Insulation
	Batt Insulation
	Building Board Insulation
	Foamed-in-Place Insulation
	Loose Fill Insulation
	Sprayed Insulation
08200	WOOD AND PLASTIC DOORS
08210	Wood Doors
	Flush Wood Doors
	Stile and Rail Wood Doors

(See also the Masterformat section numbers in Chapters 23 and 24)

SELECTED REFERENCES

1. Thallon, Rob. *Graphic Guide to Interior Details for Builders and Designers*. Newtown, Connecticut, The Taunton Press, 1996.

Profusely illustrated, clearly written, and encyclopedic in scope, this book offers complete guidance on interior finishing of Wood Light Frame buildings.

2. Dietz, Albert G. H. *Dwelling House Construction* (5th ed.). Cambridge, Massachusetts, M.I.T. Press, 1991.

This classic text has extensive chapters with clear illustrations concerning chimneys and fireplaces, insulation, wallboard, lath and plaster, and interior finish carpentry.

KEY TERMS AND CONCEPTS

roughing in
DWV pipes
supply pipes
gas pipes
hydronic heating system
thermal insulation
glass fiber batt
cellulose
rock wool
polyicynene foam
polystyrene foam
polyurethane foam
polyisocyanurate foam

R-value
raised-heel roof truss
radiant barrier
vapor retarder
vapor barrier
air-to-air heat exchanger
airtight drywall approach (ADA)
plaster
lath
gypsum board
veneer plaster
flue
hearth

damper
smoke shelf
underfire
ash dump
millwork
finish carpentry
Z-brace door
panel door
flush door
solid core
hollow core
prehung door
casing

split jamb
baseboard
nail set
baluster
wall stringer
rise
run
riser
tread
nosing
landing
underlayment

REVIEW QUESTIONS

1. List the sequence of operations required to complete the interior of a Wood Light Frame building and explain the logic of the order in which these operations occur.

2. What are some alternative ways of insulating the walls of a Wood Light Frame building to R-values beyond the range normally possible with ordinary 2 × 4 (38 × 89 mm) studs?

3. Why are plaster and gypsum board so popular as interior wall finishes in wood frame buildings? List as many reasons as you can.

4. What is the level of humidity in a building at the time installation of the interior wall finishes is completed? Why? What should be done about this, and why?

5. Summarize the most important things to keep in mind when designing a stair.

EXERCISES

1. Design and detail a fireplace for a building that you are designing, using the information provided on page 224 to work out the exact dimensions, and the information in Chapters 8 and 9 to help in detailing the masonry.

2. Design and detail a stairway for a building that you are designing, using the information provided on pages 234 and 235 to calculate the dimensions.

3. Visit a wood frame building that you admire. Make a list of the interior finish materials and components, including species of wood, where possible. How does each material and component contribute to the overall feeling of the building? How do they relate to one another?

4. Make measured drawings of millwork details in an older building that you admire. Analyze each detail to discover its logic. What woods were used and how were they sawn? How were they finished?

8

BRICK MASONRY

- **History**
- **Mortar**
- **Brick Masonry**
  Molding of Bricks
  Firing of Bricks
  Brick Sizes
  Choosing Bricks
  Laying Bricks
  Spanning Openings in Brick Walls
  Reinforced Brick Masonry

Flemish Bond brickwork combines simply and directly with cut limestone lintels and sills in this townhouse on Boston's Beacon Hill. (*Photo by the author*)

Masonry is the simplest of building techniques: The mason stacks pieces of material (bricks, stones, concrete blocks, called collectively *masonry units*) atop one another to make walls. But it is also the richest and most varied, with its endless selection of colors and textures. And because the pieces of which it is made are small, masonry can take any shape, from a planar wall to a sinuous surface that defies the distinction of roof from wall.

Masonry is the material of earth, taken from the earth and comfortably at home in foundations, pavings, and walls that grow directly from the earth. But with modern techniques of reinforcing, masonry can rise many stories from the earth, and in the form of arches and vaults, masonry can take wing and fly across space.

The most ancient of our building techniques, masonry remains labor intensive, requiring the patient skills of experienced and meticulous artisans to achieve a satisfactory result. But is has kept pace with the times and remains highly competitive technically and economically with other systems of structure and enclosure, the more so because one mason can produce in one operation a completely finished, insulated, loadbearing wall, ready for use.

Masonry is durable. The designer can select masonry materials that are scarcely affected by water, air, or fire, ones with brilliant colors that will not fade, ones that will stand up to heavy wear and abuse, and make from them a building that will last for generations.

Masonry is a material of the small entrepreneur. One can set out to build a building of bricks with no more tools than a trowel, a shovel, a hammer, a measuring rule, a level, a square of scrap plywood, and a piece of string. Yet many masons can work together, aided by mechanized handling of materials, to put up projects as large as the human mind can conceive.

HISTORY

Masonry began spontaneously in the creation of low walls from stones or pieces of caked mud taken from dried puddles. *Mortar* was originally the mud smeared into the joints of the rising wall to impart stability and weathertightness. Where stone lay readily at hand, it was preferred to bricks; where stone was unavailable, *bricks* were made from local clays and silts. Changes came with the passing millennia: People learned to quarry, cut, and dress stone with increasing precision. Fires built against mud brick walls brought a knowledge of the advantages of burned brick, leading to the invention of the kiln. *Masons* learned the simple art of turning limestone into lime, and lime mortar gradually replaced mud in the joints of masonry.

By the fourth millennium B.C., the peoples of Mesopotamia were building palaces and temples of stone

FIGURE 8.1

The Parthenon, constructed of marble, has stood on the Acropolis in Athens for more than 24 centuries. (*Photo by James Austin, Cambridge, England*)

and sun-dried brick. In the third millennium, the Egyptians erected the first of their stone temples and pyramids. In the last centuries prior to the birth of Christ, the Greeks perfected their temples of limestone and marble (Figure 8.1), and control of the western world passed to the Romans, who made the first large-scale use of masonry arches and roof vaults in their basilicas, baths, palaces, and aqueducts. Medieval civilizations in both Europe and the Islamic world brought masonry vaulting to a very high plane of development. The Islamic craftsmen built magnificent palaces, markets, and mosques of brick and often faced them with brightly glazed clay tiles. The Europeans directed their efforts toward fortresses and cathedrals of stone, culminating in the pointed vaults and flying buttresses of the great Gothic churches (Figures 8.2, 8.3). In Central America, South America, and Asia, other civilizations were carrying on a simultaneous evolution of building techniques in cut stone.

During the Industrial Revolution in Europe and North America, machines were developed that quarried and worked stone, molded bricks, and sped the transportation of these heavy materials to the building site. Sophisticated mathematics were applied for the first time to analysis of the structure of masonry arches and to the art of stonecutting. Portland cement mortar came into widespread use, enabling the construction of masonry buildings of greater strength and durability.

In the late 19th century, masonry began to lose its primacy among the materials of construction. The very tall buildings of the central cities required frames of iron or steel to replace the thick masonry bearing walls that had limited the heights to which one could build. Reinforced concrete, poured rapidly and economically into simple forms made of wood, began to replace brick and stone masonry in foundations and

FIGURE 8.2

Construction in ashlar limestone of the magnificent Gothic cathedral at Chartres, France, was begun in 1194 A.D. and was not finished until several centuries later. Seen here are the flying buttresses that resist the lateral thrusts of the stone roof vaulting.

(*Photo by James Austin, Cambridge, England*)

FIGURE 8.3
The Gothic cathedrals were roofed with lofty vaults of stone blocks. The ambulatory roof at Bourges (built 1195–1275) evidences the skill of the medieval French masons in constructing vaulting to cover even a curving floor plan. (*Photo by James Austin, Cambridge, England*)

FIGURE 8.4
Despite the steady mechanization of construction operations in general, masonry construction in brick, concrete block, and stone is still based on simple tools and the highly skilled hands that use them. (*Courtesy of International Masonry Institute*)

walls. The heavy masonry vault was supplanted by lighter floor and roof structures of steel and concrete.

The 19th-century invention of the hollow concrete block helped to avert the extinction of masonry as a craft. The concrete block was much cheaper than cut stone and required much less labor to lay than brick. It could be combined with brick or stone facings to make lower-cost walls that were still satisfactory in appearance. The brick cavity wall, an early-19th-century British invention, also contributed to the survival of masonry, for it produced a warmer, more watertight wall that was later to adapt easily to the introduction of thermal insulation when appropriate insulating materials became available in the middle of the 20th century.

Other 20th-century contributions to masonry construction include the development of techniques for steel-reinforced masonry, high-strength mortars, masonry units (both bricks and concrete blocks) that are higher in structural strength, and masonry units of many types that reduce the amount of labor required for masonry construction.

If this book had been written as recently as a century and a quarter ago, it would have had to devote little space to materials of construction other than masonry and wood. Because other materials of construction were so late in developing, most of the great works of architecture in the world, and many of the best-developed vernacular architectures, are built of masonry. We live amid a rich heritage of masonry buildings—there is scarcely a town in the world that is without a number of beautiful examples from which the serious student of masonry architecture can learn.

MORTAR

Mortar is as much a part of masonry as the masonry units themselves. Mortar serves to cushion the masonry units,

> . . . and the smothered incandescence of the kiln: in the fabulous heat, mineral and chemical treasure baking on mere clay, to issue in all the hues of the rainbow, all the shapes of imagination that never yield to time . . . these great ovens would cast a spell upon me as I listened to the subdued roar deep within.
>
> **Frank Lloyd Wright,**
> *In the Nature of Materials*

giving them full bearing against one another despite their surface irregularities. Mortar seals between the units to keep water and wind from penetrating; it adheres the units to one another to bond them into a monolithic structural unit; and, inevitably, it is important to the appearance of the finished masonry wall.

The most characteristic type of mortar is made of portland cement, hydrated lime, an inert *aggregate* (sand), and water. The sand must be clean and must be screened to eliminate particles that are too coarse or too fine; ASTM specification C144 establishes standards for mortar sand. The portland cement, produced to comply with ASTM specification C150, is the bonding agent in the mortar, but a mortar made only with portland cement is "harsh" and does not flow well on the *trowel* or under the brick, so *lime* is added to impart smoothness and workability. Lime is produced by burning limestone or seashells (calcium carbonate) in a kiln to drive off carbon dioxide and leave *quicklime* (calcium oxide). The quicklime is then slaked by allowing it to absorb as much water as it will

hold, resulting in the formation of calcium hydroxide, called *slaked lime* or *hydrated lime*. The slaking process, which releases large quantities of heat, is usually carried out in the factory. The hydrated lime is subsequently dried, ground, and bagged for shipment. ASTM specification C207 governs the production of lime. Until the late 19th and early 20th centuries, mortar was made without portland cement, and the lime itself was the bonding agent; it hardened by absorbing carbon dioxide from the air to become calcium carbonate, a very slow and uneven process.

Prepackaged *masonry cements* are also widely used for making mortar. Most are proprietary formulations that contain admixtures intended to contribute to the workability of the mortar. The formulations vary from one manufacturer to another, but all must comply with ASTM C91. Two colors of masonry cement are commonly available: *light,* which cures to about the same light gray color as ordinary concrete blocks, and *dark,* which cures to dark gray. Other colors are easily produced by the mason, either by adding pigments to the mortar at the time of mixing or by purchasing dry mortar mix that has been custom colored at the factory. Mortar mix can be obtained in shades ranging from pure white to pure black, including all the colors of the spectrum. Because mortar makes up a considerable fraction of the exposed surface area of a brick wall, typically about 20 percent, the color of the mortar is extremely important in the appearance of a brick wall and is almost as important in the appearance of stone or concrete masonry walls. Small sample walls are often constructed before a major building goes under construction to view and compare different brick and mortar combinations and make a final selection.

Mortar composition is specified in ASTM C270. Four basic mortar types are defined, as summarized in Figures 8.5 and 8.6. Type N mortar is

Mortar Type	Description	Construction Suitability	Minimum Average Compressive Strength at 28 days
M	High-strength mortar	Masonry subjected to high lateral or compressive loads or severe frost action; Masonry below grade	2500 psi (17.25 MPa)
S	Medium high-strength mortar	Masonry requiring high flexural bond strength but subjected only to normal compressive loads	1800 psi (12.40 MPa)
N	Medium strength mortar	General use above grade	750 psi (5.17 MPa)
O	Medium low-strength mortar	Nonloadbearing interior walls and partitions	350 psi (2.40 MPa)

FIGURE 8.5
Mortar types as defined by ASTM C270.

used for most purposes. Types M and S are suitable for higher-strength structural walls and for severe weather exposures. Type O, the most economical, is used only in nonload-bearing interior work.

In order to achieve a workability equivalent to conventional portland cement–lime mortars, masonry cement mortars are formulated with air-entraining admixtures that result in a high air content in the cured mortar. This reduces the bond strength between the mortar and the masonry unit to about half that of conventional mortar, which means that the flexural and shear strength of the wall is reduced and the wall is more permeable to water. For these reasons, only conventional cement–

lime mortars should be specified for masonry work that requires high strength and low permeability. For greater convenience, *mortar cement,* which consists of premixed portland cement and lime with only limited air entrainment, may be used in making cement–lime mortar.

Portland cement mortar cures by hydration, not by drying: A complex set of chemical reactions take up water and combine it with the constituents of the cement and lime to create a dense, strong, crystalline structure that binds the sand particles together. Mortar that has been mixed but not yet used can become too stiff for use, either by drying out or by commencing its hydration. If the mortar was mixed less than 90 min-

utes prior to its stiffening, it has merely dried and can safely be retempered with water to make it workable again. If the unused mortar is more than $2\frac{1}{2}$ hours old, it must be discarded because it has already begun to hydrate and cannot be retempered without reducing its final strength. On large masonry projects, an *extended-life admixture* is sometimes included in the mortar. This allows the mortar to be mixed in large batches and kept for as long as 72 hours before it must be discarded.

Most masonry units should be laid in a dry condition, but to prevent premature drying of mortar, which would weaken it, masonry units that are highly absorptive of water should be dampened before laying.

FIGURE 8.6
Some typical formulas for mortar types M through O.

Mortar Type	Parts by Volume of Portland Cement	Parts by Volume of Masonry Cement	Parts by Volume of Hydrated Lime	Aggregate Measured in a Loose, Damp Condition
M	1	1 (Type II)	—	
	1	—	$\frac{1}{4}$	
S	$\frac{1}{2}$	1 (Type II)	—	
	1	—	over $\frac{1}{4}$ to $\frac{1}{2}$	Not less than $2\frac{1}{4}$ and not more than 3 times the sum of the volumes of the cements and lime used
N	—	1 (Type II)	—	
	1	—	over $\frac{1}{2}$ to $1\frac{1}{4}$	
O	—	1 (Type I or II)	—	
	1	—	over $1\frac{1}{4}$ to $1\frac{1}{2}$	

BRICK MASONRY

Among the masonry materials, brick is special in two respects: fire resistance and size. As a product of fire, it is the most resistant to building fires of any masonry unit. Its size may account for much of the love that many people instinctively feel for brick: A traditional brick is shaped and dimensioned to fit the human hand. Hand-sized bricks are less likely to crack during drying or firing than larger bricks, and they are easy for the mason to manipulate. This small unit size makes brickwork very flexible in adapting to small-scale geometries and patterns and gives a pleasing scale and texture to a brick wall or floor.

Molding of Bricks

Because of their weight and bulk, which make them expensive to ship for long distances, bricks are produced by a large number of relatively small, widely dispersed factories from a variety of local clays and shales. The raw material is dug from pits, crushed, ground, and screened to reduce it to a fine consistency. It is then tempered with water to produce a plastic clay ready for forming into bricks.

There are three major methods used today for forming bricks: the soft mud process, the dry mud process, and the stiff mud process. The oldest is the *soft mud process*, in which a relatively moist clay (20 to 30 percent water) is pressed into simple rectangular molds, either by hand or with the aid of molding machines (Figure 8.7). To keep the sticky clay from adhering to the molds, the molds may be dipped in water immediately before being filled, producing bricks with a relatively smooth, dense surface that are known as *water-struck bricks*. If the wet mold is dusted with sand just before forming the brick, *sand-struck* or *sand-mold bricks* are produced, with a matte-textured surface.

The *dry-press process* is used for clays that shrink excessively during drying. Clay mixed with a minimum of water (up to 10 percent) is pressed into steel molds by a machine working at a very high pressure.

The high-production *stiff mud process* is the one most widely used today. Clay containing 12 to 15 percent water is passed through a vacuum to remove any pockets of air, then extruded through a rectangular

FIGURE 8.7
A simple wooden mold produces seven water-struck bricks at a time. (*Photo by the author*)

FIGURE 8.8
A column of clay emerges from the die in the stiff mud process of molding bricks.
(*Courtesy of Brick Institute of America*)

die (Figures 8.8, 8.9). As the clay leaves the die, textures or thin mixtures of colored clays may be applied to its surface as desired. The rectangular column of moist clay is pushed by the pressure of extrusion across a cutting table, where automatic cutter wires slice it into bricks.

After molding by any of these three processes, the bricks are dried for one to two days in a low-temperature dryer kiln. They are then ready for transformation into their final form by a process known as *firing* or *burning*.

Firing of Bricks

Before the advent of modern kilns, bricks were most often fired by stacking them in a loose array called a *clamp*, covering the clamp with earth or clay, building a wood fire under the clamp, and maintaining the fire for a period of several days. After cooling, the clamp would be disassembled and the bricks sorted according to the degree of burning each had experienced. Bricks adjacent to the fire (*clinker bricks*) were often overburned and distorted, making them unattractive and therefore unsuitable for use in exposed brickwork. Bricks in a zone of the clamp near the fire would be fully burned but undistorted, suitable for exterior *facing bricks* with a high degree of resistance to weather. Bricks farther from the fire would be softer and would be set aside for use as a backup bricks, while some bricks from around the perimeter of the clamp would not be burned sufficiently for any purpose and would be discarded. In the days before mechanized transportation, bricks for a building were often produced from clay obtained from the building site and were burned in clamps adjacent to the work.

Today, bricks are usually burned either in a *periodic kiln* or in a *continuous tunnel kiln*. The periodic kiln is a fixed structure that is loaded with bricks, fired, cooled, and unloaded

FIGURE 8.9
Rotating groups of parallel wires cut the column of clay into individual bricks, ready for drying and firing. (*Courtesy of Brick Institute of America*)

(a)

(b)

(c)

FIGURE 8.10
**Three stages in the firing of water-struck bricks in a small factory: (*a*) Bricks stacked
on a kiln car ready for firing. The open passages between the bricks allow the hot kiln
gases to penetrate to the interior of the stack. The bed of the kiln car is made of a
refractory material that is unaffected by the heat of the kiln. The rails on which the car
runs are recessed in the floor. (*b*) The cars of bricks are rolled into the far end of this
gas-fired periodic tunnel kiln. When a firing has been completed, the large door in the
near end of the kiln is opened and the cars of bricks are rolled out on the rails that
can be seen at the lower right of the picture. (*c*) After the fired bricks have been
sorted, they are strapped into these "cubes" for shipping. (*Photos by the author*)**

(Figure 8.10). Bricks are passed continuously through a long tunnel kiln on special railcars to emerge at the far end fully burned. In either type of kiln, the first stages of burning are *water-smoking* and *dehydration,* which drive off the remaining water from the clay. The next stages are *oxidation* and *vitrification,* during which the temperature rises to 1800 to 2400 degrees Fahrenheit (1000 to 1300° C) and the clay is transformed into a ceramic material. This may be followed by a stage called *flashing,* in which the fire is regulated to create a reducing atmosphere in the kiln that develops color variations in the bricks. Finally, the bricks are cooled under controlled conditions to achieve the desired color and avoid thermal cracking. The cooled bricks are inspected, sorted, and packaged for shipment. The entire process of firing, monitored continuously to maintain product quality, takes from 40 to 150 hours. Considerable shrinkage takes place in the bricks during drying and firing; this must be taken into account when designing the molds for the brick. The higher the temperature, the greater the shrinkage and the darker the brick. Bricks are often used in a mixed range of colors, with the darker bricks inevitably being smaller than the lighter bricks. Even in bricks of uniform color, some size variation is to be expected, and bricks in general are subject to a certain amount of distortion from the firing process.

The color of a brick depends on the chemical composition of the clay or shale and the temperature and chemistry of the fire in the kiln. Higher temperatures, as noted in the previous paragraph, produce darker bricks. The iron that is prevalent in most clays turns red in an oxidizing fire and purple in a reducing fire. Other chemical elements interact in a similar way to the kiln atmosphere to make still other colors. For bright colors, the faces of the bricks can be glazed like pottery, during the normal firing or in an additional firing.

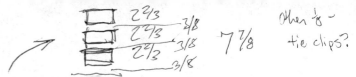

Brick Sizes

There is no truly standard brick. The nearest thing in the United States is the modular brick, dimensioned to construct walls in modules of 4 inches horizontally and 8 inches vertically, but the modular brick has not found ready acceptance in some parts of the country, and traditional sizes persist on a regional basis. Figure 8.11 shows the brick sizes that represent about 90 percent of all the bricks used in the United States. In practice, the designer, when selecting brick for a building, usually views actual samples before completing the drawings for the building, and dimensions the drawings in accordance with the size of the particular brick selected (Figure 8.22). For most bricks in the normal range of sizes, three courses of bricks plus the accompanying three mortar joints add up to a height of 8 inches (200 mm). Length dimensions must be calculated specifically for the brick selected and must include the thicknesses of the mortar joints.

The use of larger bricks can lead to substantial economies in construction. A utility brick has the same face proportions as a standard modular brick, but its in-the-wall cost per square foot is about 25 percent lower and the compressive strength of the wall is about 25 percent higher because of the smaller proportion of mortar. The designer should also consider, however, that a wall built with the larger bricks can deceive the viewer regarding the scale of the building, both because of the smaller number of courses per story and the smaller proportion of mortar to brick in the face of the wall.

Bricks may be solid, *cored, hollow,* or *frogged* (Figure 8.12). By reducing the volume and thickness of the clay, cores and frogs permit more even drying and firing of bricks, reduce fuel costs for firing, reduce shipping costs, and create bricks that are lighter and easier to handle. Hollow bricks, which may contain up to 60 percent voids, are used primarily to enable the insertion and grouting of steel reinforcing bars in single wythes of brickwork.

Custom shapes and sizes of brick are often required for buildings with special details, ornamentation, or unusual geometries (Figures 8.13, 8.14). These are readily produced by most brick manufacturers if sufficient lead time is given.

Standard Brick Sizes

Unit Name	Width	Length	Height
Modular	3½″ or 3⅝″ (90 mm)	7½″ or 7⅝″ (190 mm)	2¼″ (57 mm)
Standard	3½″ or 3⅝″ (90 mm)	8″ (200 mm)	2¼″ (57 mm)
Engineer Modular	3½″ or 3⅝″ (90 mm)	7½″ or 7⅝″ (190 mm)	2¾″ to 2¹³⁄₁₆″ (70 mm)
Engineer Standard	3½″ or 3⅝″ (90 mm)	8″ (200 mm)	2¾″ (70 mm)
Closure Modular	3½″ or 3⅝″ (90 mm)	7½″ or 7⅝″ (190 mm)	3½″ or 3⅝″ (90 mm)
Closure Standard	3½″ or 3⅝″ (90 mm)	8″ (200 mm)	3⅝″ (90 mm)
Roman	3½″ or 3⅝″ (90 mm)	11½″ or 11⅝″ (290 mm)	1⅝″ (40 mm)
Norman	3½″ or 3⅝″ (90 mm)	11½″ or 11⅝″ (290 mm)	2¼″ (57 mm)
Engineer Norman	3½″ or 3⅝″ (90 mm)	11½″ or 11⅝″ (290 mm)	2¾″ to 2¹³⁄₁₆″ (70 mm)
Utility	3½″ or 3⅝″ (90 mm)	11½″ or 11⅝″ (290 mm)	3½″ or 3⅝″ (90 mm)
King Size	3″ (75 mm)	9⅝″ (240 mm)	2⅝ or 2¾″ (70 mm)
Queen Size	3″ (75 mm)	7⅝″ or 8″ (190 mm)	2¾″ (70 mm)

FIGURE 8.11

Dimensions of bricks commonly used in North America, as established by the Brick Institute of America. This list gives an idea of the diversity of sizes and shapes available, and of the difficulty of generalizing about brick dimensions. Modular bricks are dimensioned so that three courses plus mortar joints add up to a vertical dimension of 8 inches (203 mm), and one brick length plus mortar joint has a horizontal dimension of 8 inches (203 mm). The alternative dimensions of each brick are calculated for ⅜-inch (9.5-mm) and ½-inch (12.7-mm) mortar joint thicknesses.

FIGURE 8.12
From left to right: Cored, hollow, and frogged bricks.

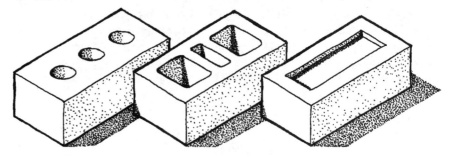

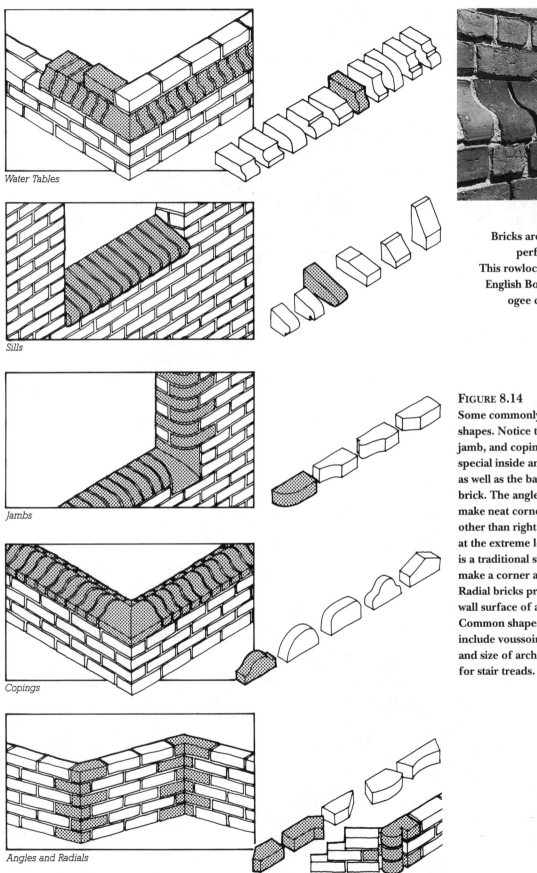

Water Tables

Sills

Jambs

Copings

Angles and Radials

FIGURE 8.13
Bricks are often custom molded to perform particular functions. This rowlock water table course in an English Bond wall was molded to an ogee curve. (*Photo by the author*)

FIGURE 8.14
Some commonly used custom brick shapes. Notice that each water table, jamb, and coping brick shape requires special inside and outside corner bricks as well as the basic rowlock or header brick. The angle bricks are needed to make neat corners in walls that meet at other than right angles. The hinge brick at the extreme lower right of the drawing is a traditional shape that can be used to make a corner at any desired angle. Radial bricks produce a smoothly curved wall surface of any specified radius. Common shapes not pictured here include voussoirs for any desired shape and size of arch and rounded-edge bricks for stair treads.

Choosing Bricks

We have already considered three important qualities that the designer must consider in choosing the bricks for a particular building: molding process, color, and size. Several other qualities are also important and are measured in accordance with standard ASTM testing procedures. ASTM C62, C216, and C652 establish three *grades* of brick based on resistance to weathering, and three *types* of facing bricks (bricks that will be exposed to view) based on the degree of uniformity in shape, dimension, texture, and color from one brick to the next (Figure 8.15). The uses of the three grades are related to a map of weathering indices prepared from National Weather Service data on winter rainfall and freezing cycles (Figure 8.16). Grade SW is recommended for use in contact with the ground, or in situations where the brickwork is likely to be saturated with water, in any of the three regions of the map. Grade MW may be used above ground in any of the regions, but SW will provide greater durability. Grade NW is intended for use in sheltered or indoor locations. Bricks used for paving of walks, drives, and patios should conform to ASTM C902 to minimize deterioration from freeze–thaw cycles.

The compressive strength of brickwork is of obvious importance in structural walls and piers, and depends on the strengths of both the brick and the mortar. Typical allowable compressive stresses for unreinforced brick walls range from 75 to 400 pounds per square inch (0.52 to 2.76 MPa).

For brickwork exposed to very high temperatures, such as the lining of a fireplace or a furnace, *firebricks* are used. These are made from special clays (*fireclays*) that produce bricks with refractory qualities. Firebricks are laid in very thin joints of fireclay mortar.

Grades for Building and Facing Bricks

Grade SW	Severe weathering
Grade MW	Moderate weathering
Grade NW	Negligible weathering

Types of Facing Bricks

Type FBX	High degree of mechanical perfection, narrow color range, minimum size variation per unit
Type FBS	Wide range of color and greater size variation per unit
Type FBA	Nonuniformity in size, color, and texture per unit

FIGURE 8.15

Grades and types of bricks as defined by ASTM C62, C216, and C652.

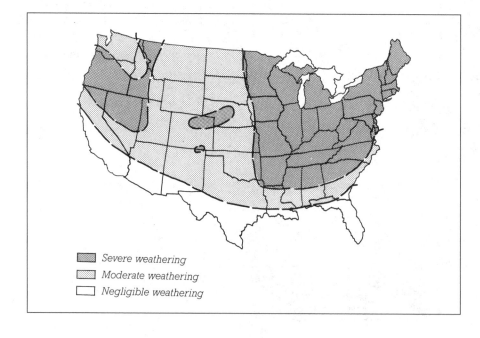

FIGURE 8.16

Weathering regions of the United States, as determined by winter rainfall and freezing cycles. Grade SW brick is recommended for exterior use in the Severe Weathering region. (*Courtesy of Brick Institute of America*)

Severe weathering
Moderate weathering
Negligible weathering

Laying Bricks

Figure 8.17 shows a basic vocabulary of bricklaying. Bricks are laid in the various positions for visual reasons, structural reasons, or both. The simplest brick wall is a single *wythe* of *stretchers*. For walls two or more wythes thick, *headers* are used to bond the wythes together into a structural unit. *Rowlock* courses are often used for caps on garden walls and for sloping sills under windows, although such caps and sills are not durable in severe climates. Architects frequently employ *soldier* courses for visual emphasis in such locations as window lintels or tops of walls.

FIGURE 8.17
Basic brickwork terminology.

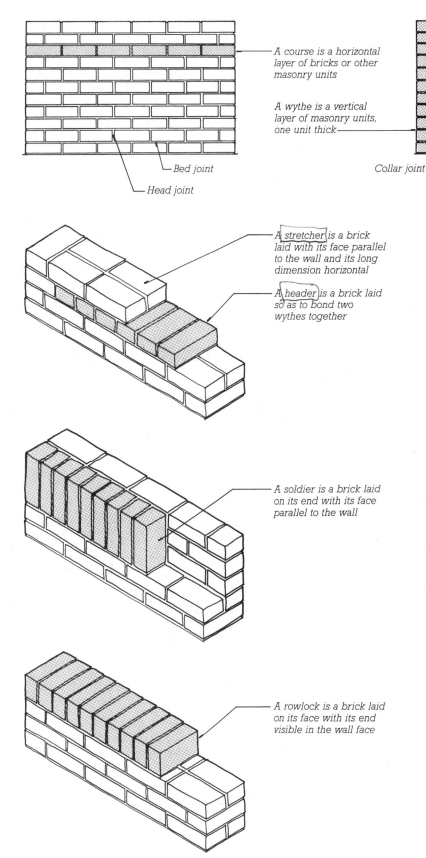

A course is a horizontal layer of bricks or other masonry units

A wythe is a vertical layer of masonry units, one unit thick

Collar joint

Bed joint

Head joint

A stretcher is a brick laid with its face parallel to the wall and its long dimension horizontal

A header is a brick laid so as to bond two wythes together

A soldier is a brick laid on its end with its face parallel to the wall

A rowlock is a brick laid on its face with its end visible in the wall face

The problem of bonding multiple wythes of brick has been solved in many ways in different regions of the world, often resulting in surface patterns that are particularly pleasing to the eye. Figures 8.18 and 8.19 show some *structural bonds* for brickwork, among which *Common Bond, Flemish Bond,* and *English Bond* are the most popular. On the exterior of buildings, the *cavity wall,* with its single outside wythe, offers the designer little excuse to use anything but Running Bond. Inside a building, safely out of the weather, one may use solid brick walls in any desired bond. For fireplaces and other very small brick constructions, however, it is often difficult to create a long enough stretch of unbroken wall to justify the use of bonded brickwork.

Running Bond consists entirely of stretchers

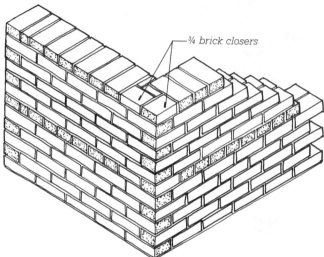

Common Bond (also known as American Bond) has a header course every sixth course. Notice how the head joints are aligned between the header and stretcher courses

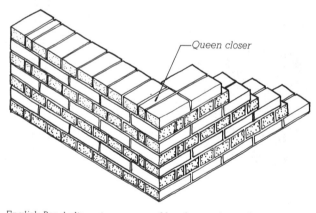

English Bond alternates courses of headers and stretchers

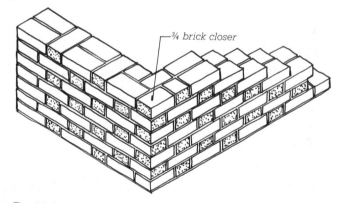

Flemish Bond alternates headers and stretchers in each course

FIGURE 8.18
Frequently used structural bonds for brick walls. Partial closer bricks are necessary at the corners to make the header courses come out even while avoiding alignments of head joints in successive courses. The mason usually cuts the closers to length with a mason's hammer, but they are sometimes cut with a diamond saw.

FIGURE 8.19
Photographs of some brick bonds. (*a*) Running Bond, (*b*) Common Bond, and (*c*) English Garden Wall Bond with Flemish header courses. In the right column, (*d*) English Bond, (*e*) Flemish Bond, and (*f*) Monk Bond, which is a Flemish Bond with two stretchers instead of one between headers. The Running Bond example shown here is from the late 18th century, with extremely thin joints, which require mortar made from very fine sand. Notice in the Common Bond wall (dating from the 1920s in this case) that the header course began to fall out of alignment with the stretcher courses, so the mason inserted a partial stretcher to make up the difference; such small variations in workmanship contribute to the visual appeal of brick walls. Flemish header courses, such as those used in the English Garden Wall Bond, are often used with bricks whose length, including mortar joint, is substantially more than twice their width; the Flemish header course avoids the thick joints between headers that would otherwise result. The Flemish Bond example is modern, and is composed of modular sand mold bricks. The Monk Bond shown here has unusually thick bed joints, approximately ¾ inch (19 mm) high; these are difficult for the mason to lay unless the consistency of the mortar is very closely controlled. (*Photos by the author*)

The process of bricklaying is summarized in Figures 8.20 and 8.21. While conceptually simple, bricklaying requires both extreme care and considerable experience to produce a satisfactory result, especially where a number of bricklayers working side by side must produce identical work on a major structure. Yet speed is essential to the economy of masonry construction. The work of a skilled mason is impressive both for its speed and for its quality. This level of expertise takes time and hard work to acquire, which is why the apprenticeship period for brickmasons is both long and demanding.

The laying of *leads* (pronounced "leeds") is relatively labor intensive. A mason's rule or a *story pole* that is

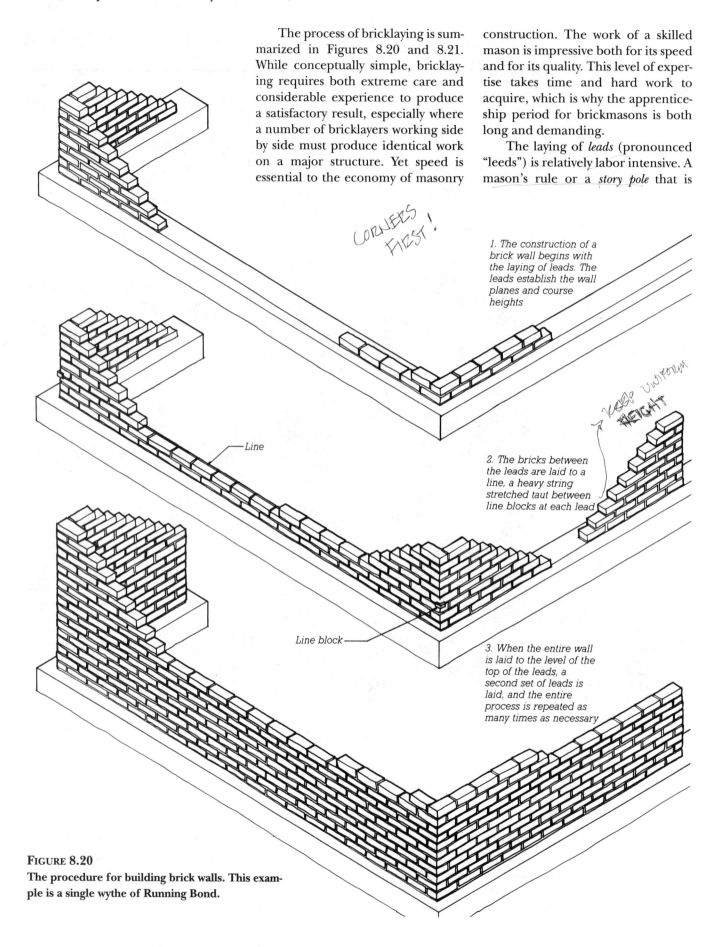

CORNERS FIRST !

1. The construction of a brick wall begins with the laying of leads. The leads establish the wall planes and course heights

KEEP UNIFORM HEIGHT

2. The bricks between the leads are laid to a line, a heavy string stretched taut between line blocks at each lead

—Line

Line block—

3. When the entire wall is laid to the level of the top of the leads, a second set of leads is laid, and the entire process is repeated as many times as necessary

FIGURE 8.20
The procedure for building brick walls. This example is a single wythe of Running Bond.

(a)

(b)

(c)

(d)

(e)

(f)

FIGURE 8.21

Laying a brick wall: (*a*) The first course of bricks for a lead is bedded in mortar, following a line marked on the foundation. (*b, c, d*) As each lead is built higher, the mason uses a spirit level to make sure that each course is level, straight, plumb, and in the same plane as the rest of the lead. A mason's rule or a story pole is also used to check the heights of the courses. (*e*) A finished lead. (*f*) A mason lays brick to a line stretched between two leads. (*Courtesy of International Masonry Institute*)

marked with the course heights is used to establish accurate course heights in the leads. The work is checked frequently with a spirit level to assure that surfaces are flat and plumb and courses are level. When the leads have been completed, a mason's line (a heavy string) is stretched between the leads, using L-shaped *line blocks* at each end to locate the end of the line precisely at the top of each course of bricks.

The laying of the infill bricks between the leads is much faster and easier because the mason needs only a trowel in one hand and a brick in the other to *lay to the line* and create a perfect wall. It follows that leads are expensive as compared to the wall surfaces between, so that where economy is important the designer should seek to minimize the number of corners in a brick structure.

Bricks may be cut as needed, either with sharp, well-directed blows of the chisel-pointed end of a mason's hammer or, for greater accuracy and more intricate shapes, with a power saw that utilizes a water-cooled diamond blade (Figure 9.24). Cutting of bricks slows the process of bricklaying considerably, however, and ordinary brick walls should be dimensioned to minimize cutting (Figure 8.22).

Mortar joints can vary in thickness from about $1/4$ inch (6.5 mm) to more than $1/2$ inch (13 mm). Thin joints work only when the bricks are identical to one another within very small tolerances and the mortar is made with a fine sand. Very thick joints require a stiff mortar that is difficult to work with. Mortar joints are usually standardized at $3/8$ inch (9.5 mm), which is easy for the mason and allows for considerable distortion and unevenness in the bricks. One-half-inch (12.7-mm) joints are also common.

The joints in brickwork are *tooled* an hour or two after laying as the mortar begins to harden, to give a neat appearance and to compact the mortar into a profile that meets the visual and weather-resistive requirements of the wall (Figures 8.23, 8.24). Outdoors, the *vee joint* and *concave joint* shed water and resist freeze–thaw damage better than the others. Indoors, a *raked* or *stripped joint* can be used if desired to accentuate the pattern of bricks in the wall and deemphasize the mortar.

After joint tooling, the face of the brick wall is swept with a soft brush to remove the dry crumbs of mortar left by the tooling process. If the mason has worked cleanly, the wall is now finished, but most brick walls are later given a final cleaning by scrubbing with muriatic acid (HCl) and rinsing with water to remove mortar stains from the faces of the bricks. Light-colored bricks can be stained by acids, and should be cleaned by other means.

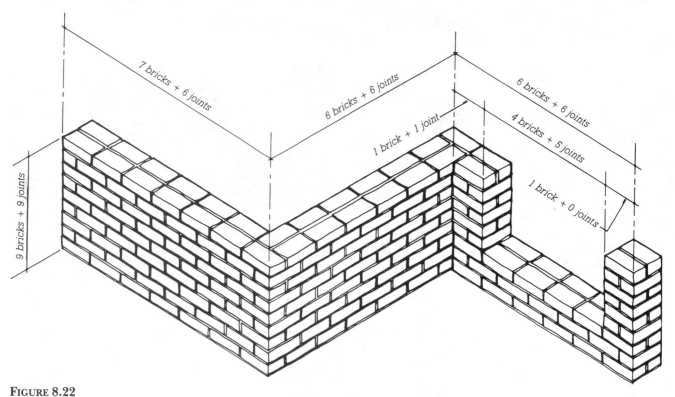

FIGURE 8.22
Dimensions for brick buildings are worked out in advance by the architect, based on the actual dimensions of the bricks and mortar joints to be used in the building. Bricks and mortar joints are carefully counted and converted to numerical dimensions for each portion of the wall.

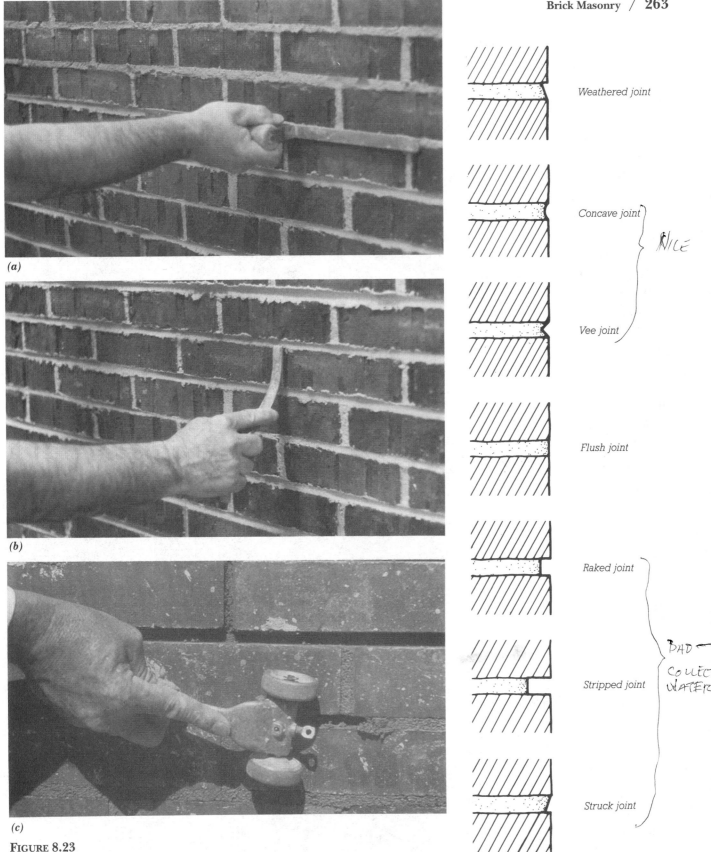

FIGURE 8.23

(*a*) Tooling horizontal joints to a concave profile. (*b*) Tooling vertical joints to a concave profile. The excess mortar squeezed out of the joints by the tooling process will be swept off with a brush, leaving a finished wall. (*c*) Raking joints with a common nail held in a skate-wheel joint raker. The head of the nail digs out the mortar to a preset depth. (*Courtesy of Brick Institute of America*)

FIGURE 8.24

Joint tooling profiles for brickwork. The concave joint and vee joint are suitable for outdoor use in severe climates.

I remember the masons on my first house. I was not much older than the apprentice whom I found choking back tears of frustration with the clumsiness of his work and the rebukes of his boss.

Nearly thirty years later, we still collaborate on sometimes difficult masonry walls, fireplaces and paving patterns. I work with bricklayers . . . whose years of learning their craft paralleled my years of trying to understand my profession. We are friends and we talk about our work like pilgrims on a journey to the same destination.

Henry Klein, Architect

Spanning Openings in Brick Walls

Brick walls must be supported above openings for windows and doors. *Lintels* of reinforced concrete, reinforced brick, or steel angles (Figures 8.25, 8.26) are all equally satisfactory from a technical standpoint. The near invisibility of the steel lintel is a source of delight to some designers but dissatisfies those who prefer that a building express visually its means of support. Wood is no longer used for lintels because of its tendency to burn, to decay, and to shrink and allow the masonry above to settle and crack.

The *corbel* is an ancient structural device of limited spanning capability, one that may be used for small openings in brick walls, for beam brackets, and for ornament (Figures 8.27–8.29). A good rule of thumb for designing corbels is that the projection of each course should not exceed half the course height; this results in a corbel angle of about 60° to the horizontal and minimizes flexural stress in the bricks.

FIGURE 8.26
Because of corbelling and arching action in the bricks, a lintel is considered to carry only the triangular area of brickwork indicated by the shaded portion of this drawing. The broken line indicates a concealed steel angle lintel.

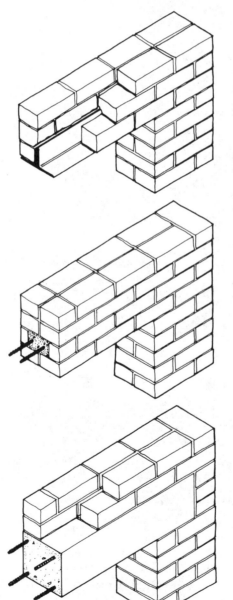

FIGURE 8.25
Three types of lintels for spanning openings in brick walls. The double-angle steel lintel (top) requires some trimming of the first courses of brick but is scarcely visible in the finished wall. The reinforced brick lintel (center) works in the same manner as a reinforced concrete beam and gives no outward clues as to what supports the bricks over the opening. The precast reinforced concrete lintel (bottom) is clearly visible. For short spans, cut stone lintels without reinforcing can be used in the same manner as the concrete lintel.

FIGURE 8.27
Corbelling has many uses in masonry construction. It is used in this example to span a door opening, and to create a bracket for support of a beam.

FIGURE 8.28
Corbelling creates a transition from the cylindrical tower to a hexagonal roof. Cut limestone is used for window sills, lintels, arch intersections, and grotesquely carved rainwater spouts. The building is the Gothic cathedral in Albi, France. (*Photo by the author*)

FIGURE 8.29
All the skills of the 19th-century mason were called into play to create the corbels and arches of this brick cornice in Boston's Back Bay. (*Photo by the author*)

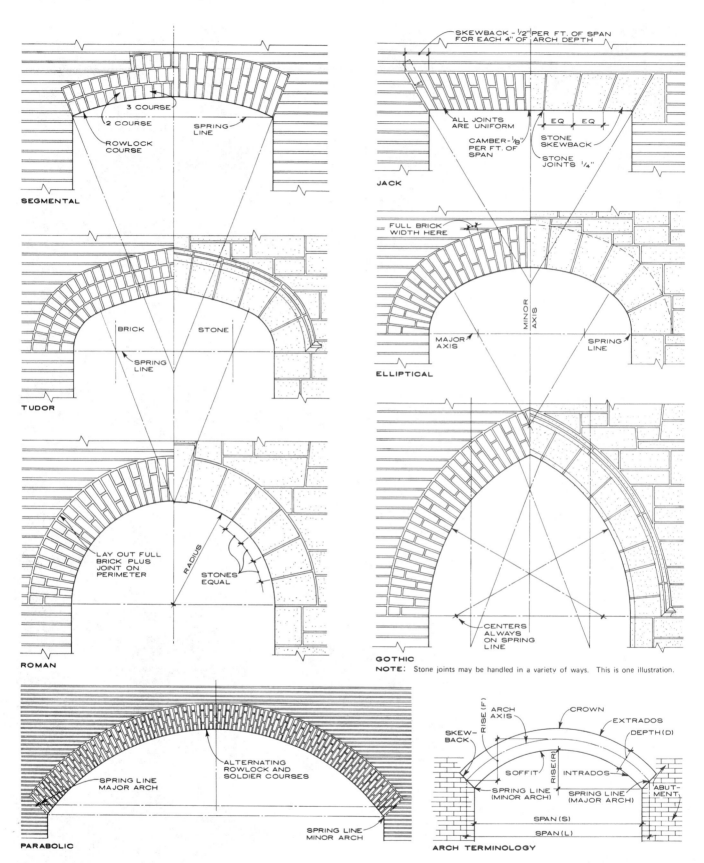

FIGURE 8.30
Arch forms and arch terminology in brick and cut stone. The spandrel is the area of wall that is bounded by the extrados of the arch. (*Reprinted by permission of John Wiley & Sons, Inc., from Ramsey/Sleeper,* Architectural Graphic Standards *(7th ed.), Robert T. Packard, A.I.A., Editor, © 1981 by John Wiley & Sons, Inc.*)

The brick *arch* is a structural form so widely used and so powerful, both structurally and symbolically, that entire books have been devoted to it (Figure 8.30). Given a *centering* of wood or steel (Figure 8.31), a mason can lay a brick arch very rapidly, although the *spandrel*, the area of flat wall that adjoins the arch, is slow to construct because of its numerous cut bricks. In an arch of *gauged brick*, each brick is rubbed to the required wedge shape on an abrasive stone, which is laborious and expensive. The *rough arch*, which depends on wedge-

(a)

(b) *(c)*

FIGURE 8.31

(*a*) **Two rough brick arches under construction, each on its wooden centering.** (*b*) **The brick locations were marked on the centering in advance to be sure that no partial bricks or unusual mortar joint thicknesses will be required to close the arch. This was done by laying the centering on its side on the floor and placing bricks around it, adjusting their positions by trial and error to achieve a uniform spacing. Then the location of each brick was marked with pencil on the curved surface of the centering.** (*c*) **The brick arches whose construction is illustrated in the previous two photographs span a fireplace room that is roofed with a brick barrel vault. The firebox is lined with firebrick and the floor is finished with quarry tiles.** (*Photos by the author*)

FIGURE 8.32
A rough brick triple rowlock arch spans a window opening. The recessing of the innermost rowlock course creates a shadow line that accentuates the arch. (*Photo by the author*)

FIGURE 8.33
A rough jack (also called a flat arch) in a wall of Flemish Bond brickwork. (*Photo by the author*)

shaped mortar joints for its curvature, is therefore much more usual in today's buildings (Figures 8.32, 8.33). A number of brick manufacturers will mold to order sets of tapered bricks for arches of any shape and span.

An arch translated along a line perpendicular to its plane produces a *barrel vault*. An arch rotated about its vertical centerline becomes a *dome*. From various intersections of these two basic roof shapes comes the infinite vocabulary of vaulted masonry construction. Brick vaults and domes, if their lateral thrusts are sufficiently tied or buttressed, are strong, stable forms. In parts of the world where labor is inexpensive, they continue to be built on an everyday basis (Figure 8.34). In North America and most of Europe, where labor is more costly, they have been replaced almost entirely by less expensive, more compact spanning elements such as beams and slabs of wood, steel, or concrete.

FIGURE 8.34
(*a*) **Masons in Mauritania, drawing on thousands of years of experience in masonry vaulting, build a dome for a patient room of a new hospital. The masonry is self-supporting throughout the process of construction; only a simple radius guide is used to maintain a constant diameter. The dome is double with an airspace between to insulate the room from the sun's heat. (*b*) The walls are buttressed with stack bond brick headers to resist the outward thrust of the domes.** (*Courtesy of ADAUA, Geneva, Switzerland*)

Reinforced Brick Masonry

Reinforced brick masonry (*RBM*) is analogous to reinforced concrete construction. The same deformed steel reinforcing bars used in concrete are placed in thickened collar joints to strengthen a brick wall or lintel. A reinforced brick wall (Figure 8.35) is created by constructing two wythes of brick 2 to 4 inches (50 to 100 mm) apart, placing the reinforcing steel in the cavity, and filling the cavity with *grout*. Grout is a mixture of portland cement, aggregate, and water. ASTM C476 specifies the proportions and qualities of grout for use in filling masonry loadbearing walls. It is important that grout be fluid enough that it will flow readily into the narrow cavity and fill it completely. The excess water in the grout that is required to achieve this fluidity is quickly absorbed by the bricks, and does not detract from the eventual strength of the grout as it would from concrete poured into formwork.

There are two methods for grouting reinforced brick walls: low-lift and high-lift. In the *low-lift method,* the masonry is constructed to a height not greater than 4 feet (1200 mm) before grouting, taking care to keep the cavity free of mortar squeezeout and droppings, which might interfere with the placement of the reinforcing and grout. To resist the hydrostatic pressure of the wet grout, the wythes are held together by galvanized steel wire *ties* laid into the bed joints and across the cavity, usually at intervals of 24 inches (600 mm) horizontally and 16 inches (400 mm) vertically. The vertical reinforcing bars are inserted into the cavity and are left projecting at least 30 bar diameters above the top of the brickwork to transfer their loads to the overlapping steel in the next lift. The cavity is then filled with grout to within 1½ inches (38 mm) of the top, and the process is repeated for the next lift. In the *high-lift method,* the wall is grouted a story at a time. The cleanliness of the cavity is assured by temporarily omitting some of the bricks in the lowest course of masonry to create *cleanout* holes. As the bricklaying progresses, the cavity is flushed periodically from above with water to drive debris down and out through the cleanouts. After the cleanouts have been filled with bricks and the mortar has cured for at least three days, the reinforcing bars are placed and grout is pumped into the cavity from above in increments not more than 4 feet (1200 mm) high. To minimize pressure on the brickwork, each increment is allowed to harden for an hour or so before the next increment is poured above it.

The low-lift method is generally easier for small work where the grout is poured by hand. Where grout pumping equipment must be rented, the high-lift method is preferred because it minimizes rental costs.

While unreinforced brick walls are adequate for many structural purposes, RBM walls are much stronger against vertical loads, flexural loads from wind or earth pressure, seismic loads, and shear loads. With RBM, it is possible to build bearing wall buildings to heights formerly possible only with steel and concrete frames, and to do so with surprisingly thin walls (Figures 8.36, 8.37). RBM is also used for brick piers that are analogous to concrete columns, and, less commonly, for structural lintels (Figure 8.25), beams, slabs, and retaining walls.

Reinforced brickwork may also be created at a smaller scale by building with hollow bricks and inserting reinforcing bars and grout into the cores. This technique is especially useful for residential construction and for single-wythe curtain wall panels (Chapter 20).

FIGURE 8.35
A reinforced brick loadbearing wall is built by installing steel reinforcing bars in a thickened collar joint, then filling the joint with portland cement grout. The cleanout holes shown here are used in the high-lift method of grouting.

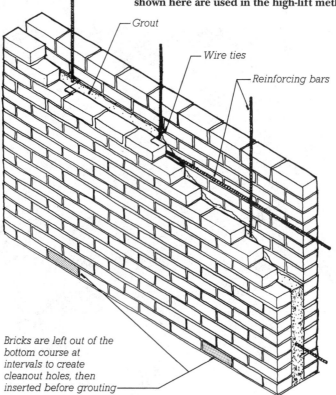

Grout

Wire ties

Reinforcing bars

Bricks are left out of the bottom course at intervals to create cleanout holes, then inserted before grouting

C.S.I./C.S.C. Masterformat Section Numbers for Brick Masonry	
04100	MORTAR AND MASONRY GROUT
04150	MASONRY ACCESSORIES
	Anchors and Tie Systems Manufactured Control Joints Joint Reinforcement
04200	UNIT MASONRY
04210	Clay Unit Masonry
04230	Reinforced Unit Masonry
04550	REFRACTORIES
04555	Flue Liners
04565	Firebrick

FIGURE 8.36
Twelve-inch (300-mm) reinforced brick walls bear the concrete floor and roof structures of a hotel. (*Photo by the author*)

FIGURE 8.37
The unreinforced brick walls of the 16-story Monadnock Building, built in Chicago in 1891, are 18 inches (460 mm) thick at the top, and 6 feet (1830 mm) thick at the base of the building. (*Architects: Burnham and Root. Photo by William T. Barnum. Courtesy of Chicago Historical Society, IChi-18292*)

FIGURE 8.38
Ornamental corbelled brickwork in an 18th-century New England chimney. Step flashings of lead sheet waterproof the junction between the chimney and the wood shingles of the roof. (*Photo by the author*)

FIGURE 8.39
In the gardens he designed at the University of Virginia, Thomas Jefferson used unreinforced serpentine walls of brick that are only a single wythe thick. The shape of the wall makes it extremely stable against overturning despite its thinness. (*Photo by Wayne Andrews*)

FIGURE 8.40
Cylindrical bays of brick with stone lintels front these Boston rowhouses. (*Photo by the author*)

FIGURE 8.41
Bricks were laid diagonally in two of the courses to create this mouse-tooth pattern. The window is spanned with a segmental arch of cut limestone. (*Photo by the author*)

FIGURE 8.42
Quoins originated long ago as cut stone blocks used to form strong corners on weak walls made of materials such as mud bricks or round fieldstones. In more recent times, quoins (pronounced "coins") have been used largely for decorative purposes. At left, cut limestone quoins and a limestone water table dress up a Common Bond brick wall. The mortar joints between quoins are finished in a protruding beaded profile to emphasize the pattern of the stones. At right, brick quoins are used to make a graceful termination of a concrete masonry wall at a garage door opening. Notice that three brick courses match perfectly to one block course. (*Photos by the author*)

FIGURE 8.43
Louis Sullivan's National Farmers' Bank in Owatonna, Minnesota, completed in 1908, rises from a red sandstone base. Enormous rowlock brick arches span the windows in the two street facades. Bands of glazed terra-cotta ornament in rich blues, greens, and browns outline the walls, and a flaring cornice of corbelled brick and terra cotta caps the building. (*Photo by Wayne Andrews*)

FIGURE 8.44
Architect Edward Larrabee Barnes constructed soldier arches of contrasting brick beneath a standing seam copper roof in this cathedral in Burlington, Vermont. (*Photo by Nick Wheeler*)

FIGURE 8.45
Frank Lloyd Wright used long, flat Roman bricks and cut limestone wall copings to emphasize the horizontality of the Robie House, built in Chicago in 1906. (*Photo by Mildred Mead. Courtesy of Chicago Historical Society, IChi-14191*)

FIGURE 8.46
The Kline Science Center at Yale University is clad with cylindrical piers of custom-made curved bricks. (*Architects: Philip Johnson and Richard Foster. Photo courtesy of John Burgee Architects with Philip Johnson*)

SELECTED REFERENCES

1. Beall, Christine. *Masonry Design and Detailing for Architects, Engineers, and Builders* (2nd ed.). New York, McGraw–Hill, 1987.

This 500-page book is the best general design reference on brick, stone, and concrete masonry.

2. Brick Institute of America. *BIA Technical Notes on Brick Construction.* McLean, Virginia, various dates.

This large ring binder contains the current set of *Technical Notes* with the latest information on every conceivable topic concerning bricks and brick masonry. (Address for ordering: 11490 Commerce Park Drive, Reston, VA 22091.)

3. Brick Institute of America. *Principles of Brick Masonry.* Reston, Virginia, 1989.

The 70 pages of this booklet present a complete curriculum in clay masonry construction for the student of building construction. (Address for ordering: See reference 2.)

4. Brick Institute of America. *Brick Shapes Guide.* Reston, Virginia, 1986.

Special shapes of brick are covered in this 24-page booklet.

KEY TERMS AND CONCEPTS

masonry unit	clamp	rowlock	concave joint
mortar	clinker bricks	soldier	raked joint
brick	facing bricks	course	stripped joint
mason	periodic kiln	bed joint	lintel
aggregate	continuous tunnel kiln	head joint	corbel
trowel	water-smoking	collar joint	arch
lime	dehydration	structural bond	centering
quicklime	oxidation	Common Bond	spandrel
slaked or hydrated lime	vitrification	Flemish Bond	gauged brick
masonry cements, light and dark	flashing	English Bond	rough arch
mortar cement	cored brick	cavity wall	barrel vault
extended-life admixture	hollow brick	leads	dome
soft mud process	frogged brick	story pole	reinforced brick masonry (RBM)
water-struck brick	firebrick	line block	grout
sand-struck or sand-mold brick	fireclay	laying to the line	low-lift method
dry-press process	wythe	tooling	high-lift method
stiff mud process	stretcher	weathered joint	cleanout
firing or burning	header	vee joint	tie

REVIEW QUESTIONS

1. How many syllables are in the word "masonry"? (*Hint:* There cannot be more syllables in a word than there are vowels. Many people, even masons and building professionals, mispronounce this word.)

2. What are the most common types of masonry units?

3. What are the molding processes used in manufacturing bricks? How do they differ from one another?

4. List the functions of mortar.

5. What are the ingredients of mortar? What is the function of each ingredient?

6. Why are mortar joints tooled? Which tooling profiles are suitable for a brick wall in a severe climate?

7. What is the function of a structural brick bond such as Common or Flemish Bond? Draw the three most popular brick bonds from memory.

EXERCISES

1. What is the exact height of a brick wall that is 44 courses high, when three courses of brick plus their three mortar joints are 8 inches (203.2 mm) high?

2. What are the inside dimensions of a window opening in a wall of modular bricks with ⅜-inch (9.5-mm) mortar joints if the opening is six and a half bricks wide and 29 courses high?

3. Obtain sand, hydrated lime, several hundred bricks, and basic bricklaying tools from a masonry supply house. Arrange for a mason to help everyone in your class learn a bit of bricklaying technique. Use lime mortar (hydrated lime, sand, and water), which hardens so slowly that it can be retempered with water and used again and again for many weeks. Lay small walls in several different structural bonds. Make simple wooden centering and construct an arch. Construct a dome about 4 feet (1.2 m) in diameter without using centering, as it is done in Figure 8.34. Dismantle what you build at the end of each day, scrape the bricks clean, stack them neatly for reuse, and retemper the mortar with water, covering it with a sheet of plastic to keep it from drying out before it is used again.

4. Design a brick fireplace for a house that you are designing. Select the size and color of brick and the color of mortar. Proportion the fireplace according to the guidelines in Chapter 7. Dimension the fireplace so that it uses only full and half bricks. Draw every brick and every mortar joint in each view of the fireplace. Use rowlocks, soldiers, corbels, and arches as desired for visual effect. How will you span the fireplace opening?

STONE AND CONCRETE MASONRY

Stone offers a wide range of expressive possibilities to the architect. In this detail of a 19th-century church, the columns to the left are made of polished granite and rest on bases of carved limestone. The limestone blocks to the right, squared and dressed by hand, have rough-pointed faces and tooth-axed edges. (*Photo by the author*)

Stone masonry and concrete masonry are similar in concept to brick masonry. Both involve the stacking of masonry units, joining them with the same mortar that is used for brick masonry. But there are important differences: Building stone must be wrested from quarries in rough blocks, then cut and carved to the shapes that we want. We cannot control the properties of stone, so we must learn to select from the bountiful assortment provided by the earth the type and color that we want, and to work with it as nature provides it to us. Concrete masonry units, like bricks, are molded to shape and size, and their properties can be closely controlled. But most concrete masonry units are much larger than bricks, and, like stone, they require slightly different techniques for laying.

STONE MASONRY

Types of Building Stone

Building stone is obtained by taking rock from the earth and reducing it to the required shapes and sizes for construction. Three types of rock are commonly quarried to produce building stone:

- *Igneous rock,* which is rock that was deposited in a molten state.

- *Sedimentary rock,* which is rock that was deposited by the action of water or wind.

- *Metamorphic rock,* which was either igneous or sedimentary rock that has been transformed by heat and pressure into a different type of rock.

Granite is the igneous rock most commonly quarried for construction in North America. It is a mosaic of mineral crystals, principally feldspar and quartz, and can be obtained in a range of colors that includes gray, black, pink, red, brown, buff, and green. Granite is nonporous, hard, strong, and durable, the most nearly permanent of building stones, suitable for use in contact with the ground or exposed to severe weathering. Its surface can be finished in any of a number of textures including a mirrorlike polish. In North America, it is quarried chiefly in the East and the upper Midwest. Various granites are also imported from abroad, chiefly from China, Finland, India, Italy, Portugal, South Africa, and Spain. Domestic granites are classified according to whether they are fine-grained, medium-grained, or coarse-grained.

Limestone and *sandstone* are the principal sedimentary rocks used in construction. Either may be found in a strongly stratified form or in deposits that show little stratification (*freestone*). Neither will accept a high polish.

Limestone is quarried throughout North America, but the major quarries for large dimension stone are in Missouri and Indiana. France, Italy, Spain, and Yugoslavia are the primary sources of imported limestone. Limestone may be composed either of calcium carbonate or of a mixture of calcium and magnesium carbonates, originally furnished in either case by the skeletons or shells of marine organisms. Its colors range from almost white through gray and buff to iron oxide red. It is a porous

FIGURE 9.1
Architect H. H. Richardson designed Austin Hall at Harvard University (1881–1884) as a showcase of stone masonry. Notice the intricate carving of the yellow Ohio sandstone capitals and arch components. The spandrels above the arches are a mosaic of two colors of Longmeadow sandstone blocks. The depth of the arches is intentionally exaggerated to impart a feeling of massiveness to the wall at the entrance to the building. (*Photo by Steve Rosenthal*)

FIGURE 9.2
The loadbearing walls of the Cistercian Abbey Church in Irving, Texas, are made of 427 rough blocks of Big Spring, Texas limestone, each 2 by 3 by 6 feet and weighing about 5000 lb (2300 kg). The stones were brought directly from the quarry to the site, without milling; drill holes are visible in many of the stones. Each stone was bedded in Type S mortar 1 inch (25 mm) thick to allow for irregularities in its horizontal faces. Darker stones are grouped in bands for visual effect. The columns were turned from limestone. The architect, Cunningham Architects, designing for a Catholic order that originated 900 years ago in Europe, wanted to build a church that would last for another 900 years.

FIGURE 9.3
The heavy timber roof of the Cistercian Abbey Church, flanked by continuous skylights, appears to float above the simple volume of the nave. (*Photos in Figures 9.2 and 9.3 by James F. Wilson*)

GRANITE NEEDS 3/8"
MARBLE 3/4"
LIMESTONE 2-3"

stone that contains considerable ground water (*quarry sap*) when quarried. While still saturated, most limestones are easy to work but are susceptible to frost damage. After seasoning in the air to evaporate the quarry sap, the stone becomes harder and is resistant to frost damage. *Travertine* is a richly patterned limestone, deposited by ancient springs, which is marblelike in its qualities.

The Indiana Limestone Institute classifies limestone into two colors, buff and gray, and four grades. *Select* grade consists of fine- to average-grained stone with a minimum of natural flaws. *Standard* grade stone may include fine- to moderately-large-grained stone with an average amount of natural flaws. *Rustic* grade may have fine- to very-coarse-grained stone with an above-average amount of natural flaws. The fourth grade, *variegated,* consists of an unselected mixture of the first three grades and permits both gray- and buff-colored stone.

Sandstone was formed in ancient times from deposits of sand (silicon dioxide). It is quarried principally in New York, Ohio, and Pennsylvania. Two of its more familiar forms are *brownstone,* widely used in wall construction, and *bluestone,* a highly stratified, durable stone especially suitable for paving and wall copings.

Slate and *marble* are the major metamorphic stones utilized in construction. Slate was formed from clay, and is a dense, hard stone with closely spaced planes of cleavage, along which it is easily split into sheets, making it useful for paving stones, roof shingles, and thin wall facings. It is quarried in Vermont, Virginia, New York, and Pennsylvania, as well as many foreign countries, in black, gray, purple, blue, green, and red.

Marble is a recrystallized form of limestone. It is easily carved and polished and occurs in white, black, and nearly every color, often with beautiful patterns of veining. The marbles

used in North America come chiefly from Alabama, Tennessee, Vermont, Georgia, Missouri, and Canada. Marbles are also imported to this continent from all over the world, primarily from Africa, China, Greece, Italy, Mexico, Portugal, Spain, and Turkey. The Marble Institute of America has established a four-step grading system for marbles, in which *Group A* includes sound marbles and stone with uniform and favorable working qualities. *Group B* marbles have working qualities somewhat less favorable than Group A, and may have some natural faults that require *sticking* (cementing together) and *waxing* (filling of voids with cements, shellac, or other materials). *Group C* marbles may have further variations in working qualities; geological flaws, voids, veins, and lines of separation are common, often requiring sticking, waxing, and the use of *liners,* which are pieces of sound stone doweled and cemented to the backs of the sheets of stone to strengthen them. *Group D* permits a maximum of variation in working qualities and a still larger proportion of natural faults and flaws. Many of the most prized, most highly colored marbles belong to this lowest group.

The stone industry is international in scope and operation. Architects and building owners select stone primarily on the basis of appearance, durability, and cost, often with little regard to national origin. The most technically advanced machinery for stoneworking, used all over the world, is designed and manufactured in Italy and Germany. A number of Italian companies have earned reputations for cutting and finishing stone to a very high standard at reasonable cost. As a result of these factors and the low cost of ocean freight relative to the value of stone, it is common to select, for example, a granite that is quarried in Finland, to have the quarry blocks shipped to Italy for cut-

ting and finishing, and to have the finished stone shipped to a North American port, where it is transferred to railcars or trucks for delivery to the building site. And because of the unique character of many domestic stones, both U.S. and Canadian quarries and mills ship millions of tons of stone to foreign countries each year, in addition to the even larger amount that is quarried, milled, and erected at home.

Quarrying and Milling of Stone

The construction industry uses stone in many different forms. *Fieldstone* is rough building stone that is obtained from riverbeds and rockstrewn fields. *Rubble stone* consists of irregular quarried fragments that have at least one good face to expose in a wall. *Dimension stone* is stone that has been quarried and cut into rectangular form; large slabs are often referred to as *cut stone,* and small rectangular blocks are called *ashlar. Flagstone* consists of thin slabs of stone that are used for flooring and paving, either rectangular or irregular in outline. Crushed and broken stone are useful in site work as freely draining fill material, base layers under concrete slabs and pavings, surfacing materials, and also as aggregates in concrete and asphalt. Stone dust and powder are used as surfacing chips and landscape materials.

The maximum sizes and minimum thicknesses of sheets of cut stone vary from one type of stone to another. Granite, the strongest stone, may be utilized in sheets as thin as $\frac{3}{8}$ inch (9.5 mm) in some applications. Marble is not generally cut thinner than $\frac{3}{4}$ inch (19 mm). In both granite and marble, somewhat greater thicknesses than these are advisable in most applications. Limestone, the weakest building stone, is never cut thinner than 2 inches (51 mm), and 3 inches (76 mm) is the preferred thickness for conventionally set stone. In a 6-inch (152-mm) thickness, limestone may be handled in sheets as large as 5 × 18 feet (1.5 × 5.5 m). A Group A marble has a maximum recommended sheet size of 5 × 7 feet (1.5 × 2.1 m).

Dimension stone, whether granite, limestone, sandstone, or marble, is cut from the quarry in blocks. The methods for doing this vary somewhat with the type of stone, and quarrying technology is changing rapidly. The most advanced machines for cutting marble and limestone in the quarry are chain saws and belt saws that utilize diamond blades (Figures 9.4, 9.5). Experiments in quarrying the much harder granite with these machines have been unsuccessful, and most granite is quarried either by drilling and blasting or by the use of a *jet burner* that combusts fuel oil with compressed air. The hot flame at the end of the jet burner lance induces local thermal stress and spalling in the granite. With repeated passes of the lance, the operator gradually excavates a deep, narrow trench to isolate each block of granite from the bedrock. Certain types of granite are also being quarried successfully with a *diamond wire* technique, in which a long, taut loop of diamond-studded wire is run at high speed against the stone to make the cut (page 515).

When preparing cut stone for a building, the stone producer works from the architect's drawings to make a set of *shop drawings* that show each individual stone in the building and how it is to be dimensioned and shaped. After these drawings have been checked by the architect, they are used to guide the work of the mill in producing the stones. Rough blocks of stone are selected in the yard, brought into the mill, and sawed into slabs. The slabs are edged, planed, planed to a molding profile, turned on a lathe, or carved, as required for each piece, and given the desired surface finish. Automated

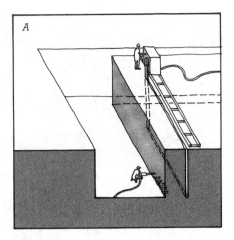

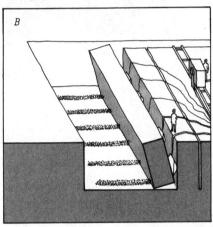

FIGURE 9.4

A typical procedure for quarrying limestone. (*a*) A diamond belt saw divides the limestone bedrock into long cuts, each about 50 feet (15 m) long and 12 feet (3.6 m) high. Multiple horizontal drill holes create a plane of weakness under the first cut. (*b*) Rubber air bags are inflated in the saw kerf to "turn the cut," breaking it free of the bedrock and tipping it over onto a prepared bed of stone chips that cushions its fall. (*c*) Quarry workers drive steel wedges into shallow drilled holes to split the cut into blocks. A front-end loader removes the blocks and stockpiles them ready for trucking to the mill.

FIGURE 9.5

(*a*) The long blade of a diamond belt saw, only the shank of which is visible here, cuts limestone full depth in one pass, using water to lubricate the saw and flush away the stone dust. The saw advances automatically on its portable rails at a maximum rate of about $2\frac{1}{2}$ inches (65 mm) per minute. A chain saw is similar to the belt saw shown here, but it uses a chain of linked, rigid teeth, while the belt saw uses a narrow, flexible belt of steel-reinforced polyethylene with diamond cutting segments. (*b*) After the cut has been turned, quarry workers split it into transportable blocks. The man at the top of the photograph is laying out the splitting pattern with a measuring tape and straightedge, working from a list of the sizes of blocks that the mill requires for the specific jobs it is working on. At the right, a worker drills shallow holes in the stone. He is followed by a second worker who places steel wedges in the holes, and a third who drives the wedges until the cut splits. Each split takes just several minutes to accomplish.

(*Photos by the author*)

equipment is often used to cut and carve repetitive pieces. Holes for *lewises* and anchors are drilled as needed (Figures 9.6, 9.7, 9.10). The finished pieces of stone are marked to correspond to their positions in the building as indicated on the shop drawings and shipped to the construction site.

Stone Masonry

Stone is used in two fundamentally different ways in buildings: It may be laid in mortar, much like bricks or concrete blocks, to make walls, arches, and vaults; or it may be mechanically attached in large sheets as a thin facing over the structural frame and walls of a building. This chapter deals only with stone masonry laid in mortar; the detailing and installation of thin facings of stone are covered in Chapters 20 and 23. There are two simple distinctions that are useful in classifying patterns of stone masonry (Figures 9.8, 9.9):

• *Rubble* masonry is composed of unsquared pieces of stone, whereas *ashlar* is made up of squared pieces.

• *Coursed* stone masonry has continuous horizontal joint lines, whereas *uncoursed* or *random* does not.

Rubble can take many forms, from rounded river-washed stones to broken pieces from a quarry. It may be either coursed or uncoursed. Ashlar masonry may be coursed or uncoursed and may be made up of blocks that are the same size or several different sizes. The terms are obviously very general in their meaning, and the reader will find some variation in usage even among people experienced in the field of stone masonry.

Rubble stonework is laid very much like brickwork, except that the irregular shapes and sizes of the stones require the mason to select

His powers continually increased, and he invented ways of hauling the stones up to the very top, where the workmen were obliged to stay all day, once they were up there. Filippo had wineshops and eating places arranged in the cupola to save the long trip down at noon. . . . He supervised the making of the bricks, lifting them out of the ovens with his own hands. He examined the stones for flaws and hastily cut model shapes with his pocket knife in a turnip or in wood to direct the men. . . .

Giorgio Vasari (1511–1574), writing of Filippo Brunelleschi (1377–1446), the architect of the great masonry dome of the Cathedral of Santa Maria del Fiore in Florence

each stone carefully to fit the available space, and occasionally to trim a stone with a mason's hammer or chisel. Ashlar stonework, though similar in many ways to brickwork, presents unique problems. The stones are often too heavy to lift manually and must be lifted and lowered into place by a hoist. This requires a means of attaching the hoisting rope to the top of the stone block so as not to interfere with the mortar joint, and several types of devices are commonly used for this purpose (Figure 9.10). Mortar joints in ashlar work are usually raked out after setting the stones, to avoid any uneven settling of

stones due to the more rapid drying and hardening of the mortar at the face of the wall. After the mortar in an ashlar wall has cured fully, the masons return to *point* the wall by filling the joints out to the face with mortar and tooling them to the desired profile. Stonework, whether ashlar or rubble, is laid with the *quarry bed* or grain of the stone running in the horizontal direction because stone is both stronger and more weather resistant in this orientation.

Some building stones, especially limestone and marble, deteriorate rapidly in the presence of acids. This restricts their outdoor use in regions whose air is heavily polluted, and it also prevents their being cleaned with acid, as is often done with bricks. Exceptional care is taken during construction to keep stonework clean: Nonstaining mortars are used, high standards of workmanship are enforced, and the work is kept covered as much as possible. Flashings in stone masonry must be of plastic or nonstaining metal. Stonework may be cleaned only with mild soap, water, and a soft brush.

From thousands of years of experience in building with stone, we have inherited a rich tradition of styles and techniques (Figures 9.11–9.19). This tradition is all the richer for its regional variations that are nurtured by the locally abundant stones: Warm, creamy Jerusalem limestone in Israel. Pristine, white Pentelic marble in Greece. Stern, gray granite and gray-streaked white Vermont marble in the northeastern United States. Joyously colorful marbles and wormy travertine in Italy. Golden limestone and gray Aswan granite in Egypt. Red and brown sandstones in New York. Cool, carveable limestone in France. Gray-black basalt and granite in Japan. Though the stone industry is now global in character, its greatest glories are often regional and local.

(a) (b) (c) (d) (e) (f)

FIGURE 9.6

Stone milling operations, showing a composite of techniques for working granite and limestone. (*a*) An over-head crane lifts a rough block of granite from the storage yard to transport it into the plant. The average block in this yard weighs 60 to 80 tons. (*b*) Two reciprocating gang saws slice blocks of limestone into slabs. The saw at the right has just completed its cuts, while the one at the left is just beginning. The intervals between the parallel blades are set by the saw operator to produce the desired thicknesses of slabs. Water cools the diamond blades of the saws and flushes away the stone dust. (*c*) A slab of granite is ground to produce a flat surface, prior to polishing operations. (*d*) If a textured thermal finish is desired on a granite slab, a propane–oxygen torch is passed across the slab under controlled conditions to cause small chips to explode off the surface. (*e*) A layout specialist, working from shop drawings for a specific building, marks a polished slab of granite for cutting. (*f*) The granite slab is squared and cut into finished pieces with a large diamond

(*g*)

(*h*)

(*i*)

(*j*)

(*k*)

(*l*)

saw that is capable of cutting 7 feet (2.1 m) per minute at a depth of 3 inches (76 mm). (*g*) Small pneumatic chisels with carbide-tipped bits are used for special details in granite. (*h*) Hand polishers are used to finish edges of granite that cannot be done by automatic machinery. (*i*) Cylindrical components of limestone are turned on a lathe. (*j*) A cylindrical column veneer is ground to true radius. (*k*) Linear shapes of Indiana limestone, a relatively soft stone, can be formed by a profiled silicon carbide blade in a planer. The piece of stone is clamped to the reciprocating bed, which passes it back and forth beneath the blade. The blade pressure and depth are controlled by the operator's left hand pressing on the wooden lever. Here the planer is making a stepped profile for a cornice. (*l*) Working free-hand with a vibrating pneumatic chisel, a carver finishes a column capital of Indiana limestone. Different shapes and sizes of interchangeable chisel bits rest in the curl of the capital. (*Photos* a, c *through* h, *and* j *courtesy of Cold Spring Granite Company; photo* i *courtesy of Indiana Limestone Institute; photos* b, k, *and* l *are by the author*)

Figure 9.7
This 9-ton Corinthian column capital was carved from a single 30-ton block of Indiana limestone. Rough cutting took 400 hours, and carving another 500. Eight of these capitals were manufactured for a new portico on an existing church. (*Architects: I. M. Pei & Partners. Photo courtesy of Indiana Limestone Institute*)

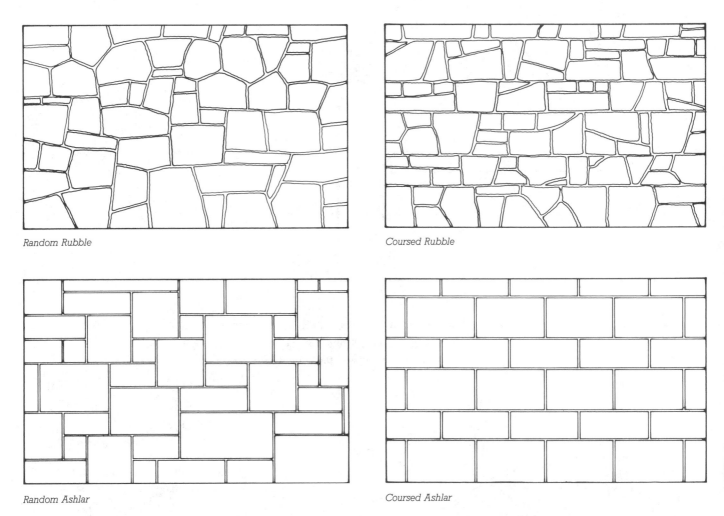

Random Rubble

Coursed Rubble

Random Ashlar

Coursed Ashlar

FIGURE 9.8
**Rubble and ashlar stone masonry,
coursed and random.**

FIGURE 9.9
Random granite rubble masonry (left) and random ashlar limestone (right). (*Photos by the author*)

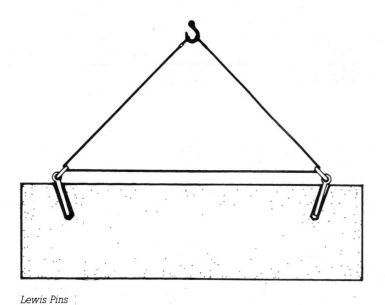

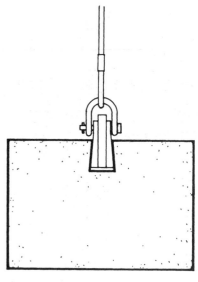

Lewis Pins

Box Lewis

FIGURE 9.10
Lewises permit the lifting and placing of blocks of building stone
without interfering with the bed joints of mortar.

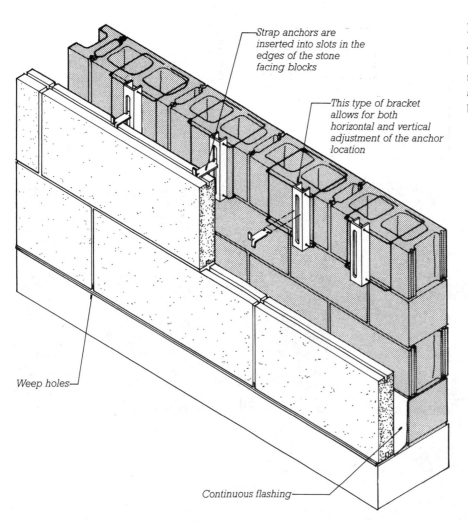

Strap anchors are
inserted into slots in the
edges of the stone
facing blocks

This type of bracket
allows for both
horizontal and vertical
adjustment of the anchor
location

Weep holes

Continuous flashing

FIGURE 9.11
A conventional method of attaching
blocks of cut stone facing to a concrete
masonry backup wall. For methods of
attaching larger panels of stone to a
building, see Chapter 20.

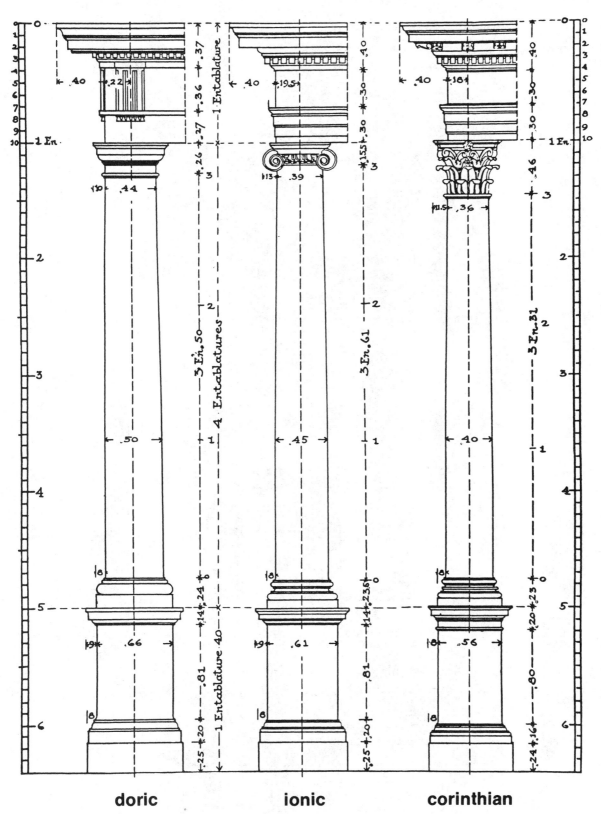

doric **ionic** **corinthian**

FIGURE 9.12
Proportioning rules for the classical
orders of architecture. (*Courtesy of*
Indiana Limestone Institute)

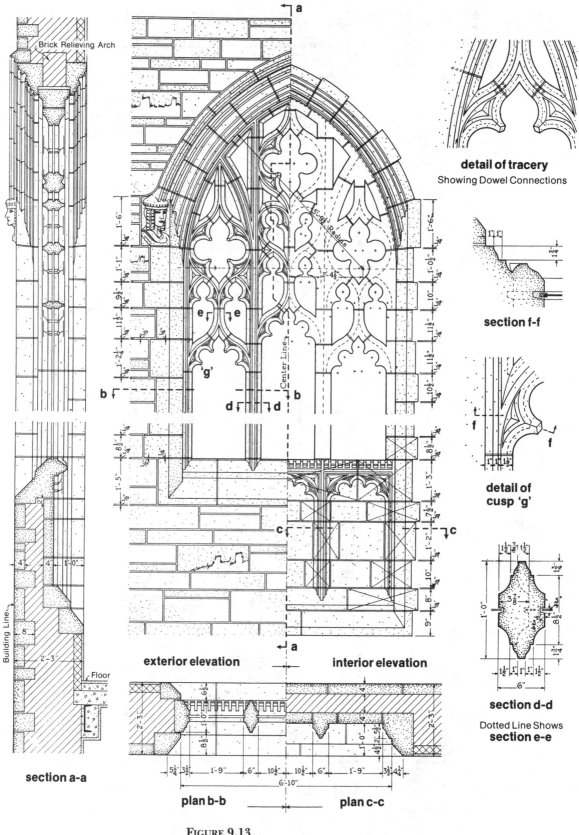

detail of tracery
Showing Dowel Connections

section f-f

**detail of
cusp 'g'**

section d-d

Dotted Line Shows
section e-e

exterior elevation

interior elevation

section a-a

plan b-b

plan c-c

FIGURE 9.13
Details of Gothic window framing and tracery in
limestone. (*Courtesy of Indiana Limestone Institute*)

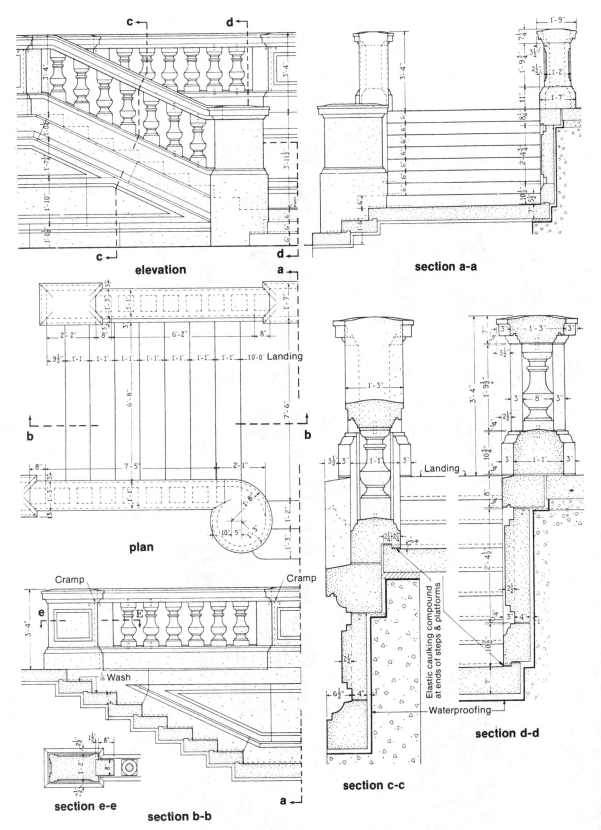

FIGURE 9.14
Details for a limestone stair and balustrade in the classical
manner. The balusters are turned in a lathe, as shown in
Figure 9.6i. (*Courtesy of Indiana Limestone Institute*)

FIGURE 9.15
The Marshall Field Wholesale Store, built in 1885 in Chicago, rested on a two-story base of red granite. Its upper walls were built of red sandstone, and its interior was framed with heavy timber. (*Architect: H. H. Richardson. Courtesy of Chicago Historical Society, IChi-01688*)

FIGURE 9.16
Granite and sandstone details from the rectory of H. H. Richardson's Trinity Church in Boston, 1872–1877. (*Photo by the author*)

FIGURE 9.17

Cut stone detailing. (*a*) Random ashlar of broken face granite on H. H. Richardson's Trinity Church in Boston (1872–1877) still shows the drill holes of its quarrying. (*b*) Dressed stonework in and around a window of the same church contrasts with the rough ashlar of the wall. (*c*) Stonework on a 19th-century apartment building grows more refined as it distances itself from the ground and meets the brick walls above. (*d*) The base of the Boston Public Library (1888–1895, McKim, Mead, and White, Architects) is constructed of pink granite, with strongly rusticated blocks between the windows. (*e*) A chapel is simply detailed in limestone. (*f*) A contemporary college library is clad in bands of limestone.

(*Architects: Shepley, Bulfinch, Richardson and Abbott. Photos by the author*)

FIGURE 9.18
Ashlar limestone loadbearing walls at the Wesleyan University Center for the Arts, designed by architects Kevin Roche, John Dinkeloo, and Associates. (*Courtesy of Indiana Limestone Institute*)

FIGURE 9.19
The East Building of the National Gallery of Art in Washington, D.C., is clad in pink marble. (*Architects: I. M. Pei & Partners. Photo by Ezra Stoller, ©ESTO*)

CONCRETE MASONRY

Concrete masonry units (*CMUs*) are manufactured in three basic forms: bricks, larger hollow units that are commonly referred to as *concrete blocks,* and, less commonly, larger solid units.

Manufacture of Concrete Masonry Units

Concrete masonry units are manufactured by vibrating a stiff concrete mixture into metal molds, then immediately turning out the wet blocks or bricks onto a rack so that the mold can be reused, at the rate of a thousand or more units per hour. The racks of concrete masonry units are cured at an accelerated rate by subjecting them to steam, either at atmospheric pressure or, for faster curing, at higher pressure (Figure 9.20). After steam curing, the units are dried to a specified moisture content and bundled on wooden pallets for shipping to the construction site.

Concrete masonry units are made in a variety of sizes and shapes (Figures 9.21, 9.22) and in different densities of concrete, some of which use cinders, pumice, or expanded lightweight aggregates rather than crushed stone or gravel. Many colors and surface textures are available, and special shapes are relatively easy to produce if there is to be a sufficient number of units produced to amortize the expense of the mold. The major ASTM standards under which concrete masonry units are manufactured are C55 (concrete bricks), C90 (hollow loadbearing units), and C145 (solid loadbearing units). ASTM C90 establishes two *grades,* two *types,* and three *weights* of hollow loadbearing concrete masonry units (concrete blocks), as shown in Figure 9.23.

Hollow concrete block masonry is generally much more economical per unit of wall area or volume than brick or stone masonry. The blocks themselves are cheaper on a volumetric basis and are made into a wall much more quickly because of their larger size (a single standard concrete block occupies the same volume as 12 modular bricks). Concrete blocks can be produced to required

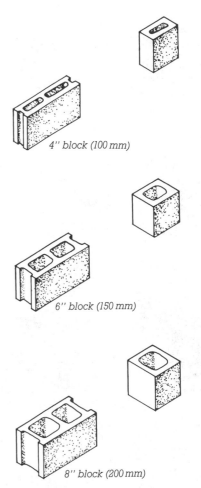

4″ block (100 mm)

6″ block (150 mm)

8″ block (200 mm)

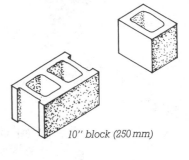

10″ block (250 mm)

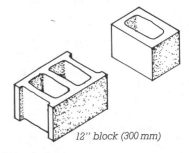

12″ block (300 mm)

FIGURE 9.21
American standard concrete blocks and half blocks. Each full block is nominally 8 inches (200 mm) high and 16 inches (400 mm) long.

FIGURE 9.20
A forklift truck loads newly molded concrete masonry units into an autoclave for steam curing. (*Courtesy of Portland Cement Association, Skokie, Illinois*)

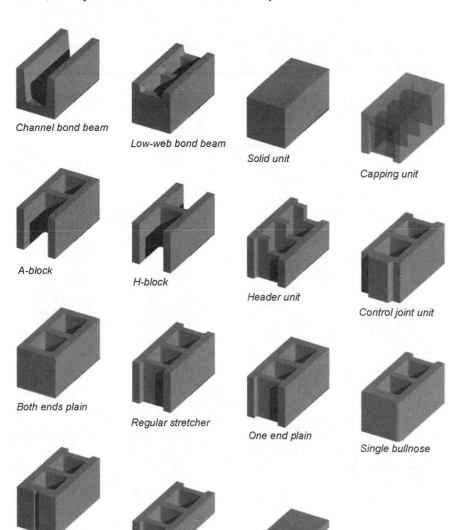

Channel bond beam

Low-web bond beam

Solid unit

Capping unit

A-block

H-block

Header unit

Control joint unit

Both ends plain

Regular stretcher

One end plain

Single bullnose

Steel sash unit

4-inch-high stretcher

Cap or paving unit

Concrete brick

FIGURE 9.22
Other concrete masonry shapes. Concrete bricks are interchangeable with modular clay bricks. Header units accept the tails of a course of headers from a brick facing. The use of control joint units is illustrated in Figure 10.17. Bond beam units have space for horizontal reinforcing bars and grout and are used to tie a wall together horizontally. They are also used for reinforced block lintels. A-blocks and H-blocks are used to build walls with vertical reinforcing bars grouted into the cores in situations where there is insufficient space to lift the blocks over the tops of the projecting bars; one such situation is a concrete masonry backup wall that is built within the frame of a building, as seen in Figure 20.1.

Grades	
Grade N	General use above and below grade
Grade S	Above grade only, in exterior walls with weather-protective coatings, or in walls not exposed to weather

Types	
Type I	Moisture-controlled units, for use where drying shrinkage of units would cause cracking
Type II	Non-moisture-controlled units

Weights	
Normal weight	Made from concrete weighing more than 125 pcf (2000 kg/m³)
Medium weight	Made from concrete weighing between 105 and 125 pcf (1680–2000 kg/m³)
Lightweight	Made from concrete weighing 105 pcf or less (1680 kg/m³)

FIGURE 9.23
Grades, types, and weights of hollow concrete masonry units, from ASTM C90.

FIGURE 9.24
Concrete blocks and bricks can be cut very accurately with a water-cooled, diamond-bladed saw. For rougher sorts of cuts, a few skillful blows from the mason's hammer will suffice. (*Courtesy of Portland Cement Association, Skokie, Illinois*)

degrees of strength, and because their hollow cores allow for the easy insertion of reinforcing steel and grout, they are widely used in masonry bearing wall construction. Where surface qualities not available in concrete masonry are desired on a masonry wall, concrete blocks are often used for the backup wythe behind a brick or stone facing. Block walls also accept plaster, stucco, or tile work directly. Single-wythe exterior concrete masonry walls, like any solid masonry walls, tend to leak in wind-driven rains. Except in dry climates, they should be painted on the outside with masonry paint, or a cavity wall construction should be used instead, as described in Chapter 10.

Although innumerable special sizes, shapes, and patterns are available, American concrete blocks are standardized around an 8-inch (203.2-mm) cubic module. The most common block is nominally $8 \times 8 \times 16$ inches ($203.2 \times 203.2 \times 406.4$ mm). The actual size of the block is $7\frac{5}{8} \times 7\frac{5}{8} \times 15\frac{5}{8}$ inches ($193.7 \times 193.7 \times$

396.9 mm), which allows for a mortar joint $\frac{3}{8}$ inch (9.5 mm) thick. This size block is designed to be lifted and laid conveniently with two hands (as compared to a brick, which is designed to be laid with one). Its double-cube proportions work well for Running Bond stretchers, for headers, and for corners. While concrete masonry units can be cut with a diamond-bladed power saw (Figure 9.24), it is more economical and produces better results if the designer lays out buildings of concrete masonry in dimensional units that correspond to the module of the block (Figure 9.25). Nominal 4-inch, 6-inch, and 12-inch block thicknesses (101.6 mm, 152.4 mm, and 304.8 mm) are also common, as is a solid concrete brick that is identical in size and proportion to a modular clay brick. A handy feature of the standard 8-inch block height is that it corresponds exactly to three courses of ordinary clay or concrete brickwork, making it easy to interweave blockwork and brickwork in composite walls.

FIGURE 9.25
Concrete masonry buildings should be dimensioned to use uncut blocks except for special circumstances.

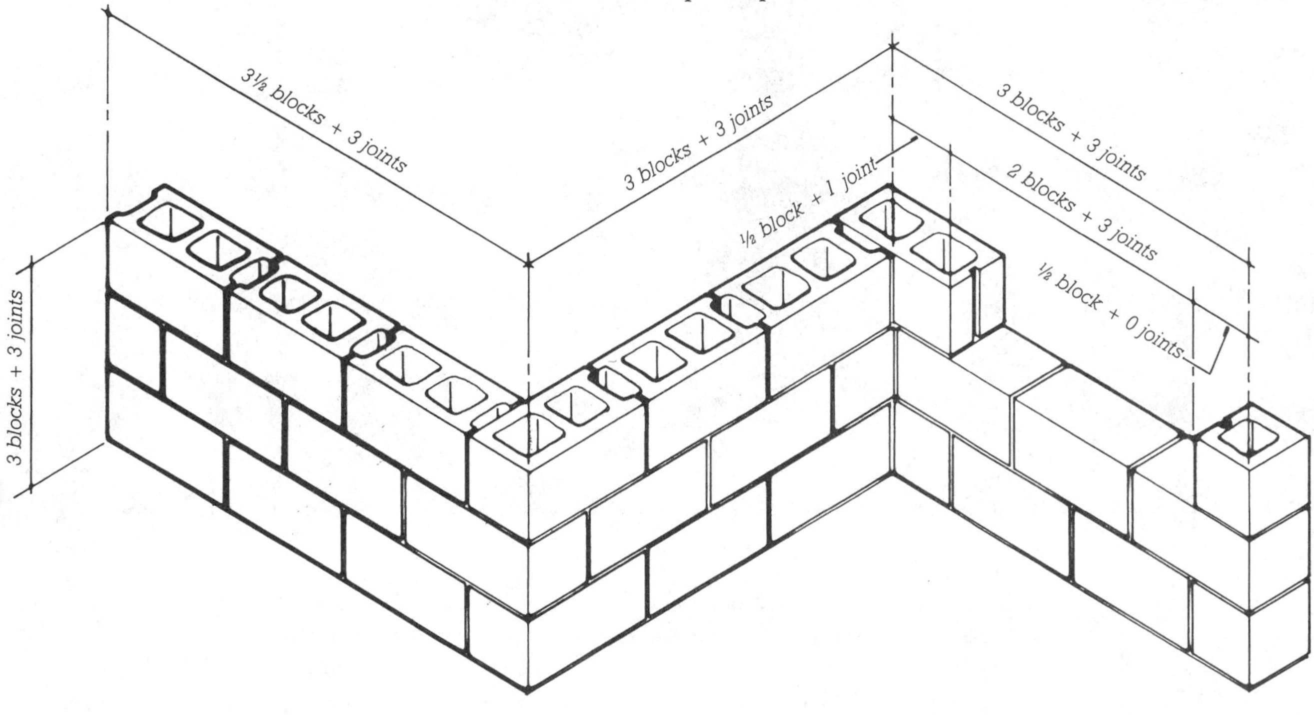

Laying Concrete Blocks

The accompanying photographic sequence (Figure 9.26) illustrates the technique for laying concrete block walls. The mortar is identical to that used in brick walls, but in most walls only the face shells of the block are mortared, with the webs left unsupported (Figure 9.26c and j).

Concrete masonry is often reinforced with steel to increase its load-bearing capacity and its resistance to cracking. Horizontal reinforcing is usually inserted in the form of welded

(a)

(b)

(c)

(d)

Figure 9.26
Laying a concrete masonry wall: (a) A bed of mortar is spread on the footing. (b) The first course of blocks for a lead is laid in the mortar. Mortar for the head joint is applied to the end of each block with the trowel before the block is laid. (c) The lead is built higher. Mortar is normally applied only to the face shells of the block and not to the webs. (d) As each new course is started on the lead, its

grids of small-diameter steel rods that are laid into the bed joints or mortar at the desired vertical intervals (Figure 9.27). If stronger horizontal reinforcing is required, bond beam blocks (Figure 9.22) or special blocks with channeled webs (Figure 9.28) allow reinforcing bars to be placed in the horizontal direction. The bars are then embedded in grout before the next course is laid. The grout is contained in the cores of the reinforced course by a strip of metal mesh that was previously laid into the bed joint beneath the course to bridge across

(e)

(f)

(g)

(h)

height is meticulously checked with either a folding rule or, as shown here, a story pole marked with the height of each course. (*e, f*) Each new course is also checked with a spirit level to be sure that it is level and plumb. Time expended in making sure the leads are accurate is amply repaid in the accuracy of the wall and the speed with which blocks can be laid between the leads. (*g*) The joints of the lead are tooled to a concave profile. (*h*) A soft brush removes mortar crumbs after tooling.   (*continued*)

(i)

(j)

(k)

(l)

FIGURE 9.26 (*continued*)

(*i*) A mason's line is held taut between the leads on line blocks. (*j*) The courses of blocks between the leads are laid rapidly, and are aligned only with the line; no story pole or spirit level is necessary. The mason has laid bed joint mortar and "buttered" the head joints for a number of blocks. (*k*) Each course of infill blocks is completed with a closer, which must be inserted between blocks that have already been laid. The head joints of the already-laid blocks are buttered. (*l*) Both ends of the closer blocks are also buttered with mortar, and the block is lowered carefully into position. Some touching up of the head joint mortar is often necessary. (*All photos courtesy of Portland Cement Association, Skokie, Illinois*)

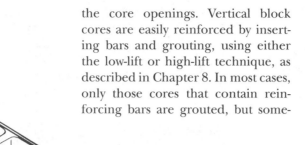

the core openings. Vertical block cores are easily reinforced by inserting bars and grouting, using either the low-lift or high-lift technique, as described in Chapter 8. In most cases, only those cores that contain reinforcing bars are grouted, but some-times all the vertical cores are filled, whether or not they contain bars, for added strength (Figure 9.29).

Lintels for concrete block walls may be made of steel angles, combinations of rolled steel shapes, reinforced concrete, or bond beam

FIGURE 9.27
Concrete masonry walls that are subjected only to moderate stresses can be reinforced horizontally with steel joint reinforcing, which is thin enough to fit into an ordinary bed joint of mortar. Vertical reinforcing is done with ordinary reinforcing bars grouted into the cores of the blocks. Horizontal joint reinforcing is available in both a "truss" pattern, as illustrated, and a "ladder" pattern. Both are equally satisfactory.

FIGURE 9.28
In this proprietary system for building more heavily reinforced concrete masonry walls, the webs are grooved to allow the insertion of horizontal reinforcing bars into the wall. The cores of the blocks are then grouted to embed the bars.
(*Courtesy of G. R. Ivany and Associates, Inc.*)

FIGURE 9.29
Grout is deposited in the cores of a reinforced concrete masonry wall using a grout pump and hose. (*Courtesy of Portland Cement Association, Skokie, Illinois*)

blocks with grouted horizontal reinforcing (Figure 9.30).

In recent years, *surface bonding* of concrete masonry walls has found application in certain low-rise buildings where the cost or availability of skilled labor is a problem. The blocks are laid without mortar, course upon course, to make a wall. Then a thin layer of a special cementitious compound containing short fibers of alkali-resistant glass is plastered on each side of the wall. This surface bonding compound, after it has cured, joins the blocks securely to one another both in tension and in compression. It serves also as a surface finish whose appearance resembles stucco.

Decorative Concrete Masonry Units

Concrete masonry units are easily and economically manufactured in an unending variety of surface patterns, textures, and colors intended for exposed use in exterior and interior walls. A few such units are diagrammed in Figure 9.33, and some of the resulting surface textures are depicted in Figures 9.32 and 9.34 through 9.36. Mold costs for producing special units are low when spread across the number of units required for medium to large buildings. Many of the textured concrete masonry units that are now considered standard originated as special designs created by architects for particular buildings.

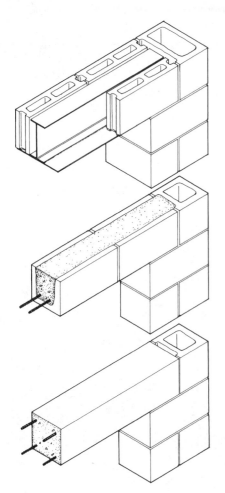

FIGURE 9.30
Lintels for opening in concrete masonry walls. At the top, a steel lintel for a broad opening is made up of a wide-flange section welded to a plate. Steel angle lintels are used for narrower openings. In the middle, a reinforced block lintel composed of bond beam units. At the bottom, a precast reinforced concrete lintel.

FIGURE 9.31
A small house by architects Clark and Menefee is cleanly detailed in ordinary concrete masonry units. A reinforced masonry retaining wall helps to tie the building visually into the site. (*Photo by Jim Rounsevell*)

FIGURE 9.32
A facade of split-face concrete masonry. (*Architects: Paderewski, Dean, Albrecht & Stevenson. Courtesy of National Concrete Masonry Association*)

Scored-face unit

Ribbed-face unit

Ribbed-face unit

FIGURE 9.33
Some decorative concrete masonry units, representative of literally hundreds of designs currently in production. The scored-face unit, if the slot in the face is filled with mortar and tooled, produces a wall that looks as if it were made entirely of half blocks. The ribbed split-face unit is produced by casting "Siamese twin" blocks joined at the ribs, then shearing them apart.

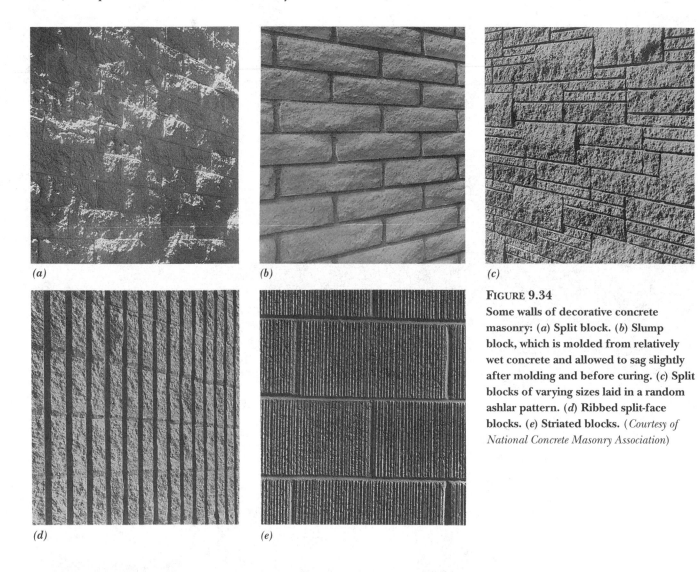

(a)

(b)

(c)

(d)

(e)

FIGURE 9.34
Some walls of decorative concrete masonry: (*a*) Split block. (*b*) Slump block, which is molded from relatively wet concrete and allowed to sag slightly after molding and before curing. (*c*) Split blocks of varying sizes laid in a random ashlar pattern. (*d*) Ribbed split-face blocks. (*e*) Striated blocks. (*Courtesy of National Concrete Masonry Association*)

FIGURE 9.35
Split-face blocks are used indoors in this high-school auditorium. (*Architects: Marcel Breuer and Associates. Courtesy of National Concrete Masonry Association*)

FIGURE 9.36
High-rise apartments constructed with reinforced bearing walls of fluted-face units.
Specially cast blocks were used to produce the curved balcony fronts. (*Architect: Paul*
Rudolph. Courtesy of National Concrete Masonry Association)

OTHER TYPES OF MASONRY UNITS

Bricks, stones, and concrete blocks are the most commonly used types of masonry units. In the past, hollow tiles of cast gypsum or fired clay were often used for partition construction (Figure 23.39), but both have been supplanted in the United States by concrete blocks, though hollow clay tiles are still widely used in other parts of the world. *Structural glazed*

FIGURE 9.37
Glass blocks used to enclose a conference room in a corporate office.
(*Courtesy of Pittsburgh Corning Corporation*)

FIGURE 9.38
Exterior use of glass blocks in the wall of a stairwell. (*Architect: Gwathmey/Siegel. Courtesy of Pittsburgh Corning Corporation*)

facing tiles of clay remain in use, especially for partitions where their durable, easily cleaned surfaces are advantageous, as in public corridors, toilet rooms, institutional kitchens, locker and shower rooms, and industrial plants (Figure 23.40). *Glass blocks,* available in many textures and in clear, heat-absorbing, and reflective glass, are enjoying a second era of popularity after nearly disappearing from the market (Figures 9.37, 9.38). *Structural terra cotta,* glazed or unglazed molded decorative units of fired clay, was widely used until the middle of the 20th century and is often seen on the facades of late-19th-century masonry buildings in the United States (Figure 8.43). Terra cotta had almost disappeared from the American materials market a few years ago, but restoration work on older buildings created a demand for it and generated a renewed desire to use it in new buildings, rescuing the last manufacturers from the brink of bankruptcy.

Autoclaved cellular concrete (ACC), though it has been manufactured and used in Europe for many years, has only recently begun to be made in North America. It is produced by mixing sand, lime, water, and a small amount of aluminum powder, and reacting these materials with steam to produce an aerated concrete that consists primarily of calcium silicate hydrates. ACC is available in blocks that are laid in mortar like other concrete masonry units. It is also made in such forms as unreinforced wall panels and steel-reinforced lintels and floor/ceiling panels. Because of its entrapped gas bubbles, which are created by the reaction of the aluminum powder with the lime, the density of ACC is similar to that of wood, and it is easily sawed, drilled, and shaped. It has moderately good thermal insulating properties. It is not as strong as normal-density concrete, but it is sufficiently strong to serve as loadbearing walls, floors, and roofs in low-rise construction. Walls of ACC are too porous to be left exposed; they are usually stuccoed on the exterior and plastered on the interior.

Masonry Wall Construction

Bricks, stones, and concrete masonry units are mixed and matched in both loadbearing and nonloadbearing walls of many types. The more important of these wall constructions are presented in the next chapter.

C.S.I./C.S.C. Masterformat Section Numbers for Stone and Concrete Masonry	
04100	**MORTAR AND MASONRY GROUT**
04150	**MASONRY ACCESSORIES**
	Anchors and Tie Systems
	Manufactured Control Joints
	Joint Reinforcement
04200	**UNIT MASONRY**
04220	**Concrete Unit Masonry**
04230	**Reinforced Unit Masonry**
04270	**Glass Unit Masonry**
04400	**STONE**
04410	**Rough Stone**
04420	**Cut Stone**
04440	**Flagstone**
04450	**Stone Veneer**
04455	**Marble**
04460	**Limestone**
04465	**Granite**
04470	**Sandstone**
04475	**Slate**
04500	**MASONRY RESTORATION AND CLEANING**

SELECTED REFERENCES

1. Indiana Limestone Institute of America, Inc. *Indiana Limestone Institute of America, Inc., Handbook.* Bedford, Indiana.

Updated frequently, this large booklet tells of the history and provenance of Indiana limestone, recommended standards and details for its use, and architectural case histories. (Address for ordering: Suite 400, Stone City Bank Building, Bedford, IN 47421.)

2. Marble Institute of America. *Dimensional Stone.* Farmington, Michigan.

Standards, specifications, and details for dimensional stonework are included in this large, periodically updated looseleaf volume. The emphasis is on marble. (Address for ordering: 33505 State Street, Farmington, MI 48024.)

3. Randall, Frank A., and William C. Panarese. *Concrete Masonry Handbook for Architects, Engineers, and Builders.* Skokie, Illinois, Portland Cement Association, 1991.

This is a clearly written, beautifully illustrated guide to every aspect of concrete masonry. (Address for ordering: 5420 Old Orchard Road, Skokie, IL 60076.)

4. National Concrete Masonry Association. *Architectural and Engineering Concrete Masonry Details for Building Construction.* Herndon, Virginia.

A treasury of typical concrete masonry details is contained in this 212-page book. (Address for ordering: NCMA, 2302 Horse Pen Road, Herndon, VA 22071.)

KEY TERMS AND CONCEPTS

igneous rock
sedimentary rock
metamorphic rock
granite
limestone
sandstone
freestone
quarry sap
travertine
brownstone

bluestone
slate
marble
sticking
waxing
liner
fieldstone
rubble stone
dimension stone
cut stone

ashlar
flagstone
jet burner
diamond wire
shop drawing
lewis
coursed
random
point
quarry bed

concrete masonry unit (CMU)
concrete block
surface bonding
structural glazed facing tiles
glass block
structural terra cotta
autoclaved cellular concrete (ACC)

REVIEW QUESTIONS

1. What are the major types of stone used in construction? How do their properties differ?

2. What sequence of operations would be used to produce rectangular slabs of polished marble from a large quarry block?

3. In what ways is the laying of stone masonry different from the laying of bricks?

4. What are the advantages of concrete masonry units over other types of masonry units?

5. How long is a wall that is made up of 22 concrete masonry units, each nominally 8 × 8 × 16 inches, joined with three-eighth-inch mortar joints?

6. How may horizontal and vertical steel reinforcement be introduced into a CMU wall?

EXERCISES

1. Design a stone masonry facade for a downtown bank building that is 32 feet wide and two stories high. Draw all the joints between stones on the elevation. Draw a detail section to show how the stones are attached to a concrete masonry backup wall.

2. Design a masonry gateway for one of the entrances to a college campus with which you are familiar. Choose whatever type of masonry you feel is appropriate, and make the fullest use you can of the decorative and structural potentials of the material. Show as much detail of the masonry on your drawings as you can.

3. Visit a masonry supply company and view all the types of concrete masonry units that they have available. What is the function of each?

4. Design a simple concrete masonry house that a student could build for his or her own use, keeping the floor area to 400 square feet (37 m²).

5. Design a decorative CMU that can be used to build a richly textured wall.

10

MASONRY LOADBEARING WALL CONSTRUCTION

- **Masonry Wall Types**
  Reinforced Masonry Walls
  Composite Masonry Walls
  Cavity Walls

- **Detailing Masonry Walls**
  Flashings and Weep Holes
  Thermal Insulation of Masonry Walls

- **Spanning Systems for Masonry Bearing Wall Construction**
  "Ordinary" Joisted Construction
  Heavy Timber or "Mill" Construction
  Steel and Concrete Decks with Masonry Bearing Walls

- **Some Special Problems of Masonry Construction**
  Expansion and Contraction
  Efflorescence
  Mortar Joint Deterioration
  Moisture Resistance of Masonry
  Cold and Hot Weather Construction

- **Masonry and the Building Codes**

- **The Uniqueness of Masonry**

America's rich fabric of 19th-century downtown buildings is made up largely of what building codes define as Ordinary construction: exterior masonry loadbearing walls with floors and roof spanned by wood joists. (*Photo by the author*)

Walls constructed of brick, stone, concrete masonry, or combinations of masonry units can be used to support roof and floor structures of wood light framing, heavy timber framing, steel, sitecast concrete, precast concrete, or masonry vaulting. Because these masonry *loadbearing walls* (usually called, simply, *bearing walls*) also serve as exterior walls and interior partitions, they are often a very economical system of construction as compared to systems that carry their structural loads on columns of wood, steel, or concrete. When used as exterior walls of buildings, masonry walls must be designed not only to support structural loads, but also to resist water penetration and the transfer of heat between indoors and outdoors.

MASONRY WALL TYPES

A masonry loadbearing wall may be classified in three different ways, according to how it is constructed:

- It may be *reinforced* or *unreinforced*.

- It may be constructed entirely of one type of masonry unit, or it may be a *composite wall*, made of two or more types of units.

- It may be a solid masonry wall or a masonry *cavity wall*.

Reinforced Masonry Walls

Loadbearing walls may be built with or without reinforcing. Unreinforced masonry walls cannot carry such high stresses as reinforced walls and are unsuitable for use in regions with high seismic risk, but such walls have been used in this country to support buildings as tall as 16 stories (Figure 8.37).

In a multistory building of masonry loadbearing wall construction, the top-floor walls support only the load of the roof. The walls on the floor below support the loads of the roof, the top floor, and the top-floor walls. Each succeeding story, counting downward from top to bottom, supports a greater load than the one above, and an unreinforced masonry bearing wall must therefore grow progressively thicker from top to bottom.

With steel reinforcing, this thickening can be reduced or eliminated entirely (Figure 8.36), with substantial savings in labor and materials, and much taller buildings may be constructed. Methods of reinforcing brick and concrete masonry walls have been described in Chapters 8 and 9, and are further illustrated in Figures 10.1, 10.2, 10.10, 10.11, and 10.12. The number, locations, and sizes of the steel reinforcing bars are determined by the structural engineer.

The engineering design of masonry loadbearing walls is governed by ACI 530/ASCE 5, a standard established jointly by the American Concrete Institute and the American Society of Civil Engineers. This document establishes quality standards for masonry units, reinforcing materials, metal ties and accessories, grout, and masonry construction. It also sets forth the procedures by which the strengths and stiffnesses of masonry structural elements are calculated.

Composite Masonry Walls

For an optimum balance between appearance and economy, solid masonry walls are often constructed with an outer wythe of stone or face brick and a backup wythe of hollow concrete masonry. The two wythes are bonded together either by steel horizontal joint reinforcing or by headers from the outer wythe that penetrate the backup wythe. Headers may penetrate completely through the backup or may interlock with courses of header blocks (Figure 9.22).

When designing any composite masonry wall, the designer should be sure that the differences in the thermal or moisture expansion characteristics of the two materials are not sufficient to cause warping or cracking of the wall. In composite loadbearing walls, the different strengths and elasticities of the two materials must be taken into account in locating the centroid of the wall and calculating the deformations it will experience under load.

Cavity Walls

Solid or composite walls are seldom used for exterior walls of buildings. Exterior masonry walls must resist water penetration and heat transfer and, for these reasons, are usually built with internal cavities. A masonry *cavity wall* consists of an inner, structural wythe and an outer wythe of masonry facing. The two are separated by a continuous airspace that is spanned only by corrosion-resistant metal ties that hold the wythes together (Figures 10.1*b, c, e, f,* 10.2, 10.3, 10.10, 10.11).

FIGURE 10.1
Masonry ties. These are only a few examples from among dozens of types available from a number of manufacturers to meet every conceivable need to tie wythes of masonry to one another or to supporting structures of wood, concrete, or steel. Corrugated ties (*a*) are widely used, but they have very little compressive strength and should be avoided. Z-ties (*b*) are simple and effective. Adjustable ties (*c*) are convenient for the masons because they allow for irregularities in course heights, but they may be insufficiently rigid for some structural purposes. The adjustable stone tie (*d*) also has excessive play unless the collar joint is filled with mortar. Ladder and truss ties (*e, f,* and *g*) are very stiff and effective. Dovetail slots (*h*) are cast into concrete walls and beams where a masonry veneer needs to be attached.

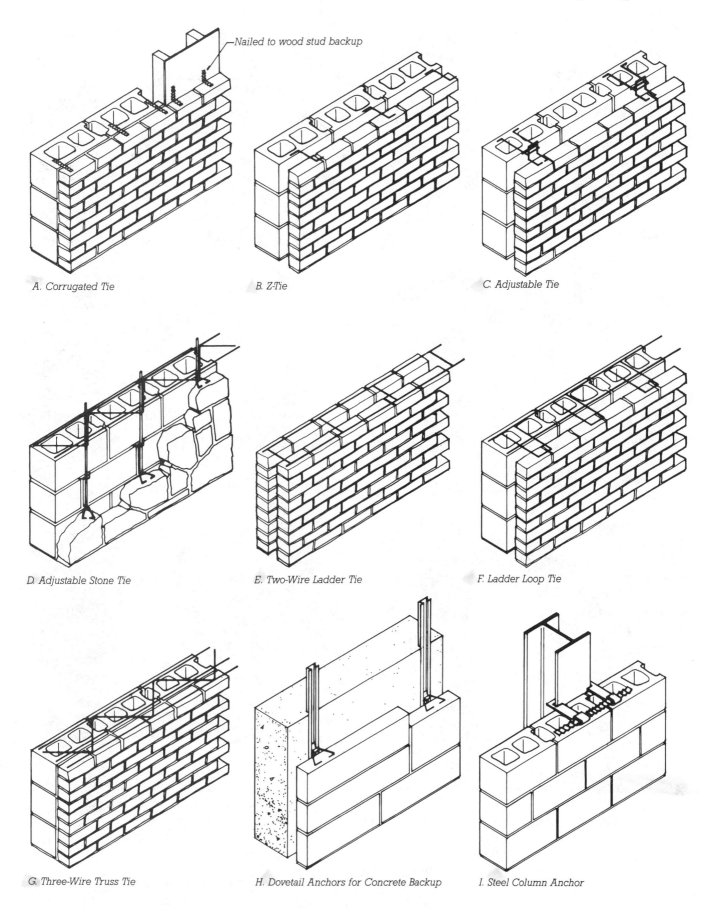

Nailed to wood stud backup

A. *Corrugated Tie*

B. *Z-Tie*

C. *Adjustable Tie*

D. *Adjustable Stone Tie*

E. *Two-Wire Ladder Tie*

F. *Ladder Loop Tie*

G. *Three-Wire Truss Tie*

H. *Dovetail Anchors for Concrete Backup*

I. *Steel Column Anchor*

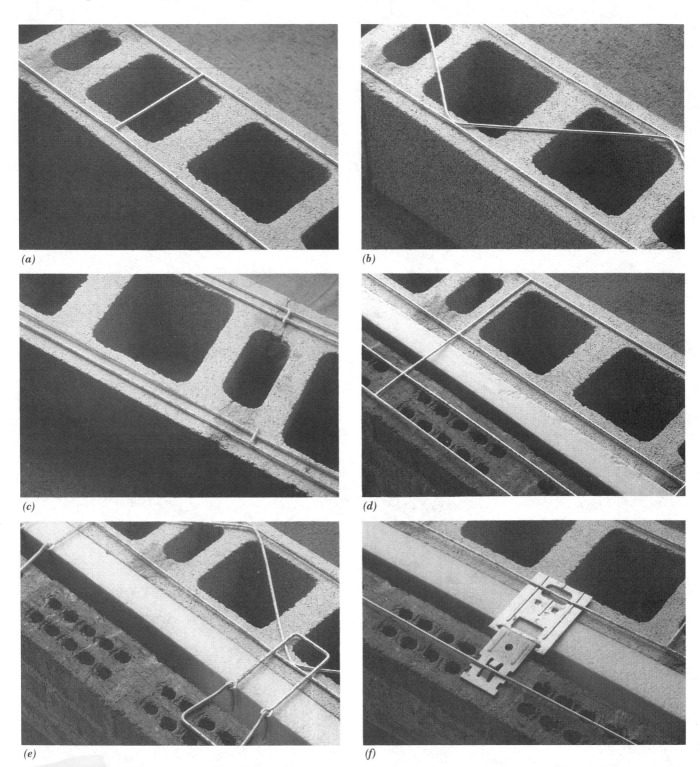

(a) (b)

(c) (d)

(e) (f)

FIGURE 10.2

Photographs of joint reinforcing and ties: (*a*) Ladder reinforcing for a single wythe of concrete masonry. (*b*) Truss reinforcing. (*c*) The double rods of seismic ladder reinforcing offer greater strength than the single rods of normal ladder reinforcing. (*d*) Ladder-type side rod reinforcing reaches across an insulated cavity to reinforce the face wythe of brick masonry and tie it to the backup wythe of concrete masonry. (*e*) Truss-type eye and pintle tie reinforcing allows for some adjustment in course heights. (*f*) Ladder-type seismic tab tie reinforcing allows the construction of the face wythe to lag behind that of the concrete masonry backup wythe, and still provides reinforcement and positive tying of the face wythe to the backup. The cavities in the last three photographs should be thicker, to provide an open cavity at least 2 inches (50 mm) wide outside the insulation boards. (*Courtesy of Dur-O-Wal Corporation*)

FOR PRELIMINARY DESIGN OF A LOADBEARING MASONRY STRUCTURE

• To estimate the size of a **reinforced brick masonry column,** add up the total roof and floor area supported by the column. A 12-inch (300-mm) column can support up to about 2000 square feet (185 m²) of area, a 16-inch (400-mm) column 3000 square feet (280 m²), a 20-inch (500-mm) column 4000 square feet (370 m²) and a 24-inch (600-mm) column 6000 square feet (560 m²). For unreinforced columns, increase these column dimensions by 30 percent.

• To estimate the size of a **reinforced concrete masonry column,** add up the total roof and floor area supported by the column. A 12-inch (300-mm) column can support up to about 1200 square feet (110 m²) of area, a 16-inch (400-mm) column 2000 square feet (185 m²), a 20-inch (500-mm) column 3500 square feet (325 m²), and a 24-inch (600-mm) column 5500 square feet (510 m²). For unreinforced columns, increase these column dimensions by 50 percent.

• To estimate the thickness of a **reinforced brick masonry loadbearing wall,** add up the total width of floor and roof decks that contribute load to a 1-foot length of wall. An 8-inch (200-mm) wall can support approximately 250 feet (75 m) of deck, a 12-inch (300-mm) wall 500 feet (150 m), a 16-inch (400-mm) wall 750 feet (230 m), and a 20-inch

(500-mm) wall 1000 feet (300 m). Increase these wall thicknesses by at least 25 percent if the walls are not reinforced.

• To estimate the thickness of a **reinforced concrete masonry loadbearing wall,** add up the total width of floor and roof decks that contribute load to a 1-foot length of wall. An 8-inch (200-mm) wall can support approximately 250 feet (75 m) of deck, a 12-inch (300-mm) wall 450 feet (140 m), and a 16-inch (400-mm) wall 600 feet (185 m). Increase these wall thicknesses by at least 25 percent if the walls are not reinforced.

These approximations are valid only for purposes of preliminary building layout, and must not be used to select final member sizes. They apply to the normal range of building occupancies such as residential, office, commercial, and institutional buildings, and parking garages. For manufacturing and storage buildings, use somewhat larger members.

For more comprehensive information on preliminary selection and layout of a structural system and sizing of structural members, see Allen, Edward, and Joseph Iano, *The Architect's Studio Companion* (2nd ed.), New York, John Wiley & Sons, Inc., 1995.

Every masonry wall is porous to some degree. Some water will find its way through even new masonry if the wall is wetted for a sustained period. Older masonry walls and walls that are imperfectly constructed will allow even more water to pass. A cavity wall prevents the water from reaching the interior of the building by interposing the cavity between the outside and inside wythes of the wall. When penetrating water reaches the cavity, it has no place to go but down. When it reaches the bottom of the cavity, it is caught by a thin, impervious membrane called a *flashing* and drained through *weep holes* to the exterior of the building.

To maintain the watertightness of a cavity wall, it is essential that no accidental bridges be created by mortar droppings that could conduct water across the cavity to the inner wythe.

The cavity should be at least 2 inches (50 mm) wide so that horizontal strips of wood can be suspended on wires in the cavity during construction to catch falling mortar and debris. The mason pulls the strips up periodically during construction to clean them, and removes them from the cavity when the wall has been completed.

DETAILING MASONRY WALLS

Flashings and Weep Holes

A flashing is a continuous sheet of impervious material that is used as a barrier against the passage of water. Two general types of flashings are used in masonry construction: *External flashings* prevent moisture

from penetrating into the masonry wall where the wall intersects the roof. *Internal flashings* (also known as *concealed flashings*) catch water that has penetrated a masonry wall and drain it through weep holes to the outdoors. Internal flashings should be placed at the bottom of the wall cavity and at every location where the cavity is interrupted: at heads of windows and doors, at window sills, at shelf angles, and over exposed spandrel beams (Figure 10.3). Internal flashings are installed by masons as they construct a wall.

The external flashing at the intersection of a flat roof and a wall parapet is usually constructed in two overlapping parts, a *base flashing* and a *counterflashing* or *cap flashing*. This makes installation easier and allows for some movement between the wall and roof components. Often the base

[handwritten note: → MUST GO IN ANY TYPE OF WALL ORDERING]

flashing is formed by the roof membrane itself. The base flashing should be turned up for a height of at least 8 inches (200 mm).

An internal flashing should be turned up 6 to 9 inches (150 to 225 mm) at the interior face of the wall cavity. It should penetrate at least 2 inches (50 mm) into the interior wythe. If the cavity is backed up by a concrete beam or wall, it may terminate in a *reglet*, a horizontal slot formed in the face of the concrete (Figure 10.3c). At the outside face of the wall, the flashing should be carried at least $\frac{3}{4}$ inch (19 mm) beyond the outside face of the wall and turned down at a 45° angle so that draining water drips free of the wall. It is a common but dangerous practice to terminate the flashing just inside the face of the wall for a better appearance; this can cause the draining water to be reabsorbed by the masonry.

An internal flashing is always accompanied by weep holes. The holes should lie immediately on top of the flashing to keep the bottom of the cavity as dry as possible. They should be installed not more than 24 inches (600 mm) apart horizontally in brick, and 32 inches (800 mm) apart in concrete masonry. The minimum diameter for a weep hole is $\frac{1}{4}$ inch (6 mm). A weep hole is created with a short piece of rope laid in the mortar joint and later pulled out (Figures 10.4, 10.6b), by a plastic or metal tube laid in the mortar, or by simply leaving a head joint unmortared. Plastic and metal weep hole accessories with insect screens are available for installation in unmortared head joints, to prevent bees or other insects from taking up residence in the cavity.

Flashings may be made of sheet metal, plastics, elastomeric compounds, or composite materials. Sheet metal flashings are the most durable and the most expensive. Copper and stainless steel are suitable metals; unsuitable ones are galvanized steel, which has a very limited

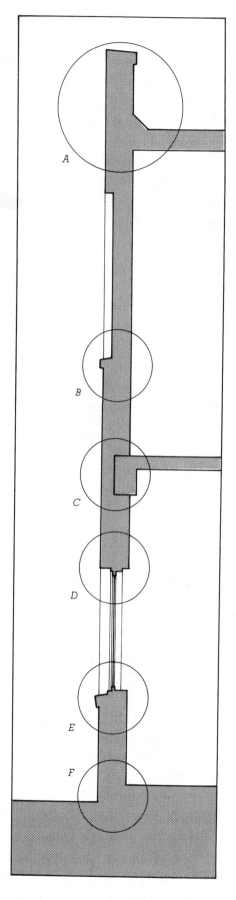

FIGURE 10.3

Typical flashings in masonry walls. The letters on the full-height wall section on this page refer to the large-scale details on the facing page. The flashings in a parapet (*a*) can be thought of as external flashings because they do not drain the cavity, and they do not usually have weep holes associated with them. The edge of the roof membrane is turned up to form its own flashing against the parapet wall, with a metal counterflashing emerging from the masonry and overlapping from above. A counterflashing is often made in two interlocking pieces that make it easier to install. The reglet used in the second shelf angle detail (*c*) is a length of formed metal or molded plastic that is attached to the inside of the formwork before the concrete is poured. The end dam shown in the center drawing prevents water trapped by the flashing from draining back into the cavity rather than out through the weep holes.

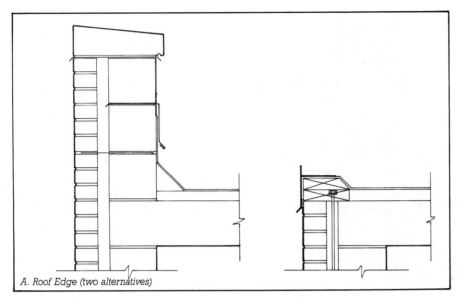

A. Roof Edge (two alternatives)

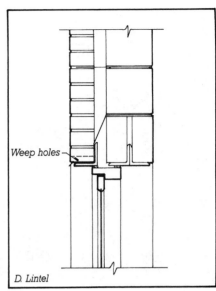

Weep holes

D. Lintel

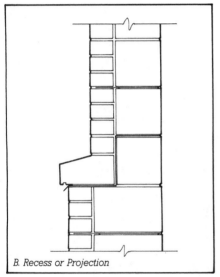

B. Recess or Projection

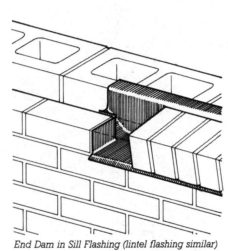

End Dam in Sill Flashing (lintel flashing similar)

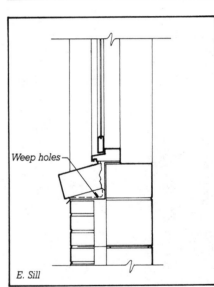

Weep holes

E. Sill

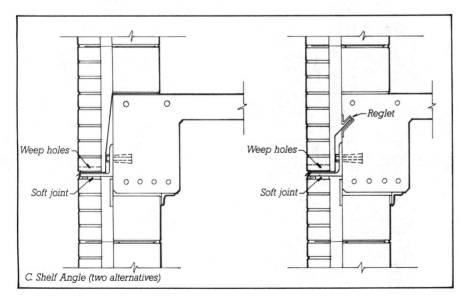

Weep holes

Reglet

Weep holes

Soft joint

Soft joint

C. Shelf Angle (two alternatives)

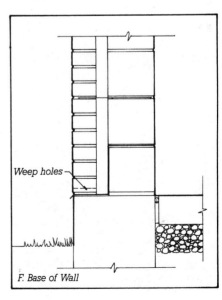

Weep holes

F. Base of Wall

FIGURE 10.4
Flashing a window sill. (*a*) The flashing sheet is turned up to form a dam at the end. Because the flashing is too flexible to project beyond the face of the wall, it should be trimmed back slightly inside the face and cemented to the top of a strip of sheet copper or stainless steel that projects beyond the face and turns down to form a drip. (*b*) Pieces of rope form weep holes in the bed joint beneath the sloping rowlock sill bricks. These will be pulled out after the mortar has stiffened. (*c*) The finished sill. (*Courtesy of Brick Institute of America*)

FIGURE 10.5
Copper flashings in a parapet. The single-ply roof membrane is turned up to self-flash under the copper counterflashing. (*Courtesy of Copper Development Association, Inc.*)

(a)

(b)

FIGURE 10.6
Flashing the base of a cavity wall. (*a*) The mason uses a mastic adhesive to seal the lap joints in the flashing at the corners. (*b*) A piece of rope forms a weep hole. The flashing projects beyond the face of the brick, but as in Figure 10.4, the projecting portion should be made of sheet metal. (*Courtesy of Brick Institute of America*)

life, and aluminum and lead, which react chemically with mortar. Plastics are the least expensive flashing materials, and some are very durable, but others (such as some PVC formulations) can deteriorate in sunlight where they emerge from the wall. Some insufficiently tested plastic flashing materials have even deteriorated within the wall. When using plastic flashings, the designer should specify only those materials that have been proven suitable for the purpose. Composite materials combine layers of material to produce flashings that are intermediate in price. Most are built around a foil or thin sheet of copper or lead, laminated with such materials as polyester film, glass fiber mesh, bitumen-coated fabric, or waterproofed kraft paper, and many are very durable.

Flashings within a wall are almost impossible to replace if they should fail in service. Even the most expensive flashing materials constitute a very small fraction of the cost of a masonry wall, so the designer should not risk using untried materials in a misguided effort to save money.

Careful supervision is necessary to be sure that flashings are properly installed. At corners and other junctions, the pieces of flashings should be lapped at least 6 inches (150 mm) and sealed with a suitable mastic (Figure 10.6). Head and sill flashings should extend well beyond the jambs of the opening and should terminate in folded end dams (Figures 10.3, 10.4*a*). End dams should also be formed at such interruptions of the wall as exposed columns and expansion or control joints.

Wall cavities should be inspected just before they are closed to be sure that the weep holes have not been obstructed by mortar droppings. Mortar-clogged weep holes have caused water leakage through masonry walls since the cavity wall was invented, and many attempts have been made to make weep holes clogproof. Some architects specify that a layer of $3/8$-inch-diameter (9 mm) gravel should be deposited in the base of the cavity, but research has shown that this is not only ineffectual, but actually tends to prevent water from flowing out even if the cavity is clean. Several proprietary systems utilize open-weave drainage matting of the type used around basement walls: One system uses a layer only at the base of the wall, and another fills the entire cavity with matting. These appear to work well, although the matting does retain slightly more water than a clean cavity.

Thermal Insulation of Masonry Walls

A solid masonry wall is a good conductor of heat, which is another way of saying that it is a poor insulator. In many hot, dry climates, the capacity of an uninsulated masonry wall to store heat and retard its passage is effective in keeping the inside of the building cool during the hot day and warm during the cold night, but in climates with sustained cold or hot seasons, measures must be taken to improve the insulating qualities of masonry walls. The introduction of an empty cavity into a wall improves its thermal insulating properties considerably, but not to a level fully sufficient for cold climates.

There are three general ways of insulating masonry walls: on the outside face, within the wall, and on the inside face. Insulation on the outside face is a relatively recent development. It is usually accomplished by means of an *exterior insulation and finish system* (*EIFS*), which consists of panels of plastic foam that are adhered to the masonry and covered

FIGURE 10.7
Insulating the cores in a concrete block wall with a dry fill insulation. The insulation is noncombustible, inorganic, nonsettling, and treated to repel water that might be present in the block cores from condensation or leakage. (*Courtesy of W. R. Grace & Co.*)

with a thin, continuous layer of polymeric stucco reinforced with glass fiber mesh. The appearance is that of a stucco building. The masonry is completely concealed, and can be of inexpensive materials and workmanship. EIFS is frequently used for insulating existing masonry buildings in cases where the exterior appearance of the masonry does not need to be retained. An advantage of EIFS is that the masonry is protected from temperature extremes, and can function effectively to stabilize the interior temperature of the building. Disadvantages are that the thin stucco coatings are usually not very resistant to denting or penetration damage, and that the EIFS is combustible. Furthermore, most exterior insulation and finish systems have no internal flashings or drainage, so that puncture damage and lapses in workmanship, especially around edges and openings, can lead to substantial moisture leakage. Figure 10.12 details a building insulated in this manner, and photographs of the installation of exterior insulation and finish systems are shown in Figure 20.23.

Insulation within the wall can take several forms. If the cavity in a wall is made sufficiently wide, the masons can insert slabs of plastic foam insulation against the inside wythe of masonry as the wall is built (Figure 10.2). The overall cavity width should be adjusted so that the net width of the airspace is at least 2 inches (50 mm), the minimum dimension that a mason can keep clean. The hollow cores of a concrete block wall can be filled with loose granular insulation (Figure 10.7) or with special molded-to-fit liners of foam plastic (Figure 10.8). Insulating the cores of concrete blocks does not retard the passage of heat through the webs of the blocks, however, and is most effective when it is coupled with an unbroken layer of insulation in the cavity or on one face of the wall.

The inside surface of a masonry wall can be insulated by adhering slabs of plastic foam to the wall and applying plaster directly to the foam.

FIGURE 10.8
This proprietary concrete masonry system is designed with polystyrene foam inserts that provide a high degree of thermal insulation. The webs of the blocks are minimal in area to reduce thermal bridging where they pass through the foam. Vertical reinforcing bars can be placed as shown, and horizontal bars can be laid in the grooved webs. The cores must be grouted where reinforcing is used. (*Courtesy of Korfil, Inc., West Brookfield, Massachusetts*)

More usual is the practice of attaching wood or metal *furring strips* to the inside of the wall with masonry nails or powder-driven metal fasteners (Figures 10.9, 10.10, 23.5–23.7). Furring strips are analogous to wood or metal studs in light frame construction and are insulated and finished in much the same manner. They may be of any desired depth to house the necessary thickness of fibrous or foam insulation. Furring strips can also solve another chronic problem of masonry construction by creating a space in which electrical wiring and plumbing can easily be concealed.

SPANNING SYSTEMS FOR MASONRY BEARING WALL CONSTRUCTION

"Ordinary" Joisted Construction

So-called *Ordinary construction*, in which the floors and roof are framed with wood joists and rafters and supported at the perimeter on masonry walls, is the fabric of which American center cities largely were built in the 19th century. It still finds use today in a small percentage of new buildings and is listed as Type 3A and Type 3B construction in the building code table in Figure 1.1. Ordinary construction is essentially balloon framing (Figure 5.2) in which the outer walls of wood are replaced with masonry bearing walls. Balloon framing of the interior loadbearing partitions is used instead of platform framing because it minimizes the sloping of floors that might be caused by wood shrinkage along the interior

lines of support. Figure 10.10 shows the essential features of Ordinary construction. Notice two very important details: the *firecut* ends of the joists, and the metal anchors used to tie the wood framing and the masonry wall together. As might be expected from the combustibility of its interior construction, building codes severely restrict the size and extent of a building that can be built of Ordinary construction (Figure 1.1).

Heavy Timber or "Mill" Construction

Heavy Timber or *"Mill" construction*, listed as Type 4 in the code table (Figure 1.1), is similar in concept to Ordinary construction, but uses heavy timbers rather than light joists, rafters, and studs, and thick timber decking rather than thin sheathing and subflooring. Because heavy timbers are slower to catch fire and burn than nominal 2-inch (38-mm) framing members, Mill construction receives more favorable treatment under the code than Ordinary construction. Mill construction is discussed in detail in Chapter 4 and is illustrated in Figures 4.5 through 4.13.

Steel and Concrete Decks with Masonry Bearing Walls

Spanning systems of structural steel, sitecast concrete, and precast concrete are frequently used in combination with masonry bearing walls. Figures 10.11 and 10.12 show representative details of two of these combinations. Depending on the degree of fire resistance of the spanning elements, these constructions may be classified under Type 1 or Type 2 in the table in Figure 1.1.

FIGURE 10.9
A worker installs foam plastic insulation on the interior of a concrete masonry wall, using a patented system of steel furring strips in which only isolated clips contact the masonry wall so as to minimize thermal conduction through the metal. The furring strips serve as a base to which interior finish panels such as gypsum board can be attached. Other furring systems for masonry walls are shown in Figures 23.5 through 23.7. (*Courtesy of W. R. Grace & Co.*)

FIGURE 10.10
Ordinary construction, shown here with a cavity wall of brick with a concrete masonry loadbearing wythe. Thermal insulation is installed between wood furring strips on the interior side. The interior of a building of Ordinary construction is framed with balloon framing of wood studs and joists. At the perimeter of the building, the joists are supported by the masonry wall. The rowlock window sill detailed here is similar to the one whose construction is shown in Figure 10.4.

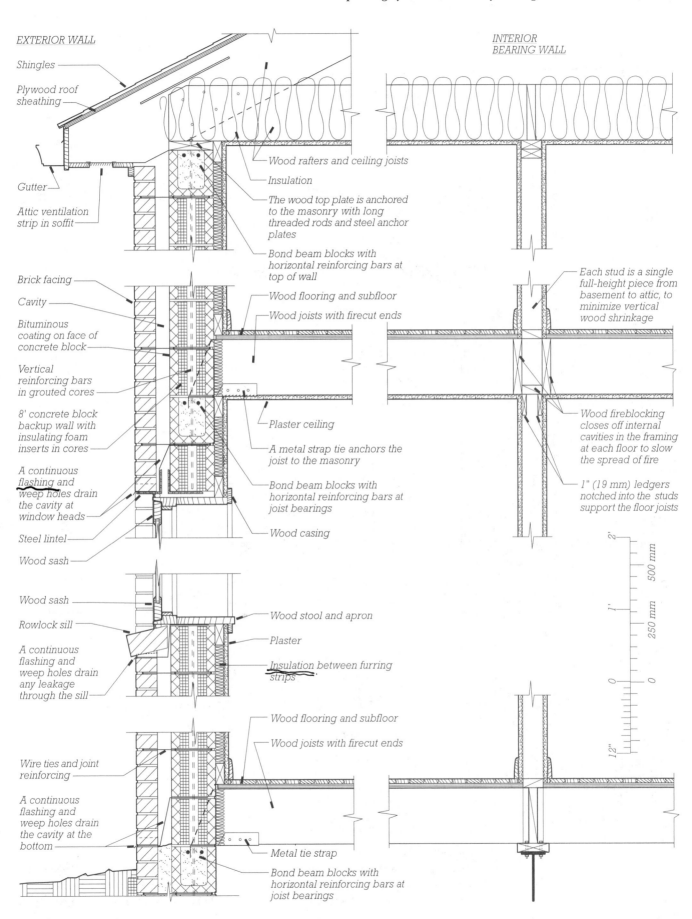

EXTERIOR WALL

- Shingles
- Plywood roof sheathing
- Gutter
- Attic ventilation strip in soffit
- Brick facing
- Cavity
- Bituminous coating on face of concrete block
- Vertical reinforcing bars in grouted cores
- 8' concrete block backup wall with insulating foam inserts in cores
- A continuous flashing and weep holes drain the cavity at window heads
- Steel lintel
- Wood sash
- Wood sash
- Rowlock sill
- A continuous flashing and weep holes drain any leakage through the sill
- Wire ties and joint reinforcing
- A continuous flashing and weep holes drain the cavity at the bottom

- Wood rafters and ceiling joists
- Insulation
- The wood top plate is anchored to the masonry with long threaded rods and steel anchor plates
- Bond beam blocks with horizontal reinforcing bars at top of wall
- Wood flooring and subfloor
- Wood joists with firecut ends
- Plaster ceiling
- A metal strap tie anchors the joist to the masonry
- Bond beam blocks with horizontal reinforcing bars at joist bearings
- Wood casing
- Wood stool and apron
- Plaster
- Insulation between furring strips
- Wood flooring and subfloor
- Wood joists with firecut ends
- Metal tie strap
- Bond beam blocks with horizontal reinforcing bars at joist bearings

INTERIOR BEARING WALL

- Each stud is a single full-height piece from basement to attic, to minimize vertical wood shrinkage
- Wood fireblocking closes off internal cavities in the framing at each floor to slow the spread of fire
- 1" (19 mm) ledgers notched into the studs support the floor joists

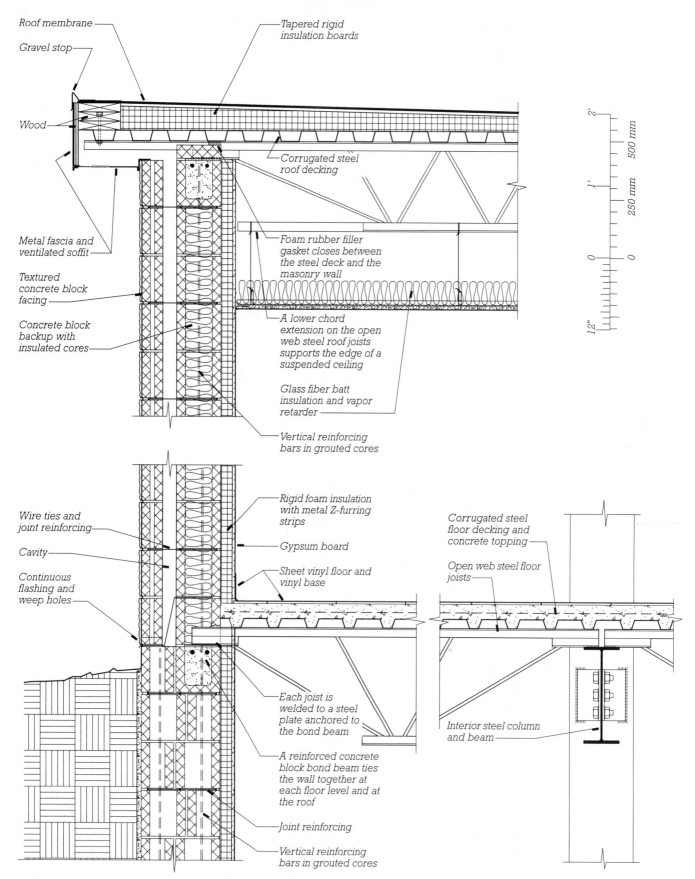

Roof membrane

Gravel stop

Wood

Metal fascia and
ventilated soffit

Textured
concrete block
facing

Concrete block
backup with
insulated cores

Tapered rigid
insulation boards

Corrugated steel
roof decking

Foam rubber filler
gasket closes between
the steel deck and the
masonry wall

A lower chord
extension on the open
web steel roof joists
supports the edge of a
suspended ceiling

Glass fiber batt
insulation and vapor
retarder

Vertical reinforcing
bars in grouted cores

Wire ties and
joint reinforcing

Cavity

Continuous
flashing and
weep holes

Rigid foam insulation
with metal Z-furring
strips

Gypsum board

Sheet vinyl floor and
vinyl base

Corrugated steel
floor decking and
concrete topping

Open web steel floor
joists

Each joist is
welded to a steel
plate anchored to
the bond beam

A reinforced concrete
block bond beam ties
the wall together at
each floor level and at
the roof

Joint reinforcing

Vertical reinforcing
bars in grouted cores

Interior steel column
and beam

FIGURE 10.11

An example of a concrete masonry exterior bearing wall with roof and floor of steel open-web joists and corrugated steel decking. With suitable fire protection for the steel components of the structure, this type of construction could be built many stories high.

326

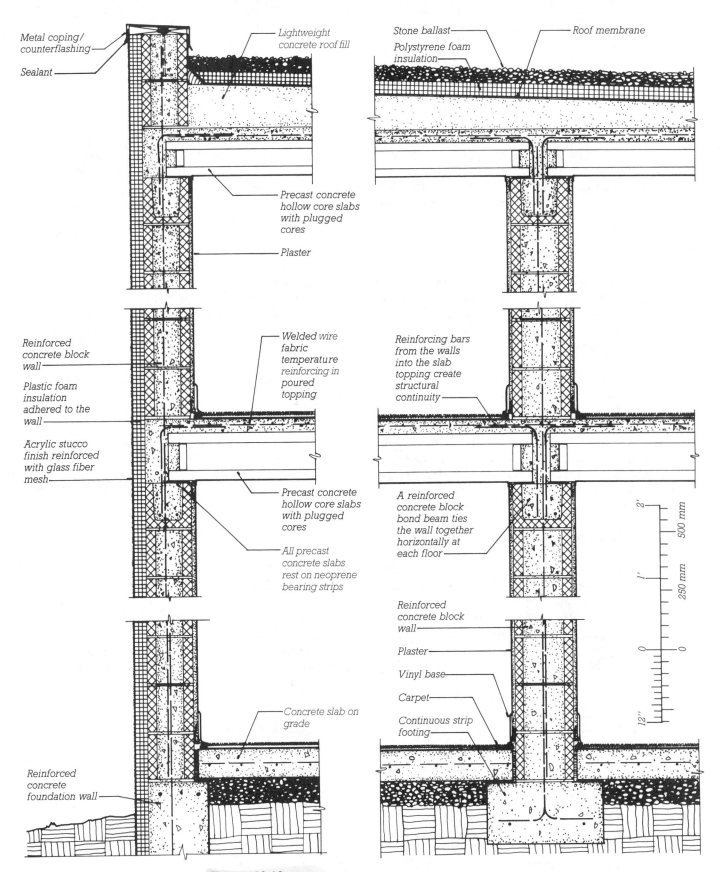

Metal coping/counterflashing

Sealant

Lightweight concrete roof fill

Stone ballast

Polystyrene foam insulation

Roof membrane

Precast concrete hollow core slabs with plugged cores

Plaster

Reinforced concrete block wall

Plastic foam insulation adhered to the wall

Acrylic stucco finish reinforced with glass fiber mesh

Welded wire fabric temperature reinforcing in poured topping

Reinforcing bars from the walls into the slab topping create structural continuity

Precast concrete hollow core slabs with plugged cores

All precast concrete slabs rest on neoprene bearing strips

A reinforced concrete block bond beam ties the wall together horizontally at each floor

Reinforced concrete block wall

Plaster

Vinyl base

Carpet

Continuous strip footing

Concrete slab on grade

Reinforced concrete foundation wall

2'   500 mm

1'   250 mm

0   0

12"

FIGURE 10.12

An example of concrete block exterior and interior bearing walls with precast concrete hollow-core slabs spanning the roof and floors. This system can be used for multistory buildings. Full wall reinforcing is shown, with EIFS applied to the outside of the building. For further details of this exterior finishing system, see Chapter 20, especially Figure 20.23.

MOVEMENT JOINTS IN BUILDINGS

Building materials and buildings are constantly in motion. Many of these motions are cyclical and never-ending: All materials shrink as they grow cooler and expand as they become warmer, each material doing so at its own characteristic rate (see Figure 10.13 and the table of coefficients of thermal expansion in the Appendix). Most porous materials grow larger when wetted by water or humid air, and smaller when they dry out, also at rates that vary from one material to another. These cyclical motions can occur on a seasonal basis (warm and moist in summer, cold and dry in winter), and they can also occur in much shorter cycles (warm days, cool nights; warm when the sun is shining on a surface, cool when a cloud covers the sun). Structural deflections under load can be very long term for dead loads and for floors supporting stored materials in a warehouse. Deflections can be medium term for snow on a roof, and very short term for walls resisting gusting winds. Some motions are onetime phenomena: Concrete and stucco shrink as they cure and dry out, while gypsum plaster expands upon curing. Concrete columns shorten slightly and concrete beams and slabs sag a bit due to plastic creep of the material during the first several years of a building's life; then they stabilize. Posttensioned slabs and beams grow measurably shorter as they take up the compression induced by the stretched steel tendons. Soil compresses under the pressure of the foundations of a new building, and then, in most cases, stops moving.

Chemical processes can cause movement in building components: If a steel reinforcing bar rusts, it expands, cracking the concrete around it. Solvent-release sealants shrink as they cure. Some plastics shrink and crack upon prolonged exposure to the sunlight. Motion can also be caused by the freezing expansion of water, as happens in the upward heaving of insufficiently deep footings during a cold winter, or in the spalling of concrete and masonry surfaces exposed to wetting and freezing.

All these motions are small in magnitude. They are largely unpreventable. But they occur in buildings of every size and every material, and if they are ignored in design and construction, they can tear the building apart, cracking brittle materials and applying forces in unanticipated ways to both structural and nonstructural building components that can cause them to fail.

We accommodate these inevitable movements in buildings in two different ways: In some cases, we strengthen a material to enable it to resist the stress that will be caused by an anticipated movement, as we do in adding shrinkage–temperature steel to a concrete slab. In most cases, however, we install *movement joints* in the fabric of the building that are designed to accommodate the movements without distress. We locate these joints in places where we anticipate maximum potential distress from expected movements. We also place them at regular intervals in large surfaces and assemblies to relieve movement-caused stresses before they can cause damage. A building that is not provided suitably with movement joints will make its own joints by cracking and spalling at points of maximum stress, creating a situation that is unsightly at best, and sometimes dangerous or catastrophic.

Types of Building Joints

The usual terminology applied to joints in buildings is confusing and often contradictory. The term "expansion joint," for example, is commonly and erroneously applied to almost any type of movement joint.

A more logical system of terminology is diagrammed in Figure A. This establishes two broad classifications,

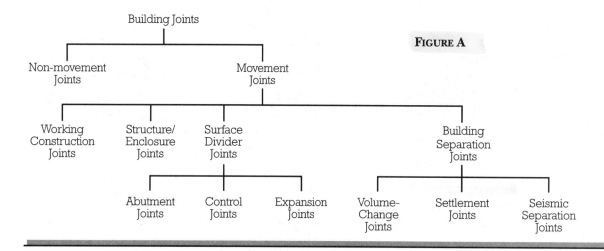

FIGURE A

movement joints and *nonmovement joints.* Nonmovement joints include most types of joints that are used to connect pieces of material in a building, joints such as the nailed connections in the wooden frame of a house, mortar joints between masonry units, welded and bolted connections in a steel frame, and joints between pours of concrete. A nonmovement joint can be made to move only by distressing one or more components of the joint, as in the pulling apart of a nailed connection, the slipping of steel members in a bolted connection, or the cracking of a weld, a mortar joint, or a concrete slab.

Movement joints are of many different kinds. What they share in common is a designed-in ability to adjust to expected amounts of motion without distress.

• The simplest movement joints are *working construction joints,* which are designed into various building materials and created in the normal process of assembling a building. An excellent example is the ordinary shingled roof, which is made up of small units of material that are applied in an overlapping pattern so that small amounts of thermal or moisture movement in the underlying roof structure or in the shingles themselves can be tolerated without distress (Figure 16.43). Other examples are wood bevel siding (Figure 6.13), which is nailed in such a way that moisture expansion and contraction are provided for; the metal clips and pans from which a sheet metal roof is assembled (Figure 16.52), which slip as necessary to allow for thermal movement; and most types of sealant joints and glazing joints (Chapters 17 and 19).

• *Structure/enclosure joints* separate structural from nonstructural elements so that they will act independently. A simple example is the sealant joint used at the top of an interior partition (Figure 23.23*a*), which assures that the partition will bear no structural load even if the floor above should sag. An important example of a structure/enclosure joint is the "soft joint" that is placed just beneath a shelf angle that supports a masonry veneer (Figures 10.3, 20.1); like the joint at the top of a partition, it prevents a nonstructural element (in this case a brick facing) from being subjected to a structural load for which it is not designed. Many other cladding attachment details shown in Chapters 19–21 are designed to allow the structural frame of the building and the exterior skin to move independently of one another, and these attachments are always associated with soft sealant joints in the skin panels. Yet another example of the structure/enclosure joint is the joint that is provided around the edge of a basement floor slab to allow for separate movement of the loadbearing wall and the slab (Figure 5.4). This type of joint is often called an *isolation joint* because it isolates

the adjacent components from each other so they can move independently.

• *Surface divider joints,* as their name implies, are used to accommodate movement in the plane of a floor, wall, ceiling, or roof. Surface divider joints can be further classified as *abutment joints, control joints,* and *expansion joints.*

• *Abutment joints* separate new construction from old construction. They are used when an existing building is altered or enlarged, to allow normal amounts of onetime movement to take place in the new materials without disturbing the original construction. If an existing brick wall is extended horizontally with new masonry, for example, a vertical abutment joint should be provided between the old work and the new, rather than trying to interlock the new courses of brick with the old. The last drawing in Figure 10.17 is an abutment joint. Abutment joints are sometimes called "construction joints" or "isolation joints."

• *Control joints* are deliberately created lines of weakness along which cracking will occur as a surface of brittle material shrinks, relieving the stresses that would otherwise cause random cracking. The regularly spaced grooves across concrete sidewalks are control joints; they serve to channel the cracking tendency of the sidewalk into an orderly pattern of straight lines rather than random jagged cracks. Elsewhere in this book, control joint designs are shown for concrete floor slabs (Figure 14.3*c*), concrete masonry walls (Figure 10.17), and plaster (Figure 23.13).

• *Expansion joints* are open seams that can close slightly to allow expansion to occur in adjacent areas of material. Expansion joints in brick walls permit the bricks to expand slightly under moist conditions (Figure 10.17). Expansion joints in aluminum curtain wall mullions (Figure 21.11) allow the elements of the wall to increase in size when warmed by sunlight.

Control joints and expansion joints should be located at geometric discontinuities such as corners, changes in the height or width of a surface, or openings. In long or large surfaces, they should also be spaced at intervals that will relieve the expected stresses in the material before those stresses rise to levels that can cause damage.

• *Building separation joints* divide a large or geometrically complex building mass into smaller, discrete structures that can move independently of one another. Building separation joints can be classified into three types:

• Large-scale effects of expansion and contraction caused by temperature and moisture are relieved by *volume-change joints.* These are generally placed at horizontal or vertical discontinuities in the massing of the building, where cracking would be most likely to occur (Figure B). They are also located at intervals of 150 to 200 feet (40 to 60 m) in very long buildings, the exact dimension depending on the nature of the materials and the rate at which dimensional changes occur.

• *Settlement joints* are designed to avoid distress caused by different rates of anticipated foundation settlement between different portions of a building, as between a high-rise tower and a connected low-rise wing, or between portions of a building that bear on different soils or have different types of foundations.

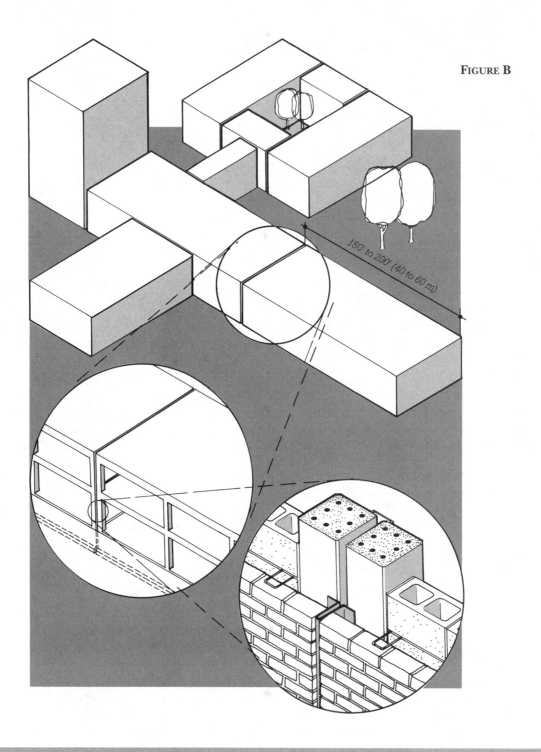

FIGURE B

150' to 200' (40 to 60 m)

• *Seismic separation joints* are used to divide a geometrically complex building into smaller units that can move independently of one another during an earthquake. (Buildings in seismic zones should also be detailed with structure/enclosure joints that permit the frame of the building to deform during an earthquake without damaging brittle cladding or partition elements.)

Building separation joints are created by constructing independent structures on either side of the plane of the joint, usually with entirely separate foundations, columns, and slabs (Figure B). Each of these independent structures is small enough and compact enough in its geometry that it is reasonable to expect that it will move as a unit in response to the forces that are expected to act upon it.

Detailing Movement Joints

The first imperative in detailing a movement joint is to determine what type or types of movement the joint must be designed to accommodate. This is not always simple. Often the same joint is called upon to perform simultaneously in several of the ways that are outlined here—as a volume-change joint, a settlement joint, and a seismic joint, for example. A joint in a composite masonry wall may serve as both an expansion joint for the brick facing and a control joint for the concrete masonry backup. Once the function or functions of a joint have been determined, the expected character and magnitude of motion can be estimated with the aid of standard technical reference works, and the joint designed accordingly.

It is important that any structural materials that would restrict movement be discontinued at a movement joint. Reinforcing bars or welded wire fabric should not extend through a control joint. Expanded metal lath is interrupted at control joints in plaster or stucco. The primary loadbearing frame of a building is interrupted at building separation joints. At the same time, it is often important to detail a movement joint so that it will maintain a critical alignment of one sort or another. Figure 10.17 shows several expansion and control joints that use interlocking masonry units or hard rubber gaskets to avoid out-of-plane movement across the joint. At control joints in concrete slabs, smooth, greased steel dowels are often inserted across the joint at close intervals at the midheight of the slab; these permit the joint to open up while assuring that the slab will remain at the same level on both

sides of the joint. The curtain wall mullion in Figure 21.11 allows for movement along one axis while maintaining alignment along the two other axes.

Joints must be designed to stop the passage of heat, air, water, light, sound, and fire. Some must carry traffic, as in the case of joints in floors or bridge pavements. All must be durable and maintainable, while simultaneously adjusting to movement and maintaining an acceptable appearance. Each joint must be detailed to allow for the expected direction and extent of movement: Some joints will have to operate only in a push–pull manner, while others can be expected to have to accommodate a shearing motion or even a twisting motion as well. The exterior joint closure is usually by means of a bellows of metal or synthetic rubber (Figure B). Some typical interior joint closures are shown in Figure 22.4. A perusal of manufacturers' catalogs classified under section 05800 of the CSI/CSC *Masterformat* system will reveal hundreds of different joint closure devices for every purpose.

Every designer of buildings must develop a sure sense of where movement joints are needed in buildings, and a feel for how to design them. This is neither quick nor easy to do, for the topic is large and complex, and authoritative reference material is widely scattered. Numerous buildings are built each year by designers who have not acquired this intuition. Many of these buildings are riddled with cracks even before they have been completed. This brief essay and the related illustrations throughout the book are intended to create an awareness of the problem of movement in buildings and to establish a logical framework that the reader can fill in with more detailed information over time.

Reference

In addition to the references listed at the ends of the chapters of this book, many of which treat movement problems, a good general reference is *Cracks, Movements and Joints in Buildings*, National Research Council of Canada, Ottawa, 1976 (NRCC 15477). (Address for ordering: Institute for Research in Construction, Ottawa, Canada K1A 0R6.) For a practical summary, see pp. 75–94 of Allen, Edward, *Architectural Detailing: Function, Constructibility, Aesthetics*, New York, John Wiley & Sons, Inc., 1993.

SOME SPECIAL PROBLEMS OF MASONRY CONSTRUCTION

Expansion and Contraction

Masonry walls expand and contract slightly in response to changes in both temperature and moisture content. Thermal movement is relatively easy to quantify (Figure 10.13). Moisture movement is more difficult: New clay masonry units tend to absorb water and expand slightly under moist conditions. New concrete masonry units usually shrink somewhat as they give off excess water following manufacture, which is the reason Type I units (moisture controlled) should be used where shrinkage must be minimized. Expansions and shrinkages in masonry materials are small as compared to the moisture movement in wood or the thermal movement in plastics or aluminum, but they must be taken into account in the design of the building by providing *surface divider joints* to avoid an excessive buildup of forces that could crack or spall the masonry.

Three different kinds of surface divider joints are used in masonry. *Expansion joints* are intentionally created slots that can close slightly to accommodate expansion of surfaces made of brick or stone masonry. *Control joints* are intentionally created cracks that can open to accommodate shrinkage in surfaces made of concrete masonry. *Abutment joints,* sometimes called *construction joints* or *isolation joints,* are placed at junctions between masonry and other materials, or between new masonry and old masonry, to accommodate differentials in movement. Figures 10.14 through 10.17 illustrate the use of movement joints in masonry walls. Movement joints are also critical in masonry facings applied over multistory structural frames of steel or concrete, to prevent fracture of the masonry when the frame deflects under load, as discussed in Chapter 20.

Joint reinforcing must be interrupted at movement joints so that it does not restrain the opening or closing of the joint. To prevent out-of-plane displacements of the wall, various kinds of vertically interlocking details are often used, as seen in Figure 10.17. Most movement joints in masonry walls are closed with elastic sealants to prevent air and water from passing through.

	in/in/°F	mm/mm/°C
Clay or shale brick masonry	0.0000036	0.0000065
Normal weight concrete masonry	0.0000052	0.0000094
Lightweight concrete masonry	0.0000043	0.0000077
Granite	0.0000047	0.0000085
Limestone	0.0000044	0.0000079
Marble	0.0000073	0.0000131
Normal weight concrete	0.0000055	0.0000099
Structural steel	0.0000065	0.0000117

FIGURE 10.13

Average coefficients of thermal expansion for some masonry materials.

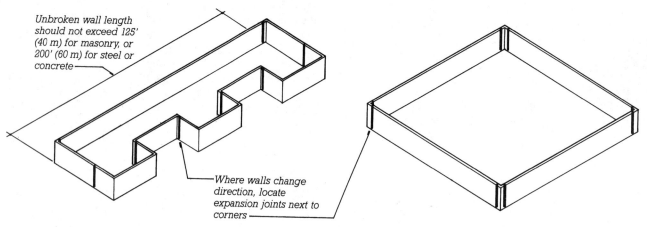

Unbroken wall length should not exceed 125' (40 m) for masonry, or 200' (60 m) for steel or concrete

Where walls change direction, locate expansion joints next to corners

FIGURE 10.14

Placing movement joints in masonry walls.

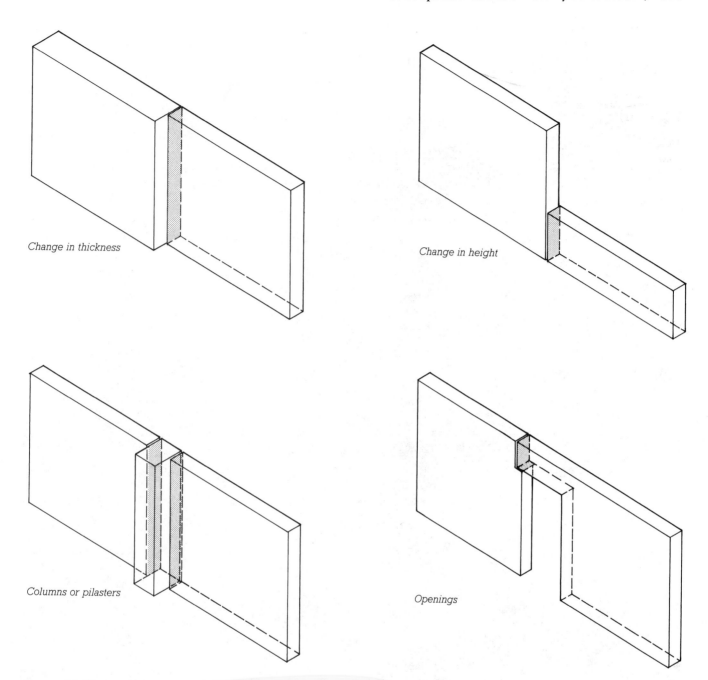

Change in thickness

Change in height

Columns or pilasters

Openings

FIGURE 10.15
Movement joints in masonry walls should be located at discontinuities in the wall, where cracks tend to form.

FIGURE 10.16
A window head and expansion joint in a brick veneer wall. The soldier course is
supported on a steel angle lintel. (*Photo by the author*)

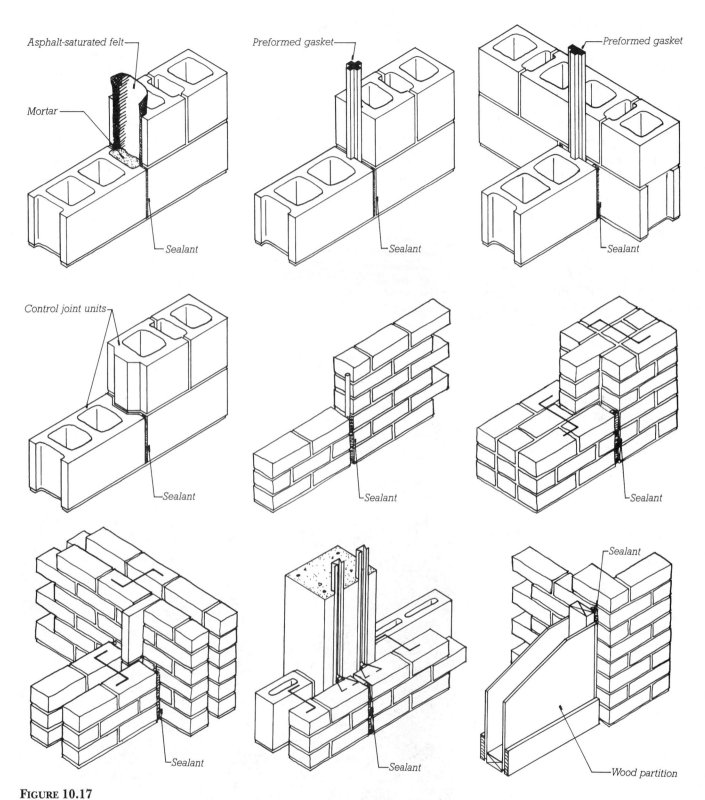

FIGURE 10.17

Some ways of making movement joints in masonry. The joints in concrete masonry walls are control joints, to control shrinkage cracking. Those in brick walls are expansion joints, to allow for moisture expansion of the bricks. The joint at the lower right is an abutment joint. Notice that many of these details interlock to prevent out-of-plane movement of the walls. Detailing of sealant joints is covered in Chapter 19.

Masonry is a massive material, taking forms permitted by the law of gravity. Our vocabulary of masonry forms was developed in buildings which became essays about gravity—great weights piled high, buttresses braced against the thrust of arched vaulting. Long after the internal steel frame relieved the need for such forms, they still retain meaning for us through their historical references and in their familiarity. The basic masonry forms have become symbols.

Michael Shellenbarger, in *Landmarks: A Tradition of Portland Masonry Architecture.* Portland, Oregon, Masonry & Ceramic Tile Institute of Oregon, 1984.

Efflorescence

Efflorescence is a fluffy powder, usually white, that sometimes appears on the surface of a wall of brick, stone, or concrete masonry (Figure 10.18). It consists of one or more water-soluble salts that were originally present either in the masonry units or in the mortar. These were brought to the surface and deposited there by water that had seeped into the masonry, dissolved the salts, then migrated to the surface and evaporated. Efflorescence can usually be avoided by choosing masonry units that have been shown by laboratory testing not to contain water-soluble salts and by using clean ingredients in the mortar. Most types of efflorescence that form soon after the completion of construction are easily removed with water and a brush. Although efflorescence is likely to reappear after such a washing, it will diminish and finally disappear with time as the salt is gradually leached out of the wall. Efflorescence that forms for the first time after a period of years is an indication that water has only recently begun to enter the wall, and is best controlled by investigating and correcting the source of leakage.

Mortar Joint Deterioration

Mortar joints are the weakest link in most masonry walls. Water running down a wall tends to accumulate in the joints, where cycles of freezing and thawing weather can gradually *spall* (split off flakes of) the mortar in an accelerating process of destruction that eventually creates water leaks and loosens the masonry units. To forestall this process as long as possible, a suitably weather-resistant mortar formulation must be used, and joints must be well filled and tightly compacted with a concave or vee tooling at the time the masonry is laid. Even with all these precautions, a masonry wall in a severe climate will show substantial joint deterioration after many years of weathering, and may require *tuck pointing*, a process of

FIGURE 10.18
Efflorescence on a wall of Flemish Bond brickwork. (*Photo by the author*)

raking and cutting out the defective mortar and replacing it with fresh mortar.

Moisture Resistance of Masonry

Most masonry materials, including mortar, are porous, and can transmit water from the outside of the wall to the inside. Water can also enter through cracks between the masonry units and the mortar. To prevent water from entering a building through a masonry wall, the designer should begin by specifying appropriate types of masonry units, mortar, and joint tooling. Cavity wall construction should be utilized rather than solid or composite wall construction. The construction process should be supervised closely to be sure that all mortar joints are free of voids, that flashings and weep holes are properly installed, and that cavities are kept clean. Masonry walls should be protected against excessive wetting of the exterior surface of the wall insofar as practical through proper roof drainage and roof overhangs. Beyond these measures, consideration may also be given to coating the wall with stucco or paint. It is important that any exterior coating be highly permeable to water vapor to avoid blistering and rupture of the coating from outward vapor migration (see pages 572–574). Masonry primer-sealers and paints based on portland cement fill the pores of the wall without obstructing the outward passage of water vapor.

Most other masonry paints are based on latex formulations and are also permeable to water vapor.

Below grade, masonry should first be *parged* (plastered on the outside) with two coats of Type M mortar to a total thickness of $\frac{1}{2}$ inch (13 mm) to seal cracks and pores. After the parging has cured and dried, it can be coated with a bituminous dampproofing compound or, if a truly watertight wall is required below grade, it can be covered with a sheet membrane or a layer of bentonite clay, as discussed in Chapter 2.

Cold and Hot Weather Construction

Mortar cannot be allowed to freeze before it has cured, or its strength and watertightness may be seriously damaged. In cold climates, special precautions are necessary if masonry work is carried on during the winter months. These include such measures as keeping masonry units and sand dry, protecting them from freezing temperatures prior to use, warming the mixing water (and sometimes the sand as well) to produce mortar at an optimum temperature for workability and curing, using a Type III (high early strength) cement to accelerate the curing of the mortar, and mixing the mortar in smaller quantities in order that it not cool excessively before it is used. The masons' workstations should be protected from wind with temporary enclosures and heated if temperatures inside the enclosures do not remain above

freezing. The finished masonry must be protected against freezing for at least 2 to 3 days after it is laid, and tops of walls should be protected from rain and snow. Chemical accelerators and so-called "antifreeze" admixtures are, in general, harmful to mortar and reinforcing steel and should not be used.

In hot weather, mortar may dry excessively before it cures. Some types of masonry units may have to be dampened before laying so that they do not absorb too much water from the mortar. It is also helpful to keep the masonry units and mortar ingredients, as well as the masons' workstations, in shade.

MASONRY AND THE BUILDING CODES

The utilization of masonry loadbearing walls in Type 1, Type 2, Type 3 (Ordinary), and Type 4 (Mill, or Heavy Timber) construction has been discussed in earlier sections of this chapter. Masonry walls are frequently constructed for interior partitions, fire separation walls, and fire walls in buildings of all construction types. Figure 10.19 gives rule-of-thumb values for fire resistance ratings of some common types of masonry walls and partitions, along with Sound Transmission Class ratings to allow comparison of the acoustical isolation capabilities of these walls with those of the partition systems shown in Chapter 23.

Wall Type	Fire Resistance (hours)	STC
4″ (100 mm) brick	1	45
8″ (200 mm) brick	4	52
10″, 12″ (250 mm, 300 mm) brick	4	59
4″ concrete block	$\frac{1}{2}$ to 1	44[a]
8″ concrete block	1 to 2	55[a]

[a] painted or plastered on both sides.

FIGURE 10.19
Rule-of-thumb Fire Resistance and Sound Transmission Class for some masonry partitions.

THE UNIQUENESS OF MASONRY

Masonry is often chosen as a material of construction for its association in people's minds with qualities of permanence and solidity and with beautiful buildings and architectural styles of the past. It is often chosen for its unique colors, textures, and patterns; for its fire resistance; and for its easy, often automatic compliance with building code requirements. Masonry is often chosen, too, because it is economical. Although it is labor intensive, it can create a high-performance, long-lasting structure and enclosure in a single operation by a single trade, bypassing the difficulties that are frequently encountered in managing the numerous trades and subcontractors needed to erect a comparable building of other materials.

Masonry, like Wood Light Frame construction, is a construction process carried out with small, relatively inexpensive tools and machines on the construction site. Unlike steel and concrete construction, it does not require (except in the case of ashlar stonework) a large and expensively equipped shop to fabricate the major materials prior to erection. It shares with sitecast concrete construction a lengthy construction schedule that requires special precautions and can encounter delays during periods of very hot, very cold, or very wet weather. But generally it does not require an extensive period of preparation and fabrication in advance of the beginning of construction because it uses standardized units and materi-

	Ultimate Compressive Strength	Density
Bricks	2000–20,000 psi (14–140 MPa)	100–140 lbs/ft³ (1600–2240 kg/m³)
Concrete masonry units	1500–6000 psi (10–41 MPa)	75–135 lbs/ft³ (1200–2160 kg/m³)
Limestone	2600–21,000 psi (18–147 MPa)	130–170 lbs/ft³ (2080–2720 kg/m³)
Sandstone	4000–28,000 psi (28–195 MPa)	140–165 lbs/ft³ (2240–2640 kg/m³)
Marble	9000–18,000 psi (62–123 MPa)	165–170 lbs/ft³ (2640–2720 kg/m³)
Granite	15,600–30,800 psi (108–212 MPa)	165–170 lbs/ft³ (2640–2720 kg/m³)

FIGURE 10.20
Ranges of strength and density for some masonry materials, to allow comparison of the properties of various bricks, blocks, and building stones. In practice, the allowable compressive stresses used in structural calculations for masonry are much lower than the values given, to take into account the strength of the mortar and a substantial factor of safety.

Material	Allowable Tensile Strength	Allowable Compressive Strength	Density	Modulus of Elasticity
Wood (average)	700 psi (4.83 MPa)	1,100 psi (7.58 MPa)	30 pcf (480 kg/m³)	1,200,000 psi (8,275 MPa)
Brick masonry (average)	0	250 psi (1.72 MPa)	120 pcf (1,920 kg/m³)	1,200,000 psi (8,275 MPa)
Steel (ASTM A36)	22,000 psi (151.69 MPa)	22,000 psi (151.69 MPa)	490 pcf (7,850 kg/m³)	29,000,000 psi (200,000 MPa)
Concrete (average)	0	1,350 psi (9.31 MPa)	145 pcf (2,320 kg/m³)	3,150,000 psi (21,720 MPa)

FIGURE 10.21
Comparative properties of four common structural materials. An allowable tensile strength of 0 is shown for brick masonry, because brick that has cracked for thermal or structural reasons has no tensile strength. In certain applications where cracking is unlikely, building codes allow the designer to assume a very low tensile strength (less than 100 psi or 1 MPa) for brickwork.

FIGURE 10.22
Some masonry paving patterns in brick and granite. All six of these examples are laid without mortar in a bed of sand. Outdoor pavings of masonry may also be laid in mortar over reinforced concrete slabs. (*Photos by the author*)

als that are put into final form as they are placed in the building.

From the beginning of human civilization, masonry has been the medium from which we have created our most nearly permanent, most carefully crafted, most highly prized buildings. It has given us the massiveness of the Egyptian pyramids, the inspirational elegance of the Parthenon, and the light-filled loftiness of the great European cathedrals, as well as the reassuring coziness of the fireplace, the brick cottage, and the walled garden. Masonry can express our highest aspirations and our deepest yearnings for a rootedness in the earth. It reflects both the tiny scale of the human hand and the boundless power of that hand to create.

FIGURE 10.23
A detail of the porch of H. H. Richardson's First Baptist Church, Boston, built in 1871.
(*Photo by the author*)

SELECTED REFERENCES

In addition to the references listed in Chapters 8 and 9, the reader is referred to *Building Code Requirements for Masonry* *Structures,* ACI 530 and ASCE 5, jointly published by the American Concrete Institute and the American Society of Civil Engineers. (Address for ordering: ACI, Customer Services Department, P.O. Box 9094, Farmington Hills, MI 48333-9094.)

KEY TERMS AND CONCEPTS

loadbearing wall	weep hole	furring strip	abutment joint
bearing wall	external flashing	Ordinary construction	construction joint
reinforced	internal or concealed flashing	firecut	isolation joint
unreinforced	base flashing	Mill construction	efflorescence
composite wall	counterflashing or cap flashing	surface divider joint	spall
cavity wall	reglet	expansion joint	tuck pointing
flashing	external insulation and finish system (EIFS)	control joint	parge

REVIEW QUESTIONS

1. Describe how a cavity wall works, and sketch its major constructional features. What aspects of cavity wall construction are most critical to its success in preventing water leakage?

2. Where should flashings be installed in a masonry wall? What is the function of the flashing in each of these locations?

3. Where should weep holes be provided? Describe the function of a weep hole and indicate several ways in which it may be constructed.

4. What are the differences between Ordinary construction and Mill construction? What features of each are related to fire resistance?

5. What types of movement joints are required in a concrete masonry wall? In a brick masonry wall? Where should these joints be located?

6. What are some ways of insulating masonry walls?

7. Why is balloon framing used rather than platform framing in Ordinary construction?

8. What precautions should be taken when constructing masonry walls in Minneapolis in the winter?

EXERCISES

1. What are the allowable height and floor area for a restaurant in a building of Mill construction (Type 4)? How do these figures change if unprotected Ordinary construction (Type 3B) is used instead? What if unprotected steel joists are substituted for the wood joists? Or precast concrete plank floors with a two-hour fire rating?

2. Examine the masonry walls of some new buildings in your area. Where have movement joints been placed in these walls? What type of joint is each? Why is it placed where it is? Do you agree with this placement?

3. Look at weep holes and flashings in these same buildings. How are they detailed? Can you improve on these details?

11

STEEL FRAME CONSTRUCTION

Ironworkers place open-web steel joists on a frame of steel wide-flange beams as a crane lowers bundles of joists from above. (*Photo by Balthazar Korab. Courtesy of Vulcraft Division of Nucor Corporation*)

Steel, strong and stiff, is a material of slender towers and soaring spans. Precise and predictable, light in proportion to its strength, it is also well suited to rapid construction, highly repetitive building frames, and architectural details that satisfy the eye with a clean, precise elegance. Among the metals, it is uniquely plentiful and inexpensive. If its weaknesses—a tendency to corrode in certain environments and a loss of strength during severe building fires—are held in check by intelligent construction measures, it offers the designer possibilities that exist in no other material.

HISTORY

Prior to the beginning of the 19th century, metals had little structural role in buildings except in connecting devices. The Greeks and Romans used hidden cramps of bronze to join blocks of stone, and architects of the Renaissance countered the thrust of masonry vaults with wrought iron chains and rods. The first all-metal structure, a cast iron bridge, was built in the late 18th century in England and still carries traffic across the Severn River more than two centuries after its construction. *Cast iron* and *wrought iron* were used increasingly for framing industrial buildings in Europe and North America in the first half of the 19th century, but their usefulness was limited by the unpredictable brittleness of cast iron and the relatively high cost of wrought iron.

Until this time, steel had been a rare and expensive material, produced only in small batches for such applications as weapons and cutlery. Plentiful, inexpensive steel first became available in the 1850s with

FIGURE 11.1
Landscape architect Joseph Paxton designed the Crystal Palace, an exposition hall of cast iron and glass, which was built in London in 1851. (*Bettmann Archive*)

FIGURE 11.2
Allied Bank Plaza, designed by architects Skidmore, Owings, and Merrill. (*Courtesy of American Institute of Steel Construction*)

the introduction of the *Bessemer process,* in which air was blown into a vessel of molten iron to burn out the impurities. By this means, a large batch of iron could be made into steel in about 20 minutes, and the structural properties of the resulting steel were vastly superior to those of cast iron. Another economical steel-making process, the open-hearth method, was developed in Europe in 1868 and was soon adopted in America. By 1889, when the Eiffel Tower was built of wrought iron in Paris (Figure 11.3), several steel frame skyscrapers had already been erected in the United States (Figure 11.4). A new material of construction had been born.

THE MATERIAL STEEL

Steel

Steel is any of a range of alloys of iron and carbon that contain less than 2 percent carbon. Ordinary structural steel, called *mild steel,* contains less than three-tenths of 1 percent carbon, plus traces of detrimental impurities such as phosphorus, sulfur, oxygen, and nitrogen, and of beneficial elements such as manganese and silicon. Ordinary cast iron, by contrast, contains 3 to 4 percent carbon and considerable quantities of impurities. Carbon content is a crucial determinant of the properties of a ferrous metal: Too much carbon makes a hard but brittle metal, while too little produces a soft, weak material. Thus, mild steel is iron whose properties have been optimized for structural purposes by controlling the amounts of carbon and other elements in the metal.

Iron is produced in a blast furnace charged with alternating layers of iron ore (oxides of iron), coke (coal whose volatile constituents have

FIGURE 11.3
Engineer Gustave Eiffel's magnificent tower of wrought iron was constructed in Paris from 1887 to 1889. (*Photo by James Austin, Cambridge, England*)

FIGURE 11.4
The Home Insurance Company Building, designed by William LeBaron Jenney and built in Chicago in 1893, was among the earliest true skyscrapers. The steel framing was fireproofed with masonry, and the exterior masonry facings were supported on the steel frame. (*Photo by William T. Barnum. Courtesy of Chicago Historical Society, IChi-18293*)

been distilled out, leaving only carbon), and crushed limestone (Figure 11.6). The coke is burned by large quantities of air forced into the bottom of the furnace to produce carbon monoxide, which reacts with the ore to reduce it to elemental iron. The limestone forms a slag with various impurities, but large amounts of carbon and other elements are inevitably incorporated into the iron. The molten iron is drawn off at the bottom of the furnace and held in a liquid state for processing into steel. The manufacture of a ton of iron requires about $1\frac{3}{4}$ tons of iron ore, $\frac{3}{4}$ ton of coke, $\frac{1}{4}$ ton of limestone, and 4 tons of air.

Today, the majority of steel for all end uses is produced by the *basic oxygen process,* in which a water-cooled lance is lowered into a container of molten iron and recycled steel scrap. A steam of pure oxygen at very high pressure is blown from the hollow lance into the metal to burn off the excess carbon and impurities. A flux of lime and fluorspar reacts with other impurities, particularly phosphorus, to form a slag that is discarded. New metallic elements may be added to the container at the end of the process to adjust the composition of the steel as desired: Manganese gives resistance to abrasion and impact, molybdenum gives strength, vanadium imparts strength and toughness, nickel and chromium give toughness and stiffness. The

FIGURE 11.5
Molten iron is poured into a crucible to begin its conversion to steel in the basic oxygen process. (*Courtesy of United States Steel Corporation*)

The gap between stone and steel-and-glass was as great as that in the evolutionary order between the crustaceans and the vertebrates.

Lewis Mumford,
The Brown Decades, New York,
Dover Publications, Inc.,
1955, pp. 130–131

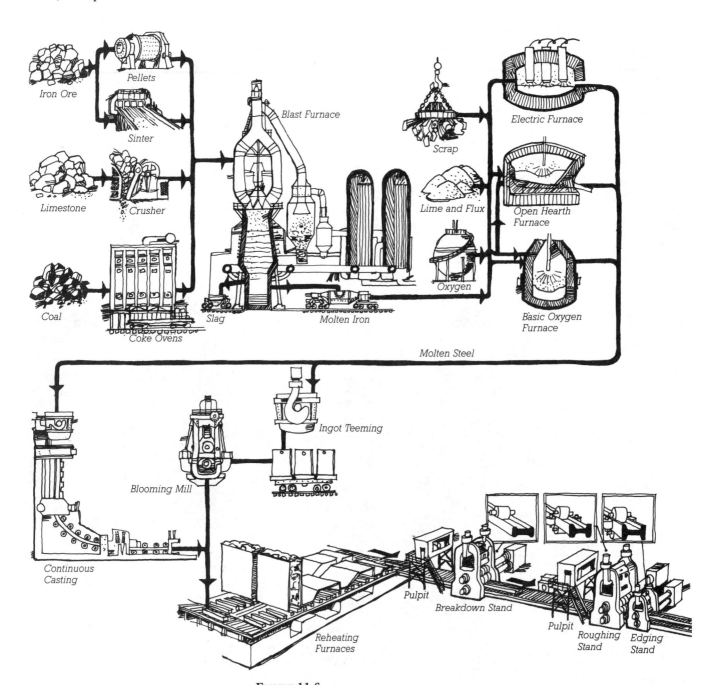

FIGURE 11.6
The steel-making process, from iron ore to structural shapes. Notice particularly the steps in the evolution of a wide-flange shape as it progresses through the various stands in the rolling mill. Most structural steel in the United States is made from steel scrap in electric furnaces. (*Adapted from "Steelmaking Flowlines," by permission of the American Iron and Steel Institute*)

entire process takes place with the aid of careful sampling and laboratory analysis techniques to assure the finished quality of the steel and takes less than an hour from start to finish.

Most structural steel for frames of buildings is produced in *electric furnaces,* largely from scrap steel, in so-called "mini-mills." These mills are miniature only in comparison to the conventional mills that they have replaced; they are housed in enormous buildings and roll structural shapes up to 40 inches (1 m) deep. The scrap comes mostly from defunct automobiles, one mini-mill alone consuming 300,000 derelict cars in an average year. Through careful metallurgical testing and control, top-quality steel is produced. It is cast continuously into beam blanks that are thick approximations of the shapes into which they will be rolled.

Production of Structural Shapes

In the *structural mill,* the hot steel blank passes through a succession of rollers that press the metal into progressively more refined approximations of the desired shape and size (Figure 11.7). The finished shape exits from the last set of rollers as a continuous length that is cut into shorter segments by a *hot saw* (Figure 11.8). These segments are cooled on a *cooling bed* (Figure 11.9). Then a *straightener* corrects any residual crookedness. Finally, each piece is labeled with its shape designation and the number of the batch of steel from which it was rolled; later, when the piece is shipped to a fabricator, it will be accompanied by a certificate that gives the chemical analysis of that particular batch, as evidence that the steel meets standard structural specifications.

The roller spacings in the structural mill are adjustable; by varying the spacings between rollers, a number of different shapes of the same nominal dimensions can be produced (Figure 11.10). This provides the architect and structural engineer with a finely graduated selection of shapes from which to select each structural member in a building and assures that little steel will be wasted through the specification of shapes that are larger than required.

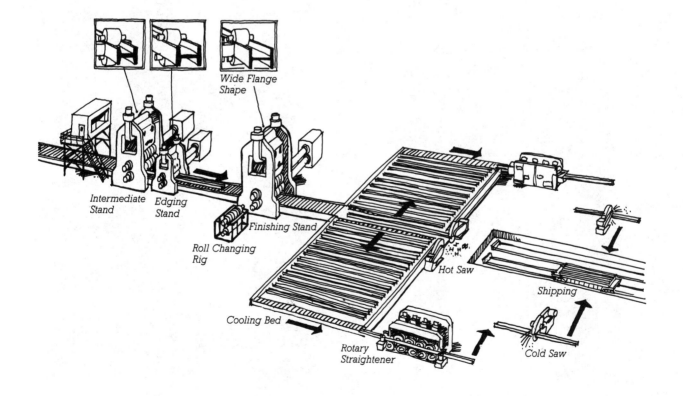

Intermediate Stand

Edging Stand

Wide Flange Shape

Roll Changing Rig

Finishing Stand

Cooling Bed

Hot Saw

Rotary Straightener

Cold Saw

Shipping

FIGURE 11.7
A glowing steel wide-flange shape
emerges from the rolls of the finishing
stand of the rolling mill. (*Photo by Mike
Engestrom. Courtesy of Nucor-Yamato Steel*)

FIGURE 11.8
A hot saw cuts pieces of wide-flange
stock from a continuous length that has
just emerged from the finishing stand in
the background. Workers in the booth
control the process. (*Courtesy of United
States Steel Corporation*)

FIGURE 11.9
Wide-flange shapes are inspected for quality on the cooling bed. The two pieces in the foreground still glow with heat. (*Photo by Mike Engestrom. Courtesy of Nucor-Yamato Steel*)

FIGURE 11.10
Examples of the standard shapes of structural steel, showing how different weights of the same section are produced by varying the spacing of the rollers in the structural mill.

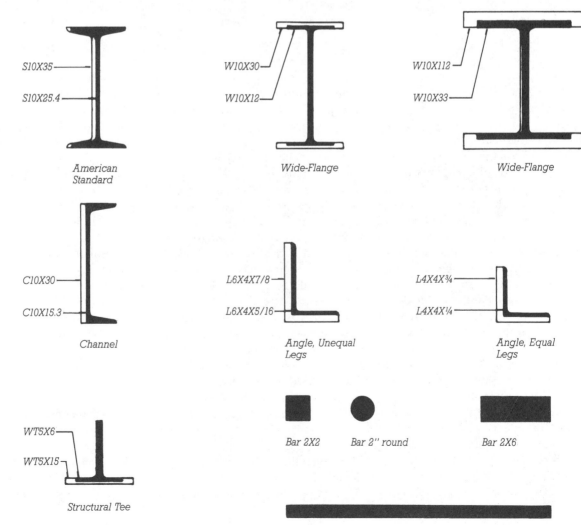

S10X35 *S10X25.4* **American Standard**	*W10X30* *W10X12* **Wide-Flange**	*W10X112* *W10X33* **Wide-Flange**
C10X30 *C10X15.3* **Channel**	*L6X4X7/8* *L6X4X5/16* **Angle, Unequal Legs**	*L4X4X¾* *L4X4X¼* **Angle, Equal Legs**
WT5X6 *WT5X15* **Structural Tee**	*Bar 2X2* *Bar 2" round*	*Bar 2X6*

Shape	Sample Designation	Explanation	Range of Available Sizes
Wide-flange	W21 × 83	W denotes a wide-flange shape. 21 is the nominal depth in inches, and 83 is the weight per foot of length in pounds.	Nominal depths from 4 to 18″ to 2″ increments, and from 18 to 36″ in 3″ increments
American Standard	S18 × 70	S denotes American Standard. 18 is the nominal depth in inches, 70 is the weight per foot of length in pounds.	Nominal depths of 3″, 4″, 5″, 6″, 7″, 8″, 10″, 12″, 15″, 18″, 20″, and 24″
Angle	L4 × 3 × ⅜	L denotes an angle. The first two numbers are the nominal depths of the two legs, and the last is the thickness of the legs.	Leg depths of 2″, 2½″, 3″, 3½″, 4″, 5″, 6″, 7″, 8″, and 9″. Leg thicknesses from ⅛″ to 1⅛″
Channel	C9 × 13.4	C denotes a channel. 9 is the nominal depth in inches, and 13.4 is the weight per foot of length in pounds.	Nominal depths of 3″, 4″, 5″, 6″, 7″, 8″, 9″, 10″, 12″, and 15″
Structural tee	WT13.5 × 47	This is a tee made by splitting a W27 × 94. It is 13.5 inches deep and weighs 47 pounds per foot of length. Tees split from American Standard shapes are designated ST rather than WT.	Nominal depths of 2 to 9″ in 1″ increments, and 10½ to 18″ in 1½″ increments

FIGURE 11.11
Commonly used steel shapes.

Wide-flange shapes are used for most beams and columns, superseding the older *American Standard (I-beam) shapes* (Figure 11.11). American Standard shapes are less efficient structurally than wide-flanges because the roller arrangement that produces them is incapable of increasing the amount of steel in the flanges without also adding steel to the web, where it does little to increase the load-carrying capacity of the member. Wide-flanges are available in a vast range of sizes and weights. The smallest available depth in the United States is a nominal 4 inches (100 mm), and the largest is 40 inches (1 m). Weights per foot range from 9 to 730 pounds (13 to 1080 kg/m), the latter for a nominal 14-inch (360-mm) shape with flanges nearly 5 inches (130 mm) thick. Some producers construct heavier wide-flange sections by welding together flange and web plates.

Wide-flanges are manufactured in two basic proportions: tall and narrow for beams, and squarish for columns and foundation piles. The accepted nomenclature for wide-flange shapes begins with a letter W, followed by the nominal depth of the shape in inches, a multiplication sign, and the weight of the shape in pounds per foot. Thus, a W12 × 26 is a wide-flange shape nominally 12 inches (305 mm) deep weighing 26 pounds per foot of length (38.5 kg/m). More information about this shape is contained in a table of dimensions and properties (Figure 11.12): Its actual depth is 12.22 inches (310.4 mm) and its flange is 6.49 inches (164.9 mm) wide. These proportions indicate that the shape is mainly intended for use as a beam or girder. Reading across the table column by column, the designer can learn everything there is to know about this section, from its thicknesses and the radii of its fillets to various quantities that are useful in computing its structural behavior under load. At the upper end of the portion of the table dealing with 12-inch (305-mm) wide-flanges, we find shapes weighing up to 336 pounds per foot (501 kg/m), with actual depths of almost 17 inches (432 mm). These heavier shapes have flanges nearly as wide as the shapes are deep, indicating that they are intended for use as columns. The United States has adopted a "soft" conversion to metric sizes, not altering the actual sizes of the steel shapes, but merely tabulating metric dimensions for them.

FIGURE 11.12

A portion of the table of dimensions and properties of wide-flange shapes from the *Manual of Steel Construction* of the American Institute of Steel Construction, reproduced by permission of the AISC. One inch equals 25.4 mm.

W SHAPES — Properties

Nominal Wt per Ft (Lb)	bf/2tf	F'y (Ksi)	d/tw	F''y (Ksi)	rT (In.)	d/Af	I (In.^{4}) X-X	S (In.^{3})	r (In.)	I (In.^{4}) Y-Y	S (In.^{3})	r (In.)	J (In.^{4})	Zx (In.^{3})	Zy (In.^{3})
336	2.3	—	9.5	—	3.71	0.43	4060	483	6.41	1190	177	3.47	243	603	274
305	2.4	—	10.0	—	3.67	0.46	3550	435	6.29	1050	159	3.42	185	537	244
279	2.7	—	10.4	—	3.64	0.49	3110	393	6.16	937	143	3.38	143	481	220
252	2.9	—	11.0	—	3.59	0.53	2720	353	6.06	828	127	3.34	108	428	196
230	3.1	—	11.7	—	3.56	0.56	2420	321	5.97	742	115	3.31	83.8	386	177
210	3.4	—	12.5	—	3.53	0.61	2140	292	5.89	664	104	3.28	64.7	348	159
190	3.7	—	13.6	—	3.50	0.65	1890	263	5.82	589	93.0	3.25	48.8	311	143
170	4.0	—	14.6	—	3.47	0.72	1650	235	5.74	517	82.3	3.22	35.6	275	126
152	4.5	—	15.8	—	3.44	0.79	1430	209	5.66	454	72.8	3.19	25.8	243	111
136	5.0	—	17.0	—	3.41	0.87	1240	186	5.58	398	64.2	3.16	18.5	214	98.0
120	5.6	—	18.5	—	3.38	0.96	1070	163	5.51	345	56.0	3.13	12.9	186	85.4
106	6.2	—	21.1	—	3.36	1.07	933	145	5.47	301	49.3	3.11	9.13	164	75.1
96	6.8	—	23.1	—	3.34	1.16	833	131	5.44	270	44.4	3.09	6.86	147	67.5
87	7.5	—	24.3	—	3.32	1.28	740	118	5.38	241	39.7	3.07	5.10	132	60.4
79	8.2	62.6	26.3	—	3.31	1.39	662	107	5.34	216	35.8	3.05	3.84	119	54.3
72	9.0	52.3	28.5	—	3.29	1.52	597	97.4	5.31	195	32.4	3.04	2.93	108	49.2
65	9.9	43.0	31.1	—	3.28	1.67	533	87.9	5.28	174	29.1	3.02	2.18	96.8	44.1
58	7.8	—	33.9	57.6	2.72	1.90	475	78.0	5.28	107	21.4	2.51	2.10	86.4	32.5
53	8.7	55.9	35.0	54.1	2.71	2.10	425	70.6	5.23	95.8	19.2	2.48	1.58	77.9	29.1
50	6.3	—	32.9	60.9	2.17	2.36	394	64.7	5.18	56.3	13.9	1.96	1.78	72.4	21.4
45	7.0	—	36.0	51.0	2.15	2.61	350	58.1	5.15	50.0	12.4	1.94	1.31	64.7	19.0
40	7.8	—	40.5	40.3	2.14	2.90	310	51.9	5.13	44.1	11.0	1.93	0.95	57.5	16.8
35	6.3	—	41.7	38.0	1.74	3.66	285	45.6	5.25	24.5	7.47	1.54	0.74	51.2	11.5
30	7.4	—	47.5	29.3	1.73	4.30	238	38.6	5.21	20.3	6.24	1.52	0.46	43.1	9.56
26	8.5	57.9	53.1	23.4	1.72	4.95	204	33.4	5.17	17.3	5.34	1.51	0.30	37.2	8.17
22	4.7	—	47.3	29.5	1.02	7.19	156	25.4	4.91	4.66	2.31	0.847	0.29	29.3	3.66
19	5.7	—	51.7	24.7	1.00	8.67	130	21.3	4.82	3.76	1.88	0.822	0.18	24.7	2.98
16	7.5	—	54.5	22.2	0.96	11.3	103	17.1	4.67	2.82	1.41	0.773	0.10	20.1	2.26
14	8.8	54.3	59.6	18.6	0.95	13.3	88.6	14.9	4.62	2.36	1.19	0.753	0.07	17.4	1.90

W SHAPES — Dimensions

Designation	Area A (In.^{2})	Depth d (In.)		Web tw (In.)		tw/2 (In.)	Flange Width bf (In.)		Flange Thickness tf (In.)		T (In.)	k (In.)	k1 (In.)
W 12x336	98.8	16.82	16 7/8	1.775	1 3/4	7/8	13.385	13 3/8	2.955	2 15/16	9 1/2	3 11/16	1 1/2
x305	89.6	16.32	16 3/8	1.625	1 5/8	13/16	13.235	13 1/4	2.705	2 11/16	9 1/2	3 7/16	1 7/16
x279	81.9	15.85	15 7/8	1.530	1 1/2	3/4	13.140	13 1/8	2.470	2 1/2	9 1/2	3 3/8	1 3/8
x252	74.1	15.41	15 3/8	1.395	1 3/8	11/16	13.005	13	2.250	2 1/4	9 1/2	2 15/16	1 5/16
x230	67.7	15.05	15	1.285	1 5/16	11/16	12.895	12 7/8	2.070	2 1/16	9 1/2	2 3/4	1 1/4
x210	61.8	14.71	14 3/4	1.180	1 3/16	5/8	12.790	12 3/4	1.900	1 7/8	9 1/2	2 5/8	1 1/4
x190	55.8	14.38	14 3/8	1.060	1 1/16	9/16	12.670	12 5/8	1.735	1 3/4	9 1/2	2 7/16	1 3/16
x170	50.0	14.03	14	0.960	15/16	1/2	12.570	12 9/16	1.560	1 9/16	9 1/2	2 1/4	1 1/4
x152	44.7	13.71	13 3/4	0.870	7/8	7/16	12.480	12 1/2	1.400	1 3/8	9 1/2	2 1/8	1 1/8
x136	39.9	13.41	13 3/8	0.790	13/16	7/16	12.400	12 3/8	1.250	1 1/4	9 1/2	1 15/16	1
x120	35.3	13.12	13 1/8	0.710	11/16	3/8	12.320	12 3/8	1.105	1 1/8	9 1/2	1 13/16	1
x106	31.2	12.89	12 7/8	0.610	5/8	5/16	12.220	12 1/4	0.990	1	9 1/2	1 11/16	15/16
x96	28.2	12.71	12 3/4	0.550	9/16	5/16	12.160	12 1/8	0.900	7/8	9 1/2	1 5/8	7/8
x87	25.6	12.53	12 1/2	0.515	1/2	1/4	12.125	12 1/8	0.810	13/16	9 1/2	1 1/2	7/8
x79	23.2	12.38	12 3/8	0.470	7/16	1/4	12.080	12 1/8	0.735	3/4	9 1/2	1 7/16	7/8
x72	21.1	12.25	12 1/4	0.430	7/16	1/4	12.040	12	0.670	11/16	9 1/2	1 3/8	7/8
x65	19.1	12.12	12 1/8	0.390	3/8	3/16	12.000	12	0.605	5/8	9 1/2	1 5/16	13/16
W 12x58	17.0	12.19	12 1/4	0.360	3/8	3/16	10.010	10	0.640	5/8	9 1/2	1 3/8	13/16
x53	15.6	12.06	12	0.345	3/8	3/16	9.995	10	0.575	9/16	9 1/2	1 1/4	13/16
W 12x50	14.7	12.19	12 1/4	0.370	3/8	3/16	8.080	8 1/8	0.640	5/8	9 1/2	1 3/8	13/16
x45	13.2	12.06	12	0.335	5/16	3/16	8.045	8	0.575	9/16	9 1/2	1 1/4	13/16
x40	11.8	11.94	12	0.295	5/16	3/16	8.005	8	0.515	1/2	9 1/2	1 1/4	3/4
W 12x35	10.3	12.50	12 1/2	0.300	5/16	3/16	6.560	6 1/2	0.520	1/2	10 1/2	1	9/16
x30	8.79	12.34	12 3/8	0.260	1/4	1/8	6.520	6 1/2	0.440	7/16	10 1/2	15/16	1/2
x26	7.65	12.22	12 1/4	0.230	1/4	1/8	6.490	6 1/2	0.380	3/8	10 1/2	7/8	1/2
W 12x22	6.48	12.31	12 3/8	0.260	1/4	1/8	4.030	4	0.425	7/16	10 1/2	7/8	1/2
x19	5.57	12.16	12 1/8	0.235	1/4	1/8	4.005	4	0.350	3/8	10 1/2	13/16	1/2
x16	4.71	11.99	12	0.220	1/4	1/8	3.990	4	0.265	1/4	10 1/2	3/4	1/2
x14	4.16	11.91	11 7/8	0.200	3/16	1/8	3.970	4	0.225	1/4	10 1/2	11/16	1/2

Steel *angles* (Figure 11.13) are extremely versatile. They can be used for very short beams supporting small loads and are frequently found playing this role as lintels spanning door and window openings in masonry construction. In steel frame buildings, they are most often seen cut into short pieces and used to connect wide-flange shapes. They also find use as diagonal braces in steel frames and as members of steel trusses, where they are paired back to back to connect conveniently to flat *gusset plates* at the joints of the truss (Figure 11.82). *Channel* sections are also used as truss members and bracing, as well as for short beams and lintels. *Tees, plates,* and *bars* all have their various roles in a steel frame building, as shown in the diagrams that accompany this text.

Steel Alloys

Mild structural steel, known by its ASTM designation of A36, is the predominant type used in steel building frames, but today's mini-mills, using scrap as their primary raw material, routinely produce a stronger steel that is designated ASTM A572 Grade 50 (with special requirements). In time, this is expected to replace ASTM A36 as the steel used for most construction. Other steels are also widely employed (Figure 11.14). With small additions of other elements to the molten metal, high-strength, low-alloy steels are produced. The increased tensile strength of these steels makes them economical for use in tension members or in columns whose cross-sectional areas are restricted by architectural considerations. But because the elastic modulus of these steels is not increased by the added elements, their use for a beam is justified only if the deflection of the beam is not the controlling factor in its design. Certain of the low-alloy steels have another useful property besides higher strength: When exposed to the atmosphere, a coating of tenacious oxide forms to protect them from further corrosion. These *weathering steels,* joined with bolts and welding electrodes of similar metallurgical composition, can be left exposed to the weather without painting. This offers significant savings in maintenance costs, and the deep, warm hue of the oxide coating can be attractive on the face of a building. In other locations where corrosion is expected

ANGLES
Equal legs and unequal legs
Properties for designing

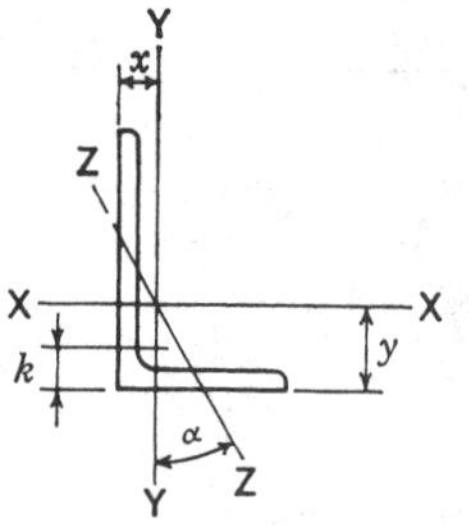

Size and Thickness	k	Weight per Foot	Area	AXIS X-X				AXIS Y-Y				AXIS Z-Z	
				I	S	r	y	I	S	r	x	r	Tan
In.	In.	Lb.	In.²	In.⁴	In.³	In.	In.	In.⁴	In.³	In.	In.	In.	α
L 4 x3 x 5/8	1 1/16	13.6	3.98	6.03	2.30	1.23	1.37	2.87	1.35	0.849	0.871	0.637	0.534
1/2	15/16	11.1	3.25	5.05	1.89	1.25	1.33	2.42	1.12	0.864	0.827	0.639	0.543
7/16	7/8	9.8	2.87	4.52	1.68	1.25	1.30	2.18	0.992	0.871	0.804	0.641	0.547
3/8	13/16	8.5	2.48	3.96	1.46	1.26	1.28	1.92	0.866	0.879	0.782	0.644	0.551
5/16	3/4	7.2	2.09	3.38	1.23	1.27	1.26	1.65	0.734	0.887	0.759	0.647	0.554
1/4	11/16	5.8	1.69	2.77	1.00	1.28	1.24	1.36	0.599	0.896	0.736	0.651	0.558
L 3 1/2x3 1/2x 1/2	7/8	11.1	3.25	3.64	1.49	1.06	1.06	3.64	1.49	1.06	1.06	0.683	1.000
7/16	13/16	9.8	2.87	3.26	1.32	1.07	1.04	3.26	1.32	1.07	1.04	0.684	1.000
3/8	3/4	8.5	2.48	2.87	1.15	1.07	1.01	2.87	1.15	1.07	1.01	0.687	1.000
5/16	11/16	7.2	2.09	2.45	0.976	1.08	0.990	2.45	0.976	1.08	0.990	0.690	1.000
1/4	5/8	5.8	1.69	2.01	0.794	1.09	0.968	2.01	0.794	1.09	0.968	0.694	1.000
L 3 1/2x3 x 1/2	15/16	10.2	3.00	3.45	1.45	1.07	1.13	2.33	1.10	0.881	0.875	0.621	0.714
7/16	7/8	9.1	2.65	3.10	1.29	1.08	1.10	2.09	0.975	0.889	0.853	0.622	0.718
3/8	13/16	7.9	2.30	2.72	1.13	1.09	1.08	1.85	0.851	0.897	0.830	0.625	0.721
5/16	3/4	6.6	1.93	2.33	0.954	1.10	1.06	1.58	0.722	0.905	0.808	0.627	0.724
1/4	11/16	5.4	1.56	1.91	0.776	1.11	1.04	1.30	0.589	0.914	0.785	0.631	0.727
L 3 1/2x2 1/2x 1/2	15/16	9.4	2.75	3.24	1.41	1.09	1.20	1.36	0.760	0.704	0.705	0.534	0.486
7/16	7/8	8.3	2.43	2.91	1.26	1.09	1.18	1.23	0.677	0.711	0.682	0.535	0.491
3/8	13/16	7.2	2.11	2.56	1.09	1.10	1.16	1.09	0.592	0.719	0.660	0.537	0.496
5/16	3/4	6.1	1.78	2.19	0.927	1.11	1.14	0.939	0.504	0.727	0.637	0.540	0.501
1/4	11/16	4.9	1.44	1.80	0.755	1.12	1.11	0.777	0.412	0.735	0.614	0.544	0.506
L 3 x3 x 1/2	13/16	9.4	2.75	2.22	1.07	0.898	0.932	2.22	1.07	0.898	0.932	0.584	1.000
7/16	3/4	8.3	2.43	1.99	0.954	0.905	0.910	1.99	0.954	0.905	0.910	0.585	1.000
3/8	11/16	7.2	2.11	1.76	0.833	0.913	0.888	1.76	0.833	0.913	0.888	0.587	1.000
5/16	5/8	6.1	1.78	1.51	0.707	0.922	0.865	1.51	0.707	0.922	0.865	0.589	1.000
1/4	9/16	4.9	1.44	1.24	0.577	0.930	0.842	1.24	0.577	0.930	0.842	0.592	1.000
3/16	1/2	3.71	1.09	0.962	0.441	0.939	0.820	0.962	0.441	0.939	0.820	0.596	1.000

Angles in shaded rows may not be readily available. Availability is subject to rolling accumulation and geographical location, and should be checked with material suppliers.

FIGURE 11.13
A portion of the table of dimensions and properties of angle shapes from the *Manual of Steel Construction* of the American Institute of Steel Construction, reproduced by permission of the AISC. One inch equals 25.4 mm.

Alloy	Yield Strength	Allowable Stress in Bending	Modulus of Elasticity
ASTM A36 — TYPICAL FOR FRAMING Carbon steel	36,000 psi (248 Mpa)	22,000 psi (152 MPa)	29,000,000 psi (200,000 MPa)
ASTM A242 High strength, low alloy; corrosion resisting	42,000–50,000 psi in 3 grades (290–345 MPa)	25,200–30,000 psi (174–207 MPa)	29,000,000 psi (200,000 MPa)
ASTM A441 High strength, low alloy; structural manganese-vanadium	40,000–50,000 psi in 3 grades (276–345 MPa)	24,000–30,000 psi (165–207 MPa)	29,000,000 psi (200,000 MPa)
ASTM A572 High strength, low alloy; columbium-vanadium steels of structural quality	42,000–65,000 psi in 4 grades (290–448 MPa)	25,200–39,000 psi (174–269 MPa)	29,000,000 psi (200,000 MPa)
ASTM A572 Gr. 50 (with special requirements)	65,000 psi (448 MPa)	39,000 psi (269 MPa)	29,000,000 psi (200,000 MPa)
ASTM A588 High strength, low alloy; corrosion resisting	42,000–50,000 psi in 3 grades (290–345 MPa)	25,200–30,000 psi (174–207MPa)	29,000,000 psi (200,000 MPa)

FIGURE 11.14
Properties of some steel alloys used in construction.

to be a problem, steel structural members are often galvanized (coated with zinc).

Open-Web Steel Joists

Among the many structural steel products fabricated from hot- and cold-rolled shapes, the most common is the *open-web steel joist*, a mass-produced truss used in closely spaced arrays to support floor and roof decks (Figure 11.15). Open-web joists are produced in three series: K Series joists are for spans up to 60 feet (18 m), and range in depth from 8 to 30 inches (200 to 760 mm). LH Series joists are designated as "Longspan" and can span as far as 96 feet (29 m). Their depths range from 18 to 48 inches (460 to 1220 mm). The DLH "Deep Longspan" Series of open-web joists are 52 to 72 inches deep (1320 to 1830 mm) and can span up to 144 feet (44 m). Most buildings utilize K Series joists that are less than 2 feet (600 mm) deep to achieve spans of up to 40 feet (13 m). The spacings between joists commonly range from 2 to 10 feet (0.6 to 3 m), depending on the magnitude of the applied loads and the spanning capability of the decking.

FIGURE 11.15
The roof of a single-story industrial building is framed with open-web steel joists supported by joist girders. The girders rest on columns of square hollow structural tubing. (*Courtesy of Vulcraft Division of Nucor Corporation*)

For Preliminary Design of a Steel Structure

- Estimate the depth of **corrugated steel roof decking** at $\frac{1}{40}$ of its span. Standard depths are 1, $1\frac{1}{2}$, 2, and 4 inches (25, 38, 50, and 100 mm).

- Estimate the overall depth of **corrugated steel floor decking plus concrete topping** at $\frac{1}{24}$ of its span. Typical overall depths range from $2\frac{1}{2}$ to 7 inches (65 to 180 mm).

- Estimate the depth of **steel open-web joists** at $\frac{1}{20}$ of their span for heavily loaded floors or widely spaced joints, and $\frac{1}{24}$ of their span for roofs, lightly loaded floors, or closely spaced joists. The spacing of joists depends on the spanning capability of the decking material. Typical joist spacings range from 2 to 10 feet (0.6 to 3.0 m). Standard joist depths are given earlier in this chapter.

- Estimate the depth of **steel beams** at $\frac{1}{20}$ of their span, and the depth of steel girders at $\frac{1}{15}$ of their span. The width of a beam or girder is usually $\frac{1}{3}$ to $\frac{1}{2}$ of its depth. For composite beams and girders, use the same ratios but apply them to the overall depth of the beam or girder, including the floor deck and concrete topping. Standard depths of steel wide-flange shapes are given earlier in this chapter.

- Estimate the depth of **triangular steel roof trusses** at $\frac{1}{4}$ to $\frac{1}{5}$ of their span. For rectangular trusses, the depth is typically $\frac{1}{8}$ to $\frac{1}{12}$ of their span.

- To estimate the size of a **steel column,** add up the total roof and floor area supported by the column. A W8 column can support up to about 3000 square feet (280 m²), and a W14 25,000 square feet (2300 m²). Very heavy W14 shapes, which are substantially larger than 14 inches in actual dimension, can support up to 50,000 square feet (4600 m²). Steel column shapes are usually square or nearly square in proportion.

These approximations are valid only for purposes of preliminary building layout, and must not be used to select final member sizes. They apply to the normal range of building occupancies such as residential, office, commercial, and institutional buildings, and parking garages. For manufacturing and storage buildings, use somewhat larger members.

For more comprehensive information on preliminary selection and layout of a structural system and sizing of structural members, see Allen, Edward, and Joseph Iano, *The Architect's Studio Companion* (2nd ed.), New York, John Wiley & Sons, Inc., 1995.

Joist girders are prefabricated steel trusses designed to carry heavy loads from the ends of open-web steel joists to columns (Figure 11.15). They range in depth from 20 to 72 inches (500 to 1800 mm). They can be used instead of wide-flange beams and girders in roof and floor structures where their greater depth is not objectionable. Open-web joists and joist girders are invariably made of high-strength steel.

Cold-Worked Steel

Steel can be formed in a cold state as well as in a hot state, by rolling or bending. Steel sheet is bent into C-shaped sections to make short-span framing members that are frequently used in partitions and exterior walls of larger buildings and in floor structures of smaller buildings (see Chapter 12). Steel sheet stock is also rolled into corrugated configurations utilized as floor and roof decking in steel-framed structures (Figures 11.60, 11.61). Heavier sheet or plate stock is cold-formed or hot-formed into square, rectangular, and round cross sections, which are then welded along the longitudinal seam to form *structural tubing.* Structural tubing is often used for columns and for members of welded steel trusses and space trusses. Tubing is especially useful for members that are subjected to torsional stresses.

Steel that is cold-rolled gains considerably in strength, through a realignment of its crystalline structure. The normal range of wide-flange shapes is too large to be cold-rolled, but cold rolling is used to produce steel rods and steel components for open-web joists, where the higher strength can be utilized to good advantage. Steel can be cold-drawn through dies to produce the very high-strength wires used in wire ropes, bridge cables, and concrete prestressing strands.

Joining Steel Members

Rivets

Steel shapes can be joined into a building frame with any of three fastening techniques—rivets, bolts, or welds—or by combinations of these. A *rivet* is a fastener consisting of a cylindrical body and a formed head, which is brought to a white heat, inserted through holes in the members to be joined, and hot-worked with a pneumatic hammer to produce a second head opposite the first (Figure 11.16). As the rivet cools, it shrinks, clamping the joined pieces together and forming a tight joint. Riveting was for many decades the predominant fastening technique

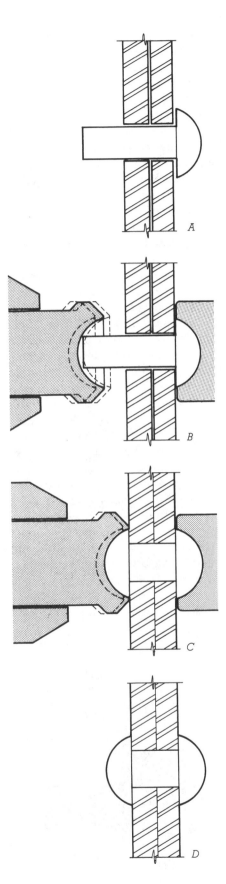

FIGURE 11.16
How riveted connections are made: (*a*) A hot steel rivet is inserted through holes in the two members to be joined. (*b, c*) Its head is placed in the cup-shaped depression of a heavy, hand-held hammer. A pneumatic hammer drives a rivet set repeatedly against the body of the rivet to form the second head. (*d*) The rivet shrinks as it cools, drawing the members tightly together.

FIGURE 11.17
An ironworker tightens high-strength bolts with a pneumatic impact wrench. (*Courtesy of Bethlehem Steel Corporation*)

used in steel frame buildings, but it has been almost entirely replaced by the less labor-intensive techniques of bolting and welding.

Bolts

The bolts commonly used in steel frame construction fall into two general categories: carbon steel bolts (ASTM A307) and high-strength bolts (ASTM A325 and A490). *Carbon steel bolts* (also called *unfinished* or *common* bolts) are similar to the ordinary machine bolts that can be purchased in hardware stores. Their installed cost is less than that of high-strength bolts, so they are used in many structural joints where their lower strength is sufficient to carry the necessary loads. They act primarily in bearing and shear.

High-strength bolts are heat treated during manufacture to develop the necessary strength. They derive their connecting ability either from their shear resistance, which is much higher than that of carbon steel bolts, or from being tightened to the point that the members they join are kept from slipping by the friction between them, producing what is known as a *friction* or *slip-critical connection*.

High-strength bolts are inserted into holes slightly larger than the shank diameter of the bolt. Washers may or may not be required, depending on the type of bolt and its specification; a washer is required in certain cases to spread the load of the bolt over a larger area of the shapes being joined. Sometimes a washer is inserted under the head or nut, whichever is turned to tighten the bolt, to prevent *galling* of the softer parent metal. The bolt is usually tightened using a pneumatic or electric *impact wrench* (Figure 11.17). If the bolt is to act only in bearing and shear, the amount of tension in the bolt is not critical; but if it is to connect by friction, it must be tightened reliably to at least 70 percent of its ultimate tensile strength. The extreme tension in the bolt produces the high clamping force necessary to allow the surfaces of the two members to transfer the load between them entirely by friction.

A major problem in friction connections is how to verify that the necessary tension has been achieved in all the bolts. This can be accomplished in any of several ways. In the *turn-of-nut method*, each bolt is tightened snug, then turned a specified additional fraction of a turn. Depending on bolt length, bolt alloy, and other factors, the additional tightening required will range from one-third of a turn to a full turn. In another method, a *load indicator washer* is placed under the head or nut of the bolt. As the bolt is tightened, the protrusions on the washer are progressively flattened in proportion to the tension in the bolt (Figure 11.18). Inspection for proper bolt tension then becomes a simple matter of inserting a feeler gauge to determine whether the protrusions have flattened sufficiently to indicate the required tension. Yet another method employs *tension control bolts* with either hex or button heads and protruding, splined ends that extend beyond the threaded portion of the body of the bolt (Figures 11.19, 11.20). The nut is tightened by a special power-driven wrench that grips both the nut and the splined end simultaneously, turning the one against the other (Figure 11.21). The splined end is formed in such a way that when the required torque has been reached, the end twists off. Installation of this type of bolt is accomplished by a single worker, while conventional high-strength bolts require a second person with a wrench to prevent the other end of the bolt assembly from turning.

A newer alternative to the high-strength bolt, the *swaged lockpin and collar fastener*, uses a special power tool to hold a boltlike lockpin under high tension while cold-forming (swaging) a steel collar around its shank end to complete the connection. As the swaging process is completed, the tail of the lockpin breaks off, furnishing visual evidence that the necessary tension has been achieved in the fastener.

Welding

Welding offers a unique and valuable capability to the structural designer: It can join the members of a steel frame as if they were a single piece. Welded connections, properly designed and executed, are stronger than the members they join in resisting both shear and moment forces. While it is possible to achieve this same performance with high-strength bolted connections, such connections are often cumbersome compared to equivalent welded joints. Bolting, on the other hand, has its own advantages: It is quick and easy for field connections that need only to resist shearing forces and can be accomplished under conditions of adverse weather or difficult physical access that would make welding impossible. Often welding and bolting are combined in the same connections to take advantage of the

unique qualities of each: Welding may be used in the fabricator's shop for its inherent economies and in the field for its structural continuity, while bolting is employed in the simpler field connections and to hold connections in alignment for welding. The choice between bolting and welding is often dictated by the designer, but such a choice may also be influenced by considerations such as the fabricator's and erector's equipment and expertise, availability of electric power, climate, and location.

Electric *arc welding* is conceptually simple. An electrical potential is established between the steel pieces to be joined and a metal *electrode* held either by a machine (as in some shop welding processes) or by a person. When the electrode is held close to the steel members, a continuous electric arc is established that generates sufficient heat to melt both a localized area of the steel members and the tip of the electrode (Figure 11.22). The molten steel from the electrode merges with that of the members to form a single puddle. The electrode is drawn slowly along the seam, leaving behind a continuous bead of metal that cools and solidifies to form a strong connection between the members. For small members, a single pass of the electrode may suffice to make the con-

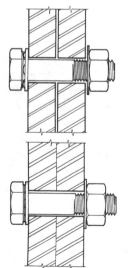

FIGURE 11.18

An untightened high-strength bolt with a load indicator washer under the head (top). The bolt and washer after tightening (bottom); notice that the protrusions on the load indicator washer have flattened.

FIGURE 11.19

A tension control bolt. (*Courtesy of Bristol Machine Company, Brea, California*)

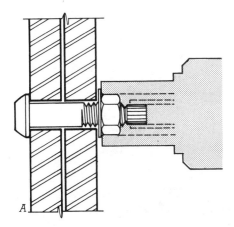

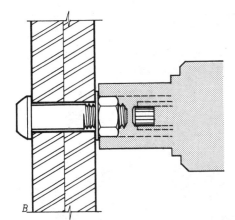

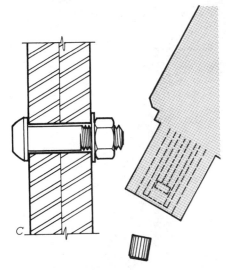

FIGURE 11.20
Tightening a tension control bolt. (*a*) The wrench holds both the nut and the splined body of the bolt, and turns them against one another to tighten the bolt. (*b*) When the required torque has been achieved, the splined end twists off in the wrench. (*c*) A plunger inside the wrench discharges the splined end into a container.

FIGURE 11.21
The compact design of the electric wrench for tightening tension control bolts makes it easy to reach bolts in tight situations. (*Courtesy of Ingersoll–Rand Corporation*)

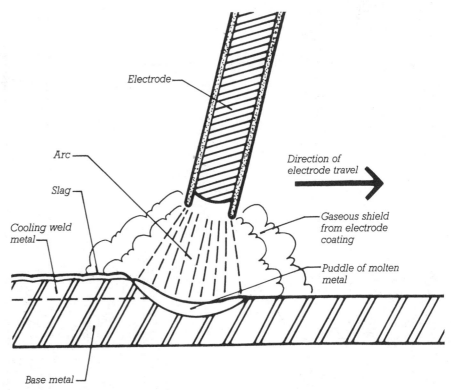

FIGURE 11.22
Close-up diagram of the electric arc welding process.

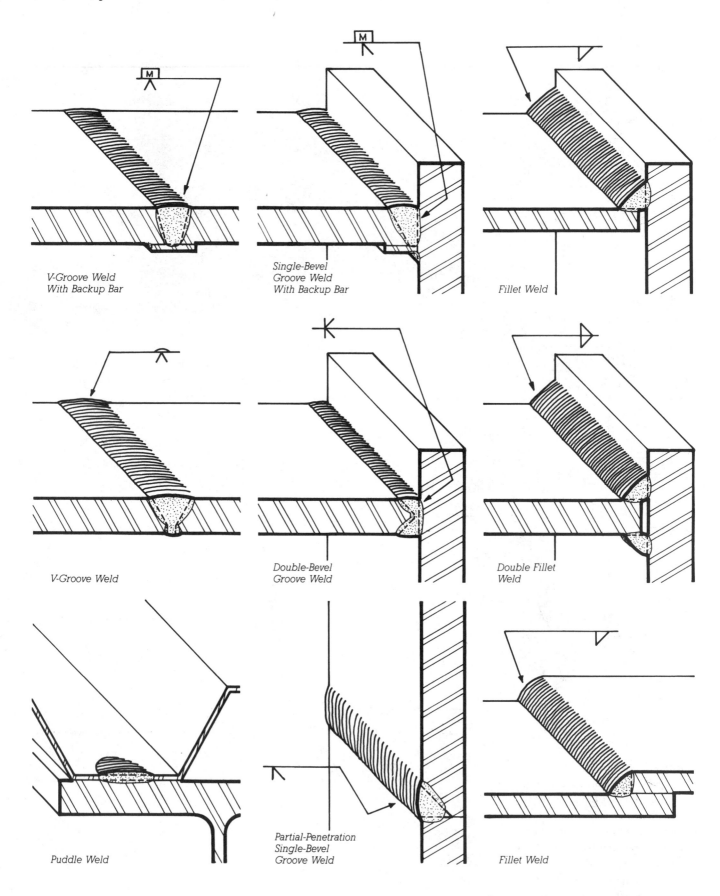

V-Groove Weld
With Backup Bar

Single-Bevel
Groove Weld
With Backup Bar

Fillet Weld

V-Groove Weld

Double-Bevel
Groove Weld

Double Fillet
Weld

Puddle Weld

Partial-Penetration
Single-Bevel
Groove Weld

Fillet Weld

FIGURE 11.23
Typical welds used in steel frame construction. Fillet welds are the most economical because they require no advance preparation of the joint, but the full-penetration groove welds are stronger. The standard symbols used here are explained in Figure 11.24.

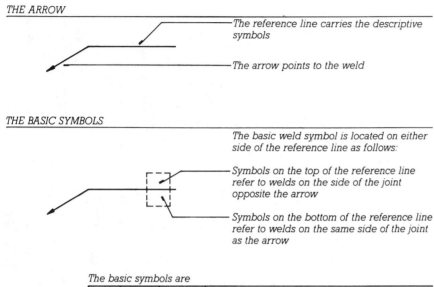

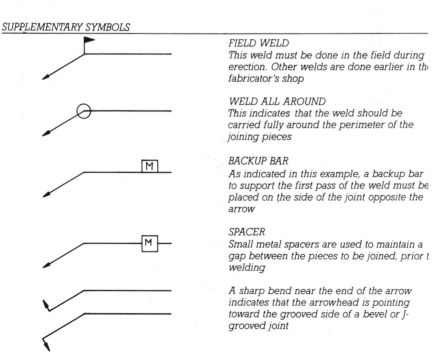

FIGURE 11.24
Standard weld symbols, as used on steel connection detail drawings.

nection. For larger members, a number of passes are made in order to build up a weld of the required depth.

In practice, welding is a complex science. The metallurgy of the structural steel and the welding electrodes must be carefully coordinated. Voltage, amperage, and polarity of the electric current are selected to achieve the right heat and penetration for the weld. Air must be kept away from the electric arc to prevent rapid oxidation of the liquid steel; this is accomplished in simple welding processes by a thick coating on the electrode that melts to create a liquid and gaseous shield around the arc, or by a core of vaporizing flux in a tubular steel electrode. It may also be done by means of a continuous flow of inert gas around the arc, or with a dry flux that is heaped over the end of the electrode as it moves across the work.

The required thickness and length of each weld are calculated by the designer to match them to the forces to be transmitted between the members. For deep welds, the edges of the members are beveled to permit access of the electrode to the full thickness of the piece (Figure 11.23). Small strips of steel called *backup bars* are welded beneath the connection prior to beginning the actual weld, to prevent the molten metal from flowing out the bottom of the groove. In some cases, *runoff bars* are required at the ends of a groove weld to facilitate the formation of a full thickness of weld metal at the edges of the member (Figure 11.46).

Workers who do structural welding are methodically trained and periodically tested to ensure that they have the required level of skill and knowledge. When an important weld is completed, it is inspected to see that it is of the required size and quality; often this involves sophisticated magnetic particle, dye penetrant, ultrasonic, or radiographic testing procedures that search for hidden voids and flaws within each weld.

Details of Steel Framing

Typical Connections

Most steel frame connections use angles, plates, or tees as transitional elements between the members being connected. A simple bolted beam-to-column-flange connection requires two angles and a number of bolts (Figures 11.25–11.27). The angles are cut to length and the holes are made in all the components prior to assembly. The angles are usually bolted to the web of the beam in the fabricator's shop. The bolts through the flange of the column are added on the construction site after the beam is in place. This type of connection, which joins only the web of the beam and not the flanges, is known as a *framed connection*. It is capable of transmitting all the vertical forces (*shear*) from a beam to a column. Because it does not connect the beam flanges to the column, it is of no value in transmitting bending moment from one to the other.

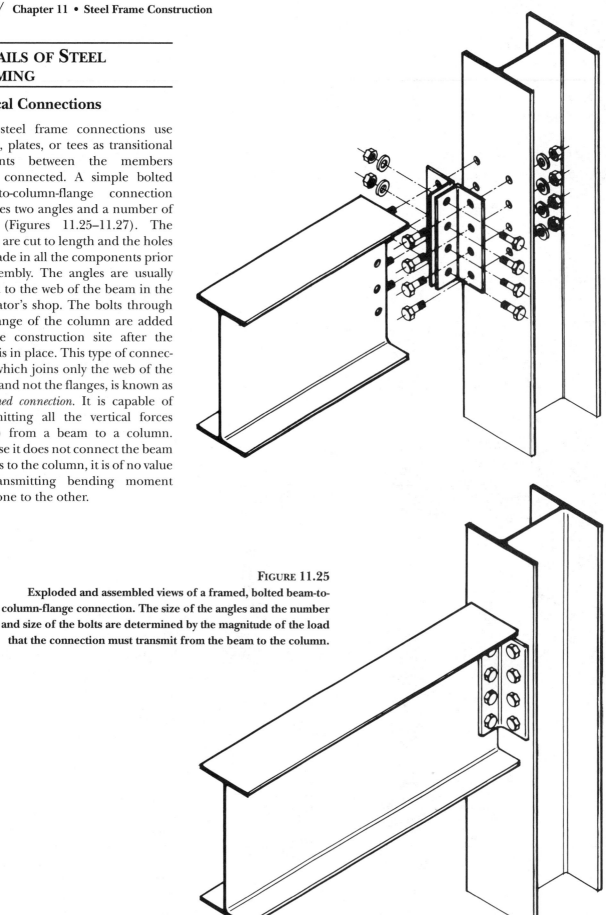

FIGURE 11.25
Exploded and assembled views of a framed, bolted beam-to-column-flange connection. The size of the angles and the number and size of the bolts are determined by the magnitude of the load that the connection must transmit from the beam to the column.

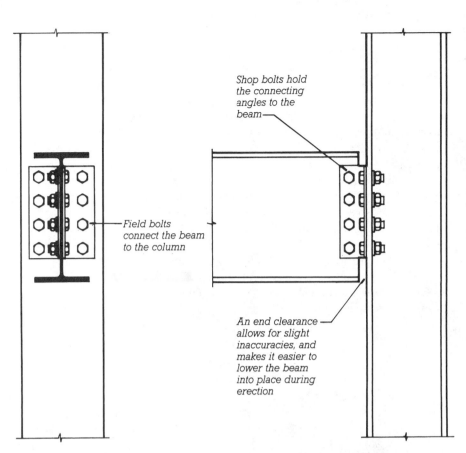

Shop bolts hold the connecting angles to the beam

Field bolts connect the beam to the column

An end clearance allows for slight inaccuracies, and makes it easier to lower the beam into place during erection

FIGURE 11.26
Two elevation views of the framed, bolted beam-to-column-flange connection shown in Figure 11.25. This is a shear connection only (AISC Type 2), because the flanges of the beam are not rigidly connected to the column.

FIGURE 11.27
A pictorial view of a framed, bolted beam-to-column-flange connection.

To produce a moment-transmitting connection, it is necessary to connect the flanges strongly across the joint by means of full-penetration welds across the beam flanges (Figures 11.28, 11.29). If the column flanges are insufficiently strong to accept the bending moment transmitted from the beam, *stiffener plates* must be installed inside the column flanges.

Shear Connections and Moment Connections

In order to understand the respective roles of shear connections and moment connections in a building frame, it is necessary to understand the means by which buildings may be made stable against the lateral forces of wind and earthquake. Three basic stabilizing mechanisms are commonly used: diagonal bracing, shear panels, and moment connections (Figure 11.30). *Diagonal bracing* works by creating stable triangular configurations within the unstable rectilinear geometry of a steel building frame. The connections within a diagonally braced frame need not transmit moments; they can behave like pins or hinges, which is another way of say-

ing that they can be *shear connections* such as the one in Figure 11.27. A special case of diagonal bracing is the *eccentrically braced frame* (Figure 11.30). Because the braces terminate in the beams, some distance away from the connections, eccentric bracing is much more elastic than conventional bracing. It is used primarily as a way of enabling a building frame to absorb large amounts of energy during an earthquake, and thus to protect itself against collapse.

Shear panels, which may be made of steel or concrete, act much like braced rectangles within the building

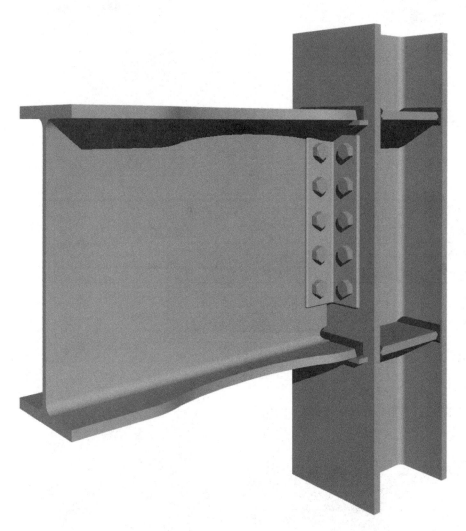

FIGURE 11.28

A welded moment connection (AISC Type 1) for joining a beam to a column flange. The bolts hold the beam in place for welding and also provide shear resistance. Small, rectangular backup bars are welded beneath the end of each beam flange to prevent the welding arc from burning through. A clearance hole (not visible in this view) is cut from the top of the beam web to permit the backup bar to pass through. A similar clearance hole that can be seen at the bottom of the beam web allows the bottom flange to be welded entirely from above for greater convenience. The groove welds develop the full strength of the flanges of the beam, allowing the connection to transmit moments between the beam and the column. If the column flanges are not stiff enough to accept the moments from the beam, stiffener plates are welded between the column flanges as shown here. The flanges of the beam are cut to a "dog bone" configuration to create a zone of the beam that is slightly weaker in bending than the welded connection itself. During a violent earthquake, the beam will deform permanently in this zone to protect the connection against failure.

frame and, like diagonal bracing, do not require moment connections.

Moment connections are capable of stabilizing a frame against lateral force without the use of either diagonal bracing or shear panels. A large number of the connections in a frame stabilized in this manner must be moment connections, but many may be shear connections.

There are two common methods of stabilizing the frame of a tall building (Figure 11.31). One is to provide a stable core area in the center of the

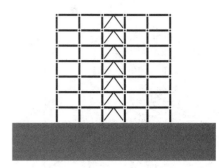
Diagonal Bracing

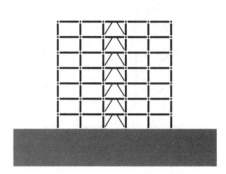
Eccentric Bracing

Moment Connections

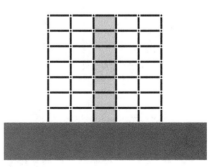
Shear Panels

FIGURE 11.29
Photograph of a moment connection similar to the one shown in Figure 11.28. The beam has just been bolted to a shear tab that is welded to the column. Next, backup bars will be welded to the column just under the beam flanges, after which the flanges will be welded to the column. (*Courtesy of American Institute of Steel Construction*)

FIGURE 11.30
Elevation views of the basic means for imparting lateral stability to a frame. Connections made with dots are shear connections only, and solid intersections indicate moment connections. Eccentric bracing is used in situations where it is advantageous for the frame to absorb seismic energy to prevent building collapse during an earthquake.

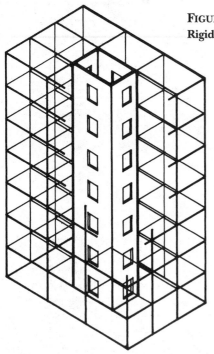

FIGURE 11.31
Rigid core versus rigid perimeter.

only, although a nominal number of moment connections may be included to provide additional rigidity or to restrict *drift,* which is a lateral displacement of the frame caused by wind or earthquake loads.

A second method of achieving stability is to make the perimeter of the building rigid, again by using diagonal braces, shear panels, or moment connections. When this is done, the entire interior of the structure can be assembled with shear connections, provided there is enough continuity in the floor system to attain *diaphragm action,* which is the rigidity possessed by a thin plate of material such as a welded steel deck with a concrete topping. Moment connections are more costly to make

building. The core, the area that contains the elevators, stairs, mechanical chases, and washrooms, is structured as a rigid tower, using diagonal braces, shear panels, or moment connections. The floors of the structure act as horizontal shear panels that make the outer bays of the building rigid by connecting them to the rigid core. The outer bays may then be structured with shear connections

Rigid Core

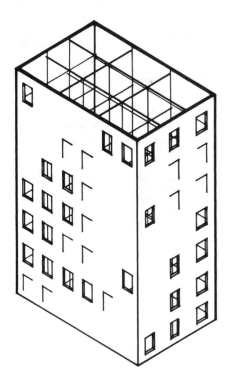

Rigid Perimeter

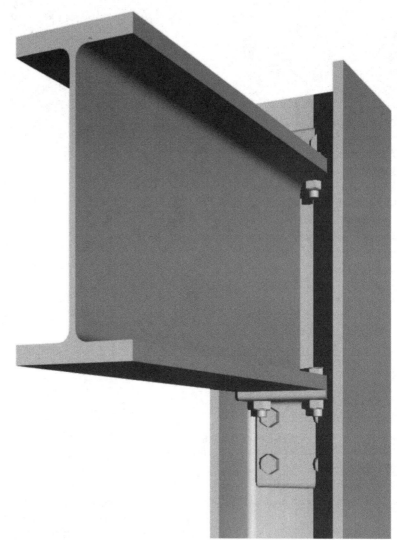

than shear connections, so the object in designing a scheme for lateral stabilization of a building is to use shear connections wherever possible, and moment connections only where necessary.

To summarize, shear connections are sufficient for most purposes in buildings that are stabilized by diagonal bracing or shear panels, and for many of the joints in buildings that are stabilized by moment connections. Moment connections may be used as the means for imparting lateral stability to a building instead of (or in addition to) diagonal bracing or shear panels. Moment connections are also utilized to connect cantilevered beams to columns (and sometimes to beams or girders).

The American Institute of Steel Construction (AISC) defines three types of Steel Frame construction, classified according to the manner in which they achieve stability against lateral forces. *AISC Type 1,* Rigid Frame construction, assumes that beam-to-column connections are sufficiently rigid that the geometric angles between members will remain virtually unchanged under loading. *AISC Type 2* construction, Simple Frame construction, assumes shear connections only and requires diagonal bracing or shear panels for lateral stability. *AISC Type 3* construction is defined as semirigid, in which the connections are not as rigid as those required for AISC Type 1 construction, but possess a dependable and

predictable moment-resisting capacity that can be used to stabilize the building.

A series of simple, fully bolted shear (AISC Type 2) connections are shown as a beginning basis for understanding steel connection details (Figures 11.25–11.27, 11.32, 11.35, 11.37). These are interspersed with a corresponding series of welded moment-resisting (AISC Type 1 and AISC Type 3) connections (Figures 11.28, 11.29, 11.33, 11.36). Welding is also widely used for making shear connections, examples of which are shown in Figures 11.34 and 11.35. A series of column connections is illustrated in Figures 11.38 through 11.41. In practice, there are a number of different ways of making any of

FIGURE 11.32
A seated beam-to-column-web connection. Although the beam flanges are connected to the column by a seat angle below and a stabilizing angle above, this is an AISC Type 2 (shear) connection, not a moment connection, because the two bolts are incapable of developing the full strength of the beam flange. This seated connection is used rather than a framed connection to connect to a column web because there is often insufficient space between the column flanges to insert a power wrench to tighten all the bolts in a framed connection.

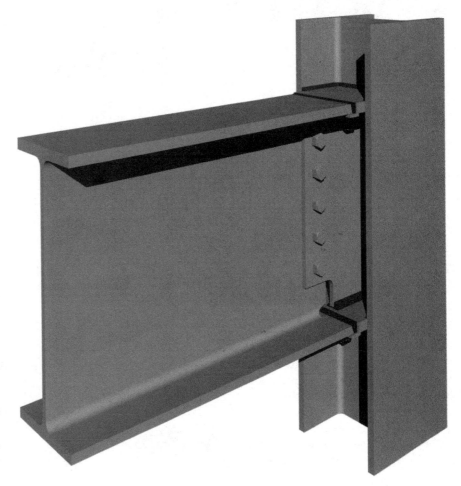

FIGURE 11.33
A welded beam-to-column-web connection (AISC Type 1). A vertical shear tab is welded to the web of the column at its centerline, and serves to receive bolts that join the column to the beam web and hold the beam in place during welding. The horizontal stiffener plates that are welded inside the column flanges are thicker than the beam flanges and extend out beyond the column flanges to reduce concentrations of stress at the welds.

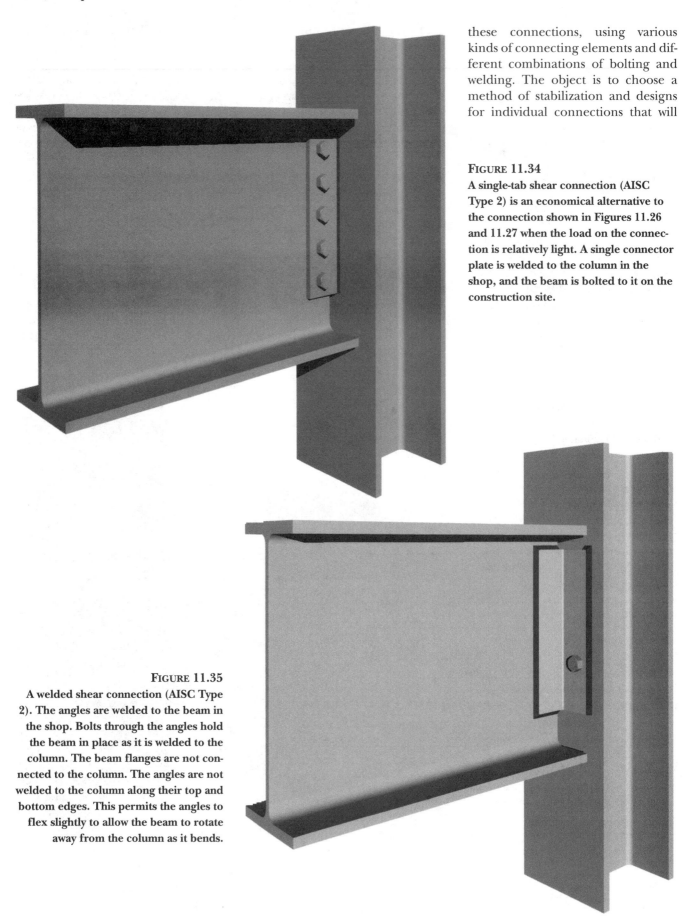

these connections, using various kinds of connecting elements and different combinations of bolting and welding. The object is to choose a method of stabilization and designs for individual connections that will

FIGURE 11.34
A single-tab shear connection (AISC Type 2) is an economical alternative to the connection shown in Figures 11.26 and 11.27 when the load on the connection is relatively light. A single connector plate is welded to the column in the shop, and the beam is bolted to it on the construction site.

FIGURE 11.35
A welded shear connection (AISC Type 2). The angles are welded to the beam in the shop. Bolts through the angles hold the beam in place as it is welded to the column. The beam flanges are not connected to the column. The angles are not welded to the column along their top and bottom edges. This permits the angles to flex slightly to allow the beam to rotate away from the column as it bends.

result in the greatest possible economy of construction for the building as a whole.

The choice of which connection to use is best left to the fabricator, who has firsthand knowledge of the safest, most erectable methods that will utilize the company's labor and equipment to the maximum efficiency. Rarely is it necessary for the structural engineer or architect to dictate a specific connection detail. Instead, the designer should concentrate on making sure that all the design loads are shown on the framing plans, and leave the design of the connections to the fabricator.

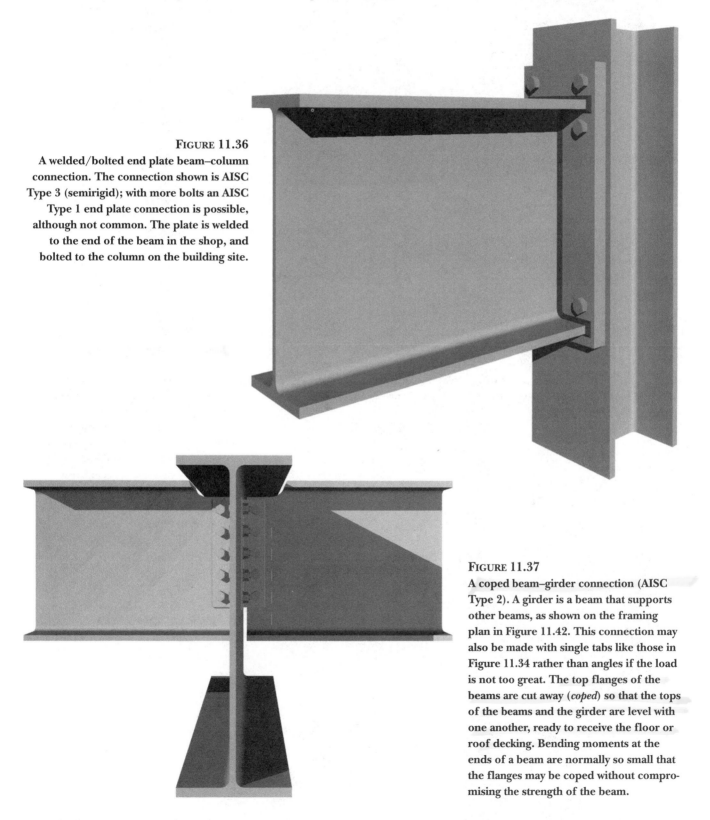

FIGURE 11.36
A welded/bolted end plate beam–column connection. The connection shown is AISC Type 3 (semirigid); with more bolts an AISC Type 1 end plate connection is possible, although not common. The plate is welded to the end of the beam in the shop, and bolted to the column on the building site.

FIGURE 11.37
A coped beam–girder connection (AISC Type 2). A girder is a beam that supports other beams, as shown on the framing plan in Figure 11.42. This connection may also be made with single tabs like those in Figure 11.34 rather than angles if the load is not too great. The top flanges of the beams are cut away (*coped*) so that the tops of the beams and the girder are level with one another, ready to receive the floor or roof decking. Bending moments at the ends of a beam are normally so small that the flanges may be coped without compromising the strength of the beam.

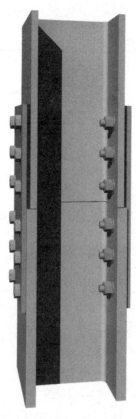

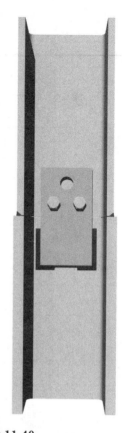

FIGURE 11.38
A bolted column–column connection for columns that are the same size. The plates are bolted to the lower section of column in the shop, and to the upper section on the site.

FIGURE 11.39
Column sizes diminish as the building rises, requiring frequent use of shim plates at connections to make up differences in flange thicknesses.

FIGURE 11.40
Column connections may be welded rather than bolted. The connector plate is welded to the lower column section in the fabricator's shop. The hole in the connector plate is used to attach a lifting line during erection. The bolts hold the column sections in alignment while the flanges are connected in the field with partial-penetration welds in bevel grooves. Partial-penetration welding allows one column to rest on the other prior to welding.

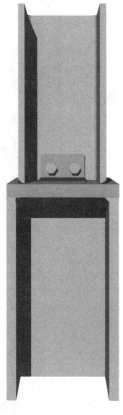

FIGURE 11.41
A welded butt plate connection is used where a column changes from one nominal size of wide-flange to another. The thick butt plate, which is welded to the lower column section in the shop, transfers the load from one section of column to the other. The bolts hold the joint in alignment while the partial-penetration weld at the base of the upper column is made on the site.

THE CONSTRUCTION PROCESS

A steel building frame begins as a rough sketch on the drafting board of an architect or engineer. As the building design process progresses, the sketch evolves through many stages of drawings and calculations to become a finished set of structural drawings (Figure 11.42). These show accurate column locations, the shapes and sizes of all the members of the frame, and all the loads of the members, but they do not give the exact length to which each member must be cut to mate with the members it joins, and they do not give details of the more routine connections of the frame. These are left to be worked out by a subsequent recipient of the drawings, the *fabricator.*

The Fabricator

The fabricator's job is to deliver steel components to the construction site ready to be assembled without further processing. This work begins with the preparation in the fabricator's shop of detailed drawings that show exactly how each piece will be made, and what its precise dimensions will be. Connections are designed to transmit the loads indicated by the engineer's drawings. Within the limits of accepted engineering practice, the fabricator is free to design the connections to be made as economically as possible, using various combinations of welding and bolting that best suit available equipment and expertise. Drawings are also prepared by the fabricator to show the general contractor exactly where and how to install foundation

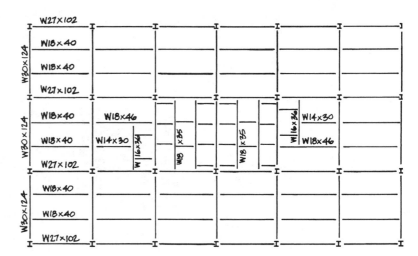

FIGURE 11.42
A typical framing plan for a multistory steel-framed building, showing size designations for beams and girders. Notice how this frame requires beam-to-column-flange connections where the W30 girders meet the columns, beam-to-column-web connections where the W27 beams meet the columns, and coped beam–girder connections where the W18 beams meet the W30 girders. The small squares in the middle of the building are openings for elevators, stairways, and mechanical shafts. An architect's or engineer's framing plan would also give dimensions between centers of columns, and would indicate the magnitudes of the loads that each joint must transfer, to enable the fabricator to design each connection.

anchor bolts to connect to the columns of the building, and to guide the erector in assembling the steel frame on the building site. When completed, the fabricator's *shop drawings* are submitted to the engineer and the architect for review and approval to be sure that they conform exactly to the intentions of the design team. Meanwhile, the fabricator places an order with a producer of steel for the stock from which the structural steel members will be fabricated. (The major beams, girders, and columns are usually ordered cut to exact length by the mill.) When the approved shop drawings, with corrections and comments, are returned by the design team, revisions are made as necessary, and full-size templates of cardboard or wood are prepared as required to assist the shop workers in laying out the various connections on the actual pieces of steel.

Plates, angles, and tees for connections are brought into the shop and cut to size and shape with oxyacetylene cutting torches, power shears, and saws. With the aid of the templates, bolt hole locations are marked. If the plates and angles are not unusually thick, the holes may be made rapidly and economically with a punching machine. In very thick stock, or in pieces that will not fit conveniently into the punching machines, holes are drilled rather than punched.

Pieces of steel stock for the beams, girders, and columns are brought into the fabricator's shop with an overhead traveling crane or conveyor system. Each is stenciled or painted with a code that tells which building it is intended for and exactly where it will go in the building. With the aid of the shop drawings, each piece is measured and marked for its exact length and for the locations of all holes, stiffeners, connectors, and other details. Cutting to length, for those members not already cut to length at the mill, is done with a power saw or an oxyacetylene cutting

torch. The ends of column sections that must bear fully on base plates or on one another are then squared and made perfectly flat by sawing, milling, or facing. In cases where the columns will be welded to one another, and for beams and girders that are to be welded, the ends of the flanges are beveled as necessary. Beam flanges are coped as required. Bolt holes are punched or drilled (Figure 11.43).

Where called for, beams and girders are *cambered* (curved slightly in an upward direction) so they will deflect into a straight line under load. Cambering may be accomplished by heating local areas of one flange of the member with a large oxyacetylene torch. As each area is heated to a cherry-red color, the metal softens, expands, and deforms to make a slight bulge in the width and thickness of the flange because the surrounding steel, which is cool, prevents the heated flange from lengthening. As the heated flange cools, the metal contracts, pulling the member into a slight bend at that

point. By repeating this process at several points along the beam, a camber of the desired shape and magnitude is produced. Cambering may also be done with a hydraulic ram that bends the beam enough to force a permanent deformation. Steel shapes can be bent to a smooth radius with a large machine that passes the shape through three rollers that flex it sufficiently to impart a permanent curvature (Figure 11.44).

As a last step in fabricating beams, girders, and columns, stiffener plates are arc welded to each piece as required, and connecting plates, angles, and tees are welded or bolted at the appropriate locations (Figure 11.45). As much connecting as possible is done in the shop, where tools are handy and access is easy. This saves time and money during erection, when tools and working conditions are less optimal and total costs per manhour are higher.

Plate girders, built-up columns, trusses, and other large components are assembled in the shop in as large

FIGURE 11.43
Punching bolt holes in a wide-flange beam. (*Courtesy of W. A. Whitney Corporation*)

The Construction Process / 373

FIGURE 11.44
Hollow steel rectangular tubing for this frame was bent into curves by the fabricator. Wide-flange shapes can also be bent. (*The Goldstein Partnership, Architects. Photo by Eliot Goldstein*)

There are 175,000 ironworkers in this country . . . and apart from our silhouettes ant-size atop a new bridge or skyscraper, we are pretty much invisible.

Mike Cherry,
On High Steel: The Education of an Ironworker,
New York, Quadrangle/
The New York Times Book Co.,
1974, p. xiii

FIGURE 11.45
Welders attach connector plates to an exceptionally heavy column section in a fabricator's shop. The twin channels bolted to the end of the column will be used to attach a lifting line for erection, after which they will be removed and reused. (*Courtesy of United States Steel Corporation*)

FIGURE 11.47
Three typical column base details. A small column with welded baseplate set on a steel leveling plate (upper left). A larger column with welded baseplate set on leveling nuts (upper right). A heavy column field welded to a loose baseplate that has been previously leveled and grouted (below).

FIGURE 11.46
Machine welding plates together to form a box column. The torches to the left preheat the metal to help avoid thermal distortions in the column. Mounds of powdered flux around the electrodes in the center indicate that this is the submerged arc process of welding. The small steel plates tack-welded onto the corners of the column at the extreme left are runoff bars, which are used to allow the welding machine to go past the end of the column to make a complete weld. These will be cut off as soon as welding is complete. (*Courtesy of United States Steel Corporation*)

units as can practically be transported to the construction site, whether by truck, railway, or barge (Figure 11.46). Intricate assemblies such as large trusses are usually preassembled in their entirety in the shop, to ensure they will go together smoothly in the field, then broken down again into transportable components.

As the members are completed, each is straightened, cleaned and painted as necessary, and inspected for quality and for conformance to the job specifications and shop drawings. The members are then taken from the shop to the fabricator's yard by crane, conveyor, trolley, or forklift, where they are organized in stacks according to the order in which they will be needed on the building site.

The Erector

Where the fabricator's job ends, the *erector's* begins. Some companies both fabricate and erect, but more often

the two operations are done by separate companies. The erector is responsible for assembling on the building site the steel components furnished by the fabricator. The erector's workers, by tradition, are called *ironworkers.*

Erecting the First Tier

Erection of a multistory steel building frame starts with assembly of the first two-story *tier* of framing. Lifting of the steel components is begun with a truck-mounted or crawler-mounted mobile crane. In accordance with the erection drawings prepared by the fabricator, the columns for the first tier, usually furnished in sections two stories high, are picked up from organized piles on the site and lowered carefully over the anchor bolts and onto the foundation, where the ironworkers bolt them down.

Foundation details for steel columns vary (Figure 11.47). Steel *baseplates,* which distribute the con-

centrated loads of the steel columns across a larger area of the concrete foundation, are shop welded to all but the largest of columns. The foundations and anchor bolts have been put in place previously by the general contractor, following the plan prepared by the fabricator. The contractor may, if requested, provide thin steel *leveling plates* that are set perfectly level and to the proper height on the bed of *grout* atop each concrete foundation. The baseplate of the column rests upon the leveling plate and is held down with the protruding anchor bolts. Alternatively, especially for larger baseplates with four anchor bolts, the leveling plate is omitted. The column is supported at the proper elevation on stacks of steel shims inserted between the baseplate and the foundation, or on leveling nuts placed beneath the baseplate on the anchor bolts. After the first tier of framing is plumbed up as described below, the baseplates are grouted and

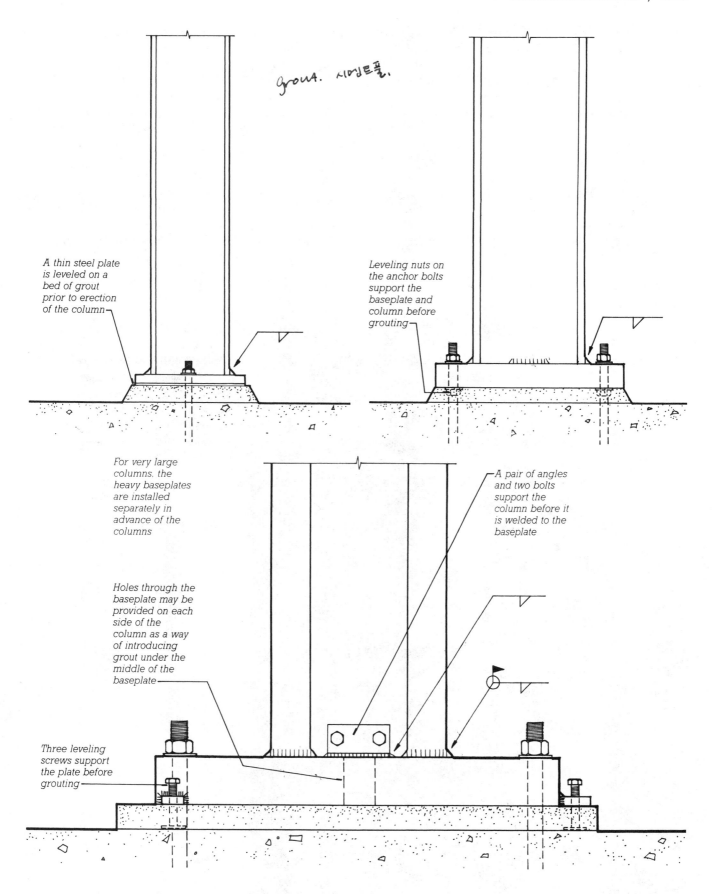

grout. 시멘트풀.

A thin steel plate is leveled on a bed of grout prior to erection of the column

Leveling nuts on the anchor bolts support the baseplate and column before grouting

For very large columns, the heavy baseplates are installed separately in advance of the columns

Holes through the baseplate may be provided on each side of the column as a way of introducing grout under the middle of the baseplate

A pair of angles and two bolts support the column before it is welded to the baseplate

Three leveling screws support the plate before grouting

the anchor bolts tightened. For very large, heavy columns, baseplates are shipped independently of the columns (Figure 11.48). Each baseplate is leveled in place with shims, wedges, or shop-attached leveling screws, then grouted prior to column placement.

After the first tier of columns has been erected, the beams and girders for the first two stories are bolted in place (Figures 11.49–11.55). The two-story tier of framing is then *plumbed up* (straightened and squared) using diagonal cables and turnbuckles, while checking the alignment with plumb bobs, transits, or laser levels. When the tier is plumb, connections are tightened, baseplates are grouted if necessary, welds are made, and permanent diagonal braces, if called for, are rigidly attached. Ironworkers scramble back and forth, up and down on the columns and beams, protected from falling by safety harnesses and life lines.

At this level, a temporary working surface of 2- or 3-inch (50- or 75-mm) wood planks or corrugated steel decking may be laid over the steel framing. Similar platforms will be placed every second story as the frame rises, unless safety nets are used instead, or the permanent floor decking is installed as erection progresses. The platforms protect workers on lower levels of the building from falling objects. They also furnish a convenient working surface for tools, materials, and derricks. Column splices are made at waist level above this platform, both as a

FIGURE 11.48
Ironworkers guide the placement of a heavy column fabricated by welding together two rolled wide-flange sections and two thick steel plates. It will be bolted to its baseplate through the holes in the small plate welded between the flanges on either side. (*Courtesy of Bethlehem Steel Corporation*)

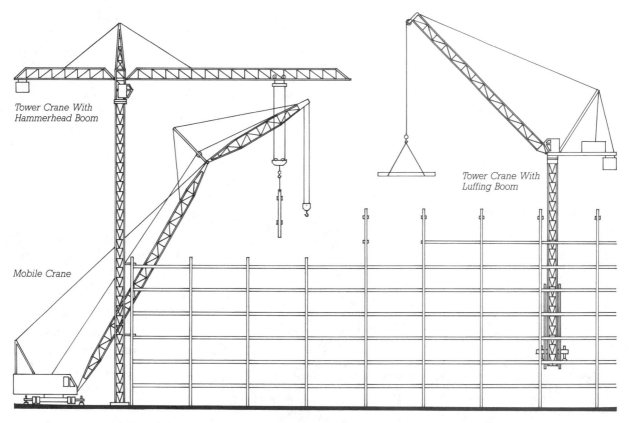

Tower Crane With Hammerhead Boom

Tower Crane With Luffing Boom

Mobile Crane

FIGURE 11.49
Two common types of tower cranes and a mobile crane. The luffing-boom crane can be used in congested situations where the movement of the hammerhead boom would be limited by obstructions. Both can be mounted on either external or internal towers. The internal tower is supported in the frame of the building, while the external tower is supported by its own foundation and braced by the building. Tower cranes climb as the building rises by means of self-contained hydraulic jacks.

FIGURE 11.50
An ironworker clips his body harness to a lifeline as he moves around a column.
(*Photo by James Digby. Courtesy of LPR Construction Company*)

matter of convenience and as a way of avoiding conflict between the column splices and the beam-to-column connections. Columns are generally fabricated in two-story lengths, a transportable size that also corresponds to the two-story spacing of the plank surfaces.

Erecting the Upper Tiers

Erection of the second tier proceeds much like that of the first. Two-story column sections are hoisted into position and connected by splice plates to the first tier of columns. The beams and columns for the two floors are set, the tier is plumbed and tightened up, and another layer of planks, decking, or safety netting is installed.

If the building is not too tall, the mobile crane will do the lifting for the entire building. For a taller building, the mobile crane does the work until it gets to the maximum height to which it can lift a climbing crane or

FIGURE 11.51
A tower crane lowers a series of beams to ironworkers. The worker to the left uses a tagline to maneuver the lowest beam into the proper orientation. (*Photo by James Digby. Courtesy of LPR Construction Company*)

> If nobody plumbed-up, all the tall buildings in our cities would lean crazily into each other, their elevators would scrape and bang against the shaft walls, and the glaziers would have to redress all the windowglass into parallelograms. . . . What leaned an inch west on the thirty-second floor is sucked back east on the thirty-fourth, and a column that refused to quit leaning south on forty-six can generally be brought over on forty-eight, and by the time the whole job is up the top is directly over the bottom.
>
> **Mike Cherry, *On High Steel: The Education of an Ironworker*, New York, Quadrangle/The New York Times Book Co., 1974, pp. 110–111**

derrick. A larger crane takes over from there, usually a *tower crane* (Figure 11.49), which builds itself an independent tower as the building rises, either alongside the building or within an elevator shaft or a vertical space temporarily left open in the frame.

As each piece of steel is lowered toward its final position in the frame, it is guided by an ironworker who holds a rope called a *tagline,* the other end of which is attached to the piece. Other ironworkers in the raising gang guide the piece by hand as soon as they can reach it, until its bolt holes align with those in the mating pieces (Figure 11.53). Sometimes crowbars or hammers must be used to pry, wedge, or drive components until they fit properly, and bolt holes may on occasion have to be reamed larger to admit bolts through slightly misaligned pieces. When an approxi-

FIGURE 11.52
Connecting a beam to a column. (*Courtesy of Bethlehem Steel Corporation*)

FIGURE 11.53
Ironworkers attach a girder to a box column. Each worker carries two wrench–drift pin combination tools in a holster on his belt, inserting the tapered drift pins into bolt holes in each connection to hold it until a few bolts can be added. Bundles of corrugated steel decking are ready to be opened and distributed over the beams to make a floor deck. (*Courtesy of Bethlehem Steel Corporation*)

FIGURE 11.54
Bolting joist girders to a column. (*Courtesy of Vulcraft Division of Nucor Corporation*)

mate alignment is achieved, tapered steel *drift pins* from the ironworker's tool belt are shoved into enough bolt holes to hold the pieces together until a few bolts can be inserted. A gang of bolters follows behind the raising gang, filling the remaining holes with bolts from leather carrying baskets, tightening them first with hand wrenches and then with impact wrenches. Field-welded connections are initially held in alignment with bolts, then welded when the frame is plumb.

The placing of the last beam at the top of the building is carried out with a degree of ceremony appropriate to the magnitude of the building. At the very least, a small evergreen tree, a national flag, or both are attached to the beam before it is

FIGURE 11.55
Welding open-web steel joists to a wide-flange beam. (*Courtesy of Vulcraft Division of Nucor Corporation*)

FIGURE 11.56
Topping out: The last beam in a steel frame is special.
(*Courtesy of United States Steel Corporation*)

lifted (Figure 11.56). For major buildings, assorted dignitaries are likely to be invited to a building-site *topping-out* party that includes music and refreshments. After the party, work goes on as usual, for although the frame is complete, the building is not. Cladding and finishing operations will continue for many months.

Floor and Roof Decking

If plank decks are used during erection of the frame, they must be replaced with permanent floor and roof decks of incombustible materials. In early steel frame buildings, shallow arches of brick or tile were often built between the beams, tied with steel tension rods, and filled over with concrete to produce level surfaces (Figure 11.58). These were both heavier, necessitating larger framing members to carry their weight, and more consumptive of labor, than the metal deck systems commonly used today.

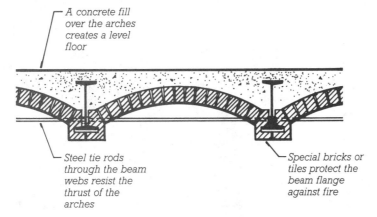

FIGURE 11.57
A 10-story steel frame nears completion. The lower floors have already been decked with corrugated steel decking.
(*Courtesy of Vulcraft Division of Nucor Corporation*)

A concrete fill over the arches creates a level floor

Steel tie rods through the beam webs resist the thrust of the arches

Special bricks or tiles protect the beam flange against fire

FIGURE 11.58
Tile or brick arch flooring is found in many older steel frame buildings.

Metal Decking

Metal decking, at its simplest, is a sheet of steel that has been corrugated to increase its stiffness. The spanning capability of the deck is determined mainly by the thickness of the sheet from which it is made and the depth and spacing of the corrugations.

Single corrugated sheets are commonly used without a concrete topping for roof decking, where concentrated loads are not expected to be great and deflection criteria are not as stringent as in floors. They are also used as permanent formwork for concrete floor decks, with a wire-fabric-reinforced concrete slab supported by the steel decking until the slab can support itself and its live loads (Figures 11.59–11.61). *Cellular decking* is manufactured by welding together two sheets, one corrugated and one flat, and can be made sufficiently stiff to support normal floor

FIGURE 11.59
Workers install corrugated steel form decking over a floor structure of open-web steel joists supported by joist girders. (*Photo by Balthazar Korab. Courtesy of Vulcraft Division of Nucor Corporation*)

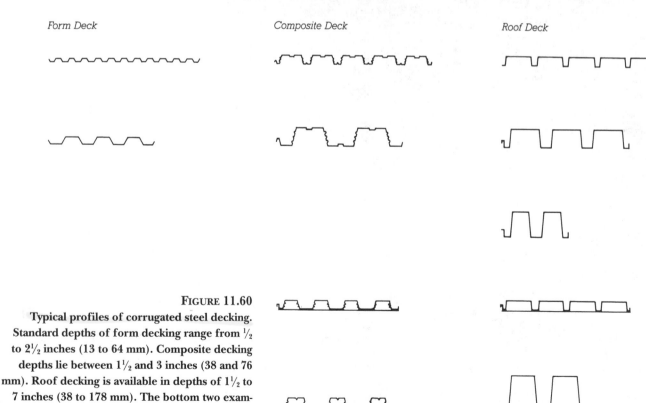

FIGURE 11.60
Typical profiles of corrugated steel decking. Standard depths of form decking range from $1/2$ to $2^1/2$ inches (13 to 64 mm). Composite decking depths lie between $1^1/2$ and 3 inches (38 and 76 mm). Roof decking is available in depths of $1^1/2$ to 7 inches (38 to 178 mm). The bottom two examples of composite and roof decking are cellular.

loads without structural assistance from the concrete fill that is poured over it to produce a level floor. Cellular decking offers the important side benefit of providing spaces for running electrical and communications wiring, as illustrated in Chapter 24. Metal decking is usually puddle welded to the joists, beams, and girders at intervals by melting through the decking to the supporting members below with a welding electrode. Self-drilling, self-tapping screws or powder-driven pins may also be used for decking attachment. If the deck is required to act as a diaphragm, the longitudinal edges of the decking panels must be connected to one another at frequent intervals with screws or welds.

Composite Construction

Composite metal decking (Figure 11.62) is designed to work together with its concrete fill to make a stiff, lightweight, economical deck. The decking itself serves as tensile reinforcing for the concrete, which is bonded to it by special rib patterns or by small steel rods or wire fabric welded to the tops of the corrugations.

FIGURE 11.61
Samples of corrugated steel decking. The second sample from the bottom achieves composite action by keying of the concrete topping to the deformations in the decking. The bottom sample has a closed end, which is used at the perimeter of the building to prevent the concrete topping from escaping during pouring. (*Courtesy of Wheeling Corrugating Company Division, Wheeling-Pittsburgh Steel Corporation*)

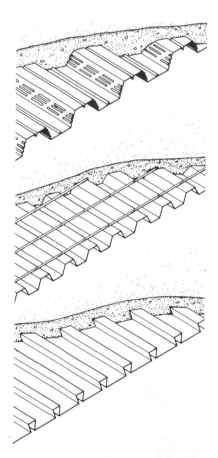

FIGURE 11.62
Composite decking acts as steel reinforcing for the concrete topping installed over it. The top example bonds to the concrete with deformed ribs, and the middle example with welded steel rods. The bottom type makes an attractive ceiling texture if left exposed and furnishes dovetail channels for the insertion of special fastening devices to hang ductwork, piping, conduits, and machinery from the ceiling.

Composite construction is often carried a step beyond the decking to include the beams of the floor. Before the concrete is poured over the metal deck, *shear studs* are welded through the decking to the top of each beam a few inches apart, using a special electric welding gun (Figures 11.63, 11.64). It would be more economical to attach the shear studs in the shop rather than in the field, but the danger of tripping up ironworkers during the erection process delays their installation until the steel decking is in place. The purpose of the studs is to create a strong shear connection between the concrete slab and the steel beam. A strip of the slab can then be assumed to act together with the top flange of the steel shape to resist compressive forces. The result of composite design is a steel member whose loadbearing capacity has been greatly enhanced at relatively low cost by taking advantage of the unused strength of the concrete topping that must be present in the construction anyway. The payoff is a stiffer, lighter, less expensive frame.

Concrete Decks

Concrete floor and roof slabs are often used in steel building frames instead of metal decking and concrete fill. Concrete may be poured in place over removable plywood forms, or it may be erected as precast planks lifted into place much like the metallic elements of the building (Figure 11.65). Precast decks are relatively light in weight and are quick to erect, even under weather conditions that would preclude the pouring of concrete, but they usually require the addition of a thin, poured-in-place concrete topping to produce a smooth floor.

FIGURE 11.63
Composite beam construction.

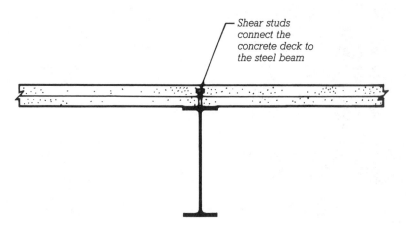

Shear studs connect the concrete deck to the steel beam

FIGURE 11.64
Pouring a concrete fill on a steel roof deck, using a concrete pump to deliver the concrete from the street below to the point of the pour. Shear studs are plainly visible over the lines of the beams below. The welded wire fabric reinforces the concrete against cracking. (*Courtesy of Schwing America, Inc.*)

FIGURE 11.65
A tower crane installs precast concrete hollow-core planks for floor decks in an apartment building. Precast concrete is also used for exterior cladding of the building. The steel framing is a design known as the *staggered truss system*, in which story-height steel trusses at alternate levels of the building support the floors. The trusses are later enclosed with interior partitions. (*Courtesy of Blakeslee Prestress, Inc.*)

Roof Decking

For roofs of low steel-framed buildings, many different types of decks are available. Corrugated metal may be used, with or without a concrete fill; many types of rigid insulation boards are capable of spanning the corrugations to provide a flat surface for the roof membrane. Some corrugated decks are finished with a weather-resistant coating that allows them to serve as the water-resistive surface of the roof. A number of different kinds of insulating deck boards are produced of such fibers as wood and glass bonded with various substances. Many of the insulating boards are designed as permanent

FIGURE 11.66
Welding truss-tee subpurlins to open-web steel joists for a roof deck. (*Courtesy of Keystone Steel and Wire*)

FIGURE 11.67
Installing insulating formboard over truss-tee subpurlins. A light wire reinforcing mesh and poured gypsum fill will be installed over the formboards. The trussed top edges of the subpurlins become embedded in the gypsum slab to form a composite deck. (*Courtesy of Keystone Steel and Wire*)

formwork for reinforced, poured slabs of gypsum or lightweight concrete; in this application, they are usually supported on steel *subpurlins* (Figures 11.66, 11.67). Heavy timber decking, or even wood joists and plywood sheathing, are also used over steel framing in situations where building codes permit combustible materials.

Corrugated steel sheets are often used for siding of industrial buildings, where they are supported on *girts,* which are horizontal zees or channels that span between the outside columns of the building (see Figure 11.80).

FIREPROOFING OF STEEL FRAMING

Building fires are not hot enough to melt steel, but are often able to weaken it sufficiently to cause structural failure (Figures 11.68, 11.69). For this reason, building codes generally limit the use of exposed steel framing to buildings of one to three stories, where escape in case of fire is rapid and where collapse of the building is unlikely to endanger people or other buildings. For taller buildings, it is necessary to protect the steel frame from heat for a length of time

sufficient that the building can be fully evacuated and the fire extinguished.

Fireproofing (fire protection might be a more accurate term) of steel framing was originally done by encasing steel beams and columns in brick masonry or poured concrete (Figures 11.70, 11.71). These heavy encasements were effective, absorbing heat into their great mass and dissipating some of it through dehydration of the mortar and concrete, but their weight added considerably to the load that the steel frame had to support and, therefore, to the weight and cost of the frame. The search for

FIGURE 11.68
An exposed steel structure following a prolonged fire in the highly combustible contents of a warehouse. (*Courtesy of National Fire Protection Association*)

FIGURE 11.69
The relationship between temperature and strength in structural steel. (*Courtesy of American Iron and Steel Institute*)

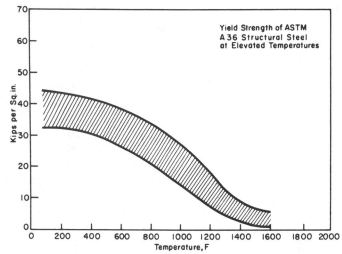

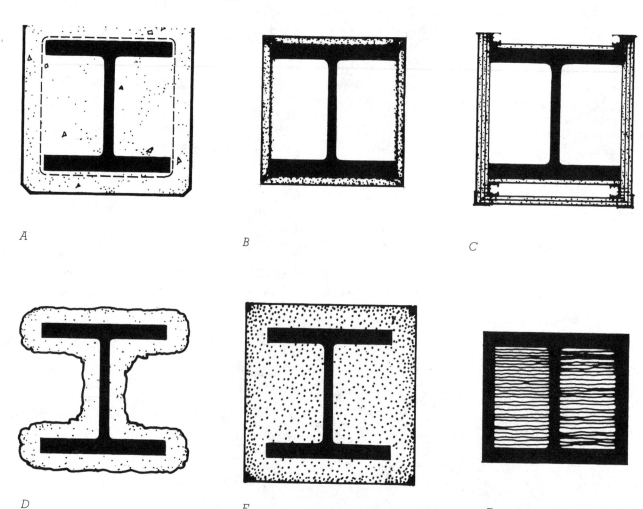

A

B

C

D

E

F

FIGURE 11.70
Some methods for fireproofing steel columns: (*a*) Encasement in reinforced concrete. (*b*) Enclosure in metal lath and plaster. (*c*) Enclosure in multiple layers of gypsum board. (*d*) Spray-on fireproofing. (*e*) Loose insulating fill inside a sheet metal enclosure. (*f*) Water-filled box column made of a wide-flange shape with added steel plates.

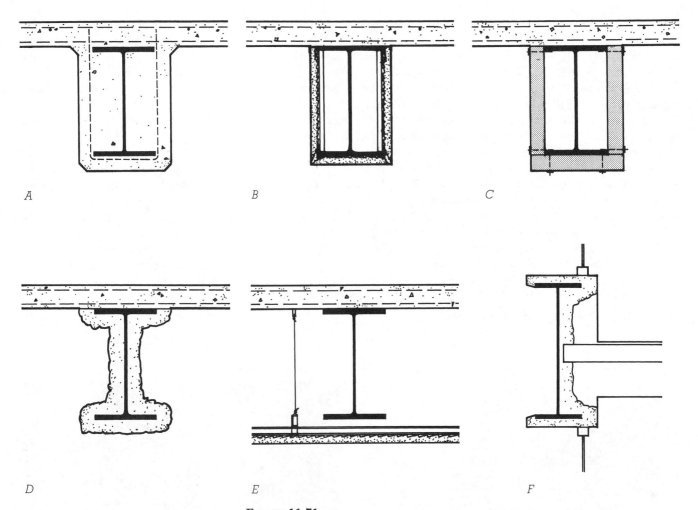

A

B

C

D

E

F

FIGURE 11.71
Some methods for fireproofing steel beams and girders: (*a*) **Encasement in reinforced concrete.** (*b*) **Enclosure in metal lath and plaster.** (*c*) **Rigid slab fireproofing.** (*d*) **Spray-on fireproofing.** (*e*) **Suspended plaster ceiling.** (*f*) **Flame-shielded exterior spandrel girder with spray-on fireproofing inside.**

lighter-weight fireproofing led first to thin enclosures of metal lath and plaster around the steel members (Figures 11.70–11.72). These derive their effectiveness from the large amounts of heat needed to dehydrate the water of crystallization from the gypsum plaster. Plasters using lightweight aggregates such as vermiculite instead of sand have come into use to further reduce the weight and to add thermal insulating properties to the plaster.

Today's designers can also choose from a group of fireproofing techniques that are lighter still. Plaster fireproofing has largely been replaced by beam and column enclosures made of boards or slabs of gypsum or other fire-resistive materials (Figures 11.70–11.75). These are fastened mechanically around the steel shapes, and, in the case of the gypsum board fireproofing, they can also serve as the finished surface on the interior of the building. Where the fireproofing material need not serve as a finished surface, spray-on materials have become the most prevalent type. These generally consist of a fiber and a binder, or of a cementitious mixture, and are sprayed over the steel to the required thickness (Figure 11.76). These products are available in densities of about 12 to 40 pounds per cubic foot (190 to 640 kg/m³). The lighter materials are fragile and unattractive, and must be covered with finish materials. The denser materials are generally more durable and attractive. All spray-on materials act primarily by insulating the steel from high temperatures for long periods of time. They are generally the least expensive form of fireproofing.

The latest generation of fireproofing techniques for steel offers new possibilities to the designer. *Intumescent mastics and paints* are coatings that allow steel structural elements to remain exposed to view in situations of low to moderate fire risk. They expand when exposed to fire to form a stable char that insulates the steel from the heat of the fire for varying lengths of time, depending on

FIGURE 11.72
Lath-and-plaster fireproofing around a steel beam. (*Courtesy of United States Gypsum Company*)

FIGURE 11.73
Gypsum board fireproofing around a steel column. The gypsum board layers are screwed to the four cold-formed steel C-channels at the corners of the column, and finished with steel corner bead and drywall compound on the corners. (*Courtesy of United States Gypsum Company*)

FIGURE 11.74
Attaching slab fireproofing made of mineral fiber to a steel column, using welded attachments. (*Courtesy of United States Gypsum Company*)

FIGURE 11.75
Slab fireproofing on a steel beam. (*Courtesy of United States Gypsum Company*)

FIGURE 11.76
Applying spray-on fireproofing to a steel beam, using a gauge to measure the depth. (*Courtesy of W. R. Grace & Co.*)

the thickness of the coating. Most intumescent coatings are available in an assortment of colors and can also serve as a base coat under ordinary paints if another color is desired.

A rather specialized technique, applicable only to steel box or tube columns exposed on the exterior of buildings, is to fill the columns with water and antifreeze (Figure 11.70*f*). Heat applied to a region of a column by a fire is dissipated throughout the column by convection in the liquid filling.

Mathematical and computer-based techniques have been developed for calculating temperatures that will be reached by steel members in various situations during a fire. These allow the designer to experiment with a variety of ways of protecting the members, including metal flame shields that allow the component to be left exposed on the exterior of a building (Figure 11.71*f*).

LONGER SPANS IN STEEL

Standard wide-flange beams are suitable for the range of structural spans normally encountered in offices, schools, hospitals, apartments, hotels, retail stores, warehouses, and other buildings in which columns may be brought to earth at intervals without obstructing the activities that take place within. For many other types of buildings—athletic buildings, certain types of industrial buildings, aircraft hangars, auditoriums, theaters, religious buildings, transportation terminals—longer spans are required than can be accomplished with wide-flange beams. A rich assortment of longer-span structural devices is available in steel for these uses.

Improved Beams

One general class of longer-span devices might be called improved beams. The *castellated beam* (Figures

11.77, 11.78) is produced by flame cutting the web of a wide-flange section along a zigzag path, then reassembling the beam by welding its two halves point to point, thus increasing its depth without increasing its weight. This greatly augments the spanning potential of the beam, provided the superimposed loads are not exceptionally heavy. For long-span beams tailored to any loading condition, *plate girders* are custom designed and fabricated. Steel plates

and angles are assembled by bolting or welding in such a way as to put the steel exactly where it is needed: The flanges often become thicker in the middle of the span where bending forces are higher, more web stiffeners are provided near the ends where shear stresses are high, and so on. Almost any depth can be manufactured as needed, and very long spans are possible, even under heavy loads (Figure 11.79). Often these members are tapered, having greater depth

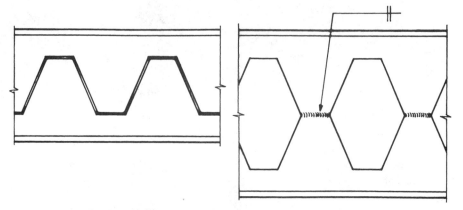

FIGURE 11.77
Manufacture of a castellated beam.

FIGURE 11.78
Castellated beams and girders frame into a wide-flange column. (*Courtesy of Castelite Steel Products, Midlothian, Texas*)

where the bending moment is largest. *Rigid frames* are easily produced by welding together steel wide-flange sections or plate girders. They may be set up in a row to roof a rectangular space (Figure 11.80) or arrayed around a vertical axis to cover a circular area. Castellated beams, plate girders, and rigid frames share the characteristic that they must be braced laterally by purlins, decking, or diagonal bracing to prevent them from buckling.

FIGURE 11.79
Erecting a welded steel plate girder. Notice how the girder is custom made with cutouts for the passage of pipes and ductwork. The section being erected is 115 feet (35 m) long, 13 feet (4 m) deep, and weights 192,000 pounds (87,000 kg).
(*Courtesy of Bethlehem Steel Corporation*)

FIGURE 11.80
The steel rigid frames of this industrial building carry steel purlins that will support the roof deck, and girts to support the corrugated steel wall cladding. The depth of each frame varies with the magnitude of the bending forces and is greatest at the eave connections, where these forces are at a maximum.
(*Courtesy of Metal Building Manufacturers Association*)

Trusses

Steel *trusses* (Figures 11.81–11.84) are generally deeper and lighter than improved beams and can span correspondingly longer distances. They can be designed to carry light or heavy loads. Earlier in this chapter, one class of steel trusses, open-web joists and joist girders, was presented. These are light members for light loadings but are capable of fairly long spans, and they are usually less expensive than custom-made trusses. Custom-made roof trusses for light loadings are most often made up of paired-angle top and bottom *chords* with paired-angle internal members. These are joined by gusset plate connectors that may be either welded or bolted. Trusses for heavier loadings, such as the transfer trusses that are used in some building frames to transmit column loads from floors

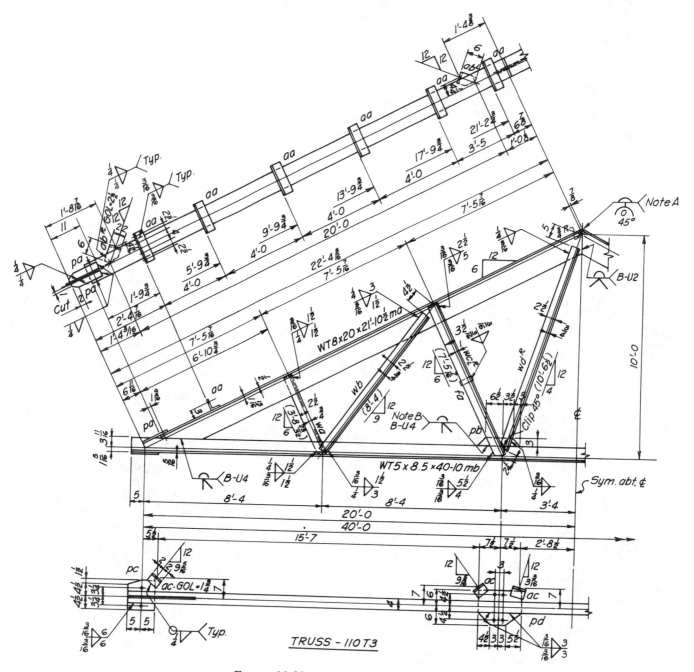

FIGURE 11.81

A fabricator's shop drawing of a welded steel roof truss made of tees and paired-angle chords. (*From* Detailing for Steel Construction, *Chicago, AISC, 1983. Reproduced by permission of the American Institute of Steel Construction*)

above across a wide meeting room or lobby in a building, can be made of wide-flange or tubular shapes. Trusses generally must be braced laterally to prevent buckling.

A steel *space truss* (more popularly called a *space frame*) is a truss made three-dimensional (Figures 11.85, 11.86). It carries its load by bending along both its axes, much like a two-way concrete slab (Chapter 13). It must be supported by columns that are spaced more or less equally in both directions.

FIGURE 11.82
Bolted steel roof trusses over a shopping mall support steel purlins that carry the corrugated steel roof deck. (*Courtesy of American Institute of Steel Construction*)

FIGURE 11.83
Ironworkers seat the end of a heavy roof
truss made of wide-flange sections.
(*Courtesy of American Institute of Steel
Construction*)

FIGURE 11.84
Tubular steel trusses support the roof of
a convention center. (*Courtesy of American
Institute of Steel Construction*)

FIGURE 11.85
Assembling a space truss. (*Courtesy of Unistrut Space-Frame Systems, GTE Products Corporation*)

FIGURE 11.86
A space truss carries the roof of a ferry terminal. (*Architects: Braccia/DeBrer/Heglund. Structural engineer: Kaiser Engineers. Photo by Barbeau Engh. Courtesy of American Institute of Steel Construction*)

FIGURE 11.87
Erecting the steel dome at Disney World.
(© *Walt Disney Productions. Photo courtesy of American Institute of Steel Construction*)

Arches

Steel *arches,* produced by bending standard wide-flange shapes or by joining plates and angles, can be made into cylindrical roof vaults or circular domes of considerable span (Figure 11.87). For greater spans still, the arches may be built of steel trusswork. Lateral thrusts are produced at the base of an arch and must be resisted by the foundations or by a tie rod.

Tensile Structures

High-tensile-strength wires of cold-drawn steel, made into cables, are the material for a fascinating variety of tentlike roofs that can span very long distances (Figures 11.88, 11.89). With *anticlastic* (saddle-shaped) curvature, cable *stays,* or other means of restraining the cable net, hanging roofs are fully rigid against wind uplift and flutter. For smaller spans, fabrics can do most of the work, supported by steel cables along the edges and at points of maximum stress.

FIGURES 11.88, 11.89
The Olympic stadium roof in Munich, Germany, is made of steel cables and transparent acrylic plastic panels. For scale, notice the worker seen through the roof at the upper left of Figure 11.89.
(*Architects: Frei Otto, Ewald Bubner, and Benisch and Partner. Photo courtesy of Institute for Lightweight Structures, Stuttgart*)

The spider web is a good inspiration for steel construction.

Frank Lloyd Wright,
"In the Cause of Architecture:
The Logic of the Plan,"
Architectural Record,
January 1928

FABRIC STRUCTURES

Fabric structures are not new: People have constructed tents since the earliest days of human civilization. But during the last several decades, new, durable fabrics and computerized methods for finding form and forces have helped to create a new construction type: a permanent, rigid, stable fabric structure that will last for 20 years or more.

Types of Fabric Structures

Fabric structures may be grouped into two general classifications, tensile and pneumatic structures (Figure A). A *tensile structure* is a membrane that is supported by masts or other rigid structural elements such as frames or arches. The membrane usually consists of a woven textile fabric, and is generally reinforced with steel cables along the main lines of stress. The fabric and cables transmit external loads to the rigid supports and ground anchors by means of tensile forces.

Pneumatic structures depend on air pressure for their stability and their capacity to carry snow and wind loads. The most common type of pneumatic structure is the *air-supported* structure, in which an airtight fabric, usually reinforced with steel cables, is held up by pressurizing the air in the inhabited space below it. The fabric and cables in an air-supported structure are stressed in tension.

Fabrics for Permanent Structures

Nearly all fabric structures are made of woven cloth that has been coated with a synthetic material. The cloth provides structural strength to resist the tensile forces in the structure, and the coating makes the fabric airtight and water resistant. The most widely used fabric is polyester cloth that has been laminated or coated with polyvinyl chloride (vinyl, PVC). Two other frequently used fabrics are based on glass fiber cloth: One is coated with polyte-

trafluorethylene (PTFE, one brand of which is known as Teflon); the other is coated with silicone. The polyester/PVC fabric is the most economical of the three, but it does not meet U.S. building code requirements for a noncombustible material and is used predominantly for smaller structures. The glass/PTFE and glass/silicone fabrics are classified as noncombustible. They also have longer life spans. Though it is more expensive, the glass/PFTE fabric remains clean longer than the other two. All three fabrics are highly resistant to such forces of deterioration as ultraviolet light, oxidation, and fungi.

Fabrics may be white, colored, or imprinted with patterns or graphics. A fabric may be totally opaque, obstructing all passage of light, or it may be translucent, allowing a controlled percentage of light to pass through.

Though a single layer of fabric has little resistance to the flow of heat, a properly designed fabric structure can achieve substantial energy savings over conventional enclosures through selective use of translucency and reflectivity. Translucency can be used to provide natural illumination, gather solar heat in the winter, and cool the space at night in the summer. A highly reflective fabric can reduce solar heat gain and conserve artificial illumination. With the addition of a second layer, a fabric liner that is suspended about a foot (300 mm) below the structural fabric, the thermal resistance of the structure can be improved. An acoustic inner liner can help to control internal sound reflection, which is especially important in air-supported structures, which tend to focus sound. Tensile structures, because of their anticlastic curvature, have little tendency to focus sound.

Tensile Structures

Tensile structures are stabilized by anticlastic curvature and prestress. *Anticlastic curvature* means that the fabric is curved simultaneously in two opposite directions. *Prestress*

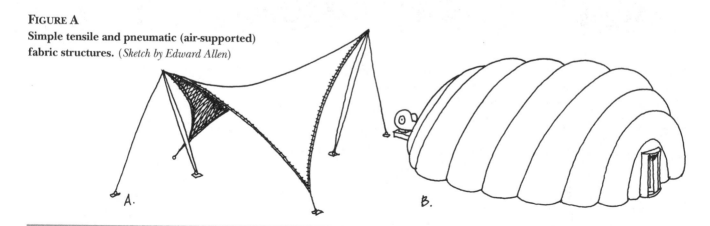

FIGURE A

Simple tensile and pneumatic (air-supported) fabric structures. (*Sketch by Edward Allen*)

A.

B.

FIGURE B
The world's largest roof structure covers the Haj Terminal in Jeddah, Saudi Arabia, a colossal airport facility that is used to facilitate the travel of vast numbers of Muslim faithful during a short period of annual pilgrimage. The roof is made up of radial shapes of fabric.
(*Architects: Skidmore, Owings & Merrill. Roof designer and structural engineer: Geiger Berger Associates. Photographs for Figures B through G are furnished by courtesy of the photographer, Horst Berger*)

FIGURE C
The fabric of the Haj Terminal is PFTE-coated glass fiber cloth.

FIGURE D
The roof canopy of the San Diego Convention Center is supported at the perimeter on the concrete frame of the building below.

FIGURE E
The San Diego Convention Center roof is raised at the middle on rigid steel struts that rest on cables supported by the concrete frame. (*Architects: Arthur Erickson Associates. Roof designer and structural engineer: Horst Berger Partners*)

FIGURE F
The form of the Denver International Airport roof combines saddle and radial shapes to echo the forms of the surrounding mountains. (*Architects: C. W. Fentriss, J. H. Bradburn & Associates. Roof designer and structural engineer: Severud Associates, Horst Berger, Principal Consultant*)

FIGURE G
Like the petals of a giant flower, tensile structures arranged in a huge circle shade the grandstand of King Fahd Stadium in Riyadh, Saudi Arabia. The masts are 58 m (190 ft) high. (*Architects: Ian Fraser, John Roberts & Partners. Roof designer and structural engineer: Geiger Berger Associates*)

is the introduction of permanent tension into the fabric in two opposing directions. Without anticlastic curvature and prestress, the fabric would flutter in the wind and destroy itself within a short time. The amount of curvature and the amount of prestressing force must both be sufficient to maintain stability under any expected wind and snow conditions. If the curvature is too flat, or if the prestressing tension is too low, excessive deflection or flutter will occur.

Two basic geometries may be used to create anticlastic curvature: One is the saddle shape (Figures D, E, G), the other the radial tent (Figures B, C). It is from combinations and variations of these geometries, as in Figure F, that all tensile structures are shaped.

The design of a tensile structure usually begins by experimenting with simple physical models. These often are based on pantyhose material or stretch fabric, either of which is easily stretched and manipulated. After a general shape has been established, a computer is used to find the exact equilibrium shape, determine the stresses in the fabric and supporting members under wind and snow loadings, and generate cutting patterns for the fabric. The design process is referred to as *form finding*, because a tensile structure cannot be made to take any arbitrary shape. Just as a hanging chain will always take a form that places its links in equilibrium with one another, a tensile structure must take a form that maintains proportionate amounts of tension in all parts of the fabric under all expected loading conditions. The designer's task is to find such a form.

A good design for a tensile structure employs short masts, so as to minimize buckling problems. The fabric generally cannot come to a peak at the mast, but must terminate in a cable ring that is attached to the mast in order to avoid high tensile stresses in the fabric. The perimeter edges of the fabric usually terminate along curving steel cables. To make a stable structure, these cables must have adequate curvature and must be anchored to foundations that offer firm resistance to uplift forces. The attachment of the fabric to the cables may be by means of sleeves sewn into the edges of the fabric or clamps that grip the fabric and pull it toward the cable.

Air-Supported Structures

Air-supported structures are pressurized by the ventilating fans that are used to heat and cool the building. The required air pressures are so low that they are scarcely discernible by people entering or leaving the building, but they are high enough (5 to 10 pounds per square foot, or 0.25 to 0.50 kPa) that they would prevent ordinary swinging doors from opening. For this reason, and because they maintain a continual seal against loss of air, revolving doors are usually used for access.

The fabric of an air-supported structure is prestressed by its internal air pressure to prevent flutter. For low-profile roof shapes, a cable net is employed to resist the high forces that result from the flat curvature. The fabric spans between the cables. The fabric and cables pull up on the foundations with a total force that is equal to the internal air pressure multiplied by the area of ground covered by the roof. The supporting elements and foundations must be designed to resist this force.

Wind causes suction forces to occur on many areas of an air-supported structure, which results in additional tension in the fabric and cables. The downward forces from wind or snow load on an air-supported structure must be resisted directly by the internal air pressure pushing upward against them. In geographic areas where snow loads may be larger than acceptable internal pressures, snow must be removed. Failure to do so has led to unplanned deflations of several air-supported roofs.

In theory, air-supported structures are not limited in span. In practice, flutter and perimeter uplift forces restrict the span of air-supported roofs to a few hundred meters, but this is sufficient to house entire football stadiums. For safety, the outer edges of most air-supported roofs terminate at a level that is well above the floor level within. Thus, if the roof should deflate because of fan failure, inadequate snow removal, or air leakage, the roof fabric will hang in suspension at a height well above the floor of the building (Figure H).

For further information, see Berger, Horst, *Light Structures; Structures of Light*, Basel, Birkhäuser, 1996.

FIGURE H
Most air-supported structures are designed so that if air pressure fails, the membrane will hang at a safe level above the heads of the occupants. (*Sketch by Edward Allen*)

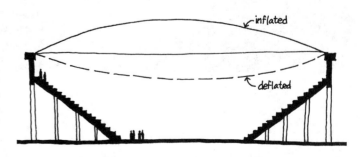

COMPOSITE COLUMNS

Columns that combine the strength of structural steel shapes and sitecast concrete through composite structural action have been used in buildings for many years. One type of composite column surrounds a steel wide-flange column with sitecast reinforced concrete. Another type consists of a steel pipe that is filled with concrete. In a third type, a wide-flange column is inserted within the pipe before the concrete is added to create a higher loadbearing capacity.

Several recent high-rise buildings use very large steel pipe columns filled with very high strength concrete to carry a major portion of both vertical and lateral loads. These columns enable reductions of as much as 50 percent in the overall quantity of steel required for the building (Figure 11.90). In one such building, a 720-foot (200-m) office tower, four 10-foot-diameter (3-m) pipe columns filled with 19,000 pounds per square inch (131 MPa) concrete carry 40 percent of the gravity loads and a large proportion of the wind loads. There is no reinforcing

or other steel inside the pipes except at certain connections that carry very heavy loads. The potential advantages of composite columns in tall buildings include reduced steel usage, greater rigidity of the building against wind forces, and simplified beam–column connections.

INDUSTRIALIZED SYSTEMS IN STEEL

Steel adapts well to industrialized systems of construction. The two most

FIGURE 11.90
A core structure of eight large composite columns, each a concrete-filled pipe 7.5 feet (2.3 m) in diameter, carries the majority of gravity and wind loads in this 44-story Seattle office building. The perimeter of the building is supported by smaller-diameter composite pipe columns. (*Courtesy of Skilling Ward Magnusson Barkshire, Inc.*)

successful prefabrication systems in the United States are probably the mobile home and the "package" industrial building. The mobile home, built largely of wood, is made possible by a rigid undercarriage welded together from rolled steel shapes. The package building is most commonly based on a structure of welded steel rigid frames supporting an enclosure of corrugated metal sheets. The mobile home is founded on steel because of steel's matchless stiffness and strength. The package building depends on steel for these qualities, for the repeatable precision with which components can be produced, and for the ease with which the relatively light steel components can be transported and assembled. It is but a short step from the usual process of steel fabrication and erection to the serial production of repetitive building components.

STEEL AND THE BUILDING CODES

Steel Frame construction appears in the typical building code tables in Figures 1.1 and 1.3 as five different construction types—1A, 1B, 2A, 2B, and 2C—the exact classification depending on the degree of fireproofing treatment applied to the various members of the frame. With a high degree of fireproofing, especially on members supporting more than one floor, unlimited building heights and areas are permitted for most types of occupancies. With no fireproofing whatsoever of steel members, building heights and areas are severely restricted, but many building types can easily meet these restrictions.

THE UNIQUENESS OF STEEL

Among the common structural materials for fire-resistant construction—masonry, concrete, and steel—steel alone has useful tensile strength, which, along with compressive strength, it possesses in great abundance (Figure 11.91). A relatively small amount of steel can do a structural job that would take a much greater amount of another material.

Thus, steel, the most dense of structural materials, is also the one that produces the lightest structures and those that span the greatest distances.

The infrastructure needed to bring steel shapes to a building site—the mines, the mills, the fabricators—is vast and complex. An elaborate sequence of advance planning and preparation activities is required for the making of a steel building frame. Once the components are on the site, however, a steel frame goes together quickly and with relatively few tools, in an erection process that is rivaled for speed and all-weather reliability only by certain precast concrete systems. With proper design and planning, steel can frame almost any shape of building, including irregular angles and curves. Ultimately, of course, steel produces only a frame. Unlike masonry or concrete, it does not lend itself easily to forming a total building enclosure except in certain industrial applications. But this is of little consequence, because steel mates easily with glass, masonry, and panel systems of enclosure, and because steel does its own job, that of carrying loads high and wide with apparent ease, so very well.

FIGURE 11.91

Steel is many times stronger and stiffer than other structural materials.

Material	Allowable Tensile Strength	Allowable Compressive Strength	Density	Modulus of Elasticity
Wood (average)	700 psi (4.83 MPa)	1,100 psi (7.58 MPa)	30 pcf (480 kg/m³)	1,200,000 psi (8,275 MPa)
Brick masonry (average)	0	250 psi (1.72 MPa)	120 pcf (1,920 kg/m³)	1,200,000 psi (8,275 MPa)
Steel (ASTM A36)	22,000 psi (151.69 MPa)	22,000 psi (151.69 MPa)	490 pcf (7,850 kg/m³)	29,000,000 psi (200,000 MPa)
Concrete (average)	0	1,350 psi (9.31 MPa)	145 pcf (2,320 kg/m³)	3,150,000 psi (21,720 MPa)

FIGURE 11.92
This elegantly detailed house in southern California is an early example of the use of structural steel at residential scale. (*Architect: Pierre Koenig, FAIA. Photo: Julius Shulman, Hon. AIA*)

FIGURE 11.93
Architect Peter Waldman utilized steel wide-flange beams and open-web steel joists for his own house in Charlottesville, Virginia.
(*Photo: Maxwell McKenzie*)

(a)

(b)

FIGURE 11.94
Two views of the Chicago Police Training Center: (*a*) Under construction. (*b*) Completed. (*Architect: Jerome R. Butler, Jr. Engineer: Louis Koncza. Photo courtesy of American Institute of Steel Construction*)

FIGURE 11.95
Architect Suzane Reatig structured the roof of a Washington, D.C., church with trusses made of steel angles. The ribs of the roof decking add a strong texture to the ceiling. (*Photo by Robert Lautman*)

FIGURE 11.96
The United Airlines Terminal at Chicago's O'Hare Airport is a high-tech wonderland
of steel framing and fritted glass. (*Architect: Murphy-Jahn. Photo by the author*)

FIGURE 11.97
Chicago is famous for its role in the development of the steel frame skyscraper (see Figure 11.4). The tallest of them is the Sears Tower, seen in the foreground of this photograph. (*Architect and engineer: Skidmore, Owings & Merrill. Photo by Chicago Convention and Tourism Bureau, Inc. Courtesy of American Institute of Steel Construction*)

C.S.I./C.S.C.
Masterformat Section Numbers for Steel Frame Construction

05100	**STRUCTURAL METAL FRAMING**
05120	Structural Steel
05150	Wire Rope
05160	Framing Systems
	Space Frames
05200	**METAL JOISTS**
05210	Steel Joists
05300	**METAL DECKING**
05310	Steel Deck
05320	Raceway Deck Systems

SELECTED REFERENCES

1. American Iron and Steel Institute. *Designing Fire Protection for Steel Beams* and *Designing Fire Protection for Steel Trusses*. Washington, D.C., 1984 and 1991, respectively.

The problem of fireproofing steel building elements is discussed, and a range of fireproofing details is illustrated in these concise booklets. (Address for ordering: P.O. Box 4327, Chestertown, MD 21690.)

2. American Institute of Steel Construction, Inc. *Manual of Steel Construction.* Chicago, updated frequently.

This is the bible of the steel construction industry in the United States. It contains detailed tables of the dimensions and properties of all the standard rolled steel sections, data on standard connections, and specifications and code information. It is available in two versions, one directed at the older allowable-stress structural design procedures and the other at the newer load-and-resistance-factor methods. (Address for ordering: P.O. Box 806276, Chicago, IL 60680-4124.)

3. Parker, Harry, and James Ambrose. *Simplified Design of Structural Steel* (6th ed.). New York, John Wiley & Sons, Inc., 1990.

This is an excellent introduction to the calculation of steel beams, columns, and connections.

4. Steel Joist Institute. *Standard Specifications, Load Tables, and Weight Tables for Steel Joists and Joist Girders*. Myrtle Beach, South Carolina (updated frequently).

Load tables, sizes, and specifications for open-web joists are given in this booklet. (Address for ordering: Suite A, 1205 48th Avenue North, Myrtle Beach, SC 29577.)

KEY TERMS AND CONCEPTS

cast iron
wrought iron
Bessemer process
steel
mild steel
basic oxygen process
electric furnace
structural mill
hot saw
cooling bed
straightener
wide-flange shapes
American Standard (I-beam) shapes
W21 × 93

angle
gusset plate
channel
tee
plate
bar
weathering steel
open-web steel joist
joist girder
structural tubing
rivet
carbon steel bolt
high-strength bolt
friction or slip-critical connection

galling
impact wrench
turn-of-nut method
load indicator washer
tension control bolt
swaged lockpin and collar fastener
arc welding
electrode
backup bar
runoff bar
framed connection
shear
stiffener plate
diagonal bracing

shear connection
eccentrically braced frame
shear panels
moment connection
drift
diaphragm action
cope
fabricator
shop drawings
camber
erector
ironworker
tier
baseplate

leveling plate
grout
plumbing up
tower crane
luffing-boom crane
hammerhead boom crane
mobile crane
tagline
drift pin
topping out
metal decking
cellular decking
composite construction
composite metal decking

shear studs
subpurlins
girts
fireproofing
intumescent mastic
castellated beam
plate girder
rigid frame
truss
chord
space truss
arch
anticlastic
stay

REVIEW QUESTIONS

1. What is the difference between iron and steel? *SMALL AMT. CARBON*

2. How are steel structural shapes produced? How are the weights and thicknesses of a shape changed? *MOSTLY IN FLANGES*

3. How does the work of the fabricator differ from that of the erector?

4. Explain the designation W21 × 68. *not exact nominal depth (inches) Weight – lbs/foot P.353*

5. How can you tell a shear connection from a moment connection? What is the role of each?

6. Why might a beam be coped?

7. What is the advantage of composite construction?

8. Explain the advantages and disadvantages of a steel building structure with respect to fire. How can the disadvantages be overcome?

9. List three different structural systems in steel that might be suitable for the roof of an athletic fieldhouse.

MAKE IT FIT

EXERCISES

1. For a simple multistory office building of your design:

a. Draw a steel framing plan for a typical floor.

b. Draw an elevation or section showing a suitable method of giving lateral stability to the building.

c. Make a preliminary determination of the approximate sizes of the decking, beams, and girders, using the information in the box on page 356.

d. Sketch details of the typical connections in the frame, using actual dimensions from the *Manual of Steel Construction* (reference 2) for the member size you have determined, and working to scale.

2. Select a method of fireproofing, and sketch typical column and beam fireproofing details for the building in Exercise 1.

3. What fire resistance ratings in hours are required for the following elements of a four-story department store with 17,500 square feet of area per floor? (The necessary information is found in Figures 1.1–1.3.)

a. Lower-floor columns
b. Floor beams
c. Roof beams
d. Walls around elevator shafts and stairways

4. Find a steel building frame under construction. Observe the connections carefully and figure out why each is detailed as it is. If possible, arrange to talk with the structural engineer of the building to discuss the design of the frame.

## 12

LIGHT GAUGE STEEL FRAME CONSTRUCTION

- **The Concept of Light Gauge Steel Frame Construction**
- **Framing Procedures**
- **Other Common Uses of Light Gauge Steel Framing**

- **Advantages and Disadvantages of Light Gauge Steel Framing**
- **Light Gauge Steel Framing and the Building Codes**
- **Finishes for Light Gauge Steel Framing**

Driving self-drilling, self-tapping screws with electric screw guns, framers add diagonal bracing straps to a wall frame made from light gauge steel studs and runner channels.
(*Courtesy of United States Gypsum Company*)

The members used in Light Gauge Steel Frame construction are manufactured by *cold forming:* Sheet steel is fed from continuous coils through machines that fold it at room temperature into long members whose shapes make them stiff and strong. Thus, these members are often referred to as *cold-formed steel components,* to differentiate them from the much heavier hot-rolled shapes that are used in structural steel framing. The term *light gauge* refers to the relative thinness (gauge) of the steel sheet from which the members are made.

THE CONCEPT OF LIGHT GAUGE STEEL FRAME CONSTRUCTION

Light Gauge Steel Frame construction is the noncombustible equivalent of Wood Light Frame construction. The external dimensions of the standard sizes of light gauge members correspond closely to the dimensions of the standard sizes of 2-inch (38-mm) framing lumber. These members are used in framing in much the same way as 2-inch wood members are used: as closely spaced studs, joists, and rafters. A Light Gauge Steel Frame building may be sheathed, insulated, wired, and finished inside and out in the same manner as a Wood Light Frame building.

The steel that is used in light gauge members is manufactured to ASTM Standard A563, and the zinc coating that protects the members from rust, to ASTM A924. For studs, joists, and rafters, the steel is formed into C-shaped (*cee*) sections (Figure 12.1). For top and bottom wall plates and for joist headers, *runner channel* sections are used. The strength and stiffness of a member depends on the shape and depth of the section and the *gauge* (thickness) of the steel sheet from which it is made. A standard range of depths and gauges is available from each manufacturer. The webs of cee members are punched at the factory to provide holes at 2 foot (600 mm) intervals; these are designed to allow wiring, piping, and bracing to pass through studs and joists without the necessity of drilling holes on the construction site.

For large projects, members may be manufactured precisely to the required lengths. Otherwise, they are furnished in standard lengths. Members may be cut to length on the construction job site with power saws or special shears. A variety of sheet metal angles, straps, plates, channels, and miscellaneous shapes are manufactured as accessories for light gauge steel construction (Figure 12.2).

Light gauge steel members are usually joined with *self-drilling, self-tapping screws,* which drill their own holes and form helical threads in the holes as they are driven. Driven

FIGURE 12.1
Typical light gauge steel framing members. To the left are the common sizes of cee studs and joists. In the center are channel studs. To the right are runner channels.

rapidly by hand-held electric or pneumatic tools, these screws are plated with cadmium or zinc to resist corrosion, and are available in an assortment of diameters and lengths to suit a full range of connection situations. Welding is often employed to assemble panels of light gauge framing that are prefabricated in a factory, and is sometimes used on the building site where particularly strong connections are needed. Other fastening techniques that are widely used include hand-held clinching devices that join members without screws or welds, and pneumatically driven pins that penetrate the members and hold by friction.

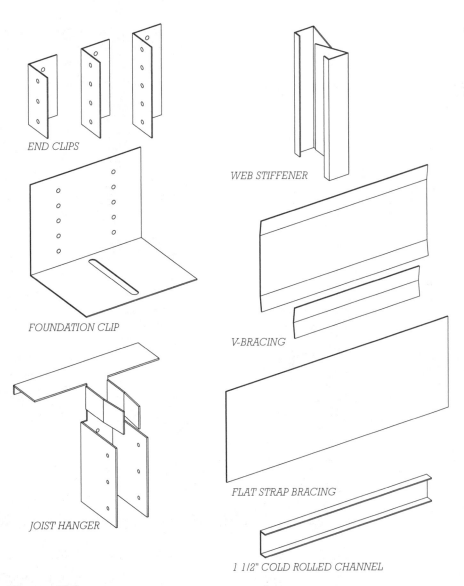

END CLIPS

FOUNDATION CLIP

JOIST HANGER

WEB STIFFENER

V-BRACING

FLAT STRAP BRACING

1 1/2" COLD ROLLED CHANNEL

FIGURE 12.2
Standard accessories for light gauge steel framing: End clips are used to join members that meet at right angles. Foundation clips attach the ground-floor platform to anchor bolts embedded in the foundation. Joist hangers connect joists to headers and trimmers around openings. The web stiffener is a two-piece assembly that is inserted inside a joist and screwed to its vertical web to help transmit wall loads vertically through the joist. The remaining accessories are used for bracing.

FRAMING PROCEDURES

The sequence of construction for a building that is framed entirely with light gauge steel members is essentially the same as that described in Chapter 5 for a building framed with nominal 2-inch wood members (Figure 12.3). Framing is usually constructed platform fashion: The ground floor is framed with steel joists. Mastic adhesive is applied to the upper edges of the joists. Wood panel subflooring is laid down and fastened to the upper flanges of the joists with screws. Steel studs are laid flat on the subfloor and joined to make wall frames. The wall frames are sheathed either with wood panels or with *gypsum sheathing panels,* which are similar to gypsum wallboard but with water-resistant paper faces and a water-resistant core formulation. The wall frames are tilted up and screwed to the floor frame. The upper-floor platform is framed, then the upper-floor walls. Finally, the ceiling and roof are framed, all in much the same way as in a wood-framed house. Prefabricated trusses of light gauge steel members that are screwed or welded together are often used to frame ceilings and roofs (Figures 12.14, 12.15). It is possible, in fact, to frame any building with light gauge steel members that can be framed with nominal 2-inch wood members. For additional fire resistance to achieve a higher construction type under the building code, floors of corrugated steel decking with a concrete topping are sometimes substituted for wood panel subflooring.

Openings in floors and walls are framed analogously to openings in Wood Light Frame construction, with doubled members around each opening and strong headers over doors and windows (Figures 12.4–12.8). Joist hangers and right-angle clips of sheet steel are used to join members around openings. Light gauge members are designed to

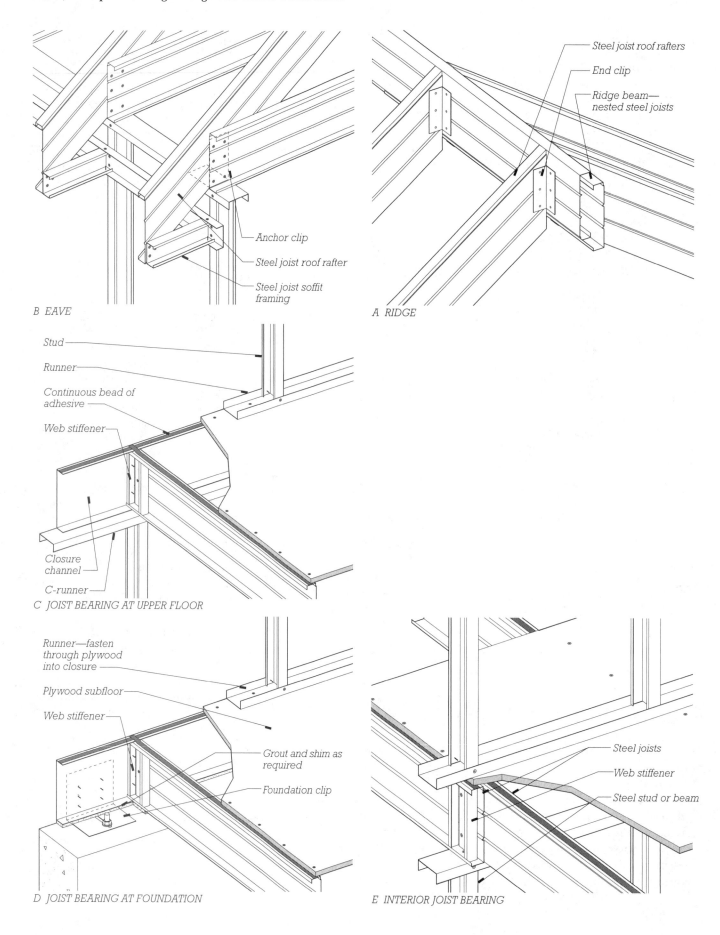

B EAVE

Anchor clip

Steel joist roof rafter

Steel joist soffit framing

A RIDGE

Steel joist roof rafters

End clip

Ridge beam— nested steel joists

Stud

Runner

Continuous bead of adhesive

Web stiffener

Closure channel

C-runner

C JOIST BEARING AT UPPER FLOOR

Runner—fasten through plywood into closure

Plywood subfloor

Web stiffener

Grout and shim as required

Foundation clip

D JOIST BEARING AT FOUNDATION

Steel joists

Web stiffener

Steel stud or beam

E INTERIOR JOIST BEARING

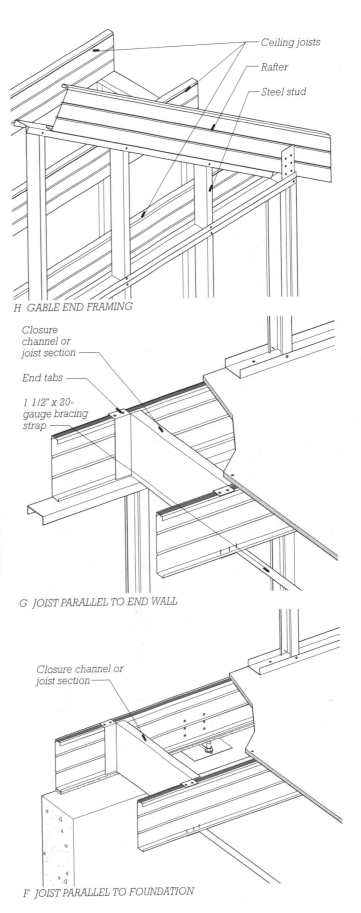

H GABLE END FRAMING

G JOIST PARALLEL TO END WALL

F JOIST PARALLEL TO FOUNDATION

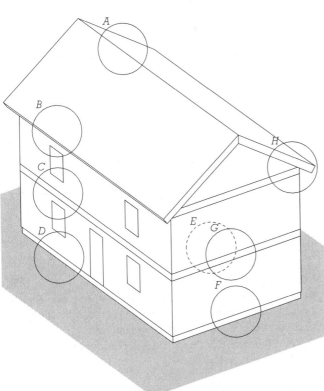

FIGURE 12.3

Typical light gauge framing details. Each detail is keyed by letter to a circle on the whole-building diagram in the center to show its location in the frame. (*a*) A pair of nested joists makes a boxlike ridge board or ridge beam. (*b*) Anchor clips are sandwiched between the ceiling joists and rafters to hold the roof framing down to the wall. (*c*) A web stiffener helps transmit vertical forces from each stud through the end of the joist to the stud in the floor below. Mastic adhesive cushions the joint between the subfloor and the steel framing. (*d*) Foundation clips anchor the entire frame to the foundation. (*e*) At interior joist bearings, joists are overlapped back to back and a web stiffener is inserted. (*f*, *g*) Short crosspieces brace the last joist at the end of the building and help transmit stud forces through to the wall below. (*h*) Like all these details, the gable end framing is directly analogous to the corresponding detail for a Wood Light Frame building as shown in Chapter 5.

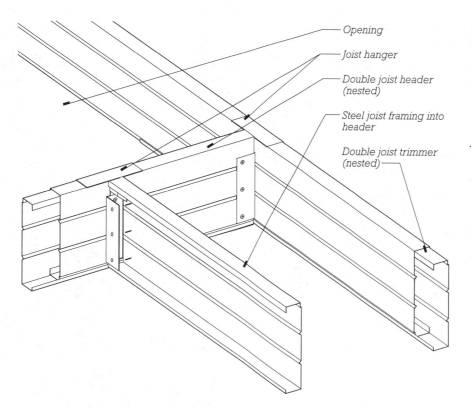

— Opening

— Joist hanger

— Double joist header (nested)

— Steel joist framing into header

— Double joist trimmer (nested)

FIGURE 12.4
Headers and trimmers for floor openings are doubled and nested to create strong, stable box members. Only one vertical flange of the joist hanger is attached to the joist; the other flange would be used instead if the web of the joist were oriented to the left rather than the right.

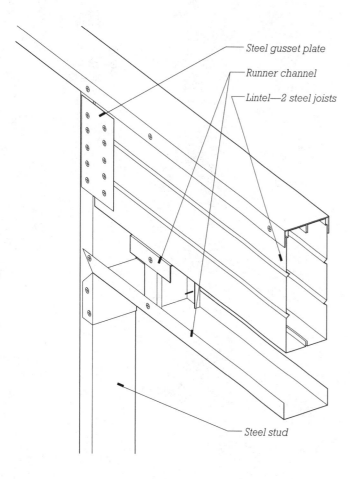

— Steel gusset plate

— Runner channel

— Lintel—2 steel joists

— Steel stud

FIGURE 12.5
A typical window or door head detail. The header is made of two joists placed with their open sides together. The top plate of the wall, which is a runner channel, continues over the top of the header. Another runner channel is cut and folded at each end to frame the top of the opening. Short studs are inserted between this channel and the header to maintain the spacing of the studs in the wall.

FIGURE 12.6
Diagonal strap braces stabilize upper-
floor wall framing for an apartment
building. (*Courtesy of United States Gypsum
Company*)

FIGURE 12.7
Temporary braces support the walls at
each level until the next floor platform has
been completed. Small cold-rolled channels pass through the web openings of the
studs; they are welded to all the studs to
help stabilize them against buckling.
(*Courtesy of Unimast Incorporated—
www.unimast.com*)

FIGURE 12.8
A detail of a window header. Because a supporting stud has been inserted under the end of the header, a large gusset plate such as the one shown in Figure 12.5 is not required. (*Courtesy of Unimast Incorporated—www.unimast.com*)

FIGURE 12.9
Ceiling joists in place for an apartment building. A brick veneer cladding has already been added to the ground floor. (*Courtesy of United States Gypsum Company*)

nest together to form a tubular configuration that is especially strong and stiff when used for a ridge board or header (Figures 12.3a, 12.4).

Because light steel members are much more prone than their wood counterparts to twist or buckle under load, somewhat more attention must be paid to their bracing and bridging. The studs in tall walls are generally braced at 4-feet (1200-mm) intervals, either with steel straps screwed to the edges of the studs or with 1½-inch (38-mm) cold-formed steel channels passed through the punched openings in the studs and welded or screwed to an angle clip at each stud (Figure 12.7). Floor joists are bridged with solid blocking between and steel straps screwed to their top and bottom edges. In locations where large vertical forces must pass through floor joists (as occurs where loadbearing studs sit on the edge of a floor platform), steel *web stiffeners* are screwed to the webs of the joists to prevent the thin web from buckling (Figure 12.3c, d, and e). Wall bracing consists of diagonal steel straps screwed to the studs (chapter opener, Figure 12.6). Permanent resistance to buckling, twisting, and lateral loads such as wind and earthquake is

FIGURE 12.10
A detail of eave framing. (*Courtesy of Unimast Incorporated—www.unimast.com*)

FIGURE 12.11
A power saw with an abrasive blade cuts quickly and precisely through steel framing members. (*Courtesy of Unimast Incorporated—www.unimast.com*)

imparted largely and very effectively by subflooring, wall sheathing, and interior finish materials.

OTHER COMMON USES OF LIGHT GAUGE STEEL FRAMING

Light gauge steel members are used to construct many components of fire-resistant buildings whose primary loadbearing structures are made of structural steel, concrete, or masonry. These components include interior walls and partitions (Chapter 23), suspended ceilings (Chapter 24), and fascias, parapets, and backup walls for such exterior claddings as masonry veneer, exterior insulation and finish system (EIFS), glass-fiber-reinforced concrete (GFRC), metal panels, and various thin stone cladding systems (Chapters 19 and 20; see also Figures 12.12 and 12.13).

In situations where noncombustibility is not an issue, metal and wood light framing are sometimes mixed in the same building. Some builders find it economical to use wood to frame exterior walls, floors, and roof, with steel framing for interior partitions. Sometimes all walls, interior and exterior, are framed with steel, and floors are framed with wood. Steel trusses made of light gauge members may be applied over wood frame walls. In such mixed uses, special care must be taken in the

FIGURE 12.12
Light gauge steel stud walls frame the exterior walls of a building whose floors and roof are framed with structural steel.
(*Courtesy of Unimast Incorporated— www.unimast.com*)

FIGURE 12.13
The straightness of steel studs is apparent in these tall walls that enclose a building framed with structural steel. (*Courtesy of Unimast Incorporated—www.unimast.com*)

FIGURE 12.14
A worker tightens the last screws to complete a connection in a light gauge steel roof truss. The truss members are held in alignment during assembly by a simple jig made of plywood and blocks of framing lumber. (*Courtesy of Unimast Incorporated—www.unimast.com*)

FIGURE 12.15
Installing steel roof trusses. (*Courtesy of Unimast Incorporated—www.unimast.com*)

FOR PRELIMINARY DESIGN OF A
LIGHT GAUGE STEEL FRAME STRUCTURE

• Estimate the depth of **rafters** on the basis of the *horizontal* (not slope) distance from the outside wall of the building to the ridge board in a gable or hip roof, and the horizontal distance between supports in a shed roof. Estimate the depth of a rafter at $1/24$ of this span, rounded up to the nearest 2-inch (50-mm) dimension.

• The depth of light gauge steel **roof trusses** is usually based on the desired roof pitch. A typical depth is $1/4$ of the width of the building, which corresponds to a $6/12$ pitch.

• Estimate the depth of light gauge steel **floor joists** at $1/20$ of the span, rounded up to the nearest 2-inch (50-mm) dimension.

• For **loadbearing studs,** estimate that a $3^5/8$-inch (92-mm) stud may be used to a maximum height of 12 feet (3.6 m), a 6-inch (150-mm) stud to 21 feet (6.4 m), and an 8-inch (100-mm) stud to 28 feet (8.5 m).

• For **exterior cladding backup walls,** estimate that a $3^5/8$-inch (92-mm) stud may be used to a maximum height of 10 feet (3.0 m), a 6-inch (150-mm) stud to 17 feet (5.2 m), and an 8-inch (100-mm) stud to 22 feet (6.7 m). For brittle cladding materials such as brick masonry, select a stud that is 2 inches (50 mm) deeper than these numbers would indicate.

All framing members are usually spaced at 24 inches (600 mm) o.c.

These approximations are valid only for purposes of preliminary building layout, and must not be used to select final member sizes. They apply to the normal range of building occupancies such as residential, office, commercial, and institutional buildings. For manufacturing and storage buildings, use somewhat larger members.

For more comprehensive information on preliminary selection and layout of structural members, see Allen, Edward, and Joseph Iano, *The Architect's Studio Companion* (2nd ed.), New York, John Wiley & Sons, Inc., 1995

details to assure that wood shrinkage will not create unforeseen stresses or damage to finish materials.

ADVANTAGES AND DISADVANTAGES OF LIGHT GAUGE STEEL FRAMING

Light gauge steel framing shares most of the advantages of wood light framing: It is versatile; flexible; requires only simple, inexpensive tools; furnishes internal cavities for utilities and thermal insulation; and accepts an extremely wide range of exterior and interior finish materials. Additionally, steel framing may be used in buildings for which noncombustible construction is required by building code, thus extending its use to larger buildings and to those whose uses require a higher degree of resistance to fire.

Steel framing members are significantly lighter in weight than the wood members to which they are structurally equivalent, an advantage that is often enhanced by spacing steel studs, joists, and rafters at 24 inches o.c. (600 mm) rather than 16 inches (400 mm). Light gauge steel joists and rafters can span slightly longer distances than nominal 2-inch wood members of the same depth. Steel members tend to be straighter and more uniform than wood members, and they are much more stable dimensionally because they are unaffected by changing humidity. Although they may corrode if exposed to moisture over an extended period of time, particularly in oceanfront locations, steel framing members cannot fall victim to termites or decay. Prices of steel framing members are relatively stable, while the price of wood, which is an agricultural commodity subject to fluctu-

ations in supply, often varies over a wide range with changing market conditions.

As compared to walls and partitions of masonry construction, equivalent walls and partitions framed with steel studs are much lighter in weight, easier to insulate, and accept electrical wiring and pipes for plumbing and heating much more readily. Steel framing, being a dry process, may be carried out in wet or cold weather conditions that would make masonry construction difficult. Masonry walls tend to be much stiffer and more resistant to the passage of sound than steel-framed walls, however.

The thermal conductivity of light gauge steel framing members is much higher than that of wood. In cold climates, steel framing members, unless detailed with *thermal breaks* such as foam plastic sheathing or insulating edge spacers between studs and sheathing, conduct heat rapidly enough to reduce substantially the thermal performance of a wall or roof. This can result in excessive energy loss and moisture condensation on interior building surfaces, with attendant mold and mildew growth and discoloration of surface finishes. Special attention must be given to designing details to block excessive heat flow in every area of the frame. At the eave of a steel-framed house, for instance, the ceiling joists readily conduct heat from the warm interior ceiling along their lengths to the cold eave unless insulating edge spacers or foam insula-

tion boards are used between the ceiling finish material and the joists.

LIGHT GAUGE STEEL FRAMING AND THE BUILDING CODES

Although light gauge steel framing members will not burn, they will lose their structural strength and stiffness rapidly if exposed to fire. They must be protected from fire in accordance with building code requirements. With suitable protection from gyp-

sum sheathing and gypsum wallboard or plaster, light gauge steel construction may be classified as high as Type 2B in the building code table shown in Figure 1.1, enabling its use in a wide variety of buildings.

The International Code Council has created a *prescriptive building code* for steel-framed one- and two-family dwellings. This code, with its structural tables and standard details, allows builders to design and construct steel-framed houses without having to employ an engineer or architect, just as they are able to do with wood light frame construction.

FIGURE 12.16
Gypsum sheathing panels have been screwed onto most of the ground-floor walls of this large commercial building.
(*Courtesy of Unimast Incorporated—www.unimast.com*)

FINISHES FOR LIGHT GAUGE STEEL FRAMING

Any exterior or interior finish material that is used in Wood Light Frame construction may be applied to a light gauge steel frame. Whereas finish materials are most often fastened to a wood frame with nails, screws must be utilized instead with a steel frame. Wood trim components are applied with special finish screws that have very small heads; these are analogous to finish nails.

C.S.I./C.S.C. Masterformat Section Numbers for Light Gauge Steel Framing	
05400	**COLD-FORMED METAL FRAMING**
05410	**Load-Bearing Metal Stud Systems**
05420	**Cold-Formed Metal Joist Systems**

FIGURE 12.17
Waferboard sheathes the walls of a house framed with light gauge steel studs, joists, and rafters. (*Courtesy of Unimast Incorporated—www.unimast.com*)

METALS IN ARCHITECTURE

Metals are dense, lustrous materials that are highly conductive of heat and electricity. They are generally *ductile,* meaning that they can be hammered thin or drawn into wires. They can be liquefied by heating and will resolidify as they cool. Most metals corrode by oxidation. Metals are the strongest building materials presently in common use.

Most metals are found in nature in the form of oxide ores. These ores are refined by processes that involve either heat and reactant materials or, in the case of aluminum, electrolysis.

Metals may be classified broadly as either *ferrous,* meaning that they consist primarily of iron, or *nonferrous.* Because iron ore is an abundant mineral and is relatively easy and economical to refine, ferrous metals tend to be much less expensive than nonferrous ones. The ferrous metals are also the strongest, but most have a tendency to rust. In general, nonferrous metals are considerably more expensive on a volumetric basis than ferrous metals, but unlike ferrous metals, most of them form thin, tenacious oxide layers that protect them from further corrosion. This makes many of the nonferrous metals valuable for finish components of buildings. Many of the nonferrous metals are also easy to work and attractive to the eye.

Modifying the Properties of Metals

A metal is seldom used in its chemically pure state. Instead, it is mixed with other elements, primarily other metals, to modify its properties for a particular purpose. Such mixtures are called *alloys.* An alloy that combines copper with a small amount of tin is known as *bronze,* and a very small, closely controlled amount of carbon mixed with iron makes *steel.* In both of these examples, the alloy is stronger and harder than the metal that is its primary ingredient. Several alloys of iron (several different steels, to be more specific) are mentioned in Chapter 11. Some of these steels have higher strengths and some form self-protecting oxide layers, all because of the influence of the alloying elements that they contain. Similarly, there are many alloys that consist primarily of aluminum; some are soft and easy to form, some are very hard and springy, some are very strong, and so on.

The properties of many metals can be changed by *heat treatment.* Steel that is heated red-hot and then plunged into cold water becomes much harder but very brittle. Steel can be *tempered* by heating it to a moderate temperature and cooling it more slowly, making it both hard and strong. Steel that is brought to a very high temperature and then cooled slowly (a process called *annealing*) will become softer, easier to work, and less brittle. Many aluminum alloys can also be heat treated to modify their characteristics.

Cold working is another way of changing the properties of a metal. When steel is beaten or rolled thinner at room temperature, its crystalline structure is changed in a way that makes it much stronger and somewhat more brittle. The highest-strength metals used in construction are the steel wires and cables that are used to prestress concrete. Their high strength (about six times that of normal structural steel) is the result of drawing the metal through smaller and smaller orifices to produce the wire, a process that subjects the metal to a high degree of cold working. Cold-rolled steel shapes with substantially higher strengths than structural steel are used as reinforcing bars and as the components of open-web joists. The effects of cold working are easily reversed by annealing. Hot rolling, which is, in effect, a self-annealing process, does not increase the strength of the metal.

To change the appearance of a metal or to protect it from oxidation, it can be coated with a thin layer of another metal. Steel is often *galvanized* by coating it with zinc to protect it against corrosion, as described below. *Electroplating* is widely used to coat metals such as chromium and cadmium onto steel to improve its appearance and protect it from corrosion. An electrolytic process is used to *anodize* aluminum, adding a thin oxide layer of controlled color and consistency to the surface of the metal. Metals are frequently finished with such nonmetallic coatings as paints, lacquers, fluoropolymers, porcelain enamel, and thermosetting powder coatings for reasons of protection and appearance.

Fabricating Metals

Metals can be shaped in many different ways. *Casting* is the process of pouring molten metal into a shaped mold; the metal retains the shape of the mold as it cools. *Rolling,* which may be done either hot or cold, forms the metal by squeezing it between shaped rollers. *Extrusion* is a process of squeezing heated but not molten metal through a shaped die to produce a long metal piece with a shaped profile. *Forging* involves heating a piece of metal until it is plastic, then beating it into shape. Forging was originally done by hand with a blacksmith's forge, hammer, and anvil, but most forging is now done with powerful hydraulic machinery that forces the metal into shaped dies. *Stamping* is the process of squeezing sheet metal between matching dies to give it a desired shape or texture. *Drawing* produces wires by pulling a metal rod through a series of progressively smaller orifices in hardened steel plates until the desired diameter is reached. These forming processes have varying effects on the strength of the resulting material: Cold drawing and cold rolling will harden and strengthen many metals. Forging imparts a grain orientation to the metal that closely follows the shape of the piece for improved structural performance. Casting tends to produce a somewhat weaker metal than most other forming processes, but it is useful for making elaborate shapes (like lavatory faucets) that could not be manufactured in any other way.

Metals can also be shaped by *machining*, which is a process of cutting unwanted material from a piece of metal to produce the desired shape. Among the most common machining operations is *milling*, in which a rotating cutting wheel is used to cut metal from a workpiece. To produce cylindrical shapes, a piece of metal is rotated against a stationary cutting tool in a *lathe*. Holes are produced by *drilling*, which is usually carried out either in a *drill press* or in a lathe. Screw threads may be produced in a hole by the use of a helical cutting tool called a *tap*, and the external threads on a steel rod are cut with a *die*. (The screw threads on mass-produced screws and bolts are formed at high speed by special rolling machines.) Grinding and polishing machines are used to create and finish flat surfaces. Sawing, shearing, and punching operations are also common methods of shaping metal components. An economical method of cutting steel of almost any thickness is with an *oxyacetylene cutting torch* that combines a slender, high-temperature gas flame with a jet of pure oxygen to burn away the metal. *Plasma cutting* with a tiny, supersonic jet of superheated gas that blows away the metal can give more precise cuts at thicknesses up to 2 inches (50 mm), and *laser cutting* gives high-quality results in thin steel plates.

Sheet metal is fabricated with its own particular set of tools. Shears are used to cut the metal sheets, and folds are made on large machines called *brakes*.

Joining Metal Components

Metal components may be joined either mechanically or by fusion. Most mechanical fastenings require drilled or punched holes for the insertion of screws, bolts, or rivets. Some small-diameter screws that are used with thin metal components are shaped and hardened so they do their own drilling and tapping as they are driven. Many sheet metal components, especially roofing sheet and ductwork, are joined primarily with interlocking, folded connections.

High-temperature fusion connections are made by *welding*, in which a gas flame or an electric arc melts the metal on both sides of the joint and mixes it with additional molten metal from a welding rod or consumable electrode. *Brazing* and *soldering* are lower-temperature processes in which the parent metal is not melted. Instead, a different metal with a lower melting point (bronze or brass in the case of brazing, and a lead–tin alloy in the most common type of solder) is melted into the heated joint and bonds to the pieces that it joins. A welded connection is generally as strong as the pieces it connects, and can be used for structural work. A soldered connection is not as strong, but it is easy to make and works well for connecting copper plumbing pipes and sheet metal roofing. As an alternative to welding or soldering, adhesives are occasionally used to join metals in certain nonstructural applications.

Some Metals That Are Common in Buildings

The ferrous metals include cast iron, wrought iron, steel, and stainless steel. **Cast iron** contains relatively large amounts of carbon and impurities; it is too brittle for most structural work. **Wrought iron** was once common; it was produced by hammering semimolten iron to produce a metal with long fibers of iron interleaved with long fibers of slag. It was much less brittle than cast iron and could therefore be used for beams, but with the introduction of economical steel-making processes, its functions were largely taken over by steel. Even the ornamental metalwork that we refer to as "wrought iron" is now made of mild steel. **Steel** is discussed in some detail in Chapter 11, and its many uses are noted throughout this book. In general, all these ferrous metals are very strong, relatively inexpensive, easy to form and machine, and must be protected from corrosion.

Stainless steels, made by alloying steel with other metals, primarily chromium and nickel, are virtually immune to corrosion. They are harder to form and machine than mild steel and are much more costly. They are available in attractive finishes that range from matte textures to a mirror polish.

Aluminum (spelled and pronounced *aluminium* in the British Commonwealth) is the nonferrous metal most used in construction. Its density is about one-third that of steel and it has moderate to high strength and stiffness, depending on which of a multitude of alloys is selected. It can be hardened by cold working, and some alloys can be heat treated. It can be hot- or cold-rolled, cast, forged, drawn, and stamped, and is particularly well adapted to extrusion (see Chapter 21). Aluminum is self-protecting, easy to machine, and has thermal and electrical conductivities that are almost as high as those of copper. It is easily made into thin foils that find wide use in insulating and vapor retarding materials. With a mirror finish, aluminum in foil or sheet form is the most highly reflective of heat and light of any architectural material. Typical uses of aluminum in buildings include roofing and flashing sheet, ductwork, curtain wall components, window and door frames, grills, ornamental railings, siding, hardware, electrical wiring, and protective coatings for other metals, chiefly steel. Aluminum powder is used in metallic paints, and aluminum oxide is used as an abrasive in sandpaper and grinding wheels.

Copper and copper alloys are widely used in construction. Copper is slightly more dense than steel and is a bright orange-red in color. When it oxidizes, it forms a self-protecting coating that is blue-green or black, depending on the contaminants in the local atmosphere. Copper is fairly strong and can be made stronger by alloying or cold working, but it is not amenable to heat treatment. It is ductile and easy to fabricate. It has the highest thermal and electrical conductivities of any metal used in construction. It may be formed by casting, drawing, extrusion, and hot or cold rolling. The primary uses of copper in buildings are roofing and flashing sheet (sometimes with a lead coating for appearance or chemical compatibility), piping and tubing, and electrical wiring. Copper is an alloying element in certain corrosion-resistant steels, and copper salts are used as wood preservatives.

Copper is the primary constituent of two versatile alloys, bronze and brass. **Bronze** is a reddish-gold metal that traditionally consists of 90 percent copper and 10 percent tin. Today, however, the term "bronze" is applied to a wide range of alloys that may also incorporate such metals as aluminum, silicon, manganese, nickel, and zinc. These various bronzes are found in buildings in the form of statuary, bells, ornamental metalwork, hardware, and weatherstripping. **Brass** is formulated of copper and zinc plus small amounts of other metals; it is usually a lighter, more yellowish color than bronze, but in contemporary usage the line between brasses and bronzes has become rather indistinct, and the various brasses occur in a wide range of colors, depending on the formulation. Brass, like bronze, is resistant to corrosion. It can be polished to a high luster. It is widely used in hinges and doorknobs, weatherstripping, ornamental metalwork, screws, bolts, nuts, and plumbing faucets (where it is usually plated with chromium). On a volumetric basis, brass, bronze, and copper are expensive metals, but they are often the most economical materials for applications that require their unique combination of functional and visual properties.

Zinc is a blue-white metal that is low in strength, relatively brittle, and moderately hard. Zinc alloy sheet is used for roofing and flashing. Alloys of zinc are also used for casting small hardware parts such as doorknobs, cabinet pulls and hinges, bathroom accessories, and components of electrical fixtures. These *die castings,* which are usually electroplated with another metal such as chromium for appearance, are not especially strong, but they are economical and they can be very finely detailed.

The most important use of zinc in construction is for *galvanizing,* the application of a zinc coating to prevent steel from rusting. The surface of the zinc coating corrodes to a self-protecting gray oxide. If the coating is accidentally scratched through to the steel beneath, zinc oxide forms in the scratch and heals the discontinuity. *Hot-dip galvanizing,* in which the steel part is dipped in molten zinc to produce a thick coating, is the most durable. Much less durable is the thin coating produced by *electrogalvanizing.*

Cadmium is similar to zinc in its properties. It is often electroplated onto steel parts to guard against corrosion, and its compounds are used as pigments.

Lead is the heaviest of the common metals, about 40 percent denser than steel. It is neither strong nor stiff, but is limp, soft, and easy to work. Lead forms an excellent barrier to the passage of sound, vibration, and electromagnetic radiation. In foil or sheet form, it is used in walls and ceilings of facilities such as recording studios to block sound transmission. It is also used in doors and partitions for X-ray rooms and nuclear installations, and in vibration-damping pads for machinery and building foundations. Lead sheet finds extensive use in flashings and sheet metal roofing. Lead is also used as a coating for copper sheet in roofing and flashing applications. Two other common building components that are made of lead are expansion anchors for fastening bolts into concrete or stone and the small divider strips (called *cames*) that separate the pieces of glass in a stained-glass window. Lead is a chemical constituent of radiation-shielding glass. Lead oxide was once the most common pigment used in paint, but because the oxide (like the parent metal) is poisonous, this use has been discontinued.

Tin is a soft, ductile, silvery metal that forms a self-protecting oxide layer. The ubiquitous "tin can" is actually made of sheet steel with a corrosion-resisting coating of tin. Tin is found in buildings primarily as a constituent of *terne metal,* an alloy of 80 percent lead and 20 percent tin that is used as a corrosion-resistant coating for steel or stainless steel roofing sheet.

Chromium is a very hard metal that can be polished to a brilliant mirror finish. It does not corrode in air. It is often electroplated onto other metals for use in ornamental metalwork, bathroom and kitchen accessories, door hardware, and plumbing and lighting fixtures. It is also a major alloying element in stainless steel and many other metals, to which it imparts hardness, strength, and corrosion resistance. Chromium compounds are used as colored pigments in paints and ceramic glazes.

Magnesium is a strong, remarkably lightweight metal (less than one-quarter the density of steel) that is much used in aircraft but is too costly for general use in buildings. It is found on the construction job site as a material for various lightweight tools and as an alloying element that increases the strength and corrosion resistance of aluminum.

Titanium is also low in density, about half the weight of steel, and very strong. It is a constituent of many alloys, and its oxide has replaced lead oxide in paint pigments.

SELECTED REFERENCE

1. American Iron and Steel Institute. *AISI Cold-Formed Steel Design Manual.* 1996. This is an engineering reference work that contains structural design tables and procedures for light gauge steel framing. (Address for ordering: USA Fulfillment/ AISI, Box 4237, Chestertown, MD 21690.)

KEY TERMS AND CONCEPTS

cold forming
cold-formed steel components
light gauge
cee
runner channel
gauge
self-drilling, self-tapping screw
gypsum sheathing panel
nest
web stiffener
thermal break
prescriptive building code

ductile
ferrous, nonferrous
alloy
heat treatment
tempering
annealing
cold working
galvanizing
electroplating
anodizing
casting
rolling

extrusion
forging
stamping
drawing
machining
milling
lathe
drilling
drill press
tap
die
oxyacetylene cutting torch

plasma cutting
laser cutting
brake
welding
brazing
soldering
die casting
hot-dip galvanizing
electrogalvanizing
cames
terne metal

REVIEW QUESTIONS

1. How are light gauge steel framing members manufactured?

2. How do the details for a house framed with light gauge steel members differ from those for a similar house with wood platform framing?

3. What special precautions should you take when detailing a steel-framed building to avoid excessive conduction of heat through the framing members?

4. If a building framed with light gauge steel members must be totally noncombustible, what materials would you use for subflooring and wall sheathing?

5. What is the advantage of a prescriptive building code for light gauge steel framing?

6. Compare the advantages and disadvantages of Wood Light Frame construction and Light Gauge Steel Frame construction.

EXERCISES

1. Convert a set of details for a Wood Light Frame house to Light Gauge Steel framing.

2. Visit a construction site where light gauge steel studs are being installed. Grasp the middle of an installed stud that has not yet been sheathed and twist it clockwise and counterclockwise. How resistant is the stud to twisting? How is this resistance increased as the building is completed?

3. On this same construction site, make sketches of how electrical wiring, electric fixture boxes, and pipes are installed in metal framing.

13

CONCRETE CONSTRUCTION

A physical sciences center at Dartmouth College, built in a highly irregular space bounded by three existing buildings, typifies the potential of reinforced concrete to make expressive, highly individual buildings. (*Architects: Shepley Bulfinch Richardson and Abbott. Photo by Ezra Stoller, ©ESTO*)

Concrete is the universal material of construction. The raw ingredients for its manufacture are readily available in every part of the globe, and concrete can be made into buildings with tools ranging from a primitive shovel to a computerized precasting plant. Concrete does not rot or burn; it is relatively low in cost; and it can be used for every building purpose, from lowly pavings to sturdy structural frames to handsome exterior claddings and interior finishes. But it has no form of its own and no useful tensile strength. Before its limitless architectural potential can be realized, the designer must learn to combine concrete skillfully with steel to bring out the best characteristics of each material and to mold and shape it to forms appropriate to its qualities.

HISTORY

The ancient Romans, while quarrying limestone for mortar, accidentally discovered a silica- and alumina-bearing mineral on the slopes of Mount Vesuvius that, when mixed with limestone and burned, produced a cement that exhibited the unique property of hardening underwater as well as in the air. This cement was also harder, stronger, much more adhesive, and cured much more quickly than the ordinary lime mortar to which they were accustomed. In time, this mortar not only became the preferred type for use in all their building projects but began also to alter the character of Roman construction. Masonry of stone or brick was used to build only the surface layers of piers, walls, and vaults, and the hollow interiors were filled entirely with large volumes of the new type of mortar (Figure 13.2). We now know that this mortar contained all the essential ingredients of modern portland cement and that the Romans were the inventors of concrete construction.

Knowledge of concrete construction was lost with the fall of the Roman Empire, not to be regained until the 18th century, when a number of English inventors began experimenting with both natural and artificially produced cements. Joseph Aspdin, in 1824, patented an artificial cement that he named *portland cement,* after English Portland limestone, whose durability as a building stone was legendary. His cement was soon in great demand, and the name portland remains in use to the present day.

Reinforced concrete, concrete that is combined with steel bars, was developed in the 1850s by several people simultaneously, including J. L. Lambot, who built several reinforced concrete boats in Paris in 1854, and an American, Thaddeus Hyatt, who made and tested a number of reinforced concrete beams. But the combination of steel and concrete did not come into widespread use until a French gardener, Joseph Monier, obtained a patent for reinforced concrete flower pots in 1867 and went on to build concrete water tanks and bridges of the new material. By the end of the 19th century, engineering design methods had been developed for structures of reinforced concrete, and a number of major structures had been built. By this time, the earliest experiments in *prestressing* (placing the reinforcing steel under tension before the structure supports a load) had also been carried out, although it remained for Eugene

FIGURE 13.1

At the time concrete is placed, it has no form of its own. This bucket of fresh concrete was filled on the ground by a transit-mix truck and hoisted to the top of the building by a crane. The worker at the right has opened the valve in the bottom of the bucket to discharge the concrete into the formwork. (*Courtesy of Portland Cement Association, Skokie, Illinois*)

438

Freyssinet in the 1920s to establish a scientific basis for the design of *prestressed concrete* structures.

CEMENT AND CONCRETE

Concrete is a rocklike material produced by mixing coarse and fine *aggregates,* portland cement, and water, and allowing the mixture to harden. Coarse aggregate is normally gravel or crushed stone, and fine aggregate is sand. Portland cement, hereafter referred to simply as *cement,* is a fine gray powder. During the hardening of concrete, considerable heat (called the *heat of hydration*) is given off as the cement combines chemically with water to form strong crystals that bind the aggregates together. During this *curing* process, especially as excess water evaporates from the concrete, concrete shrinks slightly.

In properly formulated concrete, the majority of the volume consists of coarse and fine aggregate, proportioned and graded so that the fine particles completely fill the spaces between the coarse ones (Figure 13.3). Each particle is completely coated with a paste of cement and water to join it fully to the surrounding particles.

Cement

Portland cement may be manufactured from any of a number of raw materials, provided that they are combined to yield the necessary amounts of lime, iron, silica, and alumina. Lime is commonly furnished by limestone, marble, marl, or seashells. Iron, silica, and alumina may be obtained in the form of clay or shale. The exact ingredients used

FIGURE 13.2
Hadrian's Villa, a large palace built near Rome between 125 and 135 A.D., used unreinforced concrete extensively for structures such as this dome. (*Photo by the author*)

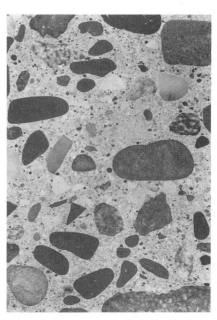

FIGURE 13.3
Photograph of a polished cross section of hardened concrete, showing the close packing of coarse and fine aggregates and the complete coating of every particle with cement paste. (*Courtesy of Portland Cement Association, Skokie, Illinois*)

depend on what is readily available, and the recipe varies widely from one geographic region to another, often including slag or flue dust from iron furnaces, chalk, sand, ore washings, bauxite, and other minerals. The selected constituents are crushed, ground, proportioned and blended, then conducted through a rotating kiln at temperatures of 2600 to 3000 degrees Fahrenheit (1400 to 1650°C) to produce *clinker* (Figures 13.4, 13.5). After cooling, the clinker is pulverized (along with a small amount of gypsum to retard the curing process) to a powder finer than flour. This powder, portland cement, is either packaged in bags or shipped in bulk. In the United States, a standard bag of cement contains 1 cubic foot of volume and weighs 94 pounds (43 kg).

The quality of cement is established by ASTM C150, which identifies eight different types of portland cement:

Type I	Normal
Type IA	Normal, air-entraining
Type II	Moderate resistance to sulfate attack
Type IIA	Moderate resistance, air-entraining
Type III	High early strength
Type IIIA	High early strength, air-entraining
Type IV	Low heat of hydration
Type V	High resistance to sulfate attack

Type I cement is used for most purposes in construction. Types II and V are used where the concrete will be in contact with water that has a high concentration of sulfates. Type III hardens more quickly than the other types and is employed in situations where a reduced curing period is desired as may be the case in cold weather, in the precasting of concrete structural elements, or when the construction schedule must be accelerated. Type IV is used in massive structures such as dams where the heat emitted by the curing concrete

STEPS IN THE MANUFACTURE OF PORTLAND CEMENT

STONE IS FIRST REDUCED TO 5-IN. SIZE, THEN 3/4-IN., AND STORED

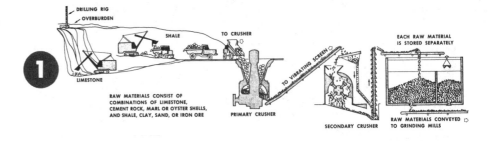

BURNING CHANGES RAW MIX CHEMICALLY INTO CEMENT CLINKER

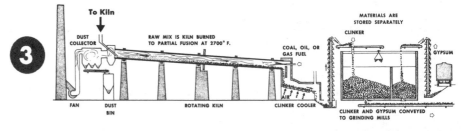

FIGURE 13.4
A rotary kiln manufacturing cement clinker. (*Courtesy of Portland Cement Association, Skokie, Illinois*)

FIGURE 13.6
A photomicrograph of air-entrained concrete shows the bubbles of entrained air (0.01 inch equals 0.25 mm). (*Courtesy of Portland Cement Association, Skokie, Illinois*)

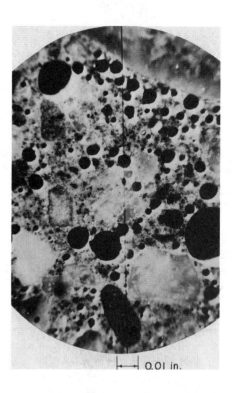

FIGURE 13.5
Steps in the manufacture of portland cement. (*Courtesy of Portland Cement Association, Skokie, Illinois*)

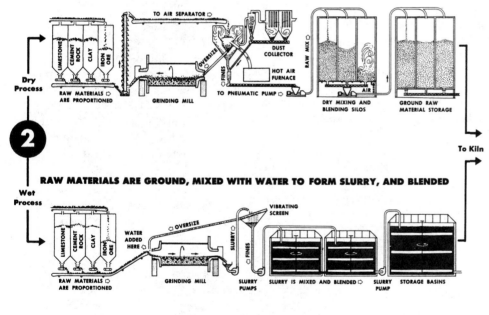

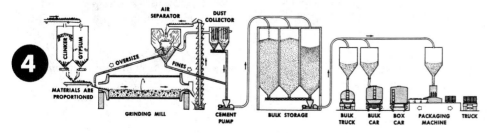

may accumulate and raise the temperature of the concrete to damaging levels.

Air-entraining cements contain ingredients that cause microscopic air bubbles to form in the concrete during mixing (Figure 13.6). These bubbles, which usually comprise 2 to 8 percent of the volume of the finished concrete, give improved workability during placement of the concrete and, more importantly, greatly increase the resistance of the cured concrete to damage caused by repeated cycles of freezing and thawing. Air-entrained concrete is commonly used for pavings and exposed architectural concrete in cold climates. With appropriate adjustments in the formulation of the mix, air-entrained concrete can achieve the same structural strength as normal concrete.

Aggregates and Water

Because aggregates make up roughly three-quarters of the volume of concrete, the structural strength of a concrete is heavily dependent on the quality of its aggregates. Aggregates

FIGURE 13.7
Taking a sample of coarse aggregate from a crusher yard for testing. (*Courtesy of Portland Cement Association, Skokie, Illinois*)

for concrete must be strong, clean, resistant to freeze–thaw deterioration, chemically stable, and properly graded for size (Figure 13.7). An aggregate that is dusty or muddy will contaminate the cement paste with inert particles that weaken it, and an aggregate that contains any of a number of chemicals from sea salt to organic compounds can cause problems ranging from corrosion of reinforcing steel to retardation of the curing process and ultimate weakening of the concrete. A number of standard ASTM laboratory tests are used to assess the various qualities of aggregates.

Size distribution of aggregate particles is important because a range of sizes must be included in each concrete mix to achieve close packing of the particles. A concrete aggregate is graded for size by passing a sample of it through a standard assortment of sieves with diminishing mesh spacings, then weighing the percentage of material that passes each sieve. This test makes it possible to compare the grading of an actual aggregate with the ideal grading for a particular concrete mixture. Size of aggregate is also significant because the largest particle in a concrete mix must be small enough to pass easily between the most closely spaced reinforcing bars and to fit easily into the formwork. In general, the maximum

aggregate size should not be greater than three-fourths of the clear spacing between bars or one-third the depth of a slab. For very thin slabs and toppings, a ³⁄₈-inch (9-mm) maximum aggregate diameter is often specified. A ³⁄₄-inch or 1¹⁄₂-inch (19-mm or 38-mm) maximum is common for most slab and structural work, but aggregate diameters up to 6 inches (150 mm) are used in dams and other massive structures. Producers of concrete aggregates use screens to sort their product for size and can furnish aggregates graded to order.

Lightweight aggregates are used instead of sand and crushed stone for various special types of concrete. *Structural lightweight aggregates* are made from minerals such as shale: The shale is crushed to the desired particle sizes, then heated in an oven to a temperature at which the shale becomes plastic in consistency. The small amount of water that occurs naturally in the shale turns to steam and "pops" the softened particles of aggregate like popcorn. Concrete made from this aggregate has a density about 80 percent of that of normal concrete, while retaining most of the strength of normal concrete. This reduces the dead weight of the components produced from structural lightweight concrete. Nonstructural lightweight concretes are made as insulating roof toppings in densities only one-fourth to one-sixth that of normal concrete. The aggregates in these concretes are usually expanded mica (*vermiculite*) or expanded volcanic glass (*perlite*), and the density of the concretes is further reduced by admixtures that entrain large amounts of air during mixing.

Mixing water for concrete must be free of harmful substances, espe-

cially organic material, clay, and salts such as chlorides and sulfates. Water that is suitable for drinking is generally suitable for concrete.

Admixtures

Ingredients other than cement, aggregates, and water are often added to concrete to alter its properties in various ways.

• *Air-entraining admixtures* may be put in the mix, if they are not already in the cement, to increase the workability of the wet concrete, reduce freeze–thaw damage, or, in larger amounts, to create very lightweight nonstructural concretes with thermal insulating properties.

• *Water-reducing admixtures* allow a reduction in the amount of mixing water while retaining the same workability, which results in a higher-strength concrete.

• *High-range water-reducing admixtures,* also known as *superplasticizers,* are organic compounds that transform a stiff concrete mix into a free-flowing liquid. They are used either to facilitate placement of concrete under difficult circumstances or to reduce the water content of a concrete mix in order to increase its strength.

• *Accelerating admixtures* cause the concrete to cure more rapidly, and *retarding admixtures* slow its curing to allow more time for working with the wet concrete.

• *Fly ash,* a fine powder that is a waste product from coal-fired power plants, increases concrete strength, decreases permeability, increases sulfate resistance, reduces temperature rise, reduces mixing water, and improves pumpability and workability of concrete.

• *Silica fume,* also known as *microsilica,* is a powder that is approximately 100 times finer than portland cement, consisting mostly of silicon dioxide. It is a byproduct of electronic chip manufacturing. When added to a concrete mix, it produces extremely high strength concrete that also has a very low permeability.

• *Blast furnace slag* is a byproduct of iron manufacture that can improve concrete workability, increase strength, reduce permeability, reduce temperature rise, and improve sulfate resistance.

• *Pozzolans* are various natural or artificial materials that react with the calcium hydroxide in wet concrete to form cementing compounds; they are used for purposes such as reducing the internal temperatures of curing concrete, reducing the reactivity of concrete with aggregates containing sulfates, or improving the workability of the concrete.

• *Workability agents* make the wet concrete easier to place in forms and finish by improving its plasticity. They include pozzolans and air-entraining admixtures, along with certain fly ashes and organic compounds.

• *Corrosion inhibitors* are used to reduce rusting of reinforcing steel in structures that are exposed to road deicing salts or other corrosion-causing chemicals.

• *Fibrous admixtures* are short fibers, usually of glass, steel, or polypropylene, that are added to a concrete mix to act as microreinforcing. Their most common use is to reduce plastic shrinkage cracking that sometimes occurs during curing of slabs. Glass fibers are also added to concrete to produce glass-fiber-reinforced concrete (GFRC), used for cladding panels (see Chapter 20).

• *Freeze protection admixtures* allow concrete to cure satisfactorily at temperatures as low as 20 degrees Fahrenheit (−7°C).

• *Extended set-control admixtures* may be used to delay the curing reaction in concrete for any desired period, even several days. The *stabilizer* component, added at the time of initial mixing, defers the onset of curing indefinitely; the *activator* component, added when desired, reinitiates the curing process.

• *Coloring agents* are dyes and pigments used to alter and control the color of concrete for building components whose appearance is important.

MAKING AND PLACING CONCRETE

Proportioning Concrete Mixes

The quality of cured concrete is measured by any of several criteria, depending on the end use of the concrete. For structural columns, beams, and slabs, compressive strength and stiffness are important. For paving and floor slabs, surface smoothness and abrasion resistance are also important. For pavings and exterior concrete walls, a high degree of weather resistance is required. Watertightness is important in concrete tanks, dams, and walls. But regardless of the criterion to which one is working, the rules for making high-quality concrete are much the same: Use clean, sound ingredients; mix them in the correct proportions; handle the wet concrete properly to avoid segregating its ingredients; and cure the concrete carefully under controlled conditions.

The design of concrete mixtures is a science that can be described here only in its broad outlines. The starting point of any mix design is to establish the desired workability characteristics of the wet concrete, the desired physical properties of the cured concrete, and the acceptable cost of the concrete, keeping in mind that there is no need to spend money to make concrete better than it needs to be for a given application. Concretes with ultimate compressive strengths as low as 2000 pounds per square inch (13.8 MPa) are satisfactory for some foundation elements. Concretes with ultimate compressive strengths of 22,000 pounds per square inch (150 MPa), produced with the aid of silica fume, fly ash, and superplasticizer admixtures, are cur-

rently being employed in the columns of some high-rise buildings, and higher strengths than this are certain to be developed in the near future. Acceptable workability is achievable at any of these strength levels.

Given a proper gradation of satisfactory aggregates, the strength of cured concrete is primarily dependent on the amount of cement in the mix and the *water–cement ratio*. While water is required as a reactant in the curing of concrete, much larger amounts of water must be added to a concrete mix than are needed for the hydration of the cement, in order to give the wet concrete the necessary fluidity and plasticity for placing and finishing. The extra water eventually evaporates from the concrete, leaving microscopic voids that impair the strength and surface qualities of the concrete (Figure 13.8). Absolute water–cement ratios by weight should be kept below 0.60 for most applications, meaning that the weight of the water in the mix should not be more

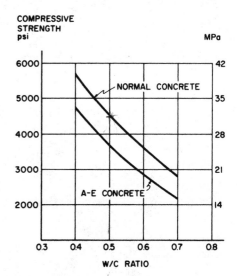

FIGURE 13.8

The effect of water–cement ratio on the strength of concrete. A-E concrete is air-entrained concrete. (*Reprinted with permission of the Portland Cement Association from* Design and Control of Concrete Mixtures)

than 60 percent of the weight of the cement. Higher water–cement ratios than this are often favored by concrete workers because they produce a fluid mixture that is easy to place in the forms, but the resulting concrete is likely to be deficient in strength and surface qualities. Low water–cement ratios make concrete that is dense and strong, but unless air-entraining or water-reducing admixtures are included in the mix, the concrete will not flow easily into the forms and will have large voids. It is important that concrete be formulated with the right quantity of water for each situation, enough to assure workability but not enough to adversely affect the final properties of the material.

Most concrete in North America is proportioned at central batch plants, using up-to-date laboratory equipment and engineering knowledge to produce concrete with the specified properties. The concrete is *transit mixed* en route in a rotating drum on the back of a truck so that it is ready to pour by the time it reaches the job site (Figures 13.9, 13.10). For very small jobs, concrete may be mixed at the job site, either in a small power-driven mixing drum or on a flat surface with shovels. For these small jobs, where the quality of the finished concrete generally does not

FIGURE 13.9
Charging a transit-mix truck with measured quantities of cement, aggregates, admixtures, and water at a central batch plant. (*Courtesy of Portland Cement Association, Skokie, Illinois*)

FIGURE 13.10
A transit-mix truck discharges its concrete, which was mixed en route in the rotating drum, into a truck-mounted concrete pump, which forces it through a hose to the point in the building at which it is being poured. (*Courtesy of Portland Cement Association, Skokie, Illinois*)

need to be precisely controlled, proportioning is usually done by rule of thumb. Typically, the dry ingredients are measured volumetrically, using a shovel as a measuring device, in proportions such as one shovel of cement to two of sand to three of gravel, with enough water to make a wet concrete that is neither soupy nor stiff.

Each load of transit-mixed concrete is delivered with a certificate from the batch plant that lists its ingredients and their proportions. As a further check on quality, a *slump test* may be performed at the time of pouring to determine if the desired degree of workability has been achieved without making the concrete too wet (Figures 13.11, 13.12). For structural concrete, standard test cylinders are also poured from each truckload. Within 48 hours of pouring, the cylinders are taken to a testing laboratory, cured for a specified period under standard conditions, and tested for compressive strength (Figure 13.13). If the laboratory results are not up to the required standard, test cores are drilled from the actual members poured from the questionable batch of concrete. If the strength of these core samples is also deficient, the contractor will be required to cut out the defective concrete and replace it. Frequently, test cylinders are also cast and cured on the construction site under the same conditions as the concrete in the forms; these may then be tested as a way of determining when the concrete is strong enough to allow removal of forms and temporary supports.

Handling and Placing Concrete

Wet concrete is not a liquid, but a slurry, an unstable mixture of solids and liquids. If wet concrete is vibrated excessively, dropped from very much of a height, or moved horizontally for any substantial distance in formwork, it is likely to *segregate*. The coarse aggregate works its way to the bottom of the form, and the water and cement paste rise to the top. The result is concrete of nonuniform and generally unsatisfactory properties. Segregation is prevented by depositing the concrete, fresh from the mixer, as close to its final position as possible. If concrete must be dropped a distance of more than 3 or 4 feet (a meter or so), it should be deposited through *dropchutes* that

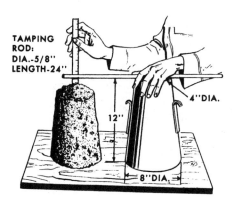

FIGURE 13.11
A slump test for concrete consistency. The hollow metal cone is filled with concrete and tamped with the rod according to a standard procedure. The cone is carefully lifted off, allowing the wet concrete to slump under its own weight. The slump in inches is measured in the manner shown. (*From the U.S. Department of Army,* Concrete, Masonry, and Brickwork)

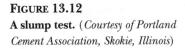

FIGURE 13.12
A slump test. (*Courtesy of Portland Cement Association, Skokie, Illinois*)

FIGURE 13.13
Inserting a standard concrete test cylinder into a structural testing machine, where it will be crushed to determine its strength. (*Courtesy of Portland Cement Association, Skokie, Illinois*)

break the fall of the concrete. If concrete must be moved a large horizontal distance to reach inaccessible areas of the formwork, it should be pumped through hoses (Figure 13.14) or conveyed in buckets or buggies, rather than pushed across or through the formwork.

Concrete must be compacted in the forms to eliminate trapped air and to fill completely around the reinforcing bars and into all the corners of the formwork. This may be done by repeatedly thrusting a rod, spade, or immersion-type vibrator into the concrete at closely spaced intervals throughout the formwork. Excessive agitation of the concrete must be avoided, however, or segregation will occur.

Curing Concrete

Because concrete cures by hydration and not by drying, it is essential that it be kept moist until its required strength is achieved. The curing reaction takes place over a very long period of time, but concrete is commonly designed to be used at the strength that it reaches after 28 days

FIGURE 13.14
Concrete being placed in a basement floor slab with the aid of a concrete pump. Concrete can be pumped for long horizontal distances, and many stories into the air. Note the soldier beams, lagging, and rakers that brace the wall of the excavation. (*Courtesy of Portland Cement Association, Skokie, Illinois*)

(4 weeks) of curing. If it is allowed to dry at any point during this time period, the strength of the cured concrete will be reduced, and its surface qualities will be adversely affected (Figure 13.15). Concrete elements cast in formwork are protected from dehydration on most of their surfaces by the formwork, but the top surfaces must be kept moist by repeatedly spraying or flooding with water, by covering with moisture-resistant sheets of paper or film, or by spraying on a *curing compound* that seals the surface of the concrete against loss of moisture. These measures are particularly important for concrete floor and paving slabs, whose large surface areas make them especially susceptible to drying. Premature drying is a particular danger when slabs are poured in hot or windy weather, which can cause a slab to crack even before it begins to cure. Temporary windbreaks may have to be erected, shade may have to be provided, and frequent fogging of the surface of the slab with a fine spray of water is required until the slab is hard enough to be covered or sprayed with curing compound.

At low temperatures, the curing reaction in concrete proceeds at a much reduced rate. If concrete reaches subfreezing temperatures while curing, the curing reaction stops completely until the temperature of the concrete rises above the freezing mark. It is important that the concrete be protected from low temperatures and especially from freezing until it is fully cured. If freshly poured concrete is covered or insulated, its heat of hydration is often sufficient to maintain an adequate temperature in the concrete even at fairly low air temperatures. Under more severe winter conditions, the ingredients of the concrete may have to be heated before mixing, and both a temporary enclosure and a temporary source of heat may have to be provided.

In very hot weather, the hydration reaction is greatly accelerated, and concrete may begin curing before there is time to place and finish it. This tendency can be controlled by using cool ingredients and, under extreme conditions, by replacing some of the mixing water with an equal quantity of crushed ice, making sure that the ice has melted fully and the concrete has been thoroughly mixed before placing. Another method of cooling concrete is to bubble liquid nitrogen through the mixture at the batch plant.

FORMWORK

Because concrete is put in place as a shapeless slurry with no physical strength, it must be shaped and supported by *formwork* until it has cured sufficiently to support itself. Formwork is usually made of wood, metal, or plastic. It is constructed as a negative of the shape intended for the concrete. Formwork for a beam or slab serves as a temporary working surface during the construction process and as the temporary means of support for reinforcing bars. Formwork must be strong enough to support the considerable weight and fluid pressure of wet concrete without excessive deflection, which often requires temporary supports that are major structures in themselves. During curing, the formwork helps to retain the necessary water of hydration in the concrete. When curing is complete, the formwork must pull away cleanly from the concrete surfaces without damage either to the concrete or to the formwork, which is usually used repeatedly as a construction project progresses. This means that the formwork should have no reentrant corners that will trap or be trapped by the concrete. Any element

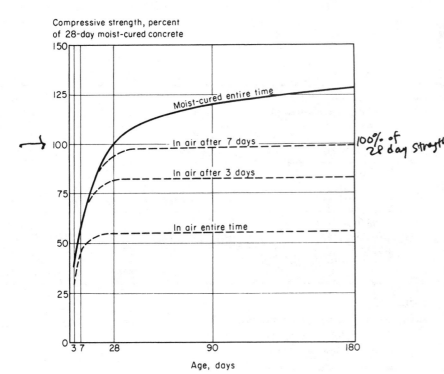

FIGURE 13.15
The growth of compressive strength in concrete over time. Moist-cured concrete is still gaining strength after 6 months, while air-dried concrete virtually stops curing altogether.
(*Courtesy of Portland Cement Association, Skokie, Illinois*)

of formwork that must be withdrawn directly from a location in which it is surrounded on four or more surfaces by concrete, such as a joist pan, must be tapered. All formwork surfaces that are in contact with concrete must be coated with a *form-release compound,* which is an oil, wax, or plastic that prevents adhesion of the concrete to the form.

The quality of the concrete surfaces can be no better than the quality of the forms in which they are cast, and the requirements for surface quality and structural strength of formwork are rigorous. Top-grade wooden boards and plastic-overlaid plywoods are frequently used to achieve high-quality surfaces. The ties and temporary framing members that support the boards or plywood are spaced closely to avoid bulging of the forms under the extreme pressure of the wet concrete.

In a sense, formwork constitutes an entire temporary building that must be erected and demolished in order to produce a second, perma-nent building of concrete. The cost of formwork is a major component of the overall cost of a concrete building frame. This cost is one of the forces that has led to the development of *precasting,* a process in which concrete is cast into permanent, reusable forms at an industrial plant, then transported in the form of fully cured structural units to the job site, where these units are hoisted into place and connected much as if they were structural steel shapes. The opposite of precasting is *sitecasting,* or *cast-in-place construction,* in which concrete is poured into forms that are erected in their final positions on the job site. In the two chapters that follow, formwork is shown for both sitecast and precast concrete.

REINFORCING

The Concept of Reinforcing

Concrete has no useful tensile strength (Figure 13.16) and was lim-ited in its structural uses until the concept of steel *reinforcing* was devel-oped. The compatibility of steel and concrete is a fortuitous accident. If the two materials had grossly differ-ent coefficients of thermal expan-sion, a reinforced concrete structure would tear itself apart during cycles of temperature variation. If they were chemically incompatible, the steel would corrode or the concrete would be degraded. If concrete did not adhere to steel, a very different and more expensive configuration of reinforcing would be necessary. But concrete and steel change dimension at nearly the same rate in response to temperature changes; steel is pro-tected from corrosion by the alkaline chemistry of concrete; and concrete bonds strongly to steel, providing a convenient means of adapting brittle concrete to structural elements that must resist tension, shear, and bend-ing, as well as compression.

The basic theory of concrete reinforcing is extremely simple: Put the steel where tension occurs in a

Material	Allowable Tensile Strength	Allowable Compressive Strength	Density	Modulus of Elasticity
Wood (average)	700 psi (4.83 MPa)	1,100 psi (7.58 MPa)	30 pcf (480 kg/m³)	1,200,000 psi (8,275 MPa)
Brick Masonry (average)	0	250 psi (1.72 MPa)	120 pcf (1,920 kg/m³)	1,200,000 psi (8,275 MPa)
Steel (ASTM A36)	22,000 psi (151.69 MPa)	22,000 psi (151.69 MPa)	490 pcf (7,850 kg/m³)	29,000,000 psi (200,000 MPa)
Concrete (average)	0	1,350 psi (9.31 MPa)	145 pcf (2,320 kg/m³)	3,150,000 psi (21,720 MPa)

FIGURE 13.16
Concrete, like masonry, has no useful tensile strength, but its compressive strength is considerable, and when com-bined with steel reinforcing, concrete can be used for every type of structure.

structural member and let the concrete resist compression. This accounts fairly precisely for the location of most of the reinforcing steel that is used in a concrete structure. But there are some important exceptions: Steel is used to resist a share of the compression in concrete columns and in beams whose height must be reduced for architectural reasons. It is also used to resist cracking that might otherwise be caused by curing shrinkage and by thermal expansion and contraction in slabs and walls.

Steel Bars for Concrete Reinforcement

Reinforcing bars for concrete construction are hot-rolled in much the same way as structural shapes. They are round in cross section, with surface ribs for better bonding to concrete (Figures 13.17, 13.18). The bars are cut to a standard length [commonly 60 feet (18.3 m) in the United States], bundled, and shipped to local fabricating shops.

Reinforcing bars are rolled in a limited number of standard diame-

FIGURE 13.17
Glowing strands of steel are shaped into reinforcing bars as they snake their way through a rolling mill. (*Courtesy of Bethlehem Steel Company*)

FIGURE 13.18
The deformations rolled into the surface of a reinforcing bar help it to bond tightly to concrete. (*Photo by the author*)

ASTM Standard Reinforcing Bars

⅛" increment.

Bar Size		Nominal Dimensions					
		Diameter		Cross-Sectional Area		Weight (mass)	
American	Metric	in.	mm	in.²	mm²	lbs/ft	kg/m
#3	#10	0.375	9.5	0.11	71	0.376	0.560
#4	#13	0.500	12.7	0.20	129	0.668	0.944
#5	#16	0.625	15.9	0.31	199	1.043	1.552
#6	#19	0.750	19.1	0.44	284	1.502	2.235
#7	#22	0.875	22.2	0.60	387	2.044	3.042
#8	#25	1.000	25.4	0.79	510	2.670	3.973
#9	#29	1.128	28.7	1.00	645	3.400	5.060
#10	#32	1.270	32.3	1.27	819	4.303	6.404
#11	#36	1.410	35.8	1.56	1006	5.313	7.907
#14	#43	1.693	43.0	2.25	1452	7.65	11.38
#18	#57	2.257	57.3	4.00	2581	13.6	20.24

FIGURE 13.19
American standard sizes of reinforcing bars. These sizes were originally established in conventional units of inches and square inches. More recently, "soft metric" designations have also been given to the bars without changing their sizes. Notice that the size designations of the bars in both systems of measurement correspond very closely to the rule-of-thumb values of ⅛ inch or 1 mm per bar size number.

ters, as shown in Figures 13.19 and 13.20. In the United States, bars are specified by a simple numbering system in which the number corresponds to the number of eighths of an inch (3.2 mm) of bar diameter. For example, a number 6 reinforcing bar is ⁶⁄₈ or ¾ inch (19 mm) in diameter, and a number 8 is ⁸⁄₈ or 1 inch (25.4 mm). Bars larger than number 8 vary slightly from these nominal diameters in order to correspond to convenient cross-sectional areas of steel. For the increasing volume of work in the United States that is carried out in SI units, a "soft" conversion of units is used: The bars are exactly the same, but a different numbering system is used, corresponding roughly to the diameter of each bar in millimeters. This avoids the expensive process of converting rolling mills to produce a slightly different set of bar sizes. In most countries out-

Metric Reinforcing Bars

Size Designation	Nominal Mass, kg/m	Nominal Dimensions	
		Diameter, mm	Cross-Sectional Area, mm²
10M	0.785	11.3	100
15M	1.570	16.0	200
20M	2.355	19.5	300
25M	3.925	25.2	500
30M	5.495	29.9	700
35M	7.850	35.7	1000
45M	11.775	43.7	1500
55M	19.625	56.4	2500

FIGURE 13.20
These "hard metric" reinforcing bar sizes are used in most countries of the world.

side the United States, a "hard metric" range of reinforcing bar sizes is standard (Figure 13.20).

In selecting reinforcing bars for a given beam or column, the structural engineer knows from calculations the required cross-sectional areas of bars. This area may be achieved with a larger number of smaller bars, or a smaller number of larger bars, in any of a number of combinations. The final bar arrangement is selected based on the physical space available in the concrete member, the required cover dimensions, the clear spacing required between bars to allow passage of the concrete aggregate, and the sizes and numbers of bars that will be most convenient to fabricate and install.

Reinforcing bars are manufactured according to ASTM Standards A615, A616, A617, and A706. They are available in grades 40, 50, and 60, corresponding to steel with yield strengths of 40,000, 50,000, and 60,000 pounds per square inch (275, 345, and 415 MPa). Grade 60 is the most readily available of the three. The higher-strength bars are useful where there is a restricted amount of space available for bars in a concrete member, and they have proved to be the most economical for vertical bars in columns.

Reinforcing bars in concrete structures that are exposed to salts such as deicing salts or those in sea water are prone to rust. Galvanized reinforcing bars and epoxy-coated reinforcing bars are often used in marine structures, highway structures, and parking garages to resist this corrosion.

Reinforcing steel is also produced in sheets or rolls of *welded wire fabric,* a grid of wires or round bars spaced 2 to 12 inches (50 to 300 mm) apart (Figure 13.22). The lighter styles of welded wire fabric resemble cattle fencing and are used to reinforce concrete slabs on grade and certain precast concrete elements. The heavier styles find use in concrete walls and structural slabs. The principal advantage of welded fire fabric over individual bars is economy of labor in placing the reinforcing, especially where a large number of small bars can be replaced by a single sheet of material. The size and spacing of the wires or bars for a particular application are specified by the structural engineer or architect of the building.

The fabrication of reinforcing steel for a concrete construction project is analogous to the fabrication of steel shapes for a Steel Frame building (Chapter 11). The fabricator receives the structural drawings for the building from the contractor and prepares shop drawings for the reinforcing bars. After the shop drawings have been checked by the engineer or architect, the fabricator sets to work cutting the reinforcing bar stock to length, making the necessary bends (Figure 13.23), and tying the fabricated bars into bundles that are tagged to indicate their destination in the building. The bundles are shipped as needed to the building site. There the bundles are opened and sorted. The bars are lifted by hand or hoisted by crane, and wired together in the forms to await pouring of the concrete. The wire has a temporary function only, which is to hold the reinforcement in position until the concrete has cured. Any transfer of load from one reinforcing bar to another in the completed building is done by the concrete. Where two bars must be spliced, they are overlapped a specified number of bar diameters, and the loads are transferred from one to the other by the surrounding concrete. The one common exception to this occurs in splices in vertical column bars, where the reinforcing is heavy and space within the column is at a premium; here the bars are often spliced end to end rather than overlapped, and

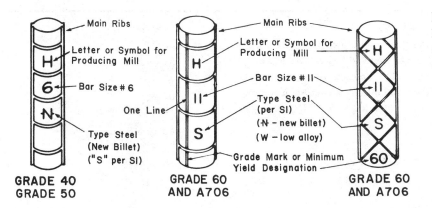

FIGURE 13.21

Examples of the identification markings rolled onto the surface of reinforcing bars. (*Courtesy of Concrete Reinforcing Steel Institute*)

Sectional Area and Weight of Welded Wire Fabric

Wire Size Number		Nominal Diameter Inches	Nominal Weight Lbs./Lin. Ft.	Area in Sq. In. Per Ft. of Width for Various Spacings						
				Center-to-Center Spacing						
Smooth	Deformed			2″	3″	4″	6″	8″	10″	12″
W20	D20	0.505	.680	1.20	.80	.60	.40	.30	.24	.20
W18	D18	0.479	.612	1.08	.72	.54	.36	.27	.216	.18
W16	D16	0.451	.544	.96	.64	.48	.32	.24	.192	.16
W14	D14	0.422	.476	.84	.56	.42	.28	.21	.168	.14
W12	D12	0.391	.408	.72	.48	.36	.24	.18	.144	.12
W11	D11	0.374	.374	.66	.44	.33	.22	.165	.132	.11
W10.5		0.366	.357	.63	.42	.315	.21	.157	.126	.105
W10	D10	0.357	.340	.60	.40	.30	.20	.15	.12	.10
W9.5		0.348	.323	.57	.38	.285	.19	.142	.114	.095
W9	D9	0.338	.306	.54	.36	.27	.18	.135	.108	.09
W8.5		0.329	.289	.51	.34	.255	.17	.127	.102	.085
W8	D8	0.319	.272	.48	.32	.24	.16	.12	.096	.08
W7.5		0.309	.255	.45	.30	.225	.15	.112	.09	.075
W7	D7	0.299	.238	.42	.28	.21	.14	.105	.084	.07
W6.5		0.288	.221	.39	.26	.195	.13	.097	.078	.065
W6	D6	0.276	.204	.36	.24	.18	.12	.09	.072	.06
W5.5		0.265	.187	.33	.22	.165	.11	.082	.066	.055
W5	D5	0.252	.170	.30	.20	.15	.10	.075	.06	.05
W4.5		0.239	.153	.27	.18	.135	.09	.067	.054	.045
W4	D4	0.226	.136	.24	.16	.12	.08	.06	.048	.04
W3.5		0.211	.119	.21	.14	.105	.07	.052	.042	.035
W3		0.195	.102	.18	.12	.09	.06	.045	.036	.03
W2.9		0.192	.099	.174	.116	.087	.058	.043	.035	.029
W2.5		0.178	.085	.15	.10	.075	.05	.037	.03	.025
W2.1		0.162	.070	.126	.084	.063	.042	.031	.025	.021
W2		0.160	.068	.12	.08	.06	.04	.03	.024	.02
W1.5		0.138	.051	.09	.06	.045	.03	.022	.018	.015
W1.4		0.134	.048	.084	.056	.042	.028	.021	.017	.014

FIGURE 13.22
Standard configurations of welded wire fabric. The heaviest "wires" are more than $\frac{1}{2}$ inch (13 mm) in diameter, making them suitable for structural slab reinforcing. (*Courtesy of Concrete Reinforcing Steel Institute*)

STANDARD HOOKS

All specific sizes recommended by CRSI below meet minimum requirements of ACI 318

RECOMMENDED END HOOKS
All Grades

D=Finished bend diameter

Bar Size	180° HOOKS			90° HOOKS
	D	A or G	J	A or G
# 3	2¼	5	3	6
# 4	3	6	4	8
# 5	3¾	7	5	10
# 6	4½	8	6	1-0
# 7	5¼	10	7	1-2
# 8	6	11	8	1-4
# 9	9½	1-3	11¾	1-7
#10	10¾	1-5	1-1¼	1-10
#11	12	1-7	1-2¾	2-0
#14	18¼	2-3	1-9¾	2-7
#18	24	3-0	2-4½	3-5

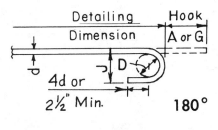

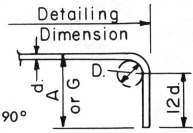

STIRRUP AND TIE HOOKS

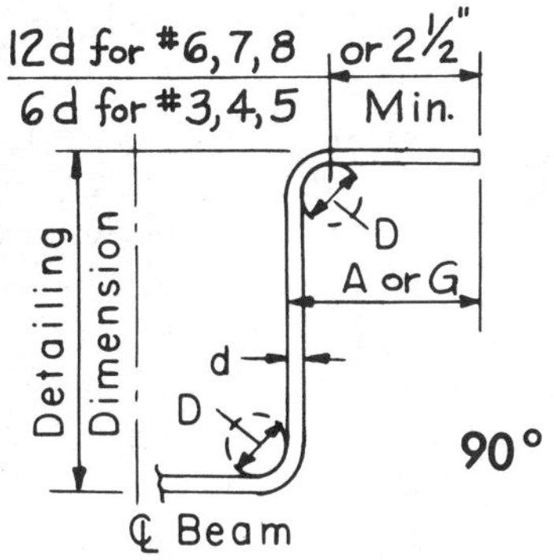

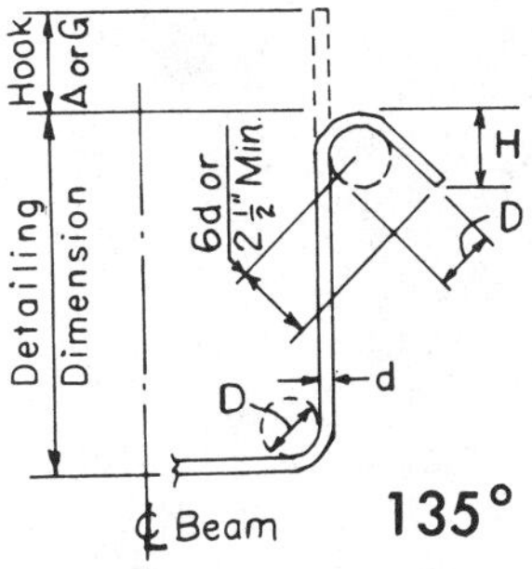

135° SEISMIC STIRRUP/TIE HOOKS

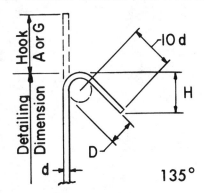

STIRRUPS
(TIES SIMILAR)

STIRRUP AND TIE HOOK DIMENSIONS
Grades 40-50-60 ksi

Bar Size	D (in.)	90° Hook	135° Hook	
		Hook A or G	Hook A or G	H Approx.
#3	1½	4	4	2½
#4	2	4½	4½	3
#5	2½	6	5½	3¾
#6	4½	1-0	7¾	4½
#7	5¼	1-2	9	5¼
#8	6	1-4	10¼	6

135° SEISMIC STIRRUP/TIE
HOOK DIMENSIONS
Grades 40-50-60 ksi

Bar Size	D (in.)	135° Hook	
		Hook A or G	H Approx.
#3	1½	5	3½
#4	2	6½	4½
#5	2½	8	5½
#6	4½	10¾	6½
#7	5¼	1-0½	7¾
#8	6	1-2¼	9

FIGURE 13.23
The bending of reinforcing bars is done according to precise standards in a fabricator's shop. (*Courtesy of Concrete Reinforcing Steel Institute*)

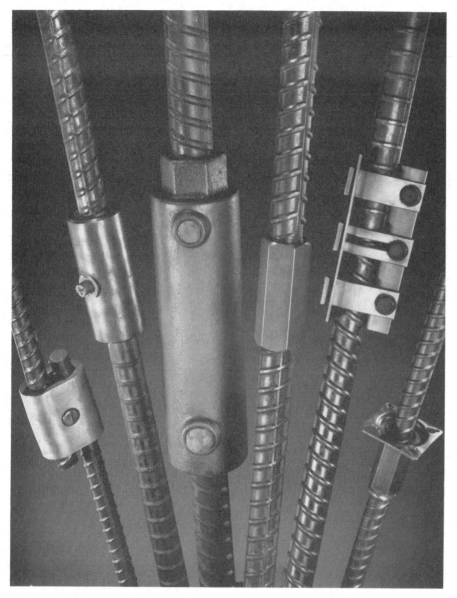

FIGURE 13.24
Some mechanical devices for splicing reinforcing bars. From left to right: A lapped, wedged connection, used primarily to connect new bars to old ones when adding to an existing structure. A welded connector, very strong and tough. A grouted sleeve connector for joining precast concrete components: One bar is threaded and screwed into a collar at one end of the sleeve, and the other bar is inserted into the remainder of the sleeve and held there with injected grout. A threaded sleeve, with both bars threaded and screwed into the ends of the sleeve. A simple clamping sleeve that serves to align compression bars in a column. A flanged coupler for splicing bars at the face of a concrete wall or beam: The coupler is screwed onto the threaded end of one bar and its flange is nailed to the inside face of the formwork. After the formwork has been stripped, the other bar is threaded and screwed through a hole in the flange and into the coupler. (*Photo courtesy of Erico, Inc.*)

loads are transferred through welds or sleevelike mechanical splicing devices (Figure 13.24).

The composite action of concrete and steel in reinforced concrete structural elements is such that the reinforcing steel is always loaded axially in tension or compression, and occasionally in shear, but never in bending. The bending stiffness of the reinforcing bars themselves is of no consequence in imparting strength to the concrete.

Reinforcing a Simple Concrete Beam

In an ideal, simply supported beam under a uniform loading, compressive forces follow a set of archlike curves that create a maximum of compressive stress in the top of the beam at midspan, with progressively lower compressive stresses toward either end. A mirrored set of curves shows the lines of tensile force, with stresses again reaching a maximum at the middle of the span (Figure 13.25). In an ideally reinforced concrete beam, steel reinforcing bars would be bent to follow these lines of tension, and the bunching of the bars at midspan would serve to carry the higher stresses at that point. But it is difficult to bend bars into these curves and to support the curved bars adequately in the formwork, so a simpler, rectilinear arrangement of reinforcing steel is substituted.

This is done with a set of bottom bars and stirrups. The *bottom bars* are placed near the bottom of the beam, leaving a specified amount of concrete below and to the sides of the rods as *cover* (Figure 13.26). The concrete cover provides a full embedment for the reinforcing bars and protects them against fire and corrosion. The bars are most heavily stressed at the midpoint of the beam span, with progressively smaller amounts of stress toward each of the supports. The differences in stress are dissipated from the bars into the concrete by means of the adhesive *bond* forces between the concrete and the steel, aided by the ribs on the surface of the bars. At the ends of the beam, some stress remains in the steel, but there is no further length of concrete into which the stress can be dissipated. This is solved by bending the ends of the bars into *hooks,* which are bends of standard dimensions that are provided for this purpose.

The bottom steel does the heavy tensile work in the beam, but lesser tensile forces remain in a diagonal orientation near the ends of the beam. These are resisted by a series of *stirrups* (Figure 13.25). The stirrups may be either open *U-stirrups* as shown or *closed stirrup-ties,* which are full rectangular loops of steel that wrap all the way around the longitudinal bars. U-stirrups are less expensive to make and install and are sufficient for many situations, but stirrup-ties are required in beams that will be subjected to torsional (twisting) forces or to high compressive forces in the top or bottom bars. In either case, the stirrups furnish vertical tensile reinforcing to resist the cracking forces that run diagonally across them. Obviously, diagonal stirrups would work more efficiently, but economy of installation dictates that stirrups should be oriented vertically.

When the simple beam of our example is formed, the bottom steel is supported at the correct cover height by steel *chairs.* In a broad beam or slab, bars are supported by

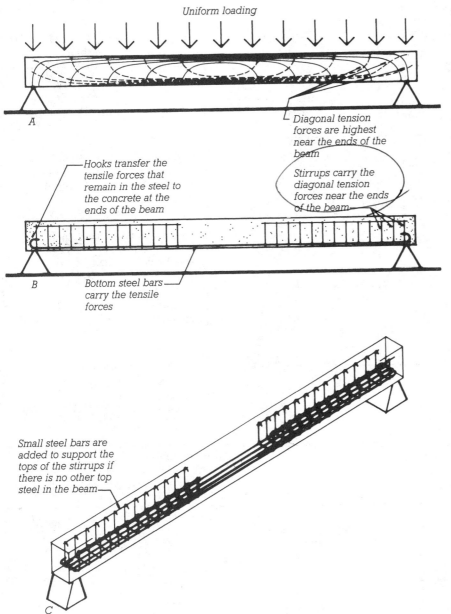

Uniform loading

A

Diagonal tension forces are highest near the ends of the beam

Stirrups carry the diagonal tension forces near the ends of the beam

Hooks transfer the tensile forces that remain in the steel to the concrete at the ends of the beam

B

Bottom steel bars carry the tensile forces

Small steel bars are added to support the tops of the stirrups if there is no other top steel in the beam

C

FIGURE 13.25
(*a*) The directions of force in a simply supported beam under a uniform loading. The solid lines represent compression, and the broken lines represent tension. Near the ends of the beam, the lines of strongest tensile force move upward diagonally through the beam. (*b*) Steel reinforcing for a simply supported beam under a uniform loading. (*c*) A three-dimensional view of the same reinforcing.

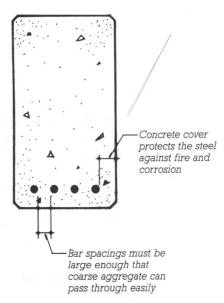

Concrete cover protects the steel against fire and corrosion

Bar spacings must be large enough that coarse aggregate can pass through easily

FIGURE 13.26
A cross section of a rectangular concrete beam showing cover and bar spacing.

SYMBOL	BAR SUPPORT ILLUSTRATION	BAR SUPPORT ILLUSTRATION PLASTIC CAPPED OR DIPPED	TYPE OF SUPPORT	SIZES
SB	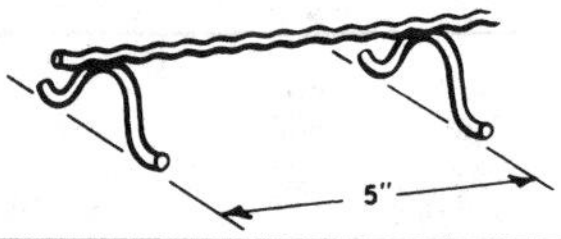 *(illustration, 5")*	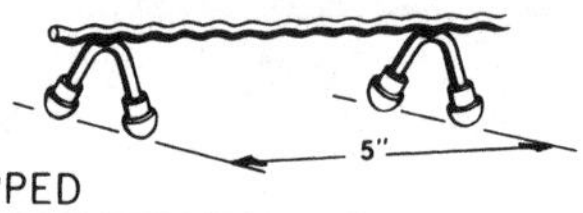 *(illustration, CAPPED, 5")*	Slab Bolster	¾, 1, 1½, and 2 inch heights in 5 ft. and 10 ft. lengths
SBU	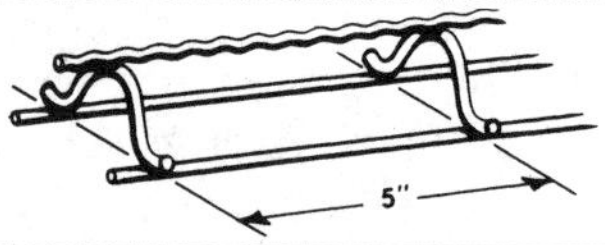 *(illustration, 5")*		Slab Bolster Upper	Same as SB
BB	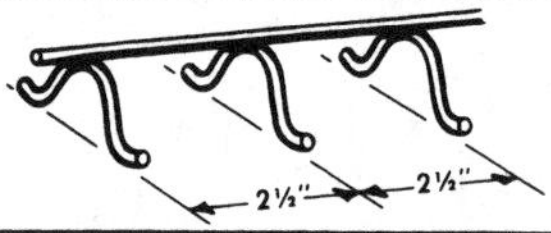 *(illustration, 2½" 2½")*	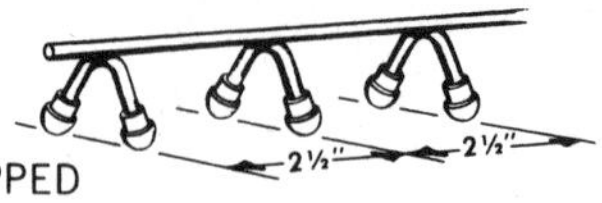 *(illustration, CAPPED, 2½" 2½")*	Beam Bolster	1, 1½, 2, over 2" to 5" heights in increments of ¼" in lengths of 5 ft.
BBU	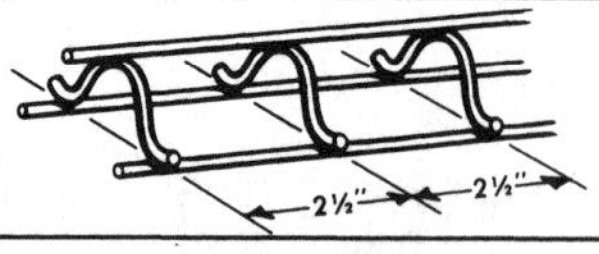 *(illustration, 2½" 2½")*		Beam Bolster Upper	Same as BB
BC	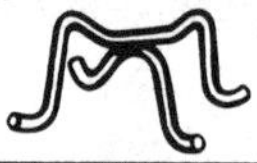 *(illustration)*	*(illustration, DIPPED)* 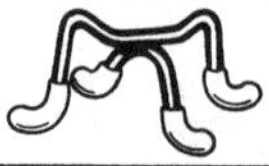	Individual Bar Chair	¾, 1, 1½, and 1¾" heights
JC	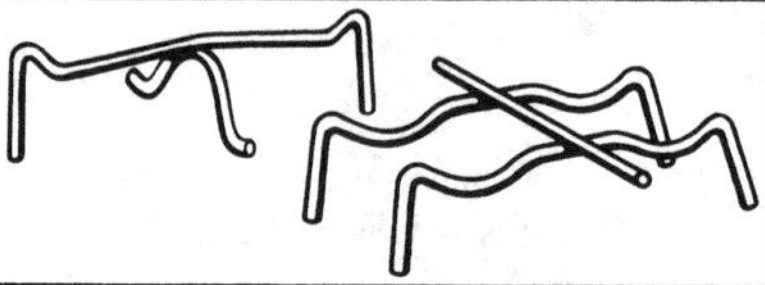 *(illustration)*	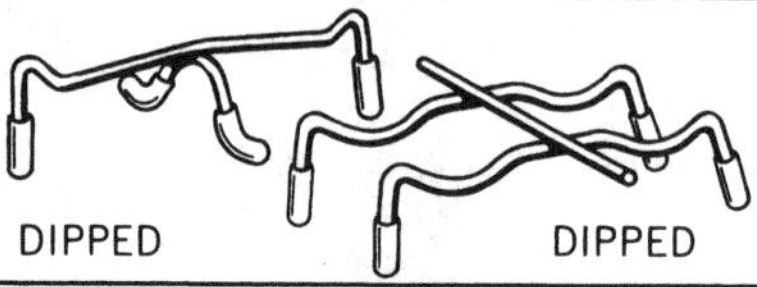 *(illustration, DIPPED DIPPED)*	Joist Chair	4, 5, and 6 inch widths and ¾, 1 and 1½ inch heights
HC	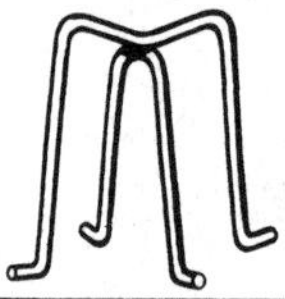 *(illustration)*	*(illustration, CAPPED)* 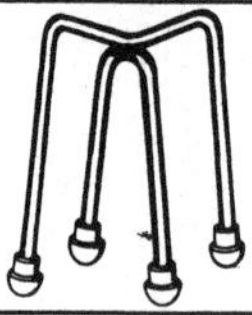	Individual High Chair	2 to 15 inch heights in increments of ¼ inch
HCM	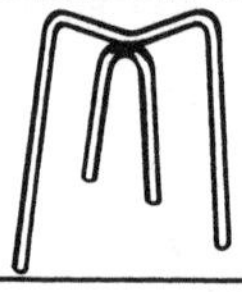 *(illustration)*		High Chair for Metal Deck	2 to 15 inch heights in increments of ¼ in.
CHC	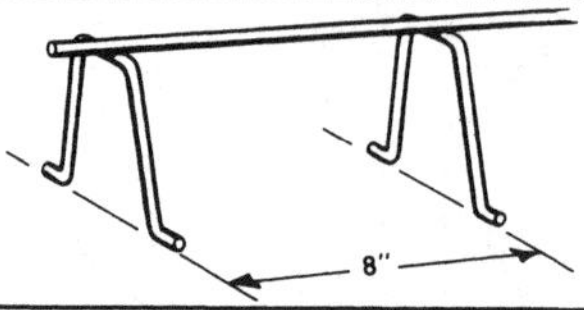 *(illustration, 8")*	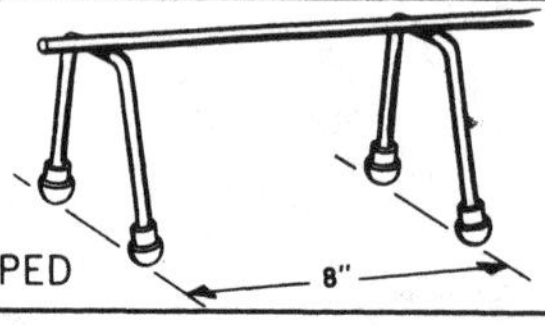 *(illustration, CAPPED, 8")*	Continuous High Chair	Same as HC in 5 foot and 10 foot lengths
CHCU	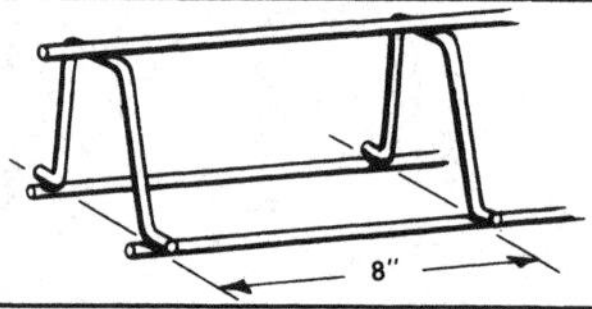 *(illustration, 8")*		Continuous High Chair Upper	Same as CHC
CHCM	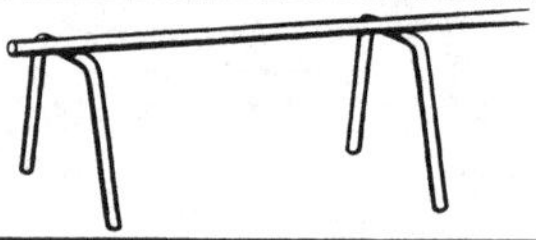 *(illustration)*		Continuous High Chair for Metal Deck	Up to 5 inch heights in increments of ¼ in.
JCU	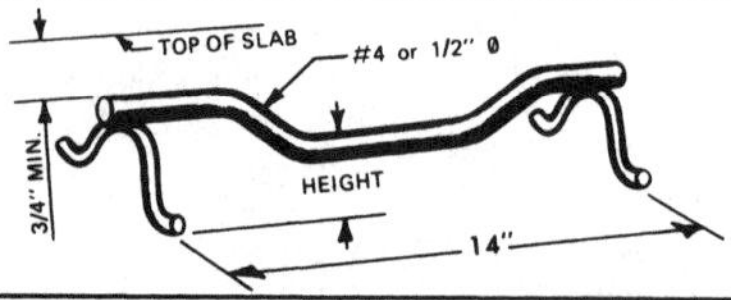 *(illustration: TOP OF SLAB, #4 or 1/2" Ø, 3/4" MIN., HEIGHT, 14")*	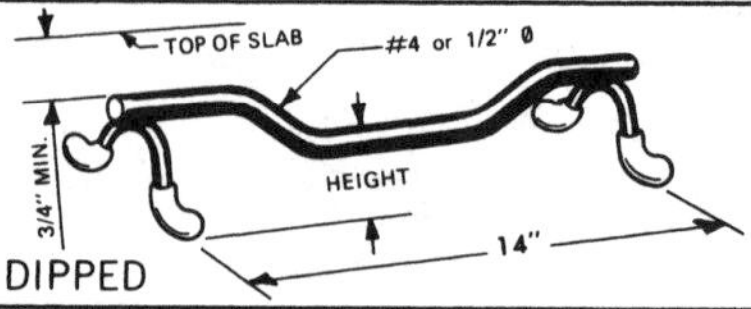 *(illustration: TOP OF SLAB, #4 or 1/2" Ø, 3/4" MIN., HEIGHT, 14", DIPPED)*	Joist Chair Upper	14" Span. Heights – 1" thru +3½" vary in ¼" increments

FIGURE 13.27
Chairs and bolsters for supporting reinforcing bars in beams and slabs. Bolsters and continuous chairs are made in long lengths for use in slabs, while chairs support only one or two bars each. (*Courtesy of Concrete Reinforcing Steel Institute*)

long chairs called *bolsters* (Figure 13.27). These accessories remain in the concrete after pouring, because although their work is then finished, there is no way to get them out. In outdoor concrete work, the feet of the chairs and bolsters sometimes

rust where they come in contact with the face of the beam or slab, unless plastic or plastic-capped steel chairs are used. Where reinforced concrete is poured in direct contact with the soil, concrete bricks or small pieces of concrete are used instead of chairs to prevent rust from forming under the feet of the chairs and spreading up into the reinforcing bars.

The stirrups in the simple beam that we have been examining are supported by wiring them to the bottom steel and by tying their tops to #3 bars (the smallest possible size) that have no function in the beam other than to keep the stirrups upright and properly spaced until the concrete has been poured and cured.

Reinforcing a Continuous Concrete Beam

Most sitecast concrete beams are not of this simple type because concrete lends itself most easily to one-piece structural frames with a high degree of structural continuity from one beam span to the next. In a continuous structure, the bottom of the beam is in tension at midspan, and the top of the beam is in tension at supporting columns or walls. This means that top bars must be provided over the supports, and bottom bars in midspan, along with the usual stirrups, as illustrated in Figure 13.28. Until a few decades ago, it was common practice to bend some of the

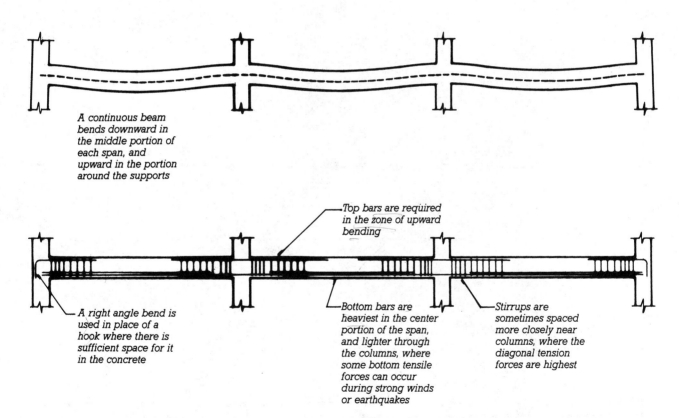

FIGURE 13.28
Reinforcing for a concrete beam that is continuous across several spans. The upper diagram shows in exaggerated form the shape taken by a continuous beam under a uniform loading; the broken line is the centerline of the beam. The lower diagram shows the arrangement of bottom steel, top steel, and stirrups conventionally used in this beam. The bottom bars are usually placed on the same level, but they are shown on two levels in this diagram to demonstrate the way in which some of the bottom steel is discontinued in the zones near the columns. There is a simple rule of thumb for determining where the bending steel must be placed in a beam: Draw an exaggerated diagram of the beam bending under load, as in the top drawing of this illustration, and put the bars as close as possible to the convex edges.

horizontal bars up or down at the points of bending reversal in continuous concrete beams so that the same bars could serve both as bottom steel at midspan and top steel over the columns, but this has largely been abandoned in favor of the simpler practice of using straight bars only.

Reinforcing Structural Concrete Slabs

The reinforcing pattern for concrete structural slabs, which may be considered as very broad beams, is similar to the reinforcing pattern in beams, except that the wide slab can usually resist the relatively weak diagonal tension forces near its supports without the aid of stirrups. Slabs are also provided with *shrinkage–temperature steel*, a lighter set of reinforcing bars set at right angles to, and on top of, the primary reinforcing in the slab in order to prevent cracking parallel to the primary reinforcing from concrete shrinkage, temperature-induced stresses, and miscellaneous forces that may occur in the building (Figure 13.29).

Two-Way Slab Action

A structural economy unique to concrete frames is easily realized through the use of *two-way action* in floor and roof slabs. Two-way slabs, which work well only for column spacing dimensions that are square or nearly square, are reinforced equally in both directions and share the bending forces equally between the two directions.

This allows two-way slabs to be somewhat shallower than one-way slabs, to use less reinforcing steel, and thus to cost less. Figure 13.30 illustrates the concept of two-way slab action. Several different two-way concrete framing systems will be shown in detail in the next chapter.

Reinforcing Concrete Columns

Columns contain two types of reinforcing: *Vertical bars* work with the concrete to share the compressive loads and to resist the tensile stresses that occur in columns when a building frame is subjected to wind or earthquake forces. *Ties* of smaller steel bars wrapped around the vertical bars help to prevent them from

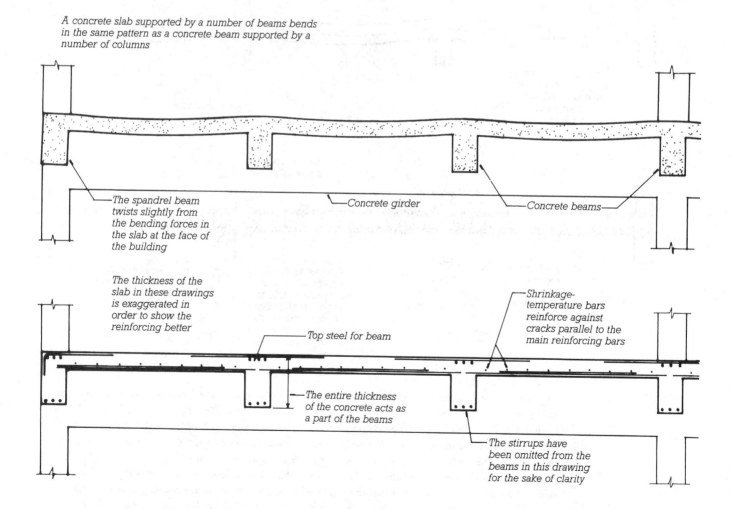

A concrete slab supported by a number of beams bends in the same pattern as a concrete beam supported by a number of columns

The spandrel beam twists slightly from the bending forces in the slab at the face of the building

Concrete girder

Concrete beams

The thickness of the slab in these drawings is exaggerated in order to show the reinforcing better

Top steel for beam

The entire thickness of the concrete acts as a part of the beams

Shrinkage-temperature bars reinforce against cracks parallel to the main reinforcing bars

The stirrups have been omitted from the beams in this drawing for the sake of clarity

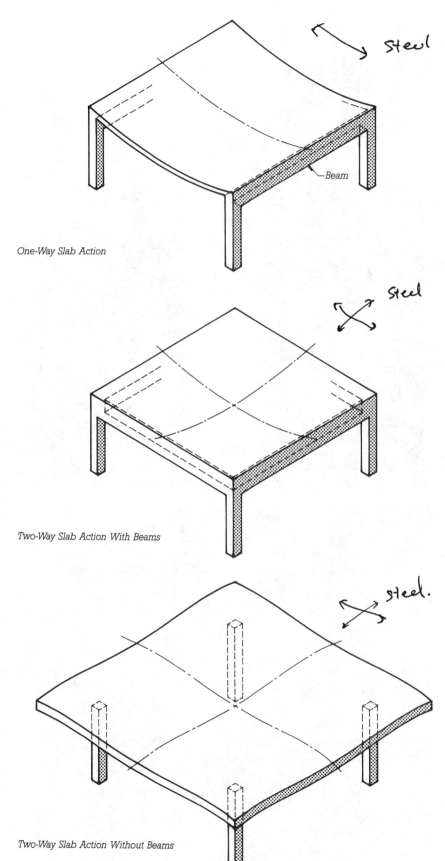

One-Way Slab Action

Two-Way Slab Action With Beams

Two-Way Slab Action Without Beams

FIGURE 13.29
Reinforcing for a one-way concrete slab. The reinforcing is similar to that for a continuous beam, except that stirrups are not usually required in the slab, and shrinkage–temperature bars must be added in the perpendicular direction. The slab does not sit on the beams; rather, the concrete around the top of a beam is part of both the beam and the slab. A concrete beam in this situation is considered to be a T-shaped member, with a portion of the slab acting together with the stem of the beam, resulting in a greater structural efficiency and reduced beam depth.

FIGURE 13.30
One-way and two-way slab action, with deflections greatly exaggerated.

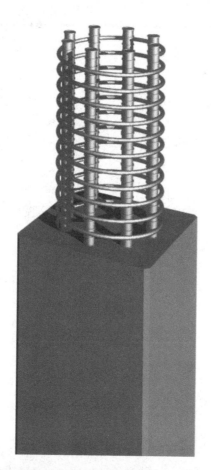

FIGURE 13.31
Reinforcing for concrete columns. To the left, a column with a rectangular arrangement of vertical bars and column ties. To the right, a circular arrangement of vertical bars with a column spiral. Either arrangement may be used in either a round or a square column.

FIGURE 13.32
Column spirals. Each double circle of vertical bars will be embedded in a single rectangular column. (*Courtesy of Concrete Reinforcing Steel Institute*)

buckling under load: Inward buckling is prevented by the concrete core of the column, and outward buckling by the ties (Figure 13.31). The vertical bars may be arranged either in a circle or in rectangular patterns. The ties may be either of two types: column ties or column spirals. *Column spirals* are shipped to the construction site as tight coils of rod that are then expanded accordion fashion to the required spacing and wired to the vertical bars (Figure 13.32). They are generally used only for square or circular arrangements of vertical bars. For rectangular arrangements of vertical bars, discrete *column ties* must be wired on one by one. Each corner bar and alternate interior bars must be contained inside a bend of rod at each column tie, so two or more column ties are often attached together at each level (Figure 13.33). A circular arrangement of vertical bars is often more economical than a rectangular one because it avoids the need to enclose bars in corners of ties. Column ties are generally more economical than spirals, so even columns with circular bar arrangements are often tied with discrete circular ties rather than spirals. The sizes and spacings of column ties and spirals are determined by the structural engineer. To minimize labor costs on the job site, the ties and vertical bars for each column are usually wired together in a horizontal position at ground level, and the finished *column cage* is lifted into its final position with a crane.

PRESTRESSING

When a beam supports a load, the compression side of the beam is squeezed slightly, and the tension side is stretched by a similar amount. In a reinforced concrete beam, the stretching tendency is resisted by the reinforcing steel but not by the brittle concrete. When the steel elongates under tension, the concrete around it cracks from the edge of the beam to the horizontal plane in the beam above which compressive forces

FIGURE 13.33
Multiple column ties at each level are arranged so that the four corner bars and alternate interior bars are contained in the corners of ties. (*Courtesy of Concrete Reinforcing Steel Institute*)

occur. This cracking is visible to the unaided eye in reinforced concrete beams that are loaded to (or beyond) their safe load-carrying capacity. In effect, over half the concrete in the beam is doing no useful work except to hold the steel in position and protect it from fire and corrosion (Figure 13.34).

If the reinforcing bars could be stretched to a high tension before the beam is loaded, and then released against the concrete that surrounds them, they would place the concrete in the vicinity of the bars in compression. If a load were subsequently put on the beam, the tension in the steel

would increase further, and the compression in the concrete surrounding the steel would diminish. But if the initial tension or *prestress* in the steel bars were of sufficient magnitude, the surrounding concrete would never be subjected to tension, and no cracking would occur. Furthermore, mathematical analysis would show that the beam is capable of carrying a much greater load with the same amounts of concrete and steel than if it were merely reinforced in the conventional manner. This is the rationale behind *prestressing*. Prestressed members are lighter than reinforced members of equivalent strength and thus,

in general, less expensive. The lighter weight also pays off by making precast, prestressed concrete members easier and cheaper to transport. For this reason, nearly all precast concrete members are prestressed.

In practice, ordinary reinforcing bars are not sufficiently strong to serve as prestressing steel. Prestressing is practical only with extremely high strength steel strands that are manufactured for the purpose.

Pretensioning

Prestressing can be accomplished in either of two ways. In *pretensioning,*

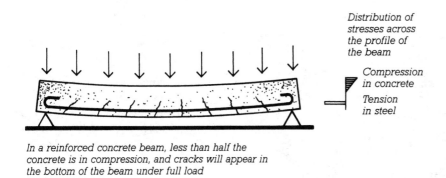

Distribution of stresses across the profile of the beam

Compression in concrete

Tension in steel

In a reinforced concrete beam, less than half the concrete is in compression, and cracks will appear in the bottom of the beam under full load

FIGURE 13.34
The rationale for prestressing concrete. In addition to the absence of cracks in the prestressed beam, the structural action is more efficient than that of a reinforced beam. The prestressed beam therefore uses less material. The small diagrams to the right indicate the distribution of stresses across the vertical cross section of each of the beams at midspan.

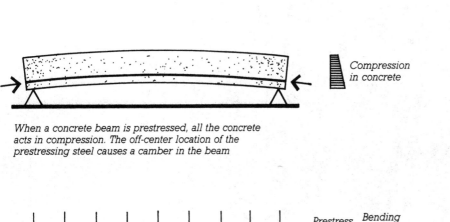

Compression in concrete

When a concrete beam is prestressed, all the concrete acts in compression. The off-center location of the prestressing steel causes a camber in the beam

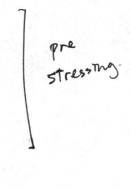

pre stressing.

Prestress   Bending stress

Under loading, the prestressed beam becomes flatter, but all the concrete still acts in compression, and no cracks appear

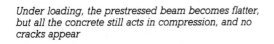

high-strength steel strands are stretched tightly between abutments in a precasting plant, and the concrete member (or, more commonly, a series of concrete members laid end to end) is cast around the stretched steel. The steel bonds to the curing concrete along its entire length. After the concrete has been cured, the steel is cut off at either end of the member. This releases the external tension and the steel recoils slightly, pulling the concrete of the member into compression. If, as is usually the case, the steel is placed as closely as possible to the tension side of the member, the member takes on a decided *camber* at the time the steel strands are cut (Figure 13.35). Much or all of this camber disappears later when the member is subjected to its full load. Because the strong abutments needed to hold the tensioned steel prior to the pouring of concrete are very difficult and expensive to construct except in a single, fixed location where many concrete members can be created within the same set of abutments, pretensioning is useful only for precast members.

Posttensioning

Posttensioning is done almost exclusively on the building site. The high-strength steel strands (called *tendons*) are covered with a steel or plastic tube to prevent them from bonding with the concrete and are not tensioned until the concrete has cured. One end of each tendon is anchored to the end of the beam or slab. A hydraulic jack is inserted between the other end of the tendon and the other end of the member. The jack applies a large tensile force to the

1. The first step in pretensioning is to stretch the steel prestressing strands tightly across the casting bed

2. Concrete is cast around the stretched strands and cured. The concrete bonds to the strands

3. When the strands are cut the concrete goes into compression and the beam takes on a camber

FIGURE 13.35
Pretensioning. Photographs of pretensioned steel strands for a beam are shown in Chapter 15.

tendon while compressing the concrete with an equal but opposite force. The stretched tendon is anchored to the second end of the member before the jack is removed (Figures 13.36–13.38). (For very long members, the tendons are jacked from both ends to ensure that frictional losses in the tubes do not prevent uniform tensioning.)

The net effect of posttensioning is identical to that of pretensioning. The difference is that abutments are not needed because the member itself provides the abutting force needed to tension the steel. When the posttensioning process is complete, the tendons may be left unbonded or, if they are in a steel tube, they may be bonded by injecting grout to fill the space between the tendons and the tube. Bonded construction is common in bridges and other heavy structures, but most posttensioning in buildings is done with unbonded tendons. These are made up of seven cold-drawn steel wires and are either 0.5 or 0.6 inch (12.7 or 15.2 mm) in diameter (Figures 13.38, 13.39). The tendon is coated with a lubricant and covered with a plastic sheath at the factory.

Even higher structural efficiencies are possible in a prestressed beam or slab if the steel strands are made to follow as closely as possible the lines of tensile force as diagrammed in Figure 13.25. In a posttensioned beam or slab, this is done by using chairs of varying heights to support the tendons along the curving line that traces the center of the tensile forces in the member. Such a tendon is referred to as being *draped*

1. In posttensioning, the concrete is not allowed to bond to the steel strands during curing

2. After the concrete has cured, the strands are tensioned with a hydraulic jack and anchored to the ends of the beam. If the strands are draped, as shown here, higher structural efficiency is possible than with straight strands

FIGURE 13.36
Posttensioning, using draped strands to more nearly approximate the flow of tensile forces in the beam.

FIGURE 13.37
Posttensioning draped tendons in a large concrete beam with a hydraulic jack. Each tendon consists of a number of individual high-strength steel strands. The bent bars projecting from the top of the beam will be embedded in the concrete slab that the beam will support, to allow them to act together as a composite structure. (*Courtesy of Portland Cement Association, Skokie, Illinois*)

FIGURE 13.38
Most beams and slabs in buildings are posttensioned with plastic-sheathed, unbonded tendons. The pump and hydraulic jack (also called a ram) are small and portable. (*Courtesy of Constructive Services, Inc., Dedham, Massachusetts*)

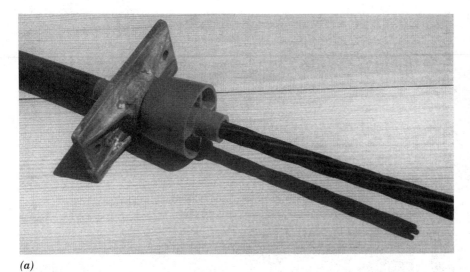

(a)

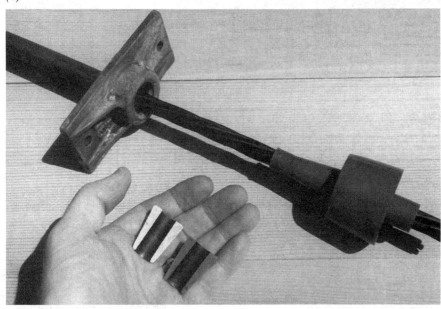

(b)

(c)

FIGURE 13.39
An end anchorage for small posttensioning tendons is ingeniously simple. (*a*) A steel anchor plate and a plastic pocket former are nailed to the inside of the concrete formwork, with the largest circular face of the pocket former placed against the vertical face of the formwork. The end of the tendon extends through a hole drilled in the formwork. (*b*) After the concrete has cured, the formwork is stripped and the pocket former is retracted, leaving a neat pocket in the edge of the slab for access to the anchor plate, which lies recessed below the surface of the concrete. Two conical wedges with sharp ridges inside are inserted around the tendon and into the conical hole in the anchor plate. (*c*) The ram presses against the wedges and draws the tendon through them until the gauge on the pump indicates that the required tension has been achieved. When the ram is withdrawn, the wedges are drawn into the conical hole, grip the tendon, and maintain the tension. After all the tendons have been tensioned, the excess length of tendon is cut off and the pocket is grouted flush with the edge of the slab. (*Photos by the author*)

Straight pretensioning strands

Depressed pretensioning strands

Harped pretensioning strands

(Figures 14.41, 14.42). Draping cannot be used in pretensioned members because the pretensioning forces are far too high, but it is possible to pull the strands up and down in the formwork to make a downward-pointing vee or flattened vee shape in each member (Figure 13.40).

Because of the high prestressing loads, there is a tendency for the concrete in a prestressed member to shorten progressively over an extended period of time in a process known as *creep*, and for the steel strands to relax slightly. Prestressing forces must be increased slightly above the theoretically correct values to accommodate these long-term movements, along with the slight curing shrinkage present in all concrete and small, short-term movements caused by elastic shortening of the concrete and frictional losses and anchorage set in posttensioned members.

Succeeding chapters will discuss prestressed concrete, both pretensioned and posttensioned, in greater detail, showing its application to various standard precast and cast-in-place systems of construction.

ACI 301

Concrete structures are built under ACI 301, *Specifications for Structural Concrete for Buildings,* a publication of the American Concrete Institute. This is a comprehensive, detailed specification that covers every aspect of concrete work: formwork, accessories, reinforcement, chairs and bolsters, concrete mixtures, handling and placing of concrete, lightweight concrete, prestressing, and the use of concrete in exposed architectural surfaces. It leaves nothing to chance. It is a standard that is familiar to architects, engineers, contractors, and building inspectors, and furnishes the basis upon which everyone works in designing and constructing a concrete building.

C.S.I./C.S.C. Masterformat Section Numbers for Concrete Construction	
03100	**CONCRETE FORMWORK**
03200	**CONCRETE REINFORCEMENT**
03210	**Reinforcing Steel**
03220	**Welded Wire Fabric**
03230	**Stressing Tendons**
03300	**CAST-IN-PLACE CONCRETE**
03400	**PRECAST CONCRETE**

SELECTED REFERENCES

1. Portland Cement Association. *Design and Control of Concrete Mixtures* (13th ed.). Skokie, Illinois, 1994.

The 15 chapters of this book summarize clearly and succinctly, with many explanatory photographs and tables, the state of current practice in making, placing, finishing, and curing concrete. (Address for ordering: 5420 Old Orchard Road, Skokie, IL 60077.)

2. Concrete Reinforcing Steel Institute. *Manual of Standard Practice* (26th ed.). Schaumburg, Illinois, 1997.

Specifications for reinforcing steel, welded wire fabric, bar supports, detailing, fabrication, and installation are standardized in this booklet. (Address for ordering: 933 North Plum Grove Road, Schaumburg, IL 60195-4758.)

3. American Concrete Institute. *ACI 301: Specifications for Structural Concrete for Buildings.* Farmington Hills, Michigan, ACI International, 1996.

This is the standard, detailed specification for every aspect of structural concrete. (Address for ordering: P.O. Box 9094, Farmington Hills, MI 48333.)

KEY TERMS AND CONCEPTS

portland cement
reinforced concrete
prestressing
prestressed concrete
concrete
aggregate
cement
heat of hydration
curing
clinker
air entraining
lightweight aggregate
structural lightweight aggregate
air-entraining admixture
water-reducing admixture
high-range water-reducing admixture
superplasticizer

accelerating admixture
retarding admixture
fly ash
silica fume
blast furnace slag
pozzolan
workability agent
corrosion inhibitor
fibrous admixture
freeze protection admixture
extended set-control admixture
stabilizer
activator
coloring agent
water–cement ratio
slump test
segregation

dropchute
curing compound
formwork
form-release compound
precasting
sitecasting
reinforcing
welded wire fabric
bottom bars
cover
bond
hook
stirrups
U-stirrups
closed stirrup-ties
chair
bolster

shrinkage–temperature steel
two-way action
vertical bars
ties
column spiral
column tie
prestress
pretensioning
camber
posttensioning
tendons
draping
creep
ACI 301

REVIEW QUESTIONS

1. What is the difference between cement and concrete?

2. List the conditions that must be met to make a satisfactory concrete mix.

3. List the precautions that should be taken to cure concrete properly. How do these change in very hot, very windy, and very cold weather?

4. What problems are likely to occur if concrete has too low a slump? Too high a slump? How can the slump be increased without increasing the water content of the concrete mixture?

5. Explain how steel reinforcing bars work in concrete.

6. Explain the role of stirrups in beams.

7. Explain the role of ties in columns.

8. What does shrinkage–temperature steel do? Where is it used?

9. Explain the differences between reinforcing and prestressing, and the relative advantages and disadvantages of each.

10. Under what circumstances would you use pretensioning, and under what circumstances would you use posttensioning?

EXERCISES

1. Design a simple concrete mixture. Mix it and pour some test cylinders for several water–cement ratios. Cure and test the cylinders. Plot a graph of concrete strength versus water–cement ratio.

2. Sketch from memory the pattern of reinforcing for a continuous concrete beam. Add notes to explain the function of each feature of the reinforcing.

3. Design, form, reinforce, and cast a small concrete beam, perhaps 6 to 12 feet (2 to 4 m) long. Get help from a teacher or professional, if necessary, in designing the beam.

4. Visit a construction site where concrete work is going on. Examine the forms, reinforcing, and concrete work. Observe how concrete is brought to the site, transported, placed, compacted, and finished. How is the concrete supported after it has been poured? For how long?

14

SITECAST CONCRETE FRAMING SYSTEMS

Boston City Hall makes bold use of sitecast concrete in its structure, facades, and interiors. Its base is faced with brick masonry. (*Architects: Kallmann, McKinnell, and Wood. Photo by Ezra Stoller, ©ESTO*)

Concrete that is cast into forms on the building site offers almost unlimited possibilities to the designer. Any shape that can be formed can be cast, with any of a limitless selection of surface textures, and the pages of books on modern architecture are filled with graphic examples of the realization of this extravagant promise. Certain types of concrete elements cannot be precast, but can only be cast on the site—foundation caissons and spread footings, slabs on grade, structural elements too large or too heavy to transport from a precasting plant, elements so irregular or special in form as to rule out precasting, slab toppings over precast floor and roof elements, and many types of structures with two-way slab action or full structural continuity from one member to another. In many cases where sitecast concrete could be replaced with precast, sitecast remains the method of choice simply because of its more massive, monolithic architectural character.

Sitecast concrete structures tend to be heavier than most other types of structures, a consideration that can lead to the selection of a precast concrete or structural steel frame instead if foundation loadings are critical. Sitecast buildings are also relatively slow to construct because each level of the building must be formed, reinforced, poured, cured, and stripped of formwork before the building can be built significantly taller. In effect, each element of a sitecast concrete building is manufactured in place, often under variable weather conditions, while the majority of the work on steel or precast concrete buildings is done in plants and shops where worker access, tooling, materials-handling equipment, and environmental conditions are generally superior to those on the job site. But the technology of sitecasting has evolved rapidly in response to its own inherent limitations, with streamlined methods of materials handling, systems of reusable formwork that can be erected and taken down almost instantaneously, extensive prefabrication of reinforcing elements, and mechanization of finishing operations, at a pace that has kept it among the construction techniques most favored by building owners, architects, and engineers.

FIGURE 14.1
Unity Temple in Oak Park, Illinois, was constructed by architect Frank Lloyd Wright in 1906. Its structure and exterior surfaces were cast in concrete, making it one of the earliest buildings in the United States to be built primarily of this material. (*Photo by John McCarthy. Courtesy of Chicago Historical Society, IChi-18291*)

CASTING A CONCRETE SLAB ON GRADE

A concrete *slab on grade* is a level surface of concrete that is supported directly by the ground for a road, a sidewalk, an airport runway, or the ground floor of a building. It usually carries little structural stress except a direct transmission of compression between its superimposed loads and the ground beneath, so it furnishes a simple example of the operations involved in the sitecasting of concrete (Figure 14.2).

To prepare for the placement of a slab on grade, topsoil is scraped away to expose the subsoil beneath. A layer of ¾-inch-diameter (19-mm) crushed stone at least 4 inches (100 mm) deep is compacted over the subsoil as a drainage layer to keep water away from the slab. A simple edge form is constructed of wood or metal around the perimeter of the area to

be poured and is coated with a form-release compound to prevent the concrete from sticking (Figure 14.3). The top edge of the form is leveled carefully. The thickness of the slab may range from 3 inches (100 mm) for a residential floor, to 6 or 8 inches (150 or 200 mm) for an industrial floor, to a foot or more (300 mm) for an airport runway, which must carry very large concentrated loads from airplane wheels and diffuse them into the ground. If the slab is to be the floor of a building, a moisture barrier (usually a heavy sheet of polyethylene plastic) is laid over the crushed stone.

The American Concrete Institute recommends that a layer of sand 1 to 3 inches thick (25 to 75 mm) be placed over the moisture barrier, as shown in Figure 14.2, to absorb excess water from the concrete and help prevent curling (warping) of the slab that can occur during curing when the top of the slab loses moisture more rapidly than the bottom.

This is a controversial recommendation. A layer of sand is never used when a structural slab is cast over an impervious formwork made of steel, plastic, or plywood, which raises the question of why it should be used over a moisture barrier. Building professionals and contractors remain divided on this issue.

A reinforcing mesh of *welded wire fabric,* cut to a size just a bit smaller than the dimensions of the slab, is laid over the moisture barrier or sand. The fabric most commonly used for lightly loaded slabs, such as those in houses, is 6×6-W1.4 × W1.4, which has a wire spacing of 6 inches (150 mm) in each direction and a wire diameter of 0.135 inch (3.43 mm); see Figure 13.22 for more information on welded wire fabric. For slabs in factories, warehouses, and airports, a fabric made of heavier wires or a grid of reinforcing bars may be used instead. The reinforcing helps protect the slab against cracking that

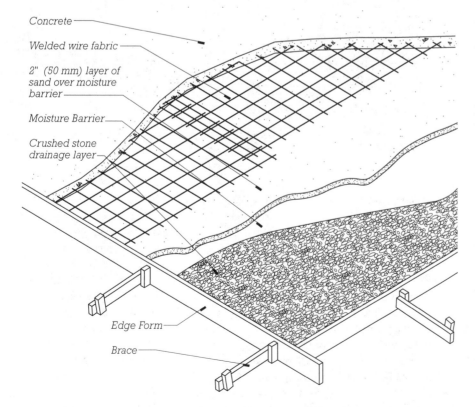

Concrete

Welded wire fabric

2" (50 mm) layer of
sand over moisture
barrier

Moisture Barrier

Crushed stone
drainage layer

Edge Form

Brace

FIGURE 14.2
The construction of a concrete slab on grade. Notice how the wire reinforcing fabric is overlapped where two sheets of fabric join.

FIGURE 14.3

Constructing and finishing a concrete slab on grade: (*a*) Attaching a proprietary slab edge form (screed joint) to a supporting stake. The profile of the screed joint forms a key to interlock two adjacent pours, and the hole knockouts allow for the placement of horizontal steel dowels to tie the pours together. (*b*) To the right, a crushed stone drainage layer for a slab on grade; to the left, a slab section ready to pour, with moisture barrier, welded wire fabric reinforcing, and edge forms in place. (*c*) This asphalt-impregnated fiber board forms a control joint when cast into the slab. The plastic cap is removed immediately after slab finishing to create a clean slot for the later insertion of an elastomeric joint sealant. (*d*) Striking off the surface of a concrete slab on grade just after pouring, using a motorized straightedging device. The motor vibrates the straightedge end to end to work the wet concrete into a level surface. (*e*) A bull float can be used for preliminary smoothing of the surface immediately after straightedging. (*f*) Hand floating brings cement paste to the surface and produces a plane surface. (*g*) Floating can be done by machine instead of by hand. (*h*) Steel troweling

might be caused by concrete shrinkage, temperature stresses, concentrated loads, frost heaves, or settlement of the ground beneath. Fibrous admixtures, which were discussed in the previous chapter, are finding increasing acceptance as a means of controlling the plastic shrinkage cracking that often takes place during curing of a slab. Fibrous reinforcing, however, is not strong enough to replace wire fabric for general crack control in slabs.

Control joints must be provided at intervals in a slab on grade. A control joint is a straight, intentional crack that is formed with a fiber-board strip (Figure 14.3c) or tooled into the surface of the slab before the concrete has hardened. Alternatively, a control joint may be created by sawing a shallow groove into the top of the slab after it has hardened. The function of a control joint is to provide a place where the forces that cause cracking can be relieved without disfiguring the slab. The reinforcing mesh is discontinued at each control joint as a further inducement for cracking to occur in this location.

In certain circumstances, it is advantageous to posttension a slab on grade, using level tendons in both directions at the midheight of the slab rather than welded wire fabric. Posttensioning is especially effective

several hours after floating produces a dense, hard, smooth surface. (*i*) A section of concrete slab on grade finished and ready for curing. The steel dowels inserted through the screed joint will connect to the sections of slab that will be poured next. (*j*) One method for damp curing a slab is to cover it with polyethylene sheeting to retain moisture inside the concrete. (*Photos* a, b, c, *and* i *courtesy of Vulcan Metal Products, Inc., Birmingham, Alabama; photos* d, e, f, g, h, *and* j *courtesy of Portland Cement Association, Skokie, Illinois*)

for slabs over unstable or inconsistent soils and for superflat floors, which are discussed below. Posttensioning makes floors more resistant to cracking under concentrated loads, minimizes cracking from other causes, eliminates the need to make control joints, and often permits the use of a thinner slab.

Pouring and Finishing the Slab on Grade

Pouring of the slab commences with the placing of concrete into the formwork. This may be done directly from the chute of a transit-mix truck, or with wheelbarrows, concrete buggies, a large crane-mounted concrete bucket, a conveyor belt, or a concrete pump and hoses; the method selected will depend on the scale of the job and the accessibility of the slab to the truck delivering the concrete. The concrete is spread by hand with shovels or rakes until the form is full, and the same tools are used to agitate the concrete slightly, especially around the edges, to eliminate air pockets. Next, using hand hooks, the concrete masons reach into the wet concrete and raise the welded wire fabric to the midheight of the slab, so that it will be able to resist tensile forces caused by forces acting either upward or downward.

The first operation in finishing the slab is to *strike off* or *straightedge* the concrete by drawing a stiff plank of wood or metal across the top edges of the formwork to achieve a level surface (Figure 14.3d). This is done with an end-to-end sawing motion that avoids tearing the projecting pieces of coarse aggregate from the surface of the wet concrete. A bulge of concrete is maintained in front of the straightedge as it progresses across the slab, so that, when a low point is encountered, concrete from the bulge will flow in to fill it. When straightedging has been completed, the top of the slab is level but rather rough. If a concrete topping will later be poured over the slab, or if a floor finish of terrazzo, stone, brick, or

quarry tile will be applied, the slab may cured without further finishing.

If a smoother surface is desired, the slab is next *floated* (Figure 14.3e–g). The masons wait until the watery sheen has evaporated from the surface, then smooth the concrete with a flat tool called a *float*. Floats may be large or small; for large slabs, rotary power floats may be used. The working surfaces of floats are made of wood or of metal with a slightly rough surface. As the float is drawn across the surface, its friction vibrates the concrete gently and brings cement paste to the surface, where it is smoothed over the coarse aggregate and into low spots. If too much floating is done, however, an excess of paste and free water rises to the surface to form puddles, and it is almost impossible to get a good finish. Experience on the part of the mason is essential to floating, as it is to all slab finishing operations, in order to know just when to begin each operation and just when to stop.

Shake-on hardeners are sometimes sprinkled over the surface of a slab between the straightedging and floating operations. These dry powders react with the concrete to form a very hard, durable surface for such heavy-wear applications as warehouses and factories.

Immediately following the floating operation, specially shaped hand tools are used to form neatly rounded edges and control joints. The floated slab has a lightly textured surface that is appropriate for outdoor walks and pavings without further finishing.

For a completely smooth, dense surface, the slab must also be *troweled*. This is done either by hand with a smooth rectangular *steel trowel* (Figure 14.3h) or with a *rotary power trowel*. Troweling is done several hours after floating, when the slab is becoming quite firm. If the concrete mason cannot reach all areas of the slab from around the edges, *knee boards* are placed on the surface of the concrete to distribute the mason's weight sufficiently that he or she can kneel on the surface without making indentations. Any marks left by the knee boards are removed by the trowel as the mason works backward across the surface from one edge to the other.

If a nonslip surface is required, a stiff-bristled janitor's broom is drawn across the surface of the slab after troweling to produce a striated texture called a *broom finish.*

When the finishing operations have been completed, the slab should be cured under damp conditions for at least a week; otherwise, its surface may crack or become dusty from premature drying. Damp curing may be accomplished by covering the slab with an absorbent material such as sawdust, earth, sand, straw, or burlap, and maintaining the material in a damp condition for the required length of time. Alternatively, an impervious sheet of plastic or waterproof paper may be drawn over the slab soon after troweling to prevent the escape of moisture from the concrete (Figure 14.3*j*). The same effect can be obtained by spraying the concrete surface with one or more applications of a liquid *curing compound,* which forms an invisible moisture barrier membrane over the slab.

No concrete floor is perfectly flat. The normal finishing process produces a surface that undulates almost imperceptibly between low and high areas that go unnoticed in everyday use. Industrial warehouses that use high-rise forklift trucks, however, require floors whose flatness is controlled to within very narrow tolerances. These *superflat floors* are specified according to an index number that corresponds to the degree of flatness that is required, and are produced using special finishing equipment and techniques. Because of its extreme accuracy, a laser-guided automatic straightedging machine (Figure 14.4) is often used in the creation of superflat floors. This device produces a slab surface that is flat to within very small tolerances, and does so at a very rapid rate, leaving the surface ready for subsequent floating and troweling operations.

CASTING A CONCRETE WALL

A reinforced concrete wall at ground level usually rests on a poured concrete strip footing (Figures 14.5–14.7). The footing is formed and poured much like a concrete slab on grade. Its cross-sectional dimensions and its reinforcing, if any, are determined by the structural engineer. A *key* is sometimes formed in the top of the footing with strips of wood that are temporarily embedded in the wet concrete. The key is a groove that forms a mechanical connection to the wall. Vertically projecting *dowels* of steel reinforcing bars are usually installed in the footing before pouring; these will later be overlapped with the vertical bars in the walls to form a strong structural connection. After pouring, the top of the footing is straightedged; no further finishing operations are required. The footing is left to cure for at least a day before the wall forms are erected.

The wall reinforcing, in one or two layers as specified by the engineer, is installed next, with the bars wired to one another at the intersections. The vertical bars are overlapped with the corresponding dowels projecting from the footing. At wall corners, L-shaped horizontal bars are installed to maintain full structural continuity between the two walls. If the wall will connect to a concrete floor or another wall at its top, rods are left projecting from the top of the formwork to form a continuous connection.

Wall forms may be custom built of lumber and plywood for each job,

FIGURE 14.4
Guided by a laser beam, the motorized straightedging device on this machine can strike off 240 square feet (22 m²) of slab surface per minute to an extremely exacting standard of flatness. The worker to the right smooths the surface with a bull float.
(*Photo by Wironen, Inc. Courtesy of Laser Screed Company, Inc., New Ipswich, New Hampshire*)

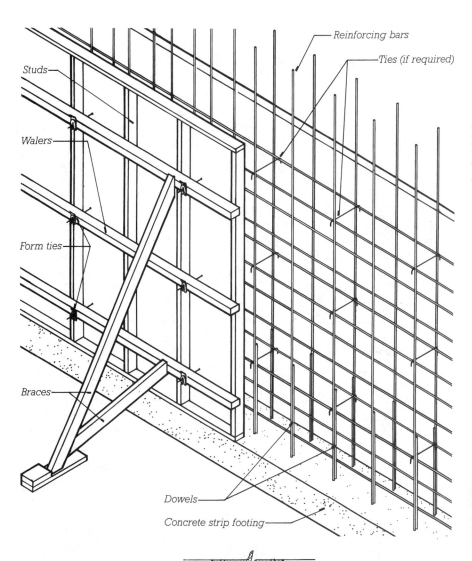

Studs

Walers

Form ties

Braces

Dowels

Concrete strip footing

Reinforcing bars

Ties (if required)

FIGURE 14.5
Formwork and reinforcing for a concrete wall. No key between the footing and the wall is shown in this example.

FIGURE 14.6
Guarded by a safety harness, a worker climbs on the reinforcing bars for a concrete wall to wire another horizontal bar in position. (*Photo courtesy of DBI/SALA, Red Wing, Minnesota*)

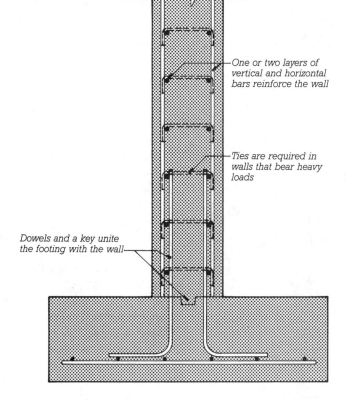

One or two layers of vertical and horizontal bars reinforce the wall

Ties are required in walls that bear heavy loads

Dowels and a key unite the footing with the wall

FIGURE 14.7
Section through a reinforced concrete wall.

but it is more usual for standard prefabricated formwork panels to be employed. The panels for one side of the form are coated with a form-release compound, set on the footing, aligned carefully, and braced. The *form ties,* which are small-diameter steel rods specially shaped to hold the formwork together under the pressure of the wet concrete, are inserted through holes provided in the formwork panels and secured to the back of the form by devices supplied with the form ties; both ties and fasteners vary in detail from one manufacturer to another (Figures 14.8, 14.9). The ties will pass straight through the concrete wall from one side to another and remain in the wall after it is poured. This may seem like an odd way to go about holding wall forms together, but the pressures of the wet, heavy concrete on the forms are so large that there is no

other economical way of dealing with them.

When the ties are in place and the reinforcing has been inspected, the formwork for the second side of the wall is erected, the walers and braces are added (Figure 14.5), and the forms are inspected to be sure that they are straight, plumb, correctly aligned, and adequately tied and braced. A surveyor's transit is used to establish the exact height to which the concrete will be poured, and this height is marked all around the inside of the forms. Pouring may then proceed.

Concrete is brought to the site, test cylinders are made, and a slump test is performed to check for the proper pouring consistency. Concrete is then transported to the top of the wall by a large crane-mounted bucket or by a concrete pump and hose. Workers standing on planks at the

top of the forms deposit the concrete in the forms, compacting it with a vibrator to eliminate air pockets (Figure 14.10). When the form has been filled and compacted up to the level that was marked inside the formwork, hand floats are used to smooth and level the top of the wall. The top of the form is then covered with a plastic sheet or canvas, and the wall is left to cure.

After a few days of curing, the bracing and walers are taken down, the connectors are removed from the ends of the form ties, and the formwork is *stripped* from the wall (Figure 14.11). This leaves the wall bristling with projecting ends of form ties. These are twisted off with heavy pliers, and the holes that they leave in the surfaces of the wall are carefully filled with grout. If required, major defects in the wall surface caused by defects in the formwork or inade-

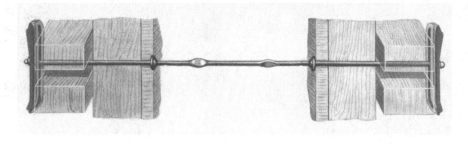

FIGURE 14.8
Detail of a form tie assembly. Two washers just inside the faces of the form maintain the correct wall thickness. Tapered, slotted wedges at the ends transmit force from the tie to the walers. The notches just inside the washers are the points at which the tie will be snapped off after the forms are stripped. (*Courtesy of Richmond Screw Anchor Co., Inc., 7214 Burns St., Fort Worth, TX 76118*)

FIGURE 14.9
Detail of a heavy-duty form tie. This assembly is tightened with special screws that engage a helix of heavy wire welded into the tie. The wire components remain in the concrete, but the screws and the plastic cone on the right are removed and reused after stripping. The purpose of the cone is to give a neatly finished hole in the exposed surface of the concrete. (*Courtesy of Richmond Screw Anchor Co., Inc., 7214 Burns St., Fort Worth, TX 76118*)

FIGURE 14.10
Compacting wet concrete after pouring, using a mechanical vibrator immersed in the concrete. (*Courtesy of Portland Cement Association, Skokie, Illinois*)

FIGURE 14.11
Three stages in the construction of a reinforced concrete wall on a strip footing: In the foreground, the reinforcing bars have been wired to the dowels that project from the footing, ready for erection of the formwork. In the center, a section of wall has been poured in steel formwork that is tied through the wall with small steel straps secured by wedges. In the background, the forms have been stripped, and some of the ties have been snapped off. (*Courtesy of Portland Cement Association, Skokie, Illinois*)

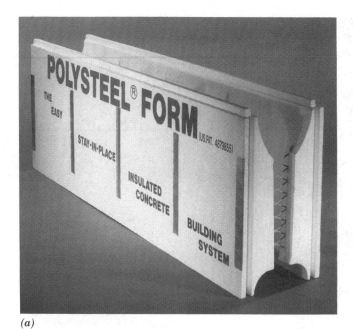

(a)

FIGURE 14.12

Forming a concrete wall with insulating concrete forms. (*a*) The forms are manufactured as interlocking blocks. The inner and outer halves of this block are tied together by steel mesh webs that connect to sheet metal strips on the inner and outer surfaces. These strips later serve to receive screws that fasten interior and exterior finish materials to the wall. (*b*) Workers stack the blocks to form all the exterior walls of a house. Openings for doors and windows are formed with dimension lumber. The worker to the right is cutting a block to length with a hand saw. (*c*) This sample wall, from which some of the foam blocks have been removed, shows that the completed wall contains a continuous core of reinforced concrete with thermal insulation inside and out. (*Courtesy of American Polysteel Forms*)

(c)

(b)

quate filling of the forms with concrete can be repaired at this time. The wall is now complete.

Insulating Concrete Forms

An alternative way of casting a concrete wall, particularly one that will be an exterior wall of a building, is to use *insulating concrete forms* that serve both to form the concrete and to remain in place permanently as thermal insulation (Figure 14.12). The forms are manufactured as interlocking hollow blocks of polystyrene foam. These weigh so little and are so accurately made that they go together almost as easily and quickly as a child's plastic building blocks. The tops of the blocks are channeled so that horizontal reinforcing bars may be laid in the top of each course, and vertical bars may be inserted into the vertical cores. The wall forms must be braced strongly to prevent them from moving during pouring. Concrete is usually deposited in the cores from the hose of a concrete pump. The full height of a wall cannot be cast in one operation because the pressure of so great a depth of wet concrete would blow out the sides of the blocks. The normal procedure is to deposit the concrete in several "lifts" of limited height, working all the way around the structure with each lift so that by the time the second lift is begun, the first lift has had an hour or two to harden somewhat, relieving the pressure at the bottom of the forms. Interior and exterior finish materials must be applied to the foam plastic faces to protect them from sunlight, mechanical damage, and fire. The thermal insulating value of the finished wall is usually sufficient to meet current code requirements.

CASTING A CONCRETE COLUMN

A column is formed and cast much like a wall, with a few important differences. The footing is usually an isolated column footing, a pile cap, or a caisson, rather than a strip footing (Figure 14.13). The dowels are sized and spaced in the footing to match the vertical bars in the column. The cage of column reinforcing is assembled with wire ties and hoisted into place over the dowels, but if space is very tight, the bars may be spliced end to end with welds or mechanical connectors (Figure 13.24). The column form can be a square

(a)

(b)

FIGURE 14.13
(a) **A column footing almost ready for pouring, but lacking dowels. The reinforcing bars are supported on pieces of concrete brick.** *(b)* **Column footings poured with projecting dowels to connect to both round and rectangular columns.** (*Photos by the author*)

FIGURE 14.14
In the foreground, a square column form tied with pairs of L-shaped steel brackets. In the background, a worker braces a round column form made of sheet steel. (*Courtesy of Ceco Corporation, Oakbrook Terrace, Illinois*)

FIGURE 14.15
Round columns may also be formed with single-use cardboard tubes. Notice the density of the steel shoring structure that is being erected to support the slab form, which will carry a very heavy load of wet concrete. (*Courtesy of Sonoco Products Company*)

box of plywood panels, a cylindrical steel or plastic tube bolted together in halves so that it can later be removed, or a waxed cardboard tube that is stripped after curing by unwinding the layers of paper that make up the tube (Figures 14.14, 14.15). Unless a rectangular column is very broad and wall-like, form ties through the concrete are not required. The vertical bars project from the top of the column to overlap or splice to the bars in the column for the story above, or they are bent over at right angles to splice into the roof structure. Where vertical bars overlap, the tops of the bars from the column below are offset (bent inward) by one bar diameter to avoid interference.

Reinforced concrete made "pilotis" possible. The house is in the air, away from the ground; the garden runs under the house, and it is also above the house, on the roof. . . . Reinforced concrete is the means which makes it possible to build all of one material. . . . Reinforced concrete brings the free plan into the house! Floors no longer have to stand simply one on top of the other. They are free. . . . Reinforced concrete revolutionizes the history of the window. Windows can run from one end of the facade to the other . . .

Le Corbusier and P. Jeanneret,
Oeuvre Complète 1910–1929,
Zurich, 1956, p. 128

ONE-WAY FLOOR AND ROOF FRAMING SYSTEMS

The One-Way Solid Slab System

A *one-way solid slab* (Figures 14.16–14.19) spans across parallel lines of support furnished by walls and/or beams. The walls and columns are poured prior to erecting the formwork for a one-way slab, but the girders and beams are nearly always formed and poured at the same time as the slab.

The girder and beam forms are erected first, then the slab forms. A form-release compound is applied to all formwork surfaces that will be in contact with concrete. The forms are supported on temporary joists and beams of metal or wood, and the temporary beams are supported on temporary *shores* (adjustable-length columns). The weight of uncured concrete that must be supported is enormous, and the temporary beams and shoring must be both numerous and strong. Formwork is, in fact, designed by a contractor's structural

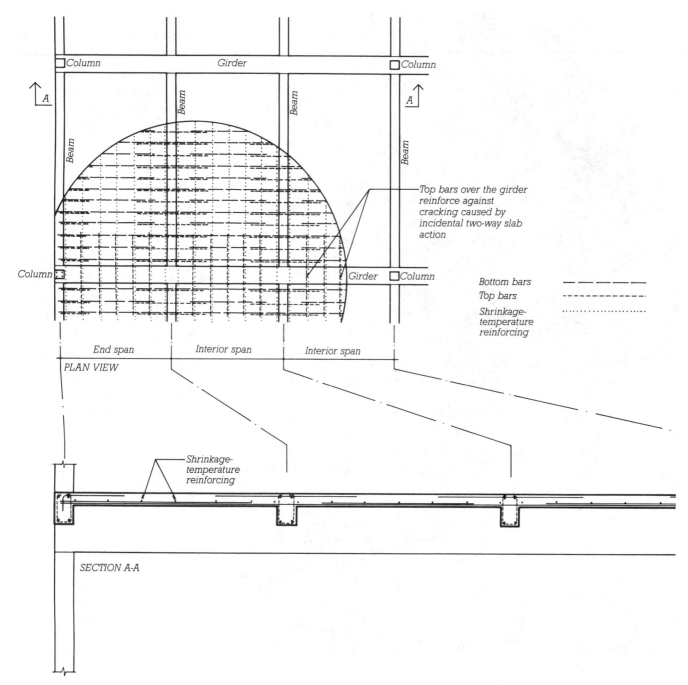

FIGURE 14.16

Plan and larger-scale section of a typical one-way solid slab system. For the sake of clarity, the girder and beam reinforcing are not shown in the plan, and the girder and column reinforcing are left out of the section. The slabs span between the beams, the beams are supported by the girders, and the girders rest on the columns.

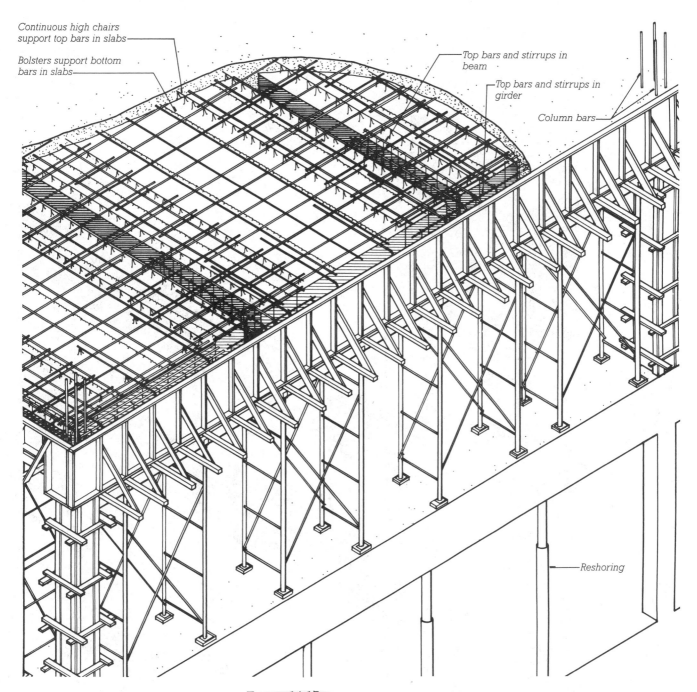

Continuous high chairs
support top bars in slabs

Bolsters support bottom
bars in slabs

Top bars and stirrups in
beam

Top bars and stirrups in
girder

Column bars

Reshoring

FIGURE 14.17
Isometric view of a one-way solid slab system under construction. The slab, beams,
and girders are created in a single pour.

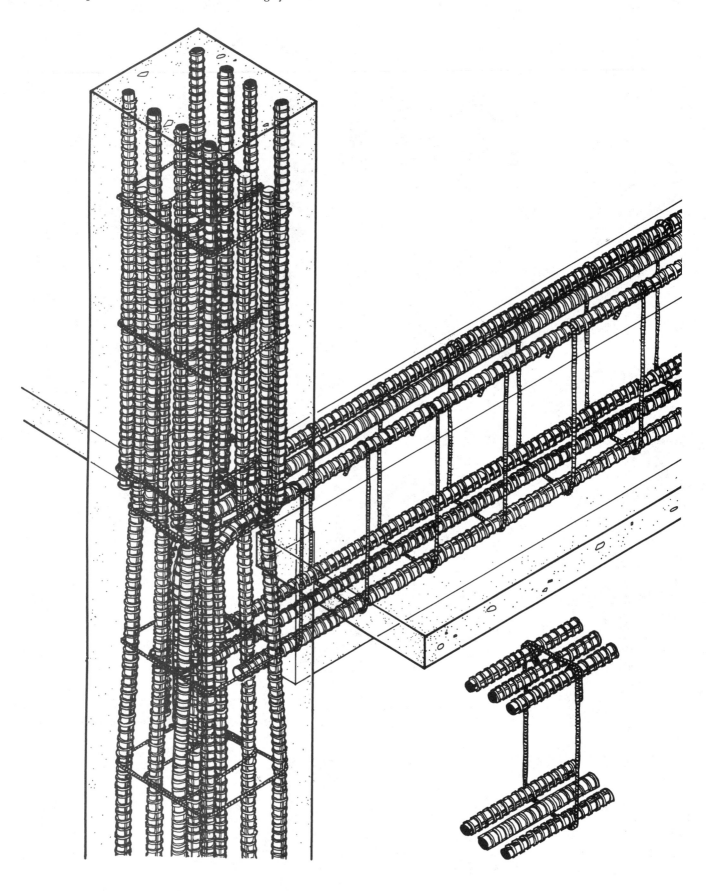

FIGURE 14.18
An example of a beam–column connection in a one-way solid slab structure, with the slab reinforcing omitted for clarity. Notice how the column bars are spliced by overlapping them just above floor level. The bars from the column below are offset at the top so that they lie just inside the bars of the column above at the splice. Structural continuity is established by running the top bars from the beam into the column. U-stirrups are shown in the beam; stirrup ties, shown in the inset detail, are often used instead.

engineers just as carefully as it would be if it were a permanent building, because a structural failure in formwork is an intolerable risk to workers and property.

Formwork is usually made so as to eliminate sharp edges on the concrete. Sharp edges of concrete tend to break off during form stripping to leave a ragged edge that is almost impossible to patch. In service, sharp edges are easily damaged by, and are potentially damaging to, people, furniture, and vehicles. Edges of concrete structures are beveled or rounded by inserting shaped strips of wood or plastic into the corners of the formwork to produce the desired profile.

In accordance with reinforcing diagrams and schedules prepared by the structural engineer, the girder and beam reinforcing—bottom bars, top bars, and stirrups—is installed in the forms, supported on chairs and bolsters to maintain the required cover of concrete. Next the slab reinforcing—bottom bars, top bars, and shrinkage–temperature reinforcing—is placed on bolsters. After the reinforcing and formwork have been

inspected, the girders, beams, and slab are poured in a single operation, with the usual sample cylinders being made for later testing. Slab depths are typically 4 to 10 inches (100 to 250 mm). The top of the slab is finished in the same manner as a slab on grade, usually to a steel trowel finish, and the slab is sealed or covered for damp curing. The only construction left projecting above the slab surface at this stage are the offset column bars, which are now ready to splice to the column bars for the floor above.

When the slab and beams have attained enough strength to support themselves safely, the formwork is stripped and the slabs and beams are *reshored* with vertical props to relieve them of loads until they have reached full strength, which will take several more weeks. Meanwhile, the formwork and the remainder of the shoring are cleaned and moved up a level above the slab and beams just poured, where the cycle of forming, reinforcing, pouring, and stripping is repeated (Figure 14.19).

Ordinarily, the most efficient and economical concrete beam is one whose depth is twice or three times its

FIGURE 14.19
A one-way solid slab building under construction. The middle floors have been reshored, the highest completed floor is still supported by its formwork, and formwork construction for the topmost floor has just begun. (*Courtesy of Portland Cement Association, Skokie, Illinois*)

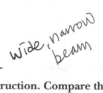

Wide, narrow beam

FIGURE 14.20
Banded slab construction. Compare the depth and breadth of the slab band in the center of the building to those of the conventional concrete beams around the perimeter. (*Courtesy of Portland Cement Association, Skokie, Illinois*)

FIGURE 14.21
This helical ramp is a special application of one-way solid slab construction. The formwork here is made of overlaid plywood for a smooth surface finish.
(*Courtesy of APA–The Engineered Wood Association*)

breadth. One-way solid slabs are often supported, however, by beams that are several times as broad as they are deep. These are called *slab bands* (Figure 14.20). Banded slab construction offers two kinds of economy: The width of the slab band reduces the span of the slab, which can result in a reduced thickness for the slab and consequent savings of concrete and reinforcing steel. Also, the reduced depth of the slab band as compared to a more conventionally proportioned concrete beam allows for reduction of the story height of the building, with attendant economies in columns, cladding, partitions, and vertical runs of piping and ductwork.

The One-Way Concrete Joist System (Ribbed Slab)

The one-way solid slab system is economical for structures in which the slab does not span very far between beams. As one-way spans increase, a progressively thicker slab is required, and, at long spans, the weight of the slab itself becomes an excessive burden, unless a substantial portion of the nonworking concrete in the lower part of the slab can be eliminated to lighten the load. This is the rationale for the *one-way concrete joist system* (Figures 14.22–14.25), also called a *ribbed slab*. The bottom steel is

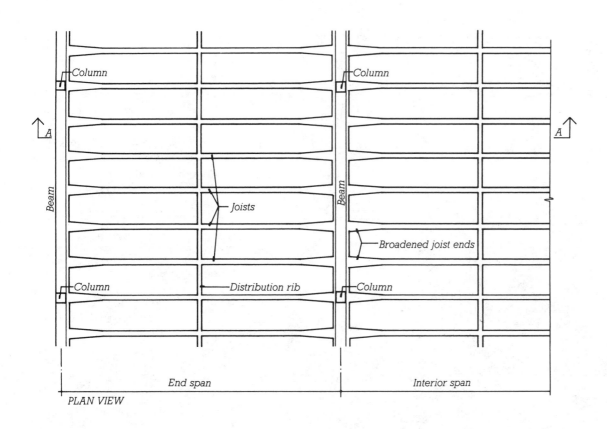

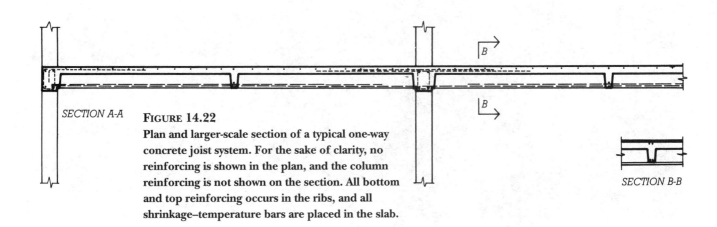

SECTION A-A

FIGURE 14.22
Plan and larger-scale section of a typical one-way concrete joist system. For the sake of clarity, no reinforcing is shown in the plan, and the column reinforcing is not shown on the section. All bottom and top reinforcing occurs in the ribs, and all shrinkage–temperature bars are placed in the slab.

SECTION B-B

FIGURE 14.23
Standard steel form dimensions for one-way concrete joist construction (1 inch equals 25.4 mm). (*Courtesy of Ceco Corporation, Oakbrook Terrace, Illinois*)

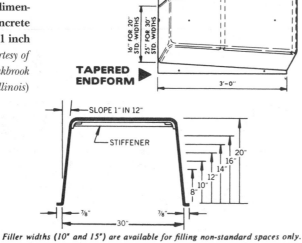

TAPERED ENDFORM ▶

SLOPE 1" IN 12"

STIFFENER

20"
16"
14"
12"
10"
8"

⅞"   30"   ⅞"

Filler widths (10" and 15") are available for filling non-standard spaces only.

SLOPE 1" IN 12"

STIFFENER

12"
10"
8"

⅞"   20"   ⅞"

FIGURE 14.24
Reinforcing being placed for a one-way concrete joist floor. Electrical conduits and boxes have been put in place, and welded wire fabric is being installed as shrinkage–temperature reinforcing. Both the tapered end pans and the square end-caps for the midspan distribution rib are clearly visible. (*Courtesy of Ceco Corporation, Oakbrook Terrace, Illinois*)

concentrated in spaced ribs or joists. The thin slab that spans across the top of the joists is reinforced only by shrinkage–temperature bars. There is little concrete in this system that is not working, with the result that a one-way concrete joist system can span considerably longer distances than a one-way solid slab. Each individual joist is reinforced as a small beam, except that stirrups are not usually used in concrete joists because of the restricted space in the narrow joist. Instead, the ends of the joists are broadened sufficiently that the concrete itself can resist the diagonal tension forces.

The joists are formed with metal or plastic *pans* supported on longitudinal strips of wood or, more usually, on a plywood deck. Pans are available in two standard widths, 20 inches (508 mm) and 30 inches (762 mm), and in depths ranging up to 20 inches (508 mm), as shown in Figure 14.23. (Larger pans are also available for forming floors designed as one-way solid slab systems.) The sides of the pans taper to allow them to drop easily out of the hardened concrete during stripping. The joist width can be varied by placing the rows of pans closer together or farther apart, with the bottom of the joist formed by the wood deck or strip of wood. The broadening of the joist ends is accomplished with standard end pans whose width tapers. A *distribution rib* is sometimes formed across the joists at midspan to distribute concentrated loads to more than one joist. After application of a form-release compound, the beam steel and joist steel are placed, the shrinkage–temperature steel is laid crosswise on bolsters over the pans, and the entire system is poured and finished (Figures 14.24, 14.25).

For greater economy of formwork, one-way concrete joists are often supported on *joist bands,* which are broad beams that are only as deep as the joists. While a deeper beam would be more efficient structurally, a joist band can be formed by the same plywood deck that supports the pans, which eliminates beam formwork entirely.

The Wide-Module Concrete Joist System

When fire resistance requirements of the building code dictate a slab thickness of 4.5 inches (115 mm) or more, the slab is capable of spanning a much greater distance than the normal space between joists in a one-way concrete joist system. This has led to the development of the *wide-module concrete joist system,* also called the *skip-joist system,* in which the joists are

FIGURE 14.25
A one-way concrete joist system after stripping of the formwork, showing broadened joist ends at the lower edge of the photograph and a distribution rib in the foreground. The dangling wires are hangers for a suspended finish ceiling (see Chapter 24). (*Courtesy of Portland Cement Association, Skokie, Illinois*)

placed 4 to 6 feet (1220 to 1830 mm) apart. The name "skip joist" arose from the original practice of achieving this wider spacing by laying strips of wood over alternate joist cavities in conventional joist pan formwork to block out the concrete. Pans are now specially produced for wide-module construction (Figures 14.26, 14.27).

Because loads are higher in wide-module joists than in conventionally spaced joists, stirrups are generally required near the ends of each joist. The width of the joist is very constricted, however, so conventional U-stirrups must be installed at an angle, or single-leg stirrups may be used instead.

FIGURE 14.26
Formwork for a wide-module concrete joist system. These pans have been placed over a flat plywood deck, which will result in joist band beams. (*Courtesy of Ceco Corporation, Oakbrook Terrace, Illinois*)

FIGURE 14.27
The underside of the finished wide-module joists, joist bands, and slab. (*Courtesy of Ceco Corporation, Oakbrook Terrace, Illinois*)

TWO-WAY FLOOR AND ROOF FRAMING SYSTEMS

The Two-Way Flat Slab and Two-Way Flat Plate Systems

Two-way concrete framing systems are generally more economical than one-way systems in buildings where the columns can be spaced in bays that are square or nearly square in proportion. A *two-way solid slab* is a system in which the slab is supported by a grid of beams running in both directions over the columns. This system is occasionally used for very heavily loaded industrial floors, but most two-way floor and roof framing systems, even for heavy loadings, are made without beams. The slab is reinforced in such a way that the varying stresses in the different zones of the slab are accommodated within a uniform thickness of concrete.

The *two-way flat slab* (Figure 14.28), a system suited to heavily loaded buildings such as storage and industrial buildings, illustrates this concept. The formwork is completely flat except for a thickening of the concrete to resist the high shear forces around the top of each column. Traditionally, this thickening was accomplished with both a *mushroom capital* and a *drop panel*, but today the capital is usually eliminated to

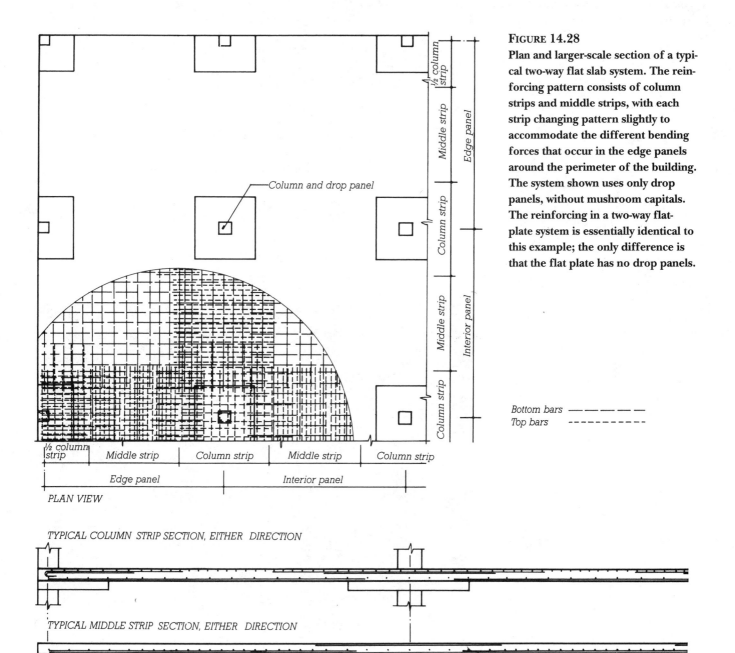

FIGURE 14.28
Plan and larger-scale section of a typical two-way flat slab system. The reinforcing pattern consists of column strips and middle strips, with each strip changing pattern slightly to accommodate the different bending forces that occur in the edge panels around the perimeter of the building. The system shown uses only drop panels, without mushroom capitals. The reinforcing in a two-way flat-plate system is essentially identical to this example; the only difference is that the flat plate has no drop panels.

Bottom bars — — — — —
Top bars - - - - - - - - -

Column and drop panel

½ column strip

Middle strip · Edge panel

Column strip

Middle strip · Interior panel

Column strip

½ column strip | Middle strip | Column strip | Middle strip | Column strip

Edge panel | Interior panel

PLAN VIEW

TYPICAL COLUMN STRIP SECTION, EITHER DIRECTION

TYPICAL MIDDLE STRIP SECTION, EITHER DIRECTION

Edge panel | Interior panel

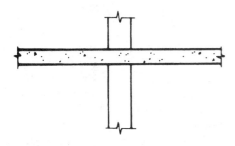

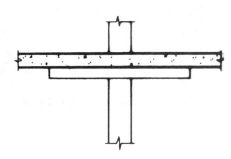

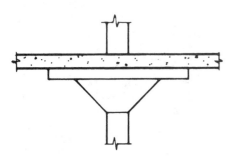

FIGURE 14.30

A two-way flat plate floor of a high-rise apartment building is poured with the aid of a concrete pump at the base of the building. The concrete is delivered up a telescoping tower to a nozzle at the end of an articulated boom. The column dowels in the center of the newly poured area demonstrate the ability of flat-plate construction to adapt to irregular patterns and spacings of columns. (*Courtesy of Schwing America, Inc.*)

FIGURE 14.29

Column capitals for two-way concrete framing systems. For slabs bearing heavy loads, shear stresses around the column are reduced by means of a mushroom capital and drop panel, or a drop panel alone. For lighter loads, no thickening of the slab is required.

FIGURE 14.31

The underside of a two-way flat plate floor. The pipes have been roughed in for an automatic sprinkler fire suppression system.

(*Courtesy of Portland Cement Association, Skokie, Illinois*)

reduce formwork cost, leaving a drop panel to do the work (Figure 14.29). Typical depths for the slab itself lie in the 6- to 12-inch (150- to 300-mm) range.

The reinforcing is laid in both directions in half-bay-wide strips of two fundamental types: *Column strips* are designed to carry the higher bending forces encountered in the zones of the slab that cross the columns. *Middle strips* have a lighter reinforcing pattern. Shrinkage–temperature steel is not needed in two-way systems because the concrete is already reinforced in both directions. The drop panel and capital (if any) have no additional reinforcing beyond that provided by the column strip; the greater thickness of concrete furnishes the required shear resistance.

In more lightly loaded buildings, such as hotels, hospitals, dormitories, and apartment buildings, the slab need not be thickened at all over the columns. This makes the formwork extremely simple and even allows some columns to be moved off the grid a bit if it will facilitate a more efficient floor plan arrangement (Figure 14.30). The completely flat ceilings of this system allow room partitions to be placed anywhere with equal ease, and the story heights of the building may be kept to an absolute minimum, which reduces the cost of exterior cladding (Figure 14.31). Typical slab depths for this *two-way flat plate system* range from 5 to 12 inches (125 to 305 mm).

The zones along the exterior edges of both the two-way flat slab system and two-way flat plate system require special attention. To take full advantage of structural continuity, the slabs should be cantilevered beyond the last row of columns a distance equal to about 30 percent of the interior span. If this cantilever is impossible, additional reinforcing must be added to the slab edges to carry the higher stresses that will result.

The omission of the drop panel in a two-way flat plate requires the insertion of extra reinforcing bars in the slab at the top of each column to resist the high shear stresses that occur in this region. Alternatively, a proprietary system of vertical steel studs may be installed in the formwork at each column head to act as stirrups and replace a much larger volume of horizontal bars (Figures 14.32, 14.33).

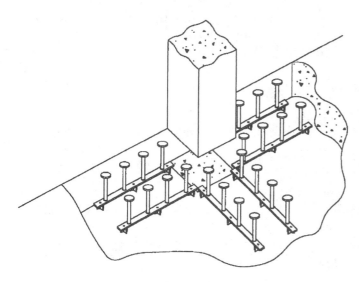

FIGURE 14.32
Shear reinforcement around columns in a two-way flat plate can be simplified by using studrails®, a proprietary system of steel studs welded to horizontal rails.

FIGURE 14.33
Studrails® nailed to the formwork around column bars, ready for installation of the top and bottom bars for the two-way flat plate. (*U.S. and Canada patents 4406103 and 1085642, respectively. Licensee: Deha, represented by Decon, 105C Atsion Road, P.O. Box 1575, Medford, NJ 08055-6675 and 35 Devon Road, Bramton, Ontario L6T 5B6*)

The Two-Way Waffle Slab System

The *waffle slab,* or *two-way concrete joist system* (Figure 14.34), is the two-way equivalent of the one-way concrete joist system. Metal or plastic pans called *domes* are used as formwork to eliminate the nonworking concrete from the slab, allowing a greater economy in longer spans. The stan- dard domes form joists 6 inches (152 mm) wide on 36-inch (914-mm) centers, or 5 inches (127 mm) wide on 24-inch (610-mm) centers, in a variety of depths up to 20 inches (500 mm), as shown in Figures 14.35 through 14.38. Special domes are also available in larger sizes. Solid concrete *heads* are created around the tops of the columns by leaving the domes out of the formwork; these serve the same function as drop pan- els in the two-way flat-slab system. If the slab cannot be cantilevered at the perimeter of the building, a perime- ter beam must be provided. Stripping of the domes is facilitated in many cases by a compressed air fitting at the top of each. This allows the dome to be popped out of the concrete with the application of a puff of com- pressed air. Waffle slabs are often

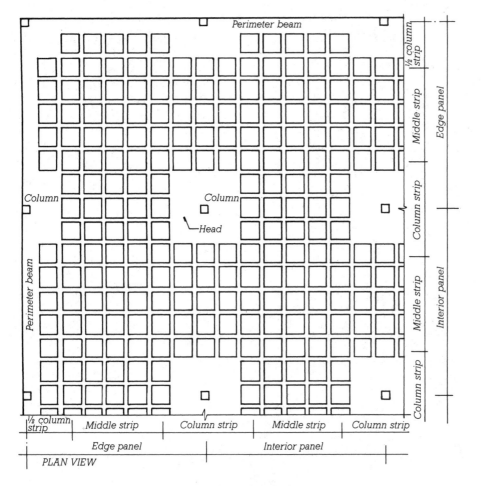

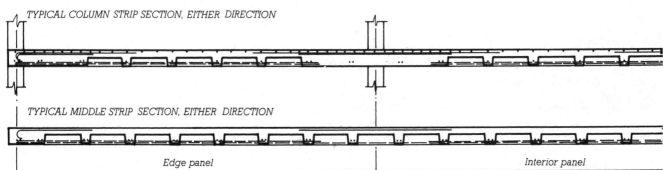

FIGURE 14.34
Plan and larger-scale section of a typ- ical two-way concrete joist system, also known as a waffle slab. For the sake of clarity, no reinforcing is shown on the plan drawing, and the section does not show the welded wire fabric that is spread over the entire form before pouring.

2'-0" MODULE
(19" x 19" Dome System)

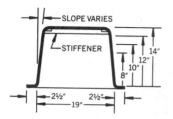

FIGURE 14.35
Standard steel dome forms for two-way concrete joist construction. A 2'-6" module, utilizing forms 24 inches square, is also available from some manufacturers (1 inch equals 25.4 mm). (*Courtesy of Ceco Corporation, Oakbrook Terrace, Illinois*)

3'-0" MODULE
(30"x 30" Dome System)

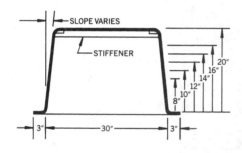

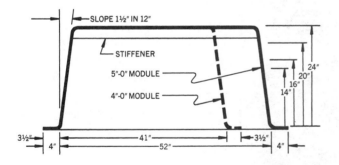

FIGURE 14.36
Steel domes being placed on a temporary plywood deck to form a two-way concrete joist floor. Pans are omitted around columns to form heads. (*Courtesy of Ceco Corporation, Oakbrook Terrace, Illinois*)

FIGURE 14.37
Plastic dome formwork being assembled for a two-way concrete joist floor. Notice the electrical conduit and junction box in the foreground, ready to be poured into the slab. (*Courtesy of Molded Fiber Glass Concrete Forms Company*)

FIGURE 14.38
Stripping plastic domes after removal of the temporary plywood deck. (*Courtesy of Molded Fiber Glass Concrete Forms Company*)

FIGURE 14.39
The underside of a two-way concrete joist floor. Notice how the joists are cantilevered for maximum structural efficiency. (*Courtesy of Ceco Corporation, Oakbrook Terrace, Illinois*)

employed in a building not only for their economy, but also for their richly coffered undersides, which can be left exposed as ceilings (Figure 14.39).

CONCRETE STAIRS

A concrete stair (Figure 14.40) is reinforced and poured as an inclined one-way solid slab with additional concrete added to make risers and treads. The underside of the form is planar. The top is built with riser forms, usually inclined to give additional tread width. The concrete is poured in one operation, and the treads are tooled to a steel trowel finish. Projecting nosings, as used in wood stair construction, are generally avoided in concrete stairs because they would have a tendency to break off. The inclined riser design is also preferable (and often mandatory) for handicapped access because it does not catch the toes of persons climbing the stairs on crutches.

SITECAST POSTTENSIONED FRAMING SYSTEMS

Posttensioning can be applied to any of the sitecast concrete framing systems. It is used in beams or girders to reduce member sizes and extend spanning capability, as well as in slabs, both one-way and two-way. Two-way flat plate structures are very commonly posttensioned, especially in cases where spans are long or zoning restrictions on the height of the building require minimal slab depths. The *tendon* layout, however, is quite different from the conventional reinforcing layout that is shown in Figure 14.28. Instead of being placed identically in both directions, the draped tendons are evenly distributed in one direction, and banded closely together over the line of columns in the other direction (Figures 14.41, 14.42). This arrangement functions better structurally in posttensioned slabs because it balances the maximum upward force from the banded tendons against the maximum downward force from the distributed tendons. It is also easier to install than distributed, draped tendons running in both directions. The same number of tendons is used in each direction. The prestressing force from the banded tendons becomes evenly distributed throughout the width of the slab within a short distance of the end anchorages because of corbelling action in the concrete.

As with any prestressed concrete framing system, both short-term and long-term losses of prestressing force must be anticipated. The short-term losses in posttensioning are caused by elastic shortening of the concrete, friction between the tendons and the

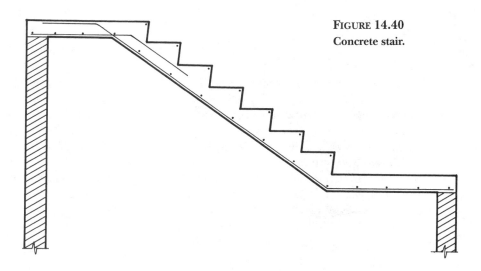

FIGURE 14.40
Concrete stair.

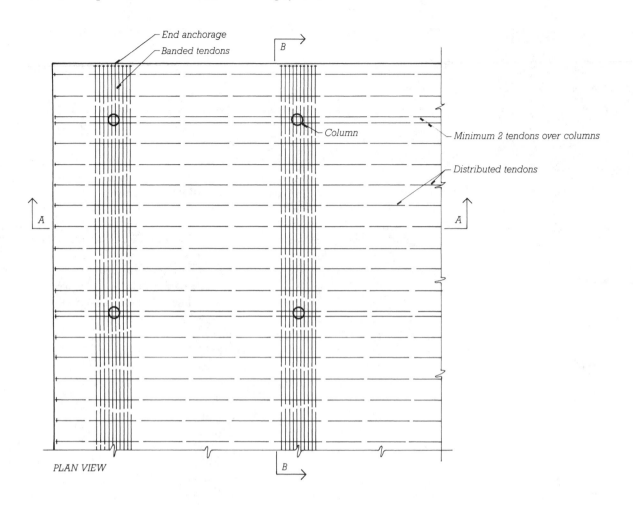

PLAN VIEW

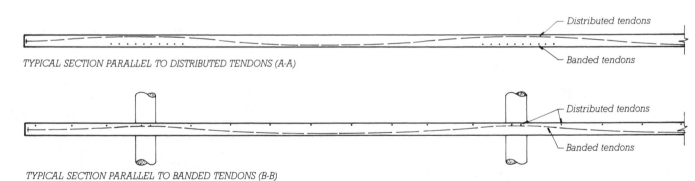

TYPICAL SECTION PARALLEL TO DISTRIBUTED TENDONS (A-A)

TYPICAL SECTION PARALLEL TO BANDED TENDONS (B-B)

FIGURE 14.41

A plan and two larger-scale sections of the tendon layout in a two-way flat plate floor with banded posttensioning. The numbers of tendons running in each of the two directions are identical, but those in one direction are concentrated into bands that run over the tops of the columns. The draping of the tendons is evident in the two section drawings. Building codes require that at least two distributed tendons run directly over each column to help reinforce against shear failure of the slab in this region. In addition to the tendons, conventional steel reinforcing is used around the columns and in midspan, but this has been omitted from these drawings for the sake of clarity.

Figure 14.42
**Banded tendons run directly through the concrete column of this flat plate floor. A
substantial amount of conventional reinforcing is used here for shear reinforcing.
Notice the end anchorage plates nailed to the vertical surface of the formwork at the
upper right; see also Figure 13.39.** (*Courtesy of Post-Tensioning Institute*)

concrete, and initial movements (*set*) in the anchorages. The long-term losses are caused by concrete shrinkage, concrete creep, and steel relaxation. The structural engineer calculates the total of these expected losses and specifies an additional amount of initial posttensioning force to compensate for them.

SELECTING A SITECAST CONCRETE FRAMING SYSTEM

Preliminary factors to be considered in the selection of a sitecast concrete framing system for a building include the following (Figures 14.43, 14.44):

1. Are the bays of the building square or nearly square? If so, a two-way system will probably be more economical than a one-way system.

2. How long are the spans? Spans less than 25 feet (7.6 m) are often accomplished most economically with a two-way flat plate system, because of the simplicity of the formwork. For longer spans, a one-way joist system or a waffle slab system may be a good choice. Posttensioning extends significantly the economical span range of any of these systems.

3. How heavy are the loads? Heavy industrial loadings are borne better by thicker slabs and larger beams than they are by light joist construction. Ordinary commercial, institutional, and residential loadings are

carried easily by flat plate or joist systems.

4. Will there be a finish ceiling beneath the slab? If not, flat plate and one-way slab construction have smooth, paintable undersides that can serve as ceilings. A waffle slab may be exposed if the rooms beneath are large and partition layouts are coordinated with the joist lines. (Precast slabs may also be appropriate—see Chapter 15.)

5. Does the lateral stability of the building against wind and seismic loads have to be provided by the rigidity of the concrete frame? Flat plate floors may not be sufficiently rigid for this purpose, which would favor a one-way system with its deeper beam-to-column connections.

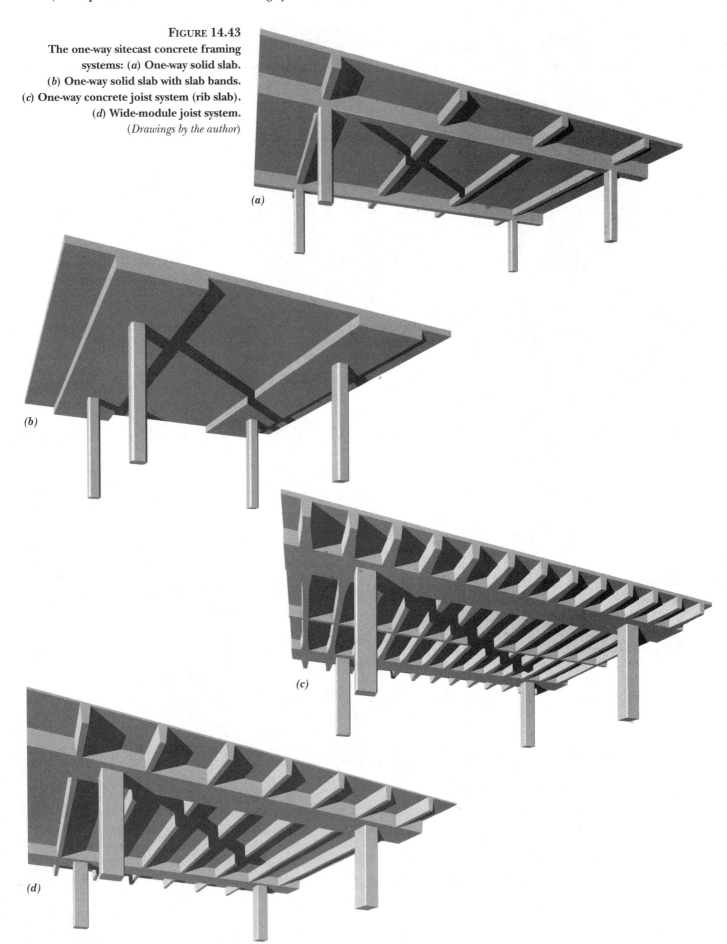

FIGURE 14.43
The one-way sitecast concrete framing
systems: (*a*) One-way solid slab.
(*b*) One-way solid slab with slab bands.
(*c*) One-way concrete joist system (rib slab).
(*d*) Wide-module joist system.
(*Drawings by the author*)

(*a*)

(*b*)

(*c*)

(*d*)

FIGURE 14.44
The two-way sitecast concrete framing
systems: (*a*) Two-way solid slab.
(*b*) Two-way flat slab. (*c*) Two-way flat plate.
(*d*) Two-way concrete joist system (waffle
slab.) (*Drawings by the author*)

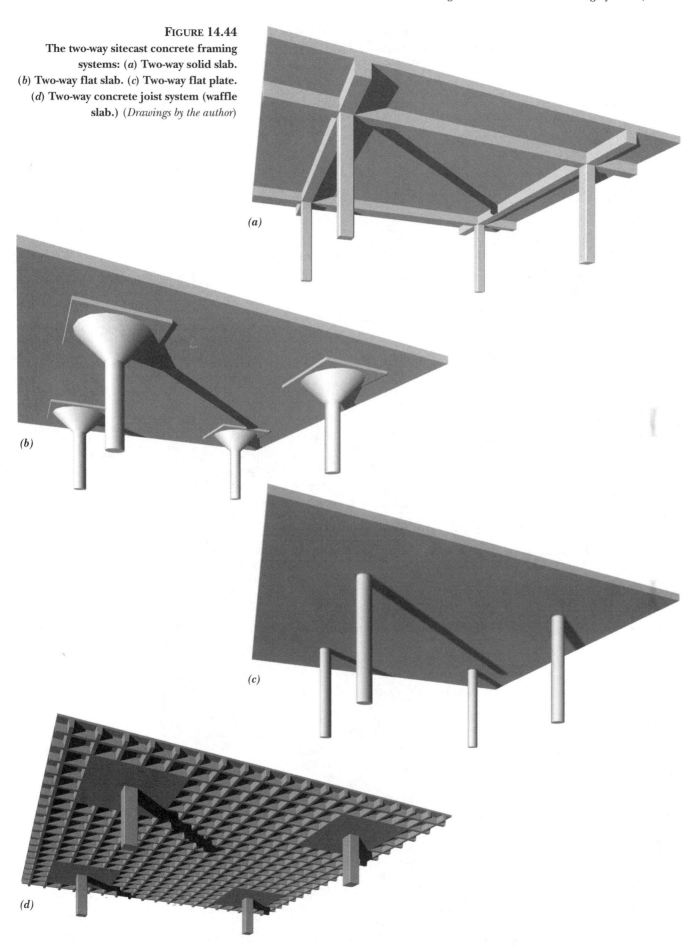

FIGURE 14.44
The two-way sitecast concrete framing
systems: (*a*) Two-way solid slab.
(*b*) Two-way flat slab. (*c*) Two-way flat plate.
(*d*) Two-way concrete joist system (waffle
slab.) (*Drawings by the author*)

(*a*)

(*b*)

(*c*)

(*d*)

FOR PRELIMINARY DESIGN OF A SITECAST CONCRETE STRUCTURE

• Estimate the depth of a **one-way solid slab** at $1/22$ of its span if it is conventionally reinforced, or $1/40$ of its span if it is posttensioned. Depths range typically from 4 to 10 inches (100 to 250 mm).

• Estimate the total depth of a **one-way concrete joist system** or **wide-module system** at $1/18$ of its span if it is conventionally reinforced, or $1/36$ of its span if it is posttensioned. For standard sizes of the pans used to form these systems, see Figure 14.23. To arrive at the total depth, a slab thickness of 3 to $4\frac{1}{2}$ inches (75 to 115 mm) must be added to the depth of the pan that is selected.

• Estimate the depth of **concrete beams** at $1/16$ of their span if they are conventionally reinforced, or $1/24$ of their span if they are posttensioned. For **concrete girders,** use ratios of $1/12$ and $1/20$, respectively.

• Estimate the depth of **two-way flat plates** and **flat slabs** at $1/30$ of their span if they are conventionally reinforced, or $1/45$ of their span if they are posttensioned. Typical depths are 5 to 12 inches (125 to 305 mm). The minimum column size for a flat plate is approximately twice the depth of the slab. The width of a drop panel for a flat slab is usually $1/3$ of the span, and the projection of the drop panel below the slab is about $1/2$ the thickness of the slab.

• Estimate the depth of a **waffle slab** at $1/24$ of its span if it is conventionally reinforced, or $1/35$ of its span if it is posttensioned. For standard sizes of the domes used to form waffle slabs, see Figure 14.35. To arrive at the total depth, a slab thickness of 3 to $4\frac{1}{2}$ inches (75 to 115 mm) must be added to the depth of the dome that is selected.

• To estimate the size of a **concrete column,** add up the total roof and floor area supported by the column. A 12-inch (300-mm) column can support up to about 2000 square feet (185 m²) of area, a 16-inch (400-mm) column 3000 square feet (280 m²), a 20-inch (500-mm) column 4000 square feet (370 m²), a 24-inch (600 mm) column 6000 square feet (560 m²), and a 28-inch (700-mm) column 8000 square feet (740 m²). These sizes are greatly influenced by the strength of the concrete used and the ratio of reinforcing steel to concrete. Columns are usually round or square.

• To estimate the thickness of a **concrete loadbearing wall,** add up the total width of floor and roof slabs that contribute load to a 1-foot length of wall. An 8-inch (200-mm) wall can support approximately 400 feet (120 m) of slab, a 10-inch (250-mm) wall 550 feet (170 m), a 12-inch (300-mm) wall 700 feet (210 m), and a 16-inch (400-mm) wall 1000 feet (300 m). These thicknesses are greatly influenced by the strength of the concrete used and the ratio of reinforcing steel to concrete.

These approximations are valid only for purposes of preliminary building layout, and must not be used to select final member sizes. They apply to the normal range of building occupancies such as residential, office, commercial, and institutional buildings, and parking garages. For manufacturing and storage buildings, use somewhat larger members.

For more comprehensive information on preliminary selection and layout of a structural system and sizing of structural members, see Allen, Edward, and Joseph Iano, *The Architect's Studio Companion* (2nd ed.), New York, John Wiley & Sons, Inc., 1995.

INNOVATIONS IN SITECAST CONCRETE CONSTRUCTION

The development of sitecast concrete construction continues along several lines. The basic materials, concrete and steel, are constantly undergoing research and development, leading to higher allowable strengths and decreased weight of the structure itself. Structural lightweight concretes are being used more widely to reduce loads still further. Shrinkage-compensating cements and admixtures have been developed for use in concrete structures that cannot be allowed to shrink during curing.

The high cost of formwork has led to many innovations. *Lift-slab construction,* used chiefly with two-way flat plate structures, eliminates most formwork by casting the slabs of a building in a stack on the ground, then using hydraulic jacks to lift the slabs up the columns to their final positions, where they are welded in place using special cast-in-place steel slab collars (Figure 14.45). For floor slabs that are cast in place, *flying formwork* is fabricated in large sections supported on deep metal trusses; the sections are moved from one floor to the next by crane, eliminating much of the labor usually expended on stripping and reerecting formwork (Figure 14.46). *Slip forming* is useful for tall walled structures such as ele-

FIGURE 14.45
Lift-slab construction in progress. The paired steel rods silhouetted against the sky to the right are part of the lifting jacks seen at the tops of the columns. (*Courtesy of Portland Cement Association, Skokie, Illinois*)

FIGURE 14.46
Flying formwork for a one-way concrete joist system being moved from one floor to the next in preparation for pouring. Stiff metal trusses allow a large area of formwork to be handled by a crane as a single piece. (*Courtesy of Molded Fiber Glass Forms Company*)

vator shafts, stairwells, and storage silos. A ring of formwork is pulled steadily upward by jacks supported on the vertical reinforcing bars, while workers add concrete and horizontal reinforcing in a continuous process. Manufacturers of concrete formwork have developed more sophisticated systems of self-climbing formwork that offer many advantages over conventional slip-forming (Figure 14.47). In *tilt-up construction* (Figure 14.48), a floor slab is cast on the ground, and reinforced concrete wall panels are poured over it in a horizontal position, then tilted into position and grouted together, thereby eliminating most of the usual wall formwork.

Shotcrete (pneumatically placed concrete) is sprayed into place from a

FIGURE 14.47
A proprietary system of <u>self-climbing formwork</u> is being used to form these sitecast concrete elevator shafts for a tall building. The top level is a working surface from which reinforcing bars are handled and the concrete is poured. The outer panels of the formwork are mounted on overhead tracks just beneath the top level. The panels can be rolled back to the outside of the perimeter walkway after each pour, allowing workers to clean the formwork and install the reinforcing for the next pour. The entire two-story apparatus raises itself a story at a time with built-in hydraulic jacks. (*Courtesy of Patent Scaffolding Company, Fort Lee, New Jersey*)

FIGURE 14.48
Tilt-up construction. The exterior wall panels were reinforced and cast flat on the floor slab. Using special lifting rings that were cast into the panels and a lifting harness that exerts equal force at each of the lifting rings, a crane tilts up each panel and places it upright on a strip foundation at the perimeter of the building. Each erected panel is braced temporarily with diagonal steel struts until the roof structure has been completed. (*Courtesy of Portland Cement Association, Skokie, Illinois*)

hose by a stream of compressed air and can be deposited without formwork even on vertical surfaces. It is used primarily for repairing damaged concrete on the faces of beams and columns and for the production of freeform structures such as swimming pools and playground structures.

The ultimate saving in formwork, of course, is realized by casting the concrete into reusable molds in a precasting plant, which is the subject of the next chapter.

Advancements in reinforcing for sitecast concrete, other than the adoption of posttensioning, include a move to higher-strength steels, and a trend toward increased prefabrication of cages and grids of reinforcing bars prior to installation in the forms. With developments in welding and fabricating machinery, welded wire fabric is pushing beyond the familiar grid of heavy wire to include complete cages of column reinforcing and entire bays of slab reinforcing.

ARCHITECTURAL CONCRETE

Concrete that is left exposed as finished interior or exterior surfaces is known as *architectural concrete*. Most formed concrete surfaces, although structurally sound, have too many blemishes and irregularities to be visually attractive. A vast amount of thought and effort has been expended to develop handsome surface finishes for concrete (Figures 14.49–14.52). *Exposed aggregate* finishes have been popular for decades. These involve the scrubbing and hosing of concrete surfaces shortly after the initial set of the concrete to remove the cement paste from the surface and leave the rough texture of the aggregate. This process is often aided by chemicals that retard the set of the cement paste; these are either sprayed on the surface of a slab or used as a coating inside formwork.

Because concrete can take on almost any texture that can be imparted to the surface of formwork, much effort has gone into developing formwork surfaces of wood, wood panel prod-

ucts, metal, plastic, and rubber to produce textures that range from almost glassy smooth to ribbed, veined, board-textured, and corrugated. After partial curing, other

FIGURE 14.49
Exposed wall surfaces of sitecast concrete. Narrow boards were used to form the walls, and form tie locations were carefully worked out in advance. (*Architect: Eduardo Catalano. Photo by Erik Leigh Simmons. Courtesy of the architect*)

steps can be taken to change the texture of concrete, including sandblasting, rubbing with abrasive stones, grinding smooth, and hammering with any of a number of types of flat, pointed, or toothed masonry hammers. Many types of pigments, dyes, paints, and sealers can be used to add color or gloss to concrete surfaces, and to give protection against weather, dirt, and wear.

Exposed wall surfaces of concrete need special attention from the designer and contractor (Figure 14.51). Chairs and bolsters need to be selected that will not create rust spots on exterior concrete surfaces. Form tie locations in exposed concrete walls should be patterned to harmonize with the layout of the walls themselves, and the holes left in the concrete surfaces by snapped-off ties

must be patched or plugged securely to prevent rusting through. Joints between pours can be concealed gracefully with recesses called *rustication strips* in the face of the concrete. The formulation of the concrete needs to be closely controlled for the next. In cold climates, air entrainment is advisable to prevent freeze–thaw damage of exterior wall surfaces.

If we were to train ourselves to draw as we build, from the bottom up ... stopping our pencil to make a mark at the joints of pouring or erecting, ornament would grow out of our love for the expression of method.

Louis I. Kahn, quoted in Vincent Scully, Jr., *Louis I. Kahn*, New York, George Braziller, 1962, p. 27

FIGURE 14.50
Exposed wall surfaces of concrete, sandblasted to expose the aggregate. Note the regular spacing of the form tie holes, which was worked out by the architect as an integral feature of the building design. (*Architect: Eduardo Catalano. Photo by Gordon H. Schenck, Jr. Courtesy of the architect*)

FIGURE 14.51
Standards specified by the architect to assure satisfactory visual quality in the exposed concrete walls of the buildings illustrated in Figures 14.49 and 14.50. (*Courtesy of Eduardo Catalano, Architect*)

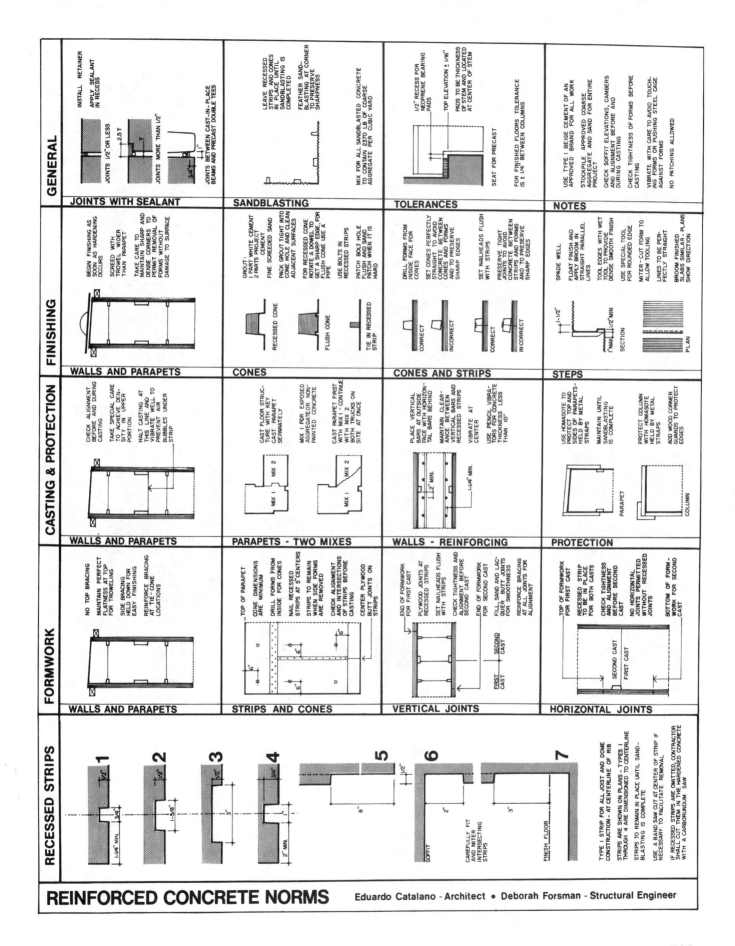

REINFORCED CONCRETE NORMS

Eduardo Catalano - Architect • Deborah Forsman - Structural Engineer

FIGURE 14.52

Close-up photographs of some surface textures for exposed concrete walls. (*a*) Concrete cast against overlaid plywood to obtain a very smooth surface shows a crazing pattern of hairline cracks. (*b*) The boat-shaped patches and rotary-sliced grain figure of A-veneered plywood formwork are mirrored faithfully in this surface. A neatly plugged form tie hole is seen at the upper left, and several lines of overspill from a higher pour have dribbled over the surface. (*c*) This exposed aggregate surface was obtained by coating the formwork with a curing retarder and scrubbing the surface of the concrete with water and a stiff brush after stripping the formwork. (*d*) The bush-hammered surface of this concrete column is framed by a smoothly formed edge. (*e, f*) Architect Paul Rudolph developed the techniques of casting concrete walls against ribbed formwork, then bush hammering the ribs to produce a very heavily textured, deeply shadowed surface. In the example to the right, the ribbed wall surface is contrasted to a board-formed slab edge, with a recessed rustication strip between. (*Photos by the author*)

CUTTING CONCRETE, STONE, AND MASONRY

It is often necessary to cut hard materials in the course of obtaining and processing construction materials, and during the construction process itself. The quarrying and milling of stone require many cutting operations. Precast concrete elements are frequently cut to length in the factory. Masonry units often need to be cut on the construction site, and masonry walls sometimes require the cutting of fastener holes and utility openings. Concrete cutting has become an industry in itself because of the need for utility openings, fastener holes, control joints, and surface grinding and texturing. Cutting and drilling are required to create new openings and remove unwanted construction during the renovation of masonry and concrete buildings. Core drilling is used to obtain laboratory test specimens of concrete, masonry, and stone. And cutting is sometimes necessary to remove incorrect work and to perform building demolition operations.

In preindustrial times, hard materials were cut with hand tools such as steel saws that employed an abrasive slurry of sand and water beneath the blade, and hardened steel drills and chisels that were driven with a heavy hammer. Wedges and explosives in drilled holes were used to split off large blocks of material. These techniques and mechanized variations of them are still used to some extent, but diamond cutting tools are rapidly taking over the bulk of tough cutting chores in the construction industry. Diamond tools are expensive in first cost, but they cut much more rapidly than other types of tools, they cut more cleanly, and they last much longer, so they are usually more economical. Furthermore, diamond tools can sometimes do things that conventional tools cannot, such as precision-sawing marble and granite into very thin sheets for floor and wall facings.

Diamonds cut hard materials efficiently because they are the hardest known material. Most of the industrial diamonds that go into cutting tools are synthetic. They are produced by subjecting graphite and a catalyst to extreme heat and pressure, then sorting and grading the small diamonds that result. Although some natural diamonds are still used in industry, synthetic diamonds are preferred for most tasks because of their more consistent behavior in use.

In the manufacture of a cutting tool, the diamonds are first embedded in a metallic bonding matrix and the mixture is formed into small cutting segments. The choice of diamonds and the exact composition of the bonding matrix are governed by the type of material that is to be cut. The cutting segments are brazed to steel cutting tools—circular saw blades, gangsaw blades, core drill cylinders—and the cutting tools are mounted in the machines that drive them. Some tools are designed to cut dry, but most are used with a spray of water that cools the blade and washes away the cut material. The accompanying illustrations (Figures A–J) show a variety of machines

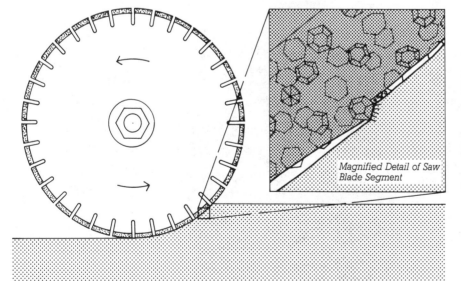

Magnified Detail of Saw Blade Segment

FIGURE A

A diamond saw blade is made up of cutting segments brazed to a steel blade core. Each cutting segment consists of diamond crystals embedded in a metallic bonding matrix. The diamonds in the cutting segment fracture chips from the material being cut. In doing so, each diamond gradually becomes chipped and worn and finally falls out of the bonding matrix altogether. The bonding matrix wears at a corresponding rate, exposing new diamonds to take over for those that have fallen out.

that use diamond cutting tools. Diamonds are also made into grinding wheels, which are used for everything from sharpening tungsten carbide tools to flattening out-of-level concrete floors and polishing granite.

Cutting tools based on materials other than diamonds are still common on the construction job site. Tungsten carbide is used for the tips of small-diameter masonry and concrete drills and for the teeth of wood-working saws. Low-cost circular saw blades composed of abrasives such as silicon carbide are useful for occasional cutting of metals, concrete, and masonry, but the cutting action is slow and the blades wear very quickly. Less precise tools such as the pneumatic jackhammer, hydraulic splitters, and the traditional sledgehammer also have their uses. But for large-volume precision cutting and drilling, there is no substitute for the industrial diamond.

FIGURE B
A hand-held pneumatically powered diamond circular saw cuts excess length from a concrete pile. (*Courtesy of Sinco Products, Inc.*)

FIGURE C
A pneumatically powered diamond circular saw slices a control joint in a concrete pavement. (*Courtesy of Partner Industrial Products, Itasca, Illinois*)

FIGURE D
Cutting a new opening in a masonry wall with a diamond circular saw. (*Courtesy of Partner Industrial Products, Itasca, Illinois*)

FIGURE E
While a conventional circular saw can cut to a depth of only about a third of its blade diameter, this edge-driven ring saw can cut hard materials to a depth of almost three-quarters of its diameter. (*Courtesy of Partner Industrial Products, Itasca, Illinois*)

FIGURE F
A hydraulically driven chain saw with diamond teeth cuts a concrete masonry wall.
(*C-150 Hydracutter, photo courtesy of Reimann & George Construction, Buffalo, New York*)

FIGURE G
A diamond core drill is used for cutting round holes and obtaining test samples of concrete or masonry. (*Courtesy of Sprague & Henwood, Inc., Scranton, Pennsylvania*)

FIGURE H
Using a technique called stitch drilling, a core drill can cut openings in shapes that are not possible with a saw. (*Courtesy of GE Superabrasives*)

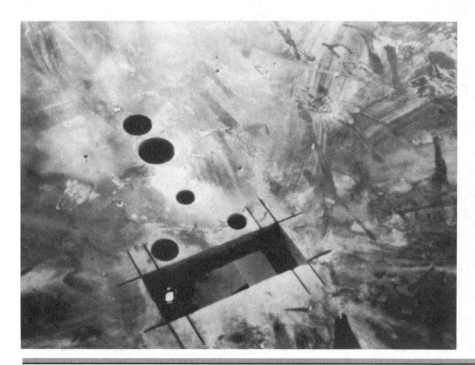

FIGURE I
Sawed and drilled openings for utility lines in a concrete floor slab. (*Courtesy of GE Superabrasives*)

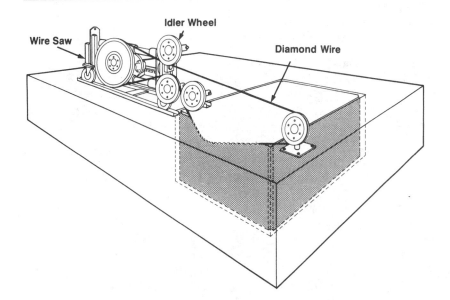

FIGURE J
Wire saws are capable of cutting depths and thicknesses of material that cannot be cut with any other kind of tool. Holes are first drilled at the corners of the cut, and the diamond cutting wire is strung through the holes and around the wheels of the sawing equipment. The equipment maintains a constant tension on the wire as it is pulled at high speed through the material. The wire gradually cuts its way out, leaving smooth, planar surfaces. The wire is actually a steel cable on which steel beads with diamonds embedded in them are strung. (*© 1988, Cutting Technologies, Inc., Cincinnati, Ohio, all rights reserved*)

FIGURE K
This hotel was created from 36 concrete grain silos originally built in 1932. The 7-inch (180-mm) walls of the silos were cut with diamond circular saw blades ranging in diameter from 24 to 42 inches (610 to 1070 mm) to create the window and door openings. (*Courtesy of GE Superabrasives*)

Basically there are two approaches to the problem of producing a good surface finish on concrete. One is to remove the cement that is the cause of the blemishes and expose the aggregate. The other is to superimpose a pattern or profile that draws attention from the blemishes.

Henry Cowan,
Science and Building: Structural and Environmental Design in the Nineteenth and Twentieth Centuries,
New York, John Wiley & Sons, Inc., 1978, p. 283

LONGER SPANS IN SITECAST CONCRETE

The ancient Romans built unreinforced concrete vaults and domes as roofs for temples, baths, palaces, and basilicas (Figure 13.2). Impressive spans were constructed, including a dome over the Pantheon in Rome, still standing, that approaches 150 feet (45 m) in diameter. Today, the arch, dome, and vault remain favorite devices for spanning long distances in concrete because of concrete's suitability to structural forms that work entirely in compression (Figures 14.53–14.55). Through folding or scalloping of vaulted forms or through the use of warped geometries such as the hyperbolic paraboloid, the required resistance to buckling can be achieved with a surprisingly thin layer of concrete, often proportionally thinner than the shell of an egg (Figure 14.55*b*).

Long-span beams and trusses are possible in concrete, including posttensioned beams and girders and reinforced deep girders analogous to steel plate girders and rigid frames. Concrete trusses and space frames are not common but are built from time to time; by definition, a truss includes strong tensile forces as well as compressive forces and is heavily dependent on steel reinforcing or prestressing.

Barrel shells and *folded plates* (Figures 14.54, 14.55*c*, 14.56) derive

FIGURE 14.53
The same wooden centering was used four times to form this concrete arch bridge.
(*Courtesy of Gang-Nail Systems, Inc.*)

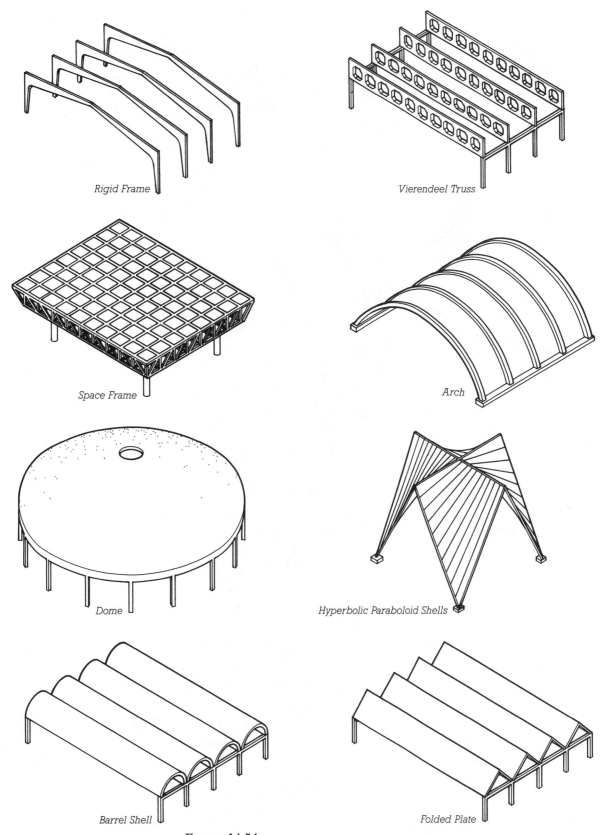

Rigid Frame

Vierendeel Truss

Space Frame

Arch

Dome

Hyperbolic Paraboloid Shells

Barrel Shell

Folded Plate

FIGURE 14.54
Examples of eight types of longer-span structures in concrete. Each is a special case of an infinite variety of forms. All can be sitecast, but the rigid frame, space frame, and Vierendeel truss are more likely to be precast for most applications.

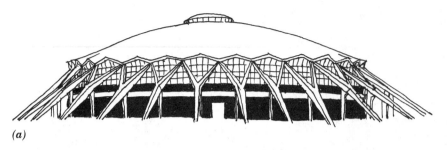

(a)

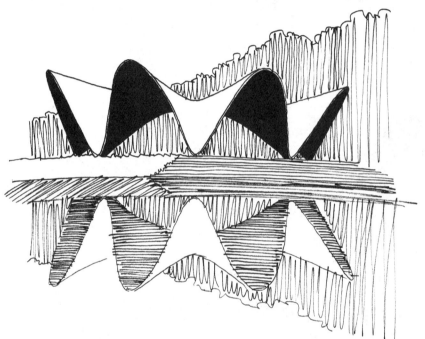

(b)

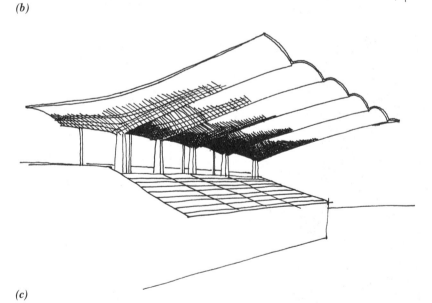

(c)

FIGURE 14.55
Three concrete shell structures by 20th-century masters of concrete engineering: (*a*) **A domed sports arena by Pier Luigi Nervi.** (*b*) **A lakeside restaurant of hyperbolic paraboloid shells by Felix Candela.** (*c*) **A racetrack grandstand roofed with cantilevered concrete barrel shells by Eduardo Torroja.** (*Drawings by the author*)

FIGURE 14.56
Flying formwork is removed from a bay of a folded-plate concrete roof for an air terminal. (*Architects: Thorshov and Cerny. Photo courtesy of APA–The Engineered Wood Association*)

SITECAST CONCRETE AND THE BUILDING CODES

Concrete structures are inherently fire resistant. When fire attacks concrete, the water of hydration is gradually driven out and the concrete loses strength, but this deterioration is slow because considerable heat is needed to raise the temperature of the mass of concrete to the point where dehydration begins, and a large additional quantity of heat is required to vaporize the water. The steel reinforcing bars or prestressing strands are buried beneath a concrete cover that affords them protection for an extended period of time. Except under unusual circumstances, such as a prolonged fire fueled by stored petroleum products, concrete structures usually survive fires with only cosmetic damage and are repaired with relative ease.

Concrete structures with adequate cover over reinforcing and adequate slab thicknesses are classified as Type 1 buildings in the tables in Figures 1.1 and 1.3. Slab thickness requirements for the several construction types are complex, depending on the type of aggregate used in the concrete and whether or not a given structural member is restrained from movement by surrounding construction. Fire resistance requirements for the highest construction types can be met in joist and waffle systems either by increasing their slab thickness beyond what is structurally necessary or by applying fireproofing materials to the lower surfaces of the floor structures.

Sitecast concrete buildings have rigid joints and, in many cases, need no additional structural elements to achieve the necessary resistance to wind and seismic forces. More restrictive seismic design provisions in the building codes have, however, increased the amount of attention paid by structural engineers to column ties and beam stirrups, particularly in the zones where beams and columns meet, to be sure that vertical

their stiffness from the folding or scalloping of a thin concrete plate to increase its rigidity without adding material. Each of these forms depends on reinforcing or posttensioning to resist the tensile forces involved.

DESIGNING ECONOMICAL SITECAST CONCRETE BUILDINGS

The cost of a concrete building frame can be broken down into the costs of the concrete, the reinforcing steel, and the formwork. Of the three, the cost of concrete is usually the least significant in North America, and the cost of formwork the most significant. Accordingly, simplification and standardization of formwork are the first requirements for an economical concrete frame. Repetitive, identical column spacings and bay sizes allow the same formwork to be used again and again without alterations. Flat plate construction is often the most economical, simply because its formwork is so straightforward. Joist band construction is usually more economical than joist construction that uses more efficiently proportioned beams, because enough is saved on formwork costs to more than compensate for the increased use of concrete and reinforcing steel in the beams. This same reasoning applies if column and beam dimensions are standardized throughout the building, even though loads may vary; the amount of reinforcing and the strengths of the concrete and reinforcing steel can be varied to meet the varying structural requirements.

FIGURE 14.57
Concrete work nears the summit of the world's tallest buildings at 450 m (1475 feet), the twin Petronas Towers in Kuala Lumpur, Malaysia. Each tower is supported by a perimeter ring of 16 cylindrical concrete columns and a concrete core structure. The columns vary in diameter from 2400 mm (8 feet) at the base of the building to 1200 mm (4 feet) at the top. For speed of construction, the floors are framed with structural steel and a composite metal deck. Concrete with strengths as high as 80 MPa (11,600 psi) was used in the columns. The architect was Cesar Pelli & Associates, Inc. The structural engineers were Thornton-Tomasetti and Rahnill Bersekutu Sdn Bhd. The U.S. partner in the joint venture team that constructed the towers was J. A. Jones Construction Co., Charlotte, North Carolina. (*Photo by Uwe Hausen, J. A. Jones, Inc.*)

bars in columns and horizontal bars in beams are adequately restrained against the unusually strong forces that can occur in these zones under seismic loadings. The joints between flat plate floors and columns may not be sufficiently rigid to brace a building of more than modest height, unless drop panels or beams are added to stiffen the slab-to-column junction.

THE UNIQUENESS OF SITECAST CONCRETE

Concrete is a shapeless material that must be given form by the designer. For economy, the designer can adopt a standard system of concrete framing. For excitement, one can invent new shapes and textures, a route taken by many of the leading practi-tioners of architecture in this century. Some have pursued its sculptural possibilities; others its surface patterns and textures, still others its structural logic. From each of these routes have come masterpieces—Le Corbusier's chapel at Ronchamp (Figure 14.60), Wright's Unity Temple (Figure 14.1), and the elegant structures of Torroja, Candela, and Nervi, examples of which are sketched in Figure 14.55.

FIGURE 14.58
The plastered surfaces of Frank Lloyd Wright's Guggenheim Museum (1943–1956) cover a helical ramp of cast-in-place concrete. (*Photo by Wayne Andrews*)

Many of these masterpieces, especially from the latter three designers, were also constructed with impressive economy. Sitecast concrete can do almost anything, be almost anything, at almost any scale, and in any type of building. It is the most potent of architectural materials and therefore the material both of spectacular architectural achievements and of dismal architectural failures. A material so malleable demands skill and restraint from those who would build with it, and a material so commonplace requires imagination if it is to rise above the mundane.

As designers, we have learned to express the internal composition of concrete by exposing its aggregates at the surface, or to show the beauties of the formwork in which it was cast, by leaving the marks of the form ties and the textures of the form boards. But we have yet to discover how to reveal in the finished structure the lovely and complex geometries of the steel bars that constitute half of the structural logic that makes concrete buildings stand up.

FIGURE 14.59
A sitecast concrete house in Lincoln, Massachusetts. (*Architects: Mary Otis Stevens and Thomas F. McNulty*)

C.S.I./C.S.C.
Masterformat Section Numbers for Sitecast Concrete
Framing Systems

03300	**CAST-IN-PLACE CONCRETE**
03310	**Structural Concrete**
03345	**Concrete Finishing**
03350	**Concrete Finishes**
03365	**Post-Tensioned Concrete**
03370	**Concrete Curing**
03430	**Structural Precast Concrete—Site Cast**
	Lift-Slab Concrete
03470	**Tilt-Up Precast Concrete—Site Cast**

FIGURE 14.60
Le Corbusier's most sculptural building in his favorite material, concrete: The chapel of Notre Dame du Haut at Ronchamp, France (1950–1955). (*Drawing by the author*)

FIGURE 14.61
The TWA Terminal at John F. Kennedy Airport, New York, 1956–1962. (*Architect: Eero Saarinen. Photo by Wayne Andrews*)

SELECTED REFERENCES

1. American Concrete Institute. *Building Code Requirements for Reinforced Concrete* (ACI Standard 318-95). Farmington Hills, Michigan, 1996.

This booklet establishes the basis for the engineering design and construction of reinforced concrete structures in the United States. (Address for ordering: ACI International, P.O. Box 9094, Farmington Hills, MI 48333.)

2. *CRSI Design Handbook* (8th ed.). Schaumburg, Illinois, Concrete Reinforcing Steel Institute, 1996.

Structural engineers working in concrete use this handbook, which is based on the ACI Code (reference 1), as their major reference. It contains examples of engineering calculation methods and hundreds of pages of tables of standard designs for reinforced concrete structural elements. (Address for ordering: CRSI, 933 North Plum Grove Road, Schaumburg, IL 60173-4758.)

3. *Placing Reinforcing Bars: Recommended Practices* (7th ed.). Schaumburg, Illinois, Concrete Reinforcing Steel Institute, 1997.

Produced as a handbook for those engaged in the business of fabricating and placing reinforcing steel, this small volume is clearly written and beautifully illustrated with diagrams and photographs of reinforcing for all the common concrete framing systems. (Address for ordering: see reference 2.)

4. Hurd, M. K. *Formwork for Concrete* (6th ed.). Farmington Hills, Michigan, American Concrete Institute, 1995.

Profusely illustrated, this book is the bible on formwork design and construction for sitecast concrete. (Address for ordering: see reference 1.)

5. *Guide to Cast-in-Place Architectural Concrete Practice*. Farmington Hills, Michigan, American Concrete Institute, 1991.

This is a comprehensive handbook on how to produce attractive surfaces in concrete. (Address for ordering: see reference 1.)

6. *Post-Tensioning Manual* (5th ed.). Phoenix, Arizona, Post-Tensioning Institute, 1990.

This heavily illustrated volume is both an excellent introduction to posttensioning for the beginner and a basic engineering manual for the expert. (Address for ordering: 1717 West Northern Avenue, Suite 218, Phoenix, AZ 85021.)

KEY TERMS AND CONCEPTS

slab on grade
welded wire fabric
control joint
strike off or straightedge
floating
float
shake-on hardener
troweling
steel trowel
rotary power trowel
knee board
broom finish
curing compound
superflat floor
key
dowels
form ties

insulating concrete forms
one-way solid slab
shores
reshoring
slab band
one-way concrete joist system (ribbed slab)
pans
distribution rib
joist band
wide-module concrete joist system
skip-joist system
two-way solid slab
two-way flat slab
mushroom capital
drop panel
column strip
middle strip

two-way flat plate system
waffle slab
two-way concrete joist system
domes
heads
tendon
set
lift-slab construction
flying formwork
slip forming
tilt-up construction
shotcrete
architectural concrete
exposed aggregate
rustication strip
barrel shell
folded plate

REVIEW QUESTIONS

1. Draw from memory a detail of a typical slab on grade, and list the steps in its production. Why can't the surface be finished in one operation, instead of waiting for hours before final finishing?

2. List the steps that are followed in forming and pouring a concrete wall.

3. Distinguish one-way concrete framing systems from two-way systems. Are steel and wood framing systems one-way or two-way? Is one-way more efficient structurally than two-way?

4. List the common one-way and two-way concrete framing systems and indicate the possibilities and limitations of each.

5. Why posttension a concrete structure rather than merely reinforce it?

EXERCISES

1. Propose a suitable reinforced concrete framing system for each of the following buildings and determine an approximate thickness for each:

a. An apartment building with a column spacing of about 16 feet (5 m) in each direction.

b. A newsprint warehouse, column spacing 20 × 22 feet (6 × 6.6 m).

c. An elementary school, column spacing 24 × 32 feet (7.3 × 9.75 m).

d. A museum, column spacing 36 × 36 feet (11 × 11 m).

e. A hotel where overall building height must be minimized in order to build as many stories as possible within a municipal height limit.

2. Look at several sitecast concrete buildings. Determine the type of framing system used in each and explain why you think it was selected. If possible, talk to the designers of the building and find out if you were right.

3. Observe a concrete building under construction. What is its framing system?

Why? What types of forms are used for its column, beams, and slabs? How is the concrete mixed? How is it raised and deposited into the forms? How is it compacted in the forms? How is it cured? Are samples taken for testing? How soon after pouring are the forms stripped? Are the forms reused? How long are the shores kept in place? Keep a diary of your observations over a period of a month or more.

PRECAST CONCRETE FRAMING SYSTEMS

Suction devices lift a precast concrete hollow-core slab from the casting bed where it was manufactured. (*Courtesy of Flexicore Co., Inc.*)

Structural *precast concrete* elements—slabs, beams, girders, columns, and wall panels—are cast and cured in industrial plants, transported to the construction job site, and erected as rigid components. Precasting offers many potential advantages over sitecasting of concrete: The production of precast elements is carried out conveniently at ground level. The mixing and pouring operations are often highly mechanized, and frequently, especially in difficult climates, they are carried out under shelter. Control of the quality of materials and workmanship is generally better than on the construction job site. The concrete is cast in permanent forms of steel, concrete, glass-fiber-reinforced plastic, or varnished wood, whose excellent surface properties are mirrored in the high-quality surfaces of the finished precast elements that they produce. The forms may be reused hundreds or thousands of times before they have to be renewed, so that formwork costs per unit of finished concrete are low. The forms are equipped to pretension the steel in the precast elements for greater structural efficiency, which translates into longer spans, lesser depths, and lower weights than for comparable reinforced concrete elements. Concrete and steel of superior strength are used in precast elements, typically 5000 pounds per square inch (35 MPa) concrete and 270,000 pounds per square inch (1860 MPa) prestressing steel.

Precast concrete elements are usually *steam cured*. Steam furnishes both heat to accelerate the curing of the concrete and moisture for full hydration. Steam curing, coupled with the use of Type III (high early strength) cement, enables a precasting plant to produce fully cured structural elements, from the laying of the prestressing strands to the removal of the finished elements from the form, on a 24-hour cycle.

When the elements produced by this expeditious technique are delivered to the construction job site, further advantages are realized: The erection process is similar to that of structural steel, but it is often faster because most precast concrete systems include a deck as an integral part of the major spanning elements, without the need for placing additional joist or decking components (Figures 15.1, 15.2). Erection is much faster than that of sitecast concrete because there is no formwork to be erected and stripped and little or no waiting for concrete to cure. And erection of precast structures can take place under some types of adverse weather conditions, such as extremely high or low temperatures, that would not permit the sitecasting of concrete.

When choosing between precast and sitecast concrete, the designer must weigh these potential advantages of precasting against some potential disadvantages. The precast structural elements, although light in weight as compared to similar elements of sitecast concrete, are nevertheless heavy and bulky to transport over the roads and hoist into place. This restricts somewhat the size and proportions of most precast elements: They can be rather long, but only as wide as the maximum legal vehicle width of 12 to 14 feet (3.66 to 4.27 m). This restricted width usually precludes utilization of the efficiencies of two-way structural action in precast slabs. And the fully three-dimensional sculptural possibilities of sitecast concrete are largely absent in precast concrete.

When choosing between precast and sitecast concrete, the designer must weigh these potential advantages of precasting against some potential disadvantages. The precast structural elements, although light in weight as compared to similar elements of sitecast concrete, are nevertheless heavy and bulky to transport over the roads and hoist into place. This restricts somewhat the size and proportions of most precast elements: They can be rather long, but only as wide as the maximum legal vehicle width of 12 to 14 feet (3.66 to 4.27 m). This restricted width usually precludes utilization of the efficiencies of two-way structural action in precast slabs. And the fully three-dimensional sculptural possibilities of sitecast concrete are largely absent in precast concrete.

Although whole walls or rooms are sometimes precast as single units in concrete, this chapter will focus on the standard precast elements that are commonly mass produced as structural components. Precast exterior curtain wall panels will be covered separately in Chapter 20.

FIGURE 15.1
A fully precast building frame under construction. A poured concrete topping will cover the hollow-core slabs and the beams to create a smooth floor and tie the precast elements together. (*Courtesy of Flexicore Co., Inc.*)

FIGURE 15.2
Workers guide two hollow-core slabs, lowered by a crane in wire rope slings, onto the precast concrete beams that will support them. (*Courtesy of Flexicore Co., Inc.*)

PRECAST, PRESTRESSED CONCRETE STRUCTURAL ELEMENTS

Precast Concrete Slabs

The most fully standardized precast concrete elements are those used for making floor and roof slabs (Figure 15.3). These may be supported by loadbearing walls of precast concrete or masonry, or by frames of steel, site-cast concrete, or precast concrete. Four kinds of precast slab elements are commonly produced: For short spans and minimum slab depths, *solid slabs* are appropriate. For longer spans, deeper elements must be used, and precast solid slabs, like their site-cast counterparts, become inefficient because they contain too much dead-weight of nonworking concrete. In *hollow-core slabs,* precast elements suitable for intermediate spans, internal longitudinal voids replace much of the nonworking concrete. For the longest spans, still deeper elements are required, and *double tees* and *single tees* eliminate still more nonworking concrete.

For most applications, precast slab elements of any of the four types are manufactured with a rough top surface. After the elements have been erected, a concrete *topping* is poured over them and finished to a smooth surface. The topping, usually 2 inches (50 mm) in thickness, bonds during curing to the rough top of the precast elements and becomes a working part of their structural action. The topping also helps the precast elements to act together as a structural unit rather than as individual planks in resisting concentrated loads and diaphragm loads, and conceals the slight differences in camber that often occur in prestressed components. Structural continuity across a number of spans can be achieved by casting reinforcing bars into the topping over the supporting beams or walls. Underfloor electrical conduits may also be embedded in the topping. (Smooth-top precast slabs are sometimes used without topping, as is discussed later in this chapter.)

Either normal-density concrete or structural lightweight concrete may be selected for use in any of the precast slab elements. The lightweight concrete, approximately 20 percent less dense than normal concrete, reduces the load on the frame and foundations of a building, but is more expensive than normal-density concrete.

There is considerable overlapping of the economical span ranges of the different kinds of precast slab elements, allowing the designer some latitude in choosing which to use in a particular situation. Solid slabs and hollow-core slabs save on overall building height in multistory structures, and their smooth undersides can be painted and used as finish ceilings in many applications. For longer spans, double tees are generally preferred to the older single-tee design because they do not need to be supported against tipping during erection.

Precast Concrete Beams, Girders, and Columns

Precast concrete beams and girders are made in several standard shapes (Figure 15.4). The projecting ledgers on *L-shaped beams* and *inverted tees* provide direct support for precast slab elements to conserve headroom in a building as compared to rectangular beams with slab elements resting on top. *AASHTO* (American Association

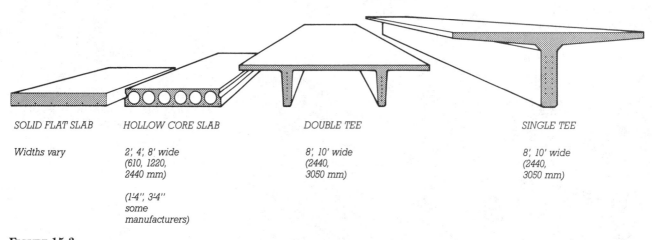

SOLID FLAT SLAB	HOLLOW CORE SLAB	DOUBLE TEE	SINGLE TEE
Widths vary	2', 4', 8' wide (610, 1220, 2440 mm)	8', 10' wide (2440, 3050 mm)	8', 10' wide (2440, 3050 mm)
	(1'-4'', 3'-4'' some manufacturers)		

FIGURE 15.3
The four major types of precast concrete slab elements. Hollow-core slabs are produced by different companies in a variety of cross-sectional patterns, using several different processes. Single tees are less commonly used than double tees because they need temporary support against tipping until they are permanently fastened in place.

FOR PRELIMINARY DESIGN OF A PRECAST CONCRETE STRUCTURE

• Estimate the depth of a **precast solid slab** at $\frac{1}{40}$ of its span. Depths typically range from $3\frac{1}{2}$ to 8 inches (90 to 200 mm).

• An 8-inch (200-mm) **precast hollow-core slab** can span approximately 25 feet (7.6 m), a 10-inch (250-mm) slab 32 feet (9.8 m), and a 12-inch (300-mm) slab 40 feet (12 m).

• Estimate the depth of **precast concrete double tees** at $\frac{1}{28}$ of their span. The most common depths of double tees are 12, 14, 16, 18, 20, 24, and 32 inches (300, 350, 400, 460, 510, 610, and 815 mm).

• A **precast concrete single tee** 36 inches (915 mm) deep spans approximately 85 feet (926 m), and a 48-inch (1220-mm) tee 105 feet (32 m).

• Estimate the depth of **precast concrete beams** and **girders** at $\frac{1}{15}$ of their span for light loadings and $\frac{1}{12}$ of their span for heavy loadings. These ratios apply to rectangular, inverted tee, and L-shaped beams. The width of a beam or girder is usually about $\frac{1}{2}$ its depth. The projecting ledgers on inverted tee and L-shaped beams are usually 6 inches (150 mm) wide and 12 inches (300 mm) deep.

• To estimate the size of a **precast concrete column,** add up the total roof and floor area supported by the column. A 10-inch (250-mm) column can support up to about 2000 square feet (185 m²) of area, a 12-inch (300-mm) column 2600 square feet (240 m²), a 16-inch (400-mm) column 4000 square feet (370 m²), and a 24-inch (600-mm) column 8000 square feet (740 m²). These values may be interpolated to columns in 2-inch (50-mm) increments of size. Columns are usually square.

These approximations are valid only for purposes of preliminary building layout, and must not be used to select final member sizes. They apply to the normal range of building occupancies such as residential, office, commercial, and institutional buildings, and parking garages. For manufacturing and storage buildings, use somewhat larger members.

For more comprehensive information on preliminary selection and layout of structural system and sizing of structural members, see Allen, Edward, and Joseph Iano, *The Architect's Studio Companion* (2nd ed.), New York, John Wiley & Sons, Inc., 1995.

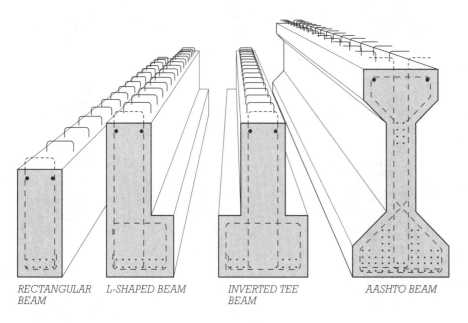

RECTANGULAR BEAM   *L-SHAPED BEAM*   *INVERTED TEE BEAM*   *AASHTO BEAM*

FIGURE 15.4
Standard precast concrete beam and girder shapes. The larger dots represent mild steel reinforcing bars, and the smaller dots represent high-strength prestressing strands. The broken lines show mild steel stirrups. Stirrups usually project above the top of the beam, as shown, to bond to the topping for composite structural action.

of State Highway and Transportation Officials) *girders* were designed originally as efficient shapes for bridge structures, but they are used sometimes in buildings as well. Precast columns are usually square or rectangular in section and may be prestressed or simply reinforced.

Precast Concrete Wall Panels

Precast solid slabs are commonly used as loadbearing wall panels in many types of low-rise and high-rise buildings. The prestressing strands are located in the vertical midplane of the wall panels to strengthen the panels against buckling and to eliminate camber. Rigid foam insulation can be cast into wall panels for thermal insulation, with suitable wire shear ties between the inner and outer wythes of concrete (Figure 15.12).

ASSEMBLY CONCEPTS FOR PRECAST CONCRETE BUILDINGS

Figure 15.5 shows a building whose precast slab elements (double tees in this example) are supported on a skeleton frame of L-shaped precast girders and precast columns. The slab elements in Figure 15.6 are sup-

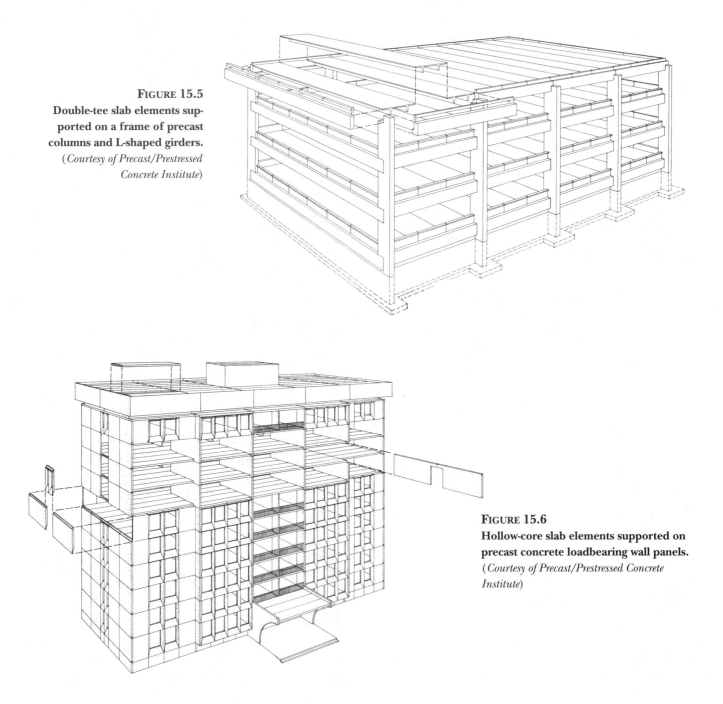

FIGURE 15.5
Double-tee slab elements supported on a frame of precast columns and L-shaped girders.
(*Courtesy of Precast/Prestressed Concrete Institute*)

FIGURE 15.6
Hollow-core slab elements supported on precast concrete loadbearing wall panels.
(*Courtesy of Precast/Prestressed Concrete Institute*)

ported on precast loadbearing wall panels. Figure 15.7 illustrates a building whose slabs are supported on a combination of wall panels and girders. These three fundamental ways of supporting precast slabs—on a precast concrete skeleton, on precast loadbearing wall panels, and on a combination of the two—occur in endless variations in buildings. The skeleton may be one bay deep or many bays; the loadbearing walls are often constructed of reinforced masonry or of any of a variety of configurations of precast concrete; the slab elements may be solid, hollow-core, or double tee, topped or untopped. One of the principal virtues of precast concrete as a structural material is that it is locally manufactured to order and is easily customized to an individual building design, usually at minimal additional cost.

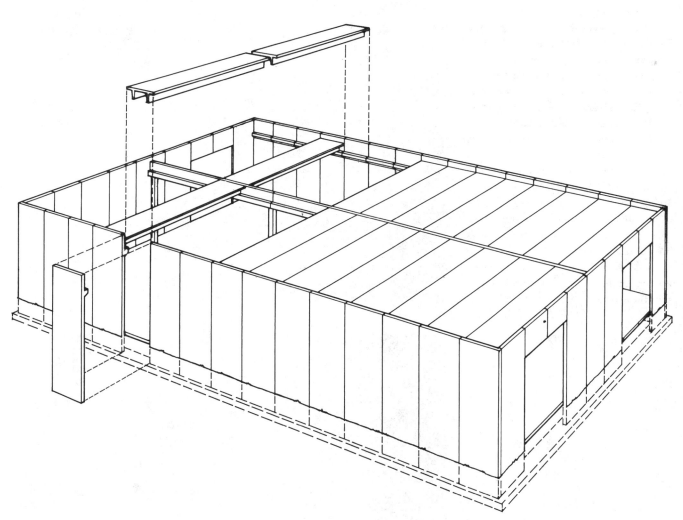

FIGURE 15.7
Double-tee slab elements supported on a perimeter of precast concrete loadbearing wall panels, and an interior structure of precast columns and inverted tee beams. (*Courtesy of Precast/Prestressed Concrete Institute*)

FIGURE 15.8
Manufacturing double-tee slabs. (*a*) A worker inserts weld plates with their V-shaped anchors of reinforcing bar in the casting bed prior to pouring the concrete. The prestressing strands and wire fabric shear reinforcement have been installed in the stems of the double tees, and a portion of the welded wire fabric for the top slab can be seen in the foreground. (*b*) The top surface of the concrete is straightedged by machine. (*c*) The next morning, following steam curing, a worker cuts the prestressing strands between bulkhead separators with an oxyacetylene cutting torch. The welded wire fabric is exposed at the ends of the elements because they will be used as untopped slabs (see Figure 15.21). (*d*) Using lifting loops cast into the ends, the slabs are lifted from the bed and will be stockpiled outside. The dapped stems will rest on inverted tee and L-shaped girders; the notched corner will fit around a column. (*e*) Loading the double tees for trucking. (*Photos by Alvin Ericson*)

(*a*)

(*b*)

THE MANUFACTURE OF PRECAST CONCRETE STRUCTURAL ELEMENTS

Casting Beds

Most precast concrete elements are produced in permanent forms called *casting beds.* Casting beds average 400 feet (125 m) in length, but extend 800 feet (250 m) or more in some plants (Figure 15.8). A cycle of precasting usually begins in the morning, as soon as the elements that were cast the previous day have been lifted from the beds. High-strength steel reinforcing strands are strung between the abutments at the extreme ends of the bed. The strands are pretensioned with hydraulic jacks, after which transverse bulkhead separators may be placed along the bed at the required intervals to divide the individual elements from one another. (For solid slabs, cored slabs, and wall panels, the bulkheads are often omitted; the cured slab is simply sawed into the required lengths before it is removed from the bed.)

When pretensioning and separator placement have been completed, mild steel reinforcing bars and welded wire fabric are placed as required. *Weld plates* and other embedments are installed. Then the concrete is placed in the bed, vibrated to eliminate voids, and struck off level. If required, the top surface is finished further with floats and trowels. Live steam or radiant heat is then applied to the concrete to accelerate its curing. Ten to twelve hours after pouring, the concrete reaches a compressive strength of 2500 to 4000 pounds per square inch (24 to 28 MPa) and has bonded to the steel strands. The next morning, after the strength of the concrete has been verified by testing cylinders in the laboratory, the exposed strands are cut between the bulkhead separators, releasing the external force in the strands and prestressing the concrete. Asymmetrically prestressed slab

(d)

(e)

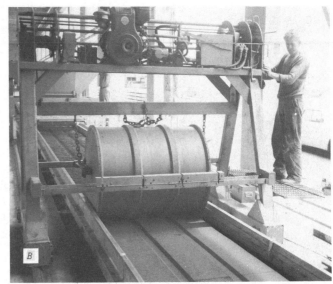

and beam elements immediately camber up from the casting bed as the prestressing force is released into them. When the elements have been separated from one another, they are hoisted off the bed and stockpiled ready for shipment, and a new cycle of casting begins.

Prestressing and Reinforcing Steel

Solid slabs, hollow-core slabs, and wall panels are cast around horizontal strands. Tees, double tees, beams, and girders are often cast around depressed or harped strands for a greater efficiency of structural action (Figures 13.40, 15.11, 15.22).

Ordinary mild steel reinforcing is also cast into prestressed concrete elements for various purposes. Beams or slabs intended to cantilever beyond their supports are given top reinforcing bars over the cantilever points. Welded wire fabric is used to reinforce the flanges of tees and double tees, and for general reinforcing of wall panels. Where stirrups are required in the stems of tees and double tees, either mild steel reinforcing bars or welded wire fabric is used. Additional reinforcing may be installed to strengthen around dapped (notched) ends and openings in panels or slabs for pipes, ducts, columns, and hatchways. Weld plates and other metal connecting devices are cast into the elements as required. Projecting steel loops are cast into many types of elements as crane attachments for lifting.

Hollow-Core Slab Production

The longitudinal voids in hollow-core slabs can be formed by any of a number of patented processes. In some processes, extrusion devices squeeze a stiff concrete mix through an extrusion die to produce the voided shape directly. This method has the disadvantage that vertical openings and weld plates cannot easily be cast in; where openings are required in extruded slabs, they must be cut out of the stiff but still wet concrete just after extrusion, or sawed after curing. Another process deposits a bottom layer of wet concrete in the casting bed, then a second layer of concrete with dry crushed stone or lightweight aggregate carefully introduced to form the voids. Special forms may be placed in the bed to make openings as required, and weld plates may be cast in by this process. After curing, the slabs are conveyed to an aggregate recovery area, where the dry stone is poured out of the voids and saved for reuse (Figure 15.9). In a third type of process, air-inflated tubes are used to form the voids. In some plants, hollow-core slabs are cast atop one another in stacks rather

FIGURE 15.9
Steps in the manufacture of Span-Deck, a proprietary hollow-core slab. (*a*) Depositing a thin bottom slab of low-slump concrete into the casting bed with a traveling hopper. (*b*) A ridged roller compacts the bottom slab and makes indentations for keying to the concrete webs. (*c*) Four prestressing strands are pulled onto the bed; these will be pretensioned with jacks. (*d*) After pretensioning, an extrusion device travels the length of the bed, forming the webs and the top of the slab and filling the cores with dry lightweight aggregate that serves as temporary support for the fresh concrete. (*e*) The next day, after curing of the slab has been completed, a saw cuts it into the required lengths. Each length is then lifted from the bed and transported by an overhead crane to the aggregate recovery area, where the dry aggregate is poured out and saved for reuse. (*f*) Finished Span-Deck slabs stockpiled, ready for transportation to the construction site. The wads of paper in the cores are to prevent concrete from filling the cores during the pouring of the topping. (*Courtesy of Blakeslee Prestress, Inc.*)

FIGURE 15.10
Workers install side forms for inverted tee beams in an outdoor casting bed. Mild steel reinforcing bars are used extensively for stirrups. (*Courtesy of Blakeslee Prestress, Inc.*)

than in a single layer and are wet cured for seven days rather than steam cured overnight.

Column Production

Precast columns are reinforced or pretensioned as directed by the architect or structural engineer. Columns are often made and shipped in multistory sections with *corbels* (Figures 15.17, 15.18) to support beams or slabs. Columns with corbels on one side or on two opposing sides are easily cast in flat beds. If corbels are required on three sides or on two adjacent sides, box forms are set atop the upper side of the column as it lies in the bed. For corbels on the fourth side, steel plate inserts are cast into the bottom of the column in the bed, to which reinforcing bars are welded after the column is removed from the bed. The corbels on the fourth side are then cast around the reinforcing bars in a separate operation.

JOINING PRECAST CONCRETE ELEMENTS

Figures 15.13 through 15.23 and 15.27 show some frequently used connection details for precast concrete construction. Bolting, welding, and

FIGURE 15.11
A precasting bed being readied for the pouring of an AASHTO girder. Side forms for the mold can be seen in the background to the right. The depressed strands are held down in the center of the beam by steel pulleys that will be left in the concrete after pouring. The bed is long enough that several girders are being cast end to end, with the depressed strands pulled up and down as required. Mild steel reinforcing bars are used for stirrups. The projecting tops of the stirrups will bond to the sitecast topping. Vertical twists of prestressing strand near the end of the girder will serve as lifting loops. (*Courtesy of Blakeslee Prestress, Inc.*)

FIGURE 15.12
A tilting table being used for the casting of a foam-insulated concrete wall panel. To the left, a concrete panel face with welded wire fabric reinforcing is being cast. Sheets of rigid foam insulation with wire ties can be seen in the center, and the welded wire fabric for the second panel face is seen at the right, with a pair of vertical bars and a pocket at the bottom being cast in as part of a system of connections. Notice the pipes at the left edge of the table for heating the mold to accelerate curing. (*Courtesy of Blakeslee Prestress, Inc.*)

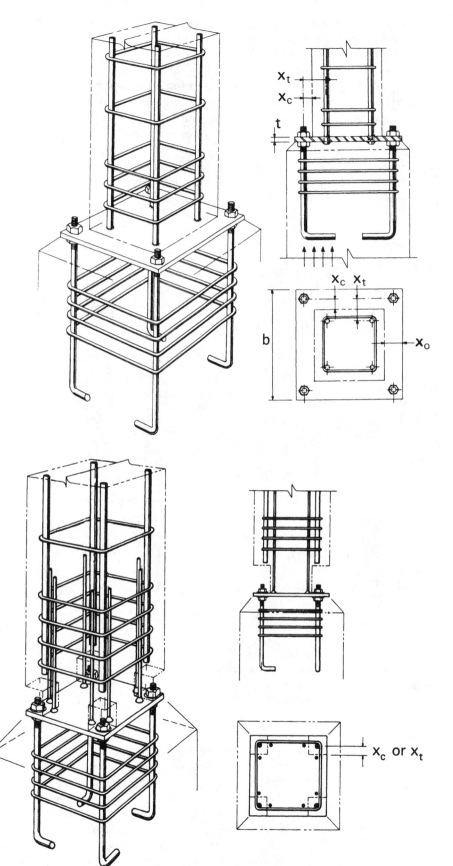

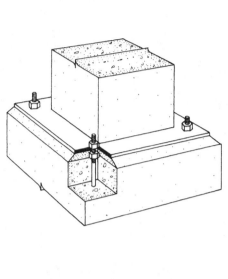

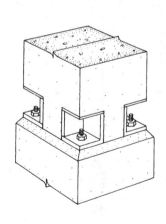

FIGURE 15.13
Some typical base details for precast concrete columns. The open corners of the lower detail are dry packed with stiff grout after the column has been aligned and bolted; this protects the metal parts of the connection from fire and corrosion. (*Courtesy of Precast/Prestressed Concrete Institute*)

grouting are all commonly employed in these connections. Connections can be posttensioned to produce continuous beam action at points of support (Figure 15.17). Exposed metal connectors that are not covered by topping are usually *dry packed* with stiff grout after being joined, to protect them from fire and corrosion.

The simplest joints in precast concrete construction are those that rely upon gravity by placing one element atop another, as is done where slab elements rest on a bearing wall or beam, or where a beam rests on the corbel of a column. *Bearing pads* are usually inserted between the concrete members at bearing points to avoid the grinding, concrete-to-concrete contact that might create points of high stress. Bearing pads also allow for expansion and contraction in the members. For solid and hollow-core slabs, these pads are strips of high-density plastic. Under elements with higher point loadings such as tees and beams, pads of synthetic rubber are used. For resistance to seismic and wind forces, the members in these simple joints must be tied together laterally. Slab elements are often joined over supports by reinforcing bars that are cast into the topping, or into the grout keys between the slabs, where no topping is used. Double tees and tees may

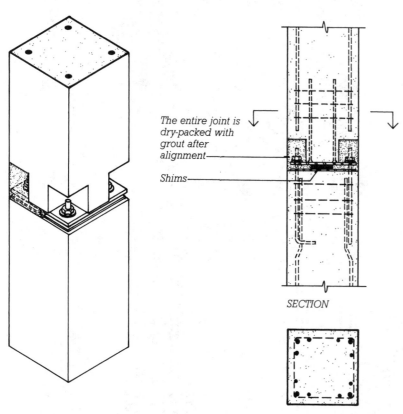

FIGURE 15.14
A bolted column-to-column connection detail. The metal shims support the upper column sections at the proper height until the grout cures.

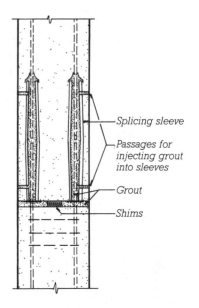

FIGURE 15.15
Section through a column-to-column connection using proprietary sleeves that are cast into the lower end of the upper column section. The sleeves are empty when the sections are assembled, except for the lower ends of the vertical bars from the upper column section, which reach down to the midheight of each sleeve. Projecting reinforcing bars from the lower column section reach into the lower halves of the sleeves. After the upper column section has been shimmed to exactly the right height, a fluid grout is injected into the sleeve to connect the reinforcing bars, and a stiff grout is dry packed between the ends of the columns.

additionally be connected by the welding together of steel plate inserts that have been embedded into the elements in the plant. The stems of these members are never welded to the supports but are left free to move on their bearing pads as the slab elements deflect under load (Figures 15.16–15.21).

Precast slab elements, especially solid slabs, occasionally require temporary shoring at midspan to help support the weight of the topping until it has cured. For construction economy, smooth-topped precast slab elements are sometimes used without topping, especially for roofs, where any unevenness between elements is bridged by rigid thermal insulation. Untopped slabs may also be used for floors that will be finished with a pad and carpet, and for parking garages. Untopped slabs require special connection details (Figures 15.18, 15.21).

Posttensioning can be used for combining large precast elements into even larger ones on the site. This is done to make very long, deep girders for bridges from precast concrete segments (Figures 15.24, 15.25), and to create tall shear walls from story-high precast panels in multistory buildings. In either case, ducts for the posttensioning tendons are placed accurately in the sections before casting, so that they will mate perfectly

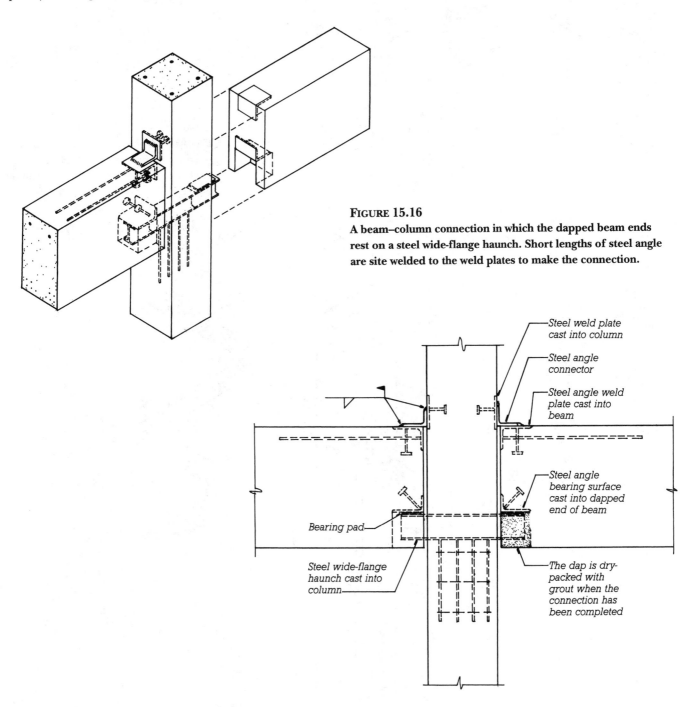

FIGURE 15.16
A beam–column connection in which the dapped beam ends rest on a steel wide-flange haunch. Short lengths of steel angle are site welded to the weld plates to make the connection.

Steel weld plate cast into column

Steel angle connector

Steel angle weld plate cast into beam

Steel angle bearing surface cast into dapped end of beam

Bearing pad

Steel wide-flange haunch cast into column

The dap is dry-packed with grout when the connection has been completed

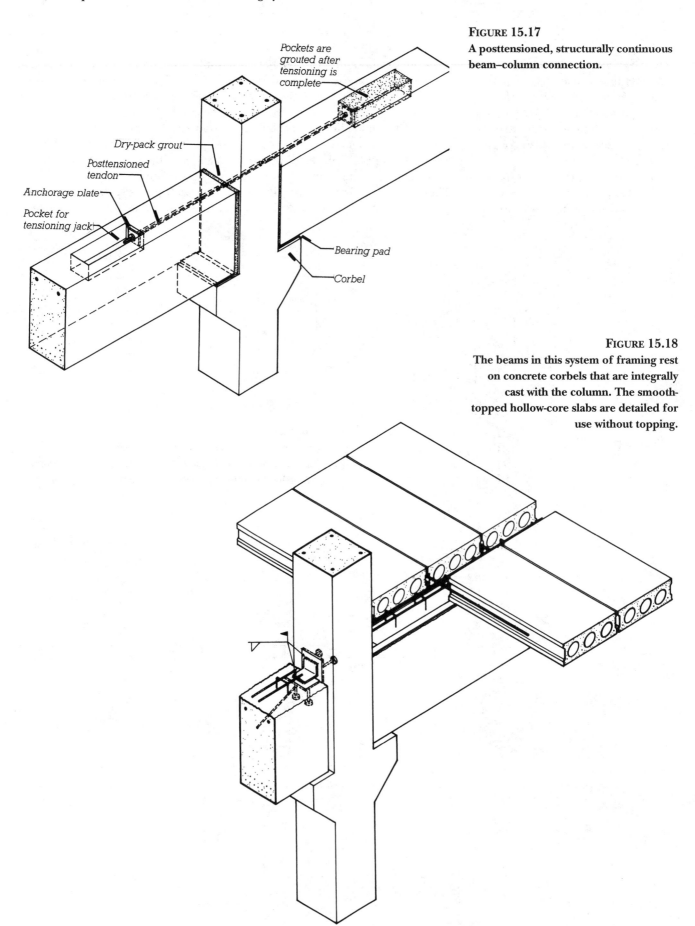

Pockets are grouted after tensioning is complete

Dry-pack grout

Posttensioned tendon

Anchorage plate

Pocket for tensioning jack

Bearing pad

Corbel

FIGURE 15.17
A posttensioned, structurally continuous beam–column connection.

FIGURE 15.18
The beams in this system of framing rest on concrete corbels that are integrally cast with the column. The smooth-topped hollow-core slabs are detailed for use without topping.

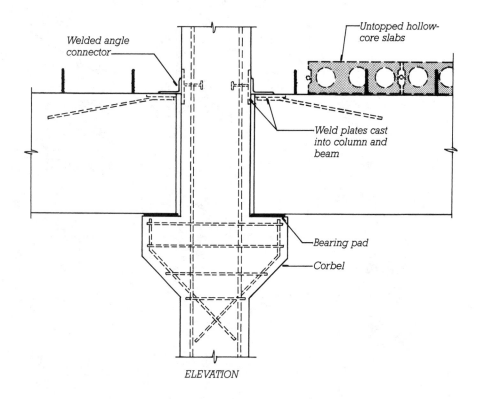

Welded angle connector

Untopped hollow-core slabs

Weld plates cast into column and beam

Bearing pad

Corbel

ELEVATION

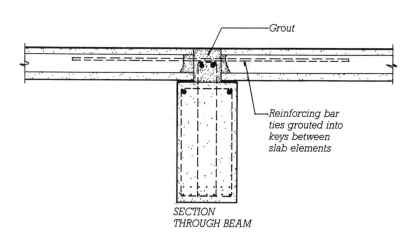

Grout

Reinforcing bar ties grouted into keys between slab elements

SECTION THROUGH BEAM

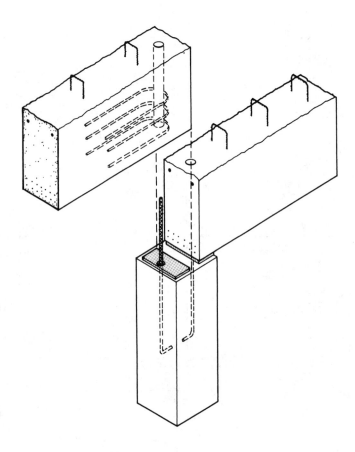

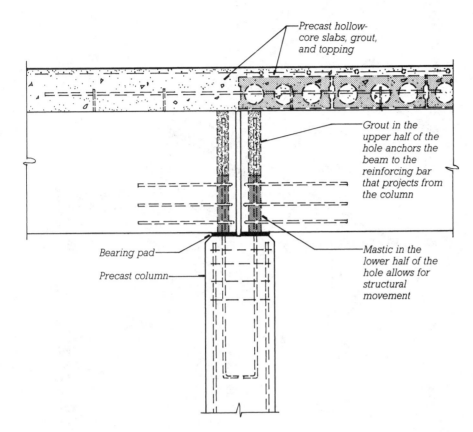

Precast hollow-core slabs, grout, and topping

Grout in the upper half of the hole anchors the beam to the reinforcing bar that projects from the column

Bearing pad

Precast column

Mastic in the lower half of the hole allows for structural movement

FIGURE 15.19
Topped hollow-core roof slabs supported on beams are joined to a column with vertical rods. The same beam–column connection can be used for floor beams resting on corbels.

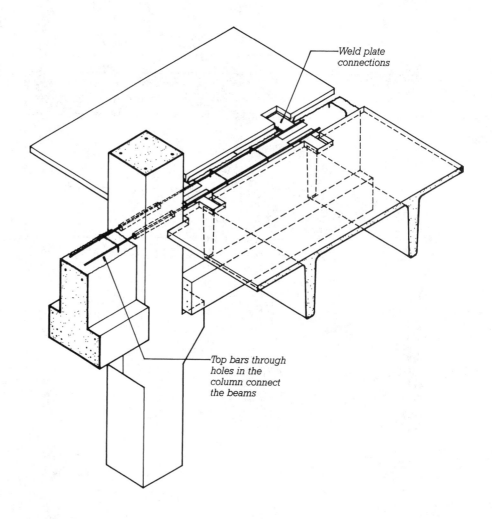

Weld plate
connections

Top bars through
holes in the
column connect
the beams

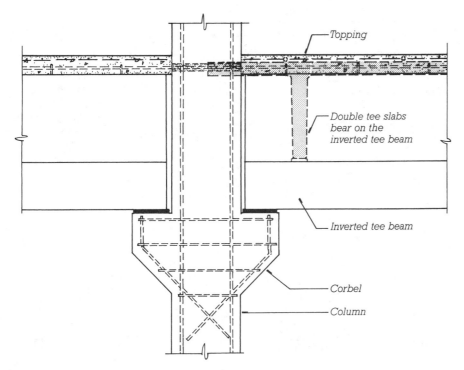

Topping

Double tee slabs bear on the inverted tee beam

Inverted tee beam

Corbel

Column

FIGURE 15.20
Topped double-tee floor slabs supported by inverted tee beams.

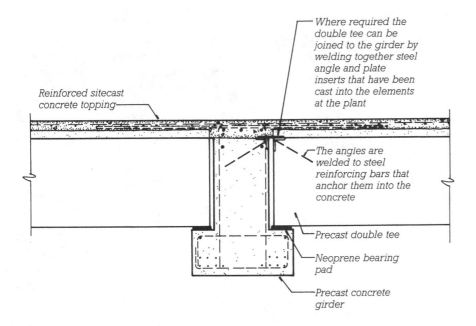

Reinforced sitecast concrete topping

Where required the double tee can be joined to the girder by welding together steel angle and plate inserts that have been cast into the elements at the plant

The angles are welded to steel reinforcing bars that anchor them into the concrete

Precast double tee

Neoprene bearing pad

Precast concrete girder

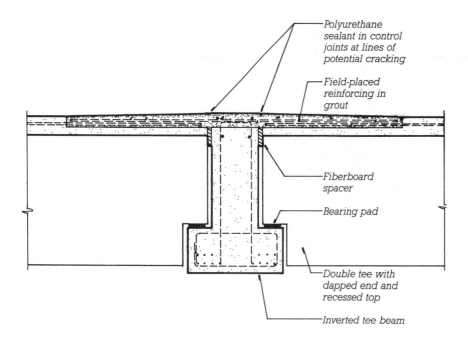

Polyurethane sealant in control joints at lines of potential cracking

Field-placed reinforcing in grout

Fiberboard spacer

Bearing pad

Double tee with dapped end and recessed top

Inverted tee beam

FIGURE 15.21
A minimum-headroom, minimum-cost floor system for parking garages uses untopped double tees. Refer to Figure 15.8 for photographs of how the ends of the tees are detailed for use in this system.

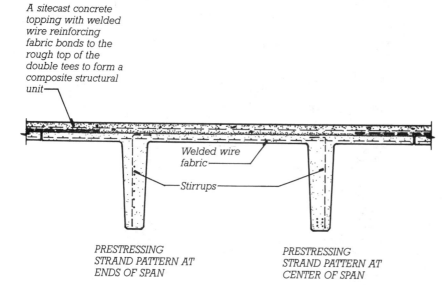

A sitecast concrete topping with welded wire reinforcing fabric bonds to the rough top of the double tees to form a composite structural unit

Welded wire fabric

Stirrups

PRESTRESSING STRAND PATTERN AT ENDS OF SPAN

PRESTRESSING STRAND PATTERN AT CENTER OF SPAN

FIGURE 15.22
A cross section through a topped double-tee slab. The prestressing strands are harped in the stems of the double tee. The stirrups are made of vertical planes of welded wire fabric.

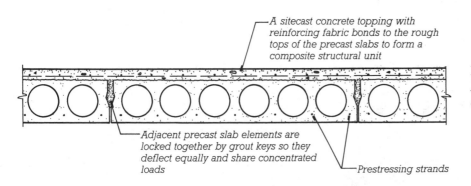

A sitecast concrete topping with reinforcing fabric bonds to the rough tops of the precast slabs to form a composite structural unit

Adjacent precast slab elements are locked together by grout keys so they deflect equally and share concentrated loads

Prestressing strands

FIGURE 15.23
A cross section through a topped hollow-core slab.

FIGURE 15.24
The Linn Cove Viaduct in Linnville, North Carolina, was built of short precast segments that were posttensioned together as they were placed to form a continuous box girder deck. A section is being placed by the derrick at the extreme right. The maximum clear span is 180 feet (55 m). (*Engineer: Figg and Muller Engineers, Inc. Photo courtesy of Precast/Prestressed Concrete Institute*)

FIGURE 15.25
The Linn Cove Viaduct was constructed with very little temporary shoring, so as to disrupt the natural landscape below as little as possible. The precast hollow box sections formed a girder that could be cantilevered for long distances during construction. The box profile is highly resistant to the torsional forces that are present in a curving beam. (*Photo courtesy of Precast/Prestressed Concrete Institute*)

FASTENING TO CONCRETE

As a concrete frame is finished, many things need to be fastened to it: exterior wall panels and facings; interior partitions; hangers for pipes, ducts, and conduits; suspended ceilings; stair railings; cabinets; machinery; and many more. Drawings *A* through *H* are examples of fastening systems that must be cast into the concrete. *A* is a familiar *anchor bolt*. *B* is a steel plate welded to a bent rod or strap anchor; this *weld plate* or *embed plate* furnishes a surface to which steel components can be welded. The steel angle in *C* has a threaded stud welded to it so that another component can be attached by bolting. *D* is an adjustable insert of malleable iron that is nailed to the formwork through the slots in the ears on either side. A special nut twists and locks into the slot to accept a bolt or threaded rod from below. A slightly different form of this type of insert is shown in the detail of a masonry shelf angle attachment in Chapter 20. *E* and *F* are two different designs of threaded inserts that are cast into the concrete. The dovetail slot in *G* is used with special anchor straps as shown to tie masonry facings to a sitecast concrete frame or wall. *H* is simply a dovetailed wood nailer strip cast into the concrete, a detail that is risky because the wood may swell and crack the concrete or shrink and become loose. *A* through *F* are heavy-duty devices that are available in

capacities sufficient to anchor heavy building components and machinery.

Details *I* through *P* depict fastening devices that are inserted in holes drilled into the cured concrete. Concrete is drilled fairly easily with carbide-tipped drill bits. *I* shows the steel post of a railing anchored into an oversize hole in a concrete slab using grout, poured lead, or epoxy; this system is also used to fasten bolts to concrete and can carry heavy loads if properly designed and installed. *J* is a plastic sleeve, *K* a wood or fiber plug, and *L* a lead sleeve; all three are inserted into drilled holes and expand to grip the sides of the hole when a screw is driven into them. *M* is a similar type of metal sleeve but has a special nail that expands the sleeve as it is driven. *N* is a special bolt with a steel sleeve over a tapered shank at the inner end. The sleeve catches against the concrete as the bolt is driven into the hole and is expanded by the taper as the bolt is tightened. *O* is a special screw and *P* is a special nail, both designed to grip tightly when inserted into drilled holes of the correct dimension. *J* through *P* are light-to-medium-duty fasteners, with the exception of *N*, which can carry rather heavy loads.

Q and *R* depict driven anchors. *Q* is the familiar concrete nail or masonry nail, made of hardened steel. If

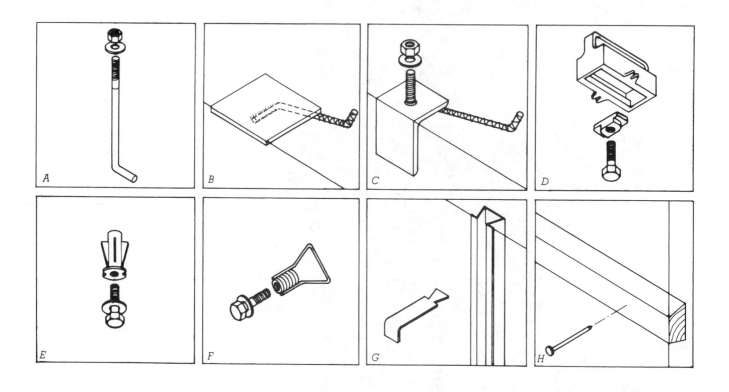

driven through a strip of wood with a few blows of a very heavy hammer, or inserted with a nail gun, it will penetrate concrete just enough to provide some shear resistance for furring strips and sleepers, but it has a tendency to loosen, particularly if driven with too many blows. Shown in *R* are three examples of *powder-driven* (often called *powder-actuated*) fasteners, which are driven into steel or concrete by an exploding cartridge of gunpowder. The first fastener is a simple pin used for attaching wood or sheet metal components to a wall or slab. The middle one is threaded to accept a nut. The eye on the fastener to the right allows a wire, such as a hanger wire for a sus-

pended ceiling, to be attached. Powder-driven fasteners are rapid to install, economical, and have a moderately high load-carrying capacity. *S* typifies devices whose perforated metal plates adhere securely with a mastic-type adhesive to surfaces of concrete or masonry. The fastener shown here has a thin sheet metal spike, over which a panel of foam plastic insulation can be impaled. The spike is then bent across the face of the insulation panel to hold it in place. Roof edge systems employing perforated plates and adhesives to fasten them to the building are illustrated in Chapter 16.

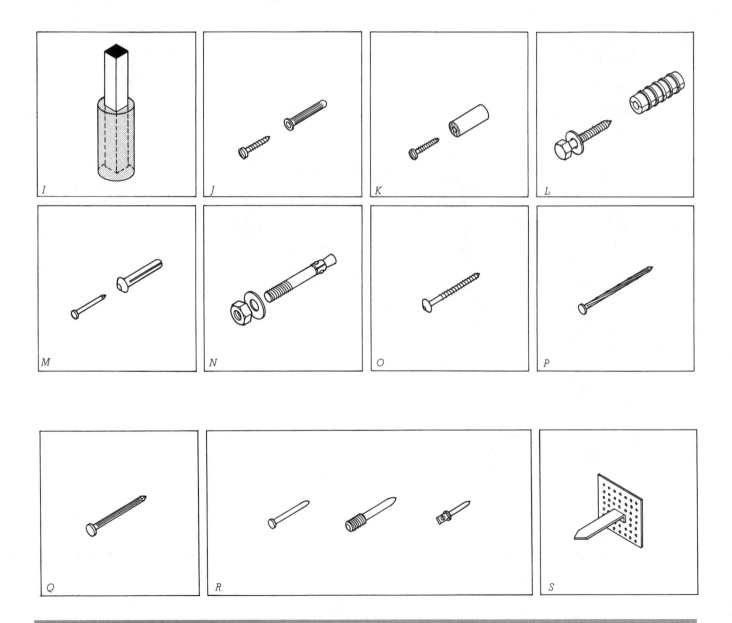

end to end when the sections are assembled on the site. After assembly, tendons are run through the ducts, horizontally in the case of girders or vertically in the case of shear walls, tensioned with portable hydraulic jacks, then grouted if required.

Joint design is the area in which precast concrete technology is developing most rapidly. New joining systems are patented each year, and, as grouts and adhesives develop further, there will be further simplifications and improvements of many kinds in precast concrete framing details.

THE CONSTRUCTION PROCESS

The construction process for precast concrete framing is directly parallel to that for steel framing. The structural drawings for the building are sent to the precasting plant, where engineers and drafters prepare shop drawings that show all the details of the individual elements and how they are to be connected. These drawings are reviewed by the engineer and architect for conformance with their design intentions and corrected as necessary. Then the production of the precast components proceeds, beginning with construction of any special molds that are required and fabrication of reinforcing cages, then continuing through cycles of casting, curing, and stockpiling as previously described. The finished elements, marked to designate their final positions in the building, are transported to the construction site as needed and placed by crane in accordance with erection drawings prepared by the precasting plant.

PRECAST CONCRETE AND THE BUILDING CODES

Precast concrete building frames and bearing wall panels can be designed to achieve any desired degree of fire resistance, depending on the amount of cover and topping specified. The building code table shown in Figure 1.3 indicates the fire resistance ratings in hours required for components of Types 1 and 2 construction. When the building type has been determined by the architect or engineer, the precaster can assist in determining how the necessary degrees of fire resistance can be achieved in each component of the building. Slab elements are readily available in 1- and 2-hour fire resistance ratings, and beams and columns in ratings ranging from 1 to 4 hours. The fire

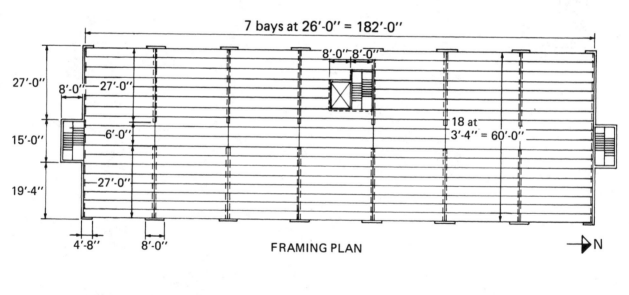

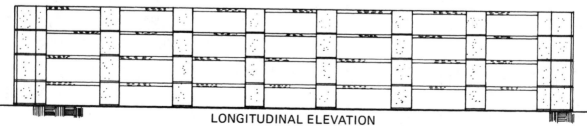

FIGURE 15.26

A framing plan and elevation of a simple four-story building made of loadbearing precast concrete wall panels and hollow-core slab elements

(*Courtesy of Precast/Prestressed Concrete Institute*)

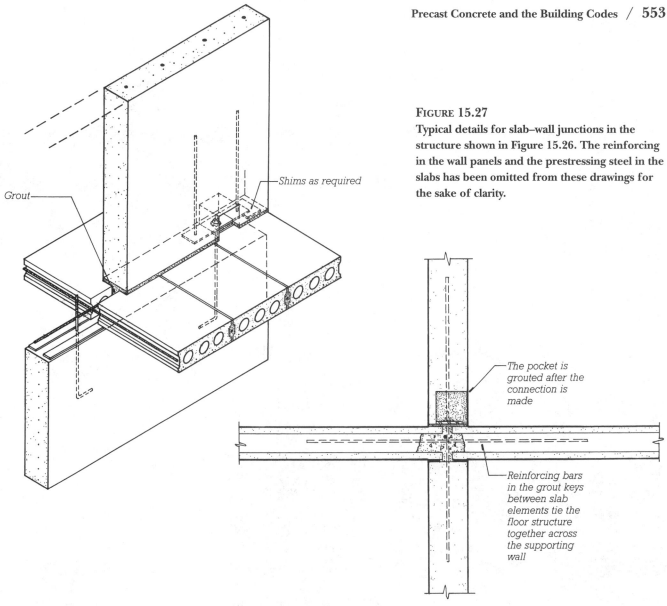

Grout

Shims as required

FIGURE 15.27
Typical details for slab–wall junctions in the structure shown in Figure 15.26. The reinforcing in the wall panels and the prestressing steel in the slabs has been omitted from these drawings for the sake of clarity.

The pocket is grouted after the connection is made

Reinforcing bars in the grout keys between slab elements tie the floor structure together across the supporting wall

FIGURE 15.28
A closer view of a building that uses precast concrete loadbearing wall panels. The rectangular pockets at the lower edges of the wall panels are for bolted connections of the type shown in Figure 15.27. Steel pipe braces are used to support the panels until all the connections have been made and the structure becomes self-stable. (*Courtesy of Blakeslee Prestress, Inc.*)

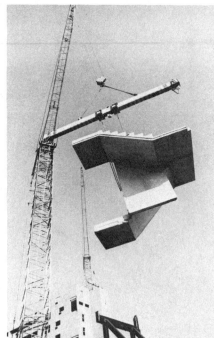

FIGURE 15.30
A crane hoists a single-piece precast stair
for a high-rise building of loadbearing
precast concrete wall panels. (*Courtesy of
Blakeslee Prestress, Inc.*)

FIGURE 15.29
Exterior loadbearing wall panels are often made of specially colored concrete, with
textured surfaces cast in. (*Courtesy of Blakeslee Prestress, Inc.*)

FIGURE 15.32
Erecting an upper-story column, using a
connection of the type illustrated in
Figure 15.14. (*Courtesy of Blakeslee
Prestress, Inc.*)

FIGURE 15.31
A precast column section is lifted from the ground by a crane.
(*Courtesy of Blakeslee Prestress, Inc.*)

resistance ratings of precast concrete slab elements may be increased by adding a topping and by increasing the thickness of the topping beyond the minimum. Solid and hollow-core slabs may achieve ratings as high as 3 hours by this means. Single and double tees require the addition of applied fireproofing material or a membrane ceiling fire protection beneath, in order to achieve a fire resistance rating higher than 2 hours.

THE UNIQUENESS OF PRECAST CONCRETE

Precast, prestressed concrete structural elements are crisp, slender in relation to span, precise, repetitive, and highly finished. They combine the rapid all-weather erection of structural steel framing with the self-fireproofing of sitecast concrete framing to offer singular functional qualities. Because precast concrete is the newest and least developed of the major framing materials for buildings, having been brought to market

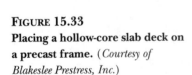

FIGURE 15.33
Placing a hollow-core slab deck on a precast frame. (*Courtesy of Blakeslee Prestress, Inc.*)

FIGURE 15.34
(*a*) **A crane lifts a column section from a flatbed truck to begin erection of a tennis stadium in New Haven, Connecticut.**
(*continued*)

(*a*)

only a few decades ago, its architectural aesthetic is just coming to maturity. Solid and hollow-core slabs have become an accepted part of our structural vocabulary in schools, hotels, apartment buildings, and hospitals, where they are ideal both functionally and economically. Engineers and architects have long been comfortable with precast concrete in longer-span building types, especially parking structures, warehouses, and industrial plants, where its awesome structural potential and efficient serial production of identical elements can be fully utilized and openly expressed. Now we are becoming increasingly successful in creating public buildings of the highest architectural quality that are built of precast concrete both inside and out (Figures 15.34–15.38). It is reasonable to expect that the most innovative of buildings in the coming years will be built of this sleek, sinewy, rapidly developing new material of construction.

(b)

(c)

(d)

FIGURE 15.34 *(continued)*
(*b*) Grouted sleeve connectors of the type shown in Figure 15.15 were used to join the precast sections of the bents that support the grandstand seating. (*c*) Stepped sections of grandstand floor await lifting. (*d*) A precast stair section is placed. (*e*) Precast hollow-core planks will be used for miscellaneous areas of floor. (*f*) The relationship of the stairs, bents, and stepped floor sections is evident in this photograph of the finished stadium. (*g*) The stadium in use, after a construction process that took only 11 months. Seating capacity is **15,000.** (*Structural engineer: Spiegel Zamecnik & Shah, Inc. Photos by Clark Broadbent. Courtesy of Blakeslee Prestress, Inc., P.O. Box 510, Branford, CT 06405, (203) 481-5306)*

(e)

(f)

FIGURE 15.35
An office building in Texas shows a carefully detailed frame of precast concrete.
(*Architects: Omniplan Architects. Photo courtesy of Precast/Prestressed Concrete Institute*)

FIGURE 15.36
Precast concrete girders span the roof of a paper mill in British Columbia.
(*Engineer: Swan Wooster Engineering Co., Ltd. Photo courtesy of Precast/Prestressed Concrete Institute*)

The Uniqueness of Precast Concrete / 561

(*Architect: Antoine Predock. Photo courtesy of Precast/Prestressed Concrete Institute*)

FIGURE 15.38
Highly customized precast concrete framing is used in this courthouse by architect **Eduardo Catalano.** (*Photo by Gordon H. Schenk, Jr. Courtesy of the architect*)

C.S.I./C.S.C.
Masterformat Section Numbers for Precast Concrete
Framing Systems

03400	**PRECAST CONCRETE**
03410	**Structural Precast Concrete—Plant Cast**
	Precast Concrete Hollow-Core Planks
	Precast Concrete Slabs
	Structural Precast Pretensioned Concrete—Plant Cast

SELECTED REFERENCES

1. Prestressed Concrete Institute. *PCI Design Handbook* (4th ed.). Chicago, 1992.

This is the major reference handbook for professionals engaged in designing precast, prestressed concrete buildings. It includes basic building assembly concepts, load tables for standard precast elements, engineering design methods, and a few suggested connection details. (Address for ordering: Precast/Prestressed Concrete Institute, 175 West Jackson Boulevard, Chicago, IL 60604.)

2. Prestressed Concrete Institute. *Design and Typical Details of Connections for Precast and Prestressed Concrete.* Chicago, 1988.

This engineering design manual includes a large collection of drawings of standard connection details. (Address for ordering: see reference 1.)

3. Prestressed Concrete Institute. *Recommended Practice for Erection of Precast Concrete.* Chicago, 1985.

The 87 pages of this manual describe and illustrate the best ways of hoisting and assembling the elements of a precast concrete building. (Address for ordering: see reference 1.)

4. Phillips, William R., and David A. Sheppard: Prestressed Concrete Institute. *Plant Cast Precast and Prestressed Concrete: A Design Guide.* New York, McGraw–Hill, 1989.

A comprehensive text on precast, prestressed concrete, this book covers production, transportation, erection, and engineering design procedures.

KEY TERMS AND CONCEPTS

precast concrete	double tee	inverted tee	corbel
steam curing	single tee	AASHTO girder	dry pack
solid slab	topping	casting bed	bearing pad
hollow-core slab	L-shaped beam	weld plate	dap

REVIEW QUESTIONS

1. In what circumstances might a designer choose a precast concrete framing system over a sitecast system? In what circumstances might a sitecast system be favored?

2. Why are precast concrete structural elements usually cured with steam?

3. Explain several methods of producing hollow-core slabs.

4. Diagram from memory several different ways of connecting precast concrete beams to columns.

5. Diagram from memory a method of connecting a pair of untopped double-tee slabs to an inverted tee beam. Then work out a similar way of connecting a double tee to an L-shaped beam, as would occur around the perimeter of the same building.

EXERCISES

1. Design a simple two-story rectangular warehouse, 90 feet by 180 feet (27 m by 54 m), using precast concrete for at least the floor and roof structure, and perhaps for the walls as well. Use the preliminary structural design information given in this chapter to help determine the column spacing, the types of elements to use, and the depths of the elements. Draw a framing plan and typical connections for the building.

2. Locate a concrete precasting plant in your area and arrange a visit to view the production process. If possible, arrive at the plant early in the morning when the strands are being cut and the elements are being lifted from the molds.

3. Learn from the management of the precasting plant where there is a precast concrete building being erected, then visit the building site. Try to trace a typical precast concrete structural element from raw materials through precasting, transporting, and erecting. Are there ways in which this process could be made more efficient? Sketch a few of the typical connections being used in the project.

16

ROOFING

- **Low-Slope Roofs**

  Roof Decks

  Thermal Insulation and Vapor Retarder

  Rigid Insulating Materials for Low-Slope Roofs

  Vapor Retarders for Low-Slope Roofs

  The Low-Slope Roof Membrane

  Ballasting and Traffic Decks

  Edge and Drainage Details for Low-Slope Roofs

  Structural Panel Metal Roofing for Low-Slope Roofs

- **Steep Roofs**

  Shingles

  Architectural Sheet Metal Roofing

- **Roofing and the Building Codes**

Architect Frank Gehry clad not only the roofs but also the walls of the entrance structure of the Weisman Art Museum at the University of Minnesota in stainless steel flat-seam sheet metal roofing. (*Photo by Don F. Wong*)

A building's roof is its first line of defense against the weather. The roof protects the interior of the building from rain, snow, and sun. The roof helps to insulate the building from extremes of heat and cold and to control the accompanying problems with condensation of water vapor. And like any front-line defender, it must itself take the brunt of the attack: A roof is subject to the most intense solar radiation of any part of a building. At midday, the sun broils a roof with radiated heat and ultraviolet light. On clear nights, a roof radiates heat to the blackness of space and becomes colder than the surrounding air. From noon to midnight of the same day, it is possible for the surface temperature of a roof to vary from near boiling to below freezing. In cold climates, snow and ice cover a roof after winter storms, and cycles of freezing and thawing gnaw at the materials of the roof. A roof is vital to the sheltering function of a building, yet it is singularly vulnerable to the destructive forces of nature.

Roofs can be covered with many different materials. These can be organized conveniently into two groups: those that work on *steep roofs* and those that work on *low-slope roofs*, roofs that are nearly flat. The distinction is important: A steep roof drains itself quickly of water, giving wind and gravity little opportunity to push or pull water through the roofing material. Therefore, steep roofs can be covered with roofing materials that are fabricated and applied in small, overlapping units—*shingles* of wood, slate, or artificial composition; *tiles* of fired clay or concrete; or even tightly wrapped bundles of reeds, leaves, or grasses (Figures 16.1, 16.2). There are several advantages to these materials: Many of them are inexpensive. The small, individual units are easy to handle and install. Repair of localized damage to the roof is easy. The effects of thermal expansion and contraction, and of movements in the structure that supports the roof, are minimized by the ability of the small roofing units to move with respect to one another. Water vapor vents itself easily from the interior of the building through the loose joints in the roofing material. And a steep roof of well-chosen materials skillfully installed can be a delight to the eye.

Low-slope roofs have none of these advantages. Water drains relatively slowly from their surfaces, and small errors in design or construction can cause them to trap puddles of standing water. Slight structural movements can tear the membrane that keeps the water out of the building. Water vapor pressure from within the building can blister and rupture the membrane. But low-slope roofs also have overriding advantages: A low-slope roof can cover a building of any horizontal dimension, whereas a steep roof becomes uneconomically tall when used on a very broad building. A building with a low-slope roof has a much simpler geometry that is often much less expensive to construct. And low-slope roofs, when appropriately detailed, can serve as balconies, decks, patios, and even landscaped parks.

FIGURE 16.1
A steep roof can be made waterproof with any of a variety of materials. This thatched roof is being constructed by fastening bundles of reeds to the roof structure in overlapping layers, in such a way that only the butts of the reeds are left exposed to the weather.

FIGURE 16.2
The finished thatched roof has gently rounded contours and a pleasing surface texture. The decorative pattern of the ridge cap is the unique signature of the thatcher who made the roof. (*Photos of thatched roofs courtesy of Warwick Cottage Enterprises, 2944 Greenhedge, Anaheim, California*)

LOW-SLOPE ROOFS

A low-slope roof (often referred to, inaccurately, as a flat roof) is defined as one whose slope is less than 3 : 12, or 25%. A low-slope roof is a complex, highly interactive assembly made up of several components. The

deck is the structural surface that supports the roof. *Thermal insulation* is installed to slow the passage of heat into and out of the building. A *vapor retarder* is essential in colder climates to prevent moisture from accumulating within the insulation. The *membrane* is the impervious sheet of material that keeps water out of the

building. *Drainage* components remove the water that runs off the membrane. Around the membrane's edges and wherever it is penetrated by pipes, vents, expansion joints, electrical conduits, or roof hatches, special *flashings* and details must be designed and installed to prevent water penetration.

Roof Decks

Previous chapters of this book have presented the types of structural decks ordinarily used under low-slope roofs: plywood or OSB panels over wood joists, solid wood decking over heavy timber framing, corrugated steel decking, panels of wood fiber bonded together with portland cement, poured gypsum over insulating formboard, sitecast concrete slab, and precast concrete slab. For a durable low-slope roof installation, it is important that the deck be adequately stiff under expected roof loadings and fully resistant to wind uplift forces. The deck must slope toward drainage points at an angle sufficient to drain reliably despite the effects of structural deflections. A slope of at least $\frac{1}{4}$ inch per foot of run (1 : 50) is recommended. To create the slope in a low-slope roof, the beams that support the deck are often sloped by shortening some of the columns. Alternatively, a tapered fill of lightweight insulating concrete or lightweight aggregate with an asphaltic binder may be poured over a dead-level structural deck, or a system of tapered boards of rigid insulation may be laid over the deck. If a roof is insufficiently sloped, puddles of water will stand for extended periods of time in the low spots that are inevitably present, leading to premature deterioration of the roofing materials in those areas. If water accumulates in low spots caused by structural deflections, progressive structural collapse becomes a possibility, with deepening puddles attracting more and more water during

rainstorms and becoming heavier and heavier, until the beams or joists become loaded to the point of failure (Figure 16.3).

If it is large in extent, the deck should be provided with enough movement joints to control the effects of expansion and contraction on the roof membrane. If the building separation joints in the structure of the building are too far apart to satisfy the requirements of the membrane, *area dividers,* which are much like building separation joints but do not extend below the surface of the roof deck, may be installed (Figure 16.29).

The roof membrane must be laid over a smooth surface. A wood deck that is to receive a roof membrane should have no large gaps or knotholes. A sitecast concrete deck should be troweled smooth, and a precast concrete plank deck, if not topped with a concrete fill, must be grouted at junctions between planks to fill the cracks and form a smooth surface. A corrugated steel deck must be covered with either wood panels or insulating boards to bridge the flutes in the deck and create a smooth surface.

It is extremely important that the deck be dry at the time roofing operations commence, to avoid later problems with water vapor trapped under the membrane. A deck should not be roofed when rain, snow, or frost is present in or on the deck material. Concrete decks and insulating fills must be fully cured and thoroughly air dried.

Thermal Insulation and Vapor Retarder

Thermal insulation for a low-slope roof may be installed in any of three positions: below the structural deck, between the deck and the membrane, or above the membrane (Figure 16.4).

Insulation Below the Deck

Below the deck, fibrous batt insulation is installed above a vapor retarder, either between framing members or on top of a suspended ceiling assembly. A ventilated airspace should be provided between the insulation and the deck to dissipate stray water vapor. Insulation in this position is relatively economical and trouble-free, but it leaves both the deck and the membrane exposed to the full range of outdoor temperature fluctuations.

Insulation Between the Deck and the Membrane

The traditional position for low-slope roof insulation is between the deck and the membrane. Insulation in this

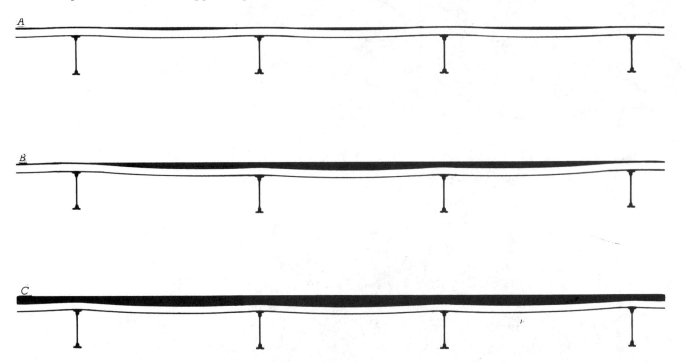

FIGURE 16.3
A low-slope roof with insufficient pitch to drain is subject to structural failure through progressive collapse, as demonstrated in this sequence of cross sections: (*a*) Water stands on the roof in puddles, its weight causing slight deflections of the roof deck between supporting beams or joists. (*b*) If heavy rainfall continues, the puddles grow and join, and the accumulating weight of the water begins to cause serious deflections in the supporting structural elements. The deflections encourage water from a broader area of the roof to run into the puddle. (*c*) As structural deflections increase, the depth of the puddle increases more and more rapidly, until the overloaded structure collapses.

position must be in the form of low-density rigid panels or lightweight concrete in order to support the membrane. The insulation protects the deck from temperature extremes and is itself protected from the weather by the membrane. But the roof membrane in this type of installation is subjected to extreme temperature variations, and any water or water vapor that may accumulate in the insulation is trapped beneath the membrane, which can lead to decay of the insulation and roof deck, and blistering and eventual rupture of the membrane from vapor pressure. (See the discussion on pages 572–574 for an explanation of insulation and vapor problems.)

Two precautions are advisable in cold climates for insulation that is located between the deck and the membrane: A vapor retarder should be installed below the insulation, and the insulation should be ventilated to allow the escape of any moisture that may accumulate there. Ventilation is accomplished through the installation of *topside vents,* one per thousand square feet (100 m²) or so, that allow water vapor to escape upward through the membrane (Figures 16.5, 16.6). Topside vents are most

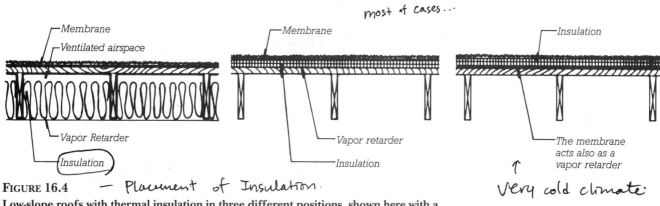

most of cases...

FIGURE 16.4 — *Placement of Insulation.*

very cold climate

Low-slope roofs with thermal insulation in three different positions, shown here with a wood joisted roof deck. At left, insulation below the deck, with a vapor retarder on the warm side of the insulation. In the center, insulation between the deck and the membrane, with a vapor retarder on the warm side of the insulation. At right, a protected membrane roof, in which the insulation is above the membrane.

FIGURE 16.5
Topside roof vents are being installed to release vapor pressure that may build up beneath a roof membrane. (*Courtesy of Manville Corporation*)

effective with a loose-laid membrane, which allows trapped moisture to work its way toward the vents from any part of the insulating layer.

Insulation Above the Membrane: The Protected Membrane Roof

Insulation above the roof membrane is a relatively new concept. It offers two major advantages: The membrane is protected from extremes of heat and cold, and the membrane is on the warm side of the insulation, where it is immune to vapor blistering problems. Because the insulation itself is exposed to water when placed above the membrane, the insulating material must be one that retains its insulating value when wet and does not decay or disintegrate. Extruded polystyrene foam board is the one material that has all these qualities (Figure 16.7). The insulating board is either laid loose or embedded in a coat of hot asphalt to adhere it to the membrane below. It is held down and protected from sunlight (which disintegrates polystyrene) by a layer of *ballast*. The ballast may consist of crushed stone, a thin concrete layer factory laminated to the upper surface of the insulating board, or interlocking concrete blocks (Figures 16.8, 16.15, 16.23). Critics of this *protected membrane roof (PMR)* system originally predicted that the membrane would disintegrate quickly because of its continual exposure to dampness trapped under and around the insulating boards, but experience of more than 20 years has shown that the membrane ages little when thus protected from sunlight and temperature extremes, despite the presence of moisture.

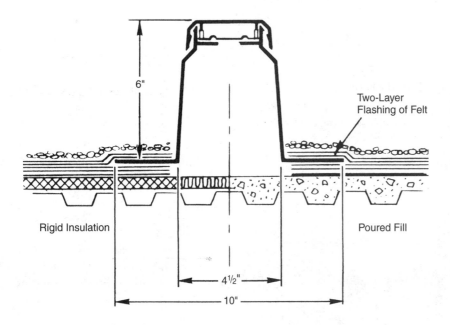

Rigid Insulation

Two-Layer Flashing of Felt

Poured Fill

6"

4½"

10"

ROOF VENT

FIGURE 16.6
This proprietary topside roof vent, made of molded plastic with a synthetic rubber valve, allows moisture vapor to escape from beneath the membrane, but closes automatically to prevent water or air from entering.
(*Courtesy of Manville Corporation*)

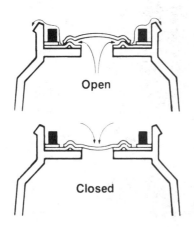

Open

Closed

FIGURE 16.7
Installing extruded polystyrene foam insulation over a roof membrane to create a protected membrane roof. (*Courtesy of Dow Chemical Company*)

FIGURE 16.8
This proprietary system of 2-inch-thick (50-mm) polystyrene foam insulation for a protected membrane roof is topped with a 3/8-inch (9-mm) layer of latex-modified concrete. The concrete protects the foam from sunlight and wear, and also ballasts it to prevent it from lifting off the roof in high winds. (*Photo courtesy of T. Clear Protected Membrane Roof System*)

THERMAL INSULATION
AND VAPOR RETARDER

Thermal Insulation

Thermal insulation is any material that is added to a building assembly for the purpose of slowing the conduction of heat through that assembly. Insulation is almost always installed in new roof and wall assemblies in North America, and often in floors and around foundations and concrete slabs on grade, anywhere that heated or cooled interior space comes in contact with unheated space, the outdoors, or the earth.

The effectiveness of a building assembly in resisting the conduction of heat is expressed in terms of its *thermal resistance,* abbreviated as R. R is expressed either in English units as square foot-hour-degree Fahrenheit per BTU or in metric units of square meter-degree Celsius per watt. The higher the *R-value,* the higher the insulating value.

Every component of a building assembly contributes in some measure to its overall thermal resistance. The amount of the contribution depends on the amount and type of material. Metals have very low R-values, and concrete and masonry materials are only slightly better. Wood has a substantially higher thermal resistance, but not nearly as high as that of an equal thickness of any of the common insulating materials. Most of the thermal resistance of any insulated building assembly is attributable to the insulating material.

In wintertime, it is warm inside a building and cold outside. The inside surface of a wall or roof assembly is warm, and the outside surface is cold. Between the two surfaces, the temperature varies between the indoor temperature and the outdoor temperature according to the thermal resistances of the various layers of the assembly. The largest temperature difference within the assembly is between the inner and outer surfaces of the insulation (Figure A).

Water Vapor and Condensation

Water exists in three different physical states, depending on its temperature and pressure: solid (ice), liquid, and vapor. *Water vapor* is an invisible gas. The air always contains water vapor. The higher the temperature of the air, the more water vapor it is capable of containing. At a given temperature, the amount of water vapor the air actually contains, divided by the maximum amount of water vapor it could contain, is the *relative humidity* of the air. Air at 50 percent relative humidity contains half as much water vapor as it could contain at the given temperature.

If a mass of air at 50 percent relative humidity is cooled, its relative humidity rises. The amount of water vapor in the air mass has not changed, but the ability of the air mass to contain water vapor has diminished because the air has become cooler. If the cooling of the air mass continues, a temperature will be reached at which the humidity is 100 percent. This temperature is known as the *dew point.* The dew point is different for

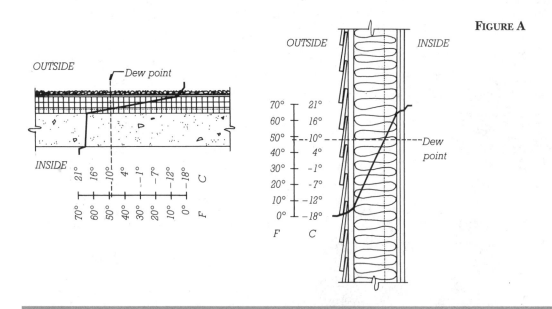

FIGURE A

every air mass. A roomful of very humid air has a high dew point, which is another way of saying that the air in the room would not have to be cooled very much before it would reach 100 percent humidity. A roomful of dry air has a low dew point; it can undergo considerable cooling before it reaches saturation.

When a mass of air is cooled below its dew point, it can no longer retain all its water vapor. Some of the vapor is converted to liquid water, usually in the form of fog droplets. The farther the air mass is cooled below its dew point, the more fog will be formed. The process of converting water vapor to liquid by cooling is called *condensation*.

Condensation takes place in buildings in many different ways. In winter, room air circulating against a cold pane of glass is cooled to below its dew point, and a fog of water droplets forms on the glass. If the air is very humid, the droplets will grow in size, then run down the glass to accumulate in puddles on the window sill. If the glass is very cold, the condensate will freeze as it forms and create patterns of ice crystals. On a hot, humid summer day, the moisture in the air in the vicinity of a cold water pipe or a cool basement wall will condense in a similar fashion.

In an insulated wall or roof assembly, condensation can become a serious problem under wintertime heating conditions. The air inside the building is at a higher temperature than the air outside and usually contains much more water vapor, especially in densely populated areas of the building or where cooking, wet industrial processes, bathing, or washing take place. Indoor air leaking through the assembly toward the outside becomes progressively cooler and reaches its dew point somewhere inside the assembly, almost always within the thermal insulation. Where air itself does not leak through the assembly, water vapor still migrates from indoors to outdoors, driven by the difference in vapor pressure between the moist indoor air and the drier air outdoors, and also reaches its dew point somewhere within the insulation. In both cases, the result is that the insulation becomes wet, and portions of it may become frozen with ice. The insulating value is lost. The materials of the roof or wall become wet and subject to rust or decay. Water may accumulate to such an extent that it runs or drips out of the assembly and spoils the finishes or contents of the building. And when the outside of the assembly is heated by warmer outdoor air or bright sunlight, the water begins to vaporize and move toward the outdoors. As it does so, its vapor pressure can raise blisters in paint films or roof membranes, blisters that can rupture like an overinflated balloon if the pressure is not relieved in time. (Most cases of peeling paint on wooden buildings are caused not by a poor painting job, but by moisture in the wood or lack of a vapor retarder on the warm side of the wall.)

Under summertime cooling conditions in hot, humid weather, the flow of water vapor through building assemblies can be reversed, as moisture moves from the warm, damp outside air toward the cooler, drier air within. This condition is not usually as severe as the winter condition because the differences in temperature and humidity between indoors and outdoors are not as great. And in most areas of North America, the cooling season is short compared to the heating season, allowing the designer to neglect the summer vapor problem and concentrate on the winter problem. But in areas of the American South, the summer problem is more severe than the winter problem, and must be solved first.

The Vapor Retarder

To prevent condensation inside building assemblies, a *vapor retarder* (often called, inexactly, a *vapor barrier*) is installed on the warmer side of the insulation layer. This is a continuous sheet, as nearly seamless as possible, of plastic sheeting, aluminum foil, kraft paper laminated with asphalt, roofing felt laminated with asphalt, troweled mastic, or some other material that is highly resistant to the passage of water vapor. The effect of the vapor retarder is to diminish the flow of air and vapor through the building assembly, preventing the moisture from reaching the point in the assembly where it would condense. For buildings in most parts of this continent, the vapor retarder should be placed on the inside of the insulation. In humid areas where warm weather cooling is the predominant problem, these positions should be reversed. In some mild climates, the vapor retarder may not be required at all.

The part of the building assembly that lies on the cooler side of the vapor retarder should be allowed to "breathe," to ventilate freely by means of attic ventilation, topside roof vents, or vapor-permeable exterior materials. This helps prevent stray moisture from becoming trapped between the vapor retarder and another near-impermeable surface.

The performance of a vapor retarder material is measured, in the United States, in *perms*. A perm is defined by ASTM Standard E96 as the passage of one grain of water vapor per hour through one square foot of material at a pressure differential of one inch of mercury between the two sides of the material. In Canada, the CSI perm is measured in terms of one nanogram per second per square meter per pascal of pressure difference. One CSI perm is equal to 57.38 U.S. perms (Figure B). For vapor retarders below low-slope roofs, the National Roofing Contractors Association of the United States recommends a perm rating approaching zero; a well-constructed vapor retarder of asphalt and felt, the best type in general use, has a rating of about 0.005 U.S. perms, and single-ply vapor retarders of foil or kraft paper range up to about 0.30,

which is satisfactory for most purposes. For walls and steep roofs, polyethylene films have U.S. perm ratings of 0.16 for a 0.002-inch thickness, and 0.08 for a 0.004-inch thickness. (For metric and CSI units, see Figure B.)

For a more complete discussion of thermal insulation and vapor retarders in buildings, consult Stein, Benjamin, and John S. Reynolds, *Mechanical and Electrical Equipment for Buildings* (8th ed.), New York, John Wiley & Sons, Inc., 1992.

	U.S. Perms	CSI Perms	**FIGURE B**
Aluminum foil			
1 mil (0.025 mm)	0.0	0.0	
Built-up roofing	0.0	0.0	
Polyethylene			
10 mil (0.25 mm)	0.03	0.0005	
4 mil (0.10 mm)	0.08	0.0014	
2 mil (0.05 mm)	0.16	0.0028	
PVC, plasticized			
4 mil (0.10 mm)	0.8–1.4	0.014–0.024	
Interior primer plus one coat flat oil paint on plaster	1.6–3.0	0.028–0.052	
Exterior oil paint, three coats on wood	0.3–1.0	0.005–0.017	
Hot melt asphalt			
2 oz/ft² (0.6 kg/m²)	0.5	0.009	
3.5 oz/ft² (1.1 kg/m²)	0.1	0.002	
Brick masonry			
4″ (100 mm) thick	0.8	0.014	
Plaster on metal lath			
³⁄₄″ (19 mm) thick	15	0.26	
Gypsum wallboard			
³⁄₈″ (9.5 mm) thick	50	0.87	
Plywood, exterior glue			
¼″ (6 mm) thick	0.7	0.012	

Rigid Insulating Materials for Low-Slope Roofs

An insulating material for low-slope roofs should have a high thermal resistance; adequate resistance to denting and gouging, moisture decay, and fire; and the ability to accept a coating of hot asphalt without melting or dissolving. No single material has all these virtues. Some rigid insulating materials commonly used on low-slope roofs in North America are listed in Figure 16.9, along with a summary of the advantages and disadvantages of each. The best choice is often a composite insulating material, which combines layers of two or more materials to exploit the best qualities of each. A composite insulating board for installation below the membrane might include a lower layer of polyisocyanurate foam for its high insulating value, and an upper layer of perlitic board for its resistance to attack by hot bitumens.

If rigid insulating boards are located below the roof membrane, they may be adhered to the deck with hot asphalt or fastened to the deck

FIGURE 16.9

A comparative summary of some rigid insulating materials for low-slope roofs.

	Composition	Advantages	Disadvantages
Cellulose fiber board	A rigid, low-density board of wood or sugar cane fibers and a binder	Economical	Lower insulating efficiency than plastic foams; susceptible to absorption of moisture
Glass fiber board	A rigid, low-density board of glass fibers and a binder	Inert; fire resistant; non-decaying; dimensionally stable; ventilates moisture freely	Lower insulating efficiency than plastic foams
Polystyrene foam board	A closed-cell rigid foam of polystyrene plastic	High insulating efficiency; resistant to moisture	Combustible; high coefficient of thermal expansion; recommended for use only in protected membrane roofs
Polyurethane foam board	A closed-cell rigid foam of polyurethane, sometimes with saturated felt facings	Very high insulating efficiency	Combustible; high coefficient of thermal expansion; best when combined with other materials to increase its resistance to fire and hot bitumens; loses some of its insulating value over time
Polyisocyanurate foam board	A closed-cell rigid foam of polyisocyanurate, sometimes with glass fiber reinforcing and saturated felt facings	Very high insulating efficiency	Best when combined with other materials to increase its resistance to fire and hot bitumens
Perlitic board	Granules of expanded volcanic glass and a binder pressed into a rigid board	Inert; fire resistant; compatible with hot bitumens, dimensionally stable	Much lower insulating efficiency than plastic foams
Lightweight concrete fill, Lightweight gypsum fill	Concretes made from very lightweight mineral aggregates (perlite or vermiculite), a cementing agent (portland cement or gypsum), and a high volume of entrained air	Can easily produce a tapered insulation layer for positive roof drainage	Much lower insulating efficiency than plastic foams; residual moisture from mixing water can cause blistering of membrane
Lightweight fill with asphaltic binder	Lightweight mineral aggregate with asphaltic binder	Can easily produce a tapered insulation layer for positive roof drainage; does not introduce mixing water under membrane	Much lower insulating efficiency than plastic foams
Composite insulating boards	Sandwich layers of foam plastic and other materials such as perlite board, glass fiber board, and saturated felt	Combine the high insulating efficiency and moisture resistance of foam plastics with the fire resistance, structural rigidity, and/or bitumen compatibility of other materials	

mechanically with nails, screws, or any of a broad selection of fasteners made especially for the purpose. Mechanical fasteners are favored by insurance companies because they are more secure against wind uplift (Figures 16.10, 16.11).

Poured lightweight gypsum and concrete *deck fill insulations* are economical. They may be applied directly to corrugated steel decking and rough concrete decks and can easily be tapered during installation to slope toward points of roof drainage. Thermal resistances per inch are not as high for these materials as for plastic foams, so greater thicknesses are required to achieve similar insulating values. Poured fill insulations contain large amounts of free water at the time they are placed. They should be cured and dried as thoroughly as possible before application of the membrane, and topside vents should be installed to allow the escape of moisture vapor from the insulation during the life of the roof.

Vapor Retarders for Low-Slope Roofs

The membrane in a protected membrane roof serves also as the vapor retarder. In other low-slope roof constructions, a separate vapor retarder is advisable except in warm, humid climates where wintertime condensation is not a problem and summertime air conditioning can cause water vapor to migrate inward through the roof.

The most common type of vapor retarder for a low-slope roof consists of two layers of asphalt-saturated roofing felt bonded together and adhered to the roof deck with hot asphalt. Proprietary vapor retarder sheets of various materials can be equally satisfactory and more economical. Polyethylene sheeting, which makes an excellent vapor retarder in many other types of construction, is seldom used in low-slope roofs because it melts at the application temperature of the hot bitumen used in most roof membranes.

FIGURE 16.10
Workers bed rigid insulation boards in strips of hot asphalt over a corrugated metal roof deck. (*Courtesy of GAF Corporation*)

FIGURE 16.11
Screws and large sheet metal washers attach insulation more securely to a metal deck. (*Courtesy of GAF Corporation*)

The vapor retarder must be located at such a point in the roof assembly that it will always be warmer than the dew point of the interior air under any conceivable condition of use. Usually this means putting the vapor retarder below the insulation, but this is not always possible. A vapor retarder installed directly over a corrugated steel deck, for example, would have to bridge across the corrugations and would be unusually vulnerable to damage until it was covered by insulation. In this case, a thin layer of roof insulation board is laid over the deck, then the vapor retarder, followed by a thicker layer of insulation board. The designer must carefully calculate the dew point location in this assembly to be sure that the vapor retarder lies below it, else condensation of moisture can occur within the lower layer of insulation.

The Low-Slope Roof Membrane

The membranes used for low-slope roofing fall into three general categories: the built-up roof membrane (BUR), the single-ply roof membrane, and the fluid-applied roof membrane.

The Built-Up Roof Membrane

A *built-up roof membrane (BUR)* is assembled in place from multiple layers of asphalt-impregnated *felt* bedded in *bitumen* (Figures 16.12–16.14). The felt fibers may be cellulose, glass, or synthetic. The felt is saturated with asphalt at the factory and delivered to the site in rolls. The bitumen is usually asphalt derived from the distillation of petroleum, but for inverted, dead-level, or very low-slope roofs, coal-tar pitch is used instead, because of its greater resistance to standing water. Both asphalt and coal-tar pitch are applied hot in order to merge with the saturant bitumens in the felt and form a single-piece membrane. The felt is laminated in overlapping layers (*plies*) to form a membrane that is two to four plies thick. The more plies used, the more durable

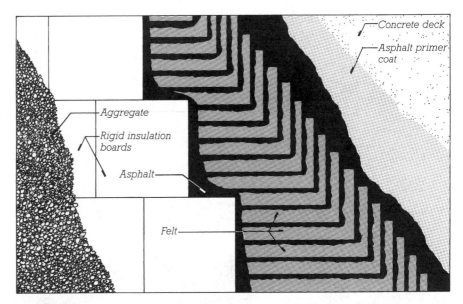

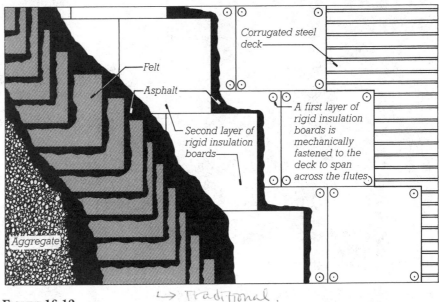

FIGURE 16.12

→ Traditional.

Two typical built-up roof constructions, as seen from above. The top diagram is a cutaway view of a protected membrane roof over a poured concrete roof deck. The membrane is made from plies of felt overlapped in such a way that it is never less than four plies thick. Rigid foam insulation boards are bedded in hot asphalt over the membrane and ballasted with stone aggregate to keep them in place and protect them from sunlight. The bottom diagram shows how rigid insulation boards are attached to a corrugated steel roof deck in two staggered layers to provide a firm, smooth base for application of the membrane. A three-ply membrane is shown. In cold climates, a vapor retarder should be installed between the layers of insulation.

3 ply - filter overlapping each other.

membrane [paint between them to stick them together

→ 3 layers of filt.

the roof. To protect the membrane from sunlight and physical wear, a layer of aggregate (crushed stone or other mineral granules) is embedded in the surface (Figures 16.15, 16.16).

Cold-applied mastics may be used in lieu of hot bitumen in built-up roof membranes. A roofing mastic is compounded of asphalt and other substances to bond to felts or to synthetic fabric reinforcing mats at ordinary ambient temperatures. The mastic may be sprayed or brushed on and hardens by the evaporation of solvents.

FIGURE 16.13
A base sheet of asphalt-saturated felt is installed over rigid insulation, using a machine that unrolls the felt and presses it into a layer of hot asphalt. (*Courtesy of Celotex Corporation*)

FIGURE 16.14
Overlapping layers of roofing felt are hot-mopped with asphalt to create a four-ply membrane. (*Courtesy of Manville Corporation*)

FIGURE 16.15
Roofers embed stone ballast in hot asphalt to hold down and protect the panels of rigid insulation in a protected membrane roof. The area of the membrane behind the wheelbarrow has not yet received its insulation. (*Courtesy of Celotex Corporation*)

FIGURE 16.16
A cutaway detail of a proprietary type of protected membrane roof shows, from bottom to top, the roof deck, the membrane, polystyrene foam insulation, a polymeric fabric that separates the ballast from the insulation, and the ballast. (*Courtesy of Dow Chemical Company*)

Single-Ply Roof Membranes

Single-ply membranes are a diverse and rapidly growing group of sheet materials that are applied to the roof in a single layer (Figures 16.17–16.22). As compared to built-up membranes, they require less on-site labor, and they are usually more elastic and therefore less prone to cracking and tearing. They are affixed to the roof by any of several means: with adhesives; by the weight of ballast; by fasteners concealed in the seams between sheets; or, if they are sufficiently flexible, with ingenious mechanical fasteners that do not penetrate the membrane (Figures 16.18–16.20).

The materials presently used for single-ply membranes fall into two general groups: *thermoplastic* and *thermosetting*. *Thermoplastic* materials may be softened by the application of heat. Sheets of thermoplastic membrane may be joined at seams by heat welding or solvent welding. Thermosetting materials cannot be softened by heat. They must be joined at seams by adhesives or pressure-sensitive tapes.

Thermoplastic membrane materials include:

• *Polyvinyl chloride (PVC)*, a compound commonly known as *vinyl*. It is relatively low in cost and very widely used. PVC sheet for roofing is 0.045 to 0.060 inch thick (1.14 to 1.5 mm). It may be laid loose, mechanically attached, adhered, or used as a protected membrane.

• *Polymer-modified bitumens*, sheets composed of bituminous materials to which such polymeric compounds as *atactic polypropylene (APP)* or *styrene-butadiene-styrene (SBS)* have been added in order to increase their flexibility, cohesion, toughness, and resistance to flow. Most are reinforced with fibers, fibrous mats, or plastic films. Thicknesses typically range from 0.040 to 0.160 inch (1.0 to 4.0 mm). Some polymer-modified bitumen sheet materials are designed to self-adhere to the roof surface.

FIGURE 16.17
Workers unfold a large single-ply roof membrane.
(*Courtesy of Carlisle SynTec Systems*)

1. Roll membrane over knobbed base plate

2. Roll and snap on white retainer clip

3. Snap and screw on threaded black cap

FIGURES 16.18, 16.19
A proprietary nonpenetrating attachment system for a single-ply roof membrane.
(*Courtesy of Carlisle SynTec Systems*)

FIGURE 16.20
Another proprietary nonpenetrating attachment system folds the membrane into a continuous slot, where it is held by a synthetic rubber spline that is inserted with a wheeled tool. (*Courtesy of Firestone*)

FIGURE 16.21
Roofers bond a composite polymer-modified bitumen membrane to a concrete deck with a cold-applied adhesive. The seams will be heat-fused together. (*Courtesy of Koppers Company, Inc.*)

Others are meant to be laid loose, bedded in hot asphalt, or softened on the underside with a gas torch at the moment of application so they will adhere to the roof. Most are factory surfaced with mineral granules, metallic laminates, or elastomeric coatings to protect against ultraviolet deterioration and fire. Seams are sealed either by torching or by using hot asphalt as an adhesive (Figures 16.21, 16.22).

• *PVC alloys* and *compounded thermoplastics*, which are mixtures of PVC and other thermoplastics. The most common of these is *copolymer alloy (CPA)*. Others include *ethylene interpolymer (EIP)*, *nitrile alloy (NBP)*, and *tripolymer alloy (TPA)*.

• *Chlorinated polyethylene (CPE)*, typically reinforced with polyester fabric.

• *Polyisobutylene (PIB)*.

• *Thermoplastic olefin (TPO)*.

Thermosetting membrane materials include: [most common]

• *EPDM (ethylene propylene diene monomer)*, the most widely used material for single-ply roof membranes. It is relatively low in cost. It is a synthetic rubber manufactured in sheets from 0.030 to 0.060 inch in thickness (0.75 to 1.5 mm). It may be laid loose, adhered, mechanically fastened, or used in a protected membrane roof.

• *Chlorosulfonated polyethylene (CSPE)*, which is highly resistant to ultraviolet deterioration and can be manufactured in light, heat-reflective colors. It is used on roofs where ballasting is unacceptable for reasons of appearance or excessive slope.

• *Epichlorhydrin*.

• *Neoprene*, a high-performance synthetic rubber compound. It is applied in sheets ranging from 0.030 to 0.120 inch (0.75 to 3.0 mm) in thickness. Because it is vulnerable to attack by ultraviolet light, it is usually coated with a protective layer of CPE.

Hybrid Membranes

A polymer-modified bitumen membrane is sometimes applied over several plies of built-up roofing to create a membrane that combines the toughness of a BUR with the wearing qualities and elasticity of a modified bitumen while weighing less than a ballasted BUR. This combination is known as a *hybrid membrane* or *composite membrane*.

FIGURE 16.22

A roofer heat-fuses a seam between two sheets of an aluminum-faced single-ply membrane. The membrane in this example consists of a flexible plastic sheet laminated between layers of modified bitumen, with protective layers of polyethylene below and embossed aluminum foil above. The bitumen layers provide the bonding elements for heat fusion. The aluminum foil protects the inner plies of the membrane from the sun and reflects solar heat from the roof.
(*Courtesy of Koppers Company, Inc.*)

Fluid-Applied Membranes

Fluid-applied membranes are used primarily for domes, shells, and other complex shapes that are difficult to roof by conventional means. Such shapes are often too flat on top for shingles, but too steep on the sides for built-up roof membranes, and if doubly curved are difficult to fit with single-ply membranes. Fluid-applied membranes are applied in liquid form with a roller or spray gun, usually in several coats, and cure to form a rubbery membrane. Materials applied by this method include neoprene (with a weathering coat of chlorosulfonated polyethylene), *silicone, polyurethane, butyl rubber,* and *asphalt emulsion.*

Fluid-applied membranes are used as a waterproofing layer over sprayed-on polyurethane foam insulation in proprietary roofing systems that are appropriate for surfaces that are hard to fit with flat sheets or insulation and membrane. These systems are also a convenient means for adding thermal insulation and a new roof membrane over existing, deteriorated built-up roofs on any shape of building.

Ballasting and Traffic Decks

Most built-up roof membranes and many single-ply membranes are covered after installation with a *ballast* of loose stone aggregate or precast concrete blocks (Figures 16.15, 16.23). The ballast serves to hold the membrane down against wind uplift, and it protects the membrane from ultraviolet light and physical wear. It also contributes to the fire resistance of the roof covering.

Traffic decks are installed over flat roof membranes for walks, roof terraces, and sometimes driveways or parking surfaces. Two different details are used: In one, low blocks of plastic or concrete are set on top of the roof membrane to support the corners of heavy square paving stones or slabs with open joints (Figure 16.24). In the other, a drainage layer

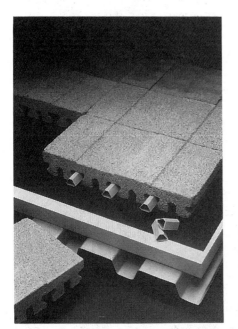

FIGURE 16.23 Ballasting.

This ballasting system uses special concrete blocks that are joined by tubular plastic splines. The grooves in the bottoms of the blocks are designed to facilitate drainage of water from the membrane. The blocks are made with lightweight aggregate so that they contribute to the thermal insulation of the roof. (*Insulating roof ballast with variable interlock feature ©National Concrete Masonry Association, 1988*)

FIGURE 16.24 Paving Block

A proprietary system for supporting stone or precast concrete paving blocks over a low-slope roof membrane permits use of a low-slope roof as an outdoor terrace. Each high-density polyethylene pedestal supports the adjacent corners of four paving blocks. The vertical spacer fins on the pedestal provide a uniform drainage space between the blocks. Matching polyethylene leveling plates (not shown) can be placed over the pedestal to compensate for irregularities in the roof surface. (*Courtesy of Envirospec, Inc., Buffalo, New York*)

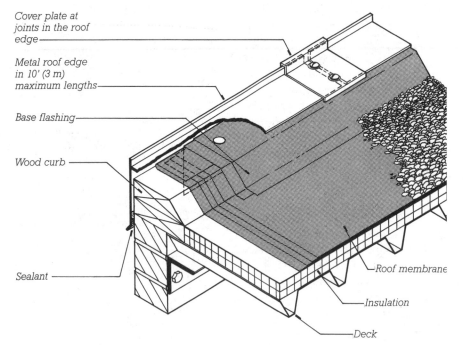

Cover plate at joints in the roof edge

Metal roof edge in 10' (3 m) maximum lengths

Base flashing

Wood curb

Sealant

Roof membrane

Insulation

Deck

FIGURE 16.25
A roof edge for a conventional built-up roof. The membrane consists of four plies of felt bedded in asphalt with a gravel ballast. The base flashing is composed of two additional plies of felt that seal the edge of the membrane and reinforce it where it bends over the curb. The curb directs water toward interior drains or scuppers rather than allowing it to spill over the edge. The exposed vertical face of the metal roof edge is called a fascia.

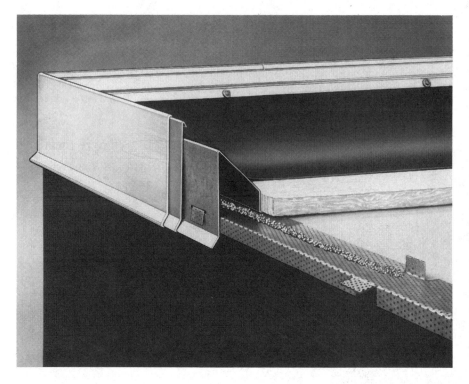

FIGURE 16.26
A proprietary roof edge system for low-slope roofs. The perforated metal strip is fastened to the roof with a mastic adhesive that oozes through the perforations to create a tighter bond. When the adhesive has hardened, a galvanized steel curb is fastened in place with the tabs of perforated metal, and an aluminum roof edge is hooked on, with a backup piece at end joints as shown to prevent leakage. Lastly, the roof edge and the membrane are locked in place simultaneously by installing a clamping strip that engages the hook on the top of the aluminum roof edge. The clamping strip is held in place by screws that pass through the edge of the membrane into the galvanized curb, as seen at the top of the photograph. (*Product of W. P. Hickman Company, Asheville, North Carolina*)

of gravel or no-fines concrete (a very porous concrete whose aggregate consists solely of a single size of coarse stone) is leveled over the membrane, and open-joined paving blocks are installed on top. In either detail, water falls through the joints in the paving and is caught and drained away by the membrane below. Notice that the membrane is not pierced in either detail.

Edge and Drainage Details for Low-Slope Roofs

Some typical details of low-slope roofs are presented in Figures 16.25 through 16.34. All are shown with built-up roof membranes, but details for single-ply membranes are similar in principle.

Structural Panel Metal Roofing for Low-Slope Roofs

Manufacturers of prefabricated metal building systems have developed pro-

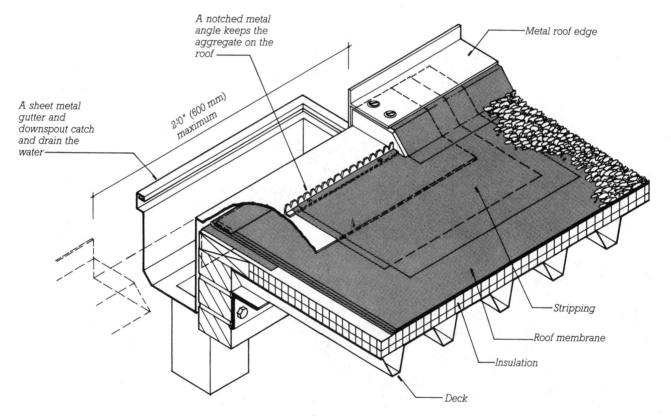

A notched metal angle keeps the aggregate on the roof

2'0" (600 mm) maximum

A sheet metal gutter and downspout catch and drain the water

Metal roof edge

Stripping

Roof membrane

Insulation

Deck

FIGURE 16.27
Detail of a scupper. The curb is discontinued to allow water to spill off the roof into a gutter and downspout. Additional layers of felt, called stripping, seal around the sheet metal components. Most roofs use interior drains (Figure 16.32) as their primary means of drainage rather than scuppers.

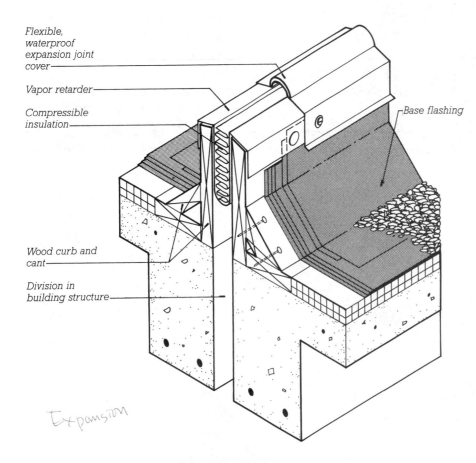

Flexible, waterproof expansion joint cover

Vapor retarder

Compressible insulation

Base flashing

Wood curb and cant

Division in building structure

Expansion

FIGURE 16.28
A building separation joint in a low-slope roof. Large differential movements between the adjoining parts of the structure can be tolerated with this type of joint because of the ability of the flexible joint cover to adjust to movement without tearing. High curbs keep standing water away from the edge of the membrane, which is sealed with a two-ply base flashing.

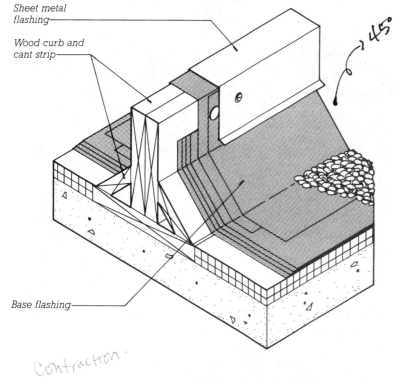

Sheet metal flashing

Wood curb and cant strip

45

Base flashing

Contraction.

FIGURE 16.29
An area divider is designed only to allow for movement in the membrane itself, not the entire structure. It is used to subdivide a very large membrane to allow for thermal movement.

← Area Divider Joint.

○ For membrane movement.

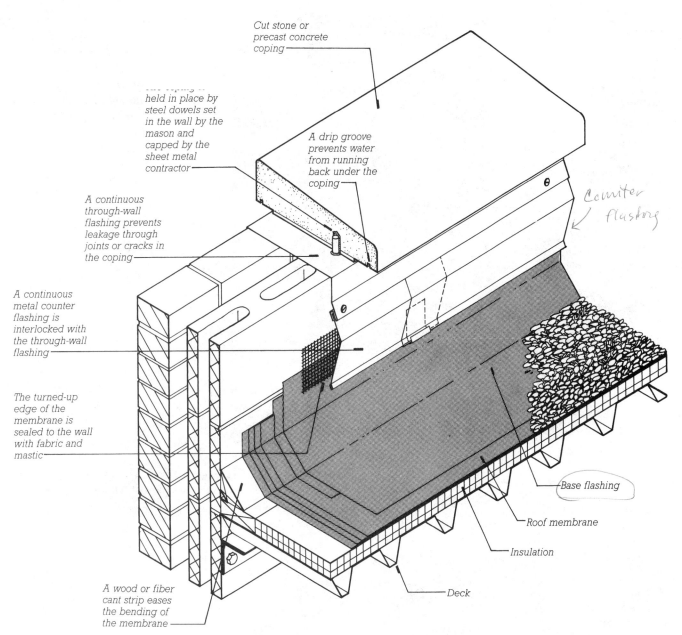

Cut stone or precast concrete coping

the coping is held in place by steel dowels set in the wall by the mason and capped by the sheet metal contractor

A drip groove prevents water from running back under the coping

A continuous through-wall flashing prevents leakage through joints or cracks in the coping

A continuous metal counter flashing is interlocked with the through-wall flashing

Counter flashing

The turned-up edge of the membrane is sealed to the wall with fabric and mastic

Base flashing

Roof membrane

Insulation

A wood or fiber cant strip eases the bending of the membrane

Deck

FIGURE 16.30
A conventional parapet design.

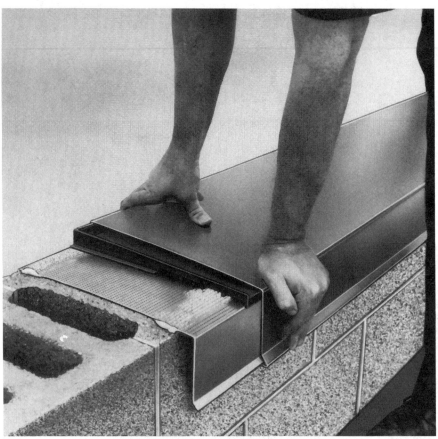

FIGURE 16.31
A proprietary parapet coping system. The perforated metal channel is fastened to the masonry with a mastic adhesive. Sections of metal coping are snapped over the channel, with a special pan element beneath the joints to drain leakage. (*Product of W. P. Hickman Company, Asheville, North Carolina*)

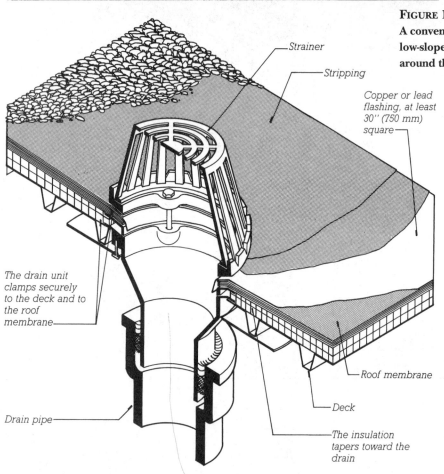

Strainer

Stripping

Copper or lead flashing, at least 30″ (750 mm) square

Back-up Drains

The drain unit clamps securely to the deck and to the roof membrane

Roof membrane

Deck

The insulation tapers toward the drain

Drain pipe

FIGURE 16.32
A conventional cast iron interior roof drain for a low-slope roof. Two plies of felt stripping seal around the sheet metal flashing.

FIGURE 16.33
A proprietary single-piece roof drain made of molded plastic. (*Product of W. P. Hickman Company, Asheville, North Carolina*)

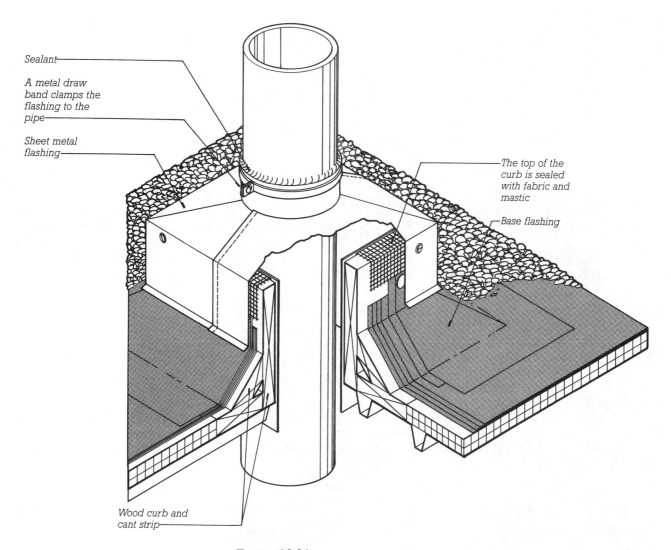

Sealant

A metal draw band clamps the flashing to the pipe

Sheet metal flashing

The top of the curb is sealed with fabric and mastic

Base flashing

Wood curb and cant strip

FIGURE 16.34
A roof penetration for a plumbing vent stack. Notice how this and all the previous edge and penetration details for a flat roof use the curb, cant strip, and stripping to keep standing water away from the edge of the membrane.

prietary systems of metal roofing panels that can be used as low-slope roofs at pitches as shallow as $\frac{1}{4}$ in 12 (1:48). These can be applied not only to prefabricated metal buildings but to buildings built of any materials. They are often employed as replacement roofs (Figures 16.35–16.37). They are called *structural* roofs because the folded shape of the metal roofing gives it sufficient stiffness that it can support itself and normal snow loads between purlins without the need for a structural deck beneath. This name also distinguishes it from *architectural* metal roofing, the traditional forms of metal roofing that are not self-supporting; these are utilized largely on steep roofs, and are described later in this chapter.

FIGURE 16.35
As a first step in reroofing an existing building with a structural metal roof, steel Z-purlins are erected over the old roof on tubular metal posts. (*Courtesy of Metal Building Manufacturers Association*)

FIGURE 16.36
A proprietary metal clip is used to fasten the metal roofing sheets to the Z-purlins while allowing for thermal movement in the sheets. The plastic foam collar minimizes thermal bridging. (*Courtesy of Metal Building Manufacturers Association*)

FIGURE 16.37
The completed structural metal roof has a slope of only 1 : 48. The numerous penetrations for plumbing vents, air vents, and ductwork are typical of low-slope roofs. (*Courtesy of Metal Building Manufacturers Association*)

STEEP ROOFS

Roofs with a pitch of 3 : 12 (25%) or greater are referred to as *steep roofs*. Roof coverings for steep roofs fall into three general categories: *thatch*, *shingles*, and *architectural sheet metal*. *Thatch*, an attractive and effective roofing consisting of bundles of reeds, grasses, or leaves (Figures 16.1, 16.2), is highly labor intensive and is rarely used today. Shingle and sheet metal roofs of many types are common to every type of building, and range in price from the most economical of roof coverings to the most expensive.

The insulation and vapor retarder in most steep roofs are installed below the roof sheathing or deck; typical details of this practice are shown in Chapters 6 and 7. Where the underside of the deck is to be left exposed as a finish surface, a vapor retarder and rigid insulation panels are applied above the deck, just below the roofing. A layer of plywood or OSB is then nailed over the insulation panels as a *nail base* for fastening the shingles or sheet metal, or special composite insulation panels with an integral nail base layer can be used.

Shingles

The word "shingles" is used here in a generic sense to include wood shingles and shakes, asphalt shingles, slates, clay tiles, and concrete tiles. What these materials share in common is that they are applied to the roof in small units and in overlapping layers with staggered vertical joints.

Wood shingles are thin, tapered slabs of wood sawn from short pieces of tree trunk with the grain of the wood running approximately parallel to the face of the shingle (Figures 16.38, 16.56). *Shakes* are split from the wood, rather than sawn, and exhibit a much rougher face texture than wood shingles (Figures 16.39, 16.40). Most wood singles and shakes in North America are made from Red cedar, White cedar, or Redwood because of the natural decay resistance of these woods. Wood roof coverings are moderately expensive and are not highly resistant to fire unless the shakes or shingles have been pressure treated with fire-retardant chemicals.

Asphalt shingles are die cut from heavy sheets of asphalt-impregnated felt. The sheets are faced with mineral granules that act as a wearing layer and decorative finish. The most common type of asphalt shingle, which covers approximately 90 percent of the single-family houses in North America, is 12 inches by 36 inches (305 mm by 914 mm) in size. (A met-

FIGURE 16.38
Applying red cedar shingles, in this example as reroofing over asphalt shingles. Small corrosion-resistant nails are driven near each edge at the midheight of the shingle. Each succeeding course covers the joints and nails in the course below. (*Courtesy of Red Cedar Shingle and Handsplit Shake Bureau*)

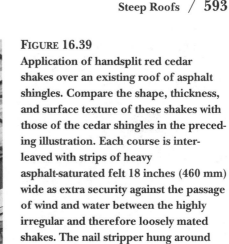

FIGURE 16.39
Application of handsplit red cedar shakes over an existing roof of asphalt shingles. Compare the shape, thickness, and surface texture of these shakes with those of the cedar shingles in the preceding illustration. Each course is interleaved with strips of heavy asphalt-saturated felt 18 inches (460 mm) wide as extra security against the passage of wind and water between the highly irregular and therefore loosely mated shakes. The nail stripper hung around the roofer's neck speeds his work by holding the nails and aligning them with points down, ready for driving. (*Courtesy of Red Cedar Shingle and Handsplit Shake Bureau*)

FIGURE 16.40
Shake application over a new roof deck using air-driven, heavy-duty staplers for greater speed. The strips of asphalt-saturated felt have all been placed in advance with their lower edges unfastened. Each course of shakes is laid out, slipped up under its felt strip, then quickly fastened by roofers walking across the roof and inserting staples as fast as they can pull the trigger. (*Courtesy of Senco Products, Inc.*)

ric shingle 337 mm × 1000 mm is also widely marketed.) Most felts are based on glass fibers, but some still retain the older cellulose composition. Each shingle is slotted twice to produce a roof that looks as though it were made of smaller shingles (Figures 16.41–16.44). Many other shingle styles are also available, including thicker shingles that are laminated from several layers of material. Asphalt shingles are inexpensive to buy, quick to install, moderately fire resistant, and have an expected lifetime of 15 to 25 years, depending on their exact composition.

The same sheet material from which asphalt shingles are cut is also manufactured in rolls 3 feet (900 mm) wide as *asphalt roll roofing*. Roll roofing is very inexpensive and is used primarily on storage and agricultural buildings. Its chief drawbacks are that thermal expansion of the roofing or shrinkage of the wood deck can cause unsightly ridges to form in the roofing and that thermal contraction can tear it.

Slate for roofing is delivered to the site split, trimmed to size, and

FIGURE 16.41

Installing asphalt shingles. To give a finer visual scale to the roof, the slots make each shingle appear as if it were three smaller shingles when the roof is finished. Many different patterns of asphalt shingles are available, including ones that do not have slots. (*Photo by the author*)

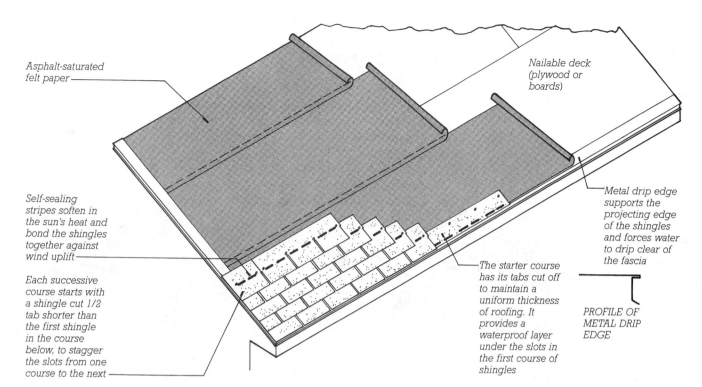

Asphalt-saturated
felt paper

Nailable deck
(plywood or
boards)

Self-sealing
stripes soften in
the sun's heat and
bond the shingles
together against
wind uplift

Each successive
course starts with
a shingle cut 1/2
tab shorter than
the first shingle
in the course
below, to stagger
the slots from one
course to the next

Metal drip edge
supports the
projecting edge
of the shingles
and forces water
to drip clear of
the fascia

The starter course
has its tabs cut off
to maintain a
uniform thickness
of roofing. It
provides a
waterproof layer
under the slots in
the first course of
shingles

PROFILE OF
METAL DRIP
EDGE

FIGURE 16.42
Starting an asphalt shingle roof. Most
building codes require the installation of
an ice-and-water barrier beneath the shin-
gles along the eave in regions with cold
winters; its function is to prevent the entry
of standing water that might be created by
ice dams. The most effective form of bar-
rier is a 3-foot-wide (900-mm) strip of
modified bitumen sheet that replaces the
lowest course of asphalt-saturated felt
paper. The bitumen self-seals around the
shanks of the roofing nails as they are
driven through it.

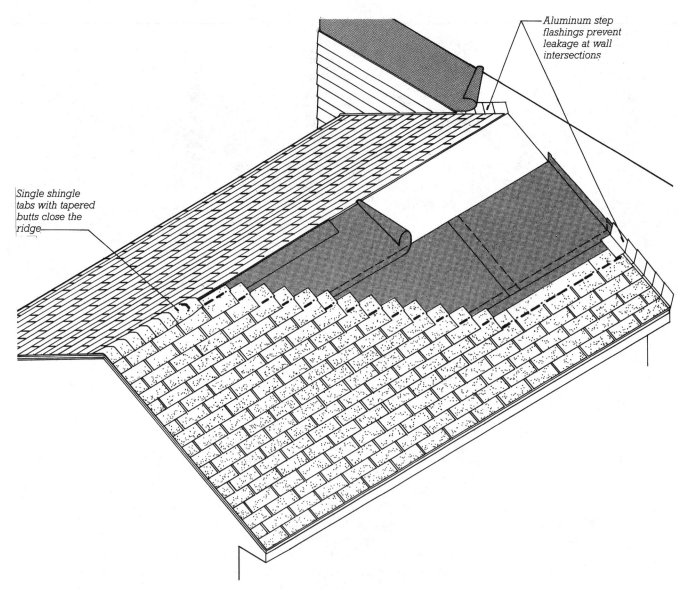

Aluminum step flashings prevent leakage at wall intersections

Single shingle tabs with tapered butts close the ridge

FIGURE 16.43
Completing an asphalt shingle roof. A metal attic ventilation strip is often substituted for the shingle tabs on the ridge.

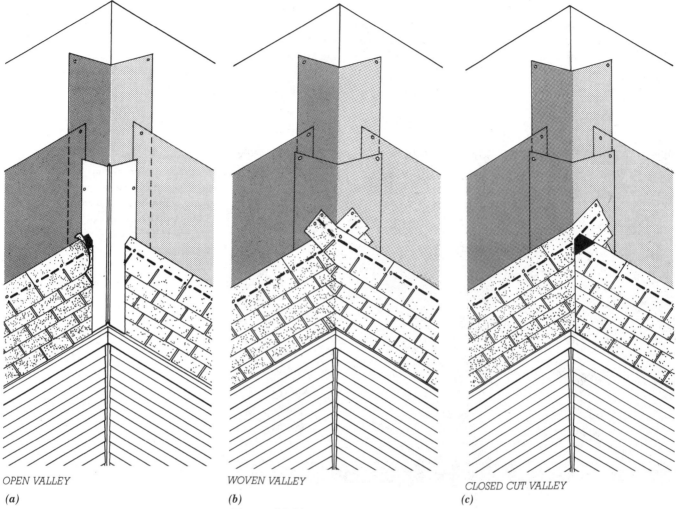

OPEN VALLEY

(a)

WOVEN VALLEY

(b)

CLOSED CUT VALLEY

(c)

FIGURE 16.44
Three alternative methods of making a valley in an asphalt shingle roof. (*a*) The open valley uses a sheet metal flashing; the ridge in the middle of the flashing helps prevent water that is coming off one slope from washing up under the shingles on the opposite slope. The woven valley (*b*) and cut valley (*c*) are favorites of roofing contractors because they require no sheet metal. The solid black areas on shingles in the open and closed cut valleys indicate areas to which asphaltic roofing cement is applied to adhere shingles to each other.

punched or drilled for nailing (Figures 16.45, 16.46). It forms a fire-resistant, long-lasting roof that is suitable for buildings of the finest materials. It is relatively costly.

Clay tiles have been used on roofs for thousands of years. It is said that the tapered barrel tiles traditional to the Mediterranean region (similar to the mission tiles in Figure 16.47)

were originally formed on the thighs of the tilemakers. Many other patterns of clay tiles are now available, both glazed and unglazed. *Concrete tiles* are generally less expensive than

FIGURE 16.45
Splitting slate for roofing. The thin slates in the background will next be trimmed square and to dimension, after which nail holes will be punched in them.
(*Photo by Flournoy. Courtesy of Buckingham-Virginia Slate Corporation*)

FIGURE 16.46
A slate roof during installation. (*Courtesy of Buckingham-Virginia Slate Corporation*)

clay tiles and are available in many of the same patterns. Tile roofs in general are heavy, durable, highly resistant to fire, and relatively expensive in first cost.

Each type of shingle, slate, or tile must be laid on a roof deck that slopes sufficiently to assure leakproof performance. Minimum slopes for each material are specified by the manufacturer. Slopes greater than the minimum should be used in locations where water is likely to be driven up the roof surface by heavy storm winds.

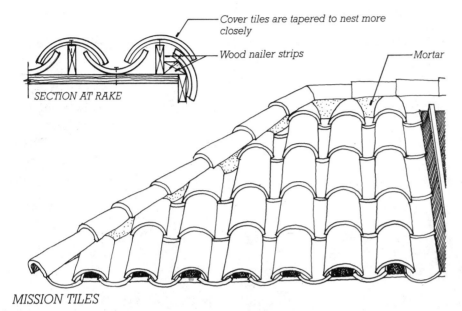

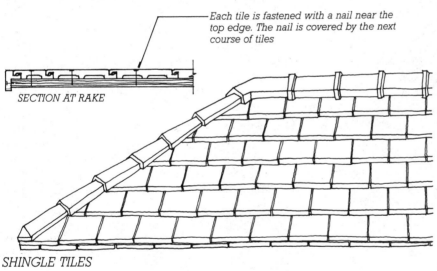

MISSION TILES

SHINGLE TILES

FIGURE 16.47
Two styles of clay tile roofs. The mission tile has very ancient origins.

FIGURE 16.48
Fall-protection devices are increasingly used by workers on steep roofs. (*Courtesy of DBI/SALA, Red Wing, Minnesota*)

Architectural Sheet Metal Roofing

Sheets of lead and copper have been used for roofing since ancient times. Both metals form self-protecting oxide layers and last for many decades. They are installed in small sheets using ingenious systems of joining and fastening to maintain watertightness at the seams (Figures 16.49–16.53). The seams, especially *standing seams* and *batten seams,* create a strong visual pattern that can be manipulated by the designer to emphasize the qualities of the roof shape. Lead roofs oxidize in time to a white color. Copper turns a beautiful blue-green in clean air, and a dignified black in an industrial atmosphere; various chemical treatments can be used to obtain and preserve the desired color. *Lead-coated copper* sheet is sometimes used to combine the greater strength of copper with the gray-white color of lead and to eliminate the staining of wall materials by oxides of copper. *Terne* roofing is made of steel or stainless steel sheet that is coated with an alloy of lead and tin. Terne-coated stainless steel is self-protecting and weathers to a soft gray color. A terne coating over plain steel must be painted to protect it against pinhole corrosion. Aluminum, zinc alloys, and stainless steel can also be used for roofing in much the same manner as copper, lead, and terne; stainless steel roofing is pictured in the opening illustration of this chapter. Architectural sheet metal roofs of any of these materials are relatively high in first cost but they can be expected to last for many decades.

Architectural panel roofs are made of long sheets of aluminum, aluminized steel, galvanized steel, or other corrosion-resistant metals

FIGURE 16.49
Standing-seam copper roofs. (*Designer: Emil Hanslin. Courtesy of Copper Development Association, Inc.*)

FIGURE 16.50
An automatic roll seamer, moving under its own power, locks standing seams in a copper roof. A cleat is just visible at the lower right. (*Courtesy of Copper Development Association, Inc.*)

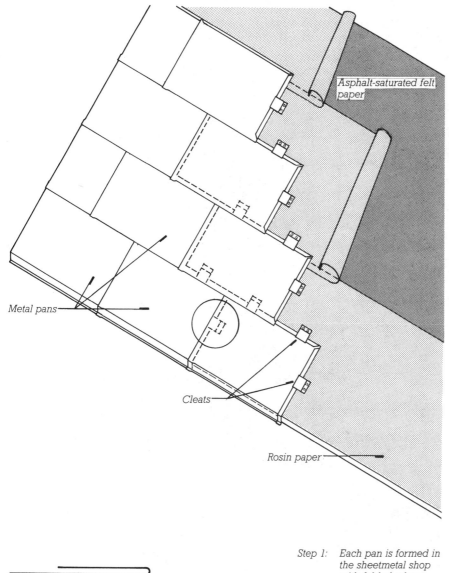

Metal pans

Cleats

Rosin paper

FIGURE 16.51
Installing a flat-seam metal roof. The three diagrams at the bottom of the illustration show the three steps in creating the seam, viewed in cross section. The cleats, which fasten the roofing to the deck, are completely concealed when the roof is finished.

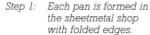

Step 1: Each pan is formed in the sheetmetal shop with folded edges.

Step 2: Sheet metal cleats interlock with the folded edges and are nailed to the deck. The cleat is folded back over the nail head to protect the pan.

Step 3: The next pan is interlocked with the first. When all pans are in place, the edges are beaten flat and soldered or sealed.

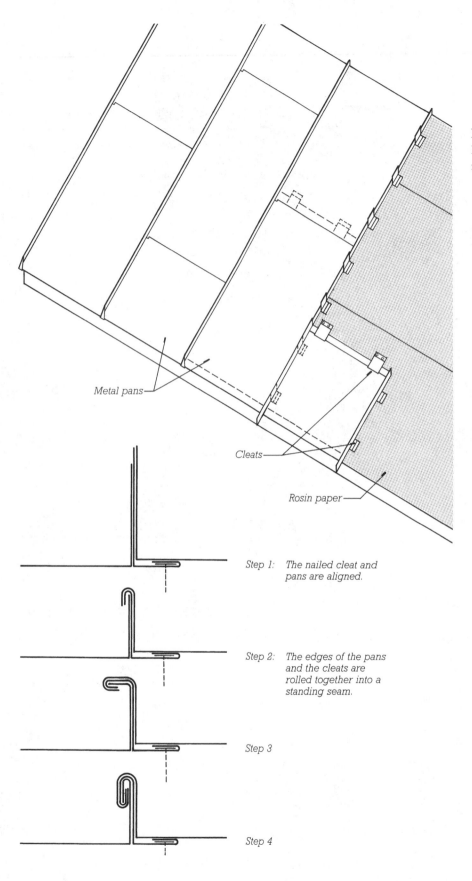

FIGURE 16.52
Installing an architectural standing-seam metal roof.

Metal pans

Cleats

Rosin paper

Step 1: The nailed cleat and pans are aligned.

Step 2: The edges of the pans and the cleats are rolled together into a standing seam.

Step 3

Step 4

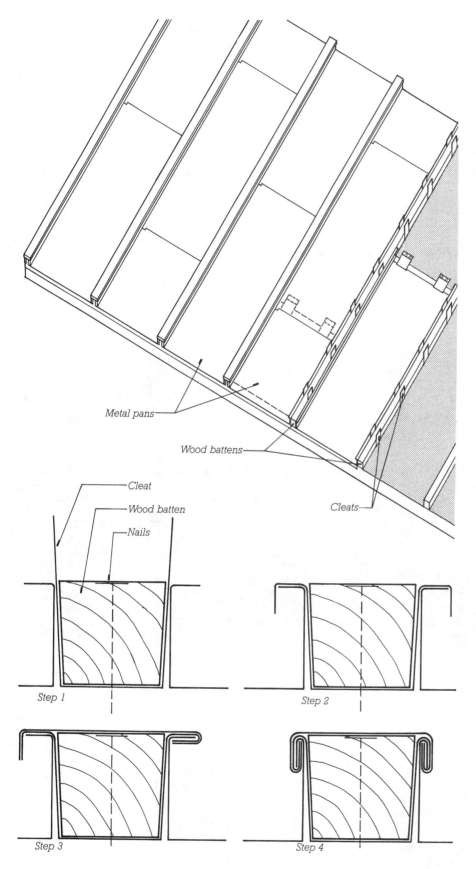

Metal pans

Wood battens

Cleat

Wood batten

Nails

Cleats

Step 1

Step 2

Step 3

Step 4

FIGURE 16.53
Installing a batten-seam metal roof. The battens are tapered in cross section to allow for expansion of the roofing metal.

(Figure 16.54). Many are coated with long-lasting polymeric coatings in various colors. The sheets are produced with raised edge seams that interlock in various ways to exclude water. The seams also include means of attachment to the roof deck. Architectural panel roofs are less costly than traditional forms of sheet metal roofing.

The same metal should be used for every component of a sheet metal roof, including the fasteners and flashings. If this is impossible, metals of similar galvanic activity should be used. Where strongly dissimilar metals touch in the presence of rainwater, which is generally acidic, galvanic action causes rapid corrosion. Figure 16.55 will be helpful to the designer in avoiding destructive combinations of metals.

ROOFING AND THE BUILDING CODES

Manufacturing standards and installation procedures for roofing materials are specified by all the model building codes. Building codes also regulate the type of roofing that may be used on a building, based on a required level of resistance to flame spread and fire penetration as measured by ASTM procedure E108. Roofing materials are grouped into four classes:

• *Class A roof coverings* are effective against severe fire exposure. They include slate, concrete tiles, clay tiles, most asphalt shingles, most built-up and single-ply roofs, and other materials certified as Class A by approved testing agencies. They may be used on any building in any type of construction.

• *Class B roof coverings* are effective against moderate fire exposure, and include many built-up and single-ply roofs, sheet metal roofings, and some composition shingles. These are the minimum class that may be used on buildings of Type 1 construction under the BOCA Code as defined in the table in Figure 1.1.

FIGURE 16.54
The raised seams in the architectural panel metal roofing accentuate the complex roof geometries of the International Center in Brattleboro, Vermont. (*Architect: William A. Hall Partnership.* *Photo by Stanley Jesudowich*)

FIGURE 16.55
A galvanic series for metals used in buildings. Each metal is corroded by all those that follow it in the list. The wider the separation between two metals on the list, the more severe the corrosion is likely to be. Some families of alloys, such as stainless steels, are difficult to place with certainty because some alloys within the family behave differently from others. To be certain, the designer should consult with the manufacturer of any metal product before installing it in contact with dissimilar metals in an outdoor environment.

Aluminum
Zinc and galvanized steel
Chromium
Steel
Stainless steel
Cadmium
Nickel
Tin
Lead
Brass
Bronze
Copper

The roof plays a primal role in our lives. The most primitive buildings are nothing but a roof. If the roof is hidden, if its presence cannot be felt around the building, or if it cannot be used, then people will lack a fundamental sense of shelter.

Christopher Alexander et al,
A Pattern Language,
New York, Oxford University
Press, 1977, p. 570

FIGURE 16.56
A house is both roofed and sided with red cedar shingles to feature its sculptural qualities. (*Architect: William Isley. Photo by Paul Harper. Courtesy of Red Cedar Shingle and Handsplit Share Bureau*)

FIGURE 16.57
A standing-seam metal roof with beautifully detailed overhangs, designed by architects **Kallmann and McKinnell.** (*Photo by Steve Rosenthal*)

- *Class C roof coverings* are effective against light fire exposure. They include fire-retardant treated wood shingles and shakes. These are the minimum class that may be used on Types 2, 3, and 4A construction.

- Nonclassified roof coverings such as untreated wood shingles may be used on Type 4B construction and on some agricultural, accessory, and storage buildings.

ASTM E108 applies to whole roof assemblies, including deck, insulation, membrane or shingles, and ballast, if any. Therefore, the classifications given in the preceding list should be taken only as a general guide. It is difficult to summarize with precision the classification of any particular type of shingle or membrane without knowing the other components of the assembly. The required class of roofing for a particular building may also be affected by an urban fire zone in which the building is located and by the proximity of the building to its neighbors.

C.S.I./C.S.C. Masterformat Section Numbers for Roofing	
07190	**VAPOR RETARDERS**
07200	**INSULATION**
07210	**Building Insulation**
07220	**Roof and Deck Insulation**
07300	**SHINGLES AND ROOFING TILES**
07310	**Shingles**
07320	**Roofing Tiles**
07400	**MANUFACTURED ROOFING AND SIDING**
07500	**MEMBRANE ROOFING**
07510	**Built-Up Bituminous Roofing**
07515	**Cold-Applied Bituminous Roofing**
07520	**Prepared Roll Roofing**
07530	**Single-Ply Membrane Roofing**
07540	**Fluid Applied Roofing**
07550	**Protected Membrane Roofing**
07600	**FLASHING AND SHEET METAL**
07610	**Sheet Metal Roofing**
07620	**Sheet Metal Flashing and Trim**
07630	**Sheet Metal Roofing Specialties**
07650	**Flexible Flashing**
07700	**ROOF SPECIALTIES AND ACCESSORIES**
07800	**SKYLIGHTS**

Selected References

1. National Roofing Contractors Association. *The NRCA Roofing and Waterproofing Manual* (4th ed.). Rosemont, Illinois, 1996.

These two thick, looseleaf volumes are a comprehensive guide to current U.S. practice for both low-slope and steep roofs. The treatment is exhaustive and both diagrams and text are excellent. (Address for ordering: 10255 West Higgins Road, Suite 600, Rosemont, IL 60018.)

2. Sheet Metal and Air Conditioning Contractors National Association, Inc. *Architectural Sheet Metal* (5th ed.). Chantilly, Virginia, 1993.

Architectural sheet metal roofs are detailed in this excellent reference, along with every conceivable flashing, fascia, gravel stop, and gutter for flat and shingled roofs. (Address for ordering: P.O. Box 221230, Chantilly, VA 22022-1230.)

KEY TERMS AND CONCEPTS

steep roof
low-slope roof
shingle
tile
deck
thermal insulation
vapor retarder
membrane
drainage
flashing
area divider
thermal resistance
R-value
water vapor
relative humidity
dew point
condensation
perm
topside vent
ballast
protected membrane roof (PMR)
deck fill insulation
built-up roof membrane (BUR)
felt
bitumen
ply

cold-applied mastic
thermoplastic
polyvinyl chloride (PVC)
polymer-modified bitumens
atactic polypropylene (APP)
styrene-butadiene-styrene (SBS)
PVC alloys
compounded thermoplastics
copolymer alloy (CPA)
ethylene interpolymer (EIP)
nitrile alloy (NBP)
tripolymer alloy (TPA)
chlorinated polyethylene (CPE)
polyisobutylene (PIB)
thermoplastic olefin (TPO)
thermosetting
ethylene propylene diene monomer (EPDM)
chlorosulfonated polyethylene (CSPE)
epichlorhydrin
neoprene
hybrid or composite membrane
fluid-applied membrane
silicone
polyurethane
butyl rubber
asphalt emulsion

traffic deck
parapet
fascia
scupper
cant
coping
counterflashing
stripping
structural sheet metal roofing
thatch
nail base
wood shingle
shake
asphalt shingle
asphalt roll roofing
slate
clay tiles
concrete tiles
architectural sheet metal roofing
flat seam
standing seam
batten seam
lead-coated copper
terne
architectural panel roofs
Class A, B, C roof coverings

REVIEW QUESTIONS

1. What are the major differences between a low-slope roof and a steep roof? What are the advantages and disadvantages of each type?

2. Discuss the three positions in which thermal insulation may be installed in a low-slope roof, and the advantages and disadvantages of each.

3. Explain in precise terms the function of a vapor retarder in a low-slope roof.

4. Compare a built-up roof membrane to a single-ply roof membrane.

5. What is the difference between cedar shingles and cedar shakes?

6. What metals are used for architectural sheet metal roofing? What are the strengths and drawbacks of each?

EXERCISES

1. For a low-slope-roofed university classroom building with a masonry bearing wall, steel interior frame, corrugated steel roof deck, and parapet:

a. Show two ways of achieving a 1:50 roof slope on structural bays 36 feet (911 m) square.

b. Sketch a set of details of the parapet edge, building separation joint, area divider, and roof drain for a low-slope roof system of your choice. Show insulation, vapor retarder (if any), roof membrane, and flashings.

2. Sketch a fascia detail for a low-slope roof system of your choice, assuming that the wall below is made of precast concrete panels and the roof deck of precast concrete slab elements.

3. Find a low-slope roof system being installed and take notes on the process until the roof is completed. Ask questions of the roofers, the architect, or your instructor about anything you don't understand.

4. Examine a number of existing low-slope roofs around your campus or neighborhood, looking for problems such as cracking, blistering, tearing, and leaking. Explain the reasons for each problem that you discover.

GLASS AND GLAZING

The lobby of the Messeturm office building in Frankfurt, Germany, features a very tall, curving wall of tempered glass. Vertical glass stiffeners support the glass sheets through bolted stainless steel angle connectors. The stiffeners are sustained against wind forces by four long, horizontal arches made of steel tubing. (*Photo courtesy of Murphy/Jahn, Architects*)

Glass plays many roles and takes many forms in buildings—Gothic church windows made of thousands of jewel-like pieces of colored glass; breathtaking expanses of smooth, uninterrupted glass that fill whole walls of today's buildings; Elizabethan casement windows with tiny diamond panes set in lead; skyscrapers that shimmer in facets of reflective glass mirroring the sky; cozy windows; comfortable windows; windows that bring soft, natural light; windows that frame spectacular views; windows that welcome winter sunlight to warm a room. But glass can also form windows that make privacy impossible; windows that admit a harsh, glaring light; winter-cold surfaces that chill the body and tax the heating system; windows that broil a room in summer-afternoon sunlight. Glass skillfully used in building contributes strongly to our enjoyment of architecture, but glass thoughtlessly used can make a building unattractive, uneconomical, and uncomfortable to inhabit.

HISTORY

The origins of glass are lost in prehistory. Initially a material for colored beads and small bottles, glass was first used in windows in Roman times. The largest known piece of Roman glass, a crudely cast sheet used for a window in a public bath at Pompeii, was nearly 3 feet by 4 feet (800 mm by 1100 mm) in size.

By the 10th century A.D., the Venetian island of Murano had become the major center of glass-making, producing *crown glass* and *cylinder glass* for windows. Both the crown and cylinder processes were begun by blowing a large glass sphere. In the crown process, the heated glass sphere was adhered to an iron rod called a *punty*, opposite the blowpipe. The blowpipe was then removed, leaving a hole opposite the punty. Then the sphere was reheated, whereupon the glassworker would spin the punty rapidly, causing centrifugal force to open the sphere into a large disk, or *crown*, 30 inches (750 mm) or more in diameter (Figure 17.2). When the crown was cut into panes, one always contained the "bullseye" where the punty was attached before being cracked off. In the cylinder process, the sphere, heated to a molten state, was swung back and forth pendulum fashion on the end of the blowpipe to elongate it into a cylinder. The hemispherical ends were cut off and the remaining cylinder was slit lengthwise, reheated, opened, and flattened into a rectangular sheet of glass that was later cut into panes of any desired size (Figure 17.3). Prior to the introduc-

FIGURE 17.1
An entire wall of the Baltimore Convention Center is made of low-emissivity double glazing supported by a tubular steel substructure. Adjustable stainless steel fittings with rubber washers connect the sheets of glass to the substructure. (*Photo of Pilkington Planar System courtesy of W & W Glass Systems, Inc.*)

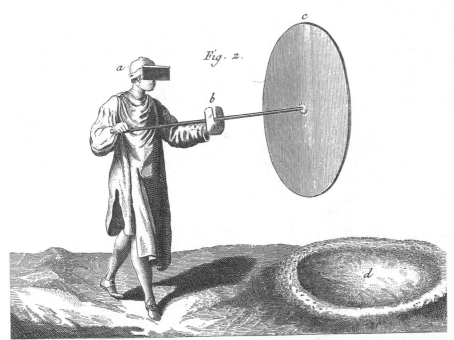

FIGURE 17.2
The glassworker in this old engraving wore a face shield (*a*) and hand shield (*b*) to protect against the heat of the large glass crown (*c*) which he had just spun on the end of a punty. After cooling, the crown was cut into small lights of window glass. (*Courtesy of the Corning Museum of Glass, Corning, New York*)

FIGURE 17.3
Making cylinder glass in the 19th century, Pittsburgh, Pennsylvania. Elongated glass bottles were blown by swinging the blowpipe back and forth in the pit in front of the furnace (center). As each bottle solidified (left), it was brought to another area where the ends were cut off to produce cylinders (right). The cylinders were reheated and flattened into sheets from which window glass was cut. (*Courtesy of the Corning Museum of Glass, Corning, New York*)

tion of modern glassmaking techniques, crown glass was favored over cylinder glass for its surface finish, which was smooth and brilliant because it was formed without contacting another material. Cylinder glass, though more economical to produce, was limited in surface quality by the texture and cleanliness of the surface on which it was flattened.

Neither crown glass nor cylinder glass was of sufficient optical quality for the fine mirrors desired by the 17th-century nobility; for this reason, *plate glass* was first produced, in France, in the late 17th century. Molten glass was cast into frames, spread into sheets by rollers, cooled, then ground flat and polished with abrasives, first on one side and then the other. The result was a costly glass of near-perfect optical quality, in sheets of unprecedentedly large size. Before long, mechanization of the grinding and polishing operations brought down the price of plate glass to a level that allowed it to be used for storefronts in both Europe and America.

In the 19th century, the cylinder process evolved into a method of drawing cylinders of molten glass vertically from a crucible, enabling the routine, economical production of cylinders 40 to 50 feet (12 to 15 m) long. In 1851, the Crystal Palace in London (Figure 11.1) was glazed with

900,000 square feet (84,000 m²) of cylinder glass supported on a cast iron structure.

In the early years of the 20th century, cylinder glass production was gradually replaced by processes that pulled flat sheets of *drawn glass* directly from a container of molten glass. Highly mechanized production lines for the grinding and polishing of plate glass were established with rough glass sheets entering the line continuously at one end and finished sheets emerging at the other.

In 1959, the English firm of Pilkington Brothers, Ltd., started production of *float glass*, which has since been licensed to other glassmakers and has become the worldwide standard, replacing both drawn glass and plate glass. In this process, a ribbon of molten glass is floated across a bath of molten tin, where it hardens before touching a solid surface (Figures 17.4–17.6). The resulting sheets of glass have parallel surfaces, high optical quality (virtually indistinguishable from plate glass), and a brilliant surface finish. Float glass has been produced in the United States since 1963, and now accounts for nearly all of domestic flat glass production.

The terminology associated with glass developed early in this long history. The term *glazing* as it applies to building refers to the installing of glass in an opening, or to the trans-

parent material (usually glass) in a glazed opening. The installer of glass is known as a *glazier*. Individual pieces of glass are known as *lights*, or often, to avoid confusion with visible light, *lites*.

THE MATERIAL GLASS

The major ingredient of glass is sand (silicon dioxide), which is mixed with soda ash (sodium hydroxide or sodium carbonate), lime, and small amounts of alumina, potassium oxide, and various elements to control color, then heated to form glass. The finished material, while seemingly crystalline and convincingly solid, is actually a supercooled liquid, for it has no fixed melting point and an open, noncrystalline microstructure.

Ordinary, clear $\frac{3}{32}$-inch (2.5-mm) glass transmits about 85 percent of the visible light incident upon it. When drawn into small fibers, glass is stronger than steel, though not as stiff. In larger pieces, the microscopic imperfections that are an inherent characteristic of glass reduce its useful strength to significantly lower levels, particularly in tension. When a surface of a sheet of glass is placed in sufficient tension, as happens when an object strikes the glass, cracks propagate from an imperfection near

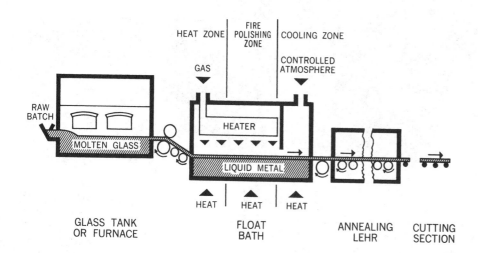

FIGURE 17.4

In the float glass process, molten glass from the furnace is floated on a bath of liquid tin to form a continuous sheet of glass. The annealing lehr cools the glass at a controlled rate to avoid internal stresses, after which it is cut into smaller sheets. (*Courtesy of PPG Industries*)

FIGURE **17.5**
The superior flatness and bright surface finish of float glass are readily seen in the reflections on the glass ribbon emerging from the annealing lehr. (*Courtesy of LOF Glass, a Libby–Owens–Ford Company*)

FIGURE **17.6**
Track-mounted cutting devices score the ribbon of cooled float glass as part of a computer-controlled cutting operation that automatically produces the glass sizes ordered by customers. (*Courtesy of PPG Industries*)

the point of maximum tension, and the glass shatters.

Thicknesses of Glass

Glass is typically manufactured in a series of thicknesses ranging from approximately $3/32$ inch (2.5 mm), which is called *single-strength*, through $1/8$ inch (3 mm), called *double-strength*, to a maximum of as much as 1 inch (25.4 mm), depending on the manufacturer. Glass thickness for a particular window is determined by the size of the glass light, the expected maximum wind loads on the glass, and the predicted breakage rate that can be accepted. For low buildings with relatively small windows, single-strength and double-strength glass are usually sufficient. For larger windows and for windows in tall buildings, where high wind velocities are experienced at higher altitudes, thicker glass is generally required, along with increased attention to how the glass is supported in its frame. It has become standard practice for architects and structural engineers to order extensive wind tunnel testing of models of tall buildings during the design process to establish the expected maximum wind pressures and suctions on the windows. Because of unavoidable manufacturing defects in the glass and the probability of subsequent damage to the glass during installation and while it is in service, a certain amount of breakage must always be anticipated in a large building. *ASTM E1300* establishes standard procedures for evaluating the structural stability and probability of breakage in glass. These are used to determine a glass thickness that will result in an acceptably low probability of breakage for a window of given dimensions, support conditions, and wind pressure.

During its manufacture, ordinary window glass is *annealed*, meaning that it is cooled slowly under controlled conditions to avoid locked-in thermal stresses that might cause it to behave unpredictably in use. But several other types of glass have come into use for particular purposes in buildings.

Tempered Glass

Tempered glass is produced by cutting annealed glass to the required sizes for use, reheating it to approximately 1200 degrees Fahrenheit (650°C), and cooling both its surfaces rapidly with blasts of air while its core cools much more slowly. This process induces permanent compressive stresses in the edges and faces of the glass, and tensile stresses in the core. The resulting glass is about four times as strong in bending as annealed glass, and much more resistant to thermal stress and impact. When tempered glass does break, the sudden release of its internal stresses reduces it instantaneously to small, square-edged granules rather than long, sharp-edged shards. These properties make tempered glass useful for windows exposed to heavy wind pressures or intense heat or cold. From the point of view of safety, the breakage characteristics of tempered glass make it especially well suited for use in and around exterior doors, where people may accidentally bump against the glass, and for floor-to-ceiling sheets of glass, which are often walked into by people who mistake them for openings in the wall. Tempered glass is also used for all-glass doors that have no frame at all (Figure 17.7), for whole walls of squash and handball courts, for hockey rink enclosures, and for basketball backboards.

Tempered glass is more costly than annealed glass. It usually has noticeable optical distortions that have been created by the tempering process. And all cutting, drilling, and edging must be done before the glass is tempered because any such operations after tempering will release the stresses in the glass and cause it to disintegrate.

Heat-Strengthened Glass

For many applications, lower-cost *heat-strengthened glass* may be used instead of tempered glass. The heat-strengthening process is similar to tempering, but the induced compressive stresses in the surface and edges are about one-third as high (typically 5000 pounds per square inch compared to 15,000 pounds per square inch for tempered glass, or 34 MPa versus 104 MPa). Heat-strengthened glass is about twice as strong in bending as annealed glass, and much more resistant to thermal stress. It usually has fewer distortions than tempered glass. Its breakage behavior is more like that of annealed glass than tempered glass.

Laminated Glass

Laminated glass is made by sandwiching a transparent *polyvinyl butyral* (*PVB*) interlayer between sheets of glass and bonding the three layers together under heat and pressure. Laminated glass is not quite as strong as annealed glass of the same thickness, but when laminated glass breaks, the soft interlayer holds the shards of glass in place rather than allowing them to fall out of the frame of the window. This makes laminated glass useful for skylights and overhead glazing, because it reduces the risk of injury to people below in case of breakage (Figures 17.8, 17.9). The PVB interlayer may be colored or patterned to produce a wide range of visual effects in laminated glass.

Laminated glass is a better barrier to the transmission of sound than solid glass. It is used to glaze windows of residences, classrooms, hospital rooms, and other rooms that must be kept quiet in the midst of noisy environments. It is especially effective when installed in two or more layers with airspaces between.

Security glass, used for drive-in banking windows and other facilities that need to be resistant to burglary, is made of multiple layers of glass and PVB, and is available in a range of thicknesses to stop any desired caliber of bullet.

Fire-Rated Glass

Glass in fire doors and fire separation walls must maintain its integrity as a barrier to the passage of smoke, heat, and flames even after it has been exposed to heat for a period of time. ASTM Standards E152, E163, and E119 establish fire endurance tests for fire door, window, and wall assemblies, respectively. Several different types of glazing are able to pass these tests when properly mounted in appropriate frames: *Wired glass* is produced by rolling a mesh of small wires into a sheet of hot glass. When wired glass breaks from thermal stress, the wires hold the fragments of glass together to act as a fire barrier. *Optical-quality ceramic* is more stable against thermal breakage than any type of glass. It looks and feels like

FIGURE 17.7
Tempered glass is used for strength and breakage safety in both the doors and the windows of this store in a downtown shopping mall. (*Photo by the author*)

FIGURE 17.8
Facets of laminated glass are borne by an elegant structure of stainless steel rods to form a skylight. (*Photo of Pilkington Planar System courtesy of W & W Glass Systems, Inc.*)

FIGURE 17.9
Laminated glass provides safety against falling glass in an overhead sloped glazing installation. (*Courtesy of PPG Industries*)

glass but does not require wires. Another type of fire-rated glazing consists of a clear, water-based polymer gel contained between two sheets of tempered glass. When the glazing is heated by a fire, the water in the gel absorbs heat until it begins to boil, at which point the polymer turns to an opaque layer that insulates against the passage of radiant heat.

Patterned Glass

Hot glass can be rolled into sheets with many different surface patterns and textures for use where light transmission is desired but vision must be obscured for privacy.

Fritted Glass

A number of producers are equipped to imprint the surface of glass with silk-screened patterns of ceramic-based paints. The paints consist primarily of pigmented glass particles that are called *frit*. After the frit has been printed onto the glass, the glass is dried and then fired in tempering furnaces, transforming the frit into a hard, permanent ceramic coating. Many colors are possible in both translucent and opaque finishes.

Typical patterns for *fritted glass* are various dot and stripe motifs (Figure 17.10), but custom-designed patterns and even text are easily reproduced. The decorative possibilities are obvious. Fritted glass is often used to control the penetration of solar light and heat into a space.

Spandrel Glass

Frits are applied in uniform layers on the interior surfaces to create special opaque glasses for covering spandrel areas (the bands of wall around the edges of floors) in glass curtain wall construction (Figure 17.11). Some

FIGURE 17.10
Fritted patterns modulate the sunlight that enters a theater lobby. (*Photo of Pilkington Planar System courtesy of W & W Glass Systems, Inc.*)

spandrel glasses are made as similar as possible in exterior appearance to the glass that will be used for the windows on a specific project. It is very difficult, however, even with reflective coated glass, to make the spandrels indistinguishable from the windows under all lighting conditions. Most spandrel glasses are made to contrast with the windows of the building. Factory-applied thermal insulation assemblies on the interior of the glass can be supplied by many glass companies, complete with vapor retarder. Spandrel glass is usually tempered or heat-strengthened to resist the thermal stresses that can be caused by accumulation of solar heat behind the spandrel.

Tinted and Reflective Coated Glass

Solar heat buildup can be problematic in the inhabited spaces of buildings with large areas of glass, especially during the warm part of the year. Fixed sun shading devices outside the windows are the best ways of blocking unwanted sunlight, but glass manufacturers have developed tinted and reflective glasses that are designed to reduce glare and cut down on solar heat gain.

Tinted Glass

Tinted glass is made by adding small amounts of selected chemical elements to the molten glass mixture to produce the desired hue and intensity of color in grays, bronzes, blues, greens, and golds. The visible light transmitted by commercially available tinted glasses varies from 14 percent in a very dark gray glass to 75 percent in the lightest tints, as compared to about 85 percent for clear glass. The total amount of transmitted solar heat is significantly higher than these figures, however, because the heat absorbed by the glass must go somewhere, and a substantial portion of it is transmitted to the interior of the building (Figure 17.12).

To simplify comparisons among various types of heat-reducing glasses, a quantity called the *shading coefficient* is used; it is the ratio of total solar

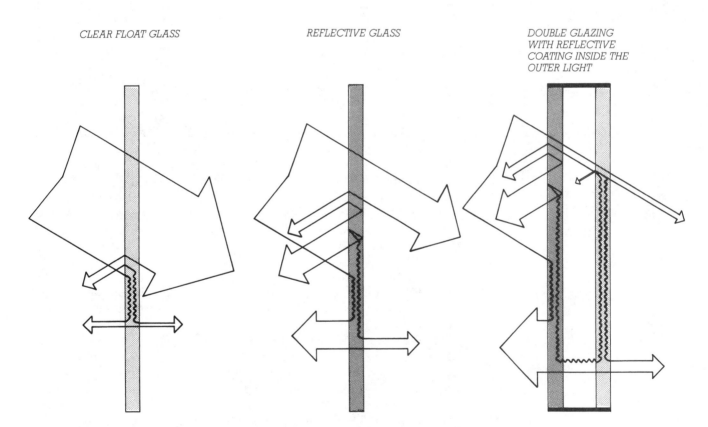

CLEAR FLOAT GLASS   REFLECTIVE GLASS   DOUBLE GLAZING WITH REFLECTIVE COATING INSIDE THE OUTER LIGHT

FIGURE 17.12
A schematic representation of the effect of three different glazing assemblies on incoming sunlight. Outdoors is to the left. The relative widths of the arrows indicate the relative percentages of the incoming light transmitted, reflected, and absorbed. In clear float glass, to the left, most of the light is transmitted, with small quantities reflected, absorbed, and reradiated as heat. Reflective coated glass, center, bounces a large proportion of the light back to the outdoors, and also absorbs and reradiates a significant amount. In double glazing, many different combinations of types of glass are possible; the one shown to the right of this diagram utilizes glass with an inside reflective coating for the outer light.

Imagine a city iridescent by day, luminous by night, imperishable! Buildings, shimmering fabrics, woven of rich glass; glass all clear or part opaque and part clear, patterned in color or stamped to harmonize with the metal tracery that is to hold it all together, the metal tracery to be, in itself, a thing of delicate beauty consistent with slender steel construction . . .

Frank Lloyd Wright, in
Architectural Record, April 1928

heat transmission through a particular glass to total solar heat transmission through double-strength clear glass. Shading coefficients for tinted glasses range from about 0.50 to 0.75, indicating that they transmit from one-half to three-quarters of the incident solar energy that would be transmitted by double-strength clear glass, whose shading coefficient is established as 1.0. *Visible transmittance* measures the transparency of glass to visible light rather than to solar heat; values range downward from about 0.9 for clear glass.

These two coefficients are combined to determine the *glazing luminous efficacy* (K_e), an important measure of the energy-conserving potential of glass. K_e is defined as the visible transmittance divided by the shading coefficient, and is tabulated for various types of tinted and reflective glass in manufacturers' catalogs. A high K_e means that a great deal of solar heat is blocked by the glass while most daylight is still allowed to enter the building. A low K_e indicates that the glass blocks similar fractions of both heat and light. Green and blue tinted glasses tend to have high K_e values, meaning that they are useful for admitting large quantities of natural light while keeping out large amounts of heat. Bronze, gold, and gray tints tend to have low K_e values.

Reflective Coated Glass

To achieve lower shading coefficients, thin, durable films of metal or metal oxide can be deposited on a surface of either clear or tinted glass sheets under closely controlled conditions. Depending on the composition of the film, the film side of the glass may be turned either toward the inside of the building or toward the outside. The film is thin enough to see through, but turns away a substantial proportion of the incident solar energy before it can enter the building. Shading coefficients for reflective coated glasses vary from about 0.31 to 0.70, depending on the density of the metallic coating.

Reflective coated glasses appear as mirrors from the outside on a bright day and are often chosen by architects for this property alone (Figure 17.13). At night, with lights on inside the building, they appear as dark but transparent glass.

Mirrored glass has the lowest K_e values of any glass. This means that, while it blocks passage of most solar heat, it also blocks most visible light, which results in a building with a dark interior.

The sunlight reflected by a building of reflective coated glass can be helpful in some circumstances by lighting an otherwise dark urban street space. It can also create problems in other circumstances by bouncing solar heat and glare into neighboring buildings and into the eyes of pedestrians and motorists.

Insulating Glass

Window glass is an extremely poor thermal insulator. A single sheet of glass conducts heat about five times as fast as an inch (25 mm) of polystyrene foam insulation, and twenty times as fast as a well-insulated wall. A second sheet of glass applied to a window with an airspace between the two sheets (*double glazing*) cuts this rate of heat loss in half, and a third sheet with its additional airspace (*triple glazing*) reduces the rate of heat loss to about a third of the rate through a single sheet. Even a triple-glazed window, however, still loses heat about six times as fast as the wall in which it is placed.

The practice of doubling or tripling layers of glass in windows has long been used in cold climates, but was frequently attended by problems of moisture condensation and frost inside the outer sheet of glass. In recent decades, the factory production of hermetically sealed double-glazed or triple-glazed window units, with dry air inserted in the space between, has virtually eliminated these problems. Until a few years ago, for small lights of double glass, the

FIGURE 17.13
Reflective coated windows with subtly different reflective coated glass spandrels. (*Architects: Paul Rudolph and 3D International. Photo courtesy of PPG Industries*)

Who when he first saw the sand and ashes . . . would have imagined that in this shapeless lump lay concealed so many conveniences in life . . . by some such fortuitous liquefaction was mankind taught to procure a body at once in a high degree solid and transparent; which might admit the light of the sun, and exclude the violence of the wind; which might extend the sight of the philosopher to new ranges of existence . . .

Dr. Samuel Johnson, writer and lexicographer, 1709–1784

edges of the two sheets were simply fused together (Figure 17.14). However, this detail is seldom used now because the fused glass edge is very conductive of heat. For lights of any size, and for any number of thicknesses of glass, a hollow metal spacer bar (also called a *spline*) is inserted between the edges of the sheets of glass, and the edges are closed with an organic sealant compound. A small amount of a chemical drying agent is left inside the spacer bar to remove any residual moisture from the trapped air. The air or gas is always inserted at atmospheric pressure to avoid structural pressures on the glass. The thickness of the airspace is less critical to the insulating value of the glass than the mere presence of an airspace: From $\frac{3}{8}$ inch (9 mm) up to about an inch (25 mm) of thickness, the insulating value of the airspace increases somewhat, but above that thickness little additional benefit is gained. A standard overall thickness for large lights of double glazing is 1 inch (25.4 mm), which results in an airspace $\frac{1}{2}$ inch (13 mm) thick if $\frac{1}{4}$-inch (6-mm) glass is used.

For slightly improved thermal performance, stainless steel, which is less conductive of heat, may be used instead of aluminum for the spacer bar, and a sealant material may be placed between the glass and the spacer bar to act as insulation.

A low-conductivity gas filling between the sheets of glass in a double glazing unit adds 12 to 18 percent to its thermal efficiency, depending on the type of gas used and the thickness of the gas-filled space. Argon and krypton, both of which are components of air, are the gases most commonly used.

The thermal performance of double glazing can be improved substantially by the use of a *low-emissivity* (*low-e*) *coating* on one or both of the sheets of glass. A low-emissivity coating is an ultrathin, transparent metallic coating that reflects selected wavelengths of light and heat radia-

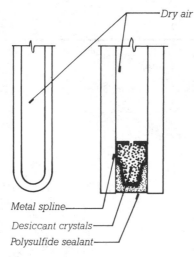

FIGURE 17.14
Two methods of sealing the edge of double glazing: fused glass edges, to the left, and a metal spline and organic sealant, to the right. Desiccant crystals in the spline (spacer bar) absorb any residual moisture in the airspace.

tion. A coating of the type used to improve the wintertime insulating performance of a window is selectively reflective within the broad spectrum of infrared wavelengths, allowing most of the shorter-wavelength infrared radiation that is contained in sunlight to pass through, but reflecting back to the indoor space most of the longer-wavelength infrared radiation that is given off by warm objects and human bodies inside the building. Typically, a double-glazed window with a low-emissivity coating on one glass surface is equivalent thermally to a triple-glazed window. Accordingly, triple-glazed windows, which are heavy and cumbersome, have largely disappeared from the market in favor of double-glazed windows with low-emissivity coatings. The performance of this type of glazing can be further enhanced by a low-conductivity gas filling. Some window manufacturers also add a very thin, stretched membrane of transparent plastic parallel to the sheets of glass in the center of

the air- or gas-space as a virtually weightless third glazing to further improve insulating performance.

Low-e coating technology has also been adapted to such uses as glass that reduces solar heat gain so as to reduce the energy consumed for cooling in a hot climate.

Glass That Changes Its Properties

New to the market are several types of glass with variable properties. *Thermochromic glass* becomes darker when it is warmed by the sun. *Photochromic glass* becomes darker when exposed to bright light. Thus, both types are potentially valuable as passive devices to reduce cooling loads in buildings. *Electrochromic glass* changes from transparent to fully opaque with the passage of a small electric current, a property that makes it useful both for privacy and for energy conservation. *Photovoltaic glass* is coated with a thin film of amorphous silicone that generates electricity from sunlight. It allows a building with a large glazed area to create at least some of the electric energy that it uses for lights and machinery.

Plastic Glazing Sheets

Transparent plastic sheets are often used instead of glass for certain specialized glazing applications. The two most common types of plastic glazing materials are *acrylic* and *polycarbonate*. Both are more expensive than annealed glass. Both have very high coefficients of thermal expansion, which cause them not merely to expand and contract with temperature changes, but also to bow visibly toward the warm side when subjected to high indoor–outdoor temperature differentials. This, in turn, requires that plastic sheet materials be installed in their frames with relatively expensive glazing details that allow for plenty of linear movement and rotation. Both polycarbonate and acrylic are soft and easily scratched, although more scratch-resistant formulations are available. Despite these problems, plastic glazing sheets find use in many situations where glass is inappropriate: The plastics can be cut to shapes with inside corners (L-shapes and T-shapes, for example) that are likely to crack if cut from glass. They can be bent easily to fit in curved frames. They can be heat-formed into domed glazing for skylights. And the plastics, especially polycarbonate, which is literally impossible to break under ordinary conditions, are widely used for windows in buildings where vandalism is a problem.

Translucent but nontransparent plastic sheets reinforced with glass fibers are also used in buildings. Corrugated sheets of this material are used for industrial skylights and residential patio roofs. Thin, flat sheets of a special formulation with a high translucency to solar energy are used for skylights and low-cost solar collector glazing.

GLAZING

Glazing Small Lights

Small lights of glass are subjected neither to large wind force stresses nor to large amounts of thermal expansion and contraction. They may be glazed by very simple means (Figure 17.15). In traditional wood sash, the glass is first held in place by small metal *glazier's points,* and then sealed on the outside with *putty,* a simple compound of linseed oil and pigment which gradually hardens by oxidation of the oil. Putty must be protected from the weather by subsequent painting. For factory-glazed sash, improved putties or *glazing compounds* are employed. These are generally more adhesive and more elastic than traditional putty, and are formulated so as not to harden and become brittle with age.

Glazing Large Lights

Large lights of glass (those over 6 square feet or 0.6 m² in area) require more care in glazing. Wind load stresses on each light of glass are higher, and the glass must span farther between its supporting edges. Any irregularities in the frame of the window may result in distortion of the glass, highly concentrated pressures on small areas of the glass, or glass-to-metal contact, any of which can lead to abrasion or fracture of the glass. In large lights, thermal expansion and contraction can also cause stresses to build up in the glass.

The design objectives for a large-light glazing system are:

1. To support the weight of the glass in such a way that the glass is not subjected to intense or abnormal stress patterns.

2. To support the glass against wind pressure and suction.

3. To isolate the glass from the effects of structural deflections in the frame of the building and in the smaller framework of *mullions* that supports the glass.

4. To allow for independent expansion and contraction of both glass and frame without damage to either.

5. To avoid contact of the glass with the frame of the window or with any other material that could abrade or stress the glass.

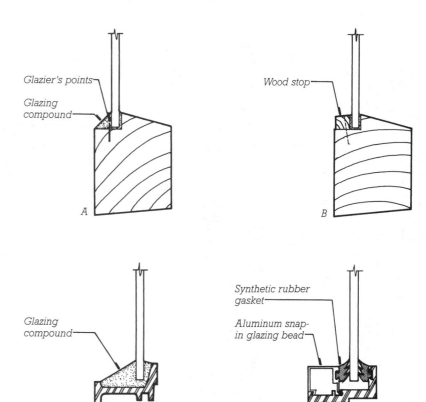

FIGURE 17.15
Alternative methods of single glazing small lights of glass. Glass is traditionally mounted in wood sash using glazier's points and glazing compound (*a*), or a wood stop nailed to the sash (*b*). Metal sash was once glazed in the manner shown in (*c*), but most metal sash is now glazed with snap-in beads and synthetic rubber gaskets, as exemplified in (*d*).

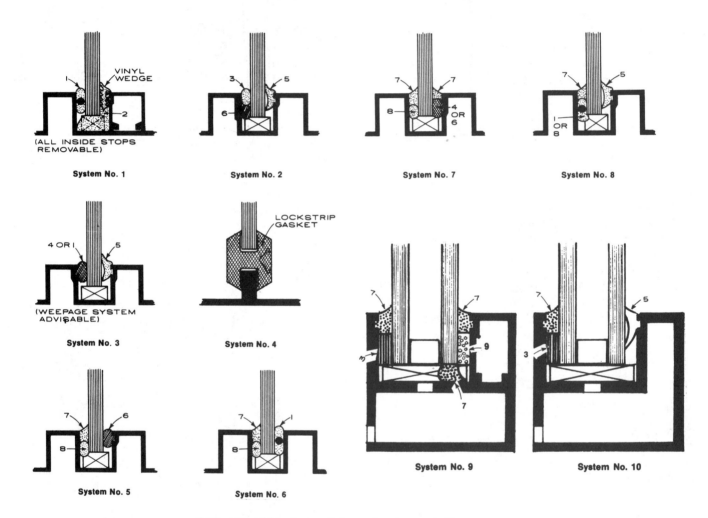

Outdoor side at left in all cases. System numbers have no significance as to order of preference.

KEY TO MATERIALS

1 Pre-shimmed butyl or polyisobutylene tape
2 Acrylic or butyl sealant
3 Butyl or polyisobutylene tape
4 Cellular neoprene
5 Dense neoprene roll-in gasket
6 Dense neoprene gasket
7 High range sealants (polysulfide, silicone or acrylic sealant) (see Note)
8 Dense neoprene roll-in rod
9 Cellular Tape

Note: Choice of cap sealant depends on opening size and other conditions; consult sealant mfr.

FIGURE 17.16

An assortment of typical large-light glazing details, with the outdoor side to the left. (*Reprinted with permission from AAMA Curtain Wall Design Guide Manual*)

The weight of a large light of glass is supported in the frame by synthetic rubber *setting blocks,* normally two per light, located at the quarter points of the bottom edge of the light. For support against wind loads, a specified amount of *bite* (depth of grip on the edge of the glass) is provided by the supporting *mullions.* If the bite is too little, the glass may pop out under wind loading; if too much, the glass may not be able to deflect enough under heavy wind loads without being stressed at the edge. The mullions, of course, must be stiff enough to transmit the wind loads from the glass to the frame of the building without deflecting so far as to overstress the glass. The resilient glazing material used to seal the glass-to-mullion joint must be of sufficient dimension and elasticity to allow for any anticipated thermal movement and for possible irregularities in the mullions.

The glazing materials that are most commonly used between the mullions and the glass include *wet glazing components* and *dry glazing components.* The wet components are mastic sealants and glazing compounds. The dry components are rubber or elastomeric gaskets. Wet glazing, with good workmanship, is more effective in sealing against penetration of water and air. Dry glazing is faster, easier, and less dependent on workmanship than wet glazing. The two types are often used in combination to utilize the best properties of each.

In the large-light glazing systems shown in Figure 17.16, rubber setting blocks are indicated by the rectangles with X's in them under the lower edge of the glass. Systems 1, 2, 3, 6, 9, and 10 use *preformed solid tape sealant,* a thick ribbon of very sticky synthetic that is adhered by pressure to the glass and the mullions. Several differ-

ent formulatio lene) are usually name *polybutene.* Th an extremely strong ho and stays plastic indefinit for movement in the glazing Systems 1, 2, 3, 8, and 10 utilize or *roll-in gaskets,* preformed strip elastomeric material that are simply pushed into the gap between the glass and the mullion on the interior side to wedge the assembly tightly together and seal against air leakage. Systems 5 through 10 seal the outside gap with a wet glazing material. System 4 comprises a two-piece *lockstrip gasket* that is a completely self-contained glazing system. Figures 17.17 through 17.19 show applications of lockstrip gasket glazing.

The properties that solid tape sealants, mastic sealants, and compression gaskets have in common are that they possess the required degree of resiliency, they can be installed in

FIGURE 17.17
Installing a lockstrip gasket on an aluminum mullion. (*Courtesy of Standard Products Company*)

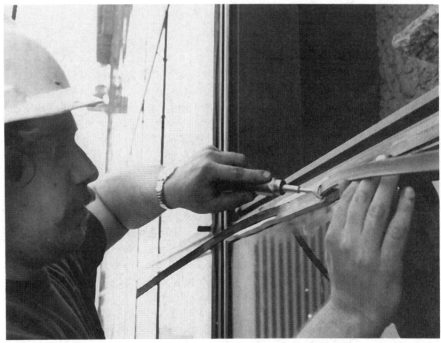

FIGURE 17.18
Inserting the lockstrip to expand the gasket and seal it against the glass.
(*Courtesy of Standard Products Company*)

...s (butyl, polyisobuty-
...lumped under the
...s material exerts
...d on the glass
...ely to allow
...r system.
wedge
...

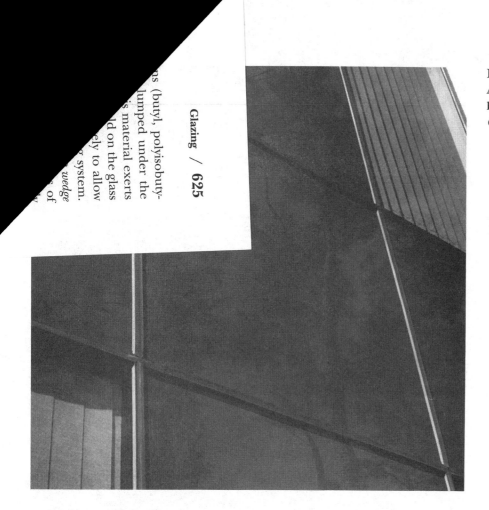

FIGURE 17.19
A finished lockstrip gasket glazing instal-
lation. (*Courtesy of Standard Products
Company*)

the thickness required to cushion the glass against all expected movements, and they form a watertight seal against both glass and frame. To guard against possible leakage and moisture condensation, however, *weep holes* should be provided to drain water from the horizontal mullions, as seen directly below the bottom edge of the glass in Systems 9 and 10 in Figure 17.16.

Advanced Glazing Systems

In their quest to design ever more minimal buildings, architects have encouraged the development of several systems of glazing that seem, in varying degrees, to defy gravity. In *butt-joint glazing*, the head and sill of the glass sheets are supported conventionally in metal frames, but the vertical mullions are eliminated, the vertical joints being made by the injection of a colorless silicone sealant between the vertical edges. This gives a strong effect of unbroken horizontal bands of glass wrapping continuously around the building (Figures 17.21, 17.22). In *structural silicone flush glazing*, the metal mullions lie entirely inside the glass, with the glass adhered to the mullions by silicone sealant. This allows the outside skin of the building to be completely flush, unbroken by protruding mullions (Figures 17.23–17.26). Notice in Figure 17.24 that the critical silicone work is done in a factory, not on a construction site. The lights of glass are transported to the site already adhered to the small aluminum channels that will bind them to the mullions.

FIGURE 17.20
Glaziers install lights of reflective glass that weigh as much as 125 pounds (57 kg) each in a spire atop an office building. The glass is attached to the hoisting rope by means of suction cups, as seen in front of the worker at the lower left. (*John Burgee Architects with Philip Johnson. Photo courtesy of PPG Industries*)

FIGURE 17.21
Mullionless butt-joint glazing uses only a bead of colorless silicone sealant at the vertical joints in the glass. The glass in this example is single glazing ³/₄ inch (19 mm) thick.
(*Courtesy of LOF Glass, a Libby–Owens–Ford Company*)

FIGURE 17.22
Another mullionless butt-joint glazing installation, as seen from the outside.
(*Architects: Neuhaus & Taylor. Photo courtesy of PPG Industries*)

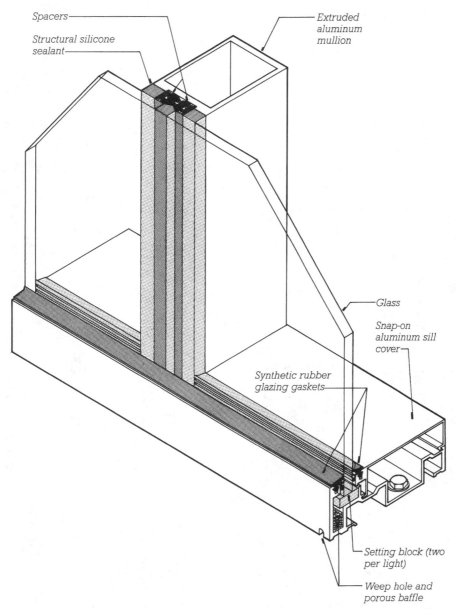

Spacers

Structural silicone sealant

Extruded aluminum mullion

Glass

Snap-on aluminum sill cover

Synthetic rubber glazing gaskets

Setting block (two per light)

Weep hole and porous baffle

FIGURE 17.23
Horizontal strip windows that need to appear mullionless only from the exterior can be created by adhering the glass to interior mullions with structural silicone sealant. The sill and head are conventionally glazed, using snap-on aluminum covers to hold the interior glazing gaskets. Either single glazing, as shown, or double glazing can be used with this type of system. (*Copied by permission from PPG EFG System 401 details. Courtesy of PPG Industries*)

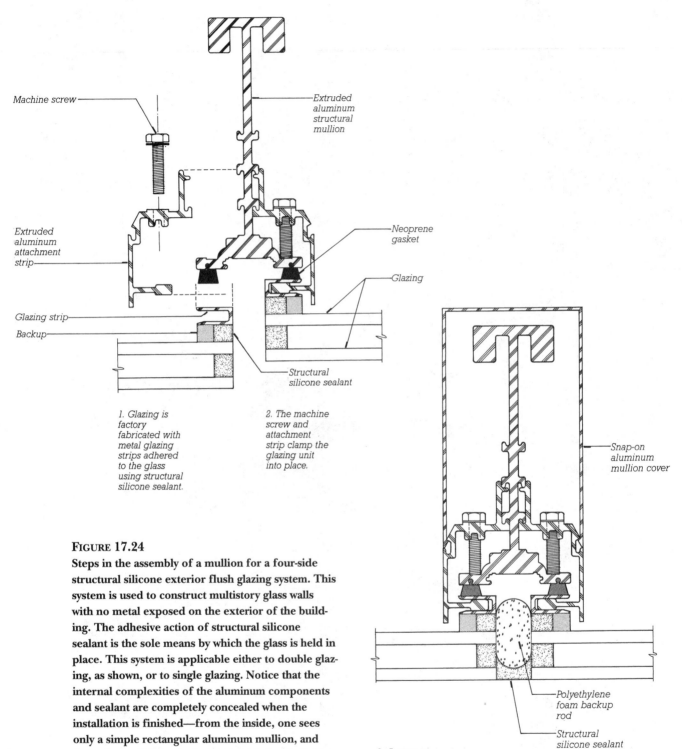

Machine screw

Extruded
aluminum
structural
mullion

Extruded
aluminum
attachment
strip

Neoprene
gasket

Glazing

Glazing strip

Backup

Structural
silicone sealant

*1. Glazing is
factory
fabricated with
metal glazing
strips adhered
to the glass
using structural
silicone sealant.*

*2. The machine
screw and
attachment
strip clamp the
glazing unit
into place.*

Snap-on
aluminum
mullion cover

Polyethylene
foam backup
rod

Structural
silicone sealant

*3. A snap-on
mullion cover
and an exterior
sealant joint
complete the
assembly.*

FIGURE 17.24
Steps in the assembly of a mullion for a four-side
structural silicone exterior flush glazing system. This
system is used to construct multistory glass walls
with no metal exposed on the exterior of the build-
ing. The adhesive action of structural silicone
sealant is the sole means by which the glass is held in
place. This system is applicable either to double glaz-
ing, as shown, or to single glazing. Notice that the
internal complexities of the aluminum components
and sealant are completely concealed when the
installation is finished—from the inside, one sees
only a simple rectangular aluminum mullion, and
from the outside, only glass and a thin bead of sili-
cone sealant. (*Copied by permission from PPG EFG*
System 712 details. Courtesy of PPG Industries)

FIGURE 17.25

Reflective glass mounted with a four-side structural silicone glazing system shows no metal on the exterior of this Texas office building, only thin lines of sealant. (*Architects: Haldeman, Miller, Bregman & Haman. Photo courtesy of PPG Industries*)

FIGURE 17.26

Structural spacer glazing is a patented s[...] flush glazing that provides a more positi[...] attachment of the double glazing unit to [...] building. The glass is fastened to the mu[...] with an aluminum pressure plate that eng[...] slot in the spacer strip between the sheet[...] glass. The desiccant required to remove [...] moisture from the airspace is mixed with [...] butyl sealant material in this system.

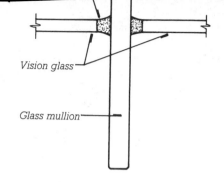

FIGURE 17.28
A typical detail of a glass mullion in a suspended glazing assembly.

the racetrack in this suspended
(Photo courtesy of PPG Industries)

entry of solar heat into buildings, to the point that they might be perceived as encouraging the designer to pay little attention to window size and orientation. But there is a growing body of buildings that are characterized by different fenestration schemes for the different sides of the building, each designed to create an optimal flow of heat into and out of the building for that orientation, and each making creative use of the available types of glass to help accomplish this objective. The results, as measured in occupant comfort and energy savings, are generally impressive, and the aesthetic possibilities are intriguing.

This last statement could apply equally well to the role of glass in admitting light to a building. Electric lighting is often the major consumer of energy in a commercial building, especially when the heat generated by the lights must be removed from the building by a cooling system. Daylight shining through windows and skylights, distributed throughout a space by reflecting and diffusing surfaces, can reduce or eliminate the need for electric lights under many

circumstances and is often more pleasant than artificial illumination. Low-cost computer models make it easy to predict the levels of *daylighting* that can be achieved with alternative designs, and more and more architects and engineers are becoming expert in this field.

GLASS AND THE BUILDING CODES

Building codes concern themselves with several functional aspects of glass: its role in providing natural light in habitable rooms; its breakage safety; its safety in preventing the spread of fire through a building; and its role in determining the energy consumption of a building.

The BOCA Code, while permitting the use of artificial illumination alone in many types of buildings, encourages the use of natural light and requires that each room have a glass area equal to at least 8 percent of its floor area. Minimum ceiling heights are established in the same section of the code, in part to assist in

Winter dining rooms and bathrooms should have a southwestern exposure, for the reason that they need the evening light, and also because the setting sun, facing them in all its splendor but with abated heat, lends a gentler warmth to that quarter in the evening. Bedrooms and libraries ought to have an eastern exposure, because their purposes require the morning light. . . . Dining rooms for Spring and Autumn to the east; for when the windows face that quarter, the sun, as he goes on his career from over against them to the west, leaves such rooms at the proper temperature at the time when it is customary to use them.

Marcus Vitruvius Pollio, Roman architect, 1st century B.C.

FIGURE 17.29
Leaded diamond-pane windows of handmade rolled glass in an Elizabethan English bay window. (*Photo by the author*)

the propagation of natural light by successive reflections and diffusions within the room.

Breakage safety is regulated in skylights and overhead glazing, in glass located in or near doors, and in large sheets of glass that might be mistaken for clear openings in a wall. Laminated glass and plastic glazing sheets, because they will not drop out of the skylight if broken, are the only skylight glazings that are permitted without restriction under most codes. In and near doors, and in windows having a glazed area in excess of 9 square feet (0.9 m^2) whose lowest edge is within 18 inches (450 mm) of the floor, tempered glass, laminated glass, or plastic glazing materials must be used to prevent the disfiguring and sometimes fatal accidents that might occur if annealed glass were used in such situations.

Fire-rated glass must be used for openings in required fire doors and fire separation walls. The maximum areas of glazed openings in these locations are regulated by all building codes. Most codes also require that windows aligned above one another in certain types of buildings over three stories in height be separated vertically by fire-resistive spandrels of a specified minimum height. The intent of this provision is to restrict the spread of fire from one floor of a building to the floors above. If a glass spandrel is used under this type of code provision, it must be backed up inside with a material such as masonry that offers the necessary fire resistance.

FIGURE 17.31
Limestone and metal mullions for a Gothic church window. (*Courtesy of Indiana Limestone Institute*)

FIGURE 17.30
Discrete windows in an office building. (*Architects: Skidmore, Owings and Merrill. Photo courtesy of PPG Industries*)

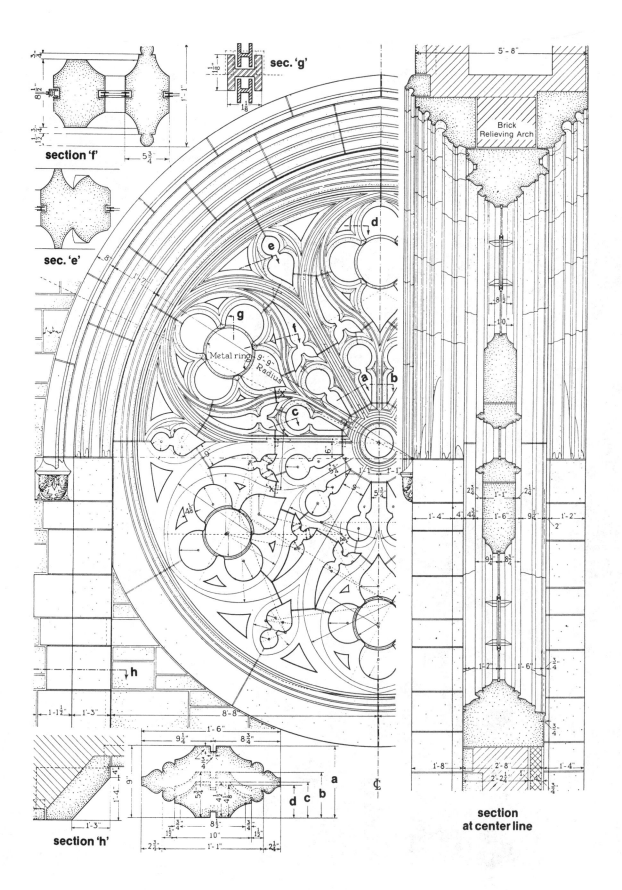

section 'f'

sec. 'g'

sec. 'e'

Metal ring 9'-9" Radius

Brick Relieving Arch

section 'h'

section at center line

Code provisions relating to glass and energy consumption are often as simple as specifying a required thermal resistance for windows in certain types of buildings. Many codes require, in effect, that double glazing or storm windows, as a minimum, be installed in residential buildings. Maximum glazed area, expressed as a percentage of overall wall area, is often specified by code. Future codes are likely to require that an analysis be performed by the designer to show that the overall performance of glass in a building has been optimized with respect to heat loss, heat gain, and daylighting.

C.S.I./C.S.C. Masterformat Section Numbers for Glass	
08800	**GLAZING**
08810	**Glass**
08840	**Plastic Glazing**
08850	**Glazing Accessories**

FIGURE 17.33
Reflective coated glass in a hotel complex. (*Architect: Welton Becket Associates. Photo courtesy of LOF Glass, a Libby–Owens–Ford Company*)

FIGURE 17.32
Leaded stained glass in the Robie House, Chicago, 1906.
(*Architect: Frank Lloyd Wright. Photo by the author*)

SELECTED REFERENCES

1. The most current information on glass will be found in manufacturers' catalogs. If you have access to *Sweet's Catalog File*, nearly every glass manufacturer's catalog is included in Section 08810.

2. Flat Glass Marketing Association. *FGMA Glazing Manual*. Topeka, Kansas.

This handbook, updated frequently, summarizes current practice in the use of glass in buildings. (Address for ordering: 3310 Harrison, Topeka, KN 66611.)

3. Amstock, Joseph S. *Handbook of Glass in Construction*. New York, McGraw–Hill, 1998.

This is a comprehensive reference work on the use of glass in buildings.

KEY TERMS AND CONCEPTS

crown glass
cylinder glass
punty
plate glass
drawn glass
float glass
glazing
glazier
lights (or lites) of glass
single-strength, double-strength
ASTM E1300
annealed glass
tempered glass
heat-strengthened glass
laminated glass
polyvinyl butyral (PVB)
security glass
fire-rated glass
wired glass

optical-quality ceramic
patterned glass
frit
fritted glass
spandrel glass
tinted glass
shading coefficient
visible transmittance
glazing luminous efficacy (K_e)
reflective coated glass
double glazing, triple glazing
spline
low-emissivity (low-e) coating
thermochromic glass
photochromic glass
electrochromic glass
photovoltaic glass
acrylic
polycarbonate

glazier's points
putty
glazing compound
setting blocks
bite
mullion
wet glazing components
dry glazing components
preformed solid tape sealant
polybutene tape
wedge or roll-in gasket
lockstrip gasket
weep hole
butt-joint glazing
structural silicone flush glazing
suspended glazing system
glass mullion system
daylighting

1. What are the advantages of float glass over drawn glass? Over plate glass?

2. Name two situations in which you might use each of the following types of glass: (a) tempered glass; (b) laminated glass; (c) wired glass; (d) patterned glass; (e) reflective glass; (f) polycarbonate plastic glazing sheet.

3. What are the design objectives for a large-light glazing system?

4. Discuss the role of glass that faces each of the principal directions of the compass in adding solar heat to an air-conditioned office building in summer. How should windows be treated on each of these facades to minimize summertime solar heat gain?

5. In what ways does a typical building code regulate the use of glass, and why?

EXERCISES

1. Examine the ways in which glass is mounted in several actual buildings and sketch a detail of each. Explain why each detail was used in its situation, and why you agree or disagree with the detail used.

2. Find a book on passive solar heating of houses and, in it, a table of solar heat gains for your area. For windows facing each of the four directions of the compass, plot a curve that represents the solar heat coming through a square foot of clear glass on an average day in each month of the year. Which window orientation maximizes wintertime heat and minimizes summertime heat? Is there an undesirable orientation that maximizes summertime heat and minimizes wintertime heat?

WINDOWS AND DOORS

The mullion patterns of steel-framed windows are matched to the rusticated joints of the concrete cladding on the Arizona State University School of Architecture.
(*Architect: Hillier Group. Photo © Jeff Goldberg/ESTO*)

Doors and windows are very special components of walls. Doors permit people, goods, and, in some cases, vehicles to pass through walls. Windows allow for simultaneous control of the passage of light, air, and visual images through walls. Windows and doors, in addition to these important functions, play a large role in establishing the character and personality of a building, much as our eyes, nose, and mouth play a large role in our personal appearance. At the same time, windows and doors are the most complex, expensive, and potentially troublesome parts of a wall. The experienced designer exercises great care and wisdom in selecting, installing, and maintaining them to achieve satisfactory results.

WINDOWS

The word *window* is thought to have originated in an old English expression that means "wind eye." The earliest windows in buildings were open holes through which smoke could escape and fresh air could enter. Devices were soon added to the holes to give greater control: hanging skins, mats, or fabric to regulate airflow; shutters for shading and to keep out burglars; translucent membranes of oiled paper or cloth, and eventually of glass, to admit light while preventing the passage of air, water, and snow. When a translucent membrane was eventually mounted in a moving *sash,* light and air could be controlled independently of one another. With the addition of woven insect screens, windows permitted air movement while keeping out mosquitoes and flies. Further improvements followed one upon another as the centuries moved by. A typical window today is an intricate, sophisticated mechanism with many layers of controls: curtains, shade or blind, sash, glazings, insulating airspace, low-emissivity coatings, insect screen, weatherstripping, and perhaps a storm sash or shutters.

Windows were formerly made on the construction site by highly skilled carpenters, but nearly all are produced now in factories. The reasons for factory production are higher efficiency, lower cost, and, most importantly, better quality. Windows need to be made to a very high standard of precision if they are to operate easily and maintain a high degree of weathertightness over a period of many years. In cold climates especially, a loosely fitted window with single glass and a frame that is highly conductive of heat will significantly increase heating fuel consumption for a building, cause noticeable discomfort to the people in the building, and create large quantities of condensate that will stain and decay materials in and around the windows.

A *prime window* is a window that is made to be installed permanently in a building. A *storm window* is a removable auxiliary unit that is added seasonally to a prime window to improve its thermal performance. A *combination window,* which is an alternative to a storm window, is an auxiliary unit that incorporates both glass and insect screening; a portion of the glass is mounted in a sash that can be opened in summer to allow ventilation through the screening. A combination window is normally left in place year-round. Some windows are designed and manufactured specifi-

cally as *replacement windows* that install easily in the openings left by deteriorated windows that have been removed from older buildings.

Types of Windows

Figure 18.1 illustrates in diagrammatic form the window types used most commonly in residential buildings, and Figure 18.2 shows additional types that are found largely in commercial and institutional buildings. *Fixed windows* are the least expensive and the least likely to leak air or water because they have no operable components. *Single-hung* and *double-hung windows* have one or two moving *sashes,* which are the frames in which the glass is mounted. The sashes slide up and down in tracks in the frame of the window. In older windows, the sashes were held in position by cords and counterweights, but today's double-hung windows usually rely on a system of springs to counterbalance the weight of the sashes. A *sliding window* is essentially a single-hung window on its side, and shares with single-hung and double-hung windows the advantage that tracks in the frame hold the sashes securely along two opposite sides. This inherently stable construction allows single-hung, double-hung, and sliding windows to be designed in an almost unlimited range of sizes and proportions. It also allows the sashes to be more lightly built than those in *projected windows,* a category that includes principally casement windows, awning windows, hopper windows, tilt/turn windows, and inswinging windows. All projected windows have sashes that rotate outward or inward from their frames and therefore must have enough structural stiffness to resist wind loads while being supported only at two corners.

With the exception of the rare triple-hung window, no window with sashes that slide can be opened to

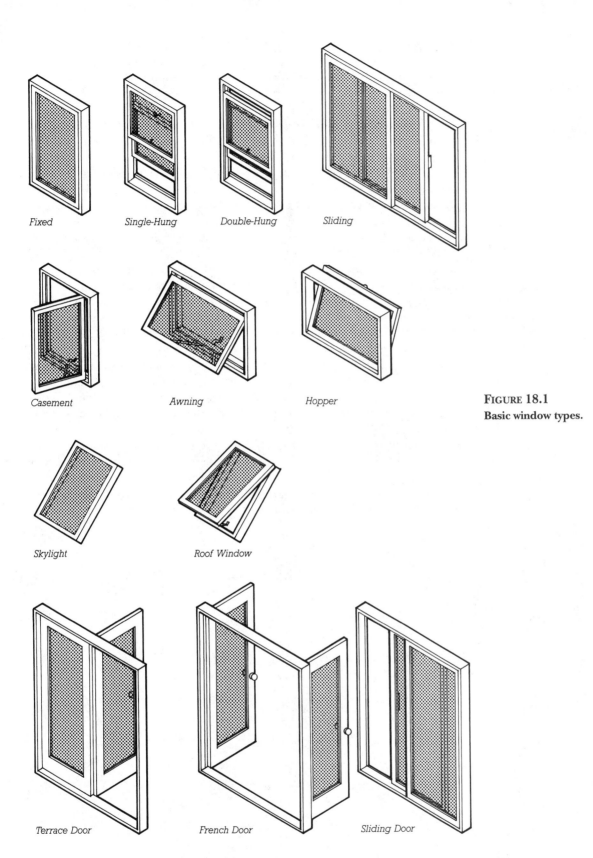

Fixed Single-Hung Double-Hung Sliding

Casement Awning Hopper

FIGURE 18.1
Basic window types.

Skylight Roof Window

Terrace Door French Door Sliding Door

Top-Hinged Inswinging

Side-Hinged Inswinging

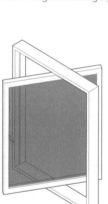

Pivoting

more than half of its total area. By contrast, all projected windows can be opened to their full area. *Casement windows* are helpful in catching passing breezes and inducing ventilation through the building; they are generally narrow in width but can be joined to one another and to sashes of fixed glass to fill wider openings. *Awning windows* can be broad but are not usually very tall. Awning windows have the advantages of protecting an open window from water during a rainstorm and of lending themselves to a building-block approach to the design of window walls (Figure 18.3). *Hopper windows* are more common in commercial buildings than in residential. Like awning windows, they will admit little or no rainwater if left open during a rainstorm. *Tilt/turn windows* have elaborate but concealed hardware that allows each window to be operated either as a casement or as a hopper.

Projected windows are usually provided with synthetic rubber weatherstripping that seals by compression around all the edges of the sash when it is closed. Single-hung, double-hung, and sliding windows generally must rely on brush-type weatherstripping because it does not exert as much friction against a sliding sash as elastomeric materials. Brush-type materials do not seal as tightly as compression weatherstripping, and they are also subject to more wear than elastomeric weatherstripping over the life of the window. As a result, projected windows are generally somewhat more resistant to air leakage than windows that slide in their frames.

Windows in roofs are specially constructed and flashed for watertightness and may be either fixed (skylights) or openable (roof windows). Large glass doors may slide in tracks or swing open on hinges. The hinged *French door* opens fully and, with its arms flung wide, is a more welcoming type of door than the sliding door, but it cannot be used to regulate airflow through the room unless it is fitted with a doorstop that can hold it securely in any open position. The French door, with its seven separate edges that must be carefully fitted and weatherstripped, is prone

FIGURE 18.2
Additional window types that are used mainly in larger buildings.

to air leakage along these edges. The recently developed *terrace door*, with only one operating door, minimizes this problem but, like the sliding door, it can open to only half its area.

Window types that are used almost exclusively in commercial and institutional work include horizontally and vertically *pivoting windows, side-hinged inswinging,* and *top-hinged inswinging windows* (Figure 18.2). These types all allow for inside washing of exterior glass surfaces. Because they project inward rather than outward, they are much less subject to the damage that could occur if, say, casement windows were opened into the high winds that swirl around tall buildings. To further limit wind damage, windows in tall buildings are often fitted with simple devices that limit the extent to which they can be opened. In most tall buildings, if the windows operate at all, they are meant to be used for ventilation only if the building's air conditioning system is inoperative.

Insect screens may be mounted only on the interior side of the sash in casement and awning windows. They are usually mounted on the outside in other window types. Sliding patio doors and terrace doors have an exterior sliding screen, while French doors require a pair of hinged screen doors on the exterior. Pivoting win-

FIGURE 18.3
Awning and fixed windows in coordinated sizes offer the architect the possibility of creating patterned walls of glass. (*Photo courtesy of Marvin Windows and Doors*)

dows cannot be fitted with insect screens.

Glass tends to attract and hold dust and dirt on both inside and outside surfaces, and must be washed at intervals if it is to remain transparent and attractive. Inside surfaces are usually easy for window washers to reach. Outside surfaces are often hard to reach, requiring ladders, scaffolding, or window-washing platforms that hang from the top of the building on cables. Accordingly, most

operable windows are designed to allow personnel to wash the outside surface of glass while standing inside the building. Casement and awning windows are usually hinged in such a way that there is sufficient space between the hinged edge of the sash and the frame when the window is open to allow access to the outer surface of glass. Double-hung and sliding windows are often designed to allow each sash to be rotated out of its track for cleaning (Figure 18.11).

Window Frames

The traditional frame material for windows is wood, but aluminum, steel, plastics, and combinations of these four materials have also come into widespread use. Wood is a fairly good thermal insulator and if free of knots is easily worked into sash, but in service it shrinks and swells with changing moisture content and requires repainting every few years. When wetted by weather, leakage, or

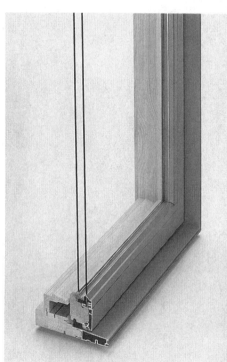

FIGURE 18.4
Cutaway sample of an aluminum-clad, wood-framed window. (*Photo courtesy of Marvin Windows and Doors*)

FIGURE 18.5
A pair of double-hung wood-framed windows in a dwelling.
(*Photo courtesy of Marvin Windows and Doors*)

condensate, wood frames are subject to decay. Knot-free wood is becoming increasingly rare and expensive, for which reason wood products made of short lengths finger-jointed together, oriented wood strands, or veneers are now used more frequently in window frames than solid lumber. These substitute materials, while functionally very satisfactory, are not attractive to the eye, so they are usually covered (*clad*) with wood veneer or a cladding of plastic or aluminum (Figures 18.4–18.6). Clad windows, at the time of this writing, account for about three-quarters of the market for wood-framed windows.

Aluminum is relatively inexpensive and is easy to form and join as a sash material. It requires no periodic repainting. But aluminum conducts heat so well that unless an aluminum frame is *thermally broken* with plastic or synthetic rubber components to interrupt the flow of heat through the metal, condensate and sometimes even frost will form on interior window frame surfaces during cold winter weather (Figure 18.7; see also Chapter 21). The majority of commercial and institutional windows and many residential windows are framed with aluminum (Figures 18.8, 18.9). For appearance, aluminum frames are usually anodized or permanently coated, as described in Chapter 21.

Plastic window frames, which are capturing a constantly increasing pro-

FIGURE 18.6
Large double-hung wood windows and a triangular fixed window bring sunlight and views. (*Photo courtesy of Marvin Windows and Doors*)

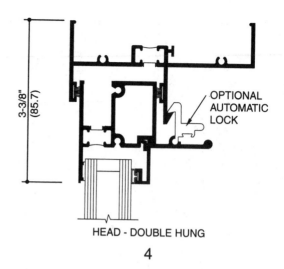

3-3/8" (85.7)

OPTIONAL AUTOMATIC LOCK

HEAD - DOUBLE HUNG

4

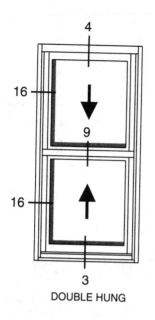

4

16

9

16

3

DOUBLE HUNG

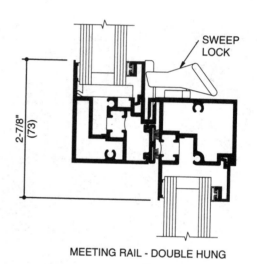

2-7/8" (73)

SWEEP LOCK

MEETING RAIL - DOUBLE HUNG

9

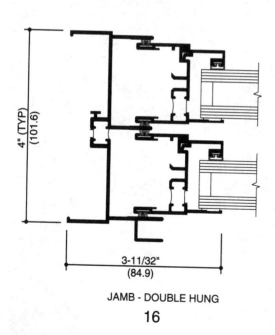

4" (TYP) (101.6)

3-11/32" (84.9)

JAMB - DOUBLE HUNG

16

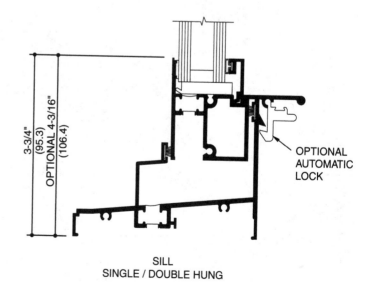

3-3/4" (95.3)

OPTIONAL 4-3/16" (106.4)

OPTIONAL AUTOMATIC LOCK

SILL
SINGLE / DOUBLE HUNG

3

FIGURE 18.7
The details of this commercial-grade double-hung window are keyed to the numbers on the small elevation view at the upper left. Cast and debridged plastic thermal breaks separate the outdoor and indoor portions of all the sash and frame extrusions. Pile weather-stripping seals against air leaks at all the interfaces between sashes and frame. For help in understanding the complexities of aluminum extrusions, see Chapter 21.
(*Courtesy of Kawneer Company, Inc.*)

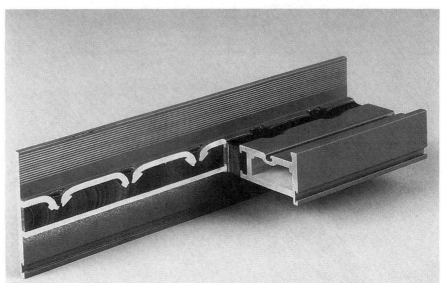

(b)

FIGURE 18.8
(*a*) Aluminum double-hung and sliding window units. (*b*) A cutaway demonstration model of a cast and debridged plastic thermal break in an aluminum window frame. In this proprietary design, the aluminum is "lanced" into the plastic at intervals to bond the two materials securely together. (*Courtesy of Kawneer Company, Inc.*)

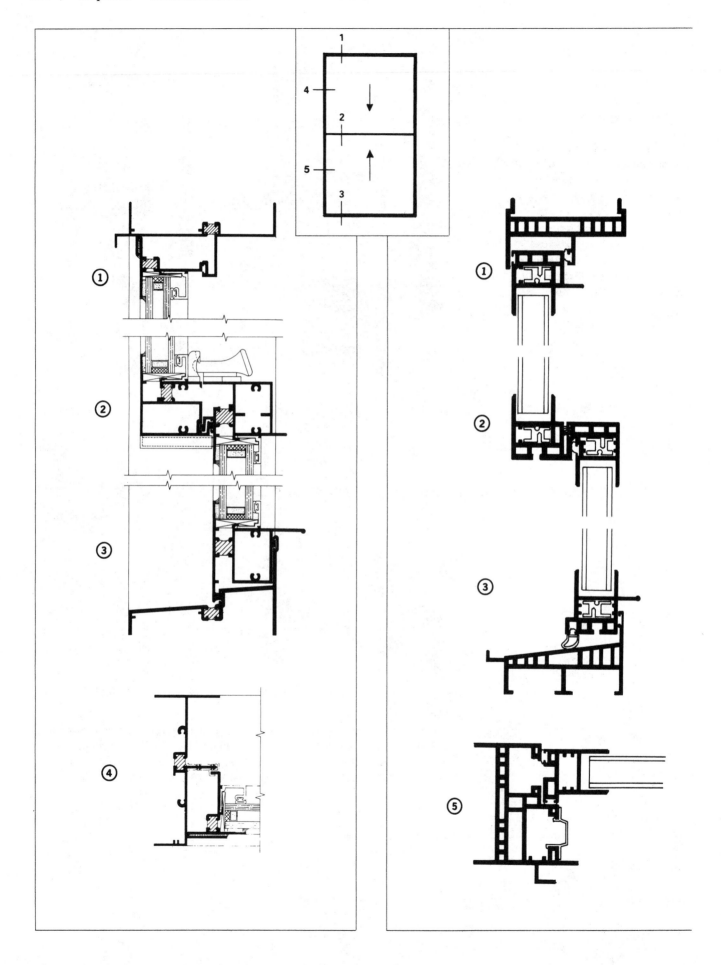

portion of the market for residential windows, never need painting and are fairly good thermal insulators. The most common material for plastic window frames is polyvinyl chloride (PVC, vinyl) that is formulated with a high proportion of inert filler material so as to minimize thermal expansion and contraction. Some typical PVC window details are shown in Figures 18.9 through 18.11.

The chief advantage of steel as a frame material for windows is its strength, which permits steel sash sections to be much more slender than those of wood and aluminum (Figures 18.12–18.16). Usually, steel windows are permanently coated to present a pleasing appearance and prevent corrosion. If they are not, they will need repainting at intervals. Steel is sufficiently less conductive of heat than aluminum that it is unlikely to form condensation under most weather conditions. Thus, steel window frames are rarely provided with thermal breaks. They are, however, more conductive of heat than wood or plastic frames.

Glass-fiber-reinforced plastics (*GFRPs*) are new to the window market. GFRP frame sections are produced by a process of *pultrusion:* Continuous bundles of glass fiber are pulled through a bath of plastic resin, usually polyester, and then through a shaped die in which the resin hardens. The resulting sash pieces are strong, stiff, and relatively low in thermal expansion. Like PVC, they are fairly good thermal insulators.

In earlier times, because of the difficulty of manufacturing large lights of glass, windows were necessarily divided into small lights by *muntins,* thin wooden bars into which

FIGURE 18.9
Comparative details of a single-hung residential window with an aluminum frame (left) and a double-hung residential window with a PVC plastic frame (right). The small inset drawing at the top center of the illustration shows an elevation view of the window with numbers that are keyed to the detail sections below. The comparatively thick sections of plastic are indicative of the greater stiffness of aluminum as a material. The diagonally hatched areas of the aluminum details are plastic thermal breaks that interrupt the flow of heat through the highly conductive metal frame. The inherently low thermal conductivity of the PVC and the multichambered construction eliminate the need for thermal breaks in the plastic window. Compare the thickness of the aluminum in this residential window to that of the aluminum in the commercial window in Figure 18.7. (*American Architectural Manufacturers Association Window Selection Guide—1988. Courtesy of AAMA*)

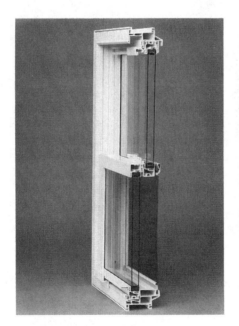

FIGURE 18.10
Cutaway sample of a plastic double-hung window with double glazing and an external half-screen. (*Courtesy of Vinyl Building Products, Inc.*)

FIGURE 18.11
For ease of washing the exterior surfaces of the glass, the sashes of this plastic window can be unlocked from the frame and tilted inward. (*Courtesy of Vinyl Building Products, Inc.*)

the glass was mounted within each sash. A typical double-hung window had its upper sash and lower sash each divided into six lights, and was referred to as a "six over six." Muntin arrangements changed with changing architectural styles and improvements in glass manufacture. Today's windows, glazed with large, virtually flawless lights of float glass, need no muntins at all, but many building owners and designers prefer the look of traditional muntined windows.

This desire for muntins is greatly complicated by the necessity of using double glazing. Some manufacturers offer the option of individual small lights of double glazing held in deep muntins. This is relatively expensive and tends to look heavy. The least expensive option utilizes grids of imitation muntin bars, made of wood or plastic, that are clipped into each sash against the interior surface of the glass. These are designed to remove easily for washing the glass.

Other alternatives are imitation muntin grids between the sheets of glass, which are not very convincing replicas of the real thing, and grids, either removable or permanently bonded to the glass, on both the outside and inside faces of the window. Another option is to use a prime window with authentic divided lights of single glazing, and to increase its thermal performance with a storm sash. This looks the best of all the options from the

FIGURE 18.12
These samples of hot-rolled steel window frame sections demonstrate a range of permanent finishes. The nearest sample includes a snap-in aluminum bead for holding the glass in place. (*Steel windows by Hope's; photography by David Moog*)

FIGURE 18.13
Cutaway sample of a steel-framed window with aluminum glazing beads and permanent finish. The details of this window are shown in Figure 18.15. (*Steel windows by Hope's; photography by David Moog*)

FIGURE 18.14
With fire-rated glazing such as wired glass, as shown here, a steel window can be fire rated. This window features an awning sash below a fixed light. (*Steel windows by Hope's; photography by David Moog*)

FIGURE 18.15
Manufacturer's details of a steel window in a masonry wall. The details are keyed by number to the elevation views at the bottom. Notice the bent steel anchors, shown with broken lines, that fasten the window unit to the masonry. (*Steel windows by Hope's*)

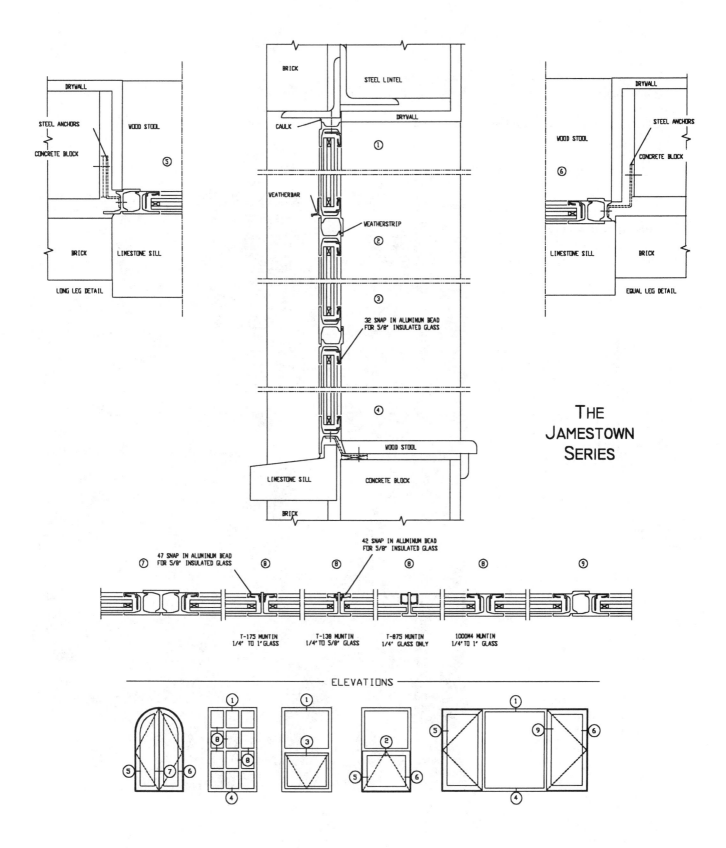

LONG LEG DETAIL

EQUAL LEG DETAIL

THE
JAMESTOWN
SERIES

47 SNAP IN ALUMINUM BEAD
FOR 5/8" INSULATED GLASS

42 SNAP IN ALUMINUM BEAD
FOR 5/8" INSULATED GLASS

T-175 MUNTIN
1/4" TO 1" GLASS

T-138 MUNTIN
1/4" TO 5/8" GLASS

T-075 MUNTIN
1/4" GLASS ONLY

1000M4 MUNTIN
1/4" TO 1" GLASS

ELEVATIONS

inside, but reflections in the storm sash largely obscure the muntins from the outside.

Glazing

A number of glazing options are available for residential windows. *Single glazing* is acceptable only in the mildest climates because of its low resistance to heat flow and the likeli-hood that moisture will condense on its interior surface in cool weather. Sealed *double glazing* or single glazing with storm windows is the minimum acceptable glazing under most build-ing codes. Removable glazing panels and storm windows must be taken down periodically and cleaned, which is a nuisance that can be avoided by using sealed double or triple glazing. The most usual form of *triple glazing* today consists of two outer sheets of glass with a thin, virtu-ally weightless, highly transparent plastic film stretched in the middle of the airspace. More than 90 percent of all residential windows sold today in North America have two or more lay-ers of glazing. Double glazing with a *low-emissivity* (*low-e*) *coating* on one glass surface performs at least as well as triple glazing.

FIGURE 18.16
The narrow sight lines of steel windows and doors are evident in this photograph. (*Steel windows by Hope's; photography by David Moog*)

PLASTICS IN BUILDING CONSTRUCTION

The first plastic was formulated more than a century ago. The major development of plastic materials has taken place since 1930, during which time plastics have come increasingly into use in buildings. Presently, the construction industry in the United States uses more than 10 billion pounds (5 billion kg) of plastics each year in hundreds of applications.

Plastics in this context may be loosely defined as synthetically produced giant molecules (polymers or copolymers) made up of large numbers of small, repetitive chemical units. Most plastics are based on carbon chemistry (Figure A), except for the silicones, which are based on silicon (Figure B). The various *synthetic rubber* compounds are usually considered as a different class of materials from plastics, although chemically they are similar; they are often referred to as *elastomers*. Plastics and elastomers are manufactured largely from organic molecules obtained from oil, natural gas, and coal.

A *polymer* is composed of many identical chemical units or *monomers*. Polyvinyl chloride (PVC), for example, is a polymer that is produced by polymerizing vinyl chloride monomers into a long chemical chain (Figure C). A *copolymer* consists of repeating patterns of two or more monomers. High-impact polystyrene is a copolymer made up of both polystyrene and polybutadiene (Figure D).

The molecular structure of a polymer or copolymer bears an important relationship to its physical properties.

A high-density polyethylene molecule, for instance, is a long, single chain containing up to 200,000 carbon atoms (Figure E). Low-density polyethylene has a branching molecular structure that does not pack together so readily as the single chain (Figure F).

There are two broad classes of plastics. *Thermoplastic* plastics consist generally of linear molecules, and may be softened by reheating at any time after their manufacture. Upon cooling, they regain their original properties. *Thermosetting* plastics have a molecular structure that is strongly cross-linked in three dimensions. They cannot be remelted after manufacture. Thermosetting plastics are generally harder and stronger than thermoplastics.

Many *modifiers* are added to various polymers to change their properties or reduce their cost. *Plasticizers* are organic compounds that impart flexibility and softness to otherwise brittle plastics. *Stabilizers* are added to resist deterioration of polymers from the effects of sunlight, heat, oxygen, and electromagnetic radiation. *Fillers* are inexpensive, nonreacting materials such as talc or marble dust that are added to reduce cost or to improve toughness or heat resistivity. *Extenders* are waxes or oils that add bulk to the plastic at low cost. *Reinforcing* fibers of glass, metal, carbon, or minerals can increase strength, impact resistance, stiffness, abrasion resistance, hardness, and other mechanical properties. *Flame retarders* are often introduced into plastics that are destined for interior use

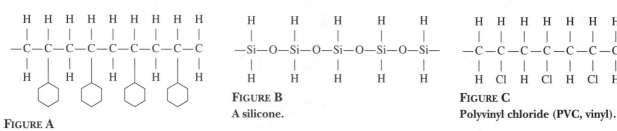

FIGURE A
Polystyrene.

FIGURE B
A silicone.

FIGURE C
Polyvinyl chloride (PVC, vinyl).

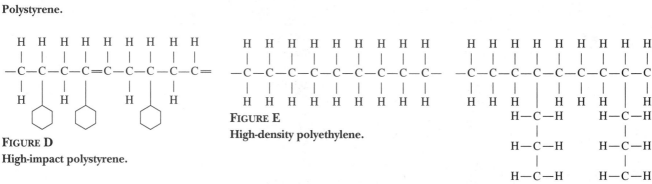

FIGURE D
High-impact polystyrene.

FIGURE E
High-density polyethylene.

FIGURE F
Low-density polyethylene.

in buildings. Color can be added to plastics with dyes or pigments. Some examples of extremely useful modifiers are the carbon black that is often added to polyethylene as a stabilizer to improve its resistance to sunlight, the lead carbonate that is used to stabilize PVC products for use outdoors, the di-isooctyl plasticizer that converts brittle PVC to a stretchable, rubbery compound, and the chopped glass fibers that reinforce polyester to make it suitable for use in boat hulls and building components.

The plastics used in buildings range from dense solids such as those used in floor tiles to the lightweight cellular foams used for thermal insulation. They range from the soft, pliable sheets used for roofing membranes and flashings to the hard, rigid plastics used for plumbing pipes. Glazing sheets are made from highly transparent plastics, while most plastics manufactured for other purposes are opaque. People readily recognize the solid plastic materials used in construction, but most tend not to realize that liquid plastics are major ingredients of many protective coatings. Plastics show up also in the form of *composites,* in which they are teamed with nonplastic materials: *laminates* consisting of paper and melamine formaldehyde, used for countertops and facings; *sandwiches* such as the foamcore plywood panels used as cladding for Heavy Timber Frame buildings; and mixes of plastics with particulate materials, such as polyester concrete (stone aggregates cemented with a polyester binder rather than portland cement) and particleboard (wood chips and phenolic resins).

Plastics are given form through any of an almost endless list of processes. *Extrusion* manufactures long, shaped sections by forcing the plastic through a shaped die. *Pultrusion,* used for certain fiber-reinforced products, is much the same as extrusion except that the section is pulled through the die rather than pushed. A host of *molding* processes cast plastic into shaped cavities to give it form. Many foam plastics can be foamed directly into the mold, which they expand to fill. Polyethylene film is produced by *film blowing,* in which air is pumped into a small extruded tube of plastic to expand and stretch it into a very thin walled tube many feet in diameter. The large tube is then slit lengthwise, folded, and rolled for distribution. Films and sheets can be made by casting plastic onto a chilled roller or a chilled, moving belt. Some plastic sheet products and many plastic laminates are produced by *calendaring,* a process of pressing a material or a sandwich of materials first through hot rollers, then cold rollers.

Thermoplastic sheets may be further formed by heating them and pressing them against shaped dies. The pressing force may be furnished mechanically by a matching die, or by air pressure. If compressed air pushes the plastic into the die, the process is called *blow forming.* In *vacuum forming,* a pump draws the air from between the heated plastic sheet and the die, and atmospheric pressure does the rest.

Many plastics are amenable to machining processes—sawing, drilling, milling, planing, turning, sanding—like those used to shape wood or metal. Nylon and acrylics are often given final form in this way. Various plastics can also be joined with adhesives or with heat or solvents that soften the two surfaces that are to be joined so they can be pressed together to reharden as a single piece.

Plastics can be joined mechanically with screws or bolts. Many plastic products are designed with ingenious snap-together features so they can be joined without fasteners.

Some Synthetic Rubbers Used in Construction

Chlorinated Polyethylene, Chlorosulfonated Polyethylene (Hypalon)	Roof membranes
Ethylene-Propylene-Diene Monomer (EPDM)	Roof membranes, flashings
Isobutylene-Isoprene Copolymer (Butyl rubber)	Flashings, waterproofing
Polychloroprene (Neoprene)	Gaskets, roof membranes
Polyisobutylene (PIB)	Roof membranes
Polysiloxane (Silicone rubber)	Sealants, adhesives, coatings, roof membranes
Polyurethanes	Sealants
Sodium Polysulfide (Polysulfide, Thiokol)	Sealants

Some Plastics Used in Construction

Chemical Name (common names and tradenames in parentheses)	Thermo-plastic or Thermo-setting	Characteristics	Some Uses in Buildings
Acrylonitrile-Butadiene-Styrene Terpolymer (ABS)	tp	Hard, dimensionally stable	Plumbing pipes, shower stalls
Alkyd	ts	Tough, transparent, weather resistant	Coatings
Epoxy	ts	Tough, hard, high adhesion	Adhesives, coatings, binders
Melamine Formaldehyde	ts	Hard, very durable	Plastic laminates
Phenol Formaldehyde (phenolic, Bakelite)	ts	Hard, very durable	Coatings, laminates, insulating foams, electrical boxes
Polyamides (Nylon)	tp	Tough, strong, elastic, machinable	Membranes for air-supported and tensile roofs, hardware for doors and cabinets
Polybutene	tp	Sticky, soft, weather resistant	Sealant tapes, sealants
Polycarbonate (Lexan)	tp	Hard, transparent, unbreakable	Glazing sheet, lighting fixtures, door sills
Polyesters (Mylar, Dacron)	ts	Durable, combustible	Fabrics for roofing membranes and geotextiles, matrix for glass-fiber-reinforced plastics
Polyethylene (polythene)	tp	Tough, flexible, impermeable	Vapor retarders, moisture barriers, piping, tarpaulins
Polyisocyanurate	ts	Heat resistant, fire retardant	Insulating foams
Polymethyl Methacrylate (acrylic, Plexiglas, Lucite)	tp	Tough, transparent, machinable	Glazing sheet, dome skylights, lighting fixtures, illuminated signs, coatings, adhesives
Polypropylene	tp	Tough, hard, does not fatigue	Crack control fibers in concrete, one-piece molded hinges
Polystyrene (Styrofoam, bead foam)	tp	Foam is excellent thermal insulator, solid is transparent	Insulating foams, glazing sheet
Polytetrafluorethylene (PTFE, Teflon)	tp	Heat resistant, low friction	Sliding joint bearings, pipe thread tape
Polyurethanes	ts or tp	Vary widely with form of plastic, from low density foams to hard, dense coatings to synthetic rubbers	Coatings, sealants, insulating foams, adhesives
Polyvinyl Chloride (PVC, vinyl), Polyvinyl Acetate (PVA, vinyl)	tp	Vary widely with form of plastic, from flexible rubberlike sheet to hard, rigid solid	Pipes, conduits, ductwork, siding, gutters, window frames, flooring sheets and tiles, coatings, wallcoverings, geotextiles, roof membranes
Polyvinyl Fluoride (PVF, Tedlar)	tp	Tough, inert	Exterior coatings
Polyvinylidene Fluoride (PVF, Kynar, fluoropolymer)	tp	Crystalline, translucent	Exterior coatings
Silicones	ts	Weather resistant, low surface tension	Masonry water repellants
Urea Formaldehyde (UF)	ts	Strong, rigid, stable	Coatings, plywood adhesives, binders for wood panel products, insulating foams

Taken as a group, plastics exhibit some common advantages and some common problems when used as building materials. Among the advantages, plastics in general are low in density, often cheaper than other materials that will do the same job, have good surface and decorative qualities, and, because their molecules are made to order for each end use, they can often offer a better solution to a building problem than more traditional materials. Plastics generally are little affected by water or biological decay, and they do not corrode galvanically. They tend to have low thermal and electrical conductivity. Many have high strength-to-weight ratios. Most plastics are essentially impermeable to water and water vapor. Many are very tough and resistant to abrasion.

Their disadvantages are also numerous. All plastics can be destroyed by fire, and many give off toxic combustion products. Some burn very rapidly, but others are slow-burning, self-extinguishing, or do not ignite at all, so careful selection of polymers and modifiers can be crucial in building applications. Flame spread ratings, smoke developed ratings, and toxicity of combustion products should be checked for each use of plastics within a building.

Plastics have very high coefficients of thermal expansion as compared to other building materials, and require close attention to the detailing of control joints, expansion joints, and other devices for accommodating volume change. Plastics tend not to be very stiff. They deflect considerably more under load than most conventional materials. Taken together with their combustibility, this severely limits their application as primary structural materials. Many also creep under prolonged loading, especially at elevated temperatures. In strength, plastics vary from fiber-reinforced composites that are as strong as many metals (but not nearly as stiff) to cellular foams than can be crushed easily between two fingers.

Plastics tend to degrade in the outdoor environment. They are especially susceptible to attack by the ultraviolet component of sunlight, oxygen, and ozone. In many plastic products (acrylic glazing sheets and some synthetic rubbers, for example), these problems have largely been surmounted by adjustments to the chemistry of the material that have resulted from a prolonged program of testing and research, but it is usually a mistake for a designer to use an untried plastic material in an exposed location without consulting with its manufacturer.

The accompanying tables list the plastics and synthetic rubbers most commonly used in construction.

Selected References

1. Dietz, Albert G. H. *Plastics for Architects and Builders.* Cambridge, Massachusetts, M.I.T. Press, 1969.

Despite its age, this deservedly famous little book is still the best introduction to the subject for building professionals.

2. Hornbostel, Caleb. *Construction Materials* (2nd ed.). New York, John Wiley & Sons, Inc., 1991.

The section on plastics of this monumental book gives excellent summaries of the plastics most used in buildings.

Figure 18.17 shows the thermal properties of various glazing options. These values apply only to the center area of each light of glass. The edges of the light transmit heat much more rapidly because of the solid strips of material used to space the sheets of glass and hold them together. For this reason, whole-window values for heat transmission that take into account heat transmission through the *edge spacers,* sash, and frame are the only accurate way of comparing the thermal efficiencies of different windows. Higher whole-window thermal efficiency is achieved by using thermally efficient frames and less conductive spacers around the perimeter of lights of multiple glazing.

In warm climates, the dominant thermal problem in windows is likely to be excessive admission of solar heat to the interior of the building through windows on the east, west, and south sides of the building. Tinted glass can reduce the transmission of solar heat somewhat and make the building easier to cool. More effective, however, are low-emissivity coatings that are designed specifically to reflect the wavelengths of sunlight that contain the most

FIGURE 18.17
Center-of-glass thermal insulating values for various types of glazing.

heat, while admitting useful amounts of light to the building and providing good viewing conditions to the outdoors.

Safety Considerations in Windows

In order to prevent accidental breakage and injuries, building codes require that glass within doors, and large lights that are near enough to the floor or to doors that they might be mistaken for open doorways, must be breakage-resistant material. Tempered glass is most often used for this purpose, but laminated glass and plastic glazing sheets are also permitted.

In residences, most building codes require that at least one window in each bedroom must open to an aperture large enough to permit occupants of the bedroom to escape through it and firefighters to enter through it. A typical requirement is that the clear opening must be at least 5.7 square feet (0.53 m²), the clear width must be at least 20 inches (510 mm), the clear height must be at least 24 inches (610 mm), and the sill may be no higher than 44 inches (1.12 m) above the floor.

Casement and awning windows should not be used adjacent to porches or walkways, where someone might be injured by running into the projecting sash. Similarly, inswinging windows should not be used in corridors unless they are above head level.

Window Testing and Standards

The designer's task in selecting windows is facilitated by programs of testing that allow objective comparisons of the thermal and structural performances of windows of different types and different manufacturers. Performance-based standards for wood and clad windows are published by the National Wood Window and Door Association (NWWDA), for aluminum and plastic windows by the American Architectural Manufacturers Association (AAMA), and for polyvinyl chloride windows by the American Society for Testing and Materials (ASTM). Each of these standards establishes *performance grades* of windows. Within each grade, a design wind pressure, water resistance, structural performance, air infiltration, and operating force are specified. The NWWDA ratings for wood windows, for example, establish six performance grades corresponding to wind pressures of 15, 20, 25, 30, 35, and 40 pounds per square foot (718 to 1914 Pa). Within each grade, a window must survive several tests: When subjected alternately to pressure and suction of a specified intensity, there must be no glass breakage or permanent deformation of the frame. When subjected to the design wind pressure and drenched on its outside surface with water at the rate of 8 inches (200 mm) of rain per hour, no water may penetrate beyond the interior face of the window. Under a specified air pressure, air leakage through the window may not exceed a specified amount. The locked window must not open in response to prying with various hand tools, and, when unlocked, the amount of force needed to open the window must not exceed a specified amount. Similar standards are established for alu-

Thermal Insulating Values, Center of Glass

	U		R	
	U.S.	S.I.	U.S.	S.I.
Single glass	1.11	6.29	0.90	0.16
Double glass or Single glass plus storm window	0.50	2.84	2.00	0.35
Double glass with argon fill	0.46	2.61	2.17	0.38
Double glass with low-e coating on one surface	0.34	1.93	2.94	0.52
Double glass with argon fill and low-e coating on one surface	0.28	1.59	3.57	0.62
Triple glass	0.34	1.93	2.94	0.52
Double glass, polyester film in middle of airspace, argon fill, low-e coating on two interior surfaces	0.17	0.96	5.88	1.03

minum and plastic windows by AAMA and ASTM. The designer may select a grade of window to correspond to the severity of exposure to weather of the building site and the level of performance that the building owner is willing to pay for. For an inexpensive residence in a sheltered location, a Grade 15 window, the lowest classification, is likely to suffice. For a tall institutional building, a Grade 35 or 40 window is likely to be a better match for the more demanding environmental conditions in which it must perform.

With regard to energy efficiency, the National Fenestration Rating Council (NFRC) sponsors a program of testing and labeling that is based on the performance of the whole window, not just the glass. At the present time, the standard label that is affixed to each window shows values for thermal insulating value (U-value), solar heat gain coefficient, and visible light transmittance (Figure 18.18). Condensation performance, long-term energy performance, and other factors are likely to be added to this label in the future.

Installing Windows

Some catalog pages for windows are reproduced in Figures 18.19 through 18.21 to give an idea of the information on window configurations that is available to the designer. The most important dimensions given in catalogs are those of the *rough opening* and *masonry opening*. The rough opening height and width are the dimensions of the hole that must be left in a framed wall for installation of the window. They are slightly larger than the corresponding outside dimensions of the window unit itself, to allow the installer to locate and level the unit accurately. The masonry opening dimensions indicate the size of the hole that should be provided if the window is mounted in a masonry wall.

Most factory-made windows are extremely easy to install, often requiring only a few minutes per window. Figure 6.9 shows a typical procedure for installing a wood replacement window in existing construction. Windows that are framed or clad in aluminum or plastic are usually provided with a continuous flange around the perimeter of the window unit. When the unit is pushed into the rough opening from the outside, the flange bears against the sheathing along all four edges. After the unit has been located and plumbed (made level and square) in the opening, it is attached to the frame by means of nails driven through the flanges. The flanges are later concealed by the exterior cladding or trim.

Methods for anchoring window units into masonry walls vary widely, from nailing the unit to wood strips that have been fastened inside the masonry with bolts or powder-driven fasteners, to attaching the unit to steel clips that have been laid into the mortar joints of the masonry. The window manufacturer's recommendations should be followed in each case.

Energy Rating Factors	Ratings		Product Description
	Residential	Nonresidential	
U-Factor Determined in Accordance with NFRC 100	0.40	0.38	Model 1000 Casement Low-e = 0.2 0.5" gap Argon Filled
Solar Heat Gain Coefficient Determined in Accordance with NFRC 200	0.65	0.66	
Visible Light Transmittance Determined in Accordance with NFRC 300 & 301	0.71	0.71	

NFRC ratings are determined for a fixed set of environmental conditions and specific product sizes and may not be appropriate for directly determining seasonal energy performance. For additional information contact:

FIGURE 18.18
A typical certification label that is affixed to a window unit so that buyers may compare energy efficiencies. (*Courtesy of National Fenestration Rating Council, Inc.*)

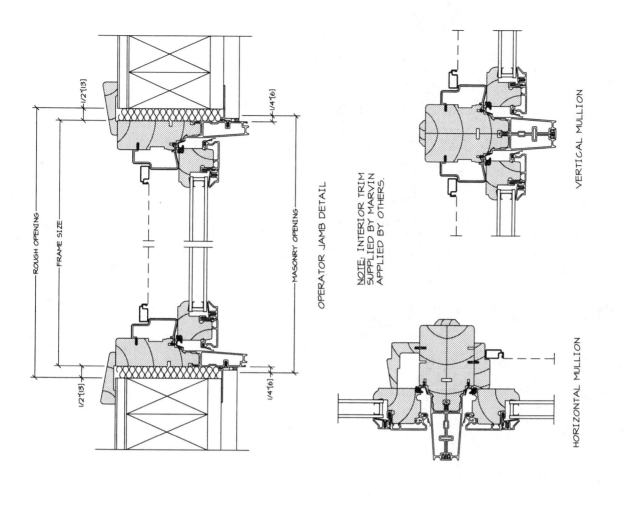

ROUGH OPENING

FRAME SIZE

1/2"[13]

1/2"[13]

1/4"[6]

MASONRY OPENING

1/4"[6]

OPERATOR JAMB DETAIL

<u>NOTE</u>: INTERIOR TRIM
SUPPLIED BY MARVIN
APPLIED BY OTHERS.

VERTICAL MULLION

HORIZONTAL MULLION

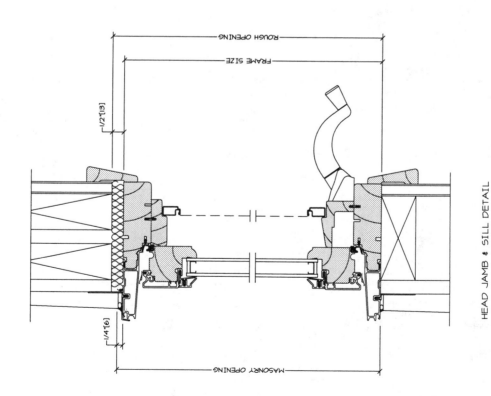

ROUGH OPENING

FRAME SIZE

1/2"[13]

1/4"[6]

MASONRY OPENING

HEAD JAMB & SILL DETAIL

FIGURE 18.19

Manufacturer's catalog details for an aluminum-clad, wood-framed casement window with double glazing and an interior insect screen. (*Courtesy of Marvin Windows and Doors*)

661

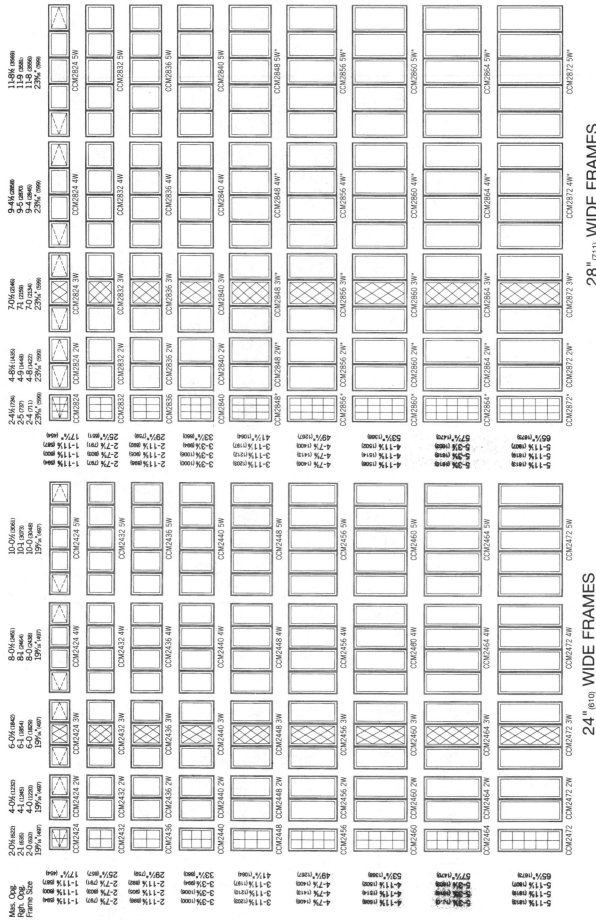

28" (711) WIDE FRAMES

24" (610) WIDE FRAMES

*THESE WINDOWS MEET NATIONAL EGRESS CODES FOR FIRE EVACUATION. LOCAL CODES MAY DIFFER.
NOTE: ANY UNIT IN THESE COMBINATIONS CAN BE OPERATING OR STATIONARY.
OPERATING SASH ARE VIEWED FROM THE EXTERIOR.

Not to scale.

FIGURE 18.20

One of several pages in a manufacturer's catalog that show stock sizes and configurations of aluminum-clad wood casement windows. (*Courtesy of Marvin Windows and Doors*)

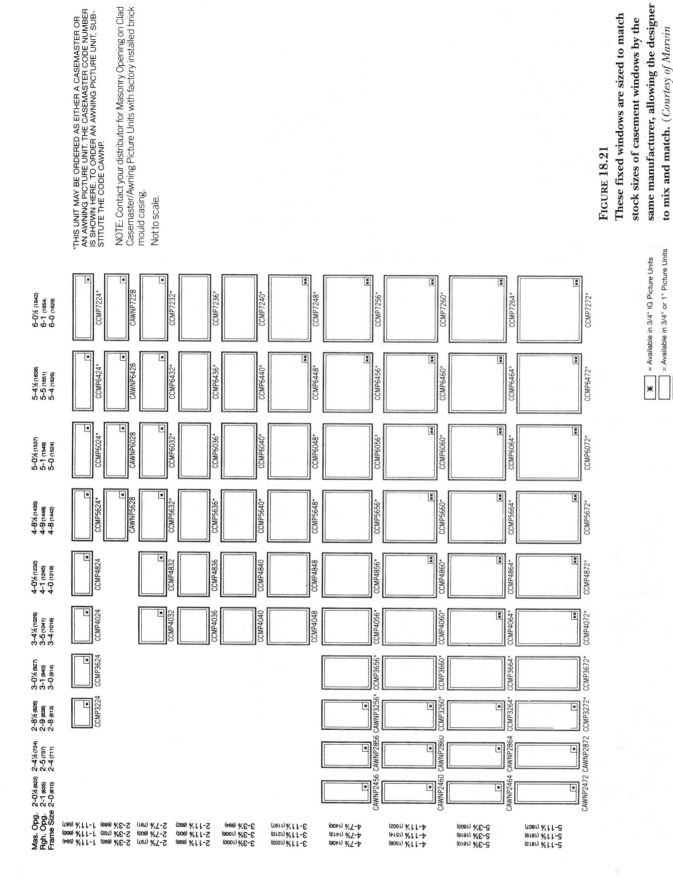

FIGURE 18.21
These fixed windows are sized to match stock sizes of casement windows by the same manufacturer, allowing the designer to mix and match. (*Courtesy of Marvin Windows and Doors*)

DOORS

Doors fall into two general categories, exterior and interior. Weather resistance is usually the most important factor in choosing exterior doors, whereas fire resistance and resistance to the passage of sound are most important in interior doors. Many different modes of door operation are possible (Figure 18.22). There are numerous types of exterior doors: solid entrance doors, entrance doors that contain glass, storefront doors that are mostly or entirely made of glass, storm doors, screen doors, vehicular doors for residential garages and industrial use, revolving doors, and sloping cellar doors, just to name a few. Interior doors come in dozens of additional types. To simplify our discussion, we will focus on swinging doors for both residential and commercial use.

Wood Doors

At one time, nearly all doors were made of wood (Figure 18.23). In simple buildings, doors made of planks and Z-bracing were once common. In more finished buildings, *stile-and-rail* doors gave a more sophisticated appearance while avoiding the worst problems of moisture expansion and contraction to which plank doors are subject (Figure 7.16).

In recent decades, while stile-and-rail doors have persisted in use in higher-quality buildings, *flush* wood doors have captured the majority of the market, chiefly because they are easier to manufacture and therefore cost less. For exterior use in small buildings, and for interior use as well in institutional and commercial buildings, flush doors are constructed with a *solid core* of wood blocks or wood composite material

(Figure 7.17). Interior doors in residences often have a *hollow core*. They consist of two veneered wood faces that are bonded to an open grid of spacers made of paperboard or wood. The perimeters of the faces are glued to wood edge strips. Flush doors with wood faces are available with a solid *mineral core* that qualifies them as fire doors.

Entrance doors must be well constructed and tightly weatherstripped if they are not to leak air and water. Properly installed and finished wood panel or solid-core doors are excellent for exterior residential use (Figures 6.10, 6.11). Pressed sheet metal doors and molded GFRP doors, usually embossed to resemble wood stile-and-rail doors, are becoming popular as an alternative to wood exterior residential doors. Their cores are filled with insulating plastic foam. Their thermal performance is superior to that of wood doors. They do not suffer from moisture expansion and contraction as wood doors do. They are often furnished *prehung*, meaning that they are already mounted on hinges in a surrounding frame, complete with weatherstripping, ready to install by merely nailing the frame into the wall. Wood doors can also be purchased prehung, although many are still hung and weatherstripped on the building site. The major disadvantage of metal and plastic exterior doors is that they do not have the satisfying appearance, feel, or sound of a wood door.

Residential entrance doors almost always swing inward and are mounted on the interior side of the door frame. For improved wintertime thermal performance of the entrance, a *storm door* is often mounted on the outside of the same frame, swinging outward. The storm door usually includes at least one large panel of tempered glass. In

FIGURE 18.22
Some modes of door operation.

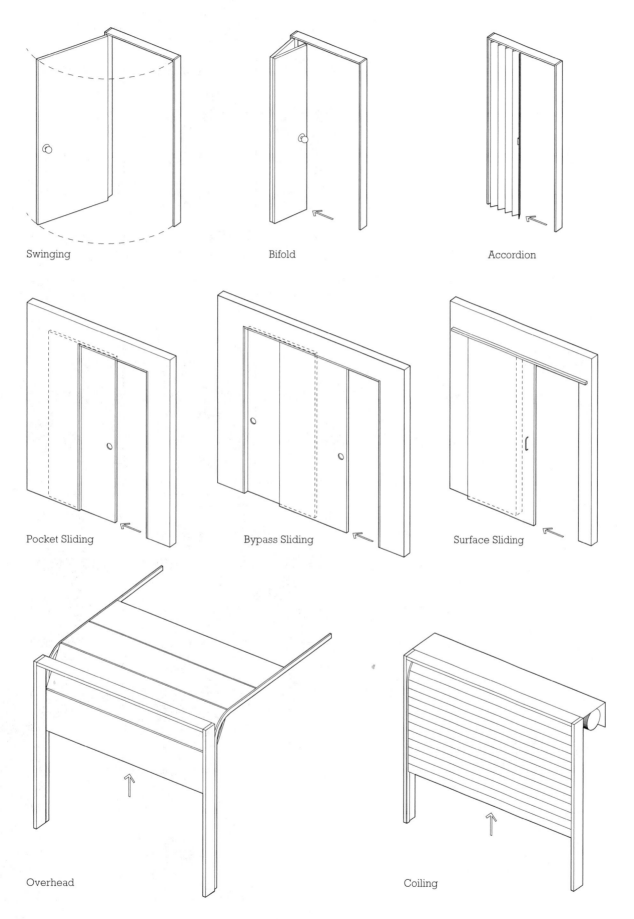

Swinging

Bifold

Accordion

Pocket Sliding

Bypass Sliding

Surface Sliding

Overhead

Coiling

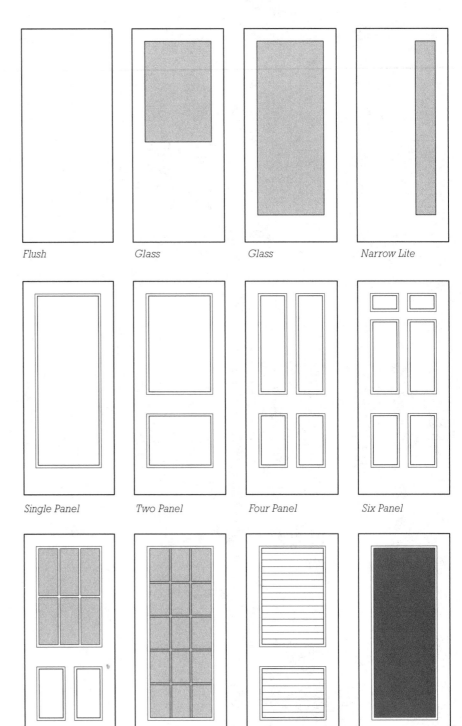

FIGURE 18.23
Some typical configurations for wood doors. The top row consists of flush doors. The middle row is made up of stile-and-rail doors.

summer, a *screen door* may be substituted for the storm door. A *combination door*, which has easily interchangeable screen and storm panels, is more convenient than separate screen and storm doors.

Steel Flush Doors

Flush doors with faces of painted sheet steel are the most common type in nonresidential buildings (Figure 18.24). For economy, interior steel doors in many situations are permitted to have hollow cores. Solid-core doors are required for exterior use and in situations that demand increased fire resistance, more

rugged construction, or better acoustical privacy between rooms.

Fire Doors

Fire doors have a noncombustible mineral core and are rated according to the period of time for which they are able to resist fire as defined by Underwriters Laboratories Standard 108. A *1½-hour door* is required by most codes for an enclosed exit stairway. Three-quarter hour or 20-minute doors are required between rooms and corridors. Three- and 4-hour doors are often required for openings in fire walls that separate uses within a building or that sepa-

rate buildings from one another. A standardized label is permanently affixed to the hinge edge of each fire door at the time of manufacture to designate its degree of fire resistance.

Except for one- and two-family residences and very small commercial buildings, any door along an egress route in a building must swing in the direction of egress travel, must be self-closing, and must be equipped with hardware that opens the door automatically if people press against it in the direction of egress travel. The amount of glass in fire doors is restricted according to the fire classification of the door, from no glass at all in the 4-hour classification to full

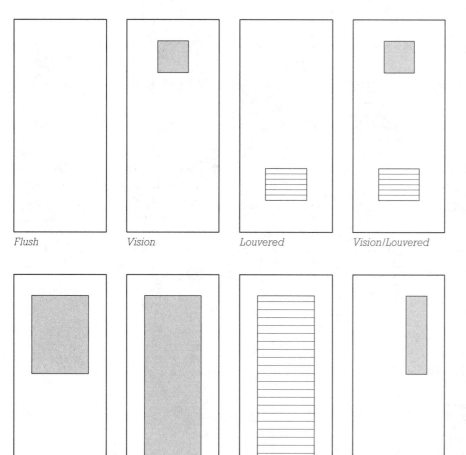

FIGURE 18.24
Some typical configurations for steel doors.

Flush   *Vision*   *Louvered*   *Vision/Louvered*

Glass   *Full Glass*   *Full Louvered*   *Narrow Lite*

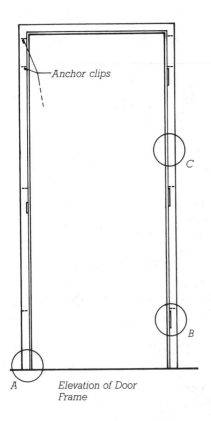

—Anchor clips

C

B

A Elevation of Door Frame

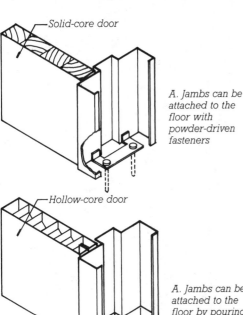

—Solid-core door

A. Jambs can be attached to the floor with powder-driven fasteners

—Hollow-core door

A. Jambs can be attached to the floor by pouring floor topping concrete around the door frame

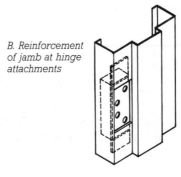

B. Reinforcement of jamb at hinge attachments

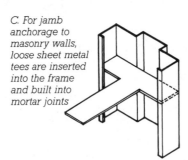

C. For jamb anchorage to masonry walls, loose sheet metal tees are inserted into the frame and built into mortar joints

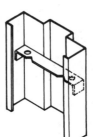

C. For jamb anchorage to steel studs, sheet metal zees are factory-welded to the jambs to receive screws driven through the studs

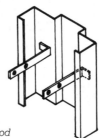

C. For jamb anchorage to wire truss studs, notches key to the vertical members of the studs, and holes provide for tie wires

C. Jambs are anchored to wood studs by nailing through holes in the jamb inserts

glass in the lowest classification. The glass in a fire door must be wired glass or some other type of fire-rated glass to prevent it from falling out of the opening if it breaks. Restricted areas of ventilating louvers are permitted in addition to glass in $1\frac{1}{2}$ hour, 1-hour, $\frac{3}{4}$-hour, and 20-minute doors.

Metal doors are usually hinged to hollow steel door frames, although aluminum frames are permitted for doors classified at $\frac{3}{4}$-hour and below. The frames for the higher-rated doors must be filled solid with mortar

FIGURE 18.25
Details of hollow steel door frames. The lettered circles on the elevation at the upper left correspond to the details on the rest of the page.

or gypsum plaster. A number of different types of anchors are available for mounting door frames to partitions of different materials (Figure 18.25).

There are many types of special-purpose doors. Among the most common are X-ray shielding doors, which contain a layer of lead foil; electric field shielding doors, with an internal layer of metal mesh that is electrically grounded through the hinges; heavily insulated cold storage doors; and bank vault doors.

C.S.I./C.S.C.
Masterformat Section Numbers for Doors and Windows

08100	**METAL DOORS AND FRAMES**
08200	**WOOD AND PLASTIC DOORS**
08250	**DOOR OPENING ASSEMBLIES**
08255	**Packaged Steel Door Assemblies**
08260	**Packaged Wood Door Assemblies**
08265	**Packaged Plastic Door Assemblies**
08300	**SPECIAL DOORS**
08400	**ENTRANCES AND STOREFRONTS**
08500	**METAL WINDOWS**
08510	**Steel Windows**
08520	**Aluminum Windows**
08600	**WOOD AND PLASTIC WINDOWS**
08610	**Wood Windows**
	Metal Clad Wood Windows
	Plastic Clad Wood Windows
08630	**Plastic Windows**
08650	**SPECIAL WINDOWS**
08655	**Roof Windows**
08670	**Storm Windows**

FIGURE 18.26
Steel and glass door and sidelights.
(*Steel windows by Hope's; photography by David Moog*)

FIGURE 18.27
The Blanchard Road Alliance Church in Wheaton, Illinois, silhouettes laminated wood trusses against a wall of vinyl-clad wood-framed fixed windows. (*Architect: Walter C. Carlson Associates. Photo courtesy of Andersen Windows, Inc. Andersen is a registered trademark of Andersen Corporation, © 1997. All rights reserved*)

SELECTED REFERENCES

1. Carmody, John, Stephen Selkowitz, and Lisa Heschong. *Residential Windows: A Guide to New Technologies and Energy Performance.* New York, W. W. Norton & Company, 1996.

This book is a clearly written, well-illustrated introduction to considerations of energy efficiency in windows.

2. American Architectural Manufacturers Association. *Window Selection Guide* (AAMA WSG.1-95). Palatine, Illinois, AAMA, 1995.

In this booklet, AAMA sets forth and explains the standards under which alu-

minum and plastic windows are manufactured and specified. (Address for ordering: AAMA, 1540 East Dundee Road, Suite 310, Palatine, IL 60067.)

KEY TERMS AND CONCEPTS

window
sash
prime window
storm window
combination window
replacement window
fixed window
single-hung window
double-hung window
sliding window
projected window

casement window
awning window
hopper window
tilt/turn window
skylight
roof window
French door
terrace door
pivoting window
side-hinged inswinging window
top-hinged inswinging window

clad
thermal break
glass-fiber-reinforced plastic (GFRP)
muntin
single, double, triple glazing
low-emissivity (low-e) coating
edge spacer
performance grade
rough opening
masonry opening
stile-and-rail door

flush door
solid core
hollow core
mineral core
prehung door
storm door
screen door
combination door
fire door
$1\frac{1}{2}$-hour door

REVIEW QUESTIONS

1. List in detail the primary functional requirements for windows in each of the following situations:

a. A residential bathroom in an urban apartment building.

b. A jail cell.

c. A display window in a department store.

d. A teller's window in a drive-in banking facility.

e. A bedroom in Nome, Alaska.

f. A living room in Hilo, Hawaii.

2. Select a type of window operation for each of the following situations:

a. A window that can be left open in the rain.

b. A window that must induce the maximum possible ventilation from passing breezes.

c. A window in a high-rise office building.

d. A window to frame an expansive view of distant mountains.

e. A window that can be operated as either a casement or a hopper.

3. Compare the advantages and disadvantages of wood, plastic-clad wood, PVC, aluminum, and steel as window frame materials.

4. A well-insulated residential wall has a U-value of 0.05 and an R-value of 20 (in U.S. units). Compare the heat loss per square foot of the worst- and best-performing window glazings with this wall.

5. Select a type of door for each of the following situations:

a. Your bedroom closet.

b. A front door of a house.

c. A front door of a department store.

d. A door between the industrial arts automotive shop and the cafeteria in a high school.

e. A door on a warehouse loading dock.

EXERCISES

1. Obtain a copy of the building code that applies to the area in which you currently live. What fire resistance ratings are required for doors in the following situations:

a. A door between a hotel room and a public corridor?

b. A door in an egress stair enclosure?

c. A door between an iron foundry and an office building?

d. A door between a single-family residence and its attached garage?

2. Obtain catalogs from several window manufacturers. From them, select a set of windows for a one-room wilderness cabin of your own design.

3. Examine closely the windows in the room in which you are now sitting. What type of glazing do they have? What type of frame? How do they operate? How are they weatherstripped? Do these windows make sense to you in terms of today's energy efficiency requirements and your own feelings about the room? How would you change them?

DESIGNING CLADDING SYSTEMS

The beautifully detailed cladding of the Hypolux Bank includes both granite blocks and metal panels. (*Architect: Richard Meier. Photo © Scott Frances/ESTO*)

Cladding typifies a paradox of building: Those parts of a building that are exposed to our view are also those that are exposed to wear and weathering. The *cladding* (nonloadbearing exterior wall enclosure) is the most visible part of a building, one to which architects devote a great deal of time to achieve the desired visual effect. It is also the part of the building that is most subject to attack by natural forces that can spoil its appearance. It is the part of the building that must defend the interior spaces against invasion by water, wind, sunlight, heat and cold, and all the other forces of nature. Its design is an intricate process that merges art, science, and craft to solve a very long and difficult list of problems.

FIGURE 19.1
A steel-framed Chicago office building during the installation of its aluminum, stainless steel, and glass curtain wall cladding. Notice the diagonal wind braces in the steel frame. (*Architects: Kohn Pedersen Fox/Perkins & Will. Photo by Architectural Camera. Courtesy of American Institute of Steel Construction*)

THE DESIGN REQUIREMENTS FOR CLADDING

Primary Functions of Cladding

The major purpose of cladding is to separate the indoor environment of a building from the outdoors in such a way that indoor environmental conditions can be maintained at levels suitable for the building's intended use. This translates into a number of separate and diverse functional requirements.

Keeping Water Out

Cladding must prevent the entry of rain, snow, and ice into a building. This requirement is complicated by the fact that water on the face of a building is often driven by wind at high velocities and high air pressures, not just in a downward direction, but in every direction, even upward. Water problems are especially acute on tall buildings, which rise to altitudes where wind velocities are much higher than at ground level, and which present a large profile to the wind. Here enormous amounts of water must be drained from a windward building face during a heavy rainstorm, and the water, pushed by wind, tends to accumulate in crevices and against projecting mullions, where it will readily penetrate the smallest crack or hole and enter the building. We will devote a considerable portion of this chapter to methods for keeping water out.

Preventing Air Leakage

The cladding of a building must prevent the unintended passage of air between indoors and outdoors. At a gross scale, this is necessary to regulate air velocities within the building. Smaller air leaks are harmful because they waste conditioned air, carry water through the wall, allow moisture vapor to condense inside the wall, and allow noise to penetrate the building from outside. Sealants, gaskets, and air barrier membranes of various types are all used to prevent air leakage through cladding.

Controlling Light

The cladding of a building must control the passage of light, especially sunlight. Sunlight is heat that may be welcome or unwelcome. Sunlight is visible light, useful for illumination but bothersome if it causes glare within a building. Sunlight includes destructive ultraviolet wavelengths that must be kept off human skin and away from interior materials that will fade or disintegrate. Cladding systems sometimes include external shading devices to keep light and solar heat away from windows. The glass in windows is often selected to control light and heat, as discussed in Chapter 17. Interior shades, blinds, and curtains may be added for further control.

Controlling the Radiation of Heat

Beyond its role in regulating the flow of radiant heat from the sun, the cladding of a building should also maintain its interior surfaces at temperatures that will not cause radiant discomfort. A very cold interior surface will make nearby people feel chilly, even if the air in the building is warmed to a comfortable level, and a hot surface or direct sunlight in summer can cause overheating of the body despite the coolness of the interior air. External sun shading devices, adequate thermal insulation, and appropriate selection of glass are potential strategies in controlling heat radiation.

Controlling the Conduction of Heat

The cladding of a building must resist to the required degree the conduction of heat into and out of the building. This requires not merely a satisfactory overall resistance of the wall to the passage of heat, but the avoidance of *thermal bridges,* wall components such as metal framing members that are highly conductive of heat and therefore likely to cause localized condensation on interior surfaces. Thermal insulation, appro-

FIGURE 19.2
This very early curtain wall, on Chicago's Reliance Building, built in 1894–1895, has spandrels constructed of white terracotta tiles. (*Architect: Charles Atwood, of Daniel H. Burnham and Company. Photo by William T. Barnum. Courtesy of Chicago Historical Society, IChi-18294*)

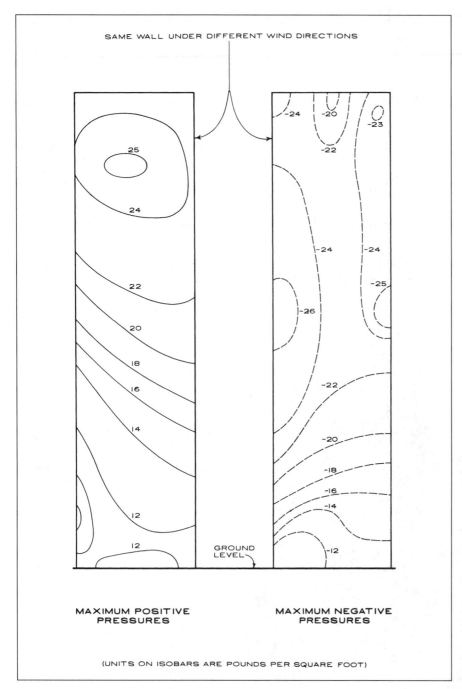

SAME WALL UNDER DIFFERENT WIND DIRECTIONS

MAXIMUM POSITIVE PRESSURES

MAXIMUM NEGATIVE PRESSURES

(UNITS ON ISOBARS ARE POUNDS PER SQUARE FOOT)

FIGURE 19.3

An example of expected positive and negative wind pressures on the cladding of a tall building, shown here in elevation, as predicted by wind tunnel testing. The building in this case is 64 stories tall and triangular in plan. Notice the high negative pressures (suctions) that can occur on the upper regions of the facade. The wind pressures on a building are dependent on many factors, including the shape of the building, its orientation, site topography, wind direction, and surrounding buildings. Each building must be modeled and tested individually to determine the pressures it is expected to undergo. (*Reprinted with permission from AAMA Curtain Wall Design Guide Manual*)

priate glazing, and thermal breaks are used to control heat conduction through cladding, as we will observe in the two chapters that follow.

Controlling Sound

Cladding serves to isolate the inside of a building from noises outside, and vice versa. Noise isolation is best achieved by walls that are airtight, massive, and resilient. The required degree of noise isolation varies from one building to another, depending on the noise levels and noise tolerances of the inside and outside environments. Cladding for a hospital near a major airport requires a high level of noise isolation. Cladding for a commercial office in a suburban office park need not perform to as high a standard.

Secondary Functions of Cladding

The fulfillment of the primary functional requirements of cladding leads unavoidably to a secondary but still important set of requirements.

Resisting Wind Forces

The cladding of a building must be adequately strong and stiff to sustain the pressures and suctions that will be placed upon it by wind. For low buildings, which are exposed to relatively slow and predictable winds, this requirement is fairly easily met. The upper reaches of taller buildings are beset by much faster winds whose directions and velocities are often determined by aerodynamic effects from surrounding buildings. High suction forces can occur on some portions of the cladding, especially near corners of the building (Figure 19.3).

Controlling Water Vapor

The cladding of a building must retard the passage of water vapor. In the heat of summer or the cold of winter, vapor moving through a wall assembly is likely to condense inside the assembly and cause problems of staining, lost insulating value, and corrosion. The cladding must be constructed with adequate levels of thermal insulation and a suitable vapor retarder to prevent condensation of moisture insofar as possible. Where condensation is inevitable, the condensate must be captured and drained to the outdoors.

Adjusting to Movement

Several different kinds of forces are always at work throughout a building, tugging and pushing both the frame and the cladding: thermal expansion and contraction, moisture expansion and contraction, and structural deflections. These forces must be anticipated and allowed for in designing a system of building cladding.

Thermal Expansion and Contraction

The cladding of a building has to accommodate thermal expansion and contraction at several levels: Indoor/outdoor temperature differences can cause warping of cladding panels due to differential expansion and contraction of their inside and outside faces (Figure 19.4). The cladding as a whole, exposed to outdoor temperature variations, grows and shrinks constantly with respect to the frame of the building, which is usually protected by the cladding from temperature extremes. And the building frame itself will expand and contract to some extent, especially between the time the cladding is installed and the time the building is first occupied and its indoor temperature is controlled.

Moisture Expansion and Contraction

Masonry and concrete cladding materials must accommodate their own expansion and contraction caused by varying moisture content. Bricks and building stone generally expand slightly after they are installed. Concrete blocks and precast concrete shrink slightly after installation in a building, as their curing is completed and excess moisture is given off. These movements are small but can

accumulate to significant and potentially troublesome quantities in long or tall panels of masonry or concrete.

Structural Movements Cladding of every type must adjust to movements in the frame of the building. Building foundations may settle unevenly, causing distortions of the frame. Gravity forces shorten columns and deflect beams and girders to which cladding is attached. Wind and earth-

quake forces push laterally on building frames and wrack panels attached to the faces. Long-term creep causes significant shortenings of concrete columns and sagging of concrete beams during the first year or two of a building's life.

If building movements from temperature differences, moisture differences, structural stresses, and creep are allowed to be transmitted between the frame and the cladding,

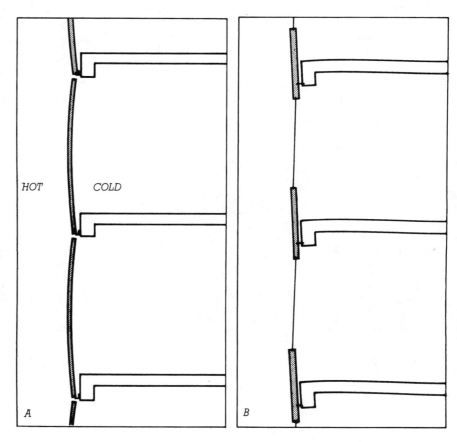

FIGURE 19.4

Distortions of curtain wall panels, illustrated in cross section. (a) Bowing caused by greater thermal expansion of the outside skin of the panels than of the inside skin under hot summertime conditions. (b) Twisting of spandrel beams because of the weight of the curtain wall.

unexpected things may happen. Cladding components may be subjected to forces for which they were not designed, which can result in broken glass, buckled cladding, sealant failures, and broken cladding attachments (Figure 19.5). In extreme cases, the building frame may end up supported by the cladding, rather than the other way around, or pieces of cladding may fall off the building. A number of provisions for dealing with movements are evident in the details of cladding systems in the two chapters that follow.

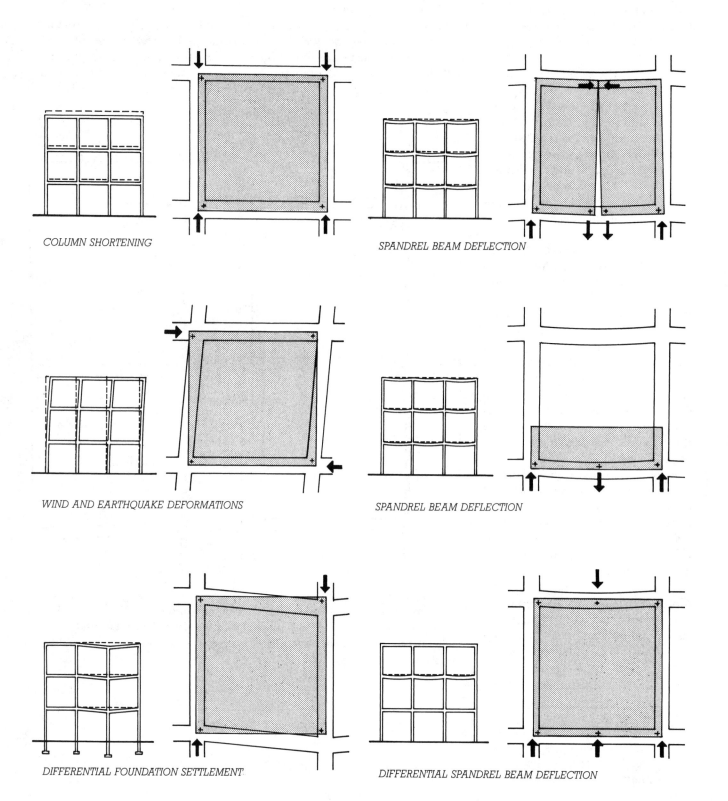

COLUMN SHORTENING

SPANDREL BEAM DEFLECTION

WIND AND EARTHQUAKE DEFORMATIONS

SPANDREL BEAM DEFLECTION

DIFFERENTIAL FOUNDATION SETTLEMENT

DIFFERENTIAL SPANDREL BEAM DEFLECTION

Resisting Fire

The cladding of a building can interact in several ways with building fires. This has resulted in a number of building code provisions relating to the construction of building cladding, as summarized in the last paragraph of this chapter.

Weathering Gracefully

To maintain the visual quality of a building, its cladding must weather gracefully. The inevitable dirt and grime should accumulate evenly, without streaking or splotching, and functional provisions must be made for maintenance operations such as glass and sealant replacement and for periodic cleaning, including scaffolding supports and safety attachment points for window washers. The cladding must resist oxidation and ultraviolet degradation of organic materials, corrosion of metallic components, and freeze–thaw damage of stone, brick, concrete, concrete block, and tile.

Installation Requirements for Cladding

Cladding should be easy to install. There should be secure places for the installers to stand, preferably on the floors of the building rather than on scaffolding outside. There must be built-in adjustment mechanisms in all the fastenings to allow for the inaccuracies that are normally present in the structural frame of the building

and the cladding components themselves. There must be dimensional clearances provided to allow the cladding components to be inserted without binding against adjacent components. And, most importantly, there must be forgiving features that allow for a lifetime of trouble-free cladding function despite all the lapses in workmanship that inevitably occur—features such as backup air barriers and drainage channels to get rid of moisture that has leaked through a faulty sealant joint, or generous edge clearances that keep a sheet of glass from contacting the hard material of the frame even if the glass is installed slightly crookedly.

CONCEPTUAL APPROACHES TO WATERTIGHTNESS IN CLADDING

In order for water to penetrate a wall, three conditions must be satisfied simultaneously:

1. There must be water present at the outer face of the wall.

2. There must be an opening through which the water can move.

3. There must be a force to move the water through the opening.

If any one of these conditions is not satisfied, the wall will not leak. This suggests that there are three concep-

tual approaches to making a wall watertight:

1. We can try to keep water completely away from the wall. This is impossible, however, on any but the smallest of buildings. A very broad overhang can keep a single-story wall dry under most conditions, but on a taller building we must assume that the wall will get wet.

2. We can try to eliminate the openings in a wall. We can build very carefully, sealing every seam in the wall with sealant or gaskets, attempting to eliminate every hole and crack.

This approach, which is called the *barrier wall* approach, works fairly well if done well, but it has inherent problems. In a wall made up of sealant jointed components, the joints are unlikely to be perfect. If an edge of a component is a bit damp, dirty, or oily, sealant may not stick to it. If the worker applying the sealant is insufficiently skilled or has to reach a bit too far to finish a joint, he or she may fail to fill the joint completely. Even if the joints are all made perfectly, building movements can tear the sealant or pull it loose. Because the sealant is on the outside of the building, it is exposed to the full destructive forces of sun, wind, water, and ice, and may fail prematurely from weathering. And whatever the cause of sealant failure, because the sealant joint is on the outside face of the wall, it is difficult to reach for inspection and repair.

In response to these problems, cladding designers often employ a strategy of *internal drainage* or *secondary defense*, which accepts the uncertainties of external sealant joints. It deals with them by providing internal drainage channels within the cladding to carry away any leakage, and backup sealant joints to the inside of the drainage channels. The ordinary masonry cavity wall facing exemplifies this strategy: The cavity, flashings, and weep holes constitute an internal drainage system for any moisture that finds its way through

FIGURE 19.5

Forces on curtain wall panels caused by movements in the frame of the building, illustrated in elevation. In each of the six examples, the drawing to the left shows the movement in the overall frame of the building, and the larger-scale drawing to the right shows its consequences on the curtain wall panels (shaded in gray) covering one bay of the building. Points of attachment between the panels and the frame are shown as crosses. The black arrows indicate forces on the wall panels caused by movement in the structure. The magnitude of the structural movements is exaggerated for clarity, and some inadvisable attachment schemes are shown to demonstrate their consequences. Forces such as these, if not taken into account in the design of the frame and cladding, can result in glass breakage, panel failures, and failure of the attachments between the panels and the frame.

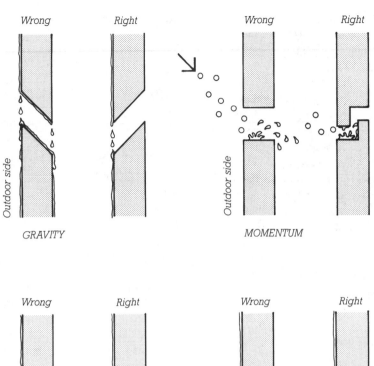

GRAVITY

MOMENTUM

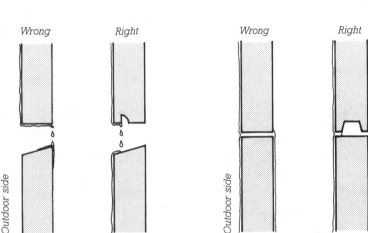

SURFACE TENSION

CAPILLARY ACTION

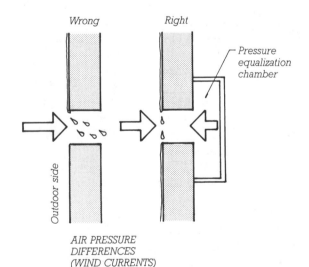

AIR PRESSURE
DIFFERENCES
(WIND CURRENTS)

Pressure
equalization
chamber

FIGURE 19.6

Five forces that can move water through an opening in a wall, illustrated in cross section with outdoors to the left. Each pair of drawings shows first a horizontal joint between curtain wall panels in which a force is causing water leakage through the wall, then an alternative design for the joint that neutralizes this force. Leakage caused by gravity is avoided by sloping internal surfaces of joints toward the outside; the slope is called a *wash*. Momentum leakage can be prevented with a simple labyrinth as shown. A drip groove and a capillary break are shown here as means for stopping leakage from surface tension and capillary action, which are closely related forces. Air pressure differences between the outside and inside of the joint will result in air currents that can transport water through the joint. This is prevented by closing the area behind the joint with a pressure equalization chamber (PEC) as shown. When wind strikes the face of the building, a slight movement of air through the joint raises the pressure in the PEC until it is equal to the pressure outside the wall, after which all air movement ceases. Each joint in an exterior wall, window, or door must be designed to neutralize all five of these forces.

the facing bricks. Internal drainage systems are an important component of every metal and glass curtain wall system on the market.

3. We can try to eliminate or neutralize all the forces that can move water through the wall. These forces are five in number: gravity, momentum, surface tension, capillary action, and wind currents (Figure 19.6).

Gravity is a factor in pulling water through a wall only if the wall contains an inclined plane that slopes into, rather than out of, the building. It is usually a simple matter to detail cladding so that no such inclined planes exist, through sometimes a loose gasket or an errant bead of sealant can create one despite the best efforts of the designer.

Momentum is the horizontal component of the energy of a raindrop that is falling at an angle toward the face of a building. It can drive water through a wall only if there is a suitably oriented slot or hole that goes completely through the wall. Momentum is easily neutralized by applying a cover to each joint in the wall or by designing each joint as a simple *labyrinth*.

Surface tension of water that causes it to adhere to the underside of a cladding component can allow water to be drawn into the building. The provision of a simple *drip* groove or flange on any underside surface to which water might adhere will eliminate the problem.

Capillary action is the surface tension effect that pulls water through any opening that can be bridged by a water drop. It is the primary force that transports water through the pores of a masonry wall. It can be eliminated as a factor in the entry of water through a wall by making each of the openings in a wall wider than a drop of water can bridge, or if this is not feasible or desirable, by providing a concealed *capillary break,* a widening somewhere inside the opening. In porous materials such as brick, capil-

lary action can be counteracted by applying an invisible coating of silicone-based water repellant, which destroys the adhesive force between water and the walls of the pores in the brick.

The solutions described in the four preceding paragraphs leave only *wind currents* remaining as the major force that must be neutralized if water is to be kept from penetrating through an opening in a wall. Wind is the force most difficult to deal with in designing a wall for watertightness. We can neutralize it by employing the *rainscreen principle.*

The Rainscreen Principle

The generic solution to the wind current problem is to allow wind pressure differences between the outside and the inside of the cladding to neutralize themselves through a concept known as the rainscreen principle. The implementation of this principle in what is sometimes called *pressure-equalized wall design* involves the creation of an airtight plane, the *air barrier,* at the interior side of the cladding, protected from direct exposure to the outdoors by a loose-fitting, unsealed, labyrinth-jointed layer known as the *rainscreen.* Between the rainscreen and the air barrier is a space known as the *pressure equalization chamber* (*PEC*). As wind pressures on the cladding build up and fluctuate, small currents of air pass back and forth through each unsealed joint in the rainscreen, just enough to equalize the pressure in the PEC with the pressure outside the joint (Figure 19.6). These currents are far too weak to carry water with them. A small flaw in the air barrier, such as a sealant bead that has pulled away from one side of the joint, is unlikely to cause a water leak because the volume of air that can pass through the flaw is still relatively small and is probably insufficient to carry water. By contrast, any flaw, no matter how small, in an exter-

nal sealant joint without an air barrier behind it will cause a water leak, because the sealant joint itself is wetted (Figure 19.7).

Because wind pressures across the face of a building may vary considerably at any given moment between one area of the face and another, the PEC should be divided into compartments not more than two stories high and a bay or two wide. If this is not done, large volumes of air may rush through the joints in higher-pressure areas of the face and flow across the air chamber to lower-pressure areas, carrying water with them as they go.

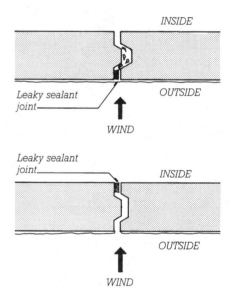

FIGURE 19.7
Leakage through a defective sealant joint between curtain wall panels, shown in plan view. In the upper example, the sealant joint is at the outside face of the panels, where it is wetted during a storm. Even a small current of air passing through the defective joint carries water with it. In the lower example, with the defective sealant installed on the inside of the panels, the volume of air leakage through the joint is insufficient to transport water through the joint, and no water penetrates.

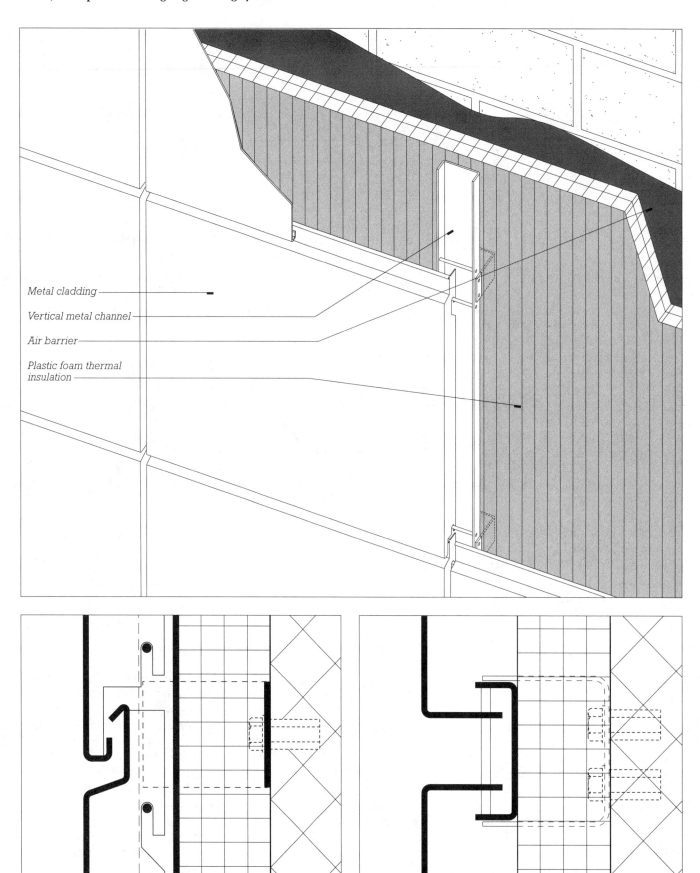

Metal cladding

Vertical metal channel

Air barrier

Plastic foam thermal
insulation

SECTION THROUGH HORIZONTAL JOINT

PLAN OF VERTICAL JOINT

A Rainscreen Cladding Design

Figure 19.8 depicts a cladding design that embodies the rainscreen principle in a very pure form. No sealants or gaskets are used. The metal rainscreen panels do not touch one another, but are separated by generous dimensions that preclude capillary movement of water, provide installation clearances, and allow for expansion and contraction. All four edges of each panel are shaped so as to create labyrinth joints. The forces of surface tension and gravity are counteracted by sloping surfaces and drips. Installation is simple and forgiving: Metal U-shaped clips are bolted to the backup wall, which is coated with an airtight mastic to create an air barrier. Rigid insulation panels are adhered to the wall, allowing the U-shaped clips to project through. Vertical metal channels are bolted to the clips. Finally, the metal panels that make up the rainscreen are hung on horizontal rods that are supported by the channels. The space between the metal rainscreen panels and the insulation acts as a pressure equalization chamber (PEC).

When wind drives rain against this wall, small quantities of air flow through the open joints in the rainscreen until the pressure in the PEC equals the pressure outside. These air flows are insufficient in volume or velocity to carry water with them. The vertical channels divide the PEC into narrow vertical compartments. Horizontal metal angles should be installed between the channels at two-story intervals to act as barriers to complete the compartmentation of the PEC.

The Rainscreen Principle at Smaller Scale

The rainscreen principle may also be applied at small scale to guide us in many aspects of exterior detailing of buildings. Figure 19.9 demonstrates how it is implemented in the placement of weatherstrip in a window sill detail. In the correct detail, the weatherstrip, whose function is to act as an air barrier, is placed to the inside of the lower rail of the sash. The open joint under the sash rail, which is provided with a capillary break, acts as the PEC. Unless the weatherstrip is grossly defective, water cannot be blown through the joint by air pressure differentials. Notice how the other forces that could transport water through the joint are counteracted: A slope (called by architects a *wash*) on the sill prevents gravity from pulling water in. The groove in the lower edge of the sash that acts as a capillary break also acts as a drip to counteract surface tension. The L-shaped joint between the sash and the sill acts as a labyrinth to prevent entry by momentum.

In the incorrect detail, the weatherstrip is wetted by rain. Any minor flaw in the weatherstrip will allow water to be blown through the joint.

There are relatively few constructed buildings that rely completely on the rainscreen principle for watertightness. But there are very few contemporary cladding details that do not employ the rainscreen principle as an important part of their defense against water penetration. Once again referring to the familiar example of the masonry cavity wall, the brick facing wythe is the rainscreen, the backup wall is the air barrier, and the cavity, pressurized through the weep holes, is the pressure equalization chamber.

FIGURE 19.8
The rainscreen in this exemplary cladding system is made up of formed metal panels. The design team included Wallace, Floyd Associates, Inc.; Bechtel/Parsons Brinckerhoff; Stull and Lee, Inc.; Gannett Fleming/URS/TAMS Consultants; and the Massachusetts Highway Department.

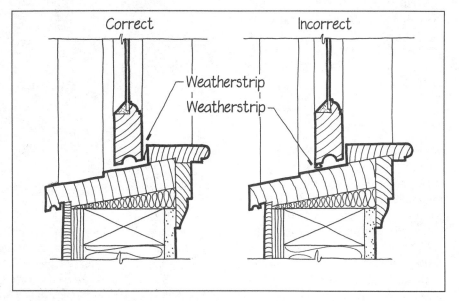

FIGURE 19.9
Applying the rainscreen principle to the detailing of the sill of a double-hung window.
(*From Allen, Edward,* Architectural Detailing: Function, Constructibility, Aesthetics, *New York, John Wiley & Sons, Inc., 1993, reproduced by permission of the publisher*)

SEALANT JOINTS IN CLADDING

Most cladding systems require sealant joints. Systems that do not use sealants as water barriers in the face of the wall generally use them to seal joints in the air barrier behind the face. The role of a *sealant* is to fill the joints between cladding components, preventing the flow of air and/or water, while still allowing reasonable dimensional tolerances for assembly and reasonable amounts of subsequent movement between the components. Sealant joint widths are usually ⅜ to ¾ inch (9 to 19 mm), but can be as small as ¼ inch (6 mm), and sometimes range up to an inch (25 mm) or more.

Sealants are often used to seal joints between panels of stone or precast concrete in a curtain wall (Figures 20.8, 20.13), to seal the joint beneath the shelf angle in a brick curtain wall (Figure 20.1), and to seal joints between dissimilar materials, such as where a metal-and-glass cladding system ends against a masonry wall (Figure 21.11, details 6, 9, and 9a). Specially formulated sealants are used to seal between lights of glass and the frames that support them (Figure 17.16), and even to prevent the passage of sound around the edges of interior partitions (Figures 23.23, 23.24, 23.35, 23.38).

Sealant Materials

Gunnable Sealant Materials

Gunnable sealant materials are viscous, sticky liquids (*mastics*) that are injected into the joints of a building with a *sealant gun* (Figure 19.10). They cure within the joint to become rubberlike materials that adhere to the surrounding surfaces and seal the joint against the passage of air and water. Gunnable sealants can be grouped conveniently in three categories according to the amount of

change in joint size that each can withstand safely after curing:

• *Low-range sealants*, also called *caulks*, are materials with very limited elongation (stretching and squeezing) capabilities, up to plus or minus 5 percent of the width of the joint. They are used mainly for filling minor cracks in secondary joints,

especially as a preparation for painting. Most caulks cure by evaporation of water or an organic solvent, and shrink substantially as they do so. None is used for sealing of joints in building cladding.

• *Medium-range sealants* are materials such as butyl rubber or acrylic that have safe elongations in the plus or

FIGURE 19.10

Applying polysulfide, a high-range gunnable sealant, to a joint between exposed-aggregate precast concrete curtain wall panels, using a sealant gun. The operator moves the gun slowly so that a bulge of sealant is maintained just ahead of the nozzle. This exerts enough pressure on the sealant that it fully penetrates the joint. Following application, the operator will return to smooth and compress the wet sealant into the joint with a convex tool, much as a mason tools the mortar joints between masonry units. (*Courtesy of Morton Thiokol, Inc., Morton Chemical Division*)

minus 5 to 10 percent range. They are used in building cladding for sealing of nonworking joints (joints that are fastened together mechanically as well as being filled with sealant). Because these sealants cure by the evaporation of water or an organic solvent, they undergo some shrinkage during curing.

• *High-range sealants* can safely sustain elongations up to plus or minus 50 percent. They include various *polysulfides,* which are usually site mixed from two components to effect a chemical cure; *polyurethanes,* which may also cure from a two-component reaction, or else from reacting with moisture vapor from the air, depending on the formulation; and *silicones,* which cure by reacting with moisture vapor from the air. None of these sealants shrinks upon curing because none relies upon the evaporation of water or a solvent to effect a cure. All adhere tenaciously to the sides of properly prepared joints. All are highly resilient, rubberlike materials that return to their original size and shape after being stretched or compressed, and all are durable for 20 years or more if properly formulated and installed. Sealants for the working joints in cladding systems are selected from among this group.

Solid Sealant Materials

In addition to the gunnable sealants, several types of solid materials are used for sealing seams in building cladding:

• *Gaskets* are strips of various fully cured elastomeric (rubberlike) materials, manufactured in a number of different configurations and sizes for different purposes (Figure 19.11). They are either compressed into a joint to seal tightly against the surfaces on either side or inserted in the joint loose, then expanded with a *lockstrip* insert as illustrated in Figures 17.17 through 17.19.

• *Preformed cellular tape sealant* is a strip of polyurethane sponge material

that has been impregnated with a mastic sealant. It is delivered to the construction site in an airtight wrapper, compressed to one-fifth or one-sixth of its original volume. When a strip is inserted, it expands to fill the joint and the sealant material cures with moisture from the air to form a watertight seam.

• *Preformed solid tape sealants* are used only in lap joints, as in the mounting of glass in a metal frame or the overlapping of two thin sheets of metal at a cladding seam (Figure 19.11). They are thick, sticky ribbons of poly-

butene or polyisobutylene that adhere to both sides of the joint to seal and cushion the junction.

Sealant Joint Design

Figures 19.12 through 19.14 show the major principles that need to be kept in mind while designing a gunnable sealant joint. For a joint between materials with high coefficients of expansion, the time of year when the sealant is to be installed must be taken into account when specifying the size of the joint and the type of

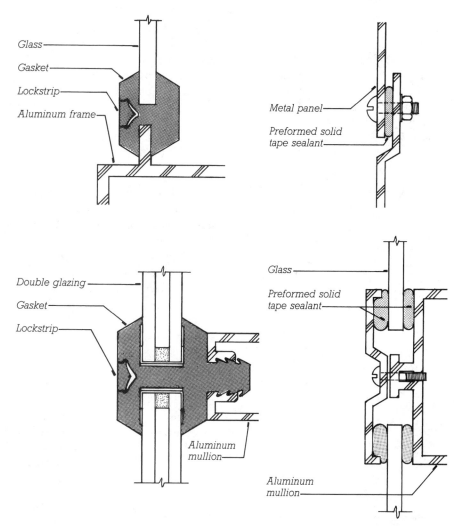

FIGURE 19.11
Some solid sealant materials. At the left, two examples of lockstrip gaskets. At the right, preformed solid tape sealants.

sealant; sealant installed in cold weather will have to stretch very little during its lifetime but will have to compress a great deal in summer, as the materials around it expand and crowd together. Sealant installed in hot weather will have to compress very little but will be greatly stretched in winter.

Installation procedures are also critical to the success of gunnable sealant joints in a cladding system. Each joint must be carefully cleaned of oil, dirt, oxide, moisture, or concrete form-release compound. If it is necessary to improve adhesion between the sealant and the substrate, the edges of the joint are *primed* with a suitable coating. Then the backer rod is inserted. The sealant is extruded into the joint from the nozzle of a sealant gun, filling the joint completely. Lastly, the sealant is mechanically tooled, much

as a masonry mortar joint is tooled, to compress the sealant material firmly against the sides of the joint and the backer rod. The tooling also gives the desired surface profile to the sealant.

Gasket sealants have generally proved to be less sensitive to installation problems than gunnable sealants. For this reason, they are widely used in proprietary cladding systems.

BASIC CONCEPTS OF BUILDING CLADDING SYSTEMS

The Loadbearing Wall

Until late in the 19th century, nearly all large buildings were built with loadbearing exterior walls. These walls supported a substantial portion of the floor and roof loads of the

Sealant

Backup rod

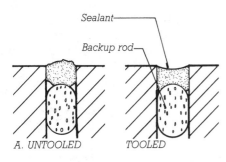
A. UNTOOLED   TOOLED

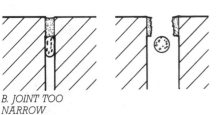
B. JOINT TOO NARROW

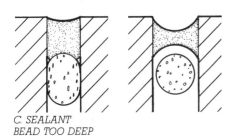
C. SEALANT BEAD TOO DEEP

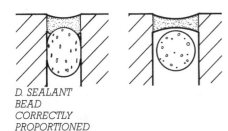
D. SEALANT BEAD CORRECTLY PROPORTIONED

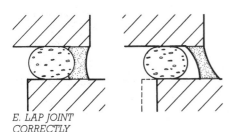
E. LAP JOINT CORRECTLY PROPORTIONED

FIGURE 19.12
Good and bad examples of sealant joint design. (*a*) **This properly proportioned joint is shown both untooled and tooled. The untooled sealant fails to penetrate completely around the backer rod and does not adhere fully to the sides of the joint.** (*b*) **A narrow joint may cause the sealant to elongate beyond its capacity when the panels on either side contract, as shown to the right.** (*c*) **If the sealant bead is too deep, sealant is wasted, and the four edges of the sealant bead are stressed excessively when the joint enlarges.** (*d*) **A correctly proportioned sealant bead. The backer rod, made of a spongy material that does not stick to the sealant, is inserted into the joint to maintain the desired depth. The width is calculated so that the expected elongation will not exceed the safe range of the sealant, and the depth is between $^1/_8$ and $^3/_8$ inch (3 and 9.5 mm).** (*e*) **A correctly proportioned lap joint. The width of the joint (the distance between the panels) should be twice the depth of the sealant bead, and also twice the expected movement in the joint.**

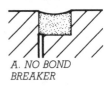
A. NO BOND BREAKER

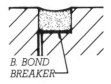
B. BOND BREAKER

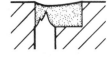

FIGURE 19.13
In three-sided joints, the sealant is likely to tear unless a nonadhering plastic bond breaker strip is placed in the joint before the sealant.

building as well as acting to separate the indoor environment from the outdoor. In noncombustible buildings, these walls were built of brick or stone masonry. Functionally, such walls had several inherent problems. They were poor thermal insulators. They tended to develop water leaks. They were heavy, which required large foundations and limited their height to a few stories.

The loadbearing wall has been brought up to date with higher-strength masonry and concrete and with the addition of components such as thermal insulating materials, cavities and flashings to make the wall more resistant to water penetration, and steel reinforcing to allow the wall to be thinner and lighter and to strengthen it against wind and seismic loads. Loadbearing masonry and concrete exterior walls are often attractive and economical for low- and medium-rise buildings. Loadbearing exterior walls are illustrated and discussed in more detail in Chapters 8, 10, and 14.

The Curtain Wall

The first steel-framed skyscrapers, built late in the 19th century, introduced the concept of the *curtain wall*, an exterior cladding supported at each story by the steel or concrete frame, rather than bearing its own load to the foundations. The earliest curtain walls were constructed of masonry (Figure 19.2). The principal advantage of the curtain wall is that, because it bears no vertical load, it can be thin and light in weight regardless of the height of the building, as compared to a masonry loadbearing wall, which becomes prohibitively thick and weighty at the base of a very tall building. The name "curtain wall" derives from the idea that the wall is thin and "hangs" like a curtain on the structural frame. (Most curtain wall panels do not actually hang in tension from the frame, but are supported from the bottom at each floor level.)

Curtain walls may be constructed of any noncombustible material that is suitable for exposure to the weather. They may be either *constructed in place* or *prefabricated*. In the next chapter, we will examine curtain walls that are made of masonry and concrete. In Chapter 21, we will look at curtain walls that are made of metal and glass. In both chapters, we will see that some types of walls are constructed in place and some are prefabricated, but all are supported by the frame of the building.

TESTING CLADDING SYSTEMS

For any new curtain wall design, it is advisable to build and test a full-scale section of wall to determine its resistance to air infiltration and water infiltration and its structural performance under heavy wind loadings. There are several outdoor laboratories in North America that are equipped to conduct these tests. A specimen of the cladding, often two stories high and a bay wide, is constructed as the exterior wall of a chamber that can be pressurized or evacuated by a calibrated blower system.

The specimen is tested first for air infiltration, using ASTM method E283, in which it is subjected to a static air pressure that corresponds to the pressure that will be created by the anticipated maximum wind velocity in the vicinity of the building. Air that leaks through the wall is carefully measured, and the rate of leakage is compared to specified standards.

A static test for water penetration is next, using ASTM E331: The wall is subjected to a static air pressure while being wetted uniformly across its sur-

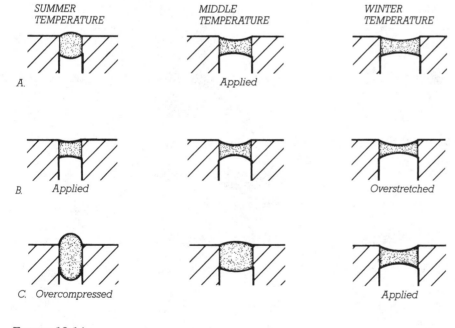

FIGURE 19.14
Sealants are best applied at temperatures that are neither excessively hot nor excessively cold. If cold- or hot-weather application of sealants is anticipated, the joints should be proportioned to minimize overstretching or overcompression. Row *A* shows the behavior of a sealant bead applied at a middle temperature. Rows *B* and *C* show beads applied at summer and winter temperatures, respectively.

face at a rate of 5 gallons per hour per square foot (204 l/h/m²). Points of water leakage are noted, and leaking water is carefully collected and measured. A dynamic water penetration test may also be performed in accordance with AAMA TM-1, using an aircraft engine and propeller to drive water against the wall.

The structural performance of the wall is tested using ASTM E330, in which the calibrated blower subjects the wall specimen to air pressures and suctions as high as 50 percent over the specified wind load, and the deflections of the structural members in the wall are measured. Thermal and sound transmission tests may also be performed, using ASTM C236 and C90, respectively.

While all these tests yield numerical results, it is also important that the behavior of the specimen be observed closely during each test so that specific problems with the design, materials, detailing, and installation can be identified and corrected. Most cladding specimens fail one or more of the tests for air and water leakage on the first attempt. By observing the sources of leakage during the test, it is usually possible to make modifications to the flashings, sealants, weep holes, or other components of the design so that the modified specimen will pass the test in a subsequent attempt. These modifications are then incorporated into the final details and specifications for the actual building.

After testing has been completed and final design adjustments have been made, production of the wall components begins, and deliveries to the site can commence as soon as the frame is ready to receive the cladding.

Curtain wall systems require careful inspection during installation to ensure that there are no defects in workmanship. Even seemingly small imperfections in assembly can lead to large and expensive problems later. It is advisable to test installed portions of curtain wall for water leakage on the actual building, using AAMA

Standard FC-1. This involves directing water at the joints in the wall with a hose that has a specified nozzle, and following specified procedures to isolate the causes of any leaks.

CLADDING AND THE BUILDING CODES

The major impact of building codes on the design of building cladding is in the areas of structural strength and fire resistance. Strength requirements relate to the strength and stiffness of the cladding itself and to the adequacy of its attachments to the building frame, with special reference to wind and seismic loadings.

Fire requirements are concerned with the combustibility of the cladding materials, the fire resistance ratings and vertical dimensions of parapets and spandrels, the fire resistance ratings of exterior walls facing other buildings that are near enough to raise questions of fire spread from one building to the other, and the closing off (*firestopping*) of any vertical passages in the cladding that are more than one story in height. The space inside column covers must be firestopped at each floor. The space between curtain wall panels and the edges of floors must also be firestopped, using a steel plate and grout, metal lath and plaster, or mineral wool *safing* (Figure 19.15).

FIGURE 19.15
Safing is a high-temperature, highly fire-resistant mineral batt material, which is inserted between a curtain wall panel and the edge of the floor slab to block the passage of fire from one floor to the next. It is seen here behind a metal-and-glass curtain wall with insulated spandrel panels. The safing is held in place by metal clips such as the one seen in the foreground. (*Courtesy of United States Gypsum Company*)

SELECTED REFERENCES

1. Latta, J. K., National Research Council Canada. *Roofs, Windows, and Walls for the Canadian Climate.* Ottawa, 1979 (NRCC #13487).

2. National Research Council Canada. *Construction Details for Airtightness.* Ottawa, 1980 (NRCC #18291).

3. National Research Council Canada. *Cracks, Movements, and Joints in Buildings.* Ottawa, 1976 (NRCC #15477).

The Division of Building Research of the National Research Council Canada has done pioneering work in theorizing about cladding design and performing tests and field observations to back up the theory. These three books treat their subjects completely and understandably, without jargon or higher mathematics. (Address for ordering: Institute for Research in Construction, National Research Council Canada, Ottawa, Ontario K1A 0R6, Canada.)

4. Anderson, J. M., and J. R. Gill. *Rainscreen Cladding: A Guide to Design Principles and Practice.* London, Butterworths, 1988.

This clear, succinct summary of rainscreen cladding principles includes an extensive bibliography on the subject.

5. Panek, Julian R., and John Philip Cook. *Construction Sealants and Adhesives* (3rd ed.). New York, John Wiley & Sons, Inc., 1992.

This book offers detailed information on selecting, detailing, and applying sealants and adhesives to all types of construction materials.

KEY TERMS AND CONCEPTS

cladding
thermal bridge
barrier wall
internal drainage
gravity
momentum
labyrinth
surface tension
drip
capillary action
capillary break
wind currents

rainscreen principle
pressure-equalized wall design
air barrier
rainscreen
pressure equalization chamber (PEC)
wash
sealant
gunnable sealant
mastic
sealant gun
caulk
polysulfide

polyurethane
silicone
gasket
lockstrip
preformed cellular tape sealant
preformed solid tape sealant
curtain wall
constructed in place
prefabricated
firestopping
safing

REVIEW QUESTIONS

1. Why is it so difficult to make cladding watertight?

2. List the functions that cladding performs and list one or two ways in which each of these functions is typically satisfied in a cladding design.

3. Explain with a series of simple sketches the principles of sealant joint design. List several sealant materials suitable for use in the joints that you have shown.

4. What are the forces that can move water through a joint in an exterior wall? How can each of these forces be neutralized?

EXERCISES

1. Examine the cladding of a building with which you are familiar. Look especially for features that have to do with insulation, condensation, drainage, and movement. Sketch a detail of how this cladding is installed and how it works. You will probably have to guess at some of the hidden features, but try to produce a complete, plausible detail. Add explanatory notes to make everything clear.

2. Work out a way to add a rainscreen window with fixed double glazing to the cladding system shown in Figure 19.8.

3. Prepare a sample sealant joint, using backer rod and silicone sealant obtained from a hardware store or building materials supplier. Apply the sealant to two parallel pieces of quarry tile or glass that are taped together with a spacer between. After the sealant has had time to cure (a week or so), remove the tape and spacer and test the joint by pulling and twisting it to find out how elastic it is and how well the sealant has adhered to the substrate.

CLADDING WITH MASONRY AND CONCRETE

Architects Thompson Ventulett Stainback & Associates created a bold pattern of precast concrete sunshades and spandrel panels for the facades of the United Parcel Service Headquarters Building in Atlanta, Georgia. (*Photo by Brian Gassel/TVS & Associates*)

Buildings framed with structural steel or reinforced concrete are often clad with brick masonry, stone masonry, cut stone panels, or precast concrete. These substantial materials, though in fact they are supported by the underlying frame, impart a sense of solidity and permanence to a building. But thin facings of masonry or concrete do not behave the same as solid, loadbearing walls. When mounted on a frame, these brittle materials must adjust to the movements of the frame and maintain weathertightness despite being applied in a layer only a few inches thick. Careful detailing and good construction practices are required to make this possible.

MASONRY VENEER CURTAIN WALLS

Figure 20.1 shows in a series of steps how a brick *masonry veneer* (a single wythe of brick masonry separated by a cavity from a structural backup wall) may be applied to a reinforced concrete frame. The veneer may also be made of ashlar or rubble stonework. The veneer wythe is erected brick by brick or stone by stone with conven-

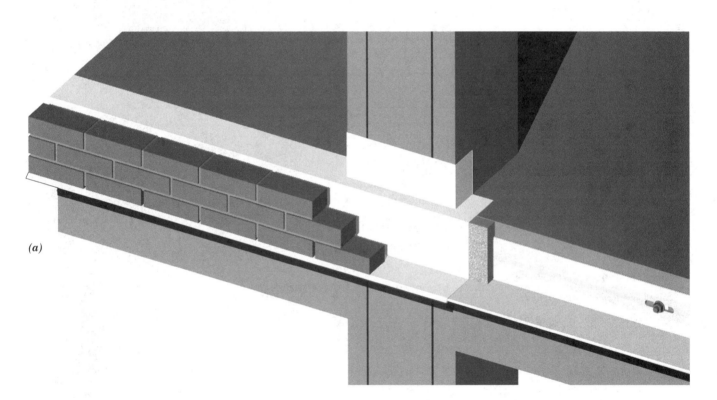

(a)

FIGURE 20.1

An example of a construction sequence for a brick veneer curtain wall supported by a reinforced concrete frame. (*a*) Before the concrete frame of the building was cast, inserts were put into the formwork to create attachments for the brick veneer, including wedge anchors along the line of each shelf angle and two vertical dovetail slots in each column. To begin installation of the brick veneer, a steel shelf angle is bolted to each spandrel beam, using wedge inserts as depicted in Figure 20.2. A slab of polystyrene foam insulation is placed over the upright leg of the shelf angle, and a continuous flashing is installed over the shelf angle, the foam, and the edge of the floor slab. The flashing is also wrapped around the base of the column. The first course of brickwork is laid directly on the shelf angle

and flashing, without a bed joint of mortar. Every third head joint in this first course is left open to form a weep hole. Three courses of brickwork bring the veneer up to the level of the floor slab. (*b*) The first course of the concrete masonry backup wall is laid on the edge of the floor slab. Vertical reinforcing bars are grouted into the hollow cores of the backup wall at intervals specified by the structural engineer. An asphaltic coating is applied to the outside of the backup wall to act as an air and moisture barrier. Three more courses of brickwork bring the veneer up to the level of the first course of concrete blocks. A slab of polystyrene foam insulation is placed against the concrete masonry wall. A combination joint reinforcing and masonry tie assembly made of galvanized heavy steel wires is

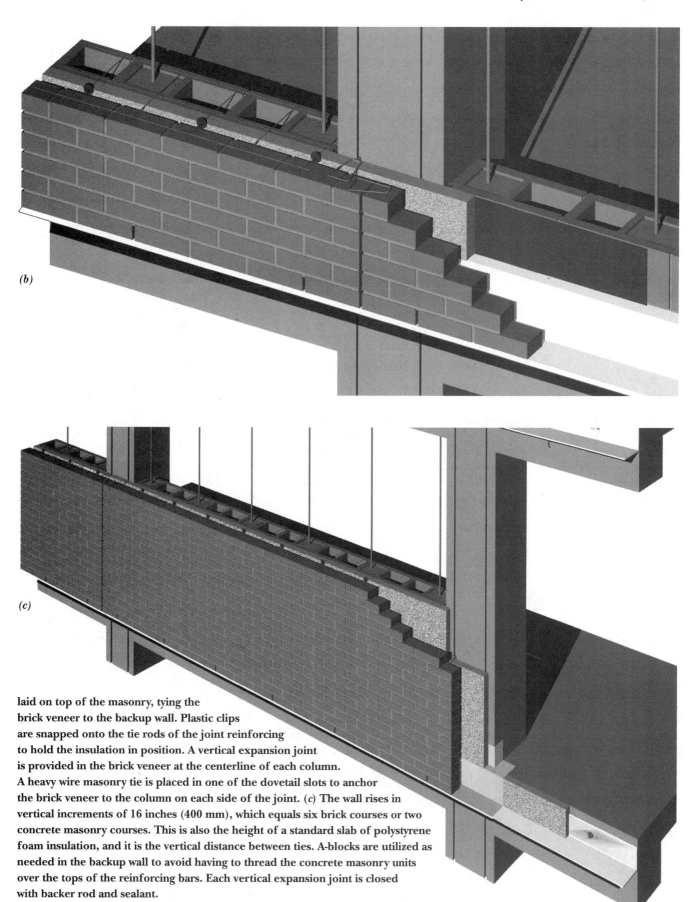

(b)

(c)

laid on top of the masonry, tying the
brick veneer to the backup wall. Plastic clips
are snapped onto the tie rods of the joint reinforcing
to hold the insulation in position. A vertical expansion joint
is provided in the brick veneer at the centerline of each column.
A heavy wire masonry tie is placed in one of the dovetail slots to anchor
the brick veneer to the column on each side of the joint. (*c*) The wall rises in
vertical increments of 16 inches (400 mm), which equals six brick courses or two
concrete masonry courses. This is also the height of a standard slab of polystyrene
foam insulation, and it is the vertical distance between ties. A-blocks are utilized as
needed in the backup wall to avoid having to thread the concrete masonry units
over the tops of the reinforcing bars. Each vertical expansion joint is closed
with backer rod and sealant.

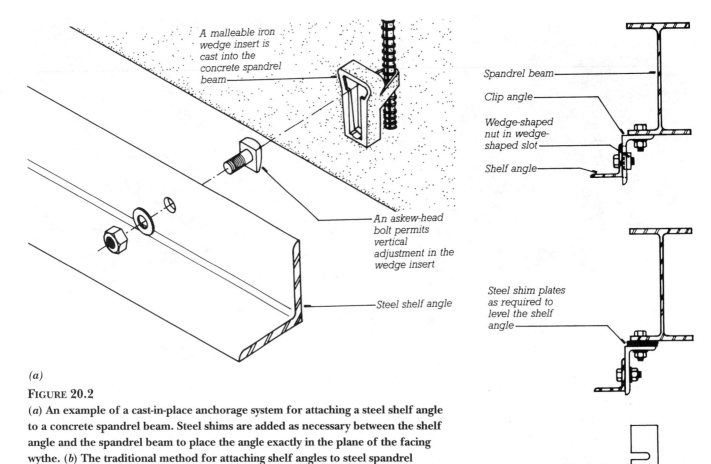

A malleable iron wedge insert is cast into the concrete spandrel beam

An askew-head bolt permits vertical adjustment in the wedge insert

Steel shelf angle

(a)

Spandrel beam

Clip angle

Wedge-shaped nut in wedge-shaped slot

Shelf angle

Steel shim plates as required to level the shelf angle

Detail of shim plate

(b)

FIGURE 20.2

(a) An example of a cast-in-place anchorage system for attaching a steel shelf angle to a concrete spandrel beam. Steel shims are added as necessary between the shelf angle and the spandrel beam to place the angle exactly in the plane of the facing wythe. (b) The traditional method for attaching shelf angles to steel spandrel beams uses steel clip angles with shim plates as needed to make up for dimensional inaccuracies in the components.

tional mortar, starting from a steel *shelf angle* that is attached to the structural frame at each floor (Figure 20.2). The construction process and details are essentially the same as for a masonry cavity wall of a single-story building, but there are some crucial differences: To accommodate movements in the frame of the building without applying a structural load to the masonry veneer, there must be a *soft joint* beneath each shelf angle (Figure 20.3). This must be dimensioned to absorb the maximum anticipated sum of column creep, brick expansion, spandrel beam deflection, and to include a dimensional tolerance to allow for construction inaccuracies, while not exceeding the maximum safe compressibility of the sealant. Masonry curtain walls also must be divided vertically by movement joints to allow the frame and

the masonry cladding to expand and contract independently of one another.

A backup wall of light-gauge steel studs covered with water-resistant sheathing panels of gypsum or cementitious materials is often

thought of as being interchangeable with a concrete masonry backup wall for a masonry facing. The stud wall even has certain advantages over masonry in its lighter weight, its greater ability to contain thermal insulation and electrical wiring, and

FIGURE 20.3

A complete detail section of the brick veneer wall shows how the top of the backup wall is fastened to the underside of the spandrel beam with a series of structural steel restraint clips that brace the top of the wall against wind loads but allow the spandrel beam to deflect under load. Two lines of backer rods and sealant along the columns and across the top of the backup wall make the backup wall airtight. A soft joint of sealant beneath the shelf angle permits the spandrel beam to deflect without applying force to the brick veneer. The brick ties nearest the underside of the shelf angle are anchored to dovetail slots in the spandrel beam. An additional plastic clip on each wire tie in the center of the cavity acts as a drip to prevent water from clinging to the tie and running toward the backup wall. The interior of the building is finished with gypsum veneer plaster mounted on steel furring channels. (*Drawing from Allen, Edward,* Architectural Detailing: Function, Constructibility, Aesthetics, *New York, John Wiley & Sons, Inc., 1993, reproduced by permission of the publisher*)

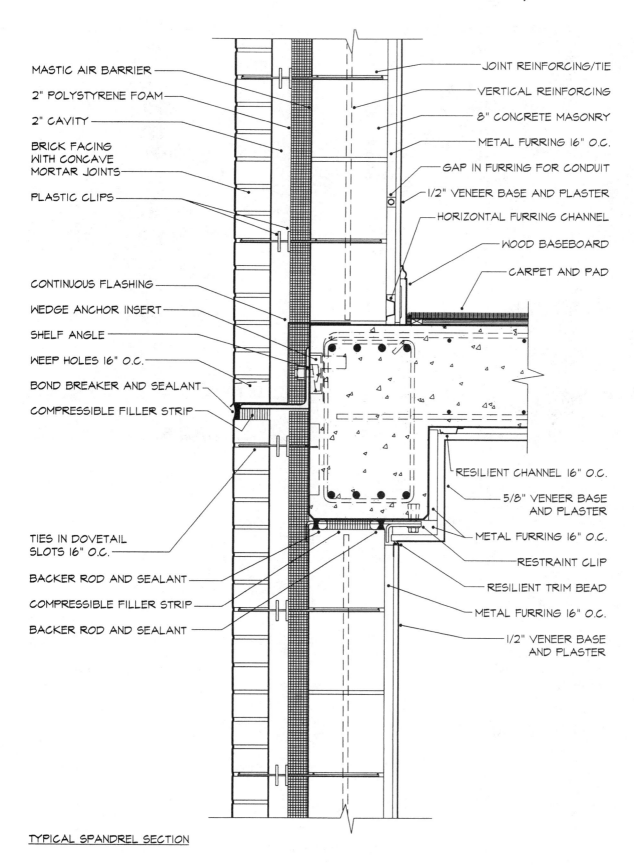

MASTIC AIR BARRIER

2" POLYSTYRENE FOAM

2" CAVITY

BRICK FACING
WITH CONCAVE
MORTAR JOINTS

PLASTIC CLIPS

CONTINUOUS FLASHING

WEDGE ANCHOR INSERT

SHELF ANGLE

WEEP HOLES 16" O.C.

BOND BREAKER AND SEALANT

COMPRESSIBLE FILLER STRIP

TIES IN DOVETAIL
SLOTS 16" O.C.

BACKER ROD AND SEALANT

COMPRESSIBLE FILLER STRIP

BACKER ROD AND SEALANT

JOINT REINFORCING/TIE

VERTICAL REINFORCING

8" CONCRETE MASONRY

METAL FURRING 16" O.C.

GAP IN FURRING FOR CONDUIT

1/2" VENEER BASE AND PLASTER

HORIZONTAL FURRING CHANNEL

WOOD BASEBOARD

CARPET AND PAD

RESILIENT CHANNEL 16" O.C.

5/8" VENEER BASE
AND PLASTER

METAL FURRING 16" O.C.

RESTRAINT CLIP

RESILIENT TRIM BEAD

METAL FURRING 16" O.C.

1/2" VENEER BASE
AND PLASTER

TYPICAL SPANDREL SECTION

(a)

(b)

FIGURE 20.4
(*a*) **A carefully detailed brick curtain wall by architects Kallman, McKinnell and Wood covers the steel frame of Hauser Hall at Harvard University. Notice the expansion joint near the corner.** (*b*) **At the base of Hauser Hall, the facing wythe is made of limestone blocks. The backup wall consists of steel studs and gypsum sheathing panels.** (*Photos by Steve Rosenthal*)

its greater receptivity to a variety of interior finish materials. However, the steel studs and their fastenings are inherently more flexible than a concrete masonry wall and may deflect enough under maximum wind pressures to cause cracking of brittle masonry veneers. Such cracking often leads to water leakage. Furthermore, if there is leakage through the facing layer of the wall because of cracking, porous masonry, or poor workmanship, the steel studs and fasteners are susceptible to corrosion, and the gypsum sheathing panels are subject to water deterioration. A concrete masonry backup wall is usually stiffer than the veneer that it supports, so that the veneer is unlikely to crack under wind loadings. A concrete masonry backup wall

also maintains its structural integrity despite prolonged periods of wetting. For these reasons, a concrete masonry backup wall is generally preferable to a wall of steel studs. If a steel stud backup system is selected, the studs, masonry ties, and fasteners should be sized very conservatively so as to be stiff enough against wind loadings that the veneer material will not crack. The sheathing material and fasteners should be selected for their durability under damp conditions. Each metal tie that connects the masonry veneer to the studs should be attached directly to a stud with at least two corrosion-resistant screws. The wall must be detailed carefully to keep leakage away from the backup components. Constant inspection is required during con-

struction to be certain that the details are faithfully executed and the cavity is kept clean so that it will drain freely.

The structural frame of a building is never absolutely flat or plumb. Thus, the attachment system for the shelf angles must allow for adjustments so that the masonry veneer may be constructed in a precisely vertical plane with level courses. Figure 20.2 shows how this is usually done for both concrete and steel frames. The attachment system in Figure 20.5, which is designed to suspend a masonry veneer spandrel wall over a continuous band of windows (Figure 20.6), also provides for free adjustment of the shelf angle location.

The flashing above the shelf angle should project beyond the face of the masonry by an inch (25 mm)

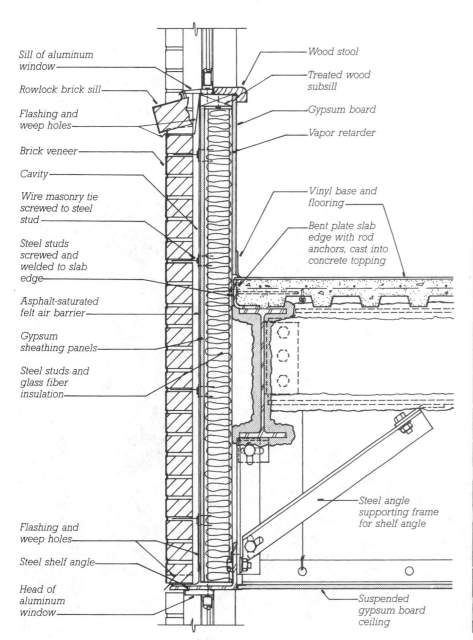

Sill of aluminum window

Rowlock brick sill

Flashing and weep holes

Brick veneer

Cavity

Wire masonry tie screwed to steel stud

Steel studs screwed and welded to slab edge

Asphalt-saturated felt air barrier

Gypsum sheathing panels

Steel studs and glass fiber insulation

Flashing and weep holes

Steel shelf angle

Head of aluminum window

Wood stool

Treated wood subsill

Gypsum board

Vapor retarder

Vinyl base and flooring

Bent plate slab edge with rod anchors, cast into concrete topping

Steel angle supporting frame for shelf angle

Suspended gypsum board ceiling

FIGURE 20.5

A detail section of a brick curtain wall that is supported below the level of the spandrel beam on a frame made of steel angles. The supporting frame becomes necessary when continuous horizontal bands of windows are to be installed between brick spandrels. All the connections in the supporting frame are made with bolts in slotted holes, to allow for exact alignment of the shelf angle. After the frame has been aligned and before the masonry work begins, the connections are welded to prevent slippage. Shelf angle constructions for masonry curtain walls require careful engineering to accommodate expected loads and structural deflections.

FIGURE 20.6
The detail shown in Figure 20.5 allows the construction of brick spandrels between continuous horizontal bands of glass. (*Photo by the author*)

or so, and should be bent downward at a 45° angle to form a drip. In this way, it is able to conduct water that has leaked into the cavity back to the outdoors and to drain it safely away from the wall. If a flexible plastic or composite flashing is used, it should be cemented to a strip of sheet metal flashing over the shelf angle, with the sheet metal forming the projecting drip.

Many architects wish to maintain the fiction that a masonry veneer is actually a solid masonry wall. They find the soft joint and projecting flashing to be objectionable in this regard. To conceal these elements, they use a brick with a face lip that hangs down over the shelf angle to conceal it from view, and they do not allow the flashing to project out of the wall. The color of the sealant that is used in the soft joint is matched as closely as possible to the color of the mortar. Unfortunately, the use of lipped bricks and recessed flashings is very risky. The recessed flashing allows water to accumulate around the toe of the angle and to cause the angle to rust. Freeze–thaw action and the expansion of the steel as it rusts are likely to cause the lips to spall off of the bricks. Eventually, the deterioration along the line of the shelf angle becomes unsightly. Worse yet, the stability of the veneer may be endangered by failure of the corroded shelf angle. A better strategy for the conscientious architect is to

FIGURE 20.7
Fabrication and installation of a brick panel curtain wall. (*a*) Masons construct the panels in a factory, using conventional bricks and mortar. Both horizontal and vertical reinforcing are used, the vertical bars being grouted into the hollow cores of the bricks. (*b*) Brick panels are stored to await shipment, complete with thermal insulation. The welded metal brackets are for attachment to the building; the structural strength of the panel comes primarily from the reinforced masonry, not the brackets. (*c*) A crane lifts a parapet panel to its final position. (*d*) Corners can be constructed as single panels.
(*Courtesy of Vet-O-Vitz Masonry Systems, Inc., Brunswick, Ohio*)

(*a*)

(*b*)

(*c*)

(*d*)

find a way to express visually the presence of the shelf angle, flashing, and soft joint, and to make them positive features of the building facade. A soldier course or cut stone sill above each shelf angle is a good start toward a frank expression of a constructional necessity.

Prefabricated Brick Panel Curtain Walls

Figure 20.7 shows the use of prefabricated reinforced brick panels for cladding. Masons construct the panels while working comfortably at ground level in a factory. Horizontal reinforcing may be laid into the mortar joints or grouted into channel-shaped bricks. Vertical reinforcing bars are placed in grouted cavities of hollow bricks. These panels are self-rigid; they need no structural backup and can be fastened to the building in much the same way as precast concrete panels. A steel stud backup wall is required to carry thermal insulation, electrical wiring, and an interior finish layer, but has no structural role.

STONE CURTAIN WALLS

Chapter 9 illustrates conventionally set stone facing systems that tie relatively small blocks of cut stone set in mortar to a concrete masonry backup wall. Slabs of stone that are larger in surface area may be fastened to framed buildings in several different ways.

Stone Panels Mounted on a Steel Subframe

Figure 20.8 shows a system for mounting stone panels on a steel subframe. The vertical members of the subframe are erected first. They are designed to transmit gravity and wind loads from the stone slabs to the frame of the building. The horizontal members are aluminum shapes that engage slots in the upper and lower edges of each panel to attach them firmly to the building. They are added as the installation of the stone

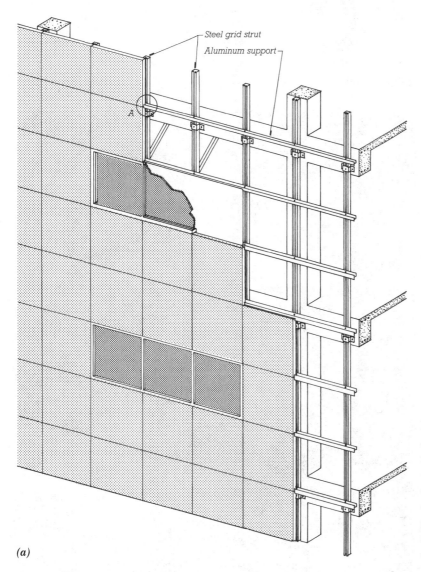

(a)

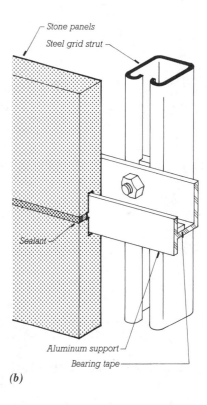

(b)

FIGURE 20.8
A subframe of vertical steel struts supports a facing of stone panels by means of horizontal metal clips that engage slots in the upper and lower edges of the panels. In order to avoid corrosion and staining problems, the steel struts should be galvanized, and the clips should be made of a nonferrous metal (usually aluminum) that is chemically compatible with the type of stone that is used.

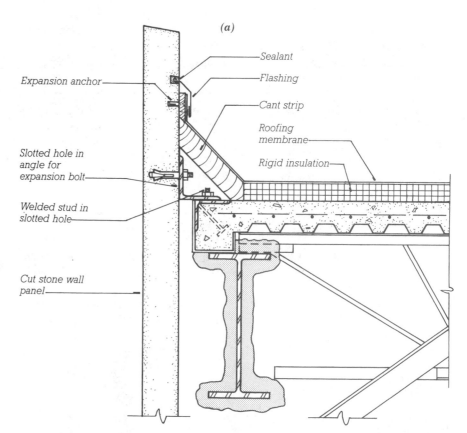

(a)

— Sealant

— Flashing

— Cant strip

Roofing membrane

Rigid insulation

Expansion anchor

Slotted hole in angle for expansion bolt

Welded stud in slotted hole

Cut stone wall panel

FIGURE 20.9

(a) Parapet and *(b)* spandrel details for a stone panel curtain wall made of limestone, marble, or granite. The broken lines indicate the outline of the interior finish and thermal insulation components, which are not shown. Each support plate holds edges of two adjacent wall panels, which are pocketed as shown to rest on the plate. The plate should be made of a noncorroding metal. The vertical joints between panels are closed with a backer rod and sealant.

FIGURE 20.10

A granite panel curtain wall of the type illustrated in Figure 20.9 wraps around the corner of a Boston office building. The upper-floor windows have not yet been installed, but the window frames have been mounted in two of the middle floors, and the lower floors have been glazed. (*Architect: Hugh Stubbins and Associates. Photo by the author*)

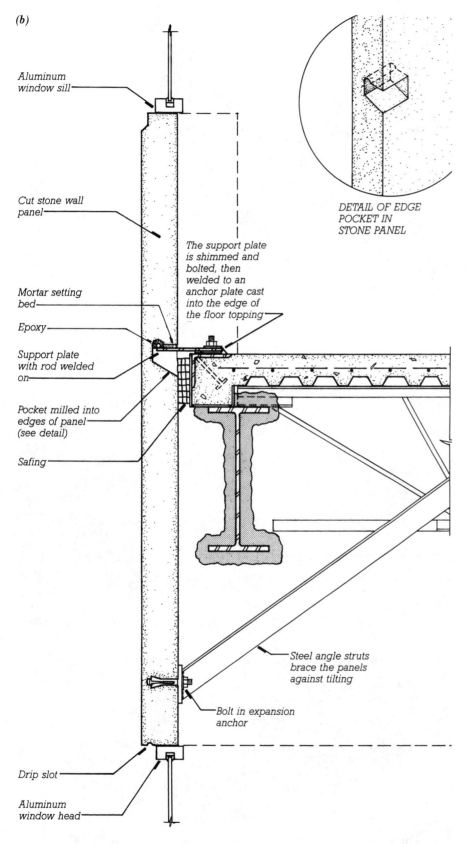

(b)

Aluminum window sill

Cut stone wall panel

Mortar setting bed

Epoxy

Support plate with rod welded on

Pocket milled into edges of panel (see detail)

Safing

The support plate is shimmed and bolted, then welded to an anchor plate cast into the edge of the floor topping

Steel angle struts brace the panels against tilting

Bolt in expansion anchor

Drip slot

Aluminum window head

DETAIL OF EDGE POCKET IN STONE PANEL

panels progresses. Backer rods and sealant fill the spaces between the panels, allowing for a considerable range of movement. A nonstructural backup wall, usually made of steel studs and gypsum sheathing panels, is constructed within the frame of the building but is not attached to the subframe. Its function is to provide an air barrier, to house thermal insulation batts and electrical wiring, and to support the interior wall finish layer, which is usually plaster or gypsum board.

A weakness of this system is its dependence on the integrity of the sealant joints. If a sealant joint should leak, water may accumulate in the slots in the tops of the stone panels, and freeze–thaw deterioration may ensue.

Monolithic Stone Cladding Panels

Figures 20.9 and 20.10 illustrate the use of monolithic panels of stone that are fastened directly to the frame of the building. The weight of each panel is transferred to two steel support plates by means of edge pockets that are cut into both sides of each panel at the stone mill. Each panel is stabilized by a pair of steel angle struts that are bolted to the stone with expansion anchors in drilled holes. Joints are closed with backer rods and sealant, and a nonstructural backup wall is required.

FIGURE 20.11
A steel truss system for stone cladding.
(*a*) Masons in a factory attach thin sheets
of stone to welded steel trusses. The vertical joints are closed with backer rods
and sealant. (*b*) The fabricated spandrel
panel is lifted onto a truck using a crane.
The metal clips that are just visible along
the top and bottom edges of the panel
engage slots in the edges of the sheets of
stone to hold the stone securely to the
truss. The steel angle clips at the two
upper corners of the truss will support
the panel on brackets welded to the steel
columns of the building frame. (*c*) The
panel is installed. (*Courtesy of
International Masonry Institute*)

(*a*)

(*b*)

(*c*)

Stone Cladding on Steel Trusses

Sheets of stone can be combined into large prefabricated panels by mounting them on structural steel trusses (Figure 20.11). Each truss is designed to carry both wind loads and the dead load of the stone to steel connection brackets that transfer these loads to the frame of the building. Sealant joints and a nonstructural backup wall finish the installation.

Posttensioned Limestone Spandrel Panels

Thick blocks of limestone may be joined with adhesives into long spandrel panels and posttensioned with high-strength steel tendons so that the assembly is self-supporting between columns (Figure 20.12). This is a relatively costly type of panel because of its use of comparatively large quantities of stone per unit area of cladding.

Very Thin Stone Facings

Extremely thin sheets of stone [as thin as $\frac{1}{4}$ inch (6.5 mm) for granite] may be stiffened with a structural backing such as a metal honeycomb and mounted as spandrel panels in an aluminum mullion system such as those pictured in Chapter 21. Very thin sheets of stone may also be used as facings for precast concrete curtain wall panels.

When specifying the thickness of stone for any exterior cladding application, the designer should work closely with the stone supplier and also consult the relevant standards of the building stone industry. Stone that has been sliced thinner than industry standards has been the cause of a number of failures of cladding systems.

FIGURE 20.12
Thicker blocks of Indiana limestone may be posttensioned together to make spandrel panels that span from column to column but require little steel. The posttensioning tendon is threaded through matching holes that are drilled in the individual stones prior to assembly.

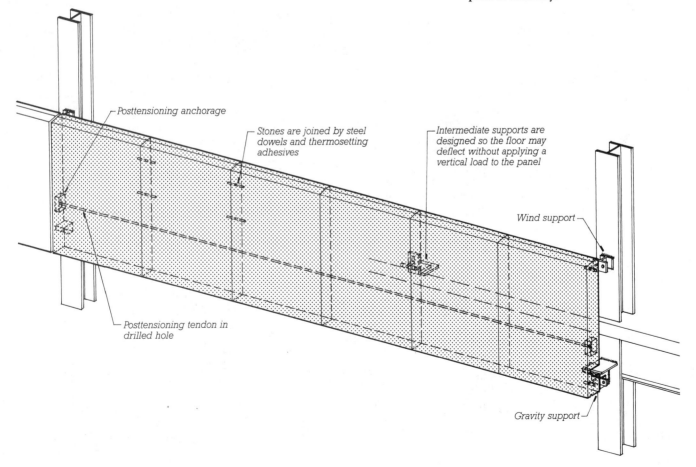

Posttensioning anchorage

Stones are joined by steel dowels and thermosetting adhesives

Intermediate supports are designed so the floor may deflect without applying a vertical load to the panel

Wind support

Posttensioning tendon in drilled hole

Gravity support

PRECAST CONCRETE CURTAIN WALLS

Curtain wall panels made of precast concrete with conventional reinforcing or prestressing (Figures 20.13–20.18) are simple in concept, but they require close attention to matters of surface finish, mold design, thermal insulation, attachment to the building frame, and sufficient strength and rigidity in the building frame to support the weight of the panels.

The factory production of concrete cladding panels makes it possible to utilize very high quality molds and a variety of surface finishes, from glassy smooth to rough, exposed aggregates. Ceramic tiles, thin bricks, and thin stone facings may be attached to precast concrete panels. Thermal insulation may be sandwiched into the panel (Figures 20.17, 20.18), may be affixed to the back of the panel, or may be provided in a nonstructural backup wall that is constructed on site. Reinforcing or prestressing must be designed to resist wind, gravity, and seismic forces, and

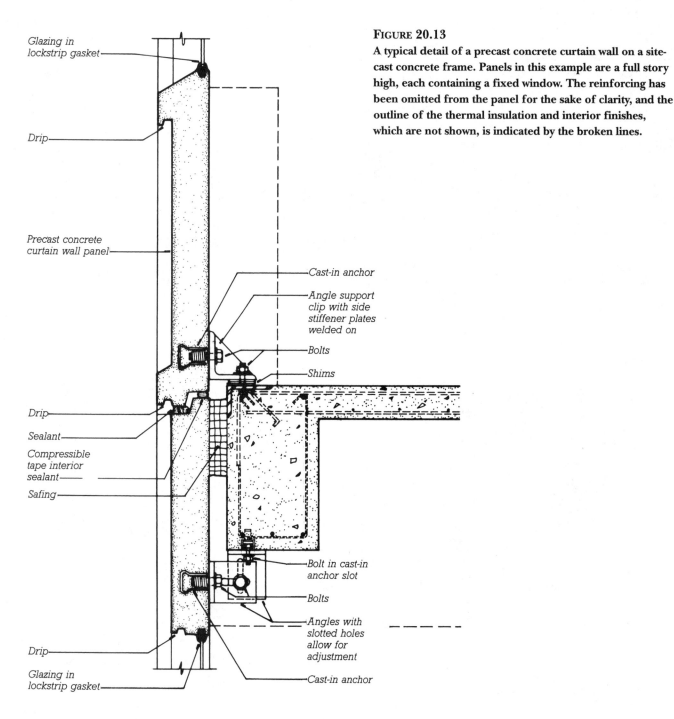

FIGURE 20.13
A typical detail of a precast concrete curtain wall on a site-cast concrete frame. Panels in this example are a full story high, each containing a fixed window. The reinforcing has been omitted from the panel for the sake of clarity, and the outline of the thermal insulation and interior finishes, which are not shown, is indicated by the broken lines.

Glazing in lockstrip gasket

Drip

Precast concrete curtain wall panel

Cast-in anchor

Angle support clip with side stiffener plates welded on

Bolts

Shims

Drip

Sealant

Compressible tape interior sealant

Safing

Bolt in cast-in anchor slot

Bolts

Angles with slotted holes allow for adjustment

Drip

Glazing in lockstrip gasket

Cast-in anchor

FIGURE 20.14
Workers install a precast concrete curtain wall panel. (*Architects and engineers: Andersen–Nichols Company, Inc. Photo by the author*)

FIGURE 20.15
A Chicago hotel is clad in handsomely detailed precast concrete panels. (*Architects: Solomon Cordwell Buenz Associates. Photo by Hedrich–Blessing*)

to control cracking of the panel. Attachments must transfer all these forces to the building frame while allowing for installation adjustment and for relative movements of the frame and the cladding.

Glass-Fiber-Reinforced Concrete (GFRC) Curtain Walls

Glass-fiber-reinforced concrete (GFRC) is a relatively new cladding material that has several advantages over conventional precast concrete panels. Its admixture of short, alkali-resistant glass fibers furnishes enough tensile strength that no steel reinforcing is required. Panel thicknesses and weights are about one-quarter of those for conventional precast concrete panels, which saves money on shipping, makes the panels easier to handle, and allows the use of lighter attachment hardware. The light weight of the cladding also allows the loadbearing frame of the building to be lighter and less expensive. GFRC can be molded into three-dimensional forms with intricate detail and an extensive range of colors and textures (Figures 20.19, 20.20). The panels may be self-stiffened with GFRC ribs, but the usual practice is to attach a welded frame made of light-gauge steel studs to the back of each GFRC facing in the factory. The attachment is made by means of thin steel rod anchors

FIGURE 20.16
Horizontal bands of smooth and textured precast concrete create the facade pattern of this suburban office building.
(*Architect: ADD, Inc. Courtesy of Precast/Prestressed Concrete Institute*)

FIGURE 20.17
Manufacturing Corewall®, a proprietary
foam-core precast concrete wall panel.
Rollers apply a ribbed texture to the outside of a panel that includes a layer of
foam plastic insulation sandwiched
between layers of concrete. (*Courtesy of
Butler Manufacturing Co.*)

FIGURE 20.18
A completed panel is lifted
from the casting bed. (*Courtesy
of Butler Manufacturing Co.*)

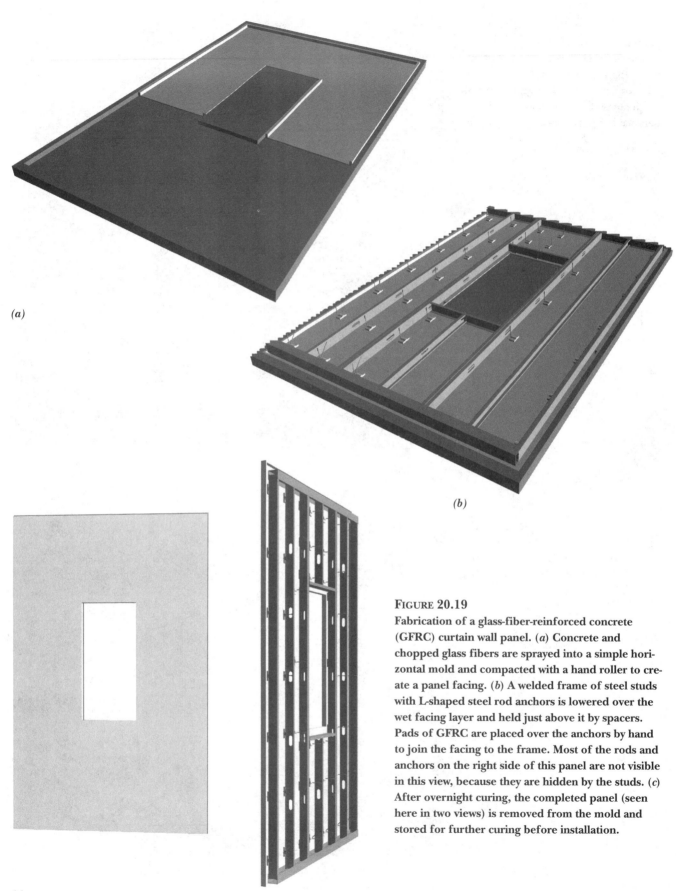

(a)

(b)

(c)

FIGURE 20.19
Fabrication of a glass-fiber-reinforced concrete (GFRC) curtain wall panel. (*a*) Concrete and chopped glass fibers are sprayed into a simple horizontal mold and compacted with a hand roller to create a panel facing. (*b*) A welded frame of steel studs with L-shaped steel rod anchors is lowered over the wet facing layer and held just above it by spacers. Pads of GFRC are placed over the anchors by hand to join the facing to the frame. Most of the rods and anchors on the right side of this panel are not visible in this view, because they are hidden by the studs. (*c*) After overnight curing, the completed panel (seen here in two views) is removed from the mold and stored for further curing before installation.

(a)

(b)

FIGURE 20.20
Fabrication of a GFRC curtain wall panel. (*a*) A special gun deposits a layer of sand–cement slurry simultaneously with 1.5-inch (38-mm) lengths of alkali-resistant glass fiber reinforcing. Three layers are usually required to make up the full thickness of the panel facing; each is compacted with a small hand roller before the next layer is applied. The overall thickness is usually ½ inch (13 mm). (*b*) After the GFRC facing layer has been completed, the steel frame is lowered over it and the operator hand applies pads of wet GFRC over the rod anchors to bond the frame to the GFRC facing. (*Courtesy of Precast/Prestressed Concrete Institute*)

that flex slightly as needed to permit relative movement between the facing and the frame. Figure 20.21 shows typical ways of attaching metal-framed GFRC panels to the building. The edges of the GFRC facing, which is usually only about $\frac{1}{2}$ inch (13 mm) thick, are flanged as shown in Figure 20.22 so that backer rods and sealant may be inserted between panels.

EXTERIOR INSULATION AND FINISH SYSTEM

An *exterior insulation and finish system* (*EIFS*) consists of a layer of plastic foam insulation that is adhered or mechanically fastened to a backup wall, a reinforcing mesh that is applied to the outer surface of the foam by embedment in a base coat of a stuccolike material, and an exterior finish coat of a similar stuccolike material that is troweled over the reinforced base coat. Most exterior insulation and finish systems are constructed in place over a backup wall made either of concrete masonry or of steel studs and water-resistant

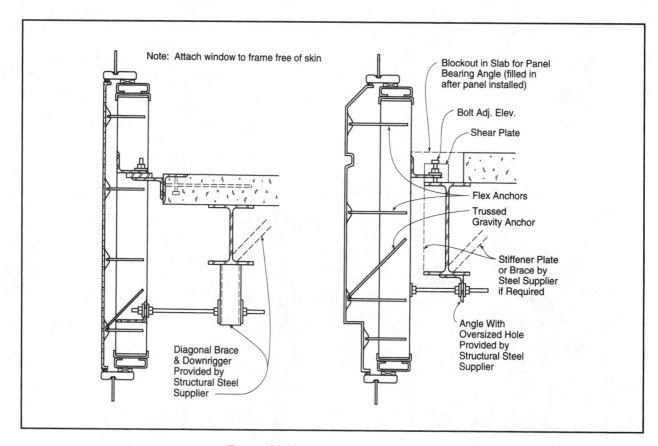

Note: Attach window to frame free of skin

Diagonal Brace & Downrigger Provided by Structural Steel Supplier

Blockout in Slab for Panel Bearing Angle (filled in after panel installed)

Bolt Adj. Elev.

Shear Plate

Flex Anchors

Trussed Gravity Anchor

Stiffener Plate or Brace by Steel Supplier if Required

Angle With Oversized Hole Provided by Structural Steel Supplier

FIGURE 20.21
Typical connections of GFRC panels to a steel frame. The lower connection in each case is a threaded rod that can flex as necessary as the height of the upper connection is adjusted with shims.
(*Courtesy of Precast/Prestressed Concrete Institute*)

sheathing (Figures 20.23, 20.24), but the system also adapts readily to prefabrication (Figure 20.25). EIFS also finds wide use over wood light framing for small commercial and residential buildings.

There are two generic types of exterior insulation and finish systems, polymer based and polymer modified. *Polymer-based systems* use a very low density expanded polystyrene bead foam insulation, a glass fiber reinforcing mesh embedded in a base coat that is formulated primarily from either portland cement or acrylic polymer, and a finish coat that consists of texture granules in an acrylic polymer vehicle. The foam insulation is adhered to the backup wall.

Polymer-modified systems use a slightly higher density, extruded polystyrene foam insulation rather than expanded bead foam. The foam panels are mechanically attached to the backup wall with metal or plastic screws (plastic screws minimize thermal bridging through the insulation). A metal reinforcing mesh is embedded in a relatively thick port-

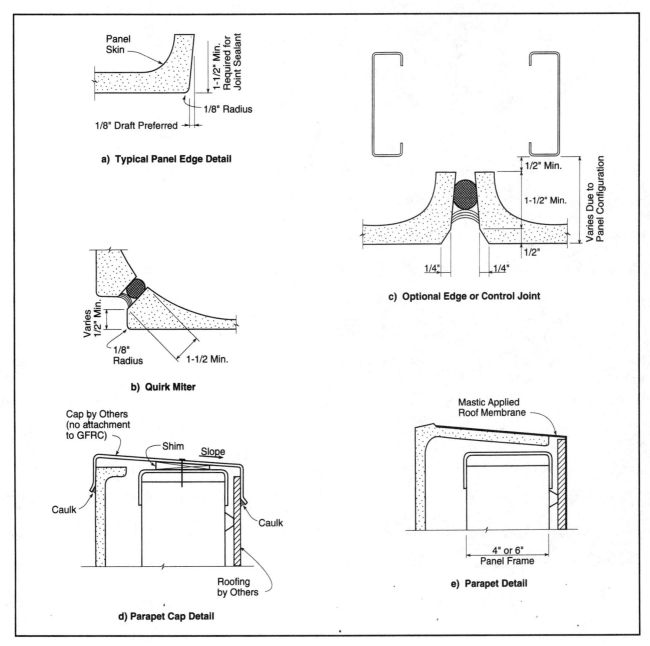

FIGURE 20.22
Typical edge details for GFRC cladding panels. (*Courtesy of Precast/Prestressed Concrete Institute*)

land cement base coat, and the finish coat is formulated of portland cement with acrylic modifiers. The net result is that polymer-modified systems are more durable (and more expensive) than polymer-based systems. They are more susceptible to shrinkage cracking during curing, but much less susceptible to denting or puncture. Polymer-based systems, on the other hand, have a very thin coating that is more elastic, less prone to crack, but relatively easy to dent or puncture when applied to areas of a building that are within reach of passersby or vehicles.

EIFS is an unusually versatile type of cladding, being applicable to buildings from single-family residences of wood or masonry construction to the largest buildings of noncombustible construction. It is used both for new construction and for refacing and insulating existing buildings. The insulating foam layer may be up to 4 inches (100 mm) thick, and there is little or no thermal bridging. The finish layer may be applied in a range of colors and textures. In appearance, EIFS is virtually

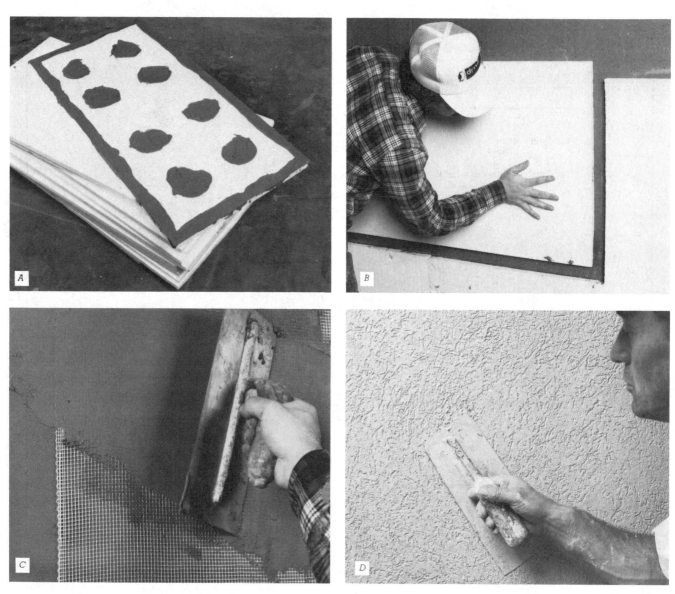

FIGURE 20.23
Four steps in installing an exterior insulation and finish system (EIFS) over a building with walls of masonry or solid sheathing. (*a*) A panel of foam is daubed with polymer-modified portland cement mortar. The foam may be as thick as required to achieve the desired thermal performance. (*b*) The foam panel is pressed into place, where it is held permanently by the daubs of mortar. (*c*) A thin base coat of polymer-modified stucco is applied to the surface of the foam panels, with an embedded mesh of glass fiber to act as reinforcing. (*d*) After the base coat has hardened, a finish coat in any desired color is troweled on. (*Courtesy of Dryvit® System, Inc.*)

FIGURE 20.24
A new bank building clad in EIFS.
(*Architect: Paul Thoryk. Photo by John Bare. Courtesy of Dryvit® System, Inc.*)

FIGURE 20.25
EIFS cladding can be shop fabricated and erected in panel form. (*a*) Steel studs are welded together to make panel frames. (*b*) Rigid sheathing is screwed to the panel frames and finished with EIFS as shown in Figure 20.23. (*c*) The finished panels are bolted to the frame of the building. (*Courtesy of Dryvit® System, Inc.*)

indistinguishable from conventional stucco.

A weakness of conventional EIFS of either type is that it is designed as a barrier system. It has no system of internal drainage that will prevent damage to the backup material if water leakage occurs at joints or through damaged areas. There have been numerous cases of extensive water damage in EIFS-faced buildings that have experienced leakage through poor detailing or faulty sealant joints, especially around windows and doors. In response to this problem, EIFS producers market rainscreen external insulation and finish systems that utilize a layer of drainage matting to create a pressure equalization chamber (PEC) behind the foam insulation (Figure 20.26). The PEC drains to plastic flashings above wall openings and at the base of the wall.

Another weakness of polymer-based EIFS is the ease with which it is dented or punctured; this may be overcome by specifying a polymer-modified system or a specially reinforced polymer-based system in areas subject to damage. Damaged spots are easily and unobtrusively patched.

FIGURE 20.26
A mockup demonstrates the features of a proprietary rainscreen EIFS system. From interior to exterior, the layers are an asphalt-saturated felt air and moisture barrier, a drainage mat composed of plastic fibers, plastic foam insulation, reinforcing mesh, base coat, and finish coat. The open structure of the drainage mat creates a pressure equalization chamber (PEC). A continuous plastic flashing under a gap at the bottom of the wall admits air to pressurize the PEC and drains any leakage out to the face of the wall. (*Photograph of Senergy CD System courtesy of Senergy, Cranston, Rhode Island*)

ings, the cavities are interrupted by framing and attachment components that are likely to splatter draining water onto the building frame and backup wall, where it can cause serious damage.

There have been many isolated efforts to design true rainscreen details for stone and concrete cladding systems, and a number of successful projects have been constructed. So far, however, no standard rainscreen details have emerged for these materials. This is an area to which trade associations and researchers should direct a great deal of effort, because reliably watertight details that are not heavily dependent on good workmanship and maintenance would save tens of millions of dollars each year in repair and reconstruction costs.

FUTURE DIRECTIONS IN MASONRY AND STONE CLADDING

The alert reader will have noticed that, with the exception of the first and last systems, brick veneer and rainscreen EIFS, the cladding systems shown in this chapter are detailed as barrier systems, not rainscreen systems. This means that they are entirely dependent on good installation and careful maintenance if they are to remain watertight. If a sealant joint fails or a stone cracks in any of these systems, there is nothing to keep water from getting behind the cladding. There is also no well-organized system of secondary drainage in these systems: Although most have cavities behind their fac-

C.S.I./C.S.C. Masterformat Section Numbers for Masonry and Concrete Cladding	
03400	**PRECAST CONCRETE**
03450	**Architectural Precast Concrete—Plant Cast Faced Architectural Precast Concrete Glass Fiber Reinforced Precast Concrete**
04200	**UNIT MASONRY**
04235	**Preassembled Masonry Panel Systems**
04400	**STONE**
04450	**Stone Veneer**
07400	**MANUFACTURED ROOFING AND SIDING**
07410	**Manufactured Roof and Wall Panels**
07420	**Composite Panels**
07440	**Faced Panels**
07450	**Glass Fiber Reinforced Cementitious Panels**

SELECTED REFERENCES

1. Brick Institute of America. *Technical Notes on Brick Construction, Nos. 18, 18A, 27, 28A, 28B*. Reston, Virginia, various dates.

These detailed pamphlets cover every aspect of brick veneer cladding systems. (Address for ordering: 11490 Commerce Park Drive, Reston, VA 22091.)

2. All the references on stone and concrete masonry listed at the end of Chapter 9 are also relevant to this chapter.

3. Precast/Prestressed Concrete Institute. *Architectural Precast Concrete* (2nd ed.). Chicago, 1989.

This well-illustrated, hardbound book covers all aspects of the design, manufacture, and installation of precast concrete curtain walls. Also available from the same source is *Architectural Precast Concrete—Color and Texture Selection Guide* (1992), an extensive set of full-color plates of finishes for precast concrete panels. (Address for ordering: Precast/Prestressed Concrete Institute, 175 West Jackson Boulevard, Chicago, IL 60604.)

4. Precast/Prestressed Concrete Institute. *GFRC: Recommended Practice for Glass Fiber Reinforced Concrete Panels* (3rd ed.). Chicago, 1993.

This 100-page booklet is a clear, complete guide to the design and manufacture of GFRC cladding systems. (Address for ordering: see reference 3.)

5. Sands, Herman. *Wall Systems: Analysis by Detail*. New York, McGraw–Hill, 1986.

This is a volume of case studies of cladding systems, illustrated comprehensively with detailed drawings, many of them in three dimensions.

KEY TERMS AND CONCEPTS

masonry veneer
shelf angle
soft joint
prefabricated brick panel curtain wall
stone cladding on steel subframe

monolithic stone cladding panel
stone cladding on steel truss
posttensioned limestone spandrel panel
precast concrete curtain wall
glass-fiber-reinforced concrete (GFRC)

exterior insulation and finish system (EIFS)
polymer-based system
polymer-modified system
rainscreen EIFS

REVIEW QUESTIONS

1. List all the common ways of attaching stone cladding to a building. Make a simple sketch to explain each system.

2. Working from memory, sketch all the details of a brick veneer wall over a concrete frame.

3. What are some options of surface finishes for precast concrete cladding panels?

4. Describe the process of producing GFRC panels, illustrating your account with simple sketches.

5. Name two types of EIFS. Name two ways of applying EIFS to a building.

EXERCISES

1. Design and detail a brick veneer cladding for a multistory building that you are designing. Rather than trying to conceal the flashings and soft joints, work out a way of expressing them boldly as part of the architecture of the building.

2. Visit one or more buildings under construction that are being clad with masonry, concrete, GFRC, or EIFS. Make sketches of how the materials are detailed, especially how they are anchored to the building. What happens if water leaks through the cladding?

3. Adapt the brick veneer details in this chapter to installation on a building framed with structural steel.

## 21

CLADDING WITH METAL AND GLASS

- **Aluminum Extrusions**

  Thermal Breaks

  Surface Finishes for Aluminum

- **Modes of Assembly**

  Outside Glazing and Inside Glazing

- **An Outside Glazed Curtain Wall System**

- **An Inside Glazed Curtain Wall System**

- **The Rainscreen Principle in Metal-and-Glass Cladding**

- **Expansion Joints in Metal-and-Glass Walls**

- **Slope Glazing**

- **Curtain Wall Design: The Process**

Kawneer 2800 Trusswall® aluminum mullions support the tall glass walls of the lobby of the Zentrum Building at the BMW Manufacturing Corporation in Spartanburg, **South Carolina.** (*Architects: Simons/AKA. Photo courtesy of Kawneer Company, Inc.*)

The contemporary metal-and-glass curtain wall is a descendent of the cast iron and glass walls that were common features of commercial buildings in the 19th century. But today's walls are vastly more sophisticated in every respect. They are carefully isolated from the frame of the building so that they support only their own weight and the forces of the wind. They are insulated and thermally broken to maximize comfort and minimize heating and cooling costs and moisture condensation. They utilize advanced glazing and spandrel materials that offer precise control of luminous and thermal properties. They are carefully gasketed and drained to discourage water leaks. Their intricate inner features are concealed behind smooth snap-on covers. They are designed for easy, forgiving installation and maintenance. And they are made of aluminum, light and strong, in the form of sleek extrusions that glisten with permanent anodized or organic finishes.

FIGURE 21.1
The basic components of the Trusswall® mullion that is used to frame the glass walls pictured on page 716 are aluminum extrusions. Aluminum plate webs, welded to the tubular extrusions, increase the depth of the mullion and make it stiff enough to support very tall glass walls against wind loads. (*Photo courtesy of Kawneer Company, Inc.*)

ALUMINUM EXTRUSIONS

Aluminum is the metal of choice for curtain walls for three primary reasons: It protects itself against corrosion. It accepts and holds a variety of attractive surface finishes. And it can be fabricated economically into elaborately detailed shapes by means of the process of *extrusion*.

The principle of extrusion is easily visualized: It is like squeezing toothpaste from a tube. In response to the squeezing pressure, the tube extrudes a column of toothpaste that is cylindrical in shape because the orifice in the tube is round. If the shape of the orifice were changed, one could produce many other shapes of toothpaste as well—square, triangular, flat, and so on (Figure 21.2).

To manufacture an aluminum extrusion, a large, cylindrical billet of aluminum is heated to a temperature at which the metal flows under pressure but still retains its shape when it is not under pressure. The heated billet is placed in the cylinder of a large press where a piston squeezes it under enormous pressure through a *die*, a steel plate with a shaped metal orifice. The die imparts its shape to a long extruded column of aluminum that is supported on rollers, cooled, straightened as necessary, and cut into convenient lengths (Figure 21.3).

Very intricate aluminum sections can be extruded for a variety of purposes, including not only curtain wall components but door frames, window frames, handrails, grillwork, and structural shapes such as wide-flanges, channels, and angles. The high precision of the extrusion process permits it to be used for close-tolerance details such as snap-in glazing beads, snap-on mullion covers, screw slots, and screw ports (Figures 21.4–21.7). Hollow shapes may be extruded by mounting the portion of the die that forms the interior of the shape on a steel "spider" within the cylinder, around whose legs the metal flows before it passes through the orifice. The accompanying illustrations show some of the ways in which extruded aluminum details are utilized in building cladding. Extrusion dies are easily produced for custom-designed sections if there will be a long enough production run to amortize their expense.

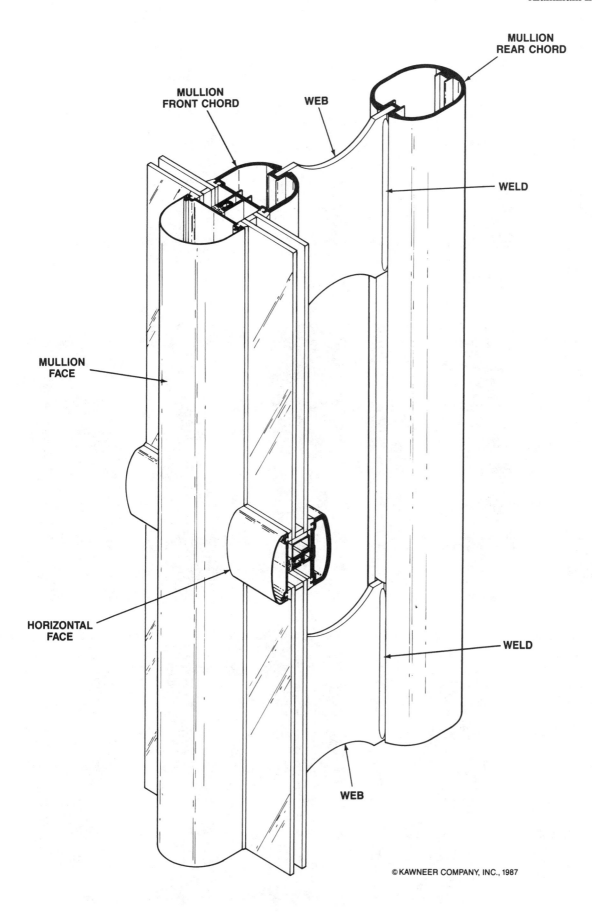

MULLION
REAR CHORD

MULLION
FRONT CHORD

WEB

WELD

MULLION
FACE

HORIZONTAL
FACE

WELD

WEB

© KAWNEER COMPANY, INC., 1987

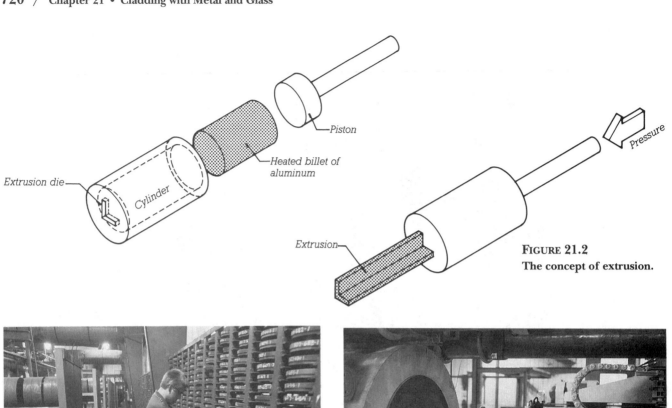

Extrusion die

Cylinder

Piston

Heated billet of aluminum

Pressure

Extrusion

FIGURE 21.2
The concept of extrusion.

(a)

(b)

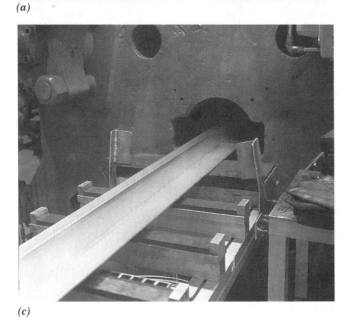

(c)

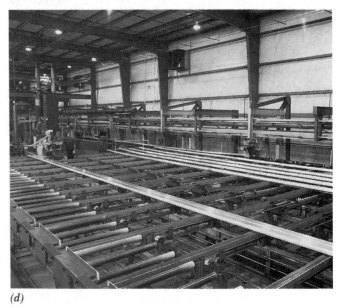

(d)

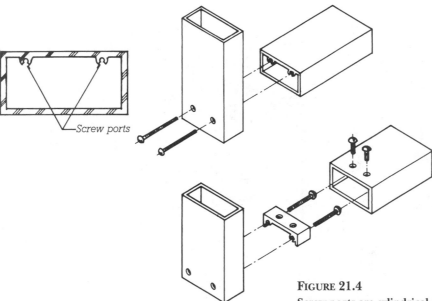

—Screw ports

FIGURE 21.3
Making aluminum extrusions for curtain wall components: (*a*) Hundreds of extrusion dies for the many components of curtain walls are organized in racks. (*b*) A heated billet of aluminum is inserted into the cylinder of the extrusion press. (*c*) An extrusion emerges from the die. (*d*) Long aluminum extrusions cool on rollers, ready for straightening and cutting to length. (*Photos courtesy of Kawneer Company, Inc.*)

FIGURE 21.4
Screw ports are cylindrical features that allow a screw to be driven parallel to the long axis of an extrusion. In the upper example, the screws pass through slightly oversized holes drilled through one box-shaped extrusion into screw ports in another box-shaped extrusion that are slightly smaller in diameter than the screws, thus engaging the screw threads tightly in the screw ports and making a snug joint. In the lower example, the screw ports are slightly larger than the outside diameter of the screws. The screws pass through the ports in a short piece of aluminum known as a shear block and engage slightly undersized holes in the vertical extrusion. Then the horizontal member is slipped over the shear block and screwed to it through drilled holes.

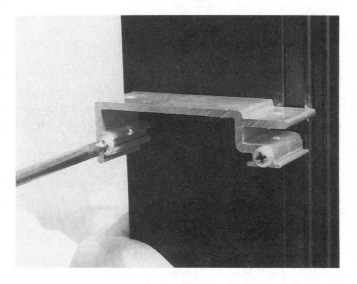

FIGURE 21.5
Self-tapping screws are driven through screw ports to fasten an extruded aluminum shear block to a vertical mullion. (*Courtesy of Kawneer Company, Inc.*)

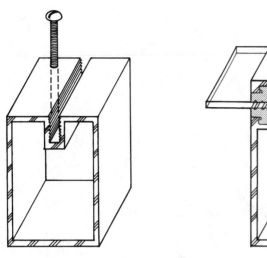

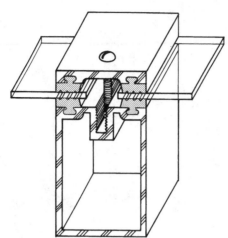

SCREW SLOT

FIGURE 21.6
Screw slots allow screws to be driven perpendicular to the long axis of an extrusion. In this example, screws pass through oversized holes in an extruded aluminum pressure plate and pull the plate down toward the screw slot. Extruded gaskets of synthetic rubber are pressed into channels in both extrusions to seal tightly against the glass.

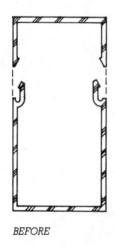

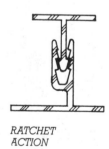

BEFORE   *AFTER*   *RATCHET ACTION*

FIGURE 21.7
Snap-on and snap-together features are commonly used in extruded aluminum curtain wall components. Assembly is accomplished simply by aligning the components and tapping firmly with a rubber mallet or squeezing with a rubber-cushioned clamp.

Thermal Breaks

Aluminum conducts heat rapidly. In very cold weather, the indoor surfaces of an aluminum member such as a window frame that passes from the outside of the building to the inside would be so cold that moisture would condense on them, and frost might even form. In very hot, humid weather, in an air-conditioned building, the outdoor surfaces of the same window frame might be cool enough to condense moisture from the air. This is why aluminum window and curtain wall members are invariably manufactured with *thermal breaks,* which are internal components of insulating material that isolate the aluminum on the interior side of the component from aluminum on the exterior side. These dramatically reduce the flow of heat through the member. Various plastics and synthetic rubbers are used in thermal breaks. There are a number of ways of creating thermal breaks; Figure 21.8 shows a *cast and debridged thermal break,* in which molten plastic is poured into a deep channel in the center of an aluminum member, where it hardens. Then the aluminum that forms the bottom of the channel is cut away in a *debridging* process that leaves only the plastic to connect the two halves of the member. Rubber or plastic gaskets, plastic strips, and plastic clips are also used as thermal breaks.

Surface Finishes for Aluminum

Aluminum, though it is a very active metal chemically, does not corrode away in service because it quickly protects itself with a thin, tenacious oxide coating that discourages further oxidation. While this oxide coating does an adequate job of protecting the aluminum, in the outdoor environment it takes on a chalky or spotty appearance that looks rather shabby.

FIGURE 21.8
This aluminum mullion extrusion is sitting on a block of dry ice in a humid room. Frost has formed on the portion of the mullion nearest to the dry ice, but the cast and debridged thermal break keeps the rest of the mullion warm enough that moisture does not condense on it. (*Courtesy of H. B. Fuller Company, St. Paul, Minnesota*)

Anodizing is a manufacturing process that produces an integral oxide coating on aluminum that is thousands of times thicker and more durable than the natural oxide film that would otherwise form. The component to be anodized is immersed in an acid bath and becomes the anode in an electrolytic process that takes oxygen from the acid and combines it with the aluminum. Color can be added to the coating by means of dyes, pigments, special electrolytes, or special aluminum alloys. The colors most frequently used in buildings are the natural aluminum color, golds, bronzes, grays, and black, but other colors are possible.

The advantages of anodized finishes are their extreme hardness and, in most colors, extreme resistance to weather and fading.

Other finishes are also used on aluminum building cladding, including organic coatings and porcelain enamels. The most widely used of the organic coatings are the *fluoropolymers,* which are based on highly inert synthetic resins that are exceptionally resistant to all forms of weathering, including ultraviolet deterioration. Fluoropolymers are applied by spray or roller in either two or three coats: a primer, a color coat, and, optionally, a clear finish coat. Fluoropolymer coatings are available in a broad spectrum of colors, including bright metallic finishes.

Powder coatings are also used on aluminum cladding components. These are manufactured with thermosetting powders that are composed of resins and pigments. The powder is electrically charged, then sprayed onto the aluminum component, which is electrically grounded so that the powder adheres to it electrostatically. The component is passed through an oven, where the powder fuses to produce a hard, resistant coating, usually in a single application. Among the advantages of powder coatings are their durability, the wide range of colors and finishes in which they are available, and their freedom from organic solvents that cause air pollution.

Modified acrylic or polyester polymers are also used in aluminum coatings. They provide finishes with very high glosses in a wide selection of colors. Polyvinyl chloride coatings, which can be applied in very heavy films to provide excellent resistance to weather and corrosion, are also common.

Mechanical finishes that include a wide range of textures may be applied to aluminum by such means as wire brushing, wheel or belt polishing, buffing, grinding, burnishing, barrel tumbling, sandblasting, blasting with steel shot or glass beads, and abrasive blasting. Chemical finishes are achieved through bright dipping, which produces mirrorlike surfaces, etching, and chemical conversion coatings such as oxides, phosphates, or chromates. Mechanical and chemical finishes may be done in preparation for the application of other types of finishes or, in some cases, may act as final finishes.

MODES OF ASSEMBLY

Metal curtain wall systems can be classified according to their degree or mode of assembly at the time of installation on the building (Figure 21.9). Many metal-and-glass curtain walls are furnished as *stick systems* whose principal components are metal mullions and rectangular panels of glass and spandrel material that are assembled in place on the building (Figure 21.9*a*). Stick systems have the advantages of low shipping bulk and a high degree of ability to adjust to unforeseen site conditions, but they must be assembled on site, under highly variable conditions, rather than in a factory with its ideal tooling, controlled environmental conditions, and lower wage rates.

The *unit system* of curtain wall installation takes full advantage of factory assembly and minimizes on-site labor, but the units require more space during shipping and more protection from damage than stick system components (Figure 21.9*b*). The *unit-and-mullion system* (Figure 21.9*c*), which is seldom used today, offers a middle ground between the stick and unit systems.

The *panel system* (Figure 21.9*d*) is made up of homogenous units that are formed from metal sheet. Its advantages and disadvantages are similar to those of the unit system, but its production involves the higher tooling costs of a custom-made die or mold, which makes it advantageous only for a building that requires a large number of identical panels.

The *column-cover-and-spandrel system* (Figure 21.9*e*) emphasizes the structural module of the building rather than creating its own grid on the facade, as the previously described systems do. A custom design must be created for each project

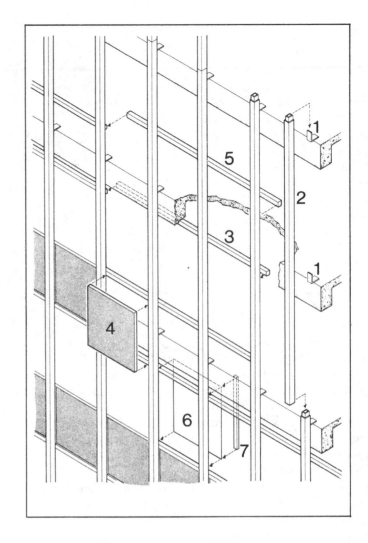

STICK SYSTEM—Schematic of typical version

1: Anchors. 2: Mullion. 3: Horizontal rail (gutter section at window head). 4: Spandrel panel (may be installed from inside building). 5: Horizontal rail (window sill section). 6: Vision glass (installed from inside building). 7: Interior mullion trim.

Other variations: Mullion and rail sections may be longer or shorter than shown. Vision glass may be set directly in recesses in framing members, may be set with applied stops, may be set in sub-frame, or may include operable sash.

(a)

because there is no standard column or floor spacing for buildings. Special care is required in detailing the spandrel panel support to ensure that the panels do not deflect when loads are applied to the spandrel beams of the building frame; otherwise, the window strips could be subjected to load-

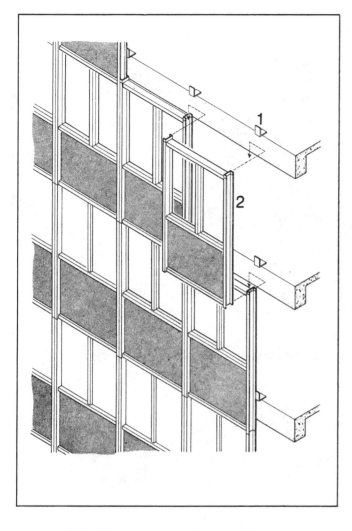

UNIT SYSTEM—Schematic of typical version

1: Anchor. 2: Pre-assembled framed unit.

Other variations: Mullion sections may be interlocking "split" type or may be channel shapes with applied inside and outside joint covers. Units may be unglazed when installed or may be pre-glazed. Spandrel panel may be either at top or bottom of unit.

(b)

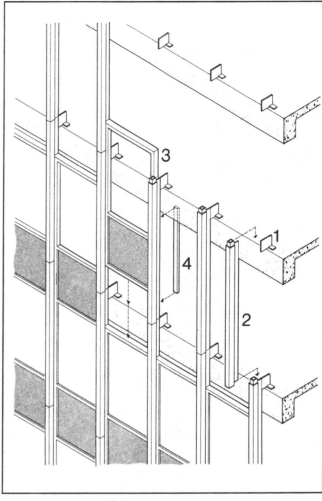

UNIT-AND-MULLION SYSTEM—Schematic of typical version

1: Anchors. 2: Mullion (either one- or two-story lengths). 3: Pre-assembled unit —lowered into place behind mullion from floor above. 4: Interior mullion trim.

Other variations: Framed units may be full-story height (as shown), either unglazed or pre-glazed, or may be separate spandrel cover units and vision glass units. Horizontal rail sections are sometimes used between units.

(c)

FIGURE 21.9   *(continued)*

Modes of assembly for curtain walls: (*a*) **Stick system.** (*b*) **Unit system.** (*c*) **Unit-and-mullion system.** (*d*) **Panel system.** (*e*) **Column-cover-and-spandrel system.** (*Courtesy of American Architectural Manufacturers Association*)

ings that could deform the mullions and crack the glass.

Outside Glazing and Inside Glazing

A metal cladding system may be designed to be *outside glazed,* which means that glass must be installed or replaced by workers standing on scaffolding or staging outside the building. Alternatively, it may be designed to be *inside glazed* by workers who stand inside the building. Inside glazing is more convenient and more economical for a tall building, but it requires a somewhat more elaborate set of extrusions. Outside glazing systems utilize a relatively simple set of extrusions and are less expensive for a building that is only one to three stories tall. Some curtain wall systems are designed so that they may be glazed from either side.

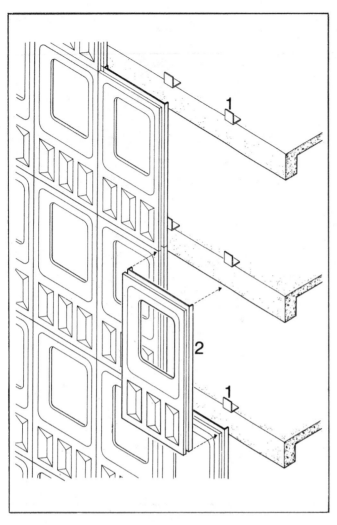

PANEL SYSTEM—Schematic of typical version

1: Anchor.   2: Panel.

Other variations: Panels may be formed sheet or castings, may be full story height (as shown) or smaller units, and may be either pre-glazed or glazed after installation.

(d)

FIGURE 21.9 (*continued*)

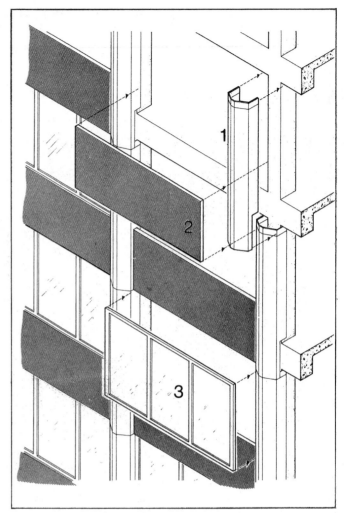

COLUMN COVER AND SPANDREL SYSTEM—Schematic of typical version

1: Column cover section.   2: Spandrel panel.   3: Glazing infill.

Other variations: Column covers may be one piece or an assembly, may be of any cross-sectional profile, and either one or two stories in height. Spandrel panel may be plain, textured or patterned. Glazing infill may be a pre-assembly, either glazed or unglazed, or be assembled in place.

(e)

FIGURE 21.10

The Kawneer 1600 System[1]® is an outside glazed stick system. Vertical mullions that run continuously from floor to floor support discontinuous horizontal mullions that are connected to them by means of shear blocks and screws. Each light of glass sits on two rubber glazing blocks in the gutter of the horizontal mullion (not shown in this drawing). The inner surface of the glass rests against extruded rubber glazing gaskets pressed into small channels in both the vertical and the horizontal mullions. An extruded aluminum pressure plate with rubber glazing gaskets is applied to both horizontal and vertical mullions to clamp the glass into place and create a weathertight seal. Each pressure plate is attached by means of screws that pass through drilled holes into an extruded screw slot. A thick rubber gasket in the screw slot acts as a thermal break. Snap-on covers conceal the screw heads and give a neat exterior appearance. A molded rubber plug at each end of each horizontal mullion contains any leakage or condensate within the horizontal mullion, from which it escapes via $^5/_{16}$-inch-diameter (8-mm) weep holes (not shown here) drilled through the pressure plate and the bottom edge of the snap-on cover. All the drawings of this system in Figures 21.10 through 21.14 are courtesy of Kawneer Company, Inc.

AN OUTSIDE GLAZED CURTAIN WALL SYSTEM

An off-the-shelf, externally glazed stick system for aluminum-and-glass curtain walls is illustrated in Figures 21.10 through 21.15. This is a system that is suitable for a low building whose walls workers can easily reach from external scaffolding.

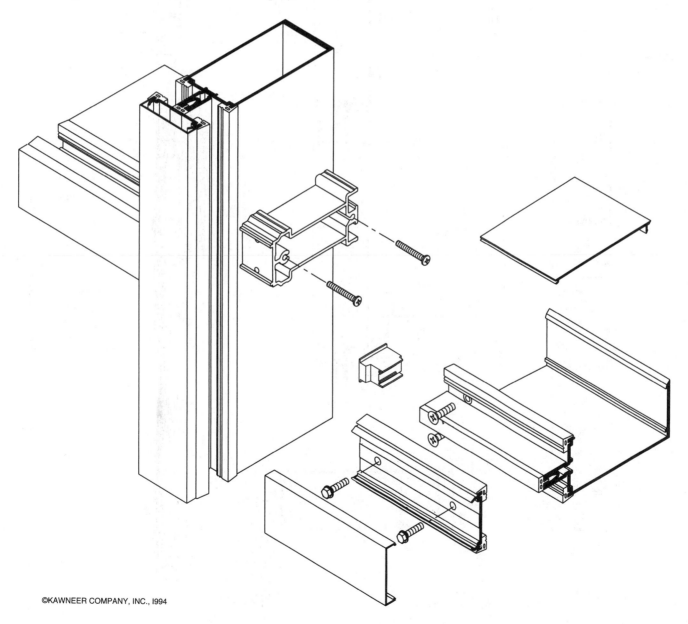

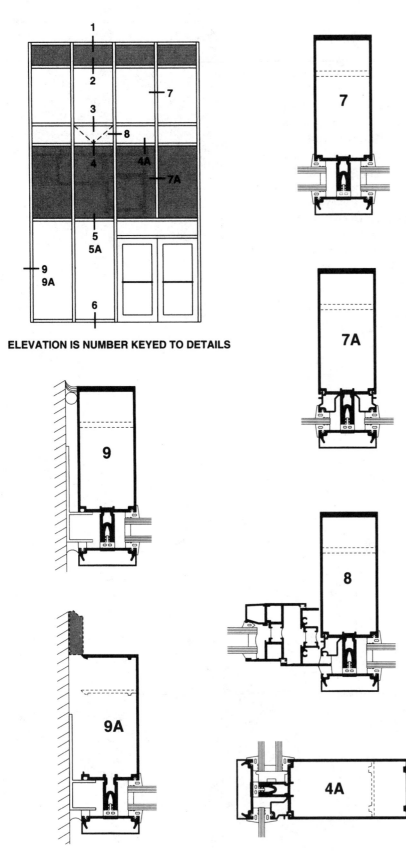

ELEVATION IS NUMBER KEYED TO DETAILS

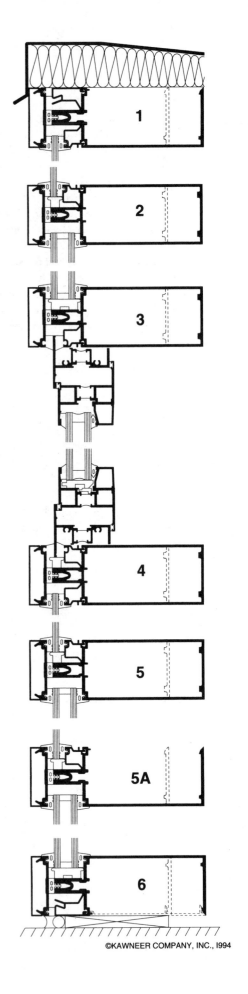

FIGURE 21.11
The manufacturer's details for the Kawneer 1600 System[1]® curtain wall are all keyed to the small elevation view at the upper left corner of this page. The details are reproduced here at one-quarter full size.

FIGURE 21.12
In this full-sized detail of the vertical mullion for the Kawneer 1600 System[1]®, we see that the basic extrusion is a rectangular box shape that is structurally stiff and presents a neat appearance inside the building. The broken lines indicate a smaller mullion (162-002) that may be used for buildings with shorter floor-to-floor heights or smaller wind loads. The extruded plastic thermal break attaches to the mullion with a projecting "pine tree" spline that is pushed into the screw slot. Holes are drilled through the thermal break for the screws (not shown) that attach the pressure plate. The four rubber glazing gaskets attach to the aluminum pieces with projecting splines.

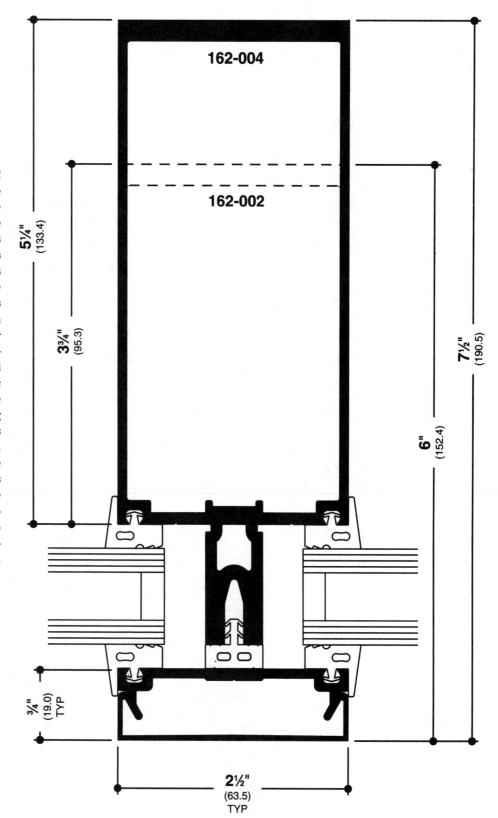

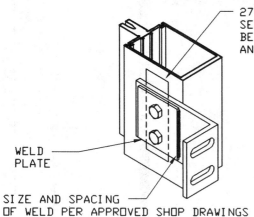

27-477
SEPARATOR
BETWEEN ANCHOR
AND MULLION

WELD
PLATE

SIZE AND SPACING
OF WELD PER APPROVED SHOP DRAWINGS

FIGURE 21.13
Vertical mullions are attached to the edges of the building's floors with the angle anchor shown in the top drawing. Sections of vertical mullion are spliced with the aluminum internal spline shown in the lower drawing. The spline is screwed to the lower section of mullion, but the upper section is free to slide, which allows for thermal expansion and contraction.

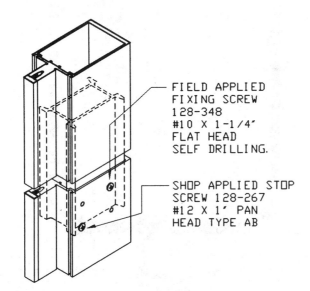

FIELD APPLIED
FIXING SCREW
128-348
#10 X 1-1/4'
FLAT HEAD
SELF DRILLING.

SHOP APPLIED STOP
SCREW 128-267
#12 X 1' PAN
HEAD TYPE AB

FIGURE 21.14
Two different horizontal mullion extrusions are available for the Kawneer 1600 System[1]®. The closed extrusion (top) is used in most situations. The open-back extrusion (bottom) allows access to the shear block on the vertical mullion for purposes of assembly; it is used for portions of the wall that butt against other materials, such as heads, sills, and end bays. A snap-on cover closes the open back of the mullion, leaving only a virtually invisible seam showing. Rubber setting blocks are shown beneath the edge of the glass in both mullions. In the upper detail, a small aluminum extrusion has been added to allow the mullion to hold a single layer of spandrel glass or vision glass rather than the standard 1-inch (25-mm) double glazing assembly. Weep holes are not shown here; they are drilled horizontally through the pressure plate just above the thermal break, and vertically through the bottom edge of the outside snap-on cover.

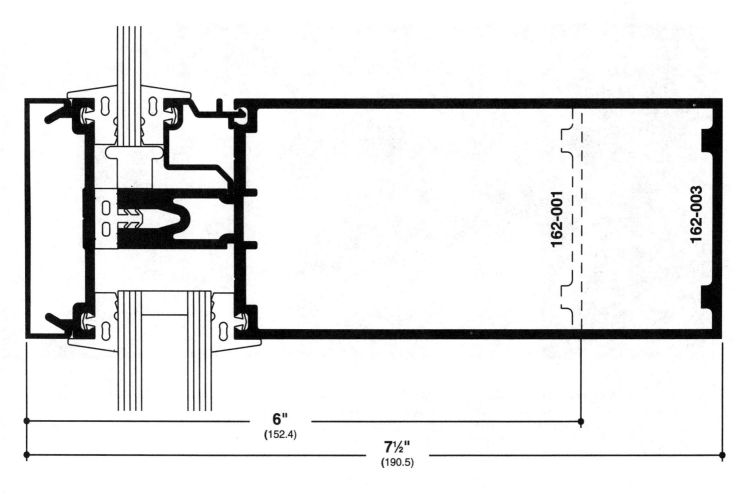

6"
(152.4)

7½"
(190.5)

162-001

162-003

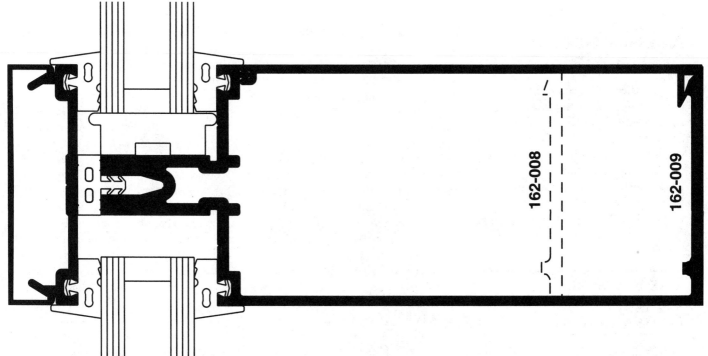

162-008

162-009

FIGURE 21.15
This building for the Harley-Davidson Motor Company in Wauwatosa,
Wisconsin, features the Kawneer 1600 System[1]®, whose details are shown in
Figures 21.10 through 21.14. (*Architect: Flad & Associates, Madison, Wisconsin.*
Photo courtesy of Kawneer Company, Inc.)

AN INSIDE GLAZED CURTAIN WALL SYSTEM

An off-the-shelf, internally glazed stick system for aluminum-and-glass curtain walls is illustrated in Figures 21.16 through 21.22. This system is suitable for use on tall buildings, because it is installed entirely by workers who are standing inside the building. Replacement of glass may also be done from within the building.

THE RAINSCREEN PRINCIPLE IN METAL-AND-GLASS CLADDING

At first glance, neither of the curtain wall systems presented in this chapter might seem to be a rainscreen design, because both use rubber gaskets to seal around the glass on the outside

of the wall as well as the inside. However, consider what would happen in either of these systems if both the inner and the outer gaskets were defective around the same light of glass: During a wind-driven rain, gravity and capillary action would be likely to draw some water past the outer gasket, into the spaces between the edges of the glass and the aluminum within the curtain wall itself. However, as we discovered in Chapter 19, even a defective inner gasket is very unlikely to permit air currents strong enough to carry water farther than this toward the interior of the building. If water has leaked in along a vertical edge of the glass, it is contained within the vertical mullion and will fall by gravity to the bottom, where it drains out through weep holes. If water accumulates in the horizontal mullion, it is prevented from running out of the ends of the mullion by rubber end plugs, which

are seen in Figure 21.10. Its only recourse is to drain to the outdoors through weep holes that are drilled horizontally through the pressure plate and vertically through the bottom edge of the external snap-on cover. Thus, the external gasket need serve only as a *deterrent seal*, essentially a rainscreen, to discourage water from entering without necessarily barring its entry altogether. The internal gasket serves as an air barrier, and the hollow spaces between the edges of the glass and the mullions act as pressure equalization chambers. The entire system functions as a rainscreen assembly. In practice, of course, the manufacturer and installer take every precaution to assure that the external gaskets are properly installed and will act as positive barriers to the passage of water, but these curtain wall systems do not depend on a perfect seal to retain their watertightness.

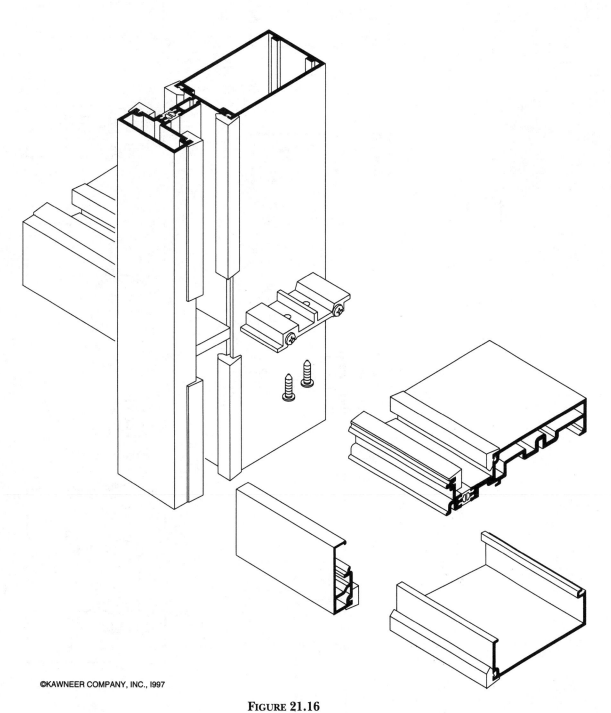

FIGURE 21.16
The Kawneer 1600 System⁴® may be glazed from either inside or outside the building. Inside glazing is usually preferred in taller buildings for its speed, safety, and convenience. All the drawings of this system in Figures 21.16 through 21.21 are courtesy of Kawneer Company, Inc.

SCALE 3" = 1' 0"

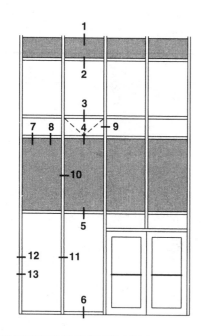

ELEVATION IS NUMBER KEYED TO DETAILS

**ALTERNATE
JAMB DETAIL**

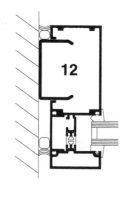

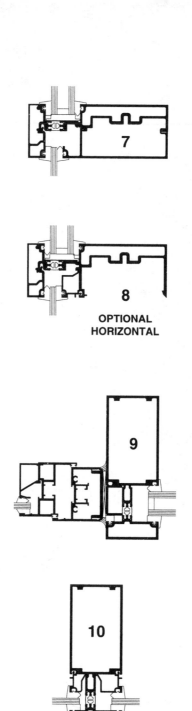

**OPTIONAL
HORIZONTAL**

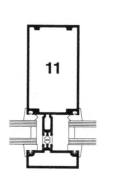

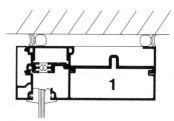

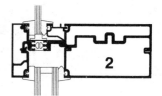

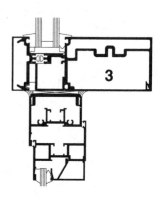

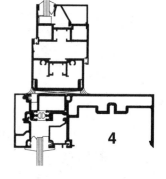

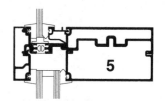

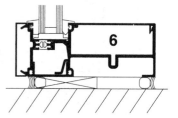

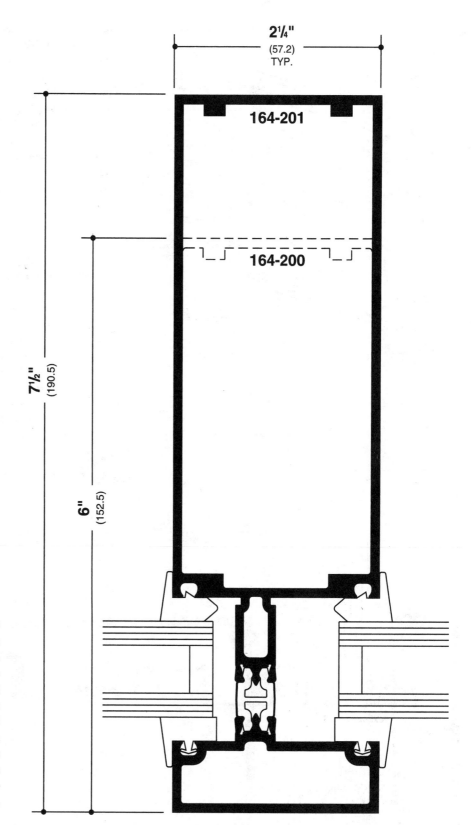

FIGURE 21.17
A complete set of details for the Kawneer 1600 System[4]®, reproduced in the manufacturer's detail book at one-quarter full size.

FIGURE 21.18
The vertical mullion for this system is rigid, with no removable components except the glazing gaskets. The thermal break is an H-shaped extrusion of rigid plastic that is crimped securely into small grooves in the inner and outer portions of the mullion. Notice that the glazing pockets are of two different depths.

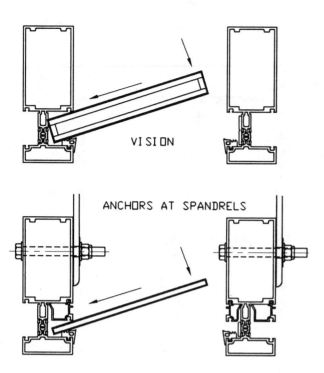

VISION

ANCHORS AT SPANDRELS

FIGURE 21.19
To glaze the Kawneer 1600 System[4]® from inside the building, rubber setting blocks are applied inside the shallow glazing pocket on the vertical mullion to cushion the edge of the glass. Rubber setting blocks are also installed in the horizontal mullion, and all the exterior glazing gaskets are inserted. Then the glass unit is pushed obliquely into the deep glazing pocket, squared away, and backed into the shallow pocket, where it rests against the setting blocks.

FIGURE 21.20
In order to make this mode of glass installation possible, the bottom half of the interior of each horizontal mullion is left off until the glass has been pushed into place. Then the bottom-half extrusion is added, and the interior glazing gaskets are installed. The interior gaskets are called glazing wedges because of their blunt wedge shape; this allows them to be installed simply by pressing them into place, wedging them between the glass and the mullion, after the glass has been installed. Notice how this shape is different from that of the exterior gaskets, which are installed before the glass.

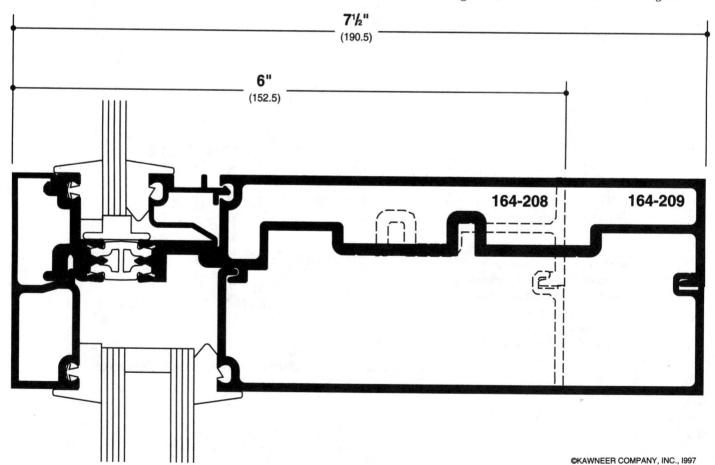

7½"
(190.5)

6"
(152.5)

164-208   164-209

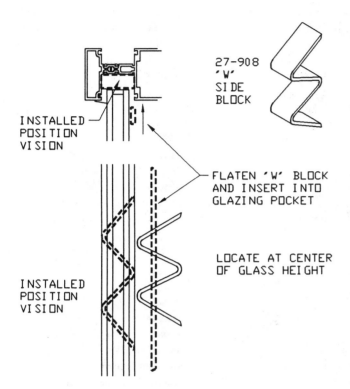

27-908 'W' SIDE BLOCK

INSTALLED POSITION VISION

FLATEN 'W' BLOCK AND INSERT INTO GLAZING POCKET

LOCATE AT CENTER OF GLASS HEIGHT

INSTALLED POSITION VISION

FIGURE 21.21
Before the glazing wedges are installed in the deep-pocket side of the vertical mullion, a W-shaped rubber block is flattened and slipped between the glass and the interior side of the mullion until it passes into the empty pocket. There it springs back into shape and serves as a side block to prevent the glass from moving too deeply into the pocket.

FIGURE 21.22
This Las Vegas casino hotel features the Kawneer 1600 System[4]®, whose details are shown in Figures 21.16 through 21.21. (*Photo courtesy of Kawneer Company, Inc.*)

EXPANSION JOINTS IN METAL-AND-GLASS WALLS

Aluminum has a relatively high coefficient of thermal expansion. The coefficient for glass is less than half as much. Because the cladding of a building is exposed to air temperature fluctuations as well as direct heating by the sun, it must be provided with expansion joints to allow thermal movement to occur without damaging the cladding or the frame of the building.

The differences in thermal movement between the glass and the aluminum are generally accommodated by very small sliding and flexing motions that occur between the glass and the gaskets in which it is mounted. Rubber blocks placed between the edge of the glass and the mullion on either side of each light prevent the glass from "walking" too far in either direction during repeated cycles of heating and cooling.

In the curtain wall systems illustrated in this chapter, vertical thermal movement in the aluminum is absorbed by telescoping joints that are provided at regular intervals in the vertical mullions (Figure 21.13). Horizontal thermal movement is accommodated by intentionally cutting horizontal aluminum components slightly short by a calculated fraction of an inch at each vertical mullion. Because the horizontal mullions are interrupted at each vertical mullion, there are many of these joints to act together in absorbing horizontal expansion and contraction.

SLOPE GLAZING

Many buildings feature glass roofs over such amenities as lobbies, restaurants, cafes, swimming pools, and garden courtyards. A glass roof presents particular problems with respect to

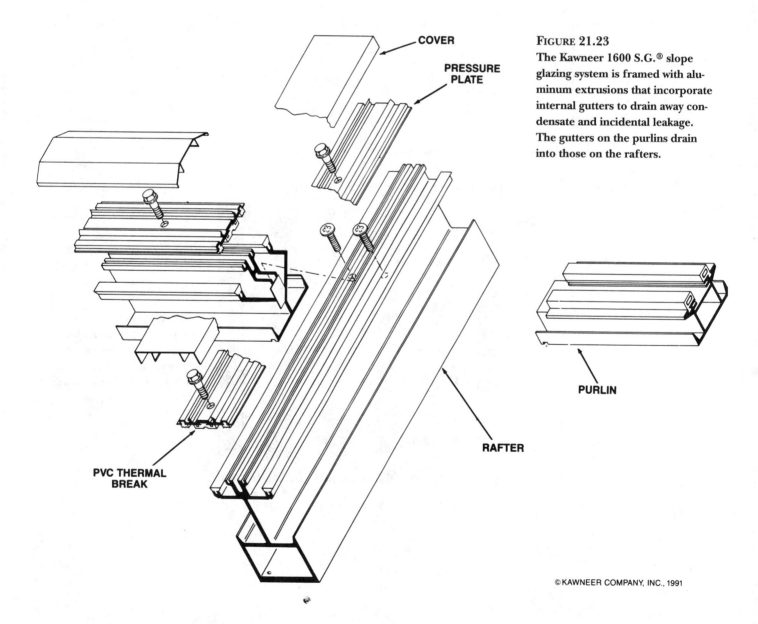

COVER

PRESSURE PLATE

PVC THERMAL BREAK

RAFTER

PURLIN

FIGURE 21.23
The Kawneer 1600 S.G.® slope glazing system is framed with aluminum extrusions that incorporate internal gutters to drain away condensate and incidental leakage. The gutters on the purlins drain into those on the rafters.

potential water leakage, because it is impossible to neutralize the force of gravity on a surface that is not vertical. Furthermore, moisture that condenses on the interior surfaces of the glass is likely to accumulate and drip onto the occupants of the space beneath. Therefore, every *slope glazing* system is designed by its manufacturer to include an internal drainage system. This system collects any water that results from leakage or condensation and drains it to the outdoors. The glass surface is sloped rather than flat because the slope enables gravity to assist in keeping water from ponding on the roof, in causing condensate to run to the lower edge of

each light of glass before dripping off, and in moving water through the drainage channels to the weep holes through which it is conducted back to the outdoors.

A proprietary slope glazing system is illustrated in Figures 21.23 and 21.24. This system is designed to adapt to a range of slopes from 15° to 60°. Water leakage is discouraged by a well-designed system of glazing gaskets, but if leakage should occur because of gasket deterioration or faulty installation, the internal drainage system will catch it along any purlin or rafter and drain it away. Moisture condensation is minimized by double glazing and thermal

breaks, but such condensate as may form under extreme weather conditions is also caught and drained by the same system of channels in the aluminum members.

FIGURE 21.24
This Orlando, Florida, hotel includes massive areas of Kawneer 1600 S.G.® slope glazing. Reflective glass is used to minimize solar overheating of the interior space. (*Photo courtesy of Kawneer Company, Inc.*)

CURTAIN WALL DESIGN: THE PROCESS

Metal curtain wall design is not undertaken by an architect alone; it is far too complex and specialized a process. For many buildings, the architect simply adopts a proprietary system that meets the design requirements. This puts the primary burden of responsibility on the manufacturer and installer of the system to see that the proper components are chosen and the wall is properly installed.

When an architect sets out to design a new system of metal cladding, as is often done for large, important buildings, other professionals are brought into the process. An independent cladding consultant can bring a large body of experience and expertise to this effort and minimize the risk to the architect. The structural engineer of the building is involved at least to the extent of understanding the weight and attachment requirements of the system. A curtain wall manufacturer is also brought into the design team early in the process.

The architect and building owner may elect either of two methods for selecting a manufacturer: One curtain wall manufacturer may be chosen on the basis of reputation and past experience. Or the architect and cladding consultant may prepare a rough design and performance specifications and submit these to several manufacturers for proposals. Each manufacturer will then submit a more detailed design and a financial proposal, and one will be selected on the basis of these proposals to proceed with the project.

The manufacturer understands better than any other member of the team the manufacturing, assembly, installation, and cost implications of a new curtain wall design. The manufacturer often does the installation as well as the manufacturing of the components, or subcontracts the installation work to companies which it certifies to be well qualified and familiar with the manufacturer's products and standards. This installation experience is also invaluable during the design process.

From conceptual drawings prepared by the architect and cladding consultant, the manufacturer prepares a more detailed set of design drawings as a basis for reaching preliminary agreement on the design. The manufacturer then prepares a very detailed set of shop drawings and installation drawings. These are checked carefully by the architect, engineer, and cladding consultant to assure compliance with design intentions and the structural capabilities of the building frame. Full-scale testing of the curtain wall, as described in Chapter 19, is usually carried out before the manufacture of a custom-designed curtain wall system is authorized to begin. The curtain wall manufacturer may wish to visit the construction site during the erection of the building frame to become familiar with the level of dimensional accuracy of the structural surfaces to which the curtain wall will be fastened.

C.S.I./C.S.C.
Masterformat Section Numbers for
Metal and Glass Cladding

08400	**ENTRANCES AND STOREFRONTS**
08410	**Aluminum Entrances and Storefronts**
08900	**GLAZED CURTAIN WALLS**
08920	**Glazed Aluminum Curtain Walls**
08960	**Slope Glazing Systems**

SELECTED REFERENCES

1. American Architectural Manufacturers Association. *Aluminum Curtain Wall Design Guide Manual.* Schaumburg, Illinois, 1996.

This publication covers its topic in exemplary fashion with clear text and beautifully prepared illustrations. The same organization also publishes a series of more specialized titles on various aspects of metal-and-glass curtain walls: rainscreen design, care and handling of aluminum components, design wind loads and wind tunnel testing, fire safety, test methods, installation procedures, finishes, and so on. (Address for ordering: 1827 Walden Office Square, Suite 104, Schaumburg, IL 60173.)

KEY TERMS AND CONCEPTS

extrusion
die
thermal break
cast and debridged thermal break
debridging
anodizing
fluoropolymer

powder coating
stick system
unit system
unit-and-mullion system
panel system
column-cover-and-spandrel system
screw slot

screw port
snap-together extrusions
outside glazed system
inside glazed system
deterrent seal
slope glazing

REVIEW QUESTIONS

1. For what reasons is aluminum the metal most frequently used in metal-and-glass cladding systems?

2. What are the relative advantages and disadvantages of a stick system and a panel system?

3. What are some characteristic features of aluminum extrusions, and what is the function of each?

4. What are the most common finishes that are applied to aluminum curtain wall components? What are the advantages of each?

5. How would you choose between an inside glazed system and an outside glazed one?

6. In what way is a slope glazing system more difficult to make watertight than a wall cladding system? How is this reflected in the details of a slope glazing system?

EXERCISES

1. Make photocopies of full-scale details of the vertical and horizontal mullions of a metal curtain wall system from a manufacturer's catalog. Paste the details on a larger sheet of paper and add notes and arrows to explain every aspect of them—the features of the extrusions, the gaskets and sealants, the glazing materials and methods, drainage, insulation, thermal breaks, rainscreen features, and so on. From your examination of the details, list the order in which the components would be assembled. Is the system glazed from inside the building or outside? How can you tell?

2. Design a coffee table that is made up of aluminum extrusions and a glass top. Design and draw complete details of the extrusions and connections. Select a surface finish for the aluminum. What type of glass will you use?

22

SELECTING INTERIOR FINISHES

- **Installation of Mechanical and Electrical Services**
- **The Sequence of Interior Finishing Operations**
- **Selecting Interior Finish Systems**

  Appearance

  Durability and Maintenance

  Acoustic Criteria

Fire Criteria

Relationship to Mechanical and
  Electrical Services

Changeability

Cost

Toxic Emissions from Interior
  Materials

- **Trends in Interior Finish Systems**

Workers complete installation of an elaborate ceiling made of gypsum board. The corners of the steplike construction have been reinforced with metal corner bead, and the joints and nail heads have been filled and sanded, ready for painting.

When a building has been roofed and at least a part of its exterior cladding has been installed, its interior is sufficiently protected from the weather that work can begin on the mechanical and electrical systems. The waste lines and water supply lines of the plumbing system are installed, as well as the pipes for an automatic sprinkler fire suppression system, if one is intended for the building. The major part of the work for the heating, ventilating, and air conditioning system is carried out, including the installation of boilers, chillers, cooling towers, pumps, fans, piping, and ductwork. Electrical, communications, and control wiring and fiber optic cables are routed through the building. Elevators and escalators are installed in the structural openings provided for them.

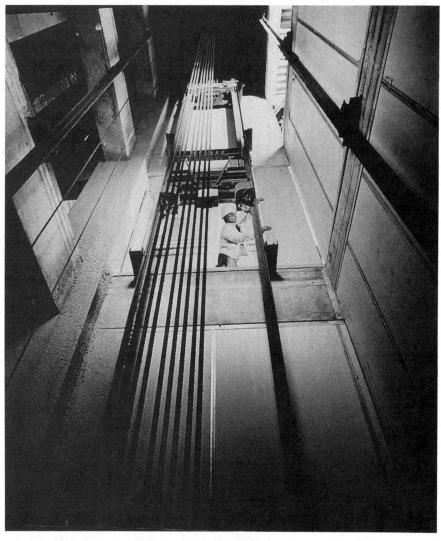

FIGURE 22.1
A worker constructs a fire-resistant wall around an elevator shaft, using gypsum panels and steel C-H studs. Chapter 23 contains more detailed information on shaft walls. (*Courtesy of United States Gypsum Company*)

INSTALLATION OF MECHANICAL AND ELECTRICAL SERVICES

The vertical runs of pipes, ducts, wires, and elevators through a multi-story building are made through vertical *shafts* whose sizes and locations were determined at the time the building was designed. Before the building is finished, each shaft will be enclosed with fire-resistive walls to prevent the vertical spread of fire (Figure 22.1). Horizontal runs of pipes, ducts, and wires are usually located just below each floor slab to keep them up out of the way. These may be left exposed in the finished building, or hidden above *suspended ceilings*. Sometimes these services, especially wiring, are concealed within a hollow floor structure, such as cellular metal decking or cellular raceways. Sometimes services are run between the structural floor deck and a raised *access flooring* system above. (For a more complete explanation of suspended ceilings, cellular floors, and access flooring, see Chapter 24.) Where several plumbing fixtures are lined up along a wall, a plumbing space is created to house the pipes.

FIGURE 22.2
Three diagrammatic plans for an actual three-story suburban office building show the principal arrangements for plumbing, communications, electricity, heating, and cooling. Heating and cooling are accomplished by means of air ducted downward through two shafts from equipment mounted on the roof.
The conditioned air from the vertical ducts is distributed around each floor by a system of horizontal ducts that run above a suspended ceiling, as shown on the plan of the intermediate floor. A row of doubled columns divides the building into two independent structures at the building separation joint, to allow for differential foundation settlement and thermal expansion and contraction. (*Courtesy of ADD Incorporated, Architects*)

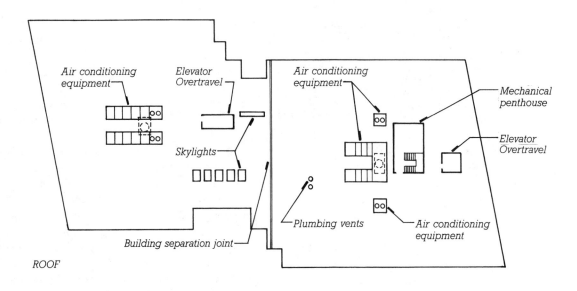

ROOF

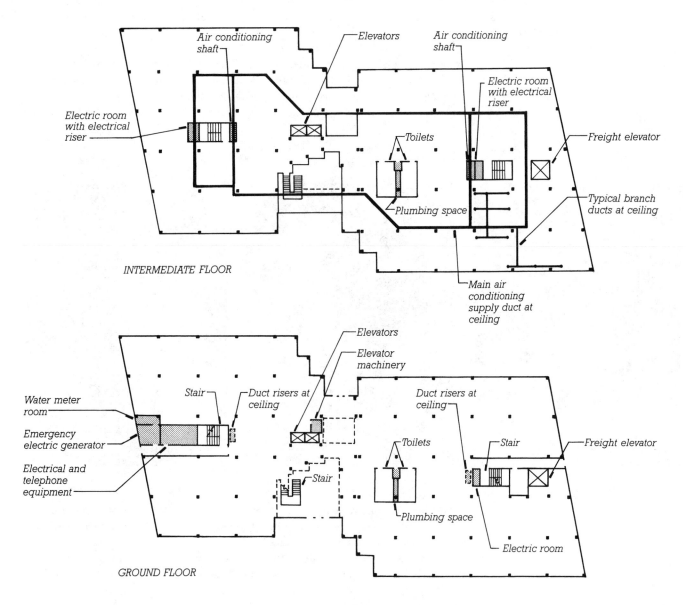

INTERMEDIATE FLOOR

GROUND FLOOR

This is done by constructing a double wall, as seen in Figure 22.3a and b.

Specific floor areas are reserved for mechanical and electrical functions in larger buildings (Figure 22.2). Distribution equipment for electrical and communications wiring and fiber optics networks is housed in special rooms or closets. Fan rooms are often provided on each floor for air handling machinery. In a large multistory building, space is set aside, usually at a basement or sub-basement level, for pumps, boilers, chillers, electrical transformers, and other heavy equipment. At the roof are penthouses for elevator machinery and such components of the mechanical systems as cooling towers and ventilating fans. In very tall buildings, one or two entire intermediate floors may be set aside for mechanical equipment, and the building is zoned vertically into groups of floors that can be reached by ducts and pipes that reach up and down from each of the mechanical floors.

The Sequence of Interior Finishing Operations

Simultaneously with the mechanical and electrical work, finishing operations are begun in a carefully ordered sequence that varies somewhat from one building to another depending on the specific requirements of each project. The first finish items to be installed are usually hanger wires for suspended ceilings, and full-height partitions and enclosures, especially those around mechanical and electrical shafts, elevator shafts, mechanical equipment rooms, and stairways. *Firestopping* is inserted around pipes, conduits, and ducts where they penetrate the floors (Figure 22.3). The full-height partitions and enclosures, firestopping, joint covers (Figure 22.4), and safing around the perimeters of the floors constitute a very important system for keeping fire from spreading vertically through the building.

When the major horizontal electrical conduits and air ducts have been installed at each ceiling, the grid for the suspended ceiling is installed on the hanger wires so that the lights and ventilating louvers can be mounted to the grid. Then, typically, the ceilings are finished, and framing for the partitions that do not penetrate the finish ceiling is installed. Electrical and communications wiring are brought down from the conduits above the ceilings to serve outlets in the partitions. The walls are finished and painted. The last major finishing operation is usually the installation of the finish floor-

(a)

(b)

(c)

FIGURE 22.3
Applying firestopping materials to floor penetrations. (*a*) Within a plumbing wall, for a large slab opening around a cast iron waste pipe, a layer of safing insulation is cut to fit and inserted by hand. Then a mastic firestopping compound is applied over the safing to make the opening airtight (*b*). (*c*) Applying firestopping compound around an electrical conduit at the base of a partition. (*Courtesy of United States Gypsum Company*)

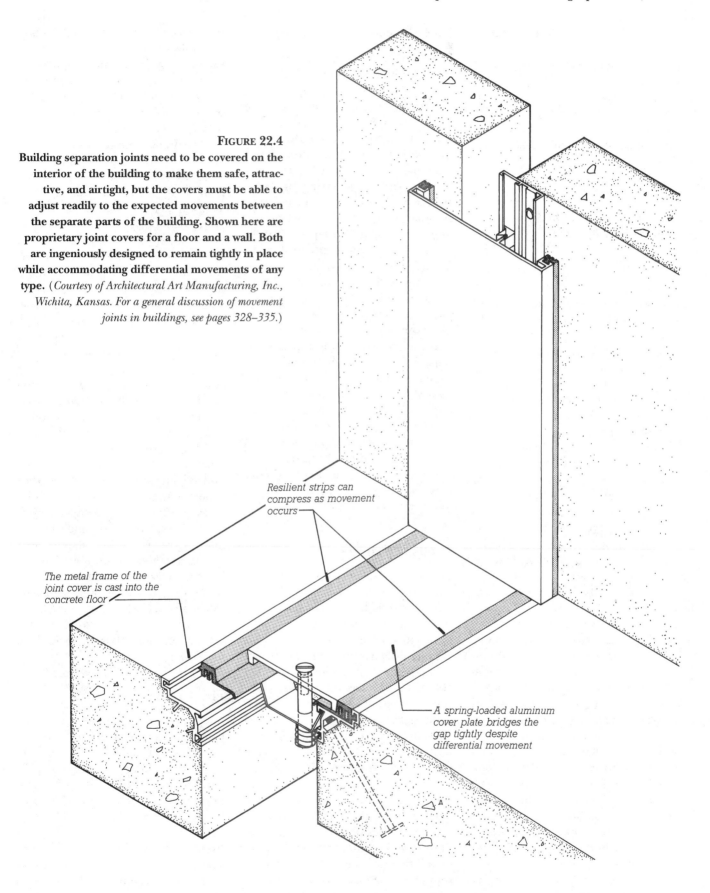

FIGURE 22.4
Building separation joints need to be covered on the interior of the building to make them safe, attractive, and airtight, but the covers must be able to adjust readily to the expected movements between the separate parts of the building. Shown here are proprietary joint covers for a floor and a wall. Both are ingeniously designed to remain tightly in place while accommodating differential movements of any type. (*Courtesy of Architectural Art Manufacturing, Inc., Wichita, Kansas. For a general discussion of movement joints in buildings, see pages 328–335.*)

Resilient strips can compress as movement occurs

The metal frame of the joint cover is cast into the concrete floor

A spring-loaded aluminum cover plate bridges the gap tightly despite differential movement

ing materials. This is delayed as long as possible, to let the other trades complete their work and get out of the building; otherwise, the floor materials could be damaged by dropped tools, spilled paint, heavy construction equipment, weld spatter, coffee stains, and construction debris ground underfoot.

SELECTING INTERIOR FINISH SYSTEMS

Appearance

A major function of interior finish components is to make the interior of the building look neat and clean by covering the rougher and less organized portions of the framing, insulation, vapor retarder, electrical wiring, ductwork, and piping. Beyond this, the architect designs the finishes to carry out a particular concept of interior space, light, color, pattern, and texture. The form and height of the ceiling, changes in floor level, interpenetrations of space from one floor to another, and the configurations of the partitions are primary factors in determining the feeling of the interior space. Light originates from windows and electric lighting fixtures, and is propagated by successive reflections off the interior surfaces of the building. Lighter-colored materials raise interior levels of illumination; darker colors and heavier textures result in a darker interior. Patterns and textures of interior finish materials are important in bringing the building down to a scale of interest that can be appreciated readily by the human eye and hand. No two buildings have the same requirements: Deep carpets and rich, polished marbles in muted tones may be chosen to give an air of affluence to a corporate lobby, brightly colored surfaces to create a happy atmosphere in a day-care center, or slick plastic and highly reflective surfaces to make a trendy ambience for the sale of designer clothing.

Durability and Maintenance

Expected levels of wear and tear must be considered carefully in selecting finishes for a building. Highly durable finishes generally cost more than shorter-lived ones and are not always required. In a courthouse, a transportation terminal, a recreation building, or a retail store, traffic is intense, and long-wearing materials are essential, but in a private office or an apartment, more economical finishes are usually adequate. Water resistance is an important attribute of finish materials in kitchens, locker and shower rooms, entrance lobbies, and some industrial buildings. In hospitals, medical offices, kitchens, food processing plants, and laboratories, finish surfaces must not trap dirt and must be easily cleaned and disinfected. Maintenance procedures and costs should be considered in selecting finishes for any building: How often will each surface be cleaned, with what type of equipment, and how much will this procedure add to the cost of owning the building? How long will each surface last, and what will it cost to replace it?

Acoustic Criteria

Interior finish materials strongly affect the quality of listening conditions and the levels of acoustic privacy inside a building. In noisy environments, interior surfaces that are highly absorptive of sound can decrease the noise intensity to a tolerable level. In lecture rooms, classrooms, meeting rooms, theaters, and concert halls, acoustically reflective and absorptive surfaces must be proportioned and placed so as to create optimum hearing conditions.

Between rooms, acoustic privacy is created by partitions that are both heavy and airtight. The acoustic isolation properties of lighter-weight partitions can be enhanced by partition details that damp the transmission of sound vibrations, using resilient mountings on one of the partition

surfaces and sound-absorbing batts of mineral wool in the interior cavity of the partition. Full-scale sample partitions of every type of material are tested for their ability to reduce the passage of sound between rooms in a procedure outlined in ASTM Standard E90. The results of this test are converted to *Sound Transmission Class* (*STC*) numbers that can be related to accepted standards of acoustic privacy. But if the cracks around the edges of a partition are not completely sealed, or if a loosely fitted door or even an unsealed electric outlet is inserted into the partition, its airtightness is compromised and the published STC value is meaningless. Similarly, a partition with a high STC is worthless if the rooms on both sides are served by a common air duct that also acts incidentally as a conduit for sound, or if the partition reaches only to a lightweight, porous suspended ceiling that allows sound to pass over the top of the partition.

Transmission of impact noise from footsteps and machinery through floor–ceiling assemblies can be a major problem. Impact noise transmission is measured by ASTM procedure E492, in which a standard machine taps on a floor above while instruments in a chamber below record sound levels. Impact noise transmission can be reduced by floor details that include soft materials which do not transmit vibration readily, such as carpeting, soft underlayment boards, or resilient underlayment matting.

Fire Criteria

A typical building code devotes many pages to provisions that control the materials and details for interior finishes in buildings. These code requirements are aimed at several important characteristics of interior finishes with respect to fire.

Combustibility

The surface burning characteristics of interior finish materials are tested in accordance with ASTM procedure

E84, also called the *Steiner Tunnel Test*. A sample of material 20 inches wide by 25 feet long (500 × 7620 mm) forms the roof of a rectangular furnace into which a controlled flame is introduced 1 foot (305 mm) from one end. The time the flame takes to spread across the face of the material from one end of the furnace to the other is recorded, along with the amount of fuel contributed by the material and the density of smoke developed. The results of this test are given in trade literature in three forms: the *flame spread rating*, which indicates the rapidity with which fire can spread across a surface of a given material; the *fuel contributed rating*, which indicates the amount of combustible substances in the material; and the *smoke developed rating*, which classifies a material according to the amount of smoke it gives off when it burns. Materials with smoke developed ratings greater than 450 are not permitted to be used inside buildings, because smoke, not heat or flame, is the primary killer in building fires. Figure 22.5 is a typical building code table that defines allowable flame spread ratings for interior finish materials for various use groups of buildings. Class I materials as defined in this table have flame spread ratings between 0 and 25, Class II between 26 and 75, and Class III between 76 and 200. (The scale of flame spread numbers is established arbitrarily by assigning a value of 0 to cement–asbestos board, and 100 to a red oak board.) Interior trim materials, if their total surface area in a room does not exceed 10 percent of the total wall and ceiling area of the room, may be of Class I, II, or III materials in any type of building.

The combustibility of flooring materials is measured for most materials by a test defined under ASTM procedure E648. For carpets, United States Department of Commerce Standard FF-1 is used. Code restrictions on floor material combustibility apply mainly to floors of exit corridors and stairways.

INTERIOR FINISH REQUIREMENTS[g]

Use Group	Required vertical exits and passageways[c]	Corridors providing exit access[i]	Rooms or enclosed spaces[a]
A-1, A-2, A-3	I	I[e]	II[b]
A-4, B, E, F, I-1, R-1, R-2	I	II	III
H	I	II	III
I-2	I[h]	I[h]	I[h]
I-3	I	I	III
M: walls	I	II	III
ceilings	I	II	II[d]
R-3	III	III	III
S, U	II	II	III

Note a. For requirements applicable to rooms and enclosed spaces, see Section 803.4.3.

Note b. Class III interior finish materials are permitted in places of assembly with a capacity of 300 persons or less.

Note c. Class III interior finish materials are permitted for wainscotting or paneling for not more than 1,000 square feet (93 m^2)of applied surface area in the grade lobby where applied directly to a noncombustible base or over furring strips applied to a noncombustible base and fireblocked as required by Section 804.0.

Note d. Class III interior finish materials are permitted in mercantile occupancies of 3,000 square feet (279 m^2) or less gross area occupied for sales purposes on the street floor only (balcony permitted).

Note e. Lobby areas shall not be less than Class II.

Note f. Where building height is over two stories, Class II shall be required.

Note g. For the classifications of interior finishes referred to herein, see Section 803.2. For interior finish requirements for exposed insulation, see Section 723.2.

Note h. Walls and ceilings shall be a minimum of Class II materials in individual rooms of not more than four persons in capacity. Where a building is equipped throughout with an automatic sprinkler system installed in accordance with Section 906.2.1, the minimum requirement for interior finish shall be Class II.

Note i. In Use Groups A, I-2 and I-3, Class II interior wall finish material shall be permitted as wainscotting extending not more than 48 inches (1219 mm) above the floor in corridors providing exit access.

FIGURE 22.5

A table of fire resistance requirements for interior finish materials, taken from the BOCA National Building Code. Class I materials are those whose flame spread ratings lie between 0 and 25. Class II flame spreads are 26 to 75, and Class III, 76 to 200.

(*Figures 22.5–22.7 are from the BOCA National Building Code/1996, © 1996, Building Officials and Code Administrators International, Inc. Published by arrangements with the author. All rights reserved. No parts of the BOCA Code may be reproduced or transmitted in any form or by any means, electronic or mechanical, including photocopying, recording or by an information storage and retrieval system without advance permission in writing from Building Officials and Code Administrators International, Inc., 4051 West Flossmoor Road, Country Club Hills, IL 60478-5795*)

Fire Resistance Ratings

A building code typically regulates the degree of fire resistance of partitions, ceilings, and floors that are used to protect the structure of the building or to separate various parts of a building from one another. Figure 22.6 is a table from the BOCA National Building Code that specifies the required *fire resistance rating* in hours of separation between different uses housed in the same building. Figure 1.3 gives required fire resistance ratings for various sorts of interior partitions, including shaft walls, exit hallways, exit stairs, dwelling unit separations, and other nonbearing partitions. These can be related to fire resistance ratings in a manufacturer's literature similar to the examples shown in Figures 1.4 through 1.6.

Fire resistance ratings for building assemblies are determined by full-scale fire endurance tests conducted in accordance with ASTM E119, which applies not only to partitions and walls, but also to beams, girders, columns, and floor–ceiling assemblies. In this test, the assembly is constructed in a large laboratory furnace and subjected to the structural load (if any) for which it is designed. The furnace is then heated according to a standard time–temperature curve, reaching 1700 degrees Fahrenheit (925°C) at 1 hour and 2000 degrees Fahrenheit (1093°C) after 4 hours. To achieve a given fire resistance rating in hours, an assembly must safely carry its design structural load for the designated period, must not develop any openings that permit the passage of flame or hot gases, and must insulate sufficiently against the heat of the fire to maintain surface temperatures on the side away from the fire within specified maximum levels. Wall and partition assemblies must

FIRERESISTANCE RATING REQUIREMENTS FOR FIRE SEPARATION ASSEMBLIES BETWEEN FIRE AREAS[a]

NP — Not Permitted
NA — Not Applicable

Use Group	A-1	A-2	A-3	A-4	A-5	B	E	F-1	F-2	H-1	H-2	H-3	H-4	I-1	I-2	I-3	M	R-1	R-2	R-3	S-1	S-2	U[b]
A-1	2	2	2	2	2	2	2	3	2	NP	4	3	2	2	3	3	2	2	2	2	3	2	3
A-2		2	2	2	2	2	2	3	2	NP	4	3	2	3	3	3	2	2	2	2	3	2	3
A-3			2	2	2	2	2	3	2	NP	4	3	2	2	3	3	2	2	2	2	3	2	3
A-4				2	2	2	2	3	2	NP	4	3	2	2	3	3	2	2	2	2	3	2	3
A-5					NA	2	2	3	2	NP	4	3	2	2	3	3	2	2	2	2	3	2	3
B						2	2	3	2	NP	4	3	2	2	3	3	2	2	2	2	3	2	3
E							2	3	2	NP	4	3	2	2	3	3	2	2	2	2	3	2	3
F-1								3	3	NP	4	3	3	3	3	3	3	3	3	3	3	3	3
F-2									2	NP	4	3	2	2	3	3	2	2	2	2	3	2	3
H-1										NP	NP	NP	NP	NP	NP	NP	NP	NP	NP	NP	NP	NP	NP
H-2											4	4	4	4	4	4	4	4	4	4	4	4	4
H-3												3	3	3	3	3	3	3	3	3	3	3	3
H-4													2	2	2	2	2	2	2	2	3	2	3
I-1														2	3	3	2	2	2	2	3	2	3
I-2															3	3	3	3	3	3	3	3	3
I-3																3	3	3	3	3	3	3	3
M																	2	2	2	2	3	2	3
R-1																		2	2	2	3	2	3
R-2																			2	2	3	2	3
R-3																				2	3	2	3
S-1																					3	3	3
S-2																						2	3
U																							3

(Left margin label: USE GROUP)

Note a. Fireresistance ratings are expressed in hours.
Note b. For requirements for private garages, see Section 407.0.

FIGURE 22.6
Requirements for fire resistance ratings in hours of fire separation assemblies between use groups, from the BOCA National Building Code.

also pass a *hose test*, in which a duplicate sample is subjected to half the rated fire exposure of the assembly, then sprayed with water from a calibrated fire nozzle for a specified period at a specified pressure in order to simulate the action of a fire hose during an actual fire. To pass this test, the assembly must not allow passage of a stream of water.

Openings in ceilings and partitions with required fire resistance ratings are restricted in size by most codes and must be protected against the passage of fire in various ways. Doors must be rated for fire resistance in accordance with a table such as that shown in Figure 22.7. Ducts that pass through rated assemblies must be equipped with sheet metal dampers that close automatically if hot gases from a fire enter the duct. Penetrations for pipes and conduits must be closed tightly with fire-resistive material.

As an example of the use of these tables, consider a multistory vocational school building of Type 2A construction that includes both a number of classrooms and a woodworking shop. For purposes of the table reproduced in Figure 22.6, the BOCA Code places classrooms in use group E (Educational) and the shop into group F-1 (Industrial, Moderate Hazard). The two uses must be separated by a wall of 3-hour construction, and the doors through this wall will also have to be rated at 3 hours (Figure 22.7). Figure 22.5 indicates that in required vertical exits (enclosed exit stairways) of this building, only Class I materials are allowed, while Class II materials are acceptable in exit corridors, and Class III materials in the individual classrooms.

Relationship to Mechanical and Electrical Services

Interior finish materials join the mechanical and electrical services of a building at the points of delivery of the services—the electrical outlets, the lighting fixtures, the ventilating diffusers and grills, the convectors, the lavatories and water closets. Beyond these points, the services may or may not be concealed by the finish materials, but spaces must be provided for the services, and the provision of these spaces is an important factor in the selection of finishes. If the service lines are to be concealed, the finish systems must provide space for them, as well as maintenance access points in such forms as access doors, panels, hatches, cover plates, or ceiling or floor components that can be lifted out to expose the lines. If service lines are to be left exposed, the architect should organize them visually and specify a sufficiently high standard of workmanship in the installation of the services that their appearance will be satisfactory.

Changeability

How often are the use patterns of a building likely to change? In a concert hall, a chapel, or a hotel, major changes will be infrequent, and fixed, unchangeable interior partitions are appropriate. Fixed finishes include many of the heavier, more expensive, more luxurious types of materials such as tile, marble, masonry, and plaster, which are considered highly desirable by many building owners. In a rental office building or a retail shopping mall, changes will be frequent; lighting and partitions should be easily and economically adjustable to new use patterns without serious delay or disruption. The likelihood of frequent change may lead the designer to select either relatively inexpensive, easily demolished construction such as gypsum wallboard partitions or relatively expensive but durable and reusable construction such as proprietary systems of modular, relocatable partitions. The func-

FIGURE 22.7
Required fire resistance ratings in hours for fire doors and other opening protective devices, from the BOCA National Building Code.

OPENING PROTECTIVE FIRE PROTECTION RATING		
Type of assembly	Required assembly rating (hour)	Minimum opening protection assembly (hour)
Fire walls and fire separation assemblies having a required fireresistance rating greater than 1 hour	4 3 2 1½	3 3 1½ 1½
Fire separation assemblies: Shaft and exit enclosure walls Other fire separation assemblies	1 1	1 ¾
Fire partitions: Exit access corridor enclosure wall Other fire partitions	1 ½ 1	⅓[a] ½[a] ¾
Note a. For testing requirements, see Section 717.1.1.		

tional and financial choices must be weighed for each building.

Cost

The cost of interior finish systems may be measured in two different ways. *First cost* is the installed cost. First cost is often of paramount importance when the construction budget is tight or the expected life of the building is short. *Life-cycle cost* is a cost figured by any of several formulas that take into account not only first cost, but the expected lifetime of the finish system, maintenance costs and fuel costs (if any) over that lifetime, replacement cost, an assumed rate of economic inflation, and the time value of money. Life-cycle cost is important to building owners who expect to retain ownership over an extended period of time. Because of higher maintenance and replacement costs, a material that is inexpensive to buy and install may be more costly over the lifetime of a building than a material that is initially more expensive.

Toxic Emissions from Interior Materials

A number of common construction materials give off substances that may be objectionable in certain interior environments. Many synthetics and wood panel products emit formaldehyde fumes for extended periods of time after the completion of construction. Solvents from paints, varnishes, and carpet adhesives often permeate the air of a new building. Airborne fibers of asbestos and glass can constitute health hazards. Some materials harbor molds and mildews whose airborne spores many people cannot tolerate. In isolated instances, stone and masonry products have proven to be sources of radon gas. Construction dust, even from chemically inert materials, can inflame respiratory passages. There is increasing pressure, both legal and societal, on building designers to select inte-

rior materials that do not create objectionable odors or endanger the health of building occupants. Compliance with this pressure is complicated by the fact that data on the toxicity of various indoor air pollutants are inconclusive. But data on emissions of pollutants from interior materials are increasingly available to designers, and it is wise to select materials that give off as small a quantity as possible of irritating or unhealthful substances.

TRENDS IN INTERIOR FINISH SYSTEMS

Interior finish systems have undergone a transformation over the last few decades. Formerly, the installation of finishes for a commercial office began with the laying up of partitions of heavy clay tiles or gypsum blocks set in mortar. These were covered with two or three coats of plaster, and joined to a three-coat plaster ceiling. The floor was commonly made of hardwood strips with a wood baseboard, or perhaps of poured terrazzo with an integral terrazzo base. Today, the same office might be framed in light metal studs and walled with gypsum board. The ceiling might be a separate assembly of lightweight, acoustically absorbent tiles, and the floor a thin layer of vinyl composition tile glued to a smooth concrete slab.

Several trends can be discerned in these changes. One trend is away from an integral, single-piece system of finishes toward a system made up of discrete components. In the old office, the walls, ceiling, and floor were all joined, and none could be changed without disrupting the others. In the new office, ceiling and floor finishes often extend uninterrupted from one side of the building to the other, and partitions can be changed at will without affecting either the ceiling or the floor. The trend toward discrete components is

epitomized by partitions made of modular, demountable, relocatable panels.

Another discernible trend leads away from heavy finish materials toward lighter ones. A partition of metal studs and gypsum board weighs a fraction as much as one of clay tiles and plaster, and a traditional terrazzo installation is many times as heavy as an equal area of vinyl composition tile. Lighter finishes dramatically reduce the dead load the structure of the building must carry. This enables the structure itself to be lighter and less expensive. Lighter finish materials reduce shipping, handling, and installation costs, and are easier to move or remove when changes are required.

"Wet" systems of interior finish, made of materials mixed with water on the building site, are steadily being replaced by "dry" ones, which are installed in rigid form. Plaster is being replaced by gypsum board and ceiling tiles in most areas of new buildings, and tile and terrazzo floors by vinyl composition tiles and carpet. "Dry" systems are quick to install and they are not heavily dependent on weather conditions. They require less skill on the part of the installer than "wet" systems, by transferring the skilled work from the job site to the factory, where it is done by machines. All these differences tend to result in a lower installed cost.

The traditional finishes, nevertheless, are far from obsolete. Gypsum board cannot rival three-coat plaster over metal lath for surface quality, durability, or design flexibility. Tile and terrazzo flooring are unsurpassed for wearing quality and appearance. In many situations, the life-cycle costs of traditional finishes compare favorably with those of lighter-weight finishes whose first cost is considerably less. And the aesthetic qualities of, for example, marble floors, wood wainscoting, and sculpted plaster ceilings, where such qualities are called for, cannot be imitated by any other material.

SELECTED REFERENCES

1. Building Officials and Code Administrators International, Inc. *The BOCA National Building Code/1996*. Country Club Hills, Illinois, 1996.

The reader is referred to Chapters 7 and 8 of this model code, which deal with fire-resistive construction requirements for interior finishing systems. (Address for ordering: 4051 West Flossmoor Road, Country Club Hills, IL 60478-5795.)

2. Allen, Edward and Joseph Iano. *The Architect's Studio Companion* (2nd ed.). New York, John Wiley & Sons, Inc., 1995.

The third section of this book gives extensive information on providing space for mechanical and electrical equipment in buildings.

KEY TERMS AND CONCEPTS

shaft
suspended ceiling
access flooring
firestopping
Sound Transmission Class (STC)

Steiner Tunnel Test
flame spread rating
fuel contributed rating
smoke developed rating
fire resistance rating

hose test
first cost
life-cycle cost

REVIEW QUESTIONS

1. Draw a flow diagram of the approximate sequence in which finishing operations are carried out on a large building of Type 2A construction.

2. List the major considerations that an architect should keep in mind while selecting interior finish materials and systems.

3. What are the two major types of fire tests conducted on interior finish systems? What measures of performance are derived from each?

4. What is the difference between first cost and life-cycle cost?

EXERCISE

1. You are designing a 31-story apartment building in a large city. What types of construction are you permitted to use under the BOCA Code? What fire resistance rating will be required for the separation between the apartment floors and the retail stores on the ground floor? What classes of finish materials can you use in the exit stairway? In the corridors to those stairways? Within the individual apartments? If a red oak board has a flame spread rating of 100, can you panel an apartment in red oak? What fire resistance ratings are required for partitions between apartments? Between an exit corridor and an apartment? What type of fire door is required between the apartment and the exit corridor? What is the required fire resistance rating for the walls around the elevator shafts?

INTERIOR WALLS
AND PARTITIONS

A plasterer applies the scratch coat of gypsum plaster to expanded metal lath. The partition is framed with open-truss wire studs. Soft-annealed galvanized steel wire is used to make all the connections in this partition. (*Courtesy of United States Gypsum Company*)

There is more to interior walls and partitions than meets the eye. Behind their simple surfaces lie assemblies of materials carefully chosen and combined to meet specific performance requirements relating to structural strength, fire resistance, durability, and acoustical isolation. A partition may be framed with steel or wood studs and faced with plaster or gypsum board. Alternatively, masons may construct it of concrete blocks or structural clay tiles. For improved appearance or durability, a partition may be faced with ceramic tiles or a masonry veneer.

TYPES OF INTERIOR WALLS

Fire Walls

A *fire wall* is a wall that forms a required separation to restrict the spread of fire through a building, and which extends continuously from the foundation to the roof. Fire walls are used to divide a building into smaller units, each of which may be considered as a separate building when calculating allowable heights and areas under a building code. A fire wall must either meet a noncombustible roof structure at the top or extend through and above the roof by a specified minimum distance, usually 32 inches (813 mm). Openings in fire walls are restricted in size and must be closed with fire doors or fire-rated glass. The required fire resistance ratings for fire walls under the BOCA Code are defined by line 2 of Figure 1.3 and by Figure 22.6.

Shaft Walls

A *shaft wall* is used to enclose a multistory opening through a building, such as an elevator shaft or a shaft for ductwork, conduits, or pipes. The required fire resistance ratings for shaft walls are defined by line 6 of Figure 1.3. Walls for elevator shafts must be able to withstand the air pressure and suction loads placed on them by the movements of the elevator cars within the shaft, and should be designed to prevent the noise of the elevator machinery from reaching other areas of the building.

Other Types of Fire-Rated Walls

Fire-rated walls are also used within a building to separate mixed occupancies (as in the case of the woodworking shop in a classroom building that was discussed in the preceding chapter), to separate tenant spaces in commercial buildings, to separate dwelling units in multiple dwellings, to separate guest rooms in hotels and dormitories, and to enclose required stairways and exitway corridors. Like a fire wall, these walls are intended to restrict the spread of fire but, unlike a fire wall, they do not necessarily extend from foundation to roof. These walls must be full-height partitions that reach from the top of one floor slab to the underside of the next; they cannot terminate at a suspended ceiling. Openings in these walls are restricted in size and must be closed with fire doors or fire-rated glass. The required fire resistance ratings for such walls are defined by lines 3, 4, 5, 7, and 8 of Figure 1.3 and by Figure 22.6.

Smoke Barriers

In certain institutional occupancies such as hospitals and prisons, where occupants are unable to leave the building in case of fire, special walls called *smoke barriers* are required. These divide floors of buildings in such a way that occupants may take refuge in case of fire by moving to the side of the smoke barrier that is away from the fire, without having to exit from the building. A smoke barrier must form a continuous, airtight assembly that fits tightly against structural slabs at top and bottom and exterior walls. Openings in smoke barrier walls must have self-closing doors of specified fire resistance ratings.

Other Nonbearing Partitions

Many of the partitions in a building neither bear a structural load nor are required as fire separation walls. These may be made of any material that meets the combustibility provisions of the code for the selected type of construction.

FRAMED PARTITION SYSTEMS

Partition Framing

Partitions that will be finished in plaster or gypsum board are usually framed with wood or metal studs (Figures 23.1, 23.2, 23.3, 23.4). Wood may be used only as allowed by combustibility requirements of the building code. Fire-retardant treated wood is allowed for partition framing in all types of construction, provided it is used as part of a partition assembly that has been tested and certified as a unit to have the required fire resistance.

Metal framing is directly analogous to wood framing and may be of *light-gauge steel studs* and *runner channels* or, for plaster partitions only, *open-truss wire studs*. Light gauge steel framing is detailed in Chapter 12. Neither the light-gauge steel stud nor the wire stud is very strong structurally until restrained from buckling by its finish surfaces of plaster or sheet material.

If plaster or gypsum board surfaces are to be applied over a masonry wall, they may be spaced away from the wall with either wood or metal *furring strips* (Figures 23.5,

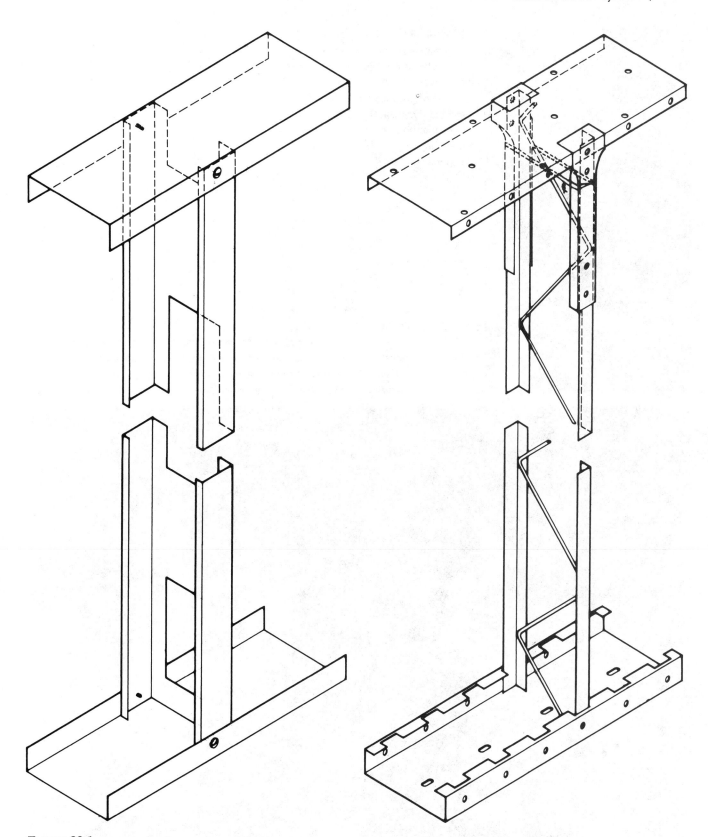

FIGURE 23.1
Partitions of light-gauge steel studs, at the left, and open-truss wire studs. The light-gauge steel studs are assembled with screws, and may be used with any type of lath or panel. The open-truss wire studs are made specifically for use with expanded metal or gypsum lath. The bottom track needs no fasteners to the studs, but metal shoes attached with wire are required at the tops of the studs. Two loops of wire are used to make a stronger connection to studs that support a door frame.

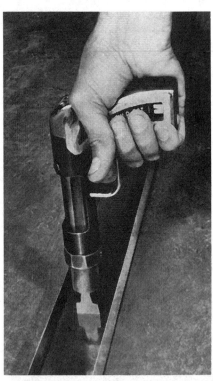

FIGURE 23.2
Attaching a runner channel to a concrete floor using powder-driven fasteners. The gun explodes a small charge of gunpowder to drive a steel pin through the metal and into the concrete to make a secure connection. (*Courtesy of United States Gypsum Company*)

FIGURE 23.3
Inserting studs into a runner channel to frame a partition of light-gauge steel studs. The cutouts in the webs of the studs provide a passage for electrical conduits. On the right side of the photograph, a stack of gypsum board awaits installation. (*Courtesy of United States Gypsum Company*)

FIGURE 23.4
Inserting an open-truss wire stud into runner track. The stud is $3\frac{1}{4}$ inches (83 mm) deep; the tip of a shoe on the right helps to give an idea of its size. (*Courtesy of United States Gypsum Company*)

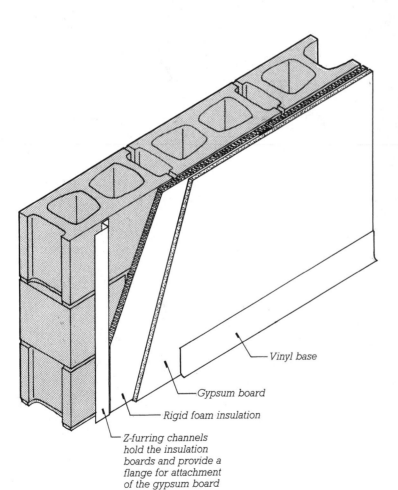

Vinyl base

Gypsum board

Rigid foam insulation

Z-furring channels hold the insulation boards and provide a flange for attachment of the gypsum board

FIGURE 23.5
A furred gypsum board finish over a concrete block wall. The Z-furring channels are attached to the masonry with powder-driven fasteners. The plastic foam insulation is tucked in behind the flange of the channel, and the gypsum board is screwed to the face of the flange. Long slots punched from the web of the channel (not visible in this drawing) help to reduce the thermal bridging effect of the Z-furring channel.

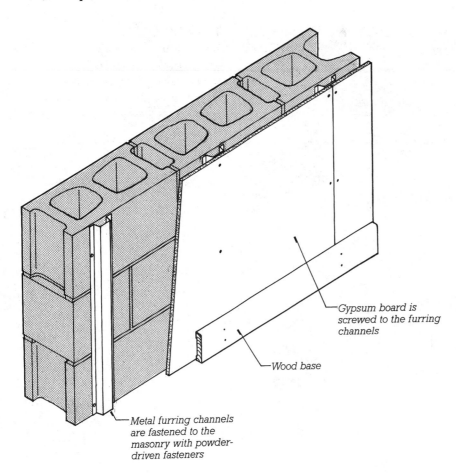

FIGURE 23.6
A furred gypsum board finish using a standard hat-shaped metal furring channel.

Gypsum board is screwed to the furring channels

Wood base

Metal furring channels are fastened to the masonry with powder-driven fasteners

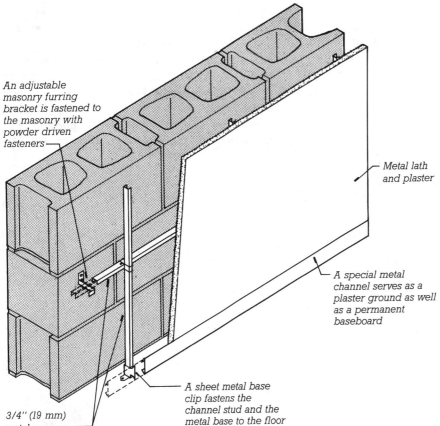

FIGURE 23.7
A furred plaster finish using adjustable furring brackets. Each bracket has a series of teeth along its upper edge so that a metal channel can be wired securely to it in any of a number of positions, allowing the lather to produce a flat wall regardless of the surface quality of the masonry.

An adjustable masonry furring bracket is fastened to the masonry with powder driven fasteners

Metal lath and plaster

A special metal channel serves as a plaster ground as well as a permanent baseboard

A sheet metal base clip fastens the channel stud and the metal base to the floor

3/4'' (19 mm) metal channels

23.6, 23.7). Furring allows for the installation of a flat wall finish over an irregular masonry surface, and provides a concealed space between the finish and the masonry for installing plumbing, wiring, and thermal insulation.

Plaster

Plaster is a generic term that refers to any of a number of cementitious substances that are applied to a surface in paste form, after which they harden into a solid material. Plaster may be applied directly to a masonry surface or, more generally, to any of a group of plaster bases known collectively as *lath* (rhymes with "math"). Plastering began in prehistoric times with the smearing of mud over masonry walls or over a mesh of woven sticks and vines to create a construction known as *wattle and daub*, the wattle being the mesh and the daub the mud. The early Egyptians and Mesopotamians invented finer, more durable plasters based on gypsum and lime. Portland cement plasters were developed in the 19th century. It is from these latter three materials—gypsum, lime, and portland cement—that the plasters used in buildings today are prepared.

Gypsum Plaster

Gypsum is an abundant mineral in nature, a crystalline hydrous calcium sulfate. It is quarried, crushed, dried, ground to a fine powder, and heated to 350 degrees Fahrenheit (175°C) in a process known as *calcining* to drive off about three-quarters of its water of hydration. The calcined gypsum, ground to a fine white powder, is known as *plaster of Paris*. When plaster of Paris is mixed with water it rehydrates and recrystallizes rapidly to return to its original, solid state. As it hardens, it gives off heat and expands slightly.

Gypsum is a major component of interior finish materials in most buildings. It has but one major disadvantage—its solubility in water.

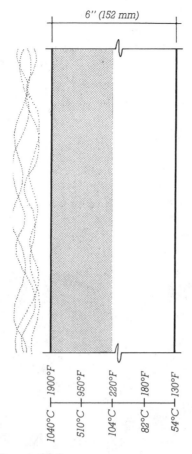

6″ (152 mm)

1040°C — 1900°F
510°C — 950°F
104°C — 220°F
82°C — 180°F
54°C — 130°F

FIGURE 23.8
The effect of fire on gypsum, based on data from Underwriters Laboratories, Inc. After a 2-hour exposure to heat following the ASTM E119 time–temperature curve, less than half the gypsum on the side toward the fire, shown here by shading, has calcined. The portions of the gypsum to the right of the line of calcination remain at temperatures below the boiling point of water.

Among its advantages are that it is durable and light in weight as compared to many other materials. It resists the passage of sound better than most materials and can be fashioned into surfaces that range from smooth to heavily textured. But above all it is inexpensive and it is highly resistant to the passage of fire.

When a gypsum building component is subjected to the intense heat of a fire, a thin surface layer is cal-

cined and gradually disintegrates. In the process, it absorbs considerable heat and gives off steam, which further cools the fire (Figure 23.8). Layer by layer, the fire works its way through the gypsum, but the process is a slow one. The uncalcined gypsum never reaches a temperature more than a few degrees above the boiling point of water, so areas behind the gypsum component are well protected from the fire's heat. Any required degree of fire resistance can be created by increasing the thickness of the gypsum as necessary. The fire resistance of gypsum can also be increased by adding lightweight aggregates to reduce its thermal conductivity and by adding reinforcing fibers to retain the calcined gypsum in place as a fire barrier.

For use in construction, calcined gypsum is carefully formulated with various admixtures to control its setting time and other properties. Gypsum plaster is made by mixing the appropriate dry plaster formulation with water and an aggregate, either fine sand or a lightweight aggregate such as vermiculite or perlite. Because of its expansion during setting, gypsum plaster is very resistant to cracking.

For certain types of fine finish coats, the plaster mix is based on hydrated lime, which has superb qualities of workability. Lime plasters, however, are always mixed with quantities of gypsum to accelerate setting and prevent cracking.

Gypsum plasters are manufactured in accordance with ASTM C28. The more common types of gypsum plasters are:

• Ordinary *gypsum plaster,* in various formulations for use with sand aggregate or lightweight aggregate and for machine application

• *Gypsum plaster with wood fiber,* for lighter weight and greater fire resistance

• *Gypsum plaster with perlite aggregate,* for lighter weight and greater fire resistance

• *Gauging plaster,* for mixing with lime putty to accelerate its set and eliminate cracking

• *High-strength basecoat plaster,* for use under high-strength finish coats such as Keenes cement

• *Keenes cement,* a high-strength gypsum plaster that produces an exceptionally strong, crack-resistant finish

• *Molding plaster,* a fast-setting, fine-textured material for molding plaster ornament and running cornices (see the sidebar on pages 770 and 771).

Lime and Portland Cement Plasters

In addition to these gypsum materials, two other types of plasters are commonly used:

• *Finish lime* is mixed with gauging plaster to make high-quality finish coat plasters.

• *Portland cement–lime plaster,* also known as *stucco,* is similar to masonry mortar. It is used where the plaster is likely to be subjected to moisture, as on exterior wall surfaces, or in commercial kitchens, industrial plants, and shower rooms. Because freshly mixed stucco is not as buttery and smooth as gypsum and lime plasters, it is not as easy to apply and finish. It shrinks slightly during curing, so it should be installed with frequent control joints to regulate cracking.

Plastering

Plaster can be applied either by machine or by hand. Machine application is essentially a spraying process (Figure 23.9). Hand application is done with two very simple tools, a *hawk* in one hand to hold a small

FIGURE 23.9
Spray applying a scratch coat of plaster to gypsum lath. (*Courtesy of United States Gypsum Company*)

quantity of plaster ready for use, and a *trowel* in the other hand to lift the plaster from the hawk, apply it to the surface, and smooth it into place (Figures 23.10, 23.18, 23.22). Plaster is transferred from the hawk to the trowel with a quick, practiced motion of both hands, and the trowel is moved up the wall or across the ceiling to spread the plaster, much as one uses a table knife to spread soft butter. After a surface is covered with plaster, it is leveled by drawing a straightedge called a *darby* across it, after which the trowel is used again to smooth the surface.

Lathing

At one time, the most common form of lath consisted of thin strips of wood nailed to wood framing with small spaces left between the strips to allow keying of the plaster. Most lath today is made either of expanded

> In lathing I was pleased to be able to send home each nail with a single blow of the hammer, and it was my ambition to transfer the plaster from the board to the wall neatly and rapidly. . . . I admired anew the economy and convenience of plastering, which so effectually shuts out the cold and takes a handsome finish . . . I had the previous winter made a small quantity of lime by burning the shells of the *Unio fluviatilis,* which our river affords . . .
>
> **Henry David Thoreau,** *Walden*

FIGURE 23.10
Applying a scratch coat over gypsum lath with a hawk (a corner of which is visible behind the plasterer's stomach) and trowel. Notice the wire clips that hold the lath to the open-truss studs, and the sheet metal clips that strengthen the end joints between panels. The end joints do not occur over studs. (*Courtesy of United States Gypsum Company*)

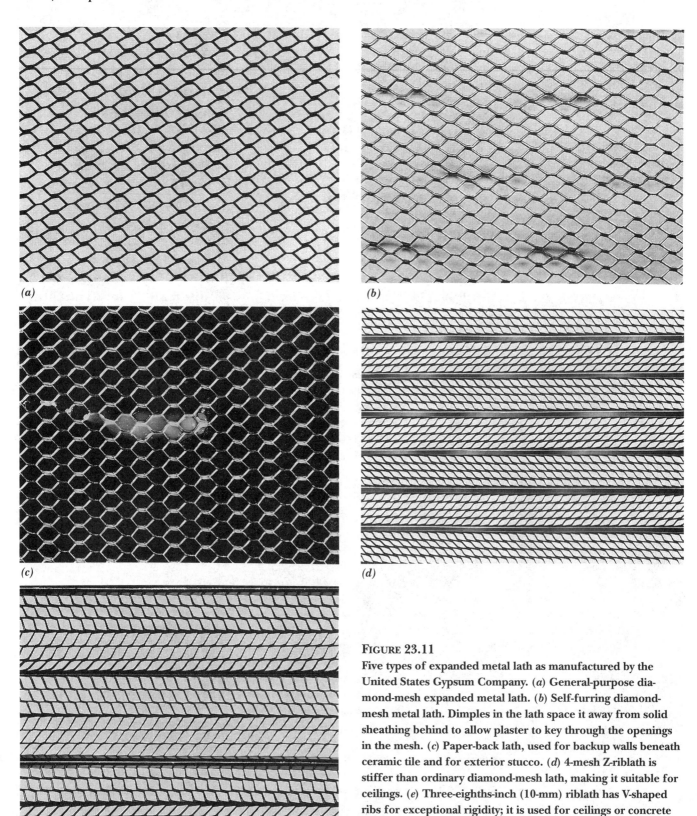

(a)

(b)

(c)

(d)

(e)

FIGURE 23.11

Five types of expanded metal lath as manufactured by the United States Gypsum Company. (*a*) General-purpose diamond-mesh expanded metal lath. (*b*) Self-furring diamond-mesh metal lath. Dimples in the lath space it away from solid sheathing behind to allow plaster to key through the openings in the mesh. (*c*) Paper-back lath, used for backup walls beneath ceramic tile and for exterior stucco. (*d*) 4-mesh Z-riblath is stiffer than ordinary diamond-mesh lath, making it suitable for ceilings. (*e*) Three-eighths-inch (10-mm) riblath has V-shaped ribs for exceptional rigidity; it is used for ceilings or concrete formwork where supports are widely spaced. (*Courtesy of United States Gypsum Company*)

metal or preformed gypsum boards. The skilled tradesperson who applies lath and trim accessories is known as a *lather.*

Expanded metal lath is made from thin sheets of steel alloy that are slit and stretched in such a way as to produce a mesh of diamond-shaped openings (Figure 23.11). It is applied either to open-truss wire studs with intermittent ties of soft steel wire, to light-gauge steel studs with self-drilling, self-tapping screws, or to wood studs with large-headed lathing nails.

Gypsum lath is made in sheets 16 inches by 48 inches (406 mm by 1220 mm) and in thicknesses of $\frac{3}{8}$ inch (9.5 mm) and $\frac{1}{2}$ inch (13 mm). It consists of sheets of hardened gypsum plaster faced with outer layers of a special absorbent paper to which

fresh plaster readily adheres, and inner layers of water-resistant paper to protect the gypsum core. Gypsum lath is attached to truss studs with special metal clips and to steel or wood studs with screws (Figure 23.12). Gypsum lath cannot be used as a base for lime plaster or portland cement stucco.

Veneer plaster base is a paper-faced gypsum board that comes in sheets 4 feet wide (1220 mm) and 8 to 14 feet long (2440 to 4270 mm). It is screw applied to wood or steel studs or nailed to wood studs, as a base for the application of *veneer plaster.*

Various lathing *trim accessories* are used at the edges of a plaster surface to make a neat, durable edge or corner (Figure 23.13). These are installed by the lather at the same time as the lath. In very long or tall

plaster surfaces, metal control joint accessories are mounted over seams in the lath at predesigned intervals to control cracking. Trim accessories are also designed to act as lines that gauge the proper thickness and plane of plaster surface. A straightedge may be run across them to level the wet plaster. In this role, the trim accessories are known collectively as *grounds.* Trim accessories are made in several different thicknesses to match the required plaster thicknesses over the different types of lath.

Trim accessories are also produced as extrusions of plastic or aluminum. The aluminum accessories and some of the plastic ones are designed for improved precision and appearance when used in innovative details for bases, edges, and shadow lines in plaster walls.

FIGURE 23.12
Installing gypsum lath over light-gauge steel studs with self-drilling, self-tapping screws. The electric screw gun disengages automatically from the screw head when the screw has been driven to the proper depth. (*Courtesy of United States Gypsum Company*)

USG Corner Beads, Trim, Control Joints

description

USG Corner Beads and Trim, made from top-quality galvanized steel, enjoy the industry's top acceptance because of their dependability and continual improvement in design. Corner beads are available in 8 and 10-ft. lengths, metal trim in 7 and 10-ft. lengths, casing beads in 7, 8 and 10-ft. lengths.

1-A Expanded Corner Bead has 2⅞″ wide expanded flanges that are easily flexed. Preferred for irregular corners. Provides increased reinforcement close to nose of bead. Made from galvanized steel or zinc alloy for exterior applications.

X-2 Corner Bead has full 3¼″ flanges easily adjusted for plaster depth on columns. Ideal for finishing corners of structural tile and rough masonry. Has perforated stiffening ribs along expanded flange.

4-A Flexible Corner Bead is an economical general purpose bead. By snipping flanges, this bead may be bent to any curved design (for archways, telephone niches, etc.).

800 Corner Bead gives ¹⁄₁₆″ grounds needed for one-coat veneer finishes. Approx. 90 keys per lin. in. provide superior bonding and strong, secure corners. The 1¼″ fine-mesh flange eliminates shadowing, is easily nailed or stapled.

900 Corner Bead is used with two-coat veneer systems, gives ³⁄₃₂″ grounds. Its 1¼″ fine-mesh flange can be either stapled or nailed. Provides superior plaster key and eliminates shadowing.

Cornerite and Striplath are strips of painted Diamond Mesh Lath used as reinforcement. **Cornerite** is bent lengthwise in the center to form a 100° angle. It should be used in all interior angles where metal lath is not lapped or carried around, over non-ferrous lath anchored to the lath, and over internal angles of masonry constructions to reduce plaster cracking. **Sizes:** 2″ x 2″ x 96″ and 3″ x 3″ x 96″. **Striplath** is a similar flat strip, used as a plaster reinforcement over joints of non-metallic lathing bases and where dissimilar bases join; also to span pipe chases. **Size:** 4″ x 96″.

USG Metal Trim comes in two styles and two grounds to provide neat edge protection for veneer finishing at cased openings and ceiling or wall intersections. All have fine-mesh expanded flanges to strengthen plaster bond and eliminate shadowing. **No. 701-A**, channel-type, and **No. 701-B**, angle edge trim, provide ³⁄₃₂″ grounds for two-coat systems. **No. 801-A**, channel-type, and **No. 801-B**, angle edge trim, provide ¹⁄₁₆″ grounds for one-coat systems. **Sizes:** for ½″ and ⅝″ IMPERIAL Gypsum Base.

USG P-1 Vinyl Trim is a channel-shaped rigid trim with flexible vinyl fins which compress on installation to provide a positive acoustical seal comparable in performance to one bead of acoustical sealant. For veneer finish partition perimeters. **Lengths:** 8, 9 and 10 ft. **Sizes:** for ½″ and ⅝″ gypsum base.

USG P-2 Vinyl Trim is a channel-shaped vinyl trim with a pressure-sensitive adhesive backing for attachment to the wall at wall-ceiling intersections. Provides positive perimeter relief in radiant heat and veneer finish systems. Allow ⅛″ to ¼″ clear space for insertion. **Length:** 10 ft.

USG Control Joint relieves stresses of expansion and contraction in large plastered areas. Made from roll-formed zinc, it is resistant to corrosion in both interior and exterior uses with gypsum or portland cement plaster. An open slot, ¼″ wide and ½″ deep, is protected with plastic tape which is removed after plastering is completed. The perforated short flanges are wire-tied to metal lath, screwed or stapled to gypsum lath. Thus the plaster is key-locked to the control joint, which not only provides plastering grounds but can also be used to create decorative panel designs. **Limitations:** Where sound and/or fire ratings are prime considerations, adequate protection must be

provided behind the control joint. USG Control Joints should not be used with magnesium oxychloride cement stuccos or stuccos containing calcium chloride additives. **Sizes and grounds: No. 50, ½″; No. 75, ¾″; No. 100, 1″** (for exterior stucco curtain walls)—10-ft. lengths.

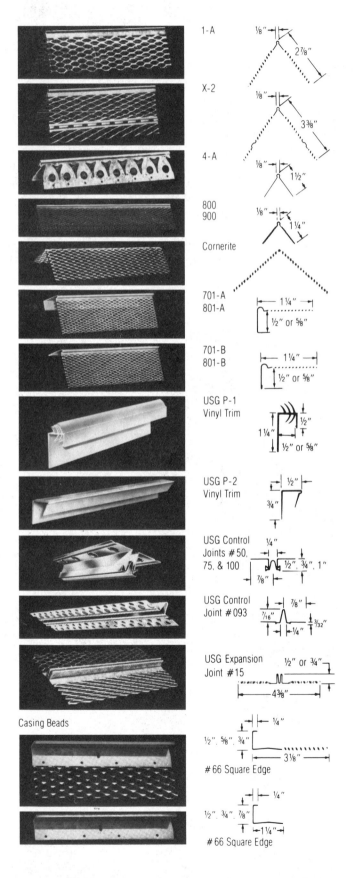

Casing Beads

FIGURE 23.13

Trim accessories for lath-and-plaster construction, as manufactured by the United States Gypsum Company. (*Courtesy of United States Gypsum Company*)

Plaster Systems

Plaster Over Expanded Metal Lath Plaster is applied over expanded metal lath in three coats (Figure 23.14). The first, called the *scratch coat,* is troweled on rather roughly and cannot be made completely flat because the uncoated lath is not very rigid under the pressure of the trowel. This first coat is scratched while still wet, using a notched darby, a broom, or a special rake, to create a rough surface to which the second coat can bond mechanically (Figure 23.15).

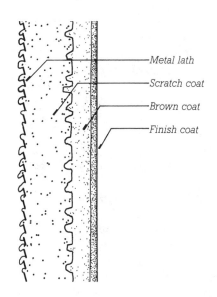

Metal lath
Scratch coat
Brown coat
Finish coat

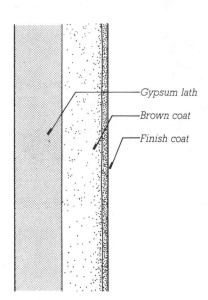

Gypsum lath
Brown coat
Finish coat

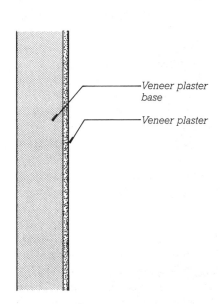

Veneer plaster base
Veneer plaster

FIGURE 23.14

Sections through the three common lath-and-plaster systems, reproduced at full scale. Metal lath (left) requires three coats of plaster; the surface of the first coat is scratched for a better bond to the second coat. Gypsum lath (middle) may be finished with three coats, or with two, as shown. Veneer plaster (right) uses only a thin coating of plaster.

FIGURE 23.15

Scratching the scratch coat while it is still wet to create a better bond to the brown coat. (*Courtesy of United States Gypsum Company*)

After the scratch coat has hardened, it works together with the lath to provide a rigid base for the second application of plaster, which is called the *brown coat*. The purpose of the brown coat is to build strength and thickness, and to present a level surface for the application of the third or *finish coat*. The level surface is produced by drawing a long straightedge across the surfaces of the lathing accessories (the grounds, edge beads, corner beads, and control joints) to strike off the wet plaster. On large, uninterrupted plaster surfaces, *plaster screeds,* intermittent spots or strips of plaster, are leveled up to the grounds in advance of brown coat plastering to serve as intermediate reference points for setting the thickness of the plaster during the striking-off operation.

The finish coat is very thin, about $1/16$ inch (1.5 mm) in thickness. It may be troweled smooth or worked into any desired texture (Figures 23.16, 23.17). The total thickness of plaster that results from this three-coat process, as measured from the face of the lath, is about $5/8$ inch (16 mm). Three-coat work over metal lath is the premium-quality plaster system, extremely strong and resistant to fire. The only disadvantage of three-coat plaster work is its cost, which can be attributed largely to the labor involved in applying the lath and the three separate coats of plaster.

Plaster Over Gypsum Lath   The best plaster work over gypsum lath is applied in three coats, but gypsum lath is sufficiently rigid that, if it is firmly mounted to the studs, only a brown coat and a finish coat need be applied. The elimination of the scratch coat has obvious economic advantages. Even with three coats of plaster, gypsum lath is often economically advantageous over metal lath because the gypsum in the lath replaces much of the plaster that would otherwise have to be mixed and applied by hand in the scratch coat. The total thickness of plaster

FIGURE 23.16
A sponge-faced float can be used to create various surface textures on plaster.
(*Courtesy of Portland Cement Association, Skokie, Illinois*)

applied over gypsum lath is $1/2$ inch (13 mm).

Plaster Applied to Masonry   Where plaster is applied directly over brick, concrete masonry, or poured concrete walls, the walls should be dampened thoroughly in advance of plastering to prevent premature dehydration of the plaster. A bonding agent may have to be applied to some types of smooth masonry surfaces to ensure good adhesion of the plaster. The number of coats of plaster required to cover a wall is determined by the degree of unevenness of the masonry surface. For the best work, three coats totaling $5/8$ inch (16 mm) should be applied, but for many walls two coats will suffice (Figure 23.18).

Veneer Plaster   Veneer plaster is the least expensive of the plaster systems and is competitive in price with gypsum board finishes in many regions. The veneer base and accessories create a very flat surface that can be finished with a layer of a specially

formulated dense plaster that is only about $1/8$ inch (3 mm) thick (Figures 23.19, 23.20, 23.21, 23.22). The plaster is applied in a "double-back" process in which a thin coat is followed immediately by a second "skim" coat that is finish troweled to the desired texture. The plaster veneer hardens and dries so rapidly that it may be painted the following day.

Stucco   Stucco is applied over galvanized metal lath, using galvanized or plastic accessories to prevent rusting in damp locations. Whereas gypsum plaster expands during hardening, and is therefore highly resistant to cracking, portland cement stucco shrinks and is prone to cracking. Stucco walls should be provided with control joints at frequent intervals to channel the shrinkage into predetermined lines rather than allowing it to cause random cracks. The curing reaction in stucco is the same as that of concrete and is much slower than that of gypsum plaster. Stucco must be kept moist for a period of at least a week before it is allowed to dry, in order to attain maximum hardness and strength through full hydration of its portland cement binder.

In exterior applications over metal or wood studs, stucco may be applied either over sheathing or without sheathing. Over sheathing, asphalt-saturated felt paper is first applied as an air and moisture barrier. Then *self-furring metal lath,* which is formed with "dimples" that hold the lath away from the surface of the wall a fraction of an inch to allow the stucco to key to the lath, is attached with nails or screws (Figure 23.11*b*). If no sheathing is used, the wall is laced tightly with strands of *line wire* a few inches apart, and paperbacked metal lath (Figure 23.11*c*) is attached to the line wire, after which stucco is applied to encase the building in a thin layer of reinforced concrete.

Stucco is usually applied in three coats, either with a hawk and trowel or by spraying (Figure 6.29). In exte-

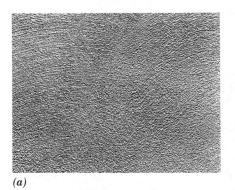 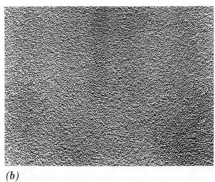 

(a) (b) (c)

FIGURE 23.17
Three different plaster surface textures from among many. (*a*) **Float finish.** (*b*) **Spray finish.** (*c*) **Texture finish.** (*Courtesy of United States Gypsum Company*)

FIGURE 23.18
Applying a finish coat of portland cement plaster over a concrete masonry partition. The block joints are visible in the base coat of plaster because of a difference in the rate of water absorption between the blocks and the mortar joints. (*Courtesy of Portland Cement Association, Skokie, Illinois*)

PLASTER ORNAMENT

Gypsum plaster, with its fine grain and even texture, has more sculptural potential than any other material used in architecture. While wet, it is easily formed with trowels and spatulas, with molds, or with templates. When dry, it is readily worked by sawing, sanding, machining, and carving. Plaster building ornament has been created for many centuries by two economical but powerful techniques, *casting* and *running*, and continues to be used in buildings of every size and every historic style.

Cast ornament is made by pouring soupy plaster into molds. The plaster hardens in a few minutes, allowing the mold to be stripped and reused. Both rigid molds and soft rubber molds are used. The rubber molds are very flexible, so that even undercut shapes can be cast without encountering difficulties in mold removal. Traditional rubber molds are created by first carving a plaster original, then brushing layers of latex over the original to build up the require wall thickness. More recently, two-component synthetic rubber compounds have replaced latex in most applications; their advantage is that they can be spread over the original in a single application rather than in layers.

Cast ornament is adhered to the brown coat of plaster in walls and ceilings with gobs of wet plaster or a mixture of plaster and glue. Once the ornament has been securely fastened in place, the finish coat of plaster is applied around it, and the plaster surfaces and adjacent pieces of ornament are merged by skillful trowel work and sanding to create a single-piece finish.

Running is used to make linear ornament such as classic cornice moldings. A rigid *blade* made of sheet metal or sheet plastic is cut to the profile of the molding. The blade is attached to a sliding wooden frame to create a *template*. The template is pushed back and forth along a guide strip mounted temporarily on the wall or ceiling while a mixture of lime putty and gauging plaster is inserted in front of the blade, which strikes it off to the desired profile. Repeated passes of the template are required to finish the molding smoothly and perfectly. These passes must be completed before the plaster begins to harden, else the setting expansion of the gypsum will cause the template to bind and spoil the plaster surface. The template may also be attached to a radius guide to produce circular moldings.

Casting and running are often used to reproduce plaster ornament during restoration of historic buildings. Rubber molds for casting can be made directly from existing ornaments, and templates for running duplicates of existing profiles are easily and cheaply produced.

New designs for plaster ornament are readily translated from the architect's paper sketches into carved plaster originals, from which rubber molds are made and duplicates cast. New profiles for moldings are quickly converted into template blades. The possibilities are almost

FIGURE A
Removing the flexible rubber mold from a cast plaster ornament.
(*Courtesy of Dovetail, Inc.*)

FIGURE B
Running a plaster cornice molding in place.
(*Courtesy of Dovetail, Inc.*)

limitless, yet few contemporary architects have chosen to explore them. This is surprising because plaster ornament is inexpensive compared to ornament carved from wood or stone, and there are few technical constraints on what can be accomplished.

Cast plaster ornament can be reinforced with short fibers of alkali-resistant glass. These greatly increase its strength and toughness and allow it to be produced in much thinner sections and much larger pieces than unreinforced plaster. This recent development has dramatically changed the economics and methods of ornamentation in plaster that is based on stock designs. A number of manufacturers issue catalogs of stock designs for ornaments made by this process. Much of the on-site assembly work for elaborate ornaments can be eliminated by combining what were formerly a number of small, thick, brittle castings into a single, larger, thinner casting that is light in weight and highly resistant to breakage. On the construction site, the lightweight castings are glued in place over gypsum board or veneer plaster base, using an ordinary mastic adhesive. The edges of the ornaments are feathered into the wall or ceiling surfaces with joint compound or veneer plaster, and the joints between pieces of ornament are smoothed over and sanded.

FIGURE C
Sculptor David Flaharty runs a circular plaster medallion on a bench in his shop.

FIGURE D
A closeup of Flaharty's blade and template. The template rides on the sledlike runner of the portion under his hand, which is called the *slipper*. The long portion of the template, called the *stock*, is a radius guide that is fastened to a pin at the end to create the circular form of the medallion.

FIGURE E
The medallion having been removed from the bench and glued in place on the ceiling, Flaharty adds cast components to complete the ornament. (*Photos in Figures C, D, and E by Brian McNeill. Courtesy of David Flaharty*)

FIGURE 23.19
Installing veneer plaster base with a screw gun.
(*Courtesy of United States Gypsum Company*)

FIGURE 23.20
Stapling a corner bead to veneer plaster base
to create a straight, durable corner. (*Courtesy
of United States Gypsum Company*)

FIGURE 23.21
Reinforcing the panel joints in veneer plaster
base with a self-adhesive glass fiber mesh tape.
A panel for access to mechanical equipment is
visible behind the installer. (*Courtesy of United
States Gypsum Company*)

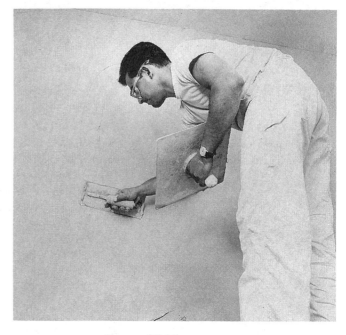

FIGURE 23.22
Applying veneer plaster with hawk and
trowel. (*Courtesy of United States Gypsum
Company*)

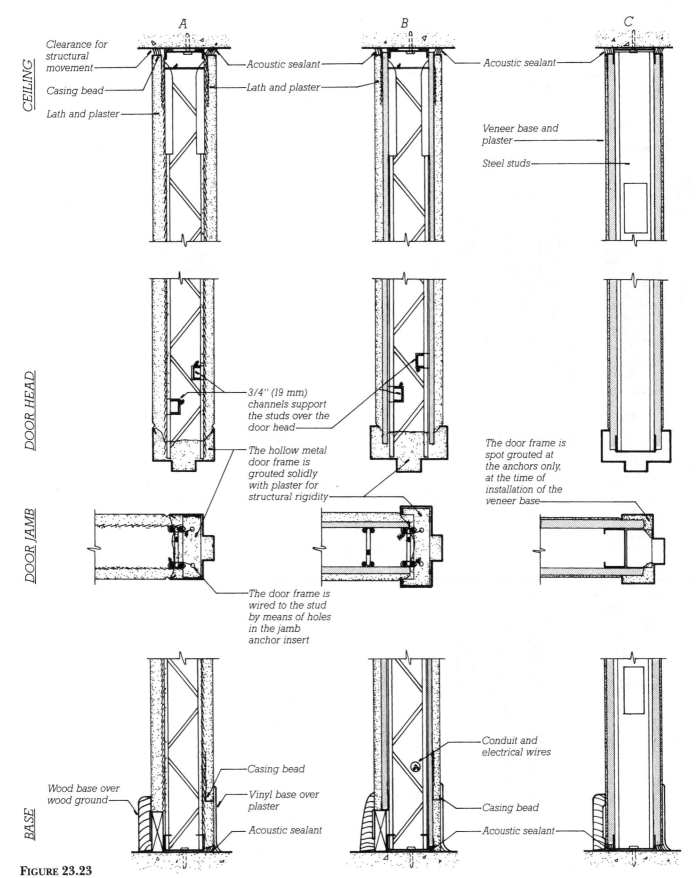

FIGURE 23.23

Three plaster partition systems. (*a*) Three coats of plaster on metal lath and open-truss wire studs, rated at 1 hour of fire resistance and STC 39. (*b*) Two coats of plaster over gypsum lath and open-truss wire studs, rated at 1 hour and STC 41. (*c*) Veneer plaster on light-gauge steel studs, STC 40, 1 hour. Notice especially the provisions for airtightness and structural movement at the top and bottom of each partition, and the methods of attachment used for hollow metal door frames. Doors weighing more than 50 pounds (23 kg) require special reinforcing details around the frames.

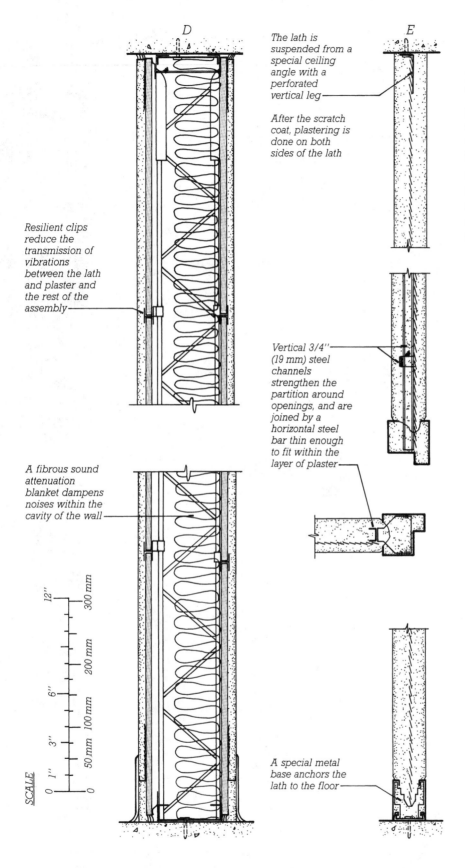

D

The lath is suspended from a special ceiling angle with a perforated vertical leg

After the scratch coat, plastering is done on both sides of the lath

Resilient clips reduce the transmission of vibrations between the lath and plaster and the rest of the assembly

A fibrous sound attenuation blanket dampens noises within the cavity of the wall

E

Vertical 3/4" (19 mm) steel channels strengthen the partition around openings, and are joined by a horizontal steel bar thin enough to fit within the layer of plaster

A special metal base anchors the lath to the floor

SCALE

12"   300 mm
6"   200 mm
       100 mm
3"   50 mm
1"
0   0

FIGURE 23.24

Two plaster partition systems. (*d*) This partition increases its STC to 51 by mounting the gypsum lath on one side with resilient metal clips and filling the hollow space in the partition with a sound attenuation blanket. (*e*) The solid plaster partition has an STC of only 38 but is used in situations where floor space must be conserved. Both these partitions are rated at 1 hour.

rior work, pigments or dyes are often added to stucco to give an integral color, and rough textures are frequently used.

Plaster Partition Systems

Several types of plaster partitions are detailed in Figures 23.23, 23.24. These diagrams show some of the ways in which the various trim accessories are used, and the precautions taken to isolate the partitions from structural or thermal movements in the loadbearing frame of the building.

Gypsum Board

Gypsum board is a prefabricated plaster sheet material that is manufactured in widths of 4 feet (1220 mm) and lengths of 8 to 14 feet (2440 to 4270 mm). It is also known as *gypsum wallboard*, *plasterboard*, and *drywall*. (The term *sheetrock* is a registered trademark of one manufacturer of gypsum board and should not be used in a generic sense.)

Gypsum board is the least expensive of all interior wall finishing materials. For this reason alone, it has found wide acceptance throughout North America as a substitute for plaster, not only in residential construction, but in buildings of every type. It retains the fire-resistive characteristics of gypsum plaster, but it is installed with less labor by semiskilled workers rather than skilled lathers and plasterers. And because it is installed largely in the form of dry materials, it eliminates some of the waiting that is associated with the curing and drying of plaster.

The core of gypsum board is formulated as a slurry of calcined gypsum, starch, water, pregenerated foam to reduce the density of the mixture, and various admixtures. This slurry is sandwiched between special paper faces and passed between sets of rollers that reduce it to the desired thickness. Within two or three minutes, the core material has hardened and bonded to the

paper faces. The board is cut to length and heated to drive off residual moisture, then bundled for shipping (Figure 23.25).

In recent years, a gypsum board that has no paper faces has come into production. The gypsum slurry from which the board is made contains an admixture of cellulose fibers that greatly strengthen the hardened material, making a paper face unnecessary.

Types of Gypsum Board

Gypsum board is manufactured in a number of different types in accordance with ASTM C36.

- *Regular gypsum board* is used for the majority of applications.

- In locations exposed to moderate amounts of moisture, a special *water-resistant gypsum board* is used, with a water-repellent paper facing and a moisture-resistant core formulation.

- For many types of fire-rated assemblies, *Type X gypsum board* is required by the building code. The core material of Type X board is reinforced with short glass fibers. In a severe fire, the fibers hold the calcined gypsum in placed to continue to act as a barrier to fire, rather than permitting it to erode or fall out.

- *Foil-backed gypsum board* can be used to eliminate the need for a separate vapor retarder in outside wall assemblies. If the back of the board faces a dead airspace at least $\frac{3}{4}$ inch (19 mm) thick, the bright foil also acts as a thermal insulator.

- *Predecorated gypsum board* is board that has been covered with paint, printed paper, or decorative plastic film. If handled carefully and installed with small nails, it needs no further finishing. This product is used in a number of systems of demountable office partitioning.

- *Coreboard* is a 1-inch-thick panel (25.4 mm) that is used for shaft walls and solid gypsum board partitions. To facilitate handling, it is fabricated in sheets 24 inches (610 mm) wide.

- *High-impact gypsum board* is a $\frac{5}{8}$-inch (15-mm) Type X panel that has a polycarbonate film bonded to its back. The film imparts considerable resistance to penetration and impact breakage, making this a useful product in buildings that are exposed to rough usage.

Gypsum board is manufactured with a variety of edge profiles, but the most common by far is the *tapered edge*, which permits sheets to be joined with a flush, invisible seam by means of subsequent joint finishing operations. Rounded and beveled edges are useful in predecorated panels, and tongue-and-groove edges serve to join coreboard panels in concealed locations.

A number of different thicknesses of board are produced:

- $\frac{1}{4}$-inch board (6.4 mm) is used as a backing board in certain sound control applications. A special $\frac{1}{4}$-inch board is also produced by some manufacturers to be used for tight-radius bends (Figure 23.30).

- $\frac{5}{16}$-inch board (8 mm) is made for manufactured housing, where weight reduction to facilitate shipping is an important consideration.

- $\frac{3}{8}$-inch board (9.5 mm) is largely used in double-layer wall finishes, but it is permitted to be used over wall studs spaced 16 inches (400 mm) apart by some building codes.

- $\frac{1}{2}$-inch board (12.7 mm) is the most common thickness. It is used for stud spacings up to 24 inches (610 mm).

- $\frac{5}{8}$-inch board (16 mm) is also limited to stud spacings of not more than 24 inches, but it is often used where additional fire resistance or structural stiffness is required.

- $\frac{3}{4}$-inch (19-mm) Type X board is produced by some manufacturers. It is used to create 2-hour partitions with a single layer of gypsum board on each face, and 4-hour partitions that include two layers on each face.

- 1-inch-thick board (25.4 mm) is made only as coreboard.

FIGURE 23.25
Sheets of gypsum board roll off the manufacturing line, trimmed and ready for packaging. (*Courtesy of United States Gypsum Company*)

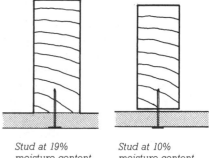

Stud at 19% moisture content   *Stud at 10% moisture content*

FIGURE 23.26
When wood studs dry and shrink during a building's first heating season, nail heads may pop through the surface of gypsum board walls.

Installing Gypsum Board

Hanging the Board Gypsum board may be installed over either light-gauge steel or wood studs, using self-drilling, self-tapping screws to fasten to steel and either screws or nails to fasten to wood. Wood studs can be troublesome with gypsum board because they usually shrink somewhat after the board is installed, which can cause nails to loosen slightly and "pop" through the finished surface of the board (Figure 23.26). *Nail popping* can be minimized by using only fully dried framing lumber, by using the shortest nail that will do the job, and by using ring-shank nails that have extra gripping power in the wood. Screws have less of a tendency to pop than nails. When screws or nails are driven into gypsum board, their heads are driven to a level slightly below the surface of the board, but not far enough to tear the paper surface.

To minimize the length of joints that must be finished, and to create the stiffest wall possible, gypsum board is usually installed with the long dimension of the boards horizontal. The longest possible boards are used, to eliminate or at least minimize end joints between boards, which are difficult to finish because the ends of boards are not tapered. Gypsum board is cut rapidly and easily by scoring one paper face with a sharp knife, snapping the brittle core along the score line with a blow from the heel of the hand, and cutting the other paper face along the fold created by the snapped core (Figure 23.27). Notches, irregular cuts, and holes for electric boxes are made with a small saw.

(a)

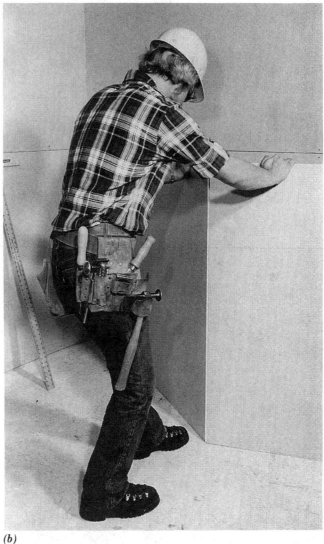

(b)

FIGURE 23.27
Cutting gypsum board: (*a*) A sharp knife and metal T-square are used to score a straight line through one paper face of the panel. (*b*) The scored board is easily "snapped," and the knife used a second time to slit the second paper face. (*Courtesy of United States Gypsum Company*)

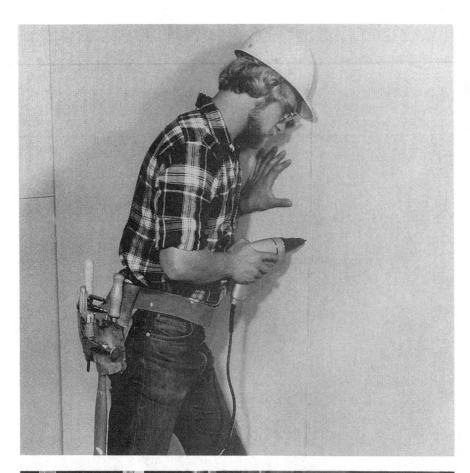

FIGURE 23.28
Attaching gypsum board to studs with a screw gun. (*Courtesy of United States Gypsum Company*)

FIGURE 23.29
Gypsum board can be curved to a large radius simply by bending it around a curving line of studs. (*Courtesy of United States Gypsum Company*)

When two or more layers of gypsum board are installed on a surface, the joints between layers are staggered in order to create a stiffer wall, and a mastic adhesive is often used to join the layers to one another. Adhesive is sometimes used between the studs and the gypsum board in single-layer installations to make a stronger joint.

Gypsum board can be curved when a design requires it. For gentle curves, the board can be bent into place dry (Figure 23.29). For somewhat sharper curves, the paper faces are moistened to decrease the stiffness of the board before it is installed. When the paper dries, the board is as stiff as before. Special high-flex $\frac{1}{4}$-inch (6.5-mm) board can be bent dry to a relatively small radius, and bent wet to an even smaller radius (Figure 23.30).

FIGURE 23.30
Tighter radii of curvature can be achieved by using $\frac{1}{4}$-inch (6.5-mm) high-flex gypsum board. (*Courtesy of Gold Bond Building Products*)

Metal trim accessories are required at exposed edges and external corners to protect the brittle board and present a neat edge (Figure 23.31).

Finishing the Joints and Fastener Holes Joints and holes in gypsum board are finished to create the appearance of a monolithic surface, indistinguishable from plaster. The finishing process is based on use of a joint compound that resembles a smooth, sticky plaster. For most purposes, a drying-type, all-purpose joint compound is used; this is a mixture of marble dust, binder, and admixtures, furnished either as a dry powder to be mixed with water or as a premixed paste. In some high-production commercial work, joint compounds that cure rapidly by chemical reaction are used to minimize waiting time between applications.

The finishing of a joint between panels of gypsum board begins with the troweling of a layer of joint compound into the tapered edge joint, and the bedding of a paper or glass fiber reinforcing tape in the compound (Figures 23.32–23.34). Compound is also troweled over the nail or screw holes. After overnight

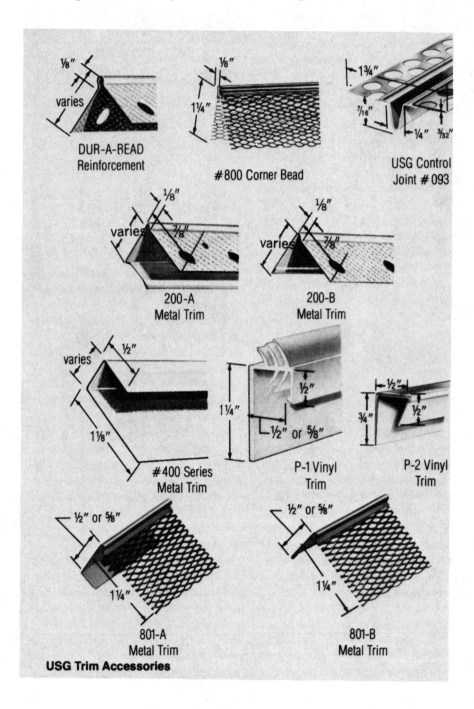

FIGURE 23.31
Accessories for gypsum board construction, as manufactured by the United States Gypsum Company. (*Courtesy of United States Gypsum Company*)

drying, a second layer of compound is applied to the joint, to bring it level with the face of the board and to fill the space left by the slight drying shrinkage of the joint compound. When this second coat is dry, the joints are lightly sanded before a very thin final coat is applied to fill any remaining voids and feather out to an invisible edge. Before painting, the wall is again sanded lightly to remove any roughness or ridges. If the finish-

ing is done properly, the painted or papered wall will show no signs at all that it is made of discrete panels of material.

Gypsum board has a smooth surface finish, but a number of spray-on textures and textured paints can be applied to give a rougher surface. Most gypsum board contractors prefer to apply a texture to ceilings; the texture masks the minor irregularities in workmanship that are likely to

occur because of the difficulty of working in the overhead position.

Gypsum Board Partition Systems

Gypsum board partition assemblies have been designed and tested to achieve fire resistance ratings up to 4 hours and acoustic performance to STC 59. A selection of these partitions is shown in Figure 23.35. Notice in these details how provisions are

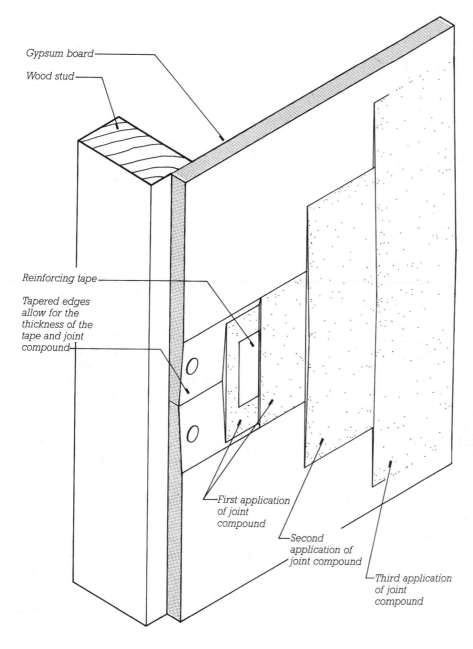

Gypsum board

Wood stud

Reinforcing tape

Tapered edges allow for the thickness of the tape and joint compound

First application of joint compound

Second application of joint compound

Third application of joint compound

FIGURE 23.32
Finishing a joint between panels of gypsum board.

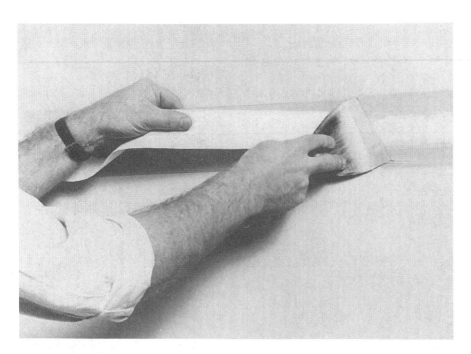

FIGURE 23.33
Applying paper joint reinforcing tape to gypsum board. (*Courtesy of United States Gypsum Company*)

FIGURE 23.34
An automatic taper simultaneously applies tape and joint compound to gypsum board joints. (*Courtesy of United States Gypsum Company*)

FIGURE 23.35
Four gypsum board partition systems. (*a*) A 1-hour partition, STC 40, using Type X gypsum board over light-gauge steel studs. (*b*) This 1-hour partition on wood studs achieves an STC of 60 to 64 through heavy laminations of gypsum board, a sound attenuation blanket, and resilient channel mounting for one face of the partition. (*c*) A 2-hour partition with an STC of 48. (*d*) A 4-hour partition, STC 58. These are representative of a large number of gypsum board partition systems in a range of fire resistance ratings and sound transmission classes.

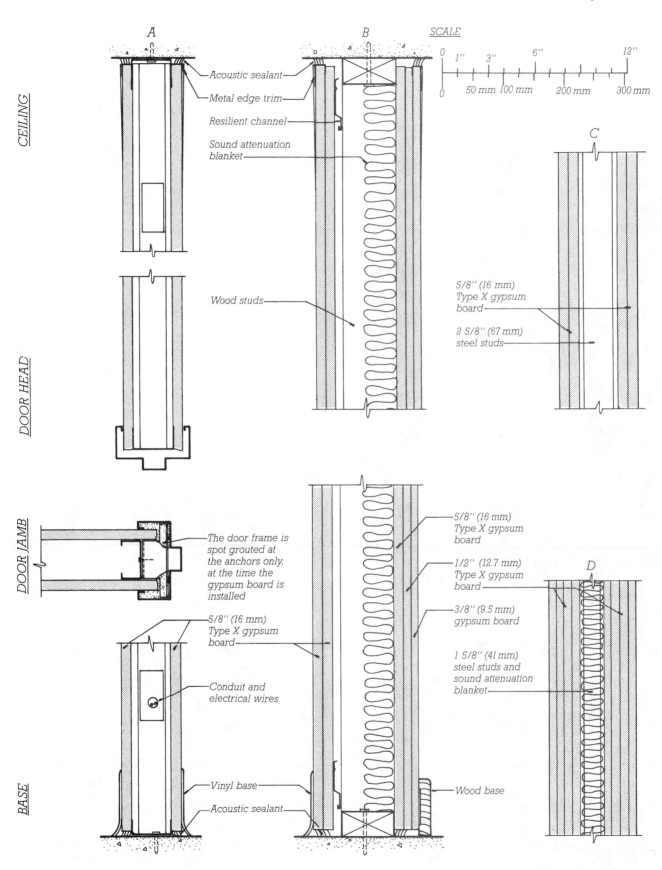

CEILING

DOOR HEAD

DOOR JAMB

BASE

A

B

SCALE

C

D

0   1''   3''   6''   12''

0   50 mm 100 mm   200 mm   300 mm

Acoustic sealant

Metal edge trim

Resilient channel

Sound attenuation blanket

Wood studs

5/8'' (16 mm) Type X gypsum board

2 5/8'' (67 mm) steel studs

The door frame is spot grouted at the anchors only, at the time the gypsum board is installed

5/8'' (16 mm) Type X gypsum board

Conduit and electrical wires

Vinyl base

Acoustic sealant

5/8'' (16 mm) Type X gypsum board

1/2'' (12.7 mm) Type X gypsum board

3/8'' (9.5 mm) gypsum board

1 5/8'' (41 mm) steel studs and sound attenuation blanket

Wood base

made to prevent the gypsum panels from being subjected to structural loadings caused by movements in the loadbearing frame of the building—structural deflections, concrete creep, moisture expansion, and temperature expansion and contraction. Notice also the use of sealant to eliminate sound transmission around the edges of the partitions.

Demountable partition systems of gypsum board have also been developed, using concealed mechanical fasteners that allow the partition to be disassembled and reassembled easily without damage to the panels (Figure 23.37). These systems are of use in buildings whose partitions must be rearranged at frequent intervals.

FIGURE 23.36
Installing a sound attenuation blanket.
(*Courtesy of United States Gypsum Company*)

FIGURE 23.37
Two photographs of relocatable partition systems. (*Courtesy of United States Gypsum Company*)

Gypsum Shaft Wall Systems

Walls around elevator shafts, stairways, and mechanical chases can be made of any masonry, lath and plaster, or gypsum panel assembly that meets fire resistance and structural requirements. Gypsum shaft wall systems offer several advantages over the alternatives: They are lighter in weight, they are installed dry, and they are erected entirely from the floor outside the shaft, with no need to erect scaffolding inside. Depending on the requirements for fire resistance rating, air pressure resistance, STC, and floor-to-ceiling height, any of a number of designs may be used. Figure 23.38 shows representative shaft wall details, and Figure 22.1 shows a shaft wall being installed.

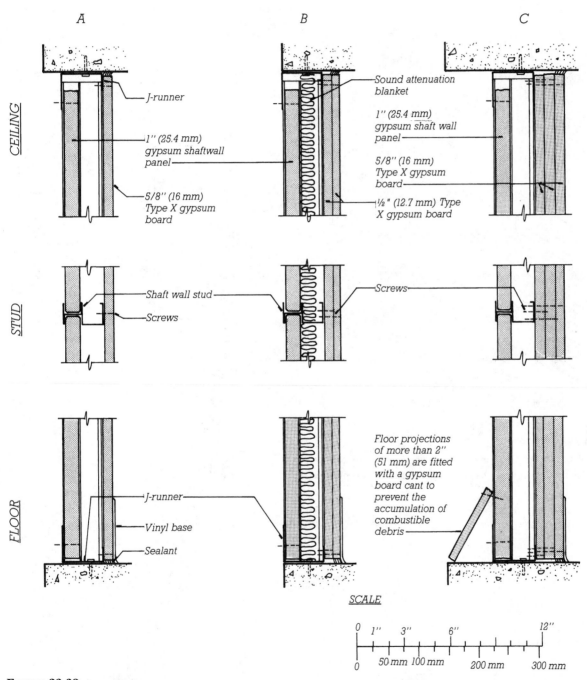

FIGURE 23.38

Gypsum shaft wall systems, all three framed with steel C–H studs. The H portion of the stud holds the 1-inch (25-mm) shaft wall panel, while the C portion accepts the screws used to attach the finish layers of gypsum board. (*a*) A 1-hour system. (*b*) A 2-hour system, STC 47. (*c*) A 3-hour system.

MASONRY PARTITION SYSTEMS

A century ago, interior partitions were often made of common brick masonry plastered on both sides. These had excellent STC and fire resistance ratings, but were labor intensive and heavy. Partition systems of hollow clay tile and hollow gypsum tile (Figure 23.39) were developed to meet these objections and continued in use until a few decades ago, but both have now become obsolete in North America, replaced by plaster, gypsum board, and concrete masonry.

Concrete masonry partitions may be plastered or faced with gypsum board but are more often left

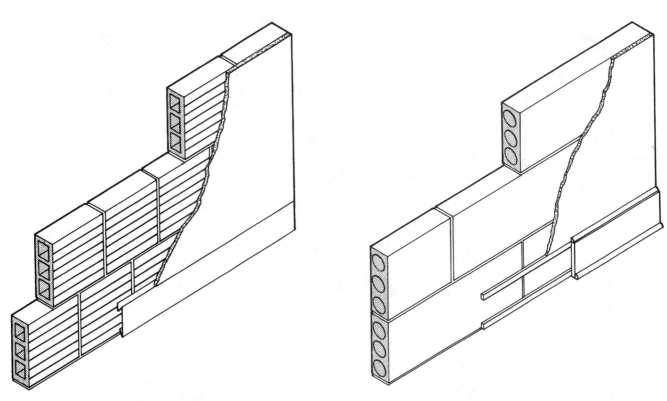

FIGURE 23.39
Obsolete partition systems often found in existing buildings: hollow clay tiles and plaster (left) and gypsum tiles and plaster (right).

exposed, either painted or unpainted. Several types of lightweight aggregate may be used to reduce the dead weight of partition blocks. Electrical wiring is relatively difficult to conceal in concrete block partitions; the electrician and the mason must coordinate their work closely, or the wiring must be mounted on the surface of the wall after the mason has finished.

Glazed structural clay tiles make very durable partitions, especially in areas with heavy wear, moisture problems, or strict sanitation requirements (Figure 23.40). The ceramic glazes are nonfading and virtually indestructible.

FIGURE 23.40
A glazed structural clay tile partition installation. The floor is finished with glazed ceramic tiles. (*Courtesy of Stark Ceramics, Inc.*)

WALL AND PARTITION FACINGS

Ceramic tile facings are often added to walls for reasons of appearance, durability, sanitation, or moisture resistance. For best quality, tile facings should be applied to a base of metal lath and portland cement mortar (Figure 23.41). Lower-cost tile facing systems eliminate the mortar base in favor of special tile base panels of fiber-reinforced lightweight concrete or, less optimally, water-resistant gypsum board. The tiles are mounted to the mortar or concrete panel base with mortar or organic adhesive, and to the gypsum with organic adhesive. After the tiles have become fully adhered, a cementitious grout of any desired color is wiped into the tile joints with a rubber trowel.

Facings of granite, limestone, marble, or slate are sometimes used in public areas of major buildings. These are usually mounted in the manner shown in Figure 23.42.

Wood wainscoting and paneling may be used in limited quantities in fire-resistant buildings. They are mounted over a backing of plaster or gypsum board to retain the fire-resistive qualities of the partition.

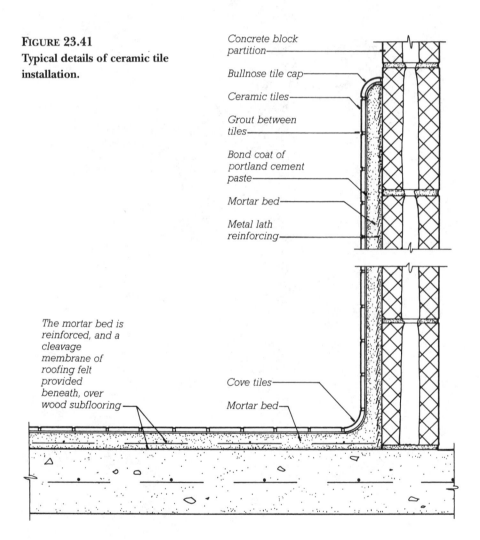

FIGURE 23.41
Typical details of ceramic tile installation.

Concrete block partition

Bullnose tile cap

Ceramic tiles

Grout between tiles

Bond coat of portland cement paste

Mortar bed

Metal lath reinforcing

The mortar bed is reinforced, and a cleavage membrane of roofing felt provided beneath, over wood subflooring

Cove tiles

Mortar bed

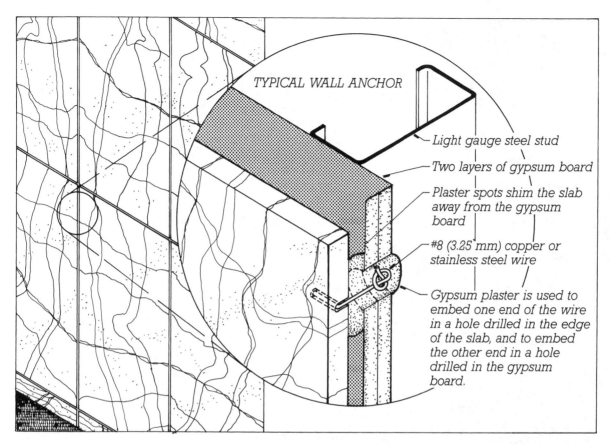

TYPICAL WALL ANCHOR

— Light gauge steel stud

— Two layers of gypsum board

— Plaster spots shim the slab away from the gypsum board

— #8 (3.25 mm) copper or stainless steel wire

— Gypsum plaster is used to embed one end of the wire in a hole drilled in the edge of the slab, and to embed the other end in a hole drilled in the gypsum board.

FIGURE 23.42
Attaching a stone facing over a backup of gypsum boards and steel studs.

C.S.I./C.S.C. Masterformat Section Numbers for Interior Walls and Partitions	
08100	**METAL DOORS AND FRAMES**
08110	**Steel Doors and Frames**
08200	**WOOD AND PLASTIC DOORS**
08210	**Wood Doors**
09100	**METAL SUPPORT SYSTEMS**
09110	**Nonloadbearing Wall Framing Systems**
09200	**LATH AND PLASTER**
09205	**Furring and Lathing**
09210	**Gypsum Plaster**
09215	**Veneer Plaster**
09220	**Portland Cement Plaster**
09250	**GYPSUM BOARD**
09300	**TILE**
09310	**Ceramic Tile**
09330	**Quarry Tile**
10600	**PARTITIONS**
10615	**Demountable Partitions**
10630	**Portable Partitions, Screens, and Panels**

FIGURE 23.43
Imaginative use of gypsum products is coordinated with concealed sources of light to create a singularly dramatic space for the D. E. Shaw & Company office and trading area in New York. (*Stephen Holl Architects. Photo by Paul Warchol*)

SELECTED REFERENCES

1. Gypsum Association. *Fire Resistance Design Manual*. Washington, D.C., updated frequently.

Fire resistance ratings and STCs are given in this booklet for a large number of wall and ceiling assemblies that use either gypsum plaster or gypsum board. (Address for ordering: 810 First Street, NE, Suite 510, Washington, DC 20002.)

2. United States Gypsum Company. *Gypsum Construction Handbook* (4th ed.). Chicago, 1992.

This manual represents manufacturers' literature at its best—more than 500 well-illustrated pages crammed with every important fact about gypsum wallboard, gypsum plaster, and associated products. (Address for ordering: P.O. Box 806278, Chicago, IL 60680-4124.)

3. Portland Cement Association. *Portland Cement Plaster (Stucco) Manual*. Skokie, Illinois, 1996.

A complete, illustrated guide to stucco. (Address for ordering: 5420 Old Orchard Road, Skokie, IL 60077.)

4. Stagg, William D., and Brian F. Pegg. *Plastering: A Craftsman's Encyclopedia*. New York, Crown Publishers, 1985.

Techniques of ornamental plastering are covered in complete detail in this 276-page reference work.

Key Terms and Concepts

fire wall
shaft wall
smoke barrier
nonbearing partition
light-gauge steel studs
runner channels
open-truss wire studs
furring strip
plaster
lath
wattle and daub
gypsum
calcining
plaster of Paris
gypsum plaster
gypsum plaster with wood fiber
gypsum plaster with perlite aggregate
gauging plaster

high-strength basecoat plaster
Keenes cement
molding plaster
finish lime
portland cement–lime plaster or stucco
hawk
trowel
darby
lather
expanded metal lath
gypsum lath
veneer plaster base
veneer plaster
trim accessories
grounds
scratch, brown, finish coats
plaster screed
self-furring metal lath

line wire
cast plaster ornament
run plaster ornament
gypsum board, gypsum wallboard,
plasterboard, drywall
regular gypsum board
water-resistant gypsum board
Type X gypsum board
foil-backed gypsum board
predecorated gypsum board
coreboard
high-impact gypsum board
tapered edge
nail popping
glazed structural clay tile
ceramic tile

Review Questions

1. What are the major types of interior walls and partitions in a larger building, such as a hospital, classroom building, apartment, or office building? How do these types differ from one another?

2. Why is gypsum used so much in interior finishes?

3. Name the coats of plaster used over expanded metal lath, and explain the role of each.

4. Under what circumstances would you specify the use of portland cement plaster? Keenes cement plaster?

5. Describe step by step how the joints between sheets of gypsum board are made invisible.

Exercises

1. Determine the construction of a number of partitions in the places where you live and work. What materials are used? What accessories? Why were these chosen for their particular situations? Sketch a detail of each partition.

2. Sketch typical details showing how the various metal lath trim accessories are used in a plaster wall.

3. Repeat Exercise 2 for gypsum board trim accessories.

4. What type of gypsum wall finish system would you specify for a major art museum? For a low-cost rental office building? Outline a complete specification of wall and partition construction for a building you are presently working on.

FINISH CEILINGS AND FLOORS

A worker lays acoustical panels in a recessed grid to create a suspended ceiling.
(*Courtesy of United States Gypsum Company*)

As the ceilings and finish floors are installed, the construction of a building is drawing to a close. The components of the mechanical and electrical systems that remain exposed are either finished or concealed, intersections of interior surfaces are neatly trimmed, and painters work their magic to reveal for the first time the interior character of the building. The architect, engineers, and municipal building officials make their last inspections and, following last-minute corrections of minor defects, the contractor turns the building over to its owner.

cation systems. And its color, texture, pattern, and shape are prominent in the overall visual impression of the room. A ceiling can be a simple, level plane, a series of sloping planes that give a sense of the roof above, a luminous surface, a richly coffered ornamental ceiling, even a frescoed plaster vault such as Michelangelo's famous ceiling in the Sistine Chapel in Rome; the possibilities are endless.

FINISH CEILINGS

Functions of Finish Ceilings

The ceiling surface is an important functional component of a room. It helps control the diffusion of light and sound about the room. It may play a role in preventing the passage of sound vertically between the rooms above and below, and horizontally between rooms on either side of a partition. Often it is designed to resist the passage of fire. Frequently, it is called upon to assist in the distribution of conditioned air, artificial light, and electrical energy. In many buildings, it must accommodate sprinkler heads for fire suppression and loudspeakers for intercommuni-

TYPES OF CEILINGS

Exposed Structural and Mechanical Components

In many buildings, it makes sense to omit a finished ceiling surface altogether and simply expose the structural and mechanical components of the floor or roof above (Figure 24.1). In industrial and agricultural build-

ings, where appearance is not of prime importance, this approach offers the advantages of economy and ease of access for maintenance. Many types of floor and roof structures are inherently attractive if left exposed, such as heavy timber beams and decking, concrete waffle slabs, and steel trusses. Other types of structures, such as concrete flat plates and precast concrete planks, have little visual interest, but can be painted and left exposed as finished ceilings in apartment buildings and hotels, which have little need for mechanical services at the ceiling; this saves money and reduces the overall height of the building. And in some buildings, the structural and mechanical elements at the ceiling, if carefully designed, installed, and painted, can create a powerful aesthetic of their own.

Exposing structural and mechanical components rather than covering them with a finished ceiling does not necessarily save money. Mechanical and structural work are not normally done in a precise, attractive fashion because they are not usually expected to be visible, and it is less expensive for workers to take only as much care in installation as is required for satisfactory functional performance. To achieve perfectly straight ductwork free of dents, steel decks without rust and weld spatter, and neat runs of electrical conduit and plumbing, the drawings and specifications for the project must tell exactly the results that are expected, and higher labor costs must be anticipated.

Tightly Attached Ceilings

Ceilings of any material may be attached tightly to wood joists, wood rafters, steel joists, or concrete slabs (Figure 24.2). Special finishing

arrangements must be worked out for any beams and girders that protrude through the plane of the ceiling and for ducts, conduits, pipes, and sprinkler heads that fall below the ceiling.

Suspended Ceilings

A ceiling that is suspended on wires some distance below the floor or roof structure can hang level and flat despite varying sizes of girders, beams, joists, and slabs above, even under a roof structure that slopes toward roof drains. Ducts, pipes, and conduits can be contained entirely in the *plenum* space between the ceiling and the structure. Lighting fixtures, sprinkler heads, loudspeakers, and fire-detection devices may be recessed into the ceiling. Such a ceiling can also serve as *membrane fire protection* for the floor or roof structure above, eliminating the need for fussy

FIGURE 24.1
Sprinkler pipes, air conditioning ductwork, electrical conduits, and lighting fixtures are exposed in a ceiling structure of painted open-web steel joists and corrugated steel decking. The office space behind has a suspended ceiling of lay-in acoustical panels with recessed lighting fixtures. The floor is of brick with a border of concrete. (*Architects: Woo and Williams. Photo by Richard Bonarrigo. Courtesy of the architects*)

FIGURE 24.2
Spraying a textured finish onto the underside of a concrete slab in a residential building, where there are no pipes, ducts, or wires to be concealed below the plane of the floor structure. (*Courtesy of United States Gypsum Company*)

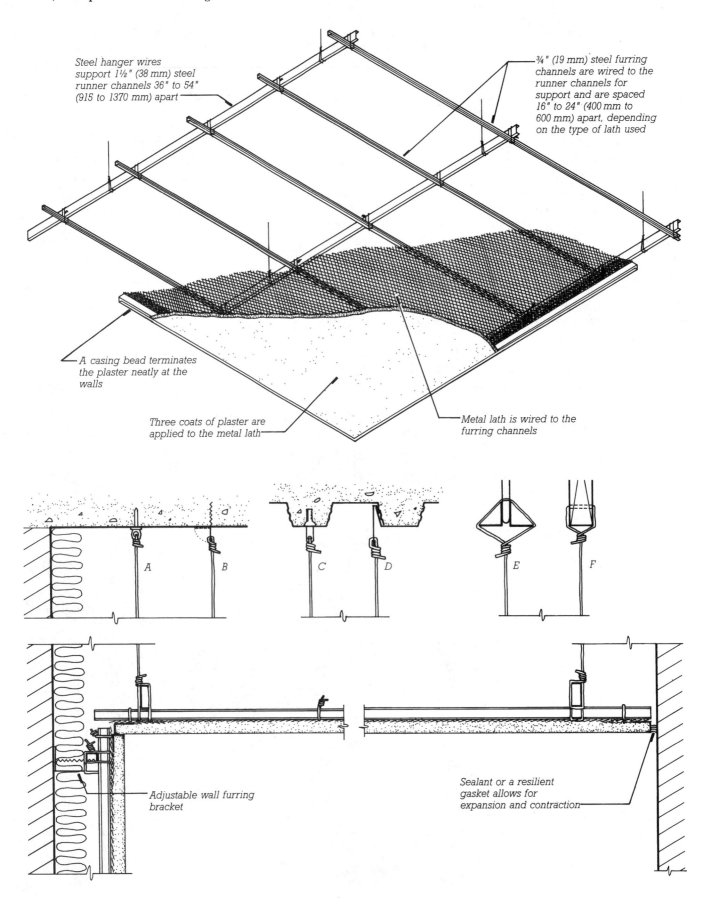

Steel hanger wires support 1½" (38 mm) steel runner channels 36" to 54" (915 to 1370 mm) apart

¾" (19 mm) steel furring channels are wired to the runner channels for support and are spaced 16" to 24" (400 mm to 600 mm) apart, depending on the type of lath used

A casing bead terminates the plaster neatly at the walls

Three coats of plaster are applied to the metal lath

Metal lath is wired to the furring channels

A

B

C

D

E

F

Adjustable wall furring bracket

Sealant or a resilient gasket allows for expansion and contraction

FIGURE 24.3

A suspended plaster ceiling on metal lath. At the top of the page is a cutaway isometric drawing, as viewed from below, of the essential components of the ceiling. Across the center of the page are details of six ways of supporting the hanger wires: (*a*) A powder-driven pin into a concrete structure. (*b*) A corrugated sheet metal tab with a hole punched in it is nailed to the formwork before the concrete is poured. When the formwork is stripped, the tab bends down and the hanger wire is threaded through the hole. (*c*) A sharp, dagger-like tab of sheet metal is driven through the corrugated metal decking before the concrete topping is poured. (*d*) A sheet metal hook is hung onto the lap joints in the metal decking. (*e*) The hanger wire is wrapped around the lower chord of an open-web steel joist. (*f*) The hanger wire is passed through a hole drilled near the bottom of a wood joist. At the bottom of the page is a section through a furred, insulated plaster wall and a suspended plaster ceiling.

individual fireproofing of steel joists, or imparting a higher fire resistance rating to wood structures. For these reasons, *suspended ceilings* have become a popular and economical feature in many types of buildings, especially office and retail structures.

Suspended ceilings can be made of almost any material; the most widely used are gypsum board, plaster, and various proprietary panels and tiles composed of noncombustible fibers. Each of these materials is supported on its own special system of small steel framing members, and the framing members are hung from the structure on heavy steel wires. Gypsum board suspended ceilings are screwed to ordinary light-gauge steel cee channels that are suspended on wires. Suspended plaster ceilings have been in use for many decades; some typical details are shown in Figure 24.3. While most suspended plaster ceilings are flat, the lather is capable of constructing ceilings that are richly sculpted, from configurations resembling highly ornamented Greek or Roman coffered ceilings to nearly any form that the contemporary designer can draw. This capability is especially useful in auditoriums, theaters, lobbies of public buildings, and other uniquely shaped rooms.

Fibrous ceiling materials are delivered to the job site as lightweight tiles or panels. Ceilings made from fibrous materials are customarily referred to as *acoustical ceilings* because most of them are highly absorptive of sound energy, unlike plaster and gypsum board, which are highly reflective of sound. The sound absorption performance of a ceiling material is measured and published in the trade literature as its *noise reduction coefficient* (*NRC*). An NRC of 0.85 would indicate that the material absorbs 85 percent of the sound that reaches it and reflects only 15 percent back into the room. NRCs for most acoustical ceiling materials range from 0.50 to 0.90, as compared

to values below 0.10 for plaster and gypsum board ceilings. This makes acoustical ceilings valuable for reducing noise levels in offices, lobbies, restaurants, retail stores, recreational spaces, and noisy industrial environments.

The lightweight, porous materials that produce high NRC ratings allow most sound energy to pass through. A ceiling made of porous materials will not furnish very good acoustic privacy between adjacent rooms unless a suitable full-height wall separates the rooms and blocks the ceiling plenum. The ability of a ceiling construction to reduce sound transmission from room to room through the plenum is measured by its *ceiling attenuation class* (*CAC*) in decibels.

Where acoustic privacy is an issue, a heavier ceiling material, such as plaster, gypsum board, or a dense type of fibrous panel, should be used. Composite ceiling panels with a highly absorbent material laminated to a dense substrate are manufactured to meet both noise reduction and sound transmission criteria simultaneously; these have high values for both NRC and CAC. The same result can be achieved by mounting acoustically absorbent tiles on a suspended ceiling of plaster or gypsum board.

The most economical acoustical ceiling systems consist of *lay-in panels* that are supported by an exposed grid (Figures 24.4, 24.5). Any panel in the ceiling can be lifted and removed for access to services in the plenum space. For a smoother appearance, a concealed grid system may be used instead. Concealed grid systems require special panels for plenum access.

Acoustical ceilings are often less costly than plaster or even gypsum board ceilings, and are available in hundreds of different designs, many of which are rated for fire resistance. Figures 24.4 and 24.5 show some basic details for acoustical ceilings,

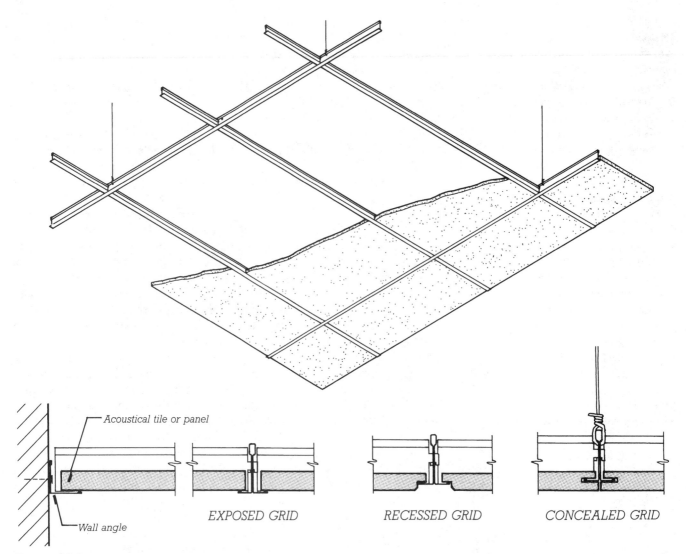

— Acoustical tile or panel

Wall angle

EXPOSED GRID   *RECESSED GRID*   *CONCEALED GRID*

FIGURE 24.4
Acoustical ceilings are supported on suspended grids of tees formed from sheet
metal. At the top of the figure is a cutaway view looking up at an acoustical ceiling of
lay-in panels. Below are sections illustrating how the grid may be exposed, recessed,
or concealed, for different visual appearances.

FIGURE 24.5
The grid for an acoustical ceiling is assembled
with a simple interlocking joint.

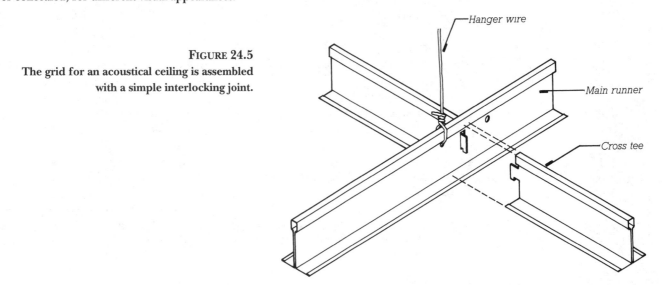

—Hanger wire

—Main runner

—Cross tee

and Figures 24.6 through 24.11 some typical examples. Figure 24.12 illustrates a suspended ceiling made of linear elements that are formed from sheet aluminum, attached to a special type of concealed grid.

Where a suspended ceiling is used as membrane fireproofing for the structure above, or where it is part of a fire-resistive assembly, penetrations of the ceiling must be detailed so as to maintain the required degree of fire resistance. Lighting fixtures must be backed up with fire-resistive material, air conditioning grills must be isolated from the ducts that feed them by means of automatic fire dampers, and any access panels pro-

FIGURE 24.6
Many acoustical ceilings are manufactured as *integrated ceiling systems* that incorporate the lighting fixtures and air conditioning outlets into the module of the grid. In this integrated ceiling, viewed from above, the hanger wires, grid, acoustical panels, fluorescent lighting fixture, and distribution boot for conditioned air have been installed. The boot will be connected to the main ductwork with a flexible oval duct. (*Courtesy of Armstrong World Industries*)

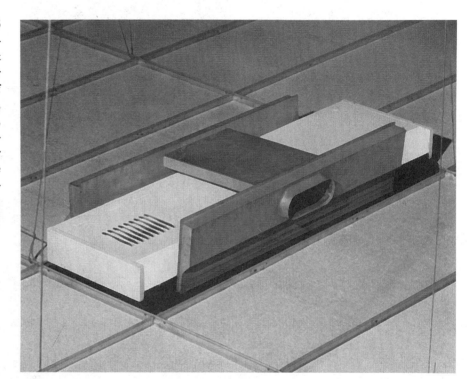

FIGURE 24.7
As viewed from below, the integrated ceiling shown in Figure 24.6 has a slot around the lighting fixture through which air is distributed from the boot above. The roughly textured acoustical panels used in this example are patterned with two cross grooves that work with the recessed grid to create the look of a ceiling composed of smaller, square panels. (*Courtesy of Armstrong World Industries*)

FIGURE 24.8
Air distribution in this integrated acoustical ceiling is through slots that occur between panels. Lighting fixtures, loudspeakers, and smoke detectors are incorporated simply and unobtrusively. (*Courtesy of United States Gypsum Company*)

FIGURE 24.9
Coffered acoustical ceilings that act as light diffusers are designed and marketed as integrated systems. (*Courtesy of Armstrong World Industries*)

FIGURES 24.10, 24.11
A variety of patterns and textures are available in acoustical ceilings. (*Courtesy of Armstrong World Industries*)

Economy has worked so great a change in our dwellings, that their ceilings are, of late years, little more than miserable naked surfaces of plaster. [A discussion of ceiling design will] possess little interest in the eye of speculating builders of the wretched houses erected about the suburbs of the metropolis, and let to unsuspecting tenants at rents usually about three times their actual value. To the student it is more important, inasmuch as a well-designed ceiling is one of the most pleasing features of a room.

Joseph Gwilt,
The Encyclopedia of Architecture,
London, 1842

FIGURE 24.12
A mirror-finish linear metal suspended ceiling, with an exposed steel space truss structure beyond. (*Architect: DeWinter and Associates. Photo courtesy of Alcan Building Products, Division of Alcan Aluminum Corporation*)

vided for maintenance of above-ceiling services must meet code requirements for fire resistance.

Interstitial Ceilings

Many hospital and laboratory buildings have extremely elaborate mechanical and electrical systems, including not just the usual air conditioning ducts, water and waste piping, and electrical and communications wiring, but also such services as fume hood ducting, fuel gas lines, compressed air lines, oxygen piping, chilled water piping, vacuum piping, and chemical waste piping. These ducts and tubes occupy a considerable volume of space in the building, often in an amount that virtually equals the inhabited volume. Furthermore, all these systems require continual maintenance and are subject to frequent change. As a consequence, many such buildings are designed with *interstitial ceilings*. An interstitial ceiling is suspended at a level that allows workers to travel freely in the plenum space, usually while walking erect, and is structured strongly enough to support safely the weight of the workers and their tools. In effect, the plenum space becomes another floor of the building, slipped in between the other floors, and the overall height of the building must be increased accordingly. Its advantage is that maintenance and updating work on the mechanical and electrical systems of the building can be carried on without interrupting the activities below. Interstitial ceilings are made of gypsum or lightweight concrete and combine the construction details of poured gypsum roof decks and suspended plaster ceilings. Figure 24.13 shows the installation of an interstitial ceiling.

FIGURE 24.13
The final steps in constructing an interstitial ceiling: The ceiling plane consists of gypsum reinforced with hexagonal steel mesh. It is framed with steel truss tee subpurlins, which are visible at the right of the picture, supported by steel wide-flange beams suspended on rods from the sitecast concrete framing above. The final layer of gypsum is being pumped onto the ceiling from the hose near the center of the picture. The wet gypsum is struck off level with the straightedge seen here hanging on the beams, and troweled to a smooth walking surface. When the gypsum has hardened, installation of the ductwork, piping, and wiring in the interstitial plenum space can begin, with workers using the gypsum ceiling as a walking surface. (*Courtesy of Keystone Steel and Wire Company*)

FINISH FLOORING

Functions of Finish Flooring

Floors have a lot to do with one's visual and tactile appreciation of a building. People sense their colors, patterns, and textures, their "feel" underfoot, and the noises they make in response to footsteps. Floors affect the acoustics of a room, contributing to a noisy quality or a hushed quality, depending on whether a hard or soft flooring material is used. Floors also interact in various ways with light: Some floor materials give mirrorlike reflections, some give diffuse reflections or none at all; dark flooring materials absorb most of the light incident upon them and contribute to the creation of a darker room, while light materials reflect most incident light and help create a brighter room.

Floors are also a major functional component of a building. They are its primary wearing surfaces, subject to water, grit, dust, and the abrasive and penetrating actions of feet and furniture. They require more cleaning and maintenance effort than any other component of a building. They must be designed to deal with problems of skid resistance, sanitation, noise reduction between floors of a building, even electrical conductivity in occupancies such as computer rooms and hospital operating rooms where static electricity would pose a threat. And like other interior finish components, floors must be selected with an eye to combustibility, fire resistance ratings, and the structural loads that they will place on the frame of the building.

The skid resistance of a flooring material is measured by its *static coefficient of friction* (*SCOF*). An SCOF of 0.50 or more is desirable to minimize accidents caused by slipping. Floor finish materials are classified for flame resistance by ASTM test procedure E648, and are grouped into Class I and Class II finishes.

Underfloor Services

Floor structures are frequently used for the distribution of electrical and communications wiring and fiber optic cables, especially in floor areas that are broad and have few fixed partitions. If the needs for electrical and telephone service are minimal and

FIGURE 24.14
A worker levels a cellular raceway over steel dome formwork for a concrete waffle slab floor structure. The concrete will be poured so that its surface is level with the tops of the access boxes, completely embedding the raceways and allowing each box to be opened merely by removing its metal cover. Electrical and communications outlets can be installed in any access box. The covers of unused access boxes will be concealed beneath vinyl composition floor tiles or carpeting. (*Courtesy of American Electric—Construction Materials Group*)

predictable, the most economical horizontal distribution system for wiring consists of conventional conduits of metal tubing that are embedded in the floor slab or concrete topping. In most commercial buildings, however, greater flexibility is required to accommodate wiring changes that will need to occur from time to time during the life of the building. There are several alternative systems for creating this flexibility. In buildings with concrete structural systems, *cellular raceways* may be cast into the floor slabs (Figure 24.14). These are sheet metal ducts that can carry many wires. Working through access boxes that rise from the top of the raceway to the surface of the slab, electricians can add or remove wiring at any time. Electrical and communications outlets can be installed in any of the access boxes. In steel-framed buildings, cellular steel decking provides the same functional advantages as cellular raceways (Figures 24.15–24.17). *Poke-through* electrical fittings (Figure 24.18) allow wiring flexibility over time without the need for raceways or cellular decking. Poke-through systems, however, require the electrician to work from the floor below the one on

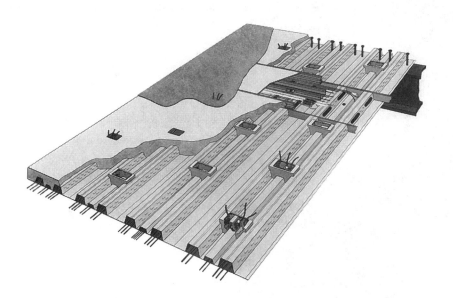

FIGURE 24.15
Cellular steel decking used for underfloor electrical and communications wiring. A transverse feeder trench, near the top of the picture, brings the wiring across the floor from the electrical risers to the cells in the deck. Boxes cast into the concrete topping give access to the cells for the installation of electrical outlets. Notice the shear studs on the steel beam at the top of the picture. (*Courtesy of H. H. Robertson Company*)

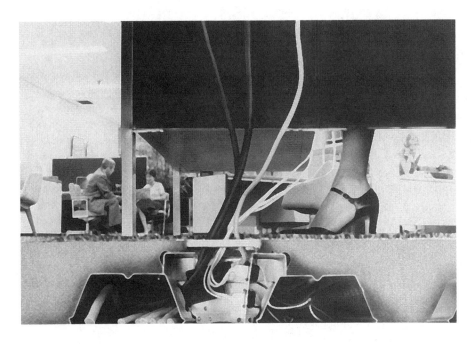

FIGURE 24.16
A sectional view of cellular steel decking, showing telephone and other low-voltage communications wiring in the cell to the left, electrical power wiring in the cell to the right, and access to these services through the box in the middle. (*Courtesy of H. H. Robertson Company*)

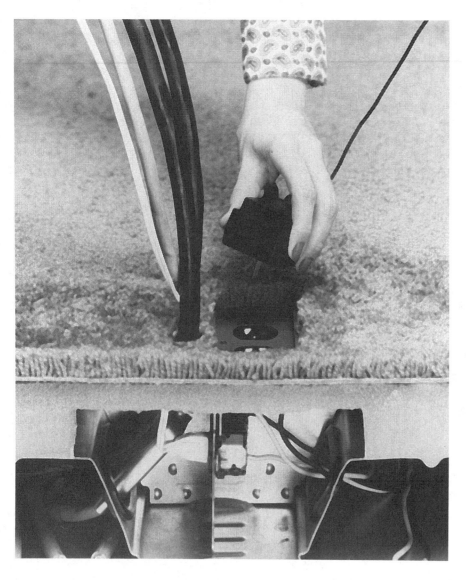

which changes are being made, which can be an inconvenience for the tenants of the lower floor.

Raised access flooring is advantageous in buildings where wiring changes are frequent and unpredictable, as in computer rooms and offices with a large number of electronic machines (Figures 24.19, 24.20). Raised access flooring has a virtually unlimited capability to meet future wiring needs, and changes in wiring are extremely easy to make. If the access floor is raised high enough, ductwork for air distribution can be run on top of the structural floor, possibly eliminating the need

FIGURE 24.17
Plugging in to a floor-mounted electric receptacle served by wires from the cells of the floor decking. (*Courtesy of H. H. Robertson Company*)

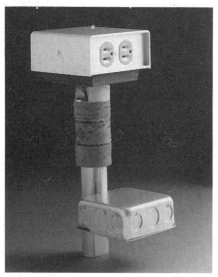

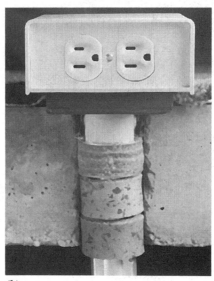

(a)   *(b)*

FIGURE 24.18
(*a*) A poke-through fitting is designed to be installed in a hole drilled through a concrete floor slab. The junction box at the bottom connects to a wiring conduit above the suspended ceiling of the floor below. Wires pass from the junction box through the vertical tube to the outlet above. (*b*) Fire-retardant gaskets seal the hole through the slab so that fire cannot pass, thus maintaining the fire resistance rating of the floor structure. If a poke-through fitting is removed in a later renovation, a special fire-resistant abandonment plug is installed to seal the hole. (*Courtesy of American Electric— Construction Materials Group*)

for a suspended ceiling below. This can be useful in buildings where the structural system is meant to be left exposed as a finish ceiling, or in older buildings with beautiful plaster or wood ceilings. Raised access flooring also works well in older buildings because its pedestal heights can be adjusted to compensate for uneven floor surfaces. Several systems for providing individualized control of air conditioning in large office buildings utilize the space below a raised access floor as a large distribution chamber for conditioned air; each workstation is provided with a small outlet diffuser in the floor surface.

FIGURE 24.19
Raised access flooring provides unlimited capacity for wiring, fiber optic cables, piping, and ductwork. The space below the flooring can serve as a plenum for air distribution. Changes in any of the underfloor systems are easily made, and wiring outlets can be installed at any point in the floor. (*Courtesy of Tate Architectural Products, Inc.*)

FIGURE 24.20
Conditioned air is supplied to this computer room through the space below the raised access flooring and is fed upward through perforated floor panels. Air is returned through slots in the suspended ceiling. (*Courtesy of Armstrong World Industries*)

Undercarpet wiring systems that use flat conductors rather than conventional round wires are appropriate in many buildings for both electrical power and communications wiring (Figures 24.21, 24.22). These are applicable to both new and retrofit projects.

Reducing Noise Transmission Through Floors

In multistory buildings, it is sometimes necessary to take precautions to reduce the amount of *impact noise* transmitted through a floor to the room below. This is particularly true of hotels and apartment buildings where people are sleeping in rooms below the rooms of others who may be awake and moving about. Impact noise is generated by footsteps or machinery and is transmitted as structureborne vibration through the material of the floor to become airborne noise in the room below.

There are several strategies for dealing with impact noise; these may be used individually or in various combinations. One is to use padded carpeting or cushioned resilient flooring to reduce the amount of impact noise that is generated. A second is to underlay the flooring material with a resilient material that is not highly conductive of impact noise. Cellulose fiber panels and nonwoven plastic filament matting are two materials marketed for this purpose. A third mechanism is to make an airtight ceiling below of a heavy, dense material such as plaster or gypsum board, and to mount this ceiling on resilient clips or on hanger wires with springs. The springs or clips absorb most of the sound energy that would otherwise travel through the structure.

Many floor–ceiling assemblies have been tested for sound transmission and rated for both *Sound Transmission Class* (*STC*), which is concerned with transmission of airborne sound, and *Impact Insulation Class* (*IIC*). These ratings will be found in trade literature concerned with various types of floor construction, and offer a ready comparison of acoustical performance.

Hard Flooring Materials

Hard finish flooring materials (concrete, stone, brick, tile, and terrazzo) are often chosen for their resistance to wear and moisture. Being rigid and

FIGURE 24.21
The flat conductors for an undercarpet wiring system lie unseen beneath the carpet, and are accessed through projecting boxes. (*Courtesy of Burndy Corporation*)

FIGURE 24.22
The flat conductors are ribbons of copper laminated between insulating layers of plastic sheet. These conductors are connected as necessary with the splicing tool shown in this photograph, and covered with a grounded metallic shield before being taped to the floor. (*Courtesy of Burndy Corporation*)

unyielding, they are not comfortable to stand upon for extended periods of time, and they contribute to a live, noisy acoustic environment. But many of these materials are so beautiful in their colors and patterns, and so durable, that they are considered among the most desirable types of flooring by designers and building owners alike.

Concrete

With a wood float finish, concrete makes an excellent finish floor for parking garages and many types of agricultural and industrial buildings. With a steel trowel finish, concrete finds its way into a vast assortment of commercial and institutional buildings, and even into homes and offices. Color can be added with a colorant admixture or a couple of coats of floor paint. Concrete's chief advantages as a finish flooring material are its low initial cost and its durability. On the minus side, extremely good workmanship is required to make an acceptable floor finish, and even the best of concrete surfaces is likely to sustain some damage and staining during construction.

Stone

Many types of building stones are used as flooring materials, in surface textures ranging from mirror-polished marble and granite to split-face slate and sandstone (Figures 24.23, 24.24). Installation is a relatively simple but highly skilled procedure of bedding the stone in mortar and filling the joints with grout

FIGURE 24.23
A slate floor in an automobile showroom. (*Photo by Bill Engdahl, Hedrich-Blessing. Courtesy of Buckingham-Virginia Slate Company*)

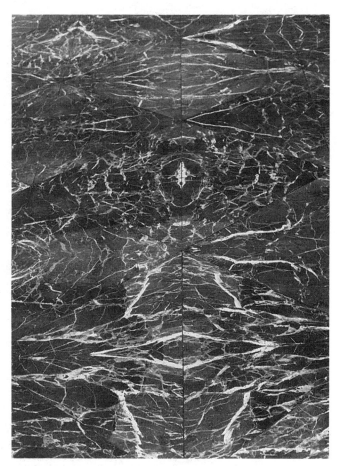

FIGURE 24.24
Flooring of matched, polished triangles of white-veined red marble gives a kaleidoscopic effect. (*Architect: The Architects Collaborative. Photograph by the author*)

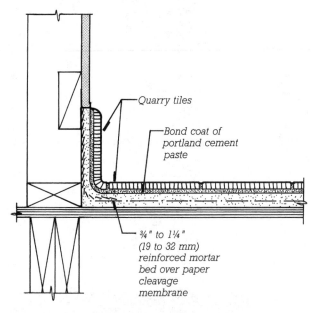

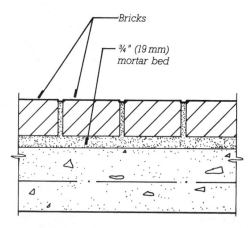

QUARRY TILE ON WOOD
SUBFLOOR

BRICKS ON CONCRETE SLAB

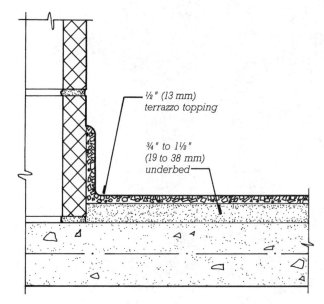

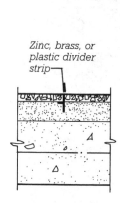

BONDED TERRAZZO

MONOLITHIC TERRAZZO

FIGURE 24.25
**Typical details of tile, brick, stone, and
terrazzo flooring. Traditional sand cushion terrazzo flooring is not illustrated.
The terrazzo systems shown here are
thinner and lighter than sand cushion terrazzo, but perform as well on a properly
engineered floor structure. Thin-set terrazzo, the lightest of the systems, uses
very small stone chips in a mortar that is
strengthened with epoxy, polyester, or
polyacrylate.**

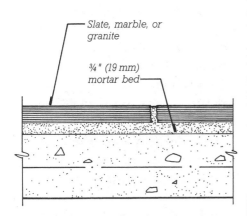

Slate, marble, or granite

¾ " (19 mm) mortar bed

STONE ON CONCRETE SLAB

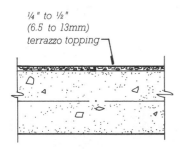

¼ " to ½ " (6.5 to 13mm) terrazzo topping

THIN SET TERRAZZO

(Figure 24.25). Most stone floorings are coated with multiple applications of a clear sealer coating, and are waxed periodically throughout the life of the building to bring out the color and figure of the stone.

Bricks and Brick Pavers

Both bricks and half-thickness bricks called *pavers* are used for finish flooring, with pavers often preferred because they add less thickness and dead weight to the floor (Figure 24.26). Bricks are usually laid with their largest surface horizontal, but they may also be laid on edge. As with stone and tile flooring, decorative joint patterns can be designed especially for each installation.

Quarry Tiles

Quarry tiles are not, as might be guessed from their name, made of stone. They are simply large, fired clay tiles, usually square but sometimes rectangular, hexagonal, octago-

FIGURE 24.26
A floor of glazed brick pavers meets a planting bed constructed of brick masonry. (*Courtesy of Stark Ceramics, Inc.*)

The brick floors, because the bricks may be made of diverse forms and of diverse colors by reason of the diversity of the chalks, will be very agreeable and beautiful to the eye. . . . The ceilings are also diversely made, because many take delight to have them of beautiful and well-wrought beams . . . these beams ought to be distant one from another one thickness and a half of the beam, because the ceilings appear thus very beautiful to the eye.

Andrea Palladio, *The Four Books of Architecture*, 1570

FIGURE 24.27
Unglazed quarry tiles used as flooring and as a facing for columns and railings.
(*Architect: Skidmore, Owings & Merrill. Interior designer: Duffy, Inc. Courtesy of American Olean Tile Company*)

nal, or other shapes (Figures 24.27, 24.28). Sizes range from about 4 inches (100 mm) to 12 inches (300 mm) square, with thicknesses ranging from $\frac{3}{8}$ inch (9 mm) to a full inch (25 mm) for some handmade tiles. Quarry tiles are available in a myriad of earth colors, as well as various kiln-applied colorations. They are usually set in a reinforced mortar bed, although in residential work they are sometimes applied directly to a wood panel subfloor with epoxy or organic adhesives. It is important that any subfloor to which tiles are glued should be exceedingly stiff; otherwise, flexing of the subfloor under changing loads will pop the tiles loose. Additional subfloor thickness and/or a stiff underlayment are advisable. An excellent underlayment material for tile is a portland cement–based backer board that is marketed by several companies.

Ceramic Tiles

Fired clay tiles that are smaller than quarry tiles are referred to collectively as *ceramic tiles*. Ceramic tiles are usually glazed. The most common shape is square, but rectangles, hexagons, circles, and more elaborate shapes are also available (Figure

FIGURE 24.28
Square and rectangular quarry tiles in contrasting colors create a pattern on the floor of a retail mall. (*Architect: Edward J. DeBartolo Corp. Courtesy of American Olean Tile Company*)

24.29). Sizes range from ½ inch (13 mm) to 4 inches (100 mm) and more. The smaller sizes of tiles are shipped from the factory adhered to large backing sheets of plastic mesh or perforated paper; the tilesetter is thus able to lay a hundred or more tiles together in a single operation, rather than as individual units.

Grout color has a strong influence on the appearance of tile surfaces, as it does for brick and stone. Many different premixed colors are available, or the tilesetter may color a grout with pigments. The use of ceramic tiles on interior wall surfaces is mentioned in the previous chapter, and typical details of ceramic tile installation on floors and walls are shown in Figure 23.41.

Terrazzo

Terrazzo is an exceptionally durable flooring. It is made by grinding and polishing a concrete that consists of marble or granite chips selected for size and color, in a matrix of colored portland cement or other binding agent. The polishing brings out the pattern and color of the stone chips, and a sealer is usually applied to further enhance the appearance of the floor (Figure 24.30). Terrazzo may be formed in place, or may be installed as factory-made tiles. For stair treads, window sills, and other large components, terrazzo is often precast.

The endless variety of colors and textures of terrazzo leads to its use in decorative flooring patterns, where the colors are separated from one another by *divider strips* of metal, plastic, or marble. The divider strips are installed in the underbed prior to placing the terrazzo, and are ground and polished flush in the same operation as the terrazzo itself.

Traditionally, terrazzo is installed over a thin bed of sand that isolates it from the structural floor slab, thus protecting it to some extent from movements in the building frame. This *sand cushion terrazzo* is heavy, however, due to its thickness, usually 2½ inches (64 mm). For greater economy and reduced thickness, the sand bed may be eliminated to produce *bonded terrazzo*, or both the sand bed and underbed may be eliminated with *monolithic terrazzo* (Figure 24.25). In any of these systems, a terrazzo baseboard can be formed and finished as an integral part of the floor, thus eliminating a dirt-catching seam where the floor meets the wall.

Wood Flooring

Wood is used in several different forms as a finish flooring material, the most common of which is *wood strip flooring*, typically made of white oak, red oak, pecan, or maple (Figures 24.31, 24.32). The strips are held tightly together and *blind nailed* by driving nails diagonally through the upper interior corners of the tongues, where they are entirely concealed from view. The entire floor is then sanded smooth, stained if desired, and finished with two coats of a varnish or other clear coating. When its

FIGURE 24.29
Ceramic tile wainscoting and flooring in a bar. (*Architect: Daughn/Salisbury, Inc. Designer: Morris Nathanson Design. Courtesy of American Olean Tile Company*)

FIGURE 24.30
A terrazzo floor in a residential entry uses divider strips and contrasting colors to create a custom floor pattern.
(*Courtesy of National Terrazzo and Mosaic Association, Inc.*)

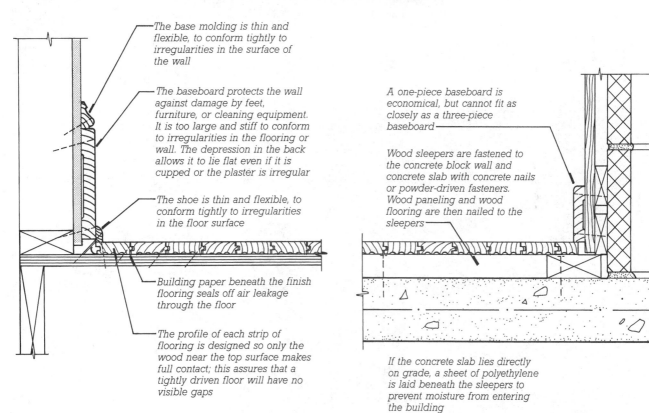

The base molding is thin and flexible, to conform tightly to irregularities in the surface of the wall

The baseboard protects the wall against damage by feet, furniture, or cleaning equipment. It is too large and stiff to conform to irregularities in the flooring or wall. The depression in the back allows it to lie flat even if it is cupped or the plaster is irregular

The shoe is thin and flexible, to conform tightly to irregularities in the floor surface

Building paper beneath the finish flooring seals off air leakage through the floor

The profile of each strip of flooring is designed so only the wood near the top surface makes full contact; this assures that a tightly driven floor will have no visible gaps

A one-piece baseboard is economical, but cannot fit as closely as a three-piece baseboard

Wood sleepers are fastened to the concrete block wall and concrete slab with concrete nails or powder-driven fasteners. Wood paneling and wood flooring are then nailed to the sleepers

If the concrete slab lies directly on grade, a sheet of polyethylene is laid beneath the sleepers to prevent moisture from entering the building

FIGURE 24.31
Details of hardwood strip flooring installation. At left, the flooring is applied to a wood joist floor, and at right, to wood sleepers over a concrete slab. The blind nailing of the flooring is shown only for the first several strips of flooring at the left. The baseboard makes a neat junction between the floor and the wall, covering up the rough edges of the wall material and the flooring. The three-piece baseboard shown at the left does this job somewhat better than the one-piece baseboard, but it is more expensive and elaborate.

surface becomes worn, the flooring can be restored to a new appearance by sanding and refinishing.

For greater economy, wood flooring is available in square-edged strips about $\frac{5}{16}$ inch (8 mm) thick that are face nailed to the subfloor. The nail holes are filled before the floor is sanded and finished. The initial appearance is essentially the same as for the thicker flooring, but the thinner floor cannot be sanded and refinished as many times, so it has a shorter lifetime. Another less costly form of wood flooring consists of factory-made, prefinished wood planks and parquet tiles. These are furnished in many different woods and patterns, and are usually fastened to the subfloor with a mastic adhesive. Most are too thin to be able to withstand subsequent sanding and refinishing.

Exceptionally long-wearing industrial wood floors are made of small blocks of wood set in adhesive with their grain oriented vertically. While this type of floor is relatively high in first cost, it is economical for heavily used floors and is sometimes chosen for use in public spaces because of the beauty of its pattern and grain.

Resilient Flooring

The oldest *resilient flooring* material is *linoleum,* a sheet material made of ground cork in a linseed oil binder over a burlap backing. Asphalt tiles were later developed as an alternative to linoleum, but most of today's resilient sheet floorings and tiles are made of vinyl (polyvinyl chloride), often in combination with mineral reinforcing fibers. Thicknesses of these *vinyl composition* floorings are on the order of $\frac{1}{8}$ inch (3 mm), slightly thinner for lighter-duty floorings, and slightly thicker if a cushioned back is added to the product. Most resilient flooring materials are glued to the concrete or wood of the structural floor (Figures 24.33–24.35). The primary advantages of resilient floorings are the wide range of available colors and patterns, a moderately high degree of durability, and low initial cost. Vinyl composition tile has the lowest installed cost of any flooring material except concrete and is used in vast quantities on the floors of residences, offices, classrooms, and retail space. Resilient tiles are usually 12 inches (305 mm) square. Sheet flooring is furnished in rolls 6 to 12 feet (1.83 to 3.66 m) wide. If skillfully made, the seams between strips of sheet flooring are virtually invisible.

Resilient floorings of rubber, cork, and other materials are also available. Each offers particular advantages of durability or appearance.

Most resilient flooring materials are so thin that they show even the slightest irregularities in the floor deck beneath. Concrete surfaces to which resilient materials will be applied must first be scraped clean of construction debris and spatters. Wood panel decks are covered with a layer of smooth *underlayment panels,* usually of hardboard, particleboard, or sanded plywood, to prepare them for resilient flooring materials. Joints between underlayment panels are offset from joints in the subfloor to eliminate soft spots. The thickness of the underlayment is chosen to make the surface of the resilient flooring level with the surfaces of flooring materials such as hardwood and ceramic tile that are used in surrounding areas of the building.

Carpet

Carpet is manufactured in fibers, styles, and patterns to meet almost any flooring requirement, indoors or out, except for rooms that need thor-

FIGURE 24.32
Oak strip flooring in a hair salon. Notice the use of exposed ducts and lighting track at the ceiling. (*Architects: Michael Rubin and Henry Smith-Miller in association with Kenneth Cohen. Courtesy of Oak Flooring Institute*)

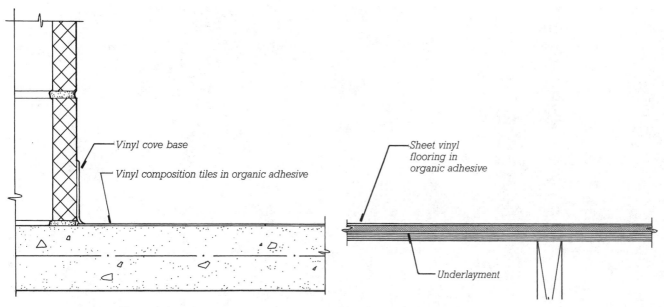

FIGURE 24.33
Typical installation details for resilient flooring. At left, vinyl composition tiles applied directly to a steel-trowel-finish concrete slab, with a vinyl cove base adhered to a concrete masonry partition. To the right, sheet vinyl flooring on underlayment and a wood joist floor structure.

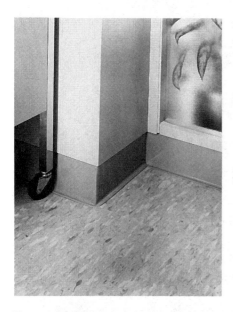

FIGURE 24.34
Vinyl composition tile and a vinyl cove base. Notice how the base is simply folded around the outside corner of the partition. (*Courtesy of Armstrong World Industries*)

FIGURE 24.35
Sheet vinyl flooring can be flash-coved to create an integral base that is easily cleaned, for use in health care facilities, kitchens, and bathrooms. The seams are welded to eliminate dirt-catching cracks. (*Courtesy of Armstrong World Industries*)

FIGURE 24.36
Wall-to-wall carpeting in a commercial installation. (*Courtesy of Armstrong World Industries*)

ough sanitation, such as hospital rooms, food processing facilities, and toilet rooms. Some carpets are tough enough to wear for years in public corridors (Figure 24.36), yet others are soft enough for intimate residential interiors. Costs of carpeting are often competitive with those of other flooring materials of similar quality, whether measured on an installed-cost or life-cycle-cost basis.

Carpets are either glued directly to the floor deck or stretched over a carpet pad and attached around the perimeter of the room by means of a *tackless strip,* a continuous length of wood, fastened to the floor, that has protruding spikes along the top to catch the backing of the carpet and hold it taut.

If carpet is laid directly over a wood panel subfloor such as plywood, the panel joints perpendicular to the floor joists should be blocked beneath to prevent movement between sheets. Tongue-and-groove plywood subflooring accomplishes the same result without blocking. Alternatively, a layer of underlayment panels may be nailed over the subfloor with its joints offset from those in the subfloor.

FLOORING THICKNESS

Thicknesses of floor finishes vary from the $\frac{1}{8}$ inch (3 mm) or less of resilient flooring to 3 inches (76 mm) or more for brick flooring. Frequently, several different types of flooring are used on different areas of the same floor. If the differences in the thicknesses of the flooring materials are not great, they can be resolved by using tapered edgings or thresholds at changes of material. Otherwise, the level of the top of the floor deck must be adjusted from one part of the building to the next, to

C.S.I./C.S.C. Masterformat Section Numbers for Ceilings and Floor Finishes	
09100	METAL SUPPORT SYSTEMS
09120	Ceiling Suspension Systems
09130	Acoustical Suspension Systems
09200	LATH AND PLASTER
09250	GYPSUM BOARD
09300	TILE
09310	Ceramic Tile
09330	Quarry Tile
09400	TERRAZZO
09500	ACOUSTICAL TREATMENT
09510	Acoustical Ceilings
09545	Special Ceiling Surfaces
09550	WOOD FLOORING
09560	Wood Strip Flooring
09565	Wood Block Flooring
09570	Wood Parquet Flooring
09600	STONE FLOORING
09610	Flagstone Flooring
09615	Marble Flooring
09620	Granite Flooring
09625	Slate Flooring
09630	UNIT MASONRY FLOORING
09635	Brick Flooring
09650	RESILIENT FLOORING
09660	Resilient Tile Flooring
09665	Resilient Sheet Flooring
09680	CARPET

bring the finish floor surfaces to the same elevation. The architect should work out the necessary level changes in advance and indicate them clearly on the construction drawings. In many cases, special structural details must be drawn to indicate how the level changes should be made. In wood framing, they can usually be accomplished either by notching the ends of the floor joists to lower the subfloor in parts of the building with thicker floor materials or by adding sheets of underlayment material of the proper thickness to areas of thinner flooring. In steel and concrete buildings, slab or topping thicknesses can change, or whole areas of structure can be raised or lowered by the necessary amount.

SELECTED REFERENCES

Because so many ceiling and flooring materials and systems are proprietary, much of the best information on ceiling and floor finishes is to be found in manufacturers' literature. Certain generic products are, however, well documented in trade association literature:

1. Ceilings and Interior Systems Construction Association. *Ceiling Systems Handbook*. Elmhurst, Illinois, updated frequently.

Aimed at construction workers, this manual covers everything pertaining to the installation of ceilings. (Address for ordering: 579 West North Avenue, Suite 301, Elmhurst, IL 60126.)

2. Tile Council of America, Inc. *Handbook for Ceramic Tile Installation*. Princeton, New Jersey, updated frequently.

Crammed into this booklet are details and specifications for ceramic tile floors and walls of every conceivable type, as applied to various kinds of underlying structures. (Address for ordering: P.O. Box 326, Princeton, NJ 08540.)

3. National Oak Flooring Manufacturers Association. *Hardwood Flooring Installation Manual* and *Hardwood Floors, Walls, and Ceilings*. Memphis, Tennessee, updated frequently.

Oak flooring grades and installation procedures are covered by these two pamphlets. (Address for ordering: 804 Sterick Building, Memphis, TN 38103.)

4. The National Terrazzo and Mosaic Association, Inc. *Terrazzo Design/Technical Data Book*. Des Plaines, Illinois, updated frequently.

Terrazzo details, specifications, colors, and patterns are treated thoroughly in this looseleaf binder of information. (Address for ordering: 3166 Des Plaines Avenue, Suite 15, Des Plaines, IL 60018.)

KEY TERMS AND CONCEPTS

plenum
membrane fire protection
suspended ceiling
hanger wire
runner channel
furring channel
acoustical ceiling
noise reduction coefficient (NRC)
ceiling attenuation class (CAC)
lay-in panels
linear metal ceiling
exposed grid
recessed grid

concealed grid
interstitial ceiling
static coefficient of friction (SCOF)
cellular raceway
poke-through
raised access flooring
undercarpet wiring system
impact noise
Sound Transmission Class (STC)
Impact Insulation Class (IIC)
brick paver
quarry tile
ceramic tile

terrazzo
divider strips
sand cushion terrazzo
bonded terrazzo
monolithic terrazzo
wood strip flooring
blind nailing
resilient flooring
linoleum
vinyl composition
underlayment panels
carpet
tackless strip

REVIEW QUESTIONS

1. List the potential range of functions of a finish ceiling, and of a finish floor.

2. What are the advantages and disadvantages of a suspended ceiling as compared to a tightly attached ceiling?

3. When designing a building with its structure and mechanical equipment left exposed at the ceiling, what sorts of precautions should you take to assure a satisfactory appearance?

4. What does underlayment do?

5. List several different approaches to the problem of running electrical and communications wiring beneath a floor of a building framed with steel or concrete.

EXERCISES

1. Visit some rooms that you happen to like in nearby buildings: classrooms, auditoriums, theaters, restaurants, bars, museums, shopping centers—both new buildings and old. For each room, list the ceiling and floor finishes used. Why was each material chosen? Sketch details of critical junctions between materials. What does each room sound like—noisy, hushed, "live"? How does this acoustical quality relate to the floor and ceiling materials? What is the quality of the illumination in the room, and what role do ceiling and floor materials play in creating this quality?

2. Look in current architectural magazines for interior photographs of buildings that appeal to you. What ceiling and floor materials are used in each? Why?

Densities and Coefficients of Thermal Expansion of Common Building Materials

Material		Density		Coefficient of Thermal Expansion	
		lb/ft³	kg/m³	in./in./°F	mm/mm/°C
Wood (seasoned)					
Douglas fir	parallel to grain	32	510	0.0000021	0.0000038
	perpendicular to grain			0.000032	0.000058
Pine	parallel to grain	26	415	0.0000030	0.0000054
	perpendicular to grain			0.000019	0.000034
Oak, red or white	parallel to grain	41–46	655–735	0.0000027	0.0000049
	perpendicular to grain			0.000030	0.000054
Masonry					
Limestone		160	2560	0.0000044	0.0000079
Granite		165	2640	0.0000047	0.0000085
Marble		165	2640	0.0000073	0.0000131
Brick (average)		100–140	1600–2240	0.0000036	0.0000065
Concrete masonry units		100–140	1600–2240	0.0000052	0.0000094
Concrete					
Normal weight concrete		145	2320	0.0000055	0.0000099
Metals					
Steel		490	7850	0.0000065	0.0000117
Stainless steel, 18–8		490	7850	0.0000099	0.0000178
Aluminum		165	2640	0.0000128	0.0000231
Copper		556	8900	0.0000093	0.0000168
Finish Materials					
Gypsum board		43–50	690–800	0.000009	0.0000162
Gypsum plaster, sand		105	1680	0.000007	0.0000126
Glass		156	2500	0.0000050	0.0000090
Acrylic glazing sheet		72	1150	0.0000410	0.0000742
Polycarbonate glazing sheet		75	1200	0.0000440	0.0000796
Polyethylene		57–61	910–970	0.000085	0.000153
Polyvinyl chloride		75–106	1200–1700	0.000040	0.000072

APPENDIX

English/Metric Conversions

English	Metric	Metric	English
1 in.	25.4 mm	1 mm	0.0394 in.
1 ft	304.8 mm	1 m	39.37 in.
1 ft	0.3048 m	1 m	3.2808 ft
1 lb	0.454 kg	1 kg	2.205 lb
1 ft^2	0.0929 m^2	1 m^2	10.76 ft^2
1 psi	6.89 kPa	1 kPa	0.145 psi
1 psi	0.00689 MPa	1 MPa	145.14 psi
1 lb/ft^2	4.884 kg/m^2	1 kg/m^2	0.205 lb/ft^2
1 lb/ft^3	16.019 kg/m^3	1 kg/m^3	0.0624 lb/ft^3
1 ft/min	0.0051 m/s	1 m/s	196.85 ft/min
1 cfm	0.0005 m^3/s	1 m^3/s	2119 cfm
1 BTU	1.055 kJ	1 kJ	0.9479 BTU
1 BTU	3.9683 kcal	1 kcal	0.252 BTU
1 BTUH	0.2928 W	1 W	3.412 BTUH

$$1 \text{ Pa} = 0.102 \text{ kg/m}^2$$
$$1 \text{ kg/m}^2 = 9.80 \text{ Pa}$$

Note: Units are converted in the text to a degree of precision consistent with the precision of the number being converted.

GLOSSARY

A

Abutment joint A surface divider joint designed to allow free movement between new and existing construction, or between differing materials.

ACC *See* **autoclaved cellular concrete.**

Access flooring A raised finish floor surface consisting entirely of small, individually removable panels beneath which wiring, ductwork, and other services may be installed.

Acoustical ceiling A ceiling of fibrous tiles that are highly absorbent of sound energy.

Acrylic A transparent plastic material widely used in sheet form for glazing windows and skylights.

ADA *See* **airtight drywall approach.**

Admixture A substance other than cement, water, and aggregates included in a concrete mixture for the purpose of altering one or more properties of the concrete.

Aggregate Inert particles, such as sand, gravel, crushed stone, or expanded minerals, in a concrete or plaster mixture.

Air-entraining cement A portland cement with an admixture that causes a controlled quantity of stable, microscopic air bubbles to form in the concrete during mixing.

Airtight drywall approach (ADA) Restricting the passage of water vapor into the insulated cavities of a light frame building by eliminating passages around and through the gypsum board interior finish.

Air-to-air heat exchanger A device that exhausts air from a building while recovering much of the heat from the exhausted air and transferring it to the incoming air.

AISC American Institute of Steel Construction.

Alloy A substance composed of two or more metals, or of a metal and a nonmetallic constituent.

Anchorage The device that fastens the end of a posttensioning tendon to the end of a concrete slab.

Anchor bolt A bolt embedded in concrete for the purpose of fastening a building frame to a concrete or masonry foundation.

Angle A structural section of steel or aluminum whose profile resembles the letter L.

Annealed Cooled under controlled conditions to minimize internal stresses.

Anodizing An electrolytic process that forms a permanent, protective oxide coating on aluminum, with or without added color.

ANSI American National Standards Institute, an organization that fosters the establishment of voluntary industrial standards.

Anticlastic Saddle-shaped, or having curvature in two opposing directions.

Apron The finish piece that covers the joint between a window stool and the wall finish below.

Arch A structural device that supports a vertical load by translating it into axial, inclined forces at its supports.

Architectural sheet metal roofing A roof covering made up of sheets of metal in a traditional shop-fabricated pattern such as standing seam, flat seam, or batten seam.

Arc welding A process of joining two pieces of metal by melting them together at their interface with a continuous electric spark and adding a controlled additional amount of molten metal from a metallic electrode.

Area divider A curb used to partition a large roof membrane into smaller areas, to allow for expansion and contraction in the deck and membrane.

Ash dump A door in the underfire of a fireplace that permits ashes from the fire to be swept into a chamber beneath, from which they may be removed at a later time.

Ashlar Squared stonework.

Asphalt A tarry brown or black mixture of hydrocarbons.

Asphalt roll roofing A continuous sheet of the same roofing material used in asphalt shingles. *See* **asphalt shingle.**

Asphalt-saturated felt A moisture-resistant sheet material, available in several different thicknesses, usually consisting of a heavy paper that has been impregnated with asphalt.

Asphalt shingle A roofing unit composed of a heavy organic or inorganic felt saturated with asphalt and faced with mineral granules.

ASTM American Society for Testing and Materials, an organization that promulgates standard methods of testing the performance of building materials and components.

Atactic polypropylene An amorphous form of polypropylene used as a modifier in modified-bitumen roofing.

Auger A helical tool for creating cylindrical holes.

Autoclaved cellular concrete (ACC) Concrete formulated so as to contain a large percentage of gas bubbles as a result of a chemical reaction that takes place in an atmosphere of steam.

Awning window A window that pivots on an axis at or near the top edge of the sash and projects toward the outdoors.

Axial In a direction parallel to the long axis of a structural member.

B

Backer rod A flexible, compressible strip of plastic foam inserted into a joint to limit the depth to which sealant can penetrate.

Backfill Earth or earthen material used to fill the excavation around a foundation; the act of filling around a foundation.

Backup, Backup wall A vertical plane of masonry, concrete, or framing used to support a thin facing such as a single wythe of brickwork.

Backup bar A small rectangular strip of steel applied beneath a joint to provide a solid base for beginning a weld between two steel structural members.

Ballast A heavy material installed over a roof membrane to prevent wind uplift and shield the membrane from sunlight.

Balloon frame A wooden building frame composed of closely spaced members nominally 2 inches (51 mm) in thickness, in which the wall members are single pieces that run from the sill to the top plates at the eave.

Baluster A small, vertical member that serves to fill the opening between a handrail and a stair or floor.

Band joist A wooden joist running perpendicular to the primary direction of the joists in a floor and closing off the floor platform at the outside face of the building.

Bar A small rolled steel shape, usually round or rectangular in cross section; a rolled steel shape used for reinforcing concrete.

Barrel shell A scalloped roof structure of reinforced concrete that spans in one direction as a barrel vault and in the other as a folded plate.

Barrel vault A segment of a cylindrical surface that spans as an arch.

Barrier wall An exterior wall of a building whose watertightness depends on its freedom from passages through the wall.

Baseboard A strip of finish material placed at the junction of a floor and a wall to create a neat intersection and to protect the wall against damage from feet, furniture, and floor-cleaning equipment.

Baseplate A steel plate inserted between a column and a foundation to spread the concentrated load of the column across a larger area of the foundation.

Batten A strip of wood or metal used to cover the crack between two adjoining boards or panels.

Batten seam A seam in a sheet metal roof that encloses a wood batten.

Bay A rectangular area of a building defined by four adjacent columns; a portion of a building that projects from a facade.

Bead A narrow line of weld metal or sealant; a strip of metal or wood used to hold a sheet of glass in place; a narrow, convex molding profile; a metal edge or corner accessory for plaster.

Beam A straight structural member, usually horizontal, that acts primarily to resist nonaxial loads.

Bearing A point at which one building element rests upon another.

Bearing block A piece of wood fastened to a column to provide support for a beam or girder.

Bearing pad A block of plastic or synthetic rubber used to cushion the point at which one precast concrete element rests upon another.

Bearing wall A wall that supports floors or roofs.

Bed *See* **casting bed.**

Bed joint The horizontal layer of mortar beneath a masonry unit.

Bedrock A solid stratum of rock.

Bending moment The moment that causes a beam or other structural member to bend.

Bending stress A compressive or tensile stress resulting from the application of a nonaxial force to a structural member.

Bent A plane of framing consisting of beams and columns joined together, often with rigid joints.

Bentonite clay An absorptive, colloidal clay that swells to several times its dry volume when saturated with water.

Bevel An end or edge that is cut at an angle other than a right angle.

Bevel siding Wood cladding boards that taper in cross section.

Billet A large cylinder or rectangular solid of material.

Bite The depth to which the edge of a piece of glass is held by its frame.

Bitumen A tarry mixture of hydrocarbons, such as asphalt or coal tar.

Blast furnace slag A byproduct of iron manufacture used as a concrete admixture.

Blind nailing Attaching boards to a frame, sheathing, or subflooring with toe nails driven through the edge of each piece so as to be completely concealed by the adjoining piece.

Blocking Pieces of wood inserted tightly between joists, studs, or rafters in a building frame to stabilize the structure, inhibit the passage of fire, provide a nailing surface for finish materials, or retain insulation.

Bloom A rectangular solid of steel formed from an ingot as an intermediate step in creating rolled steel structural shapes.

Blooming mill A set of rollers used to transform an ingot into a bloom.

Bluestone A sandstone that is gray to blue-gray in color and splits readily into thin slabs.

Board foot A unit of lumber volume, a rectangular solid nominally 12 square inches in cross-sectional area and 1 foot long.

Board siding Wood cladding made up of boards, as differentiated from shingles or manufactured wood panels.

BOCA Building Officials and Code Administrators International, Inc., an organization that publishes a model building code.

Bolster A long chair used to support reinforcing bars in a concrete slab.

Bolt A fastener consisting of a cylindrical metal body with a head at one end and a helical thread at the other, intended to be inserted through holes in adjoining pieces of material and closed with a threaded nut.

Bond In masonry, the adhesive force between mortar and masonry units, or the pattern in which masonry units are laid to tie two or more wythes together into a structural unit. In reinforced concrete, the adhesion between the surface of a reinforcing bar and the surrounding concrete.

Bond breaker A strip of material to which sealant does not adhere.

Bonded posttensioning A system of prestressing in which the tendons are grouted after stressing so as to bond them to the surrounding concrete.

Bonded terrazzo Terrazzo flooring whose underbed is poured directly upon the structural floor.

Bottom bars The reinforcing bars that lie close to the bottom of a beam or slab.

Box beam A bending member of metal or plywood whose cross section resembles a closed rectangular box.

Box girder A major spanning member of concrete or steel whose cross section is a hollow rectangle or trapezoid.

Bracing Diagonal members, either temporary or permanent, installed to stabilize a structure against lateral loads.

Brad A small finish nail.

Bridging Bracing or blocking installed between steel or wood joists at midspan to stabilize them against buckling and, in some cases, to permit adjacent joists to share loads.

British thermal unit (BTU) The quantity of heat required to raise 1 pound of water 1 degree Fahrenheit.

Broom finish A skid-resistant texture imparted to an uncured concrete surface by dragging a stiff-bristled broom across it.

Brown coat The second coat of plaster in a three-coat application.

Brownstone A brownish or reddish sandstone.

BTU *See* **British thermal unit.**

Buckling Structure failure by gross lateral deflection of a slender element under compressive stress, such as the sideward buckling of a long, slender column or the upper edge of a long, thin floor joist.

Building code A set of legal restrictions intended to assure a minimum standard of health and safety in buildings.

Building separation joint A plane along which a building is divided into separate structures that may move independently of one another.

Built-up roof (BUR) A roof membrane laminated from layers of asphalt-saturated felt or other fabric, bonded together with bitumen or pitch.

Buoyant uplift The force of water or liquefied soil that tends to raise a building foundation out of the ground.

BUR *See* **built-up roof.**

Butt The thicker end, such as the lower edge of a wood shingle or the lower end of a tree trunk; a joint between square-edged pieces; a weld between square-edged pieces of metal that lie in the same plane; a type of door hinge that attaches to the edge of the door.

Butt-joint glazing A type of glass installation in which the vertical joints between lights of glass do not meet at a mullion, but are made weathertight with a sealant.

Button head A smooth, convex bolt head with no provision for engaging a wrench.

Buttress A structural device of masonry or concrete that resists the diagonal forces from an arch or vault.

Butyl rubber A synthetic rubber compound.

C

CAC *See* **ceiling attenuation class.**

Caisson A cylindrical sitecast concrete foundation unit that penetrates through unsatisfactory soil to rest upon an underlying stratum of rock or satisfactory soil; an enclosure that permits excavation work to be carried out underwater.

Calcining The driving off of the water of hydration from gypsum by the application of heat.

Camber A slight, intentional initial curvature in a beam or slab.

Cambium The thin layer beneath the bark of a tree that manufactures cells of wood and bark.

Cantilever A beam, truss, or slab that extends beyond its last point of support.

Cant strip A strip of material with a sloping face used to ease the transition from a horizontal to a vertical surface at the edge of a membrane roof.

Capillary action The pulling of water through a small orifice or fibrous material by the adhesive force between the water and the material.

Capillary break A slot or groove intended to create an opening too large to be bridged by a drop of water, and thereby to prevent the passage of water by capillary action.

Carbide-tipped tools Drill bits, saws, and other tools with cutting edges made of an extremely hard alloy.

Carbon steel Low-carbon or mild steel.

Carpenter One who makes things of wood.

Casement window A window that pivots on an axis at or near a vertical edge of the sash.

Casing The wood finish pieces surrounding the frame of a window or door; a cylindrical steel tube used to line a drilled or driven hole in foundation work.

Castellated beam A steel wide-flange section whose web has been cut along a zigzag path and reassembled by welding in such a way as to create a deeper section.

Casting Pouring a liquid material or slurry into a mold whose form it will take as it solidifies.

Casting bed A permanent, fixed form in which precast concrete elements are produced.

Cast-in-place Concrete that is poured in its final location; sitecast.

Cast iron Iron with too high a carbon content to be classified as steel.

Caulk A low-range sealant.

Cavity wall A masonry wall that includes a continuous airspace between its outermost wythe and the remainder of the wall.

Cee A metal framing member whose cross-sectional shape resembles the letter C.

Ceiling attenuation class (CAC) An index of the ability of a ceiling construction to obstruct the passage of sound between adjacent rooms.

Cellular decking Panels made of steel sheets corrugated and welded together in such a way that hollow longitudinal cells are created within the panels.

Cellular raceway A rectangular tube cast into a concrete floor slab for the purpose of housing electrical and communications wiring.

Cellulose A complex polymeric carbohydrate of which the structural fibers in wood are composed.

Celsius A temperature scale on which the freezing point of water is established as 0 and the boiling point as 100 degrees.

Cement A substance used to adhere material together; in concrete work, the dry powder that, when it has combined chemically with the water in the mix, cements the particles of aggregate together to form concrete.

Cementitious Having cementing properties, usually with reference to inorganic substances, such as portland cement and lime.

Centering Temporary formwork for an arch, dome, or vault.

Centering shims Small blocks of synthetic rubber or plastic used to hold a sheet of glass in the center of its frame.

Ceramic tile Small, flat, thin clay tiles intended for use as wall and floor facings.

Chair A device used to support reinforcing bars.

Chamfer A flattening of a longitudinal edge of a solid member on a plane that lies at an angle of 45° to the adjoining planes.

Channel A steel or aluminum section shaped like a rectangular box with one side missing.

Chlorinated polyethylene A plastic material used in roof membranes.

Chlorosulfonated polyethylene A plastic material used in roof membranes.

Chord A top or bottom member of a truss.

C–H stud A steel wall framing member whose profile resembles a combination of the letters C and H, used to support gypsum panels in shaft walls.

Chuck A device for holding a steel wire, rod, or cable securely in place by means of steel wedges in a tapering cylinder.

Churn drill A steel tool used with an up-and-down motion to cut through rock at the bottom of a steel pipe caisson.

Cladding A material used as the exterior wall enclosure of a building.

Clamp A tool for holding two pieces of material together temporarily; unfired bricks piled in such a way that they can be fired without using a kiln.

Class A, B, C roofing Roof covering materials classified according to their resistance to fire when tested in accordance with ASTM E108. Class A is the highest, and Class C is the lowest.

Cleanout hole An opening at the base of a masonry wall through which mortar droppings and other debris can be removed prior to grouting the interior cavity of the wall.

Clear dimension, clear opening The dimension between opposing inside faces of an opening.

Climbing crane A heavy-duty lifting machine that raises itself as the building rises.

Clinker A fused mass that is an intermediate product of cement manufacture; a brick that is overburned.

Closer The last masonry unit laid in a course; a partial masonry unit used at the corner of a header course to adjust the joint spacing; a mechanical device for regulating the closing action of a door.

CLSM *See* **controlled low-strength material.**

CMU *See* **concrete masonry unit.**

Code *See* **building code.**

Cohesive soil A soil such as clay whose particles are able to adhere to one another by means of cohesive and adhesive forces.

Cold-formed steel construction Steel framing composed of members that were created by folding sheet steel at room temperature.

Cold-rolled steel Steel rolled to its final form at a temperature at which it is no longer plastic.

Cold-worked steel Steel formed at a temperature at which it is no longer plastic, as by rolling or forging at room temperature.

Collar joint The vertical mortar joint between wythes of masonry.

Collar tie A piece of wood nailed across two opposing rafters near the ridge to resist wind uplift.

Column An upright structural member acting primarily in compression.

Column cage An assembly of vertical reinforcing bars and ties for a concrete column.

Column-cover-and-spandrel system A system of cladding in which panels of material cover the columns and spandrels, with horizontal strips of windows filling the remaining portion of the wall.

Column spiral A continuous coil of steel reinforcing used to tie a concrete column.

Column tie A single loop of steel bar, usually bent into a rectangular configuration, used to prevent spreading of the vertical bars in a concrete column.

Combination door A door with interchangeable inserts of glass and insect screening, usually used as a second, exterior door and mounted in the same opening with a conventional door.

Combination window A sash that holds both insect screening and a retractable sheet of glass, mounted in the same frame with a window and used to increase its thermal resistance.

Common bolt An ordinary carbon steel bolt.

Common bond Brickwork laid with each five courses of stretchers followed by one course of headers.

Composite A material or assembly made up of two or more materials bonded together to act as a single structural unit.

Composite column An upright structural member, acting primarily in compression, that is composed of concrete and a steel structural shape, usually a wide-flange or a tube.

Composite construction Any element in which concrete and steel, other than reinforcing bars, work as a single structural unit.

Composite metal decking Corrugated steel decking manufactured in such a way that it bonds securely to the concrete floor fill to form a reinforced concrete deck.

Composite wall A masonry wall that incorporates two or more different types of masonry units, such as clay bricks and concrete blocks.

Compression A squeezing force.

Compression gasket A synthetic rubber strip that seals around a sheet of glass or a wall panel by being squeezed tightly against it.

Compressive strength The ability of a structural material to withstand squeezing forces.

Concave joint A mortar joint tooled into a curved, indented profile.

Concealed grid A suspended ceiling framework that is completely hidden by the tiles or panels it supports.

Concrete A structural material produced by mixing predetermined amounts of portland cement, aggregates, and water, and allowing this mixture to cure under controlled conditions.

Concrete block A concrete masonry unit, usually hollow, that is larger than a brick.

Concrete masonry unit (CMU) A block of hardened concrete, with or without hollow cores, designed to be laid in the same manner as a brick or stone; a concrete block.

Condensate Water formed as a result of condensation.

Condensation The process of changing from a gaseous to a liquid state, especially as applied to water.

Conduit A steel or plastic tube through which electrical wiring is run.

Continuous ridge vent A screened, water-shielded ventilation opening that runs continuously along the ridge of a gable roof.

Contractor A person or organization that undertakes a legal obligation to do construction work.

Control joint An intentional, linear discontinuity in a structure or component, designed to form a plane of weakness where cracking can occur in response to various forces so as to minimize or eliminate cracking elsewhere in the structure.

Controlled low-strength material (CLSM) A concrete that is purposely formulated to have a very low but known strength, used primarily as a backfill material.

Convector A heat exchange device that uses the heat in steam, hot water, or an electric resistance element to warm the air in a room; often called, inaccurately, a *radiator.*

Cope The removal of a portion of the flange at the end of a steel beam in order to facilitate connection to another member.

Coped connection A joint in which the end of one member is cut to match the profile of the other member.

Coping A protective cap on the top of a masonry wall.

Coping saw A handsaw with a thin, very narrow blade, used for cutting detailed shapes in the ends of wood moldings and trim.

Copolymer A large molecule composed of repeating patterns of two or more chemical units.

Corbel A spanning device in which masonry units in successive courses are cantilevered slightly over one another; a projecting bracket of masonry or concrete.

Coreboard A thick gypsum panel used primarily in shaft walls.

Corner bead A metal or plastic strip used to form a neat, durable edge at an outside corner of two walls of plaster or gypsum board.

Cornice The exterior detail at the meeting of a wall and a roof overhang; a decorative molding at the intersection of a wall and a ceiling.

Corrosion Oxidation, such as rust.

Corrosion inhibitor A concrete admixture intended to prevent oxidation of reinforcing bars.

Corrugated Pressed into a fluted or ribbed profile.

Counterflashing A flashing turned down from above to overlap another flashing turned up from below so as to shed water.

Course A horizontal layer of masonry units one unit high; a horizontal line of shingles or siding.

Coursed Laid in courses with straight bed joints.

Cove base A flexible strip of plastic or synthetic rubber used to finish the junction between resilient flooring and a wall.

Crawlspace A space that is not tall enough to stand in, located beneath the lowest floor of a building.

Creep A permanent inelastic deformation in a material due to changes in the material caused by the prolonged application of a structural stress.

Cripple stud A wood wall-framing member that is shorter than full-length studs because it is interrupted by a header or sill.

Cross-grain wood Wood incorporated into a structure in such a way that its direction of grain is perpendicular to the direction of the principal loads on the structure.

Crosslot bracing Horizontal compression members running from one side of an excavation to the other, used to support sheeting.

Crown glass Glass sheet formed by spinning an opened hollow globe of heated glass.

Cruck A framing member cut from a bent tree so as to form one-half of a rigid frame.

Cup A curl in the cross section of a board or timber caused by unequal shrinkage or expansion between one side of the board and the other.

Curing The hardening of concrete, plaster, gunnable sealant, or other wet materials. Curing can occur through evaporation of water or a solvent, hydration, polymerization, or chemical reactions of various other types, depending on the formulation of the material.

Curing compound A liquid that, when sprayed on the surface of newly placed concrete, forms a water-resistant layer to prevent premature dehydration of the concrete.

Curtain wall An exterior building wall that is supported entirely by the frame of the building, rather than being self-supporting or loadbearing.

Cylinder glass Glass sheet produced by blowing a large, elongated glass cylinder, cutting off its ends, slitting it lengthwise, and opening it into a flat rectangle.

D

Damper A flap to control or obstruct the flow of gases; specifically, a metal control flap in the throat of a fireplace or in an air duct.

Dampproofing A coating applied to the outside face of a basement wall as a barrier to the passage of water.

Dap A notch at the end of a piece of material.

Darby A stiff straightedge of wood or metal used to level the surface of wet plaster.

Daylighting Illuminating the interior of a building by natural means.

Dead load The weight of the building or building component itself.

Deadman A large and/or heavy object buried in the ground as an anchor.

Decking A material used to span across beams or joists to create a floor or roof surface.

Deformation A change in the shape of a structure or structural element caused by a load or force acting on the structure.

Depressed strand A pretensioning tendon that is pulled to the bottom of the beam at the center of the span to follow more closely the path of tensile forces in the member.

Derrick Any of a number of devices for hoisting building materials on the end of a rope or cable.

Dew point The temperature at which water will begin to condense from a mass of air of a given temperature and moisture content.

Diagonal bracing *See* **bracing.**

Diamond saw A tool with a moving chain, belt, wire, straight blade, or circular blade whose cutting action is carried out by diamonds.

Diaphragm action A bracing action that derives from the stiffness of a thin plane of material when it is loaded in a direction parallel to the plane. Diaphragms in buildings are typically floor, wall, or roof surfaces of wood panels, reinforced masonry, steel decking, or reinforced concrete.

Die An industrial tool for giving identical form to repeatedly produced or continuously generated units, such as a shaped orifice for giving form to a column of clay, a steel wire, or an aluminum extrusion; a shaped punch for making cutouts of sheet metal or paper; a mold for casting plastic or metal; a cutting tool that produces screw threads.

Die-cut Manufactured by punching from a sheet material.

Differential settlement Subsidence of the various foundation elements of a building at differing rates.

Diffuser A louver shaped so as to distribute air about a room.

Dimension lumber Lengths of wood, rectangular in cross section, sawed directly from the log.

Dimension stone Building stone cut to a rectangular shape.

Distribution rib A transverse beam at the midspan of a one-way concrete joist structure used to allow the joists to share concentrated loads.

Divider strip A strip of metal or plastic embedded in terrazzo to form control joints and decorative patterns.

Dome An arch rotated about its vertical axis to produce a structure shaped like an inverted bowl; a form used to make one of the cavities in a concrete waffle slab.

Dormer A structure protruding through the plane of a sloping roof, usually containing a window and having its own smaller roof.

Double glazing Two parallel sheets of glass with an airspace between.

Double-hung window A window with two overlapping sashes that slide vertically in tracks.

Double shear Acting to resist shear forces at two locations, such as a bolt that passes through a steel supporting angle, a beam web, and another supporting angle.

Double-strength glass Glass that is approximately $\frac{1}{8}$ inch (3 mm) in thickness.

Double tee A precast concrete slab element that resembles the letters TT in cross section.

Dovetail slot anchor A system for fastening to a concrete structure that uses metal tabs inserted into a slot that is small at the face of the concrete and larger behind.

Dowel A short cylindrical rod of wood or steel; a steel reinforcing bar that projects from a foundation to tie it to a column or wall, or from one section of a concrete slab or wall to another.

Downspout A vertical pipe for conducting water from a roof to a lower level.

Drainage Removal of water.

Draped tendon A posttensioning strand that is placed along a curving profile that approximates the path of the tensile forces in a beam.

Drawing Shaping a material by pulling it through an orifice, as in the drawing of steel wire or the drawing of a sheet of glass.

Drawn glass Glass sheet pulled directly from a container of molten glass.

Drift Lateral deflection of a building caused by wind or earthquake loads.

Drift pin A tapered steel rod used to align bolt holes in steel connections during erection.

Drip A discontinuity formed into the underside of a window sill, soffit, or wall component to force adhering drops of water to fall free of the face of the building rather than to move farther toward the interior.

Drop panel A thickening of a two-way concrete structure at the head of a column.

Dry pack A low-slump grout tamped into the space in a connection between precast concrete members.

Dry press process A method of molding slightly damp clays and shales into bricks by forcing them into molds under high pressure.

"Dry" systems Systems of construction that use little or no water during construction, as differentiated from systems such as masonry, plastering, and ceramic tile work.

Drywall *See* **gypsum board.**

Dry well An underground pit filled with broken stone or other porous material, from which rainwater from a roof drainage system can seep into the surrounding soil.

Duct A hollow conduit, commonly of sheet metal, through which air can be circulated; a tube used to establish a passage for a posttensioning tendon in a concrete structure.

DWV Drain–waste–vent pipes, the part of the plumbing system of a building that removes liquid wastes and conducts them to the sewer or sewage disposal system.

E

Eave The horizontal edge at the low side of a sloping roof.

Edge bead A strip of metal or plastic used to make a neat, durable edge where plaster or gypsum board abuts another material.

Efflorescence A powdery deposit on the face of a surface of masonry or concrete, caused by the leaching of chemical salts by water migrating from within the structure to the surface.

EIFS *See* **exterior insulation and finish system.**

Elastic Able to return to its original size and shape after removal of stress.

Elastomer A rubber or synthetic rubber.

Elastomeric Rubberlike.

Electrode A consumable steel wire or rod used to maintain an arc and furnish additional weld metal in electric arc welding.

Elevation A drawing that views a building from any of its sides; a vertical height above a reference point such as sea level.

Elongation Stretching under load; growing longer because of temperature expansion.

Enamel A glossy or semigloss paint.

End dam The turned-up end of a flashing that prevents water from running out of the end; a block inserted into the space within a horizontal aluminum mullion for the same purpose.

End nail A nail driven through the side of one piece of lumber and into the end of another.

Engineered fill Earth compacted into place in such a way that it has predictable physical properties, based on laboratory tests and specified, supervised installation procedures.

English bond Brickwork laid with alternating courses each consisting entirely of headers and stretchers.

EPDM Ethylene propylene diene monomer, a synthetic rubber material used in roofing membranes.

Erector The subcontractor who raises, connects, and plumbs up a building frame from fabricated steel or precast concrete components.

Expanded metal lath A thin sheet of metal that has been slit and stretched to transform it into a mesh used as a base for the application of plaster.

Expansion joint A surface divider joint that provides space for the surface to expand. In common usage, a building separation joint.

Exposed aggregate finish A concrete surface in which the coarse aggregate is revealed.

Exposed grid A framework for an acoustical ceiling that is visible from below after the ceiling is completed.

Extended life admixture A substance that retards the onset of the curing reaction in mortar so that the mortar may be used over a protracted period of time after mixing.

Exterior insulation and finish system (EIFS) A cladding system that consists of a thin layer of reinforced stucco applied directly to the surface of an insulating plastic foam board.

Extrados The convex surface of an arch.

Extrusion The process of squeezing a material through a shaped orifice to produce a linear element with the desired cross section; an element produced by this process.

F

Fabricator The company that prepares structural steel members for erection.

Facade An exterior face of a building.

Face brick A brick selected on the basis of appearance and durability for use in the exposed surface of a wall.

Face nail A nail driven through the side of one wood member into the side of another.

Face shell The portion of a hollow concrete masonry unit that forms the face of the wall.

Fahrenheit A temperature scale on which the boiling point of water is fixed at 212 degrees and the freezing point at 32.

Fanlight A semicircular or semielliptical window above an entrance door, often with radiating muntins that resemble a fan.

Fascia The exposed vertical face of an eave.

Felt A thin, flexible sheet material made of soft fibers pressed and bonded together. In building practice, a thick paper or a sheet of glass or plastic fibers.

Fibrous admixture Short fibers of glass, steel, or polypropylene mixed into concrete to act as reinforcement against plastic shrinkage cracking.

Fieldstone Rough building stone gathered from river beds and fields.

Figure The surface pattern of the grain of a piece of smoothly finished wood or stone.

Fillet A rounded inside intersection between two surfaces that meet at an angle.

Fillet weld A weld at the inside intersection of two metal surfaces that meet at right angles.

Finger joint A glued end connection between two pieces of wood, using an interlocking pattern of deeply cut "fingers." A finger joint creates a large surface for the glue bond, allowing it to develop the full tensile strength of the wood that it connects.

Finial An ornament at the top of a roof or spire.

Finish Exposed to view; material that is exposed to view.

Finish carpenter One who does finish carpentry.

Finish carpentry The wood components exposed to view on the interior of a building, such as window and door casings, baseboards, bookshelves, and the like.

Finish coat The final coat of plaster.

Finish floor The floor material exposed to view, as differentiated from the *subfloor*, which is the loadbearing floor surface beneath.

Finish lime A fine grade of hydrated lime used in finish coats of plaster and in ornamental plaster work.

Firebrick A brick made from special clays to withstand very high temperatures, as in a fireplace, furnace, or industrial chimney.

Firecut A sloping end cut on a wood beam or joist where it enters a masonry wall. The purpose of the firecut is to allow the wood member to rotate out of the wall without prying the wall apart, if the floor or roof structure should burn through in a fire.

Fireproofing Material used around a steel structural element to insulate it against excessive temperatures in case of fire.

Fire-rated glass Glass that is capable of retaining its integrity in an opening after being exposed to fire.

Fire resistance rating The time, in hours or fractions of an hour, that a material or assembly will resist fire exposure as determined by ASTM E119.

Fire resistant Noncombustible; slow to be damaged by fire; forming a barrier to the passage of fire.

Fire separation wall A wall required under a building code to divide two parts of a building from one another as a deterrent to the spread of fire.

Firestop A wood or masonry baffle used to close an opening between studs or joists in a balloon or platform frame in order to retard the spread of fire through the opening.

Firestopping A component or mastic installed in an opening through a floor or around the edge of a floor to retard the passage of fire.

Fire wall A wall extending from foundation to roof, required under a building code to separate two parts of a building from one another as a deterrent to the spread of fire.

Fire zone A legally designated area of a city in which construction must meet established standards of fire resistance and/or combustibility.

Firing The process of converting dry clay or shale into a ceramic material through the application of intense heat.

First cost The cost of construction.

Fixed window Glass that is immovably mounted in a wall.

Flagstone Flat stones used for paving or flooring.

Flame spread rating A measure of the rapidity with which fire will spread across the surface of a material as determined by ASTM E84.

Flange A projecting crosspiece of a wide-flange or channel profile; a projecting fin.

Flash cove A detail in which a sheet of resilient flooring is turned up at the edge and finished against the wall to create an integral baseboard.

Flashing A thin, continuous sheet of metal, plastic, rubber, or waterproof paper used to prevent the passage of water through a joint in a wall, roof, or chimney.

Flat seam A sheet metal roofing seam that is formed flat against the surface of the roof.

Flemish bond Brickwork laid with each course consisting of alternating headers and stretchers.

Flitch-sliced veneer A thin sheet of wood cut by passing a block of wood vertically against a long, sharp knife.

Float A small platform suspended on ropes from a steel building frame to permit ironworkers to work on a connection; a trowel with a slightly rough surface used in an intermediate stage of finishing a concrete slab; as a verb, to use a float for finishing concrete.

Float glass Glass sheet manufactured by cooling a layer of liquid glass on a bath of molten tin.

Flocculated A "fluffy" microstructure of clay particles in which the platelets are randomly oriented.

Flue A passage for smoke and combustion products from a furnace, stove, water heater, or fireplace.

Fluid-applied roof membrane A roof membrane applied in one or more coats of a liquid that cure to form an impervious sheet.

Fluoropolymer A highly stable organic compound used as a finish coating for building cladding.

Flush Smooth, lying in a single plane.

Flush door A door with smooth, planar faces.

Flush glazing *See* **structural silicone flush glazing.**

Flux A material added to react chemically with impurities and remove them from molten metal. Fluxes are used both in steel making and in welding. Welding fluxes serve the additional purpose of shielding the molten weld metal from the air to reduce oxidation and other undesirable effects.

Fly ash A waste product of coal-fired power plants, used as a concrete admixture.

Flying formwork Large sections of slab formwork that are moved by crane.

Fly rafter A rafter in a rake overhang.

Foil-backed gypsum board Gypsum board with aluminum foil laminated to its back surface to act as a vapor retarder and thermal insulator.

Folded plate A roof structure whose strength and stiffness derive from a pleated or folded geometry.

Footing The widened part of a foundation that spreads a load from the building across a broader area of soil.

Form deck Thin, corrugated steel sheets that serve as permanent formwork for a reinforced concrete deck.

Form-release compound A substance applied to concrete formwork to prevent concrete from adhering.

Form tie A steel or plastic rod with fasteners on each end, used to hold together the two surfaces of formwork for a concrete wall.

Formwork Temporary structures of wood, steel, or plastic that serve to give shape to poured concrete, and to support it and keep it moist as it cures.

Foundation The portion of a building that has the sole purpose of transmitting structural loads from the building into the earth.

Framed connection A shear connection between steel members made by means of steel angles or plates connecting to the web of the beam or girder.

Framing plan A diagram showing the arrangement and sizes of the structural members in a floor or roof.

Freestone Fine-grained sedimentary rock that has no planes of cleavage or sedimentation along which it is likely to split.

French door A symmetrical pair of glazed doors hinged to the jambs of a single frame and meeting at the center of the opening.

Frictional soil A soil such as sand that has little or no attraction between its particles, and derives its strength from geometric interlocking of the particles; a noncohesive soil.

Friction connection Two or more structural steel members clamped together by high-strength bolts with sufficient force that the loads on the members are transmitted between them by friction along their mating surfaces.

Frit Ground-up colored glass that is heat-fused to lights of glass to form functional or decorative patterns.

Frost line The depth in the earth to which the soil can be expected to freeze during a severe winter.

Fuel contributed rating A measure of the extent to which a building material will provide additional energy to a fire.

Furring channel A formed sheet metal furring strip.

Furring strip A length of wood or metal attached to a masonry or concrete wall to permit the attachment of finish materials to the wall using screws or nails.

G

Gable The triangular wall beneath the end of a gable roof.

Gable roof A roof consisting of two oppositely sloping planes that intersect at a level ridge.

Gable vent A screened, louvered opening in a gable, used for exhausting excess heat and humidity from an attic.

Galling Chafing of one material against another under extreme pressure.

Galvanizing The application of a zinc coating to steel as a means of preventing corrosion.

Gambrel A roof shape consisting of two superimposed levels of gable roofs with the lower level at a steeper pitch than the upper.

Gasket A dry, resilient material used to seal a joint between two rigid assemblies by being compressed between them.

Gauge A measure of the thickness of sheet material.

Gauged brick A brick that has been rubbed on an abrasive stone to reduce it to a trapezoidal shape for use in an arch.

Gauging plaster A gypsum plaster formulated for use in combination with finish lime in finish coat plaster.

General contractor A contractor who has responsibility for the overall conduct of a construction project.

Geotextile A synthetic cloth used beneath the surface of the ground to stabilize soil or promote drainage.

GFRC *See* **glass-fiber-reinforced concrete.**

Girder A horizontal beam that supports other beams; a very large beam, especially one that is built up from smaller elements.

Girt A horizontal beam that supports wall cladding between columns.

Glass block A hollow masonry unit made of glass.

Glass fiber batt A thick, fluffy, nonwoven insulating blanket of filaments spun from glass.

Glass-fiber-reinforced concrete (GFRC) Concrete with a strengthening admixture of short alkali-resistant glass fibers.

Glass mullion system A method of constructing a large glazed area by stiffening the sheets of glass with perpendicular glass ribs.

Glaze A glassy finish on a brick or tile; as a verb, to install glass.

Glazed structural clay facing tile A hollow clay block with glazed faces, usually used for constructing interior partitions.

Glazier One who installs glass.

Glazier's points Small pieces of metal driven into a wood sash to hold glass in place.

Glazing The act of installing glass; installed glass; as an adjective, referring to materials used in installing glass.

Glazing compound Any of a number of types of mastic used to bed small lights of glass in a frame.

Glazing luminous efficacy The visible transmittance of glass divided by the shading coefficient.

Glue-laminated timber A timber made up of a large number of small strips of wood glued together.

Glulam A short expression for glue-laminated timber.

Grade A classification of size or quality for an intended purpose; to classify as to size or quality.

Grade The surface of the ground; to move earth for the purpose of bringing the surface of the ground to an intended level or profile.

Grade beam A reinforced concrete beam that transmits the load from a bearing wall into spaced foundations such as pile caps or caissons.

Grain In wood, the direction of the longitudinal axes of the wood fibers, or the figure formed by the fibers.

Granite Igneous rock with visible crystals of quartz and feldspar.

Groove weld A weld made in a groove created by beveling or milling the edges of the mating pieces of metal.

Ground A strip attached to a wall or ceiling to establish the level to which plaster should be applied.

Grout A high-slump mixture of portland cement, aggregates, and water, which can be poured or pumped into cavities in concrete or masonry for the purpose of embedding reinforcing bars and/or increasing the amount of load-bearing material in a wall; a specially formulated, mortarlike material for filling under steel baseplates and around connections in precast concrete framing; a mortar used to fill joints between ceramic tiles or quarry tiles.

Gunnable sealant A sealant material that is extruded in liquid or mastic form from a sealant gun.

Gusset plate A flat steel plate used to connect the members of a truss; a stiffener plate.

Gutter A channel to collect rainwater and snowmelt at the eave of a roof.

Gypsum Hydrous calcium sulfate.

Gypsum backing board A lower-cost gypsum panel intended for use as an interior layer in multilayer constructions of gypsum board.

Gypsum board An interior facing panel consisting of a gypsum core sandwiched between paper faces. Also called *drywall, plasterboard.*

Gypsum lath Small sheets of gypsum board manufactured specifically for use as a plaster base.

Gypsum plaster Plaster whose cementing substance is gypsum.

Gypsum sheathing panel A water-resistant, gypsum-based sheet material used for exterior sheathing.

Gypsum wallboard *See* **gypsum board.**

H

Hammerhead boom crane A heavy-duty lifting device that uses a tower-mounted horizontal boom that may rotate only in a horizontal plane.

Hardboard A very dense panel product, usually with at least one smooth face, made of highly compressed wood fibers.

Hawk A square piece of sheet metal with a perpendicular handle beneath, used by a plasterer to hold a small quantity of wet plaster and transfer it to a trowel for application to a wall or ceiling.

Header Lintel; band joist; a joist that supports other joists; in steel construction, a beam that spans between girders; a brick or other masonry unit that is laid across two wythes with its end exposed in the face of the wall.

Head joint The vertical layer of mortar between ends of masonry units.

Hearth The noncombustible floor area outside a fireplace opening.

Heartwood The dead wood cells in the center region of a tree trunk.

Heat-fuse To join by softening or melting the edges with heat and pressing them together.

Heat of hydration The thermal energy given off by concrete or gypsum as it cures.

Heat-strengthened glass Glass that has been strengthened by heat treatment, though not to as great an extent as tempered glass.

Heaving The forcing upward of ground or buildings by the action of frost or pile driving.

Heavy timber construction *See* **mill construction.**

High-lift grouting A method of constructing a reinforced masonry wall in which the reinforcing bars are embedded in grout in story-high increments.

High-range sealant A sealant that is capable of a high degree of elongation without rupture.

High-strength bolt A bolt designed to connect steel members by clamping them together with sufficient force that the load is transferred between them by friction.

Hip The diagonal intersection of planes in a hip roof.

Hip roof A roof consisting of four sloping planes that intersect to form a pyramidal or elongated pyramid shape.

Hollow concrete masonry Concrete masonry units that are manufactured with open cores, such as ordinary concrete blocks.

Hollow-core door A door consisting of two face veneers separated by an airspace, with solid wood spacers around the four edges. The face veneers are usually connected by a grid of thin spacers within the airspace.

Hollow-core slab A precast concrete slab element that has internal longitudinal cavities to reduce its self-weight.

Hook A semicircular bend in the end of a reinforcing bar, made for the purpose of anchoring the end of the bar securely into the surrounding concrete.

Hopper window A window whose sash pivots on an axis along or near the sill, and that opens by tilting toward the interior of the building.

Horizontal force A force whose direction of action is horizontal or nearly horizontal. *See also* **lateral force.**

Hose test A standard laboratory test to determine the relative ability of an interior building assembly to stand up to water from a fire hose during a fire.

Hot-rolled steel Steel formed into its final shape by passing it between rollers while it is very hot.

Hydrated lime Calcium hydroxide produced by burning calcium carbonate to form calcium oxide (quicklime), then allowing the calcium oxide to combine chemically with water.

Hydration A process of combining chemically with water to form molecules or crystals that include hydroxide radicals or water of crystallization.

Hydronic heating system A system that circulates warm water through convectors to heat a building.

Hydrostatic pressure Pressure exerted by standing water.

Hygrosopic Readily absorbing and retaining moisture.

Hyperbolic paraboloid shell A concrete roof structure with a saddle shape.

I

I-beam (obsolete term) An American Standard section of hot-rolled steel.

Ice dam An obstruction along the eave of a roof caused by the refreezing of water emanating from melting snow on the roof surface above.

Igneous rock Rock formed by the solidification of magma.

I-joist A manufactured wood framing member whose cross-sectional shape resembles the letter I.

Impact Insulation Class (IIC) An index of the extent to which a floor assembly transmits impact noise from a room above to the room below.

Impact noise Noise generated by footsteps or other impacts on a floor.

Impact wrench A device for tightening bolts and nuts by means of rapidly repeated torque impulses produced by electrical or pneumatic energy.

Ingot A large block of cast metal.

Insulating glass Double or triple glazing.

Internal drainage Providing a curtain wall with hidden channels and weep holes to remove any water that may penetrate the exterior layers of the wall.

Interstitial ceiling A suspended ceiling with sufficient structural strength to support workers safely as they install and maintain mechanical and electrical installations above the ceiling.

Intrados The concave surface of an arch.

Intumescent coating A paint or mastic that expands to form a stable, insulating char when exposed to fire.

Inverted roof A membrane roof assembly in which the thermal insulation lies above the membrane.

Iron In pure form, a metallic element. In common usage, ferrous alloys other than steels, including cast iron and wrought iron.

Iron dog A heavy U-shaped staple used to tie the ends of heavy timbers together.

Ironworker A skilled laborer who erects steel building frames or places reinforcing bars in concrete construction.

Isocyanurate foam A thermosetting plastic foam with thermal insulating properties.

J

Jack A device for exerting a large force over a short distance, usually by means of screw action or hydraulic pressure.

Jack rafter A shortened rafter that joins a hip or valley rafter.

Jamb The vertical side of a door or window.

Jet burner A torch that burns fuel oil and compressed air, used in quarrying granite.

Joist One of a parallel array of light, closely spaced beams used to support a floor deck or flat roof.

Joist band A broad, shallow concrete beam that supports one-way concrete joists whose depths are identical to its own.

Joist girder A light steel truss used to support open-web steel joists.

K

Keenes cement A strong, crack-resistant gypsum plaster formulation.

Key A slot formed into a concrete surface for the purpose of interlocking with a subsequent pour of concrete; a slot at the edge of a precast member into which grout will be poured to lock it to an adjacent member; a mechanical interlocking of plaster with lath.

Kiln A furnace for firing clay or glass products; a heated chamber for seasoning wood; a furnace for manufacturing quicklime, gypsum hemihydrate, or portland cement.

Knee wall A short wall under the slope of a roof.

kPa Kilopascal, a unit of pressure equal to 1 kilonewton per square meter.

L

Labyrinth A cladding joint design in which a series of interlocking baffles prevents drops of water from penetrating the joint by momentum.

Lacquer A coating that dries extremely quickly through evaporation of a volatile solvent.

Lagging Planks placed between soldier beams to retain earth around an excavation.

Lag screw A large-diameter wood screw with a square or hexagonal head.

Laminate As a verb, to bond together in layers; as a noun, a material produced by bonding together layers of material.

Laminated glass A glazing material consisting of outer layers of glass bonded to an inner layer of transparent plastic.

Laminated veneer lumber (LVL) Wood members that are made up of thin wood veneers joined with glue.

Laminated wood *See* **glue-laminated timber.**

Landing A platform in or at either end of a stair.

Lap joint A connection in which one piece of material is placed partially over another piece before the two are fastened together.

Lateral force A force acting generally in a horizontal direction, such as wind, earthquake, or soil pressure against a foundation wall.

Lateral thrust The horizontal component of the force produced by an arch, dome, vault, or rigid frame.

Latex caulk A low-range sealant based on a synthetic latex.

Lath (rhymes with "math") A base material to which plaster is applied.

Lathe (rhymes with "bathe") A machine in which a piece of material is rotated against a sharp cutting tool to produce a shape, all of whose cross sections are circles; a machine in which a log is rotated against a long knife to peel a continuous sheet of veneer.

Lather (rhymes with "rather") One who applies lath.

Lay-in panel A finish ceiling panel that is installed merely by lowering it onto the top of the metal grid components of the ceiling.

Lead (rhymes with "bead") In masonry work, a corner or wall end accurately constructed with the aid of a spirit level to serve as a guide for placing the bricks in the remainder of the wall.

Leader (rhymes with "feeder") A vertical pipe for conducting water from a roof to a lower level.

Leaf The moving portion of a door.

Lehr A chamber in which glass is annealed.

Let-in bracing Diagonal bracing that is nailed into notches cut in the face of the studs so that it does not increase the thickness of the wall.

Level cut A saw cut that produces a level surface at the lower end of a sloping rafter when the rafter is in its final position.

Leveling plate A steel plate placed in grout on top of a concrete foundation to create a level bearing surface for the lower end of a steel column.

Lewis A device for lifting a block of stone by means of friction exerted against the sides of a hole drilled in the top of the block.

Life-cycle cost A cost that takes into account both the first cost and the costs of maintenance, replacement, fuel consumed, monetary inflation, and interest over the lifetime of the object being evaluated.

Lift-slab construction A method of building multistory sitecast concrete buildings by casting all the slabs in a stack on the ground, then lifting them up the columns with jacks and welding them in place.

Light A sheet of glass.

Light-gauge steel stud A length of thin sheet metal folded into a stiff shape and used as a wall framing member.

Lignin The natural cementing substance that binds together the cellulose in wood.

Limestone A sedimentary rock consisting of calcium carbonate, magnesium carbonate, or both.

Linear metal ceiling A finish ceiling whose exposed face is made up of long, parallel elements of sheet metal.

Liner A piece of marble doweled and cemented to the back of another sheet of marble.

Line wire Wire stretched across wall studs as a base for the application of metal mesh and stucco.

Linoleum A resilient floorcovering material composed primarily of ground cork and linseed oil on a burlap or canvas backing.

Lintel A beam that carries the load of a wall across a window or door opening.

Liquid sealant Gunnable sealant.

Lite *See* **light.**

Live load The weight of snow, people, furnishings, machines, vehicles, and goods in or on a building.

Load A weight or force acting on a structure.

Loadbearing Supporting a superimposed weight or force.

Load indicator washer A disk placed under the head or nut of a high-strength bolt to indicate sufficient tensioning of the bolt by means of the deformation of ridges on the surface of the disk.

Lockstrip gasket A synthetic rubber strip compressed around the edge of a piece of glass or a wall panel by inserting a spline (lockstrip) into a groove in the strip.

Lookout A short rafter, running perpendicular to the other rafters in the roof, which supports a rake overhang.

Louver An array of numerous sloping, closely spaced slats used to diffuse air or to prevent the entry of rainwater into a ventilating opening.

Low-e coating *See* **low-emissivity coating.**

Low-emissivity coating A surface coating for glass that permits the passage of most shortwave electromagnetic radiation (light and heat), but reflects most longer-wave radiation (heat).

Low-lift grouting A method of constructing a reinforced masonry wall in which the reinforcing bars are embedded and grouted in increments not higher than 4 feet (1200 mm).

Low-range sealant A sealant that is capable of only a slight degree of elongation prior to rupture.

Low-slope roof A roof that is pitched at an angle so near to horizontal that it must be made waterproof with a continuous membrane rather than shingles; commonly and inaccurately referred to as a flat roof.

Luffing-boom crane A heavy-duty lifting device that uses a tower-mounted boom that may rotate in any vertical plane as well as in a horizontal plane.

LVL *See* **laminated veneer lumber.**

M

Mandrel A stiff steel core placed inside the thin steel shell of a sitecast concrete pile to prevent it from collapsing during driving.

Mansard A roof shape consisting of two superimposed levels of *hip roofs* with the lower level at a steeper pitch than the upper.

Marble A metamorphic rock formed from limestone by heat and pressure.

Mason One who builds with bricks, stones, or concrete masonry units; one who works with concrete.

Masonry Brickwork, concrete block-work, and stonework.

Masonry cement Portland cement with dry admixtures designed to increase the workability of mortar.

Masonry opening The clear dimension required in a masonry wall for the installation of a specific window or door unit.

Masonry unit A brick, stone, concrete block, glass block, or hollow clay tile intended to be laid in mortar.

Masonry veneer A single wythe of masonry used as a facing over a frame of wood or metal.

Masterformat The copyrighted title of a uniform indexing system for construction specifications, as created by the Construction Specifications Institute and Construction Specifications Canada.

Mastic A viscous, doughlike, adhesive substance; any of a large number of formulations for different purposes such as sealants, adhesives, glazing compounds, or roofing cements.

Mat foundation A single concrete footing that is essentially equal in area to the area of ground covered by the building.

Medium-range sealant A sealant material that is capable of a moderate degree of elongation before rupture.

Meeting rail The wood or metal bar along which one sash of a double-hung, single-hung, or sliding window seals against the other.

Member An element of a structure such as a beam, a girder, a column, a joist, a piece of decking, a stud, or a component of a truss.

Membrane A sheet material that is impervious to water or water vapor.

Membrane fire protection A ceiling used to provide fire protection to the structural members above.

Metal lath A steel mesh used primarily as a base for the application of plaster or stucco.

Metamorphic rock A rock created by the action of heat or pressure on a sedimentary rock or soil.

Microsilica *See* **silica fume.**

Middle strip The half-span-wide zone of a two-way concrete slab or plate that lies midway between columns.

Mild steel Ordinary structural steel, containing less than three-tenths of 1 percent carbon.

Mill construction A building type with exterior masonry bearing walls and an interior framework of heavy timbers and solid timber decking.

Milling Shaping or planing by using a rotating cutting tool.

Millwork Wood interior finish components of a building, including moldings, windows, doors, cabinets, stairs, mantels, and the like.

Miter A diagonal cut at the end of a piece; the joint produced by joining two diagonally cut pieces at right angles.

Mobile home Euphemism for a portable house that is entirely factory built on a steel underframe supported by wheels.

Model building code A code that is offered by a recognized national organization as worthy of adoption by state or local governments.

Modified bitumen A natural bitumen with admixtures of synthetic compounds to enhance such properties as flexibility, plasticity, and durability.

Modular Conforming to a multiple of a fixed dimension.

Modular home Euphemism for a house assembled on the site from two or more boxlike factory-built sections.

Modulus of elasticity An index of the stiffness of a material, derived by measuring the elastic deformation of the material as it is placed under stress, and then dividing the stress by the deformation.

Moisture barrier A membrane used to prevent the migration of liquid water through a floor or wall.

Molding A strip of wood, plastic, or plaster with an ornamental profile.

Molding plaster A fast-setting gypsum plaster used for the manufacture of cast ornament.

Moment A twisting action; a torque; a force acting at a distance from a point in a structure so as to cause a tendency of the structure to rotate about that point. *See also* **bending moment, moment connection.**

Moment connection A connection between two structural members that is highly resistant to rotation between the members, as differentiated from a *shear connection*, which allows rotation.

Momentum The energy possessed by a moving body.

Monolithic Of a single massive piece.

Monolithic terrazzo A thin terrazzo topping applied to a concrete slab without an underbed.

Mortar A substance used to join masonry units, consisting of cementitious materials, fine aggregate, and water.

Mortise-and-tenon A joint in which a tonguelike protrusion (tenon) on the end of one piece is tightly fitted into a rectangular slot (mortise) in the side of the other piece.

Movement joint A line or plane along which movement is allowed to take place in a building or a surface of a building in response to such forces as moisture expansion and contraction, thermal expansion and contraction, foundation settling, and seismic forces.

MPa Megapascal, a unit of pressure or stress equal to 1 meganewton per square meter.

Mud slab A slab of weak concrete placed directly on the ground to provide a temporary working surface that is hard, level, and dry.

Mullion A vertical or horizontal bar between adjacent window or door units. A framing member in a metal-and-glass curtain wall.

Muntin A small vertical or horizontal bar between small lights of glass in a sash.

Muriatic acid Hydrochloric acid.

Mushroom capital A flaring, conical head on a concrete column.

N

Nail-base sheathing A sheathing material, such as wood boards or plywood, to which siding can be attached by nailing, as differentiated from one such as cane fiber board or plastic foam board that is too soft to hold nails.

Nail popping The loosening of nails holding gypsum board to a wall, caused by drying shrinkage of the studs.

Nail set A hardened steel punch used to drive the head of a nail to a level below the surface of the wood.

Needle beam A steel or wood beam threaded through a hole in a bearing wall and used to support the wall and its superimposed loads during underpinning of its foundation.

Needling The use of needle beams.

Neoprene Polychloroprene, a synthetic rubber.

Noise reduction coefficient (NRC) An index of the proportion of incident sound that is absorbed by a surface, expressed as a decimal fraction of 1.

Nominal dimension An approximate dimension assigned to a piece of material as a convenience in referring to the piece.

Nonaxial In a direction not parallel to the long axis of a structural member.

Nonbearing Not carrying a load.

Nonmovement joint A connection between materials or elements that is not designed to allow for movement.

Nonworking joint A joint that is not subjected to significant deformations.

Nosing The projecting forward edge of a stair tread.

NRC *See* **noise reduction coefficient.**

Nut A steel fastener with internal helical threads, used to close a bolt.

O

o.c. Abbreviation for "on center," meaning that the spacing of framing members is measured from the center of one member to the center of the next, rather than the clear spacing between members.

Ogee An S-shaped curve.

Oil-based caulk A low-range sealant made with linseed oil.

One-way action The structural action of a slab that spans between two parallel beams or bearing walls.

One-way concrete joist system A reinforced concrete framing system in which closely spaced concrete joists span between parallel beams or bearing walls.

One-way solid slab A reinforced concrete floor or roof slab that spans between parallel beams or bearing walls.

Open-truss wire stud A wall framing member in the form of a small steel truss.

Open-web steel joist A prefabricated, welded steel truss used at closely spaced intervals to support floor or roof decking.

Optical-quality ceramic A transparent, glasslike material that is used as a fire-rated glazing sheet.

Ordinary construction A building type with exterior masonry bearing walls and an interior structure of balloon framing.

Organic soil Soil containing decayed vegetable and/or animal matter; topsoil.

Oriented strand board (OSB) A building panel composed of long shreds of wood fiber oriented in specific directions and bonded together under pressure.

Oxidation Corrosion; rusting; rust.

P

Package fireplace A factory-built fireplace that is installed as a unit.

Paint A heavily pigmented coating applied to a surface for decorative and/or protective purposes.

Pan A form used to produce the cavity between joists in a one-way concrete joist system.

Panel A broad, thin piece of wood; a sheet of building material such as plywood or particleboard; a prefabricated building component that is broad and thin, such as a curtain wall panel; a region of a truss bounded by two vertical interior members.

Panel door A wood door in which one or more thin panels are held by stiles and rails.

Panic hardware A mechanical device that opens a door automatically if pressure is exerted against the device from the interior of the building.

Parallel strand lumber (PSL) Manufactured wood components that are made of wood shreds oriented parallel to the long axis of each piece and bonded together with adhesive.

Parapet The portion of an exterior wall that projects above the level of the roof.

Parging Portland cement plaster applied over masonry to make it less permeable to water.

Particleboard A building panel composed of small particles of wood bonded together under pressure.

Parting compound *See* **form-release compound.**

Partition An interior nonloadbearing wall.

Patterned glass Glass into which a texture has been rolled during manufacture.

Paver A half-thickness brick used as finish flooring.

PEC *See* **pressure equalization chamber.**

Pediment The gable end of a roof in classical architecture.

Penetrometer A device for testing the resistance of a material to penetration, usually used to make a quick, approximate determination of its compressive strength.

Performance grade A rating used to indicate the relative weather resistance of a window.

Periodic kiln A kiln that is loaded and fired in discrete batches, as differentiated from a tunnel kiln, which is operated continuously.

Perlite Expanded volcanic glass, used as a lightweight aggregate in concrete and plaster and as an insulating fill.

Perm A unit of vapor permeability.

Pier A caisson foundation unit.

Pilaster A vertical, integral stiffening rib in a masonry or concrete wall.

Pile A long, slender piece of material driven into the ground to act as an element of a foundation.

Pile cap A thick slab of reinforced concrete poured across the top of a pile cluster to cause the cluster to act as a unit in supporting a column or grade beam.

Piledriver A machine for driving piles.

Pintle cap A metal device used to transmit compressive forces between superimposed columns in Mill construction.

Pitch The slope of a roof or other plane, often expressed as inches of rise per foot of run; a dark, viscous hydrocarbon distilled from coal tar; a viscous resin found in wood.

Pitched roof A sloping roof.

Pivoting window A window that opens by rotating around its vertical centerline.

Plainsawing Sawing a log into dimension lumber without regard to the direction of the annual rings.

Plain slicing Cutting a flitch into veneers without regard to the direction of the annual rings.

Planing Smoothing the surface of a piece of wood, stone, or steel with a cutting blade.

Plaster A cementitious material, usually based on gypsum or portland cement, applied to lath or masonry in paste form, to harden into a finish surface.

Plasterboard *See* **gypsum board.**

Plaster of Paris Pure calcined gypsum.

Plaster screeds Intermittent spots or strips of plaster used to establish the level to which a large plaster surface will be finished.

Plasticity The ability to retain a shape attained by pressure deformation.

Plastics Synthetically produced giant molecules.

Plate A broad sheet of rolled metal $\frac{1}{4}$ inch (6.35 mm) or more in thickness; a two-way concrete slab; a horizontal top or bottom member in a platform frame wall structure.

Plate girder A large beam made up of steel plates, sometimes in combination with steel angles, that are welded, bolted, or riveted together.

Plate glass Glass of high optical quality produced by grinding and polishing both faces of a glass sheet.

Platform frame A wooden building frame composed of closely spaced members nominally 2 inches (51 mm) in thickness, in which the wall members do not run past the floor framing members.

Plenum The space between the ceiling of a room and the structural floor above, used as a passage for ductwork, piping, and wiring.

Plumb Vertical.

Plumb cut A saw cut that produces a vertical (plumb) surface at the lower end of a sloping rafter after the rafter is in its final position.

Plumbing up The process of making a steel building frame vertical and square.

Ply A layer, as in a layer of felt in a built-up roof membrane or a layer of veneer or veneers in plywood.

Plywood A wood panel composed of an odd number of layers of wood veneer bonded together under pressure.

Pointing The process of applying mortar to the surface of a mortar joint after the masonry has been laid, either as a means of finishing the joint or to repair a defective joint.

Poke-through fitting An electrical outlet that is installed by drilling a hole through a floor, inserting the outlet from above, and bringing in the wiring from the plenum below.

Polybutene A sticky, masticlike tape used to seal nonworking joints, especially between glass and mullions.

Polycarbonate An extremely tough, strong, usually transparent plastic used for window and skylight glazing, light fixture globes, door sills, and other applications.

Polyethylene A thermoplastic widely used in sheet form for vapor retarders, moisture barriers, and temporary construction coverings.

Polymer A large molecule composed of many identical chemical units.

Polypropylene A plastic formed by the polymerization of propylene.

Polystyrene foam A thermoplastic foam with thermal insulating properties.

Polysulfide A high-range gunnable sealant.

Polyurethane Any of a large group of synthetic resins and synthetic rubber compounds used in sealants, varnishes, insulating foams, and roof membranes.

Polyurethane foam A thermosetting foam with thermal insulating properties.

Polyvinyl chloride (PVC) A thermoplastic material widely used in construction products, including plumbing pipes, floor tiles, wall coverings, and roof membranes. Called "vinyl" for short.

Portal frame A rigid frame; two columns and a beam attached to one another with moment connections.

Portland cement The gray powder used as the binder in concrete, mortar, and stucco.

Posttensioning The compressing of the concrete in a structural member by means of tensioning high-strength steel tendons against it after the concrete has cured.

Pour To cast concrete; an increment of concrete casting carried out without interruption.

Powder coating A coating produced by applying a powder consisting of thermosetting resins and pigments, adhering it to the substrate by electrostatic attraction, and fusing it into a continuous film in an oven.

Powder-driven Inserted by a gunlike tool using energy provided by an exploding charge of gunpowder.

Precast concrete Concrete cast and cured in a position other than its final position in the structure.

Predecorated gypsum board Gypsum board finished at the factory with a decorative layer of paint, paper, or plastic.

Preformed cellular tape sealant A sealant inserted into a joint in the form of a compressed sponge impregnated with compounds that cure to form a watertight seal.

Preformed joint filler A strip of rubbery or spongelike material designed to fit snugly into a gap between two materials.

Preformed solid tape sealant A sealant inserted into a joint in the form of a flexible strip of solid material.

Prehung door A door that is hinged to its frame in a factory or shop.

Prescriptive building code A set of legal regulations that mandate specific construction details and practices rather than establish performance standards.

Pressure equalization chamber (PEC) The wind-pressurized cavity in a rainscreen wall.

Pressure-equalized wall design Curtain wall design that utilizes the *rainscreen principle*.

Pressure-treated lumber Lumber that has been impregnated with chemicals under pressure, for the purpose of retarding either decay or fire.

Prestressed concrete Concrete that has been pretensioned or posttensioned.

Prestressing Applying an initial compressive stress to a concrete structural member, either by pretensioning or by posttensioning.

Pretensioning The compressing of the concrete in a structural member by pouring the concrete for the member around stretched high-strength steel strands, curing the concrete, and releasing the external tensioning force on the strands.

Prime window A window unit that is made to be installed permanently in a building.

Priming Covering a surface with a coating that prepares it to accept another type of coating or a sealant.

Protected membrane roof A membrane roof assembly in which the thermal insulation lies above the membrane.

PSL *See* **parallel strand lumber.**

Punty A metal rod used in working with hot glass.

Purlins Beams that span across the slope of a steep roof to support the roof decking.

Putty A simple glazing compound used to seal around a small light.

PVC *See* **polyvinyl chloride.**

Pyrolitic coating A coating applied at a very high temperature.

Q

Quarry An excavation from which building stone is obtained; the act of taking stone from the ground.

Quarry bed A plane in a building stone that was horizontal before the stone was cut from the quarry.

Quarry sap Excess water found in rock at the time of its quarrying.

Quarry tile A large clay floor tile.

Quartersawn Lumber sawn in such a way that the annual rings run roughly perpendicular to the face of each piece.

Quartersliced Veneer sliced in such a way that the annual rings run roughly perpendicular to the face of each veneer.

Quoin (pronounced "coin") A corner reinforcing of cut stone or bricks in a masonry wall, usually done for decorative effect.

R

Rabbet A longitudinal groove cut at the edge of a member to receive another member; also called a *rebate*.

Radiant barrier A reflective foil placed adjacent to an airspace in roof or wall assemblies as a deterrent to the passage of infrared energy.

Raft A mat footing.

Rafter A framing member that runs up and down the slope of a steep roof.

Rail A horizontal framing piece in a panel door; a handrail.

Rainscreen principle A theory by which wall cladding is made watertight by providing wind-pressurized air chambers behind joints to eliminate air pressure differentials between the outside and the inside that might transport water through the joints.

Raised access flooring *See* **access flooring.**

Rake The sloping edge of a steep roof.

Raker A sloping brace for supporting sheeting around an excavation.

Ram A hydraulic piston device used for bending steel, tensioning steel strands in prestressed concrete, or lifting heavy loads.

Ratchet A mechanical device with sloping teeth that allows one piece to be advanced against another in small increments, but not to move in the reverse direction.

Ray A tubular cell that runs radially in a tree trunk.

RBM *See* **reinforced brick masonry.**

Reflective coated glass Glass onto which a thin layer of metal or metal oxide has been deposited to reflect light and/or heat.

Reglet A horizontal slot, inclined in cross section, into which a flashing or roof membrane may be inserted in a concrete or masonry surface.

Reinforced brick masonry (RBM) Brickwork into which steel bars have been embedded to impart tensile strength to the construction.

Reinforced concrete Concrete work into which steel bars have been embedded to impart tensile strength to the construction.

Relative humidity A percentage representing the ratio of the amount of water vapor contained in a mass of air to the amount of water it could contain under the existing conditions of temperature and pressure.

Relieved back A longitudinal groove or series of grooves cut from the back of a flat wood molding or flooring strip to minimize cupping forces and make the piece easier to fit to a flat surface.

Removable glazing panel A framed sheet of glass that can be attached to a window sash to increase its thermal insulating properties.

Replacement window A window unit that is designed to install easily in an opening left in a wall by a deteriorated window unit that has been removed.

Reshoring Inserting temporary supports under concrete beams and slabs after the formwork has been removed, to prevent overloading prior to full curing of the concrete.

Resilient clip A springy mounting device for plaster or gypsum board that helps reduce the transmission of sound vibrations through a wall or ceiling.

Resilient flooring A manufactured sheet or tile flooring made of asphalt, polyvinyl chloride, linoleum, rubber, or other elastic material.

Resin A natural or synthetic, solid or semisolid organic material of high molecular weight, used in the manufacture of paints, varnishes, and plastics.

Retaining wall A wall that resists horizontal soil pressures at an abrupt change in ground elevation.

Ridge board The board against which the tips of rafters are fastened.

Rigid frame Two columns and a beam or beams attached to one another with moment connections; a moment-resisting building frame.

Rim joist *See* **band joist.**

Rise A difference in elevation, such as the rise of a stair from one floor to the next or a rise per foot of run in a sloping roof.

Riser A single vertical increment of a stair; the vertical face between two treads in a stair; a vertical run of plumbing, wiring, or ductwork.

Rivet A structural fastener on which a second head is formed after the fastener is in place.

Rock anchor A posttensioned rod or cable inserted into a rock formation for the purpose of tying it together.

Rock wool An insulating material manufactured by forming fibers from molten rock.

Roofer One who installs roof coverings.

Roofing The material used to make a roof watertight, such as shingles, slate, tiles, sheet metal, or a roof membrane; the act of applying roofing.

Roof window An openable window designed to be installed in the sloping surface of a roof.

Rotary-sliced veneer A thin sheet of wood produced by rotating a log against a long, sharp knife blade in a lathe.

Rough arch An arch made from masonry units that are rectangular rather than wedge shaped.

Rough carpentry Framing carpentry, as distinguished from *finish carpentry*.

Roughing in The installation of mechanical, electrical, and plumbing components that will not be exposed to view in the finished building.

Rough opening The clear dimensions of the opening that must be provided in a wall frame to accept a given door or window unit.

Rowlock A brick laid on its long edge, with its end exposed in the face of the wall.

Rubble Unsquared stones.

Run Horizontal dimension in a stair or sloping roof.

Runner channel A steel member from which furring channels and lath are supported in a suspended plaster ceiling.

Running bond Brickwork consisting entirely of stretchers.

Runoff bar One of a pair of small rectangular steel bars attached temporarily at the end of a prepared groove for the purpose of permitting the groove to be filled to its very end with weld metal.

Run plaster ornament A linear molding produced by passing a profiled sheet metal or plastic template back and forth across a mass of wet plaster.

R-value A numerical measure of resistance to the flow of heat.

S

Safing Fire-resistant material inserted into a space between a curtain wall and a spandrel beam or column, to retard the passage of fire through the space.

Sand cushion terrazzo Terrazzo with an underbed that is separated from the structural floor deck by a layer of sand.

Sand-mold brick, sand-struck brick A brick made in a mold that was wetted and then dusted with sand before the clay was placed in it.

Sandstone A sedimentary rock formed from sand.

Sandwich panel A panel consisting of two outer faces of wood, metal, gypsum, or concrete bonded to a core of insulating foam.

Sapwood The living wood in the outer region of a tree trunk or branch.

Sash A frame that holds glass.

Scab A piece of framing lumber nailed to the face of another piece of lumber.

Scarf joint A glued end connection between two pieces of wood, using a sloping cut to create a large surface for the glue bond, to allow it to develop the full tensile strength of the wood that it connects.

SCOF *See* **static coefficient of friction.**

Scratch coat The first coat in a three-coat application of plaster.

Screed A strip of wood, metal, or plaster that establishes the level to which concrete or plaster will be placed.

Screw port A three-quarter circular profile in an aluminum extrusion, made to accept a screw driven parallel to the long axis of the extrusion.

Screw slot A serrated slot profile in an aluminum extrusion, made to accept screws driven at right angles to the long axis of the extrusion.

Scupper An opening through a parapet through which water can drain over the edge of a flat roof.

Sealant gun A tool for injecting sealant into a joint.

Sealer A coating used to close the pores in a surface, usually in preparation for the application of a finish coating.

Seated connection A connection in which a steel beam rests on top of a steel angle or tee that is fastened to a column or girder.

Security glass A glazing sheet with multiple laminations of glass and plastic, designed to stop bullets.

Sedimentary rock Rock formed from materials deposited as sediments, such as sand or sea shells, which form sandstone and limestone, respectively.

Segregation Separation of the constituents of wet concrete caused by excessive handling or vibration.

Seismic Relating to earthquakes.

Seismic load A load on a structure caused by movement of the earth relative to the structure during an earthquake.

Seismic separation joint A building separation joint that allows adjacent building masses to oscillate independently during an earthquake.

Self-drilling Drills its own hole.

Self-furring lath Metal lath with dimples that space the lath away from the sheathing behind to allow plaster to penetrate the lath and key to it.

Self-tapping Creates its own screw threads on the inside of a hole.

Self-weight The weight of a beam or slab.

Set To cure; to install; to recess the heads of nails; a punch for recessing the heads of nails.

Setting block A small block of synthetic rubber or lead used to support the weight of a sheet of glass at its lower edge.

Settlement joint A building separation joint that allows the foundations of adjacent building masses to settle at different rates.

Shading coefficient The ratio of total solar heat passing through a given sheet of glass to that passing through a sheet of clear double-strength glass.

Shaft An unbroken vertical passage through a multistory building, used for elevators, wiring, plumbing, ductwork, and so on.

Shaft wall A wall surrounding a shaft.

Shake A shingle split from a block of wood.

Shake-on hardener A dry powder that is dusted onto the surface of a concrete slab before troweling, to react with the concrete and produce a hard wearing surface for industrial use.

Shale A rock formed from the consolidation of clay or silt.

Shear A deformation in which planes of material slide with respect to one another.

Shear connection A connection designed to resist only the tendency of one member to slide past the other, and not to resist any tendency of the members to rotate with respect to one another.

Shear panel A wall, floor, or roof surface that acts as a deep beam to help stabilize a building against deformation by lateral forces.

Shear stud A piece of steel welded to the top of a steel beam or girder so as to become embedded in the concrete fill over the beam and cause the beam and the concrete to act as a single structural unit.

Sheathing The rough covering applied to the outside of the roof, wall, or floor framing of a light frame structure.

Shed A building or dormer with a single, sloping roof plane.

Sheeting A stiff material used to retain the soil around an excavation; a material such as polyethylene in the form of very thin, flexible sheets.

Sheet metal Flat rolled metal less than $1/4$ inch (6.35 mm) in thickness.

Shelf angle A steel angle attached to the spandrel of a building to support a masonry facing.

Shim A thin piece of material placed between two components of a building to adjust their relative positions as they are assembled; to insert shims.

Shingle A small unit of water-resistant material nailed in overlapping fashion with many other such units to render a wall or sloping roof watertight; to apply shingles.

Shiplap A board with edges rabbeted so as to overlap flush from one board to the next.

Shop drawings Detailed plans prepared by a fabricator or manufacturer to guide the shop production of such building components as cut stonework, steel or precast concrete framing, curtain wall panels, or cabinetwork.

Shoring Temporary vertical or sloping supports of steel or timber.

Shotcrete A low-slump concrete mixture that is deposited by being blown from a nozzle at high speed with a stream of compressed air.

Shrinkage–temperature steel Reinforcing bars laid at right angles to the principal bars in a one-way slab, for the purpose of preventing excessive cracking caused by drying shrinkage or temperature stresses in the concrete.

Side-hinged inswinging window A window that opens by pivoting inward on hinges at or near a vertical edge of the sash.

Sidelight A tall, narrow window alongside a door.

Siding The exterior wall finish material applied to a light frame wood structure.

Siding nail A nail with a small head, used to fasten siding to a building.

Silica fume Very finely divided silicon dioxide, used as an admixture in the formulation of very high strength, low-permeability concrete.

Silicone A polymer used for high-range sealants, roof membranes, and masonry water repellents.

Sill The strip of wood that lies immediately on top of a concrete or masonry foundation in wood frame construction; the horizontal bottom portion of a window or door; the exterior surface, usually sloped to shed water, below the bottom of a window or door.

Sill sealer A resilient, fibrous material placed between a foundation and a sill to reduce air infiltration between the outdoors and the indoors.

Single-hung window A window with two overlapping sashes, the lower of which can slide vertically in tracks, and the upper of which is fixed.

Single-ply membrane A sheet of plastic, synthetic rubber, or modified bitumen used as a roofing sheet for a low-slope roof.

Single-strength glass Glass approximately $\frac{3}{32}$ inch (2.5 mm) thick.

Single tee A precast slab element whose profile resembles the letter T.

Sitecast Concrete that is poured and cured in its final position in a building.

Skip joist *See* **wide-module concrete joist system.**

Skylight A fixed window installed in a roof.

Slab band A very broad, shallow beam used with a one-way solid slab.

Slab on grade A concrete surface lying upon, and supported directly by, the ground beneath.

Slag The mineral waste that rises to the top of molten iron or steel or to the top of a weld.

Slaked lime Calcium hydroxide.

Slate A metamorphic form of clay, easily split into thin sheets.

Sliding window A window with one fixed sash and another that moves horizontally in tracks.

Slip-critical connection A steel connection in which high-strength bolts clamp the members together with sufficient force that the load is transferred between them by friction.

Slip forming Building multistory sitecast concrete walls with forms that rise up the wall as construction progresses.

Slope glazing A system of metal and glass components used to make an inclined, transparent roof.

Slump test A test in which wet concrete or plaster is placed in a cone-shaped metal mold of specified dimensions and allowed to sag under its own weight after the cone is removed. The vertical distance between the top of the mold and the top of the slumped mixture is an index of its working consistency.

Slurry A watery mixture of insoluble materials.

Smoke developed rating An index of the toxic fumes generated by a material as it burns.

Smoke shelf The horizontal area behind the damper of a fireplace.

Soffit The undersurface of a horizontal element of a building, especially the underside of a stair or a roof overhang.

Soffit vent An opening under the eave of a roof, used to allow air to flow into the attic or the space below the roof sheathing.

Soft mud process Making bricks by pressing wet clay into molds.

Soldier A brick laid on its end, with its narrow face toward the outside of the wall.

Sole plate The horizontal piece of dimension lumber at the bottom of the studs in a wall in a light frame building.

Solid-core door A flush door with no internal cavities.

Solid slab A concrete slab, without ribs or voids, that spans between beams or bearing walls.

Solid tape sealant *See* **preformed solid tape sealant.**

Solvent A liquid that dissolves another material.

Sound Transmission Class (STC) An index of the resistance of a partition to the passage of sound.

Space frame, space truss A truss that spans with two-way action.

Span The distance between supports for a beam, girder, truss, vault, arch, or other horizontal structural device; to carry a load between supports.

Spandrel The wall area between the head of a window on one story and the sill of a window on the floor above; the area of a wall between adjacent arches.

Spandrel beam A beam that runs along the outside edge of a floor or roof.

Spandrel glass Opaque glass manufactured especially for use in spandrel panels.

Spandrel panel A curtain wall panel used in a spandrel.

Span rating The number stamped on a sheet of plywood or other wood building panel to indicate how far in inches it may span between supports.

Specifications The written instructions from an architect or engineer concerning the quality of materials and execution required for a building.

Spirit level A tool in which a bubble in an upwardly curving, cylindrical glass vial indicates whether a building element is level or not level, plumb or not plumb.

Splash block A small precast block of concrete or plastic used to divert water at the bottom of a downspout.

Spline A thin strip inserted into grooves in two mating pieces of material to hold them in alignment; a ridge or strip of material intended to lock to a mating groove; an internal piece that serves to align two hollow components to one another.

Split jamb A door frame fabricated in two interlocking halves, to be installed from the opposite sides of an opening.

Staggered truss system A steel framing system in which story-high trusses, staggered one-half bay from one story to the next, support floor decks on both their top and their bottom chords.

Stain A coating intended primarily to change the color of wood or concrete, without forming an impervious film.

Standing and running trim Door and window casings and baseboards.

Standing seam A sheet metal roofing seam that projects at right angles to the plane of the roof.

Static coefficient of friction (SCOF) A measure of the slip resistance of a flooring material.

Stay A sloping cable used to stabilize a structure.

STC *See* **Sound Transmission Class.**

Steam curing Aiding and accelerating the setting reaction of concrete by the application of steam.

Steel Iron with a controlled amount of carbon, generally less than 1.7 percent.

Steep roof A roof with sufficient slope that it may be made waterproof with shingles.

Sticking The cementing together of defects in marble slabs.

Stick system A metal curtain wall system that is largely assembled in place.

Stiffener plate A steel plate attached to a structural member to support it against heavy localized loading or stresses.

Stiff mud process A method of molding bricks in which a column of damp clay is extruded from a rectangular die and cut into bricks by fine wires.

Stile A vertical framing member in a panel door.

Stirrup A vertical loop of steel bar used to reinforce a concrete beam against diagonal tension forces.

Stirrup-tie A stirrup that forms a complete loop, as differentiated from a U-stirrup, which has an open top.

Stool The interior horizontal plane at the sill of a window.

Storm window A sash added to the outside of a window in winter to increase its thermal resistance and decrease air infiltration.

Story pole A strip of wood marked with the exact course heights of masonry for a particular building, used to make sure that all the leads are identical in height and coursing.

Straightedge To strike off the surface of a concrete slab using screeds and a straight piece of lumber or metal.

Strain Deformation under stress.

Stress Force per unit area.

Stressed-skin panel A panel consisting of two face sheets of wood, metal, or concrete bonded to perpendicular spacer strips. Often used as a term for a sandwich panel with an insulating foam core and wood panel faces.

Stretcher A brick or masonry unit laid in its most usual position, with the broadest surface of the unit horizontal and the length of the unit parallel to the surface of the wall.

Striated Textured with parallel scratches or grooves.

Stringer The sloping wood or steel member that supports the treads of a stair.

Strip flooring Wood finish flooring in the form of long, narrow tongue-and-groove boards.

Stripping Removing formwork from concrete; sealing around a roof flashing with layers of felt and bitumen.

Structural bond The interlocking pattern of masonry units used to tie two or more wythes together in a wall.

Structural glazed facing tile A hollow clay masonry unit with glazed faces.

Structural mill The portion of a steel mill that rolls structural shapes.

Structural silicone flush glazing Glass secured to the face of a building with strong, highly adhesive silicone sealant so as to eliminate the need for any metal to appear on the exterior of the building.

Structural standing-seam metal roofing Sheets of folded metal that serve both as decking and as the waterproof layer of a roof.

Structural terra cotta Molded components, often highly ornamental, made of fired clay, designed to be used in the facades of buildings.

Structural tubing Hollow steel cylindrical or rectangular shapes made to be used as structural members.

Structure/enclosure joint A connection designed to allow the structure of a building and its cladding or partitions to move independently.

Stucco Portland cement plaster used as an exterior cladding or siding material.

Stud One of an array of small, closely spaced, parallel wall framing members; a heavy steel pin.

Styrene-butadiene-styrene A copolymer of butadiene and styrene used as a modifier in polymer-modified bitumen roofing.

Subcontractor A contractor who specializes in one area of construction activity and who usually works under a general contractor.

Subfloor The loadbearing surface beneath a finish floor.

Subpurlin A very small roof framing member that spans between joists or purlins.

Substrate The base to which a coating or veneer is applied.

Substructure The occupied, below-ground portion of a building.

Sump A pit designed to collect water for removal from an excavation or basement.

Superflat floor A concrete slab finished to a high degree of flatness according to a recognized system of measurement.

Superplasticizer A concrete admixture that makes wet concrete extremely fluid without additional water.

Superstructure The above-ground portion of a building.

Supply pipe A pipe that brings clean water to a plumbing fixture.

Supporting stud A wall framing member that extends from the sole plate to the underside of a header and that supports the header.

Surface bonding Bonding a concrete masonry wall together by applying a layer of glass-fiber-reinforced stucco to both its faces.

Surface divider joint A line along which a surface may expand and/or contract without damage.

Suspended ceiling A finish ceiling that is hung on wires from the structure above.

Suspended glazing Large sheets of glass hung from clamps at their top edges so as to eliminate the need for metal mullions.

Swaged lockpin and collar fastener A boltlike device that is passed through holes in structural steel components, held in very high tension, and closed with a steel ring that is squeezed onto its protruding shank.

T

Tackless strip A wood strip with projecting points used to fasten a carpet around the edge of a room.

Tagline A rope attached to a building component to help guide it as it is lifted by a crane or derrick.

Tapered edge The longitudinal edge of a sheet of gypsum board, which is recessed to allow room for reinforcing tape and joint compound.

Tee A metal or precast concrete member with a cross section resembling the letter T.

Tempered glass Glass that has been heat-treated to increase its toughness and its resistance to breakage.

Tendon A steel strand used for prestressing a concrete member.

Tensile strength The ability of a structural material to withstand stretching forces.

Tensile stress A stress caused by stretching of a material.

Tension A stretching force; to stretch.

Tension control bolt A bolt tightened by means of a splined end that breaks off when the bolt shank has reached the required tension.

Terne An alloy of lead and tin, used to coat sheets of carbon steel or stainless steel for use as metal roofing sheet.

Terrace door A double glass door, one leaf of which is fixed, and the other hinged to the fixed leaf at the centerline of the door.

Terrazzo A finish floor material consisting of concrete with an aggregate of marble chips selected for size and color, which is ground and polished smooth after curing.

Thatch A thick roof covering of reeds, straw, grasses, or leaves.

Thermal break A section of material with a low thermal conductivity, installed between metal components to retard the passage of heat through a wall or window assembly.

Thermal bridge A component of higher thermal conductivity that conducts heat more rapidly through an insulated building assembly, such as a steel stud in an insulated stud wall.

Thermal conductivity The rate at which a material conducts heat.

Thermal insulation A material that greatly retards the passage of heat.

Thermal resistance The resistance of a material or assembly to the conduction of heat.

Thermoplastic Having the property of softening when heated and rehardening when cooled.

Thermosetting Not having the property of softening when heated.

Thrust A lateral or inclined force resulting from the structural action of an arch, vault, dome, suspension structure, or rigid frame.

Thrust block A wooden block running perpendicular to the stringers at the bottom of a stair, whose function is to hold the stringers in place.

Tie A device for holding two parts of a construction together; a structural device that acts in tension.

Tieback A tie, one end of which is anchored in the ground, with the other end used to support sheeting around an excavation.

Tie beam A reinforced concrete beam cast as part of a masonry wall, whose primary purpose is to hold the wall together, especially against seismic loads, or cast between a number of isolated foundation elements to maintain their relative positions.

Tier The portion of a multistory steel building frame supported by one set of fabricated column pieces, commonly two stories in height.

Tie rod A steel rod that acts in tension.

Tile A fired clay product that is thin in cross section as compared to a brick, either a thin, flat element (ceramic tile or quarry tile), a thin, curved element (roofing tile), or a hollow element with thin walls (flue tile, tile pipe, structural clay tile); also a thin, flat element of another material, such as an acoustical ceiling unit or a resilient floor unit.

Tilt/turn window A window that may open either as a casement window or a hopper window.

Tilt-up construction A method of constructing concrete walls in which panels are cast and cured flat on a floor slab, then tilted up into their final positions.

Timber Standing trees; a large piece of dimension lumber.

Tinted glass Glass that is colored with pigments, dyes, or other admixtures.

Toe nailing Fastening with nails driven at an angle.

Tongue-and-groove An interlocking edge detail for joining planks or panels.

Tooling The finishing of a mortar joint or sealant joint by pressing and compacting it to create a particular profile.

Top-hinged inswinging window A window that opens inward on hinges on or near its head.

Topping A thin layer of concrete cast over the top of a floor deck.

Topping-out Placing the last member in a building frame.

Top plate The horizontal member at the top of a stud wall.

Topside vent A water-protected opening through a roof membrane to relieve pressure from water vapor that may accumulate beneath the membrane.

Torque Twisting action; moment.

Torsional stress Stress resulting from the twisting of a structural member.

Tracheids The longitudinal cells in a softwood.

Traffic deck A walking surface placed on top of a roof membrane.

Transit-mixed concrete Concrete mixed in a rotating drum on the back of a truck as it is transported to the building site.

Travertine A richly patterned, marble-like form of limestone.

Tread One of the horizontal planes that make up a stair.

Tremie A large funnel with a tube attached, used to deposit concrete in deep forms or beneath water or slurry.

Trim accessories Casing beads, corner beads, expansion joints, and other devices used to finish edges and corners of a plaster wall or ceiling.

Trimmer A beam that supports a header around an opening in a floor or roof frame.

Trowel A thin, flat steel tool, either pointed or rectangular, provided with a handle and held in the hand, used to manipulate mastic, mortar, plaster, or concrete. Also, a machine whose rotating steel blades are used to finish concrete slabs; to use a trowel.

Truss A triangulated arrangement of structural members that reduces nonaxial external forces to a set of axial forces in its members. *See also* **Vierendeel truss.**

Tuck pointing The process of removing deteriorated mortar from the zone near the surface of a brick wall and inserting fresh mortar.

Tunnel kiln A kiln through which clay products are passed on railroad cars.

Turn-of-nut method A method of achieving the correct tightness in a high-strength bolt by first tightening the nut snugly, then turning it a specified additional fraction of a turn.

Two-way action Bending of a slab or deck in which bending stresses are approximately equal in the two principal directions of the structure.

Two-way concrete joist system A reinforced concrete framing system in which columns directly support an orthogonal grid of intersecting joists.

Two-way flat plate A reinforced concrete framing system in which columns directly support a two-way slab that is planar on both its surfaces.

Two-way flat slab A reinforced concrete framing system in which columns with mushroom capitals and/or drop panels directly support a two-way slab that is planar on both its surfaces.

Type X gypsum board A fiber-reinforced gypsum board, used where greater fire resistance is required.

U

Unbonded construction Posttensioned concrete construction in which the tendons are not grouted to the surrounding concrete.

Uncoursed Laid without continuous horizontal joints, random.

Undercarpet wiring system Flat, insulated electrical conductors that run under carpeting, and their associated outlet boxes and fixtures.

Undercourse A course of shingles laid beneath an exposed course of shingles at the lower edge of a wall or roof, in order to provide a waterproof layer behind the joints in the exposed course.

Underfire The floor of the firebox in a fireplace.

Underlayment A panel laid over a subfloor to create a smooth, stiff surface for application of finish flooring.

Underpinning The process of placing new foundations beneath an existing structure.

Unfinished bolt An ordinary carbon steel bolt.

Uniform settlement Subsidence of the various foundation elements of a building at the same rate, resulting in no distress to the structure of the building.

Unit-and-mullion system A curtain wall system consisting of prefabricated panel units secured with site-applied mullions.

Unit system A curtain wall system consisting entirely of prefabricated panel units.

Unreinforced Constructed without steel reinforcing bars or welded wire fabric.

Up–down construction A sequence of construction activity in which construction proceeds downward on the sublevels of a building at the same time as construction proceeds upward on the superstructure.

Upside-down roof A membrane roof assembly in which the thermal insulation lies above the membrane.

U-stirrup An open-top, U-shaped loop of steel bar, used as reinforcing against diagonal tension in a concrete beam.

V

Valley A trough formed by the intersection of two roof slopes.

Valley rafter A diagonal rafter that supports a valley.

Vapor barrier *See* **vapor retarder.**

Vapor retarder A layer of material intended to obstruct the passage of water vapor through a building assembly. Also called, less accurately, *vapor barrier.*

Varnish A slow-drying transparent coating.

Vault An arched surface. A strongly built room for such purposes as housing large electrical equipment or safeguarding money.

Vee joint A mortar joint whose profile resembles the letter V.

Veneer A thin layer, sheet, or facing.

Veneer-based lumber Linear wooden structural elements manufactured by bonding together strands or sheets of rotary-sliced wood veneer.

Veneer plaster A wall finish system in which a thin finish layer of plaster is applied over a special gypsum board base.

Veneer plaster base The special gypsum board over which veneer plaster is applied.

Vent spacer A device used to maintain a free air passage above the thermal insulation in an attic or roof.

Vermiculite Expanded mica, used as an insulating fill or a lightweight aggregate.

Vertical bar An upright reinforcing bar in a concrete column.

Vertical grain lumber Dimension lumber sawed in such a way that the annual rings run more or less perpendicular to the faces of each piece.

Vierendeel truss A truss with rectangular panels and rigid joints but no diagonal members. The members of a Vierendeel truss are subjected to strong nonaxial forces.

Vinyl *See* **polyvinyl chloride.**

Visible transmittance The fraction of light at visible wavelengths that passes through a sheet of glass.

Vitrification The process of transforming a material into a glassy substance by means of heat.

Volume change joint A building separation joint that allows for expansion and contraction of adjacent portions of a building without distress.

Voussoir A wedge-shaped element of an arch or vault.

W

Waferboard A building panel made by bonding together large, flat flakes of wood.

Waffle slab A two-way concrete joist system.

Wainscoting A wall facing, usually of wood, cut stone, or ceramic tile, that is carried only partway up a wall.

Waler A horizontal beam used to support sheeting or concrete formwork.

Wane An irregular rounding of a long edge of a piece of dimension lumber caused by cutting the lumber from too near the outside surface of the log.

Washer A steel disk with a hole in the middle, used to spread the load from a bolt, screw, or nail across a wider area of material.

Water–cement ratio A numerical index of the relative proportions by weight of water and cement in a concrete mixture.

Water-resistant gypsum board A gypsum board designed for use in locations where it may be exposed to occasional dampness.

Water smoking The process of applying heat to evaporate the last water from clay products before they are fired.

Waterstop A synthetic rubber or bentonite clay strip used to seal joints in concrete foundation walls.

Water-struck brick A brick made in a mold that was wetted before the clay was placed on it.

Water table The level at which the pressure of water in the soil is equal to atmospheric pressure; effectively, the level to which ground water will fill an excavation; a wood molding or shaped brick used to make a transition between a thicker foundation and the wall above.

Water vapor Water in its gaseous phase.

Wattle and daub Mud plaster (daub) applied to a primitive lath of woven twigs or reeds (wattle).

Waxing Filling of voids in marble slabs.

Weathered joint A mortar joint finished in a sloping, planar profile that tends to shed water to the outside of the wall.

Weathering steel A steel alloy that forms a tenacious, self-protecting rust layer when exposed to the atmosphere.

Weatherstrip A ribbon of resilient, brushlike, or springy material used to reduce air infiltration through the crack around a sash or door.

Web A cross-connecting piece, such as the portion of a wide-flange shape that is perpendicular to the flanges, or the portion of a concrete masonry unit that is perpendicular to the face shells.

Web stiffener A metal rib used to support the web of a light gauge steel joist or a structural steel girder against buckling.

Weep hole A small opening whose purpose is to permit drainage of water that accumulates inside a building component or assembly.

Weld A joint between two pieces of metal formed by fusing the pieces together by the application of intense heat, usually with the aid of additional metal melted from a rod or electrode.

Welding The process of making a weld.

Weld plate A steel plate anchored into the surface of concrete, to which another steel element can be welded.

"Wet" systems Construction systems that utilize considerable quantities of water on the construction site, such as masonry, plaster, sitecast concrete, and terrazzo.

Wide-flange shape Any of a wide range of structural steel components rolled in the shape of a letter I or H.

Wide-module concrete joist system A one-way concrete framing system with joists that are spaced more widely than those in a conventional one-way concrete joist system.

Wind brace A diagonal structural member whose function is to stabilize a frame against lateral forces.

Winder (rhymes with "reminder") A stair tread that is wider at one end than at the other.

Wind load A load on a building caused by wind pressure and/or suction.

Wind uplift Upward forces on a structure caused by negative aerodynamic pressures that result from certain wind conditions.

Wired glass Glass in which a wire mesh is embedded during manufacture.

Working construction joint A connection that is designed to allow for small amounts of relative movement between two pieces of a building assembly.

Wracking Forcing out of plumb.

Wrought iron A form of iron that is soft, tough, and fibrous in structure, containing about 0.1 percent carbon and 1 to 2 percent slag.

Wythe (rhymes with "scythe" and "tithe") A vertical layer of masonry that is one masonry unit thick.

Y

Yield strength The stress at which a material ceases to deform in a fully elastic manner and begins to deform irreversibly.

Z

Z-brace door A door made of vertical planks held together and braced on the back by three pieces of wood whose configuration resembles the letter Z.

Zero-slump concrete A concrete mixed with so little water that it does not sag when piled vertically.

Zoning ordinance A law that specifies in detail how land may be used in a municipality.

INDEX